# Modern
# Bioelectrochemistry

# Modern Bioelectrochemistry

Edited by
## Felix Gutmann
Macquarie University
North Ryde, New South Wales, Australia

and

## Hendrik Keyzer
California State University, Los Angeles
Los Angeles, California

PLENUM PRESS • NEW YORK AND LONDON

Library of Congress Cataloging in Publication Data

Main entry under title:

Modern bioelectrochemistry.

   Includes bibliographies and index.
   1. Bioelectrochemistry. I. Gutmann, Felix, 1908–    . II. Keyzer, Hendrik.
QP517.B53M63   1985                    574.19′283                    85-19337
ISBN-13: 978-1-4612-9246-3      e-ISBN-13: 978-1-4613-2105-7
DOI: 10.1007/978-1-4613-2105-7

© 1986 Plenum Press, New York

A Division of Plenum Publishing Corporation
233 Spring Street, New York, N.Y. 10013

Dedicated to

FREEMAN W. COPE

*In Memoriam*

# Contributors

**Mohammad Amin** • School of Life Sciences, Jawaharlal Nehru University, New Mehrauli Road, New Delhi-110067, India

**John O'M. Bockris** • Department of Chemistry, Texas A & M University, College Station, Texas 77843

**James D. Bond** • Science Applications, Inc., 1710 Goodridge Drive, McLean, Virginia 22102

**Phoebe K. Dea** • Department of Chemistry and Biochemistry, California State University, 5151 State University Drive, Los Angeles, California 90032

**H. P. Dhar** • Energy Research Corporation, 3 Great Pasture Road, Danbury, Connecticut 06810

**Silvia Doglia** • Dipartimento di Fisica dell'Università, Via Celoria, 16-20133 Milano, Italy, and Gruppo Nazionale di Struttura della Materia del C.N.R., Milano, Italy

**George M. Eckert** • Clinical Pharmacologist, The St. George Hospital, Kogarah, N.S.W. 2217, Australia

**Sidney W. Fox** • Institute for Molecular and Cellular Evolution, University of Miami, 521 Anastasia Avenue, Coral Gables, Florida 33134

**H. Fröhlich** • Department of Physics, Oliver Lodge Laboratory, Oxford Street, P.O. Box 147, Liverpool L69 3BX, England

**Emilio Del Giudice** • Dipartimento di Fisica dell'Università, Via Celoria, 16-20133 Milano, Italy, and Istituto Nazionale di Fisica Nucleare, Sez. di Milano, Italy

**Felix Gutmann** • School of Chemistry, Macquarie University, North Ryde, N.S.W. 2113, Australia

**M. A. Habib** • Department of Chemistry, Texas A & M University, College Station. Texas 77843

**Gerald C. Huth** • Institute for Physics, University of Southern California, 4676 Admiralty Way, Marina del Rey, California 90291

**Hendrik Keyzer** • Department of Chemistry and Biochemistry, California State University, 5151 State University Drive, Los Angeles, California 90032

**Shahed U. M. Khan** • Department of Chemistry, Duquesne University, Pittsburgh, Pennsylvania 15282

**Paavo K. J. Kinnunen** • Department of Medical Chemistry and Department of Chemistry, University of Helsinki, Siltavuorenpenger 10, SF-00170 Helsinki 17, Finland

**H. Kranck** • Laboratoire de Biophysique, Institut Polytechnique Méditerranéen, Université de Nice, Parc Valrose-06034, Nice, Cedex, France

**Gilbert N. Ling** • Department of Molecular Biology, Pennsylvania Hospital, Eight and Spruce Streets, Philadelphia, Pennsylvania 19107

**Marziale Milani** • Dipartimento di Fisica dell'Università, Via Celoria, 16-20133 Milano, Italy, and Istituto Nazionale di Fisica Nucleare, Sez. di Milano, Italy

**Eberhard Neumann** • Department of Physical and Biophysical Chemistry, University of Bielefeld, P.O. Box 8640, D-4800 Bielefeld 1, Federal Republic of Germany

**Ronald Pethig** • School of Electronic Engineering Science, University College of North Wales, Dean Street, Bangor, Gwynedd LL57 1UT, United Kingdom

**Herbert A. Pohl** • National Magnet Laboratory, Massachusetts Institute of Technology, Building NW-14, Cambridge, Massachusetts 02139

**J. Kent Pollock** • Pohl Cancer Research Laboratory, 515 Harned Avenue, Stillwater, Oklahoma 74075

**Aleksander T. Przybylski** • Institute for Molecular and Cellular Evolution, University of Miami, 521 Anastasia Avenue, Coral Gables, Florida 33134

***B. L. Reid*** • Queen Elizabeth II Research Institute for Mothers & Children, University of Sydney, Sydney, Australia

***Hiram Rivera*** • Physics Department, Oklahoma State University, Stillwater, Oklahoma 74078

***B. S. Thornton*** • Faculty of Mathematical and Computing Sciences, The New South Wales Institute of Technology, P.O. Box 123, Broadway, N.S.W. 2007, Australia

***D. Vasilescu*** • Laboratoire de Biophysique, Institut Polytechnique Méditerranéen, Université de Nice, Parc Valrose-06034, Nice, Cedex, France

***Jorma A. Virtanen*** • Department of Medical Chemistry and Department of Chemistry, University of Helsinki, Siltavuoren-penger 10, SF-00170 Helsinki 17, Finland

***Giuseppe Vitiello*** • Dipartimento di Fisica dell'Università, 84100-Salerno, Italy, and Istituto Nazionale di Fisica Nucleare, Sez. di Napoli, Italy

***S. J. Webb*** • The New South Wales Institute of Technology, Sydney, Australia

# Foreword

As stated by Buckminster Fuller in *Operation Manual for Spaceship Earth,* "Synergy is the behavior of whole systems unpredicted by separately observed behaviors of any of the system's separate parts". In a similar vein, one might define an intellectual synergy as "an improvement in our understanding of the behavior of a system unpredicted by separately acquired viewpoints of the activities of such a system". Such considerations underlie, and provide a motivation for, an interdisciplinary approach to the problem of unraveling the deeper mysteries of cellular metabolism and organization, and have led a number of pioneering spirits, many represented in the pages which follow, to consider biological systems from an electrochemical standpoint.

Now electrochemistry is itself, of course, an interdisciplinary branch of science, and there is no doubt that many were introduced to it via Bockris and Reddy's outstanding, wide-ranging and celebrated textbook *Modern Electrochemistry.* If I am to stick my neck out, and seek to define bioelectrochemistry, I would take it to refer to "the study of the mutual interactions of electrical fields and biological materials, including living systems". Its proper understanding therefore requires an up-to-date knowledge both of biochemistry and electrochemistry, and, both for experts in a single discipline and for the bioelectrochemical tyro, there is a great need for a work which will provide a background to the range of ideas which fall within its purview. Following the success of their earlier volume *Bioelectrochemistry,* Professors Gutmann and Keyzer have assembled here another outstanding range of experts and practitioners, with the aims of informing us of the background and present state of many areas of bioelectrochemistry, and of bridging the intellectual gaps which necessarily occur in such interdisciplinary endeavors.

Perhaps the most difficult problem in such a subject is to demystify the jargon of its sub-disciplines, so as to distinguish more accurately the known from the possible, and perhaps the most satisfactory means by which to do it is to be exposed to as wide a range as feasible of the relevant ideas. By the breadth and depth of its coverage, this multi-author work provides such an exposure, and if it serves to stimulate and clarify the design of new experiments and hypotheses, as I believe it surely will, then its aims will have been amply fulfilled. I am therefore delighted to have had the chance to be associated with this most exciting enterprise; what follows provides most stimulating and thought-provoking reading.

Douglas B. Kell
*Department of Botany and Microbiology*
*University College of Wales*
*Aberystwyth*
*Dyfed SY23 3DA*
*United Kingdom*

# Preface

The interdisciplinary nature of bioelectrochemistry is a root of its fascination but also a cause of many actual and potential controversies. Bioelectrochemistry forms more a part of biophysics than of biochemistry. The philosophy of the cell, or organelle, being considered as a "bag of enzymes", which presumes biological processes are on the whole governed by conventional chemical rate equations—whether in the bulk or on enzymatic sites—is assaulted in part at least by the notion that many biological processes are physical rather than chemical in essence. The stunning successes of classical biochemistry make it difficult for many of its ardent practitioners to accept ideas that, e.g., in biological energy transduction it is the physics of charge transfer which predominates, or that cellular communication involves radiative processes even if these are, most likely, associated with the microtrabecular lattices.

Bioelectrochemistry enters the scene because of the vectorial, directed flow of electric charges and charged particles and because of the critical, nay, determining, role played by electrified surfaces in the chemistry of so many biological reactions. These are electrode processes even if there are no metallic or even classically semiconducting electrodes present. Application of electrochemical ideas to biology presupposes the existence, in the biological system under consideration, of (1) an electron donating and an electron accepting surface or site, (2) an electronically conducting "external" pathway linking the electrodes, and (3) an ionically conducting "internal" pathway supplying the electro-active species to the electrodes.

However, the scope of electrochemistry includes the transfer of electrical energy by oscillatory behavior of molecules or molecular assemblies, i.e., by alternating current phenomena. In the limit, direct current and alternating current effects can become indistinguishable.

This book is concerned with the study of such processes. Apart from the first Conference Report on this field, a meeting in Pasadena in 1979, it is one of the first attempts to present the subject matter of bioelectrochemistry in an organized and coordinated fashion. It explains the several shortcomings of which we are only too well aware. However, in an extremely fast moving field of science, the desire for perfection and depth for detailed study of individual phenomena, for extensive reviews and literature surveys, is incompatible with the overriding need to maintain topicality and freshness of ideas. We welcome comment and criticism.

We wish to take this opportunity to record our indebtedness to the late *FREEMAN W. COPE* whose seminal contributions stimulated so much of what is now *Bioelectrochemistry*. Following some early work by Geissmann (*Quant. Rev. Biol.* **29**, 309 (1949)), it was Cope who first suggested that an enzyme particle could, in a way, act as an electrode.

We wish to thank the several authors of this volume for their cooperation. We greatly appreciate the encouragement and patience exhibited by the staff of Plenum Press, who at all times were most cooperative. We also thank Anne Wimberley for her unstinting help to get this book to the publisher.

Felix Gutmann
*School of Chemistry*
*Macquarie University*
*North Ryde,*
*New South Wales 2113*
*Australia*

Hendrik Keyzer
*Department of Chemistry & Biochemistry*
*California State University, Los Angeles*
*5151 State University Drive*
*Los Angeles, California 90032*

# Contents

Chapter 2.    The Origin of Cellular Electrical Potentials
                    *Gilbert N. Ling*

Chapter 3.    Electrodic Chemistry in Biology
                    *M. A. Habib and John O'M. Bockris*

Chapter 4.    Elementary Analysis of Chemical Electric Field Effects
              in Biological Macromolecules. I. Thermodynamic Foun-
              dations

*Eberhard Neumann*

## Chapter 20.   Muscular Contraction
### *Mohammad Amin*

## Chapter 21.   Transport in Plants
### *Mohammad Amin*

Chapter 22. Electrochemical Methods for the Prevention of Microbial Fouling

*H. P. Dhar*

# Fundamental Aspects of Electron Transfer at Interfaces

## Shahed U. M. Khan and John O'M. Bockris

*ABSTRACT:* The chemical and electrical implications of charge transfer are discussed. The basic differences between chemical and electrochemical reactions are highlighted. Electrochemical kinetics and its various aspects are treated in detail. Tunneling, electronic, and surface states are discussed in the context of interfaces. A current potential relation at semiconductor–solution interfaces receives attention, as do insulator–solution interfaces.

## 1. Introduction

An attempt will be made in the following pages to outline the subject of electron transfer at interfaces, on the assumption that the reader is a person familiar with general physical chemistry who has not as yet studied electrochemical kinetics.

Books on electrochemistry have been varied in the emphasis they place upon electrode kinetics. Those published before 1970 portray the subject as largely connected with the physical chemistry of ions in solution. In reality, the subject concerns the physical chemistry of ions in solutions, and the other processes which occur with charge transfer at interfaces between an ionic conductor and electronic conductors. Although the latter has been, traditionally, a metal,[1–3] studies are increasingly carried out in which the electronic conductor is a semiconductor and sometimes an insulator.[4–6]

There is a close relationship between kinetics of processes at electrodes (or, more generally, at interfaces between electron and ion conductors,

*Shahed U. M. Khan* • Department of Chemistry, Duquesne University, Pittsburgh, Pennsylvania 15282.    *John O'M. Bockris* • Department of Chemistry, Texas A & M University, College Station, Texas 77843.

respectively) and the subject of heterogeneous chemical kinetics, often dealt with under the name "catalysis." Correspondingly, part of the field of electrode kinetic is entitled "electrocatalysis." The principal thing to understand in the beginning of a study of electrode kinetics is that, whereas in the heterogeneous kinetics of reactions involving metals and gases, the principal mode of observation is to follow the rate of reaction (e.g., moles $m^{-2} s^{-1}$). The electrochemical analog is usually put in terms of a current density. This is a matter of convenience because the rate of an electrochemical reaction is often read with instruments which measure currents (amperes) and thus current densities ($A\ m^{-2}$), but it is easy to convert the meaning of the measure of amperes into number of electrons per second per unit area of electrode surface.

In chemical kinetics the foundation equation is the Arrhenius equation between the velocity of reaction $V$, and the temperature:

$$V = A \exp(-E_a/RT) \tag{1}$$

where $A$ is a number taken as independent of temperature and $E_a$ is the "energy of activation."

There is an analogous relationship in electrochemical chemistry which is of comparable status to that of the Arrhenius equation. It originated with some work by Julius Tafel in 1905, in which the so-called "overpotential"* of an electrochemical reaction and the log of the current density was related. Tafel wrote the so-called Tafel relation as

$$\eta = a \pm b \log i \tag{2}$$

where $a$ and $b$ are independent of the current density. The relationship can then be rewritten in the form

$$i = A' e^{\alpha \eta F/RT} \tag{3}$$

where $F$ is the Faraday constant (the charge upon one mole of electrons), $A'$ is a constant independent of the current density, and $\alpha$ is a constant, known as the transfer coefficient[2] and its value lies in the range $0 < \alpha < 1$. Physically $\alpha$ takes into account the fact that only a fraction of electrical potential or overpotential influences electrochemical reaction rate at the interface.

---

* The overpotential will be defined more completely later on. Its operational definition is the departure, for a certain defined rate of the electrochemical reaction, of the electrode potential from the value which it would have were the reaction to be at equilibrium at the surface concerned.

## 2. The Chemical and Electrical Implications of Charge Transfer at Interfaces

It has been stated that electrochemistry itself is divided into two branches, ionics (the physical chemistry of ions in solution) and electrodics (electron transfer at interfaces). In this article we intend to give a summary of central aspects of electron transfer at interfaces which will be helpful to biophysicists and biochemists who want to know something about the electrical properties of biological situations.

Firstly, it is quite helpful to point out the overlap in similarity between electronics (the study of electron transfer between semiconductor and vacua) and electrodics, the study of electron transfer between electron conducting materials and ionic conducting materials, e.g., a metal in solution. Electrodics is less well known because electronics, during the last half century, has become such a base to our technology, and enters into everyday life, e.g., television. Although electrodics enters into everyday life (for example, electron transfer between a metal and surrounding moisture film is the basic event in determining the stability of materials), it does not pop up, as it were, in advertisements in the supermarket, and therefore is less well known. The relevance of these remarks to the present chapter is that one can begin to understand electrodics better if one has a background (as many have) in electronics.

Another general thing to remember as we approach an understanding of electrochemical reactions is the electrical character of all interfaces. Thus, it is quite easy to show[9] that when interfaces between electron conductors and ion conductors are made, there must be a net electrical charge on the metal and an equal and opposite one on the solution phase (Figure 1).

This fact has a large implication for material science, because as material science deals predominantly with interfacial phenomena [i.e., the stability and material properties are controlled by interfaces (external or internal)], there is an electrical character about these happenings and thus they are subject to electrochemical science and electrochemical arguments. The main difference in an electrochemical (compared with a chemical) reaction, of course, is that an electronic charge transfer occurs in the electrochemical one. However, there are other differences which do not meet the eye. Electrochemical reactions always occur in two different locations. One cannot have an electrode operating in isolation in a solution. It always must be adjoined to another electrode, by an external circuit, in which electrons pass through a wire, and at this other electrode another electrochemical reaction takes place (Figure 2).

Thus, one of the differences between electrochemical and chemical reactions is that, in the first, the overall reaction takes place in a separated

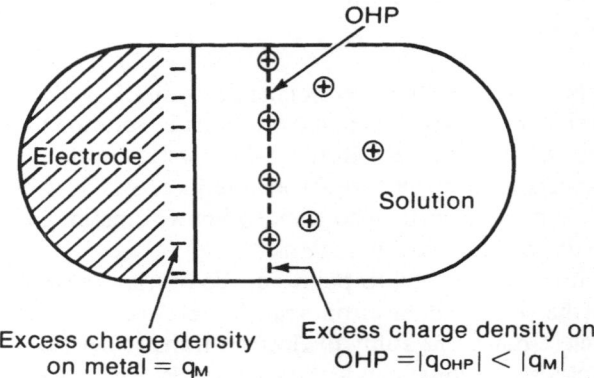

FIGURE 1. A situation where the excess-charge density on the OHP is smaller in magnitude than the charge on the metal. $|q_S| - |q_M| > |q_{OHP}|$. The remaining charge is distributed in the solution. The solvation sheaths of the ions and the water molecules on the electrode are not shown in the diagram. OHP, outer Helmholtz plane.

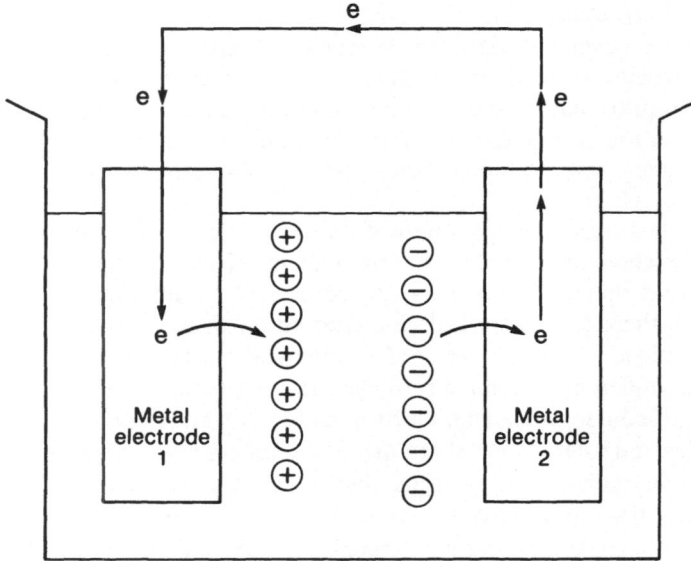

FIGURE 2. An electrochemical cell where it is shown that the electron passes through the externally connected wire.

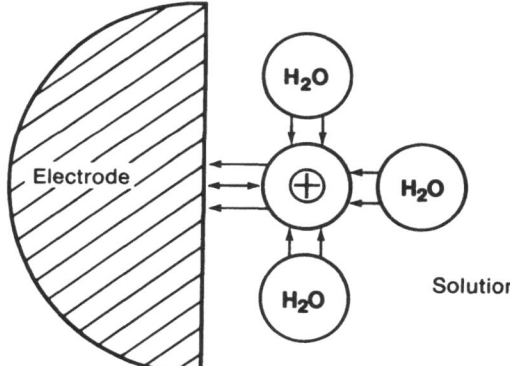

*FIGURE 3.* Collision of a reactant (e.g., a positive ion) with a negatively charged electron and the water molecules adjacent to the reacting ion.

way. If the overall reaction is the formation of hydrogen and oxygen, then at one electrode protons will undergo charge transfer to form hydrogen but at another electrode, connected at the first to the external circuit on the one hand and also through the outer conductor on the other, a different reaction will take place in which $OH^-$ ions will be discharged to form oxygen. Thus, electrochemical reactions on the whole do not take place at one spot but take place at two separated spots. Furthermore, electrochemical reactions do not take place (as chemical reactions do) by collision among the reactants (although such collisions may also occur). The essential thing is that, in an electrochemical reaction, one of the components will undergo charge transfer *with an electronic conductor*, and not necessarily with other reactants. In chemical reactions the principal point is the collision between the reactants. In electrochemical reactions the principal point is the collision between the electrode (the electron sink or source) and the reactants (Figure 3).

## 3. Energy Conversion: A Basic Difference between Chemical and Electrochemical Reactions

A fundamental difference, which has practical importance in biology, exists between the energy conversion aspects of chemical and electrochemical reactions.

In a chemical reaction energy conversion occurs by means of the production of heat. If we take, again, the chemical reaction of hydrogen and oxygen reaction in the gas phase, hydrogen collides with oxygen and a reaction occurs to form water, together with the evolution of heat, which arises from the difference in the potential energy of the hydrogen and oxygen molecules on the one hand and the water on the other. Heat is given out, i.e., there is less heat energy in the water than the sum of the heat energies in the hydrogen and the oxygen. Only a part of this heat energy is then converted to mechanical work (and perhaps also electrical energy, if that is desired), by means of the collision of these hotter product molecules with the head of a piston when the piston is moved, and work is done.

This type of energy conversion, which is called "the conversion of heat to work by a machine," is associated (as is well explained in all physical chemistry texts on thermodynamics) with the so-called Carnot efficiency factor. The *maximum* amount of work which can be converted is given by $(T_{high} - T_{low})/T_{high}$. The "high" and the "low" refer to the temperature at which the reaction begins and the temperature at which it ends. This figure often works out to somewhere in the region of 30% in practical machines such as steam engines.

Such a Carnot efficiency limitation is an important negative aspect of the normal methods which we use in energy conversion. The energy conversion which we are using at the present time (burning oil with air to give heat and conversion of this heat to mechanical power and electricity is outlined above) wastes roughly 2/3 of the total energy in a reaction by losing part of the energy as heat to the surroundings.

Electrochemical reactions can give energy conversion to electricity in a completely different way which is free of this Carnot limitation and thus of the energy loss referred to. In fact, theoretical maximum efficiencies in electrochemical reactions are in the 90% region and practical ones above 50%. Such energy conversion arises by means of a fuel cell, the principle of which is shown in Figure 4.

One of the important things to realize about electrochemical reactions—and that part of electrochemistry called electrodics (the part which deals with charge transfer at interfaces and which is the subject of this chapter)—is the wide sweep of phenomena for which such reactions have some kind of meaning. Thus, electrodics is an interdisciplinary subject which is not only in chemistry itself but also in metallurgy, engineering, biology, geology, and many other subaspects of these sciences. Very briefly to exemplify these statements we can take electro-organic syntheses as the example for chemistry. For metallurgy we have already mentioned corrosion as an important part of metallurgy, in which electron transfer at interfaces is the principal event. In engineering the applications of electrochemistry are very widespread indeed, but one which has obtained some attention from the public is the provision of power in space for the auxilliary functions in space vehicles. In biology, the conduction of electricity through nerves is a phenomenon in which the electrochemical components have been recognized for 80 years. In geology the applications are less well realized, but for example the properties of soils on which the

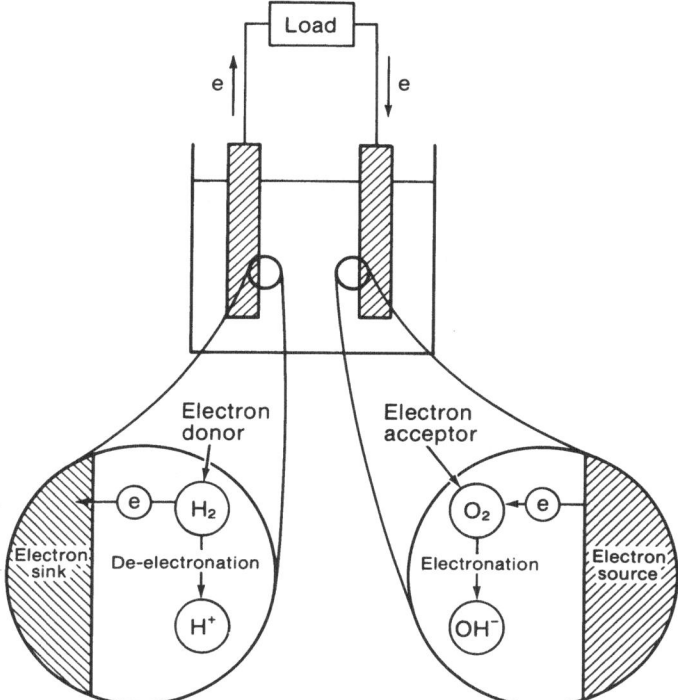

**FIGURE 4.** A schematic representation of the processes occurring in an electrochemical energy producer (fuel cell).

stability depends are dependent upon the thixotropic aspects of the soil, and this in turn depends upon the electrochemical double layers (the opposite layer of charges which occurs at interfaces between colloidal particles in the surrounding solutions), so that electrochemical considerations control thixotropy, and thus the flow of the soil under pressure. Shifting of the foundation of buildings may, therefore, be a phenomenon which requires some knowledge of electrochemistry for its understanding.

Lastly, it may be relevant to remark that consideration should be given to the idea that there are two chemistries. In the one (historically the first and the one always taught to beginners), one regards the basic mechanism of events as connected with reactions in the gas phase and based upon collisions of the reactants. This (i.e., collision between reactants) is the primary chemical event in nature, and regards all other phenomenon as derived from it.

However, another view can be taken, and that is that the electrochemical type of reaction—in which there is no collision between the reactants but the essential reaction is the electrical charge transfer between an electronic conductor (often a metal in solution) and an ionic conductor (ions in aqueous solvent)—is the prevalent reaction and reactions that take place in the gas phase are less common in nature.

Such a view is not yet prevalent. However, the prevalence of surfaces as reaction sites is gradually becoming understood and, in particular, in biology, it is clear that most reactions are interfacial.

Such thoughts belong to discussions in chemistry and are only briefly mentioned here to help develop a broader viewpoint.

## 4. Electrochemical Kinetics

### 4.1. The Equivalence of Current Density at an Interface and Reaction Rate

It has been mentioned that in heterogeneous kinetics (the province of chemistry) the typical reaction would be the catalytic combination of hydrogen and oxygen to form water on a platinum surface, and the way of measuring the rate would be to follow the change in pressure arising from the reaction, and to convert this by means of a simple phenomenological equation to a reaction rate in moles per unit area and time.

The corresponding measurement in electrochemical kinetics is the current, measured on an ammeter in the external circuit between two electrodes (Figure 4).

It is easy to relate this current density to the many familiar expression for the rate in terms of moles/cm$^2$/s. Let the current density be given by the symbol $i$ (current per unit area and time, e.g., $A/m^{-2}/s^{-1}$); then the number of coulombs which are passing in the time $t$ is given by definition as $i \cdot t$, where $i$ is in A cm$^{-2}$ and $t$ is in seconds. A constant called the Faraday (which has a value of 96,500 C/mole) is associated with every mole of charge so that if in an electrochemical reaction one passes $n$ electrons in one act of the overall reaction the number of coulombs flowing is $nF$, or $n \times 96,500$ coulombs per mole every time the reaction occurs once (in the molar sense).

If, therefore, we divide the number of coulombs which actually flow in the time $t$ by the number of electrons associated with one act of the overall reaction, we would have the number of times this reaction took place in the time $t$. If we make this $t$ one second we have

$$\frac{i}{nF} = v \tag{4}$$

where $v$ is the rate, i.e., the moles per unit area and time of the reaction.

To exemplify this, let us consider the hydrogen evolution reaction:

$$2H^+ + 2e \rightarrow H_2 \tag{5}$$

Then one act of the overall reaction is associated with *two* electrons, so that if the current density of this reaction is given by $i$, the moles $cm^{-2} s^{-1}$ for the evolution of hydrogen will be given by $i/2F$.

Thus, there is the simple phenomenological relation [that given in Eq. (4) above] for the relation of the electrochemical reaction rate given in moles $cm^{-2} s^{-1}$.

In electrochemistry we always deal in terms of current density when we want to express our rates. But it is easy, as seen, to convert to the more familiar chemical designation of moles $cm^{-2} s^{-1}$ for a heterogeneous reaction rate.

### 4.2. Two-Way Electron Transfer across an Interface

When one considers electron transfer at interfaces, one has to realize that there is always two-way traffic at any interface. In the one the electrons leave the electrode (or electronic conductor) to reach some entity in solution. In the other the reverse occurs: the electrons leave some entity in solution to reach the electronic conductor (Figure 5).

An easy way to illustrate this is in terms of the two ions of a redox couple, for example $Fe^{3+}$ and $Fe^{2+}$. When in contact with an electrode, the ion with the higher positive charge upon it (and therefore, compared with the atom, having lost a greater number of electrons), the ferric ion, tends to receive electrons from the electronic conductor and become a ferrous ion.

Such electron transport at the interface, where the electron leaves the electron conductor and goes to the solution, causes a *cathodic* electron

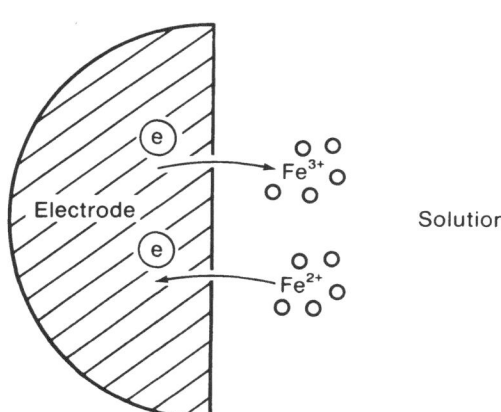

FIGURE 5. Two-way traffic of electron from the ion to the electrode and also from the electrode to the ion.

transfer, and the corresponding stream of electrons is called a *cathodic* current (Figure 6a).

Conversely, the ferrous ions tend to give up an electron to the electrode and become ferric ions. When the transfer of electrons is in this direction, from the ionic conductor to the electronic conductor, the flow of electrons occurring is called an *anodic* current (Figure 6b).

In some presentations in electrochemical books, the cathodic current is represented by the symbol $i$, on top of which is an arrow going to the right, and the anodic current by the symbol $i$ with an arrow going to the left. Thus, what one measures on the ammeter, outside the actual electrochemical system in which this two-way track is occurring, is a *net* current, the difference of the rate of electron transfer in one direction and the opposite direction.

Thus, that which is measured in the outer circuit by the ammeter can be itself net cathodic or net anodic current. In the first case what is meant is that the rate of electrons leaving the electrode for the solution is greater than the rate of electrons leaving the solution for the electrode; but in the case of a net anodic current, the electrons leaving the solution for the electrode are greater in rate than the electrons leaving the electrode for the solution.

One could, therefore, write the following equation:

$$i_{\text{net,cathodic}} = \vec{i}_c - \vec{i}_a \tag{6}$$

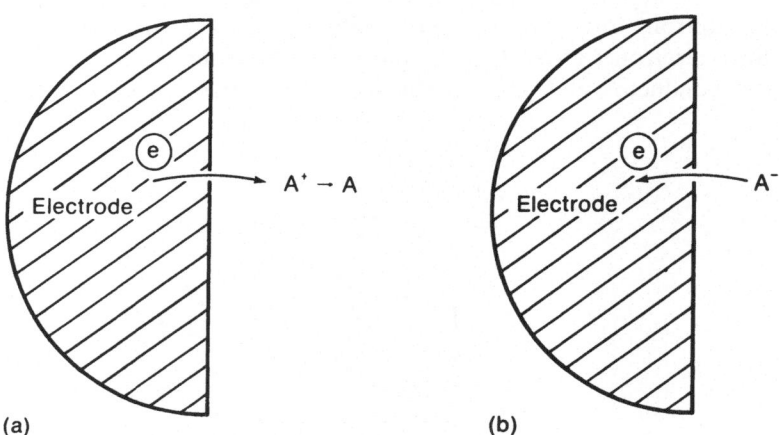

(a)                                                    (b)

*FIGURE 6.* (a) Net electron transfer to an ion $A^+$ to form A (i.e., a net cathodic electron transfer reaction). (b) Net electron transfer from an ion to the electrode (i.e., a net anodic electron transfer).

On the other hand

$$i_{\text{net anodic}} = \vec{i}_a - \vec{i}_c \tag{7}$$

Under these circumstances there must obviously be a situation where the net current is zero, and one can see that this means that the cathodic partial current and the anodic partial current are then equal, whereupon one has the condition of *equilibrium*, i.e.,

$$\vec{i}_c = \vec{i}_a \tag{8}$$

Many workers have in their mind equilibrium as a quiescent state, but in fact both in electrochemistry and of course also in ordinary chemical equilibria it is a dynamic state in which the reaction rate is equal and opposite in each direction.

A special term is given in electrochemistry to the current density which exists in equal and opposite degree at equilibrium. The term used is "the exchange current density," and the symbol $i_0$ is given to such a quantity. Thus, if one says: "The exchange current density for the deposition of protons on platinum to form hydrogen molecules is $3 \times 10^{-3}$ A cm$^{-2}$ at 25°C, it means that, when the hydrogen evolution reaction is at equilibrium on platinum, it is really occurring to an equal and opposite extent in both directions (the one forming hydrogen from protons, and the other dissolving hydrogen molecules to form protons (Figure 5). These exchange current densities are common parlance among electrochemists and they express, in fact, the rates of an electrode reaction at equilibrium. From such quantities, as will be seen (cf. the Butler–Volmer equation below) the rate at any potential can be calculated.

### 4.3. The Butler–Volmer Equation: The Rate of an Electrochemical Reaction at a Given Degree of Displacement from Equilibrium

In order to understand the variation of the electrochemical reaction rate (measured in current density) with the potential of the electrode, it is best to start by thinking of the situation at equilibrium. As explained above there is a certain potential (referred to as the thermodynamic reversible potential) at which the two reaction rates, the anodic and the cathodic, are equal in magnitude and opposite in direction.

It stands to reason, then, that if one biases the electrode in the cathodic direction (this would mean making the surface charge more negative), electron transfer towards the solution would be favored, and correspondingly electron transfer from the solution will be inhibited. As the two reaction rates have been equal at the reversible potential, when one makes the electrode potential more negative than the reversible potential, the cathodic partial current will exceed the anodic partial current and there will be a net flow of electrons from the electrode to the solution. The current will then be called "cathodic," and the word "net" will be neglected (Figure 6a).

Correspondingly, if one biases the current more in the positive direction (increases the positive charges by the net decess of the electron charge), then the transfer of electrons from the solution to the electrode will be encouraged and the transfer of electrons from the electrode to the solution will be inhibited, so that there will be a net anodic current, usually simply called "an anodic current" (Figure 6b).

It goes without saying, then, that these net anodic currents and net cathodic currents must be equal and opposite at the reversible equilibrium potential, when the net current must be zero, although there will still be a partial anodic and partial cathodic current equal in magnitude but opposite in direction and equals to the exchange current density.

This biased voltage which we have talked about, that is the *amount* one biases the potential of the electrode away from the value corresponding to the reversible to the thermodynamic potential, and the value which corresponds to any given net cathodic current— or any given net anodic current—has a name. This name is, logically, *overpotential*, the "over" referring to the fact that there has been a displacement compared with the value for the reversible thermodynamic potential.

This overpotential is an important quantity in electrochemistry—one of the central ones; and it should be understood. A historical picture of how the idea of overpotential developed and what its status is today has been given by Bockris.[8]

To deduce the mathematical relationship between the electrochemical reaction rate (cathodic or anodic) and the overpotential is beyond the scope of our present chapter and is given in books such as that by Bockris and Reddy.[9] Nevertheless, arguments can be presented which make the relationship which we shall present acceptable. Thus, if the current density at the equilibrium in the cathodic direction (electrons to the solution from the electrode) is $i_0$, then as we depart from the reversible potential by an amount $n_c$ (the cathodic overpotential) there should be a change in the reaction rate which will be related to exponential function of the change in the energy of activation of the reaction (cf. the Arrhenius equation, rate = $Ae^{-E_a/kT}$) and the analogous Tafel relation, $i = Ae^{-\beta F\eta_c/kT}$.

If the electrode potential is changed by $\eta_c$, then there will be a corresponding change in the energy of activation by an amount which should be proportional to $\beta F\eta_c$. Thus, in the theory of electricity, the energy which corresponds to potential $\eta$ is that energy times the charge associated with it. If we are considering, for a very simple electrode reaction, one electronic charge, then the energy associated with $\eta$ (per mole of electrons) is $\eta F$.

If it is true that the energy of activation is decreased by the (understandably, negative) value of $\eta_c$ for the cathodic reaction, then one must ask oneself whether one simply subtracts the value of $\eta$ from the energy of activation, i.e., whether the whole of this electrical energy is applied to reducing the energy of activation. This is where a more complex part of the argument comes in, and it will be stated at this time that the amount one does apply is somewhere in the region of about half the value $\eta F$. This fraction is given in electrochemical books by $\alpha$, with bears the name *transfer coefficient*. The value of $\alpha$ for more complex reactions is often not a half, but it is always between 0 and 1.

Thus, for the partial cathodic current, with an overpotential of $\eta_c$ the value will be

$$\vec{i} = i_0 e^{-\alpha \eta F/RT} \tag{9}$$

At the same time, of course, something corresponding will happen to the anodic current when we change the potential of the electrode from the reversible potential where the cathodic current and anodic current were both equal to each other though opposite in direction. It can be shown that the factor which applies to the *anodic* current, and by which the value in one direction at equilibrium, $i_0$, should be multiplied, to obtain the value of the current at an overpotential, $\eta_a$, is $(1 - \alpha) \eta_a F$.

Thus, the partial anodic curent for an overpotential $\eta_a$ is given by

$$\vec{i} = i_0 e^{(1 - \alpha) \eta_a F/kT} \tag{10}$$

Then, if the current is cathodic, one will write

$$i_{\text{net cathodic}} = i_0 (e^{-\alpha \eta_c F/RT} - e^{(1 - \alpha) \eta_c F/RT}) \tag{11}$$

Correspondingly, if the current is net anodic one will write

$$i_{\text{net anodic}} = i_0 e^{(1 - \alpha) \eta_a F/RT} - i_0 e^{-\alpha \eta_a F/RT} \tag{12}$$

These equations are sample forms of the so-called Butler–Volmer equation, and this is usually written

$$i = i_0 (e^{(1 - \alpha) \eta_a F/RT} - e^{-\alpha \eta_c RT}) \tag{13}$$

where $\eta$ may be either cathodic, i.e., $\eta_c$, or anodic, $\eta_a$. In electrochemistry, two more equations remain to argue out before we leave the presentation of it. When one does the arithmetic of working out the numerical values of $\alpha \eta F/RT$ and $(1 - \alpha) \eta F/RT$, then using the "halfway" value of $\alpha$, i.e., $\alpha = 1/2$

(because that is an often, but by no means always, observed value of this quantity), one finds that for $\eta = 0.05$ V, the value of the exponential is about one. From the well-known properties of exponential series, one can then expand the exponential as $\alpha\eta F/RT$, or correspondingly, $(1-\alpha)\eta F/RT$ so long as the exponential is "much less than" one. If we take this as "much less than" to mean one, when the overpotential is less than around 0.025 Volt, there can be a linear expansion of the Butler–Volmer equation, and simple algebra shows that one obtains thereby

$$(i)_{\eta < 0.025 \text{ Volt}} = i_0 \eta F/RT \tag{14}$$

Of course, the precise value of $\eta$ at which one is allowed to assume that the values of $i$ and $\eta$ are linear in this way will vary upon the details—for example, the temperature, the value of $\alpha$, and so forth—but there will be a lower value of $\eta$, in the region of tens of millivolts (mV), in which linearity between current and potential will occur.

The other side of the coin to this argument is that when the value of $\eta$ is sufficiently large so that one of the two exponential terms becomes at least ten times more than the other, it is reasonable, as an approximation, to neglect the exponential term.

In this way one gets an equation for the relation of $i$ and $\eta$ similar to the empirical equation of Tafel (cf. Section 1) because, as can be seen,

$$i = i_0 e^{(1-\alpha)\eta_a F/RT} \tag{15}$$

for the anodic reaction and correspondingly

$$i = i_0 e^{-\alpha\eta_c F/RT} \tag{16}$$

for the cathodic reaction.

If one compares these equations with the equation which has been written above as an analog of the Arrhenius equation in normal chemical kinetics, one can see that they do have the same form (for $\eta F$ is an energy), i.e., the exponential form of the Tafel equation is a special case of the Butler–Volmer equation in which one of the exponentials can be neglected compared with the other.

## 4.4. The Measurement of Potential in Electrochemical Reactions

We have already referred above to the measurement of current density in electrochemical reactions. This is a very simple measurement so long as one does not want to do it at very low times, e.g., in the microsecond range, but under the normal conditions of ordinary steady state measurements where the time may be seconds or longer, a simple measurement in some kind of ammeter which is external for the circuit will do.

The measurement of potential, and correspondingly overpotential, is, however, by no means simple, and in fact it involves quite special reasoning, and even conventions, which would herewith be presented.

First of all, it must be understood that one of the weaknesses of the subject of electrochemistry is that there is no experimental method available (or even possible) by which the absolute solution electrode p.d. can be measured. This may seem surprising, considering that electrochemists talk about potential as the center of their arguments, but it is true and not controversial, and the reasoning and details concerning the why and the wherefore can be read in a text such as that of Bockris and Reddy.[2]

Thus, one has to make a relative measurement of electrode potential, and the way one does this, in principle, is shown in Figure 7. The typical electrochemical cell consists of an electrode, which contains the interface at which the studies are to be made (i.e., the working electrode). The counterelectrode is the other electrode in which electrons can enter or leave the circuit and is generally unpictured in the experimental procedure; but there is also what is called a "reference electrode," the position of which is seen in the diagram.

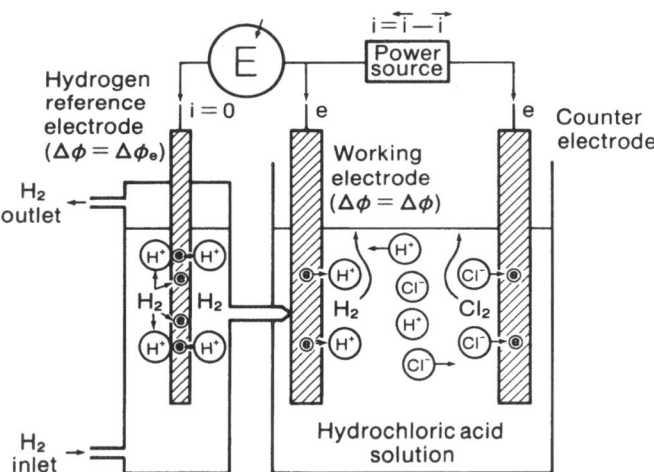

FIGURE 7. The three-electrode system required to measure electrode overpotential, i.e., $\Delta\phi - \Delta\phi_e$. The potential between the working electrode and the reference electrode when both $\Delta\phi$ and $\Delta\phi_e$ correspond to the same reaction is equal to the overpotential $\eta$. The tube joining reference electrode and working electrode is called a Luggin capillary. It helps diminish the inclusion of the illicit $IR$ drop in the measurement.

The essence of this reference electrode is that it presents a well-known and stable potential, in comparison with which the potential of the working electrode can be measured. Thus, suppose one has a reference electrode which is some arbitrary standard set up so that the conditions which lead to its reproduction become easy. In practice, two reference electrodes are chosen. The first and more fundamental is the hydrogen electrode, which is the electrode produced by bubbling hydrogen gas at one atmosphere over a highly catalytically active surface (which is in the solution of proton, $H^+$ of activity unity) for the reaction of hydrogen to equilibrate protons in solution and electrons (platinized platinum is often used) with a hydrogen gas. The potential of this electrode (it could be used as a reference) is arbitrarily taken to be zero volt.

All scales in electrochemistry are then related to this particular scale with the hydrogen electrode as zero. Another electrode which is often used as a reference electrode is the calomel electrode, which is derived by placing a pool of mercury in contact with a calomel, $Hg_2Cl_2$, in powder form, both of them being in contact with a KCl solution of a certain concentration.

If the solution is saturated with respect to KCl, the potential of this calomel electrode with respect to a hydrogen electrode is 0.242 V.

Hence, if one wants to measure the relative potential of an electrode which is passing current (for example while it is evolving hydrogen) at a certain rate $i$ (this will be on the working electrode itself, Figure 7), one then places a potential-measuring device between the working electrode and the reference electrode. This potential measuring device must have a special characteristic: it must pass only an extremely small current, usually less than $10^{-12}$ A, so that there is no question of passing significant amounts of current across the reversible electrode; otherwise it would no longer be reversible, i.e., no longer be at equilibrium. If one measures the potential of a working electrode with respect to reference electrode on the hydrogen scale, then one has to calculate what is the reversible potential of the electrode reaction concerned.

To take an example, suppose one is measuring the electrode deposition of copper, the reversible (i.e., the situation in thermodynamic equilibrium) potential for a copper ions in solution is found by experiment to be 0.337 V.

What this statement means is that if one had a piece of copper in a solution in which the cupric ion activity was unity, and one connected this electrode by means of an electronic conductor (or wire) on the one end to a hydrogen electrode in its reversible standard condition ($p_{H_2} = 1$ atm and $H^+ = 1$), then the potential measured by a voltmeter using virtually zero current placed in the electronic circuit outside would be 0.337 V. Of course, the situation would have to be similar to that shown in Figure 7, i.e., there would have to be an ionic connection between two solutions.

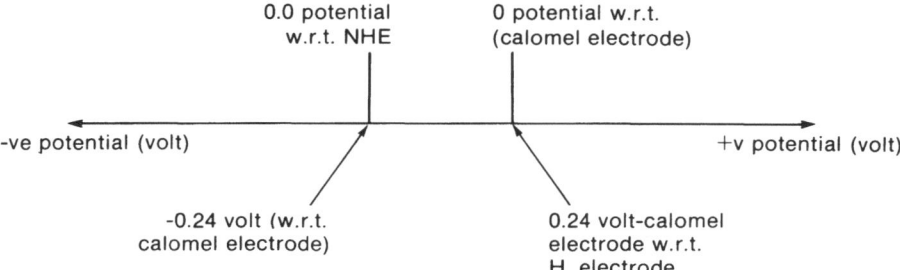

FIGURE 8. The method of calculation of electrode potential with respect to a reference electrode. The figure shows one with respect to a reference normal hydrogen electrode (NHE).

What would the overpotential be for a copper deposition if one made the current density more cathodic and departed from the reversible condition by, say, $-0.2$ V (i.e., $\eta_c = -0.2$ V)? By definition, the latter quantity would be the overpotential, but what the copper electrode would read with respect to a hydrogen reference electrode would be $+0.137$ because the reversible potential of a copper electrode is $+0.337$ on the hydrogen scale and by giving 0.2 V more negative one has made the potential now $(0.337 - 0.2$ V) or $+0.137$ V.

At first, the calculations of overpotentials are tricky because of the signs, but it is possible to make a diagram which is shown in Figure 8 and to calculate the overpotential in that way.*

### 4.5. The Electrical Control of Charge Transfer Reactions

From the above presentations, it is clear that electrochemical reactions have a unique feature which is lacking in chemical reactions. In the laboratory the velocity of an electrochemical reaction can be changed by orders of magnitude simply by changing the setting on the potential which changes the rate of the reaction concerned (cf. the Butler–Volmer equation).

To illustrate this, suppose that one has an electrode potential for which the equilibrium potential is 0.377 V on the hydrogen scale, and applies an overpotential of $-0.5$ V to the electrode, then the increase in the

---

* The potential on an electrode can also be expressed in a different scale, used by some theoretical electrochemists. This is the so-called vacuum scale of electrode potentials (cf. Bockris and Argade).[28] Thus, it has been shown by Trassatti[29] and others[5] that if one adds 4.5 V to the potential on the hydrogen scale already described, then one would obtain the value of the potential for the transfer of electrons from a vacuum level to the oxidized ion in solution to form the reduced ion of the overall reaction.

rate of reaction would be about $e^{-\beta\eta F/RT} = e^{[(-0.5)F/RT]/2} =$ $e^{-[(0.5)96,500/8.315 \times 298]/2}$, where 8.315 is the value of the gas constant, $R$, in joules and $T = 298$ K represents the absolute temperature of 25°C.

Computation shows that the value of the exponential stated here [cf. the simple form of the Butler–Volmer equation for overpotentials greater than about 0.5 V] is $10^{4.16}$, i.e., the change of overpotential by 0.5 V has increased the value of the rate by some four orders of magnitude. Large changes of this kind are, therefore, very simply induced by change of the electrodic potential, using an outside instrument. An important limitation to a calculation of this kind of higher overpotential is the so-called limiting currents which occur when the rate of electron flow to or from the inter-face—or equally the rate of transport of ions to and from the inter-face—cannot keep up with the demands for electrons at the interface. Then, as will be shown in the next section, the Butler–Volmer equation is no longer simply applicable and calculation of the above kind no longer has any simple meaning.

One of the first things one has to do in any electrode kinetic calculation, therefore, is to find out where the limiting currents occur. If the region in which the electrochemical reaction is being carried out is, say, an order of magnitude less than the limiting current, use of the Butler–Volmer equation will be a reasonable approximation. An accurate one at, say, two orders of magnitude less than the limiting current is the Butler–Volmer equation and often in the simple exponential form [Eqs. (13) and (14)]. The situation relating to the limiting current will be dealt with in the next section.

Limiting currents are important in bioelectrochemistry because bioelectrochemistry deals with electronic conduction in what may be con-sidered poor semiconductors or even insulators. Under these situations the current will often be controlled by the limiting rate of electron flow to the interface.

### 4.6. Transport Control at the Interface

The simplest aspect of transport control is that in which the control arises from the diffusion of ions in solution to the electrode. Let us consider a cathodic reaction, in which electrons are emitted from the electrode to the solution. It is clear that we must here supply ions to the interface.

In the conditions which are assumed to exist in this chapter so far, the supply of ions is plentiful, i.e., the claims of the electrode for electron trans-fer to ions are always easily met by diffusion of ions from the solution. When one part of a consecutive series of reactions is "easy" it is generally said to be in quasiequilibrium. For example, one might think of 100 ions arriving at a small patch on the electrode, 99% of them going back again, and one being used in charge transfer reaction. This would be when the situation is "easy" and diffusion under these circumstances is not rate controlling.

At the other end of the spectrum of events, when the demand by electrons is greater than can be easily supplied by diffusion, so that each ion that reaches the electrode is at once used up by the waiting electrons, the control of events will pass to transport ions in the solution and this will be no longer governed by the Butler–Volmer equation.

The reason why so much attention is devoted to the Butler–Volmer equation and to the exponential dependence of current on potential is interesting to note. First of all, many electrochemical experiments, particularly the fundamental ones, are carried out in the region where the influence of transport control from the solution side is purposely avoided. One simply calculates what the limiting current will be (see below) and then makes one's experiments in the situation when the currents examined are much less than the limiting current, so that the Butler–Volmer equation is applicable.

Sometimes this is not practical, particularly in industrial practices, where the highest possible rate of production of material is needed, and therefore industrial plants tend to work near the limiting current region. Conversely, in biological situations, it may be that, as mentioned, charge transport within the "insulator" protein will control the rate of the reaction.

Before assaying the relation between current and potential under these situations (i.e., the rate–potential relation under diffusion control), it is worthwhile deducing an expression for the limiting current when diffusion is coming from the solution.

Often the transport in solution is controlled by complex factors. It may involve turbulence or some kind of convection process. However, we shall adopt the simplifying assumption that it is only *diffusion* (e.g., not convective transport, i.e., no stirring or natural convection due to density differences near the electrode).

Hence, with diffusion control, the transport flux is given by

$$\frac{i}{nF} = -D \left( \frac{dc}{dx} \right)_{x=0} \tag{17}$$

where $D$ is the diffusion coefficient of ions in solution. The distance $x = 0$ refers to the distance at which the concentration gradient, $dc/dx$, is measured from the electrode, where $c$ is the concentration of ion in the electrolyte near the electrode. Thus, the equation tells us that if the concentration gradient near to the electrode surface is $dc/dx$, then, under diffusion control, the rate of the reaction, $i/nF$ is proportional to the diffusion gradient.

A more practical version of this equation can be given by assuming that the concentration gradient near the surface is linear. This is not an accurate assumption, but it is blessed by history and is always used. In fact, the model here, which was originated by the famous German physical chemist Nernst, is artificial in the sense that it linearizes the current density.

If one lets the concentration of the simple ion in the bulk solution be $C^0$ and the concentration of this ion just outside the electrode be $C_{x=0}$, then in the linearized model

$$\frac{i}{nF} = -D\frac{C_0 - C_{x=0}}{\delta} \tag{18}$$

Here the $\delta$ (cf. Figure 10) has been used to represent "thickness of the diffusion layer," and the actual limiting current is then obtained by putting $C_{x=0}$ equal to 0, for it can be seen that then current under diffusion will increase as $C_{x=0}$ is reduced but cannot increase further when $C_{x=0}$ become equal to zero. Under these conditions, one gets the limiting current, $i_L$, as

$$i_L = \frac{-DnFC_0}{\delta} \tag{19}$$

We can go at once to a simplifying calculation to get the value of a limiting current. Suppose we do it for the case of oxygen.

The solubility of oxygen in water around 25°C is some $2 \times 10^{-3}$ moles/liter. The unit of $C_0$ used in the above equation is in moles/cm³, so that one needs the value of $2 \times 10^{-6}$ moles/cm³. The diffusion coefficient of most gases in solution is around $10^{-5}$ cm² s$^{-1}$. The value of $\delta$ is often obtained (for unstirred solutions) by experience as 0.05 cm.

On this basis one has (for $n = 2$ and $F = 10^5$ C/mole)

$$i_L = \frac{10^{-5} \times 2 \times 10^5 \times 2 \times 10^{-6}}{5 \times 10^{-2}} = 8 \times 10^{-5} \text{ A/cm}^2 \tag{20}$$

It may be observed that one has implicitly assumed the value for partial pressure of oxygen, i.e., $p_{O_2} = 1$, and it is much more likely that the oxygen present in most solutions will correspond to that in air or about 0.2 atm. If this is so, the limiting diffusion current for oxygen diffusing to an interface should be reduced by one-fifth.

This way of finding out what is the limiting current for the access of oxygen to a surface [Eq. (17)] should then not be made in a turbulent condition. The theory of the situation under these conditions is complex but an empirical device is available for very rough order-of-magnitude calculations under stirred conditions. Under stirred condition no exact values of $\delta$ can be given because of the difficulty of defining stirring rate (except for rotating disk electrodes, for which equations do exist). For "very stirred" conditions a value of $\delta$ having an order of magnitude of $10^{-3}$ cm might be used and for extremely turbulent conditions several times less than this value could be used.

### 4.7. Potential–Current Relation under Transport Control

We have considered above the Butler–Volmer equation for the relationship between current density and potential under the situation when transport of ions in solution makes little or no difference to the rate of an electrode reaction. In order to considered the situation in which transport does control the flow we shall adopt a correspondingly simple counterassumption: electron transfer at the interface no longer has control of the electrode reaction.

The overpotential caused by electron transfer is assumed to be limitingly low. Under these conditions, the relationship between potential and current density is given by considering it to consist of the shift in the equilibrium potential due to a change in the concentration of reactants in solution away from that at equilibrium to that caused by the holdup in transport control.[2]

The fact that the situation is not quite like this in reality is not very important at the moment in order for us to get a basic relationship. Further on in the chapter we shall state, without deduction, the full relationship which takes into account both transport control and activation control.

Thinking now of the simplification to which we have just referred, we have

$$\eta_{\text{transport}} = \frac{RT}{nF} \ln \frac{C_{x=0}}{C_0} \tag{21}$$

where $\eta_{\text{transport}}$ is the overpotential due to transport. Now the concentration in this equation can be related in the case of a cathodic reaction by the equations already deduced in the consideration of what the limiting current is.

We saw that

$$i = -DnF \frac{C_0 - C_{x=0}}{\delta} \tag{22}$$

Trivial algebra transforms this to

$$\frac{C_{x=0}}{C_0} = \frac{i_L - i}{i_L} \tag{23}$$

where the value for the $i$ already deduced above has been taken into account.

Substituting this value of concentration from Eq. (21) into the expression for the overpotential in Eq. (19) one gets

$$\eta_{transport} = \frac{RT}{nF} \ln \frac{i_L - i}{i_L} \tag{24}$$

This last equation would be written in the form

$$i = i_L(1 - e^{nF\eta/RT}) \tag{25}$$

This equation gives a relation between the current density (or rate of electrode reactions) as a function of the overpotential, $\eta$, under the limiting condition when electron transfer at the interface is "easy" (nearly at equilibrium in fact) and the major difficulty in making the reaction go is diffusion to the interface from the solution.

That the reality is different (i.e., some overpotential due to electron transfer difficulty will exist) is important to realize, but an equation is easy to deduce which takes into account both kinds of overpotential which exist for electron transfer and other rate-determining reactions occurring on the electrode, and that which exists due to the transport of entities from the solution to the electrode. It is

$$i = \frac{i_0(e^{(1-\alpha)F/RT} - e^{-\alpha\eta F/RT})}{1 - (i_0/i_L) e^{-\alpha\eta F/RT}} \tag{26}$$

Some algebraic manipulation of this equation shows that it does indeed give the two limiting forms Eqs. (13) and (25) which one would expect. If the $i_L$ is much smaller than $i_0$, i.e., the transport dominates, then one reobtains the equation deduced in this section. On the other hand, if $i_0$ is very small compared with $i_L$, the Butler–Volmer equation is reobtained.

### 4.8. The General Relationship between Current and Potential at an Interface

It is easy to show that the general relationship always has the same shape, and this is a so-called S-shaped curve (Figure 9).

All electrode reactions manifest this type of shape. Somewhat like the situation with the examination of the proverbial elephant, the situation may seem different depending upon the point at which one touches reality. If for example, one measures an extremely low current density very near the reversible potential, it may seem that the relationship between current density and potential is a linear one, but in the large range of potentials one may obtain a logarithmic relationship between the current density and

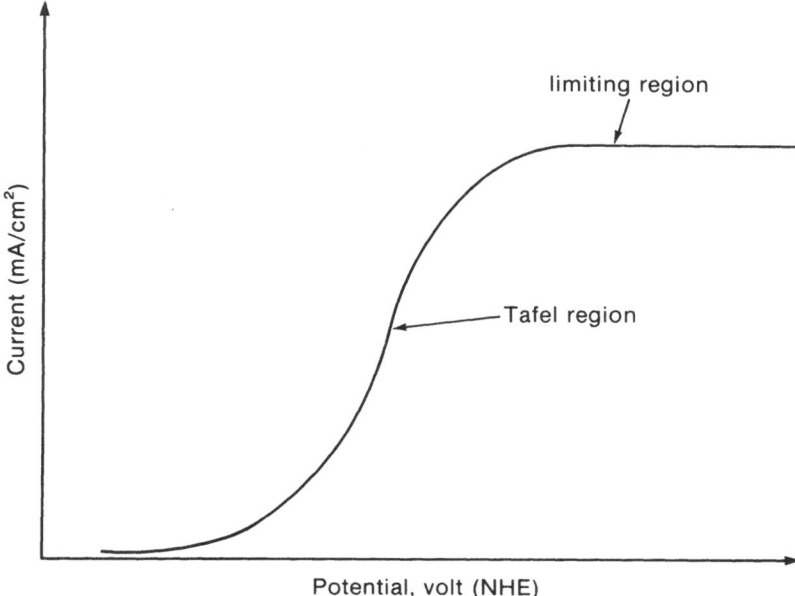

FIGURE 9. S-shape current potential dependence.

the overpotential, similar to that of the Tafel relationship. However, if one is in a situation where limiting transport factors come into control, then electrocatalysis and interfacial phenomena will be less important, and it will be diffusion, convection, or for semiconductor solution interfaces, various ways to create carriers inside the electrode, that will dominate the situation.

## 5. Phenomena Connected with Tunneling at an Interface

### 5.1. The Interfacial Barrier and Its Penetration

The first ideas about this were given by R. Gurney,[9] whose best-known work resides in a paper given early after the beginning of the quantum theory pertaining to the escape of radioactive particle from nuclei, i.e., the beginning of the quantum theory of radioactivity.

Gurney applied the thinking which he had made for the radioactive case to electrodes and considered the mechanism of electron escape, the cathodic case. Taking the clue from the situation within the nucleus, Gurney regarded the metal itself as a kind of box in which electrons were contained, and the solution as a sink for these electrons, starting some distance away from the surface of the metal (e.g., about 5 Å). The interphasial region between the metal and the solution Gurney regarded as containing a "potential energy barrier" which for reasons for simplification he regarded as a square in shape (Figure 10).

The origin of this barrier is mainly due to image potential and field drop at the interface. However, most readers will be familiar with the general idea of "an activation barrier" in chemical reactions. There, of course, is the energy of activation which controls the rate of electrode reactions and has appeared in this chapter already in the $E_a$ of the Arrhenius equation[10] quoted above. If this barrier is high, the reaction is slow because the value of the exponential function in the Arrhenius equation is small. Fast reactions are associated, then, with small energies of activation.

The physical origin of energies of activation has to be discussed individually in the case of every situation. In general, however, it can be stated that when the reactants arrive at the interface for reaction, most of them are not in an energetic state appropriate to electron transfer. Such a state, which involves rearrangement of the atoms within the molecule, and perhaps some stretching of bonds, is only attained if one waits a certain time (of the order of magnitude of $10^{-12}$–$10^{-13}$ s) while some of the molecules get into a certain condition, experience activation into a state suitable for the receipt of electrons.

Gurney then invoked the quantum mechanical phenomena of tun-

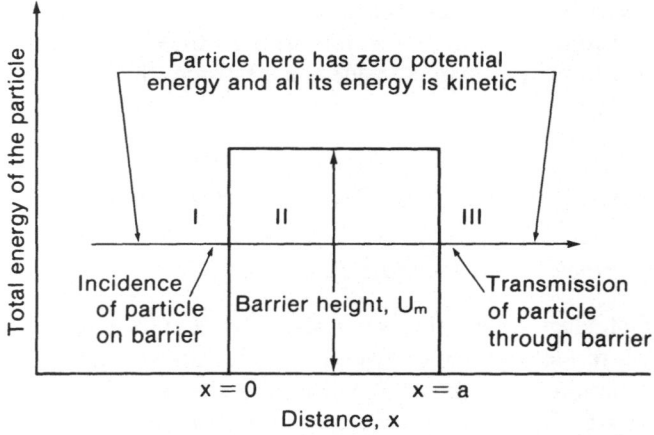

FIGURE 10. A schematic diagram of a rectangular potential energy barrier for a tunneling particle. In regions I and III, the particle has total energy equal to kinetic energy. In region II, the particle experiences a potential energy barrier of height $U_m$ and width $a$. Within the barrier, the potential energy is constant.

neling. This phenomenon allows the penetration of a barrier by a particle. Books on quantum mechanics[12,13] should be consulted for an explanation as to how this occurs, but very briefly, the phenomenon rests upon the wave character of an electron whereby when an electron strikes the barrier, its wavelike properties allow there to be a certain probability (certain density of electrons) that it will transfer through the barrier without going over it as it would have to behave as a classical particle with no wave properties.

An equation exists which gives the probability, with many simplifying assumptions, for the penetration of this barrier. This is called "the Gamow equation."

For a square-shaped barrier the probability of the penetration of the barrier can be expressed as[12,13]

$$p_T = \exp\left\{-\frac{4\pi l}{h}[2m(U-E)]^{1/2}\right\} \tag{27}$$

In this equation $E$ is the total energy of the particle including both kinetic and potential energy and $U$ is the height of the barrier potential energy at the interface, $l$ is the width of the barrier (Figure 10), $m$ is the mass of the tunnelling particle, and $h$ is Planck's constant.

Apart from the energy factors, it can be seen that this equation is much dependent upon the value of the mass of the particle, $m$. It turns out upon numerical calculation that the only particles that tunnel through the barriers with any realistic probability are the electron and the proton. The electron transfers through barriers rather easily and in electrode processes the transfer of electrons and holes through barriers is regarded by some authors as occurring with a probability of one, though this assumption is not always applicable. Sometimes in electrode kinetics, protons transfer through barriers, and then their probability of transfer is very much lower than one. Even with this rough Gamow approximation, the probability is also dependent upon the values of $E$, $U$, and $l$. Thin barriers (thickness perhaps 1 Å) are much more penetrable than a thick barrier ($l$ approximately $= 10$ Å). Barriers that are low $(U-E) < 1$ eV are much easier to penetrate than barriers that are high $(U-E) > 1$ eV.

Gurney[9] found out a condition whereby nonradiative tunneling would occur. If there is to be no radiation, it is clear that the electron must arrive in the same energy state on the other side of the barrier (in a cathodic reaction this would be inside a molecule in solution), perhaps an $H_3O^+$ ion, which is then transformed to a hydrogen atom by accepting the tunneled electron. The Gurney condition of radiationless tunnel transition is

$$E_{e,\text{ in electrode}} = E_{e,\text{ in particle solution}} \tag{28}$$

Gurney's idea was that the tunneling was possible only when this condition (28) is satisfied.

Electrode kinetic theory, including some quantum theory, is worked out on this basis.

However, the most modern view is that this equation is applicable only to a degree. The details of this have been worked out[12] and it has been found that, roughly speaking, it applies well for electrons, but for protons other equations are better used. In particular a different attitude towards transfer probability is now used in modern quantum mechanics. This involves the application of an equation called Fermi's Golden Rule,[12,13] which states that the probability of transfer of an electron between two states, in general, is given by the equation

$$T(E) = \frac{2\pi}{\hbar} |\psi_f| \, V |\psi_i|^2 \rho(E_f) \tag{29}$$

where the $\psi_i$ and $\psi_f$ are the wave functions for the particles in the initial and final state, respectively, $V$ is the perturbing potential that influences the transfer of the particle from the initial to final state, and $\rho(E_f)$ is the density of the electronic states in the final state.

This equation has occasionally been applied to electrode kinetics,[12] but its applicability is rather difficult in the numerical sense and the Gamow equation is often still used to make approximate calculations.

## 5.2. Distribution of Electronic States at the Interface

The distribution of electronic acceptor and donor states in ions in solution arises due to the stretching of bonds in the inner layer, and this mechanism becomes feasible when one recognizes the continuum nature of the energy distribution in the inner bonds of ion–solvent complex. The communication of energy from the solvent is vibronic and collisional. The ion–solvent complex must be considered as the matrix into which thermal energy is communicated from the vibrational and cage-effect collision[14] of the neighboring solvents, and thus the ion–solvent or ion–ligand bonds of the inner layer are stretched and become activated.

The distribution of electronic states in the acceptor ion in solution can be expressed as[12]

$$G(E) = \exp[-\beta(E_0 - E)/kT] \tag{30}$$

where $E_0$ is the ground state energy of the acceptor ion in solution and $\beta$ is the symmetry factor the value of ranges from $\langle \beta \rangle$ 0.

Most of the theory of electrochemistry, in its electrodic aspect, has been worked out in respect to metal–solution interfaces. Only in the last 20 years, and to a quite restricted extent, have semiconductor–interfaces been considered in detail, and here the work of Gerischer[5] is to be cited as of particular value (cf. the founding papers which were by Clark and Garrett[15] and, in the electrode kinetic form, by Green.[16,17]

Garrett and Brattain[18] were the first to attack this problem, and they relied upon the similarity to the distribution of ions in solution when they considered the distribution of electrons and holes inside semiconductors. Thus, deep inside an intrinsic semiconductor the excess charge density must be zero because of the equality of electrons and holes.

Hence

$$n^0 = p^0 \tag{31}$$

Correspondingly,

$$\rho_{\text{bulk}} = e_0\, p^0 - e_0 n^0 = 0 \tag{32}$$

where $n^0$ and $p^0$ are, respectively, the concentration of electrons and holes in the bulk of the semiconductor.

In the surface of a charged semiconductor the numbers of electrons and holes are not equal (this, of course, is just another statement of the fact that the surface is charged).

One can treat the charge density at a distance $x$ inside the semiconductor by using the Poisson–Boltzmann equation and obtain

$$\frac{d^2 V}{dx^2} = -\frac{4\pi\rho}{\varepsilon_s \varepsilon_0} \tag{33}$$

where $V$ is the potential inside the semiconductor at any distance $x$ from the interface and $\rho$ is the charge density inside the semiconductor.

$$V(x) = V_0 e^{-\kappa\chi} \tag{34}$$

where $V_0$ is the potential at the interface of the semiconductor, $\kappa$ is a constant the inverse of which is known as the Debye length, and $\chi$ is the distance from the surface, into the semiconductor, at which one expects to find the potential $V$.

Some diagrams will be helpful in the consideration of this situation, and these are given in Figure 11.

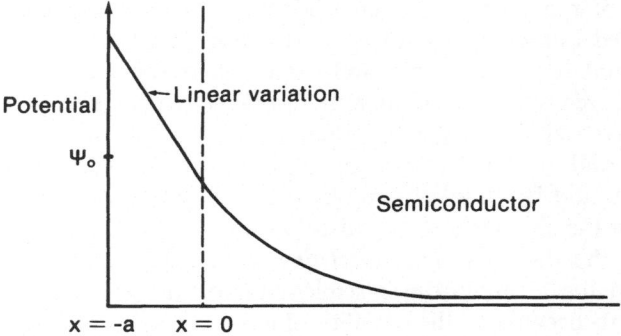

*FIGURE 11.* The variation of potential inside the semiconductor.

It is important to consider a phenomenon known as the bending of the bands which occurs near semiconductor surfaces. It has been stated that in the interior of the semiconductor there is an equality of charges, and as the charges become unequal near the surface, the top of the valency band and the bottom of the conduction band inside the semiconductor change with distance as shown in Figure 12. Thus, there is a space charge region and a potential drop inside the semiconductor. This potential drop gives rise to a region often referred to as the Schottky barrier. This is because an *n*-type

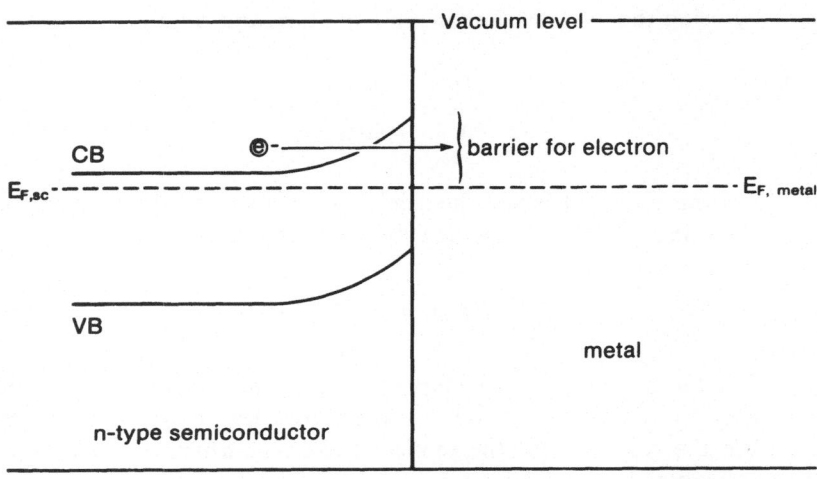

*FIGURE 12.* Schottky barrier for majority carriers (electrons) in the *n*-type semiconductor. CB, conduction band; VB, valence band.

semiconductor band inside the semiconductor moves downward with respect to surface level and produces a depletion region with respect to the majority carrier (i.e., electron for $n$-type semiconductor). Thus, this band bending acts as a barrier for the majority carrier electron (Figure 12) and hence is called a "Schottky barrier." However, for $p$-type semiconductor, the band bends upward with respect to surface level and acts as a barrier for the majority carrier hole (Figure 13).

Semiconductor–solution interfaces behave as capacitors, but the electrical capacity to which they give rise is different from the electrical capacity which is observed at a metal–solution interface. This is due to a so-called Helmholtz region at the metal–solution interface (Figure 14), and in the case of the semiconductor the space charge region inside the solid body gives rise to an extra capacitance which is less in value than the Helmholtz capacitance. The capacitance due to the space charge region inside the semiconductor is added as a capacity in series with the Helmholtz capacitance and this gives rise to an observed capacitance.

$$\frac{1}{C_{\text{obs}}} = \frac{1}{C_{\text{sc}}} + \frac{1}{C_{\text{hL}}} \tag{35}$$

If the $C_{\text{sc}}$ is small and the $C_{\text{hL}}$ is large (usually the case of semiconductors) observation will only give rise to $C_{\text{sc}}$.

Hence, the difference between a semiconductor–solution interface and a metal–solution interface is that in the latter the Helmholtz region is the main part of the double layer and in the former there are two, i.e., the

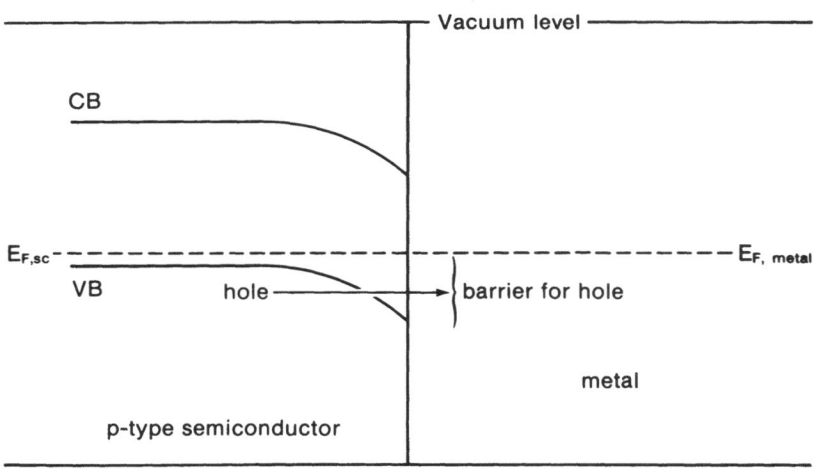

FIGURE 13. Schottky barrier for majority carriers (holes) in the $p$-type semiconductor. CB, conduction band; VB, valence band.

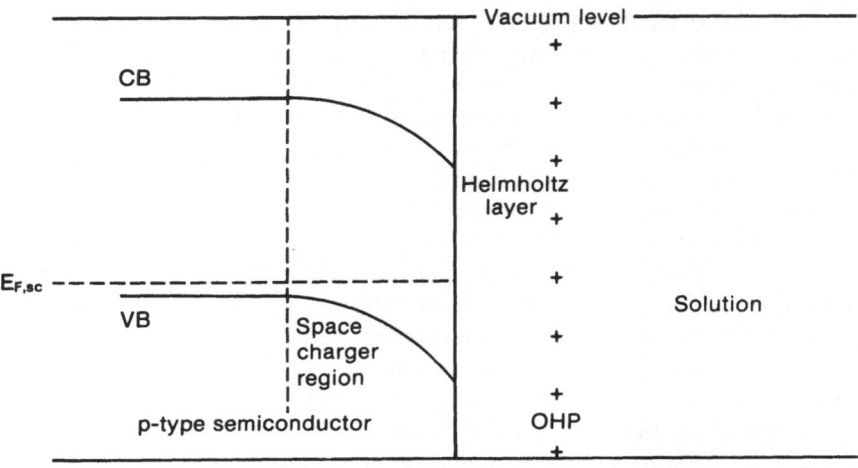

*FIGURE 14.* The space charge region and the Helhmoltz region at the semiconductor–solution interface.

space charge region inside the semiconductor and also the Helmholtz double layer at the interface.

This space charge region depends, for its characteristics, upon the electron–hole concentration of the semiconductor. Briefly, if these are low, then the space charge region is stretched out far inside the semiconductor, whereas if it is high it is a short region. The distances we are speaking about here as "high" are in the region of thousands of angstroms, and "low" means in the region of 10,000 Å.

### 5.3. The Effect of Surface States upon the Distribution of Potential at the Semiconductor–Solution Interface

The picture we have given so far, the Schottky barrier picture, of the semiconductor–solution interface is an ideal one. It certainly exists in reality in certain semiconductors, but in general semiconductors have a much more complex situation at their interfaces, which arises from the fact that at the actual surface of a semiconductor with the solution there are likely to be surface states.

These surface states arise in three main ways:

1. Intrinsic surface states arise due to "dangling bonds" for the discontinuity of the surface.
2. Specifically adsorbed ions at the semiconductor–solution interface will also form surface states and will attract electrons.

3. In the case where electron transfer results in adsorbed materials on the surface of electrodes (as in the evolution of oxygen, for example, where oxygen atoms are adsorbed upon the semiconductor–solution interface) surface states will arise from this cause.

If the semiconductor–solution interface involves simple redox cations (a $Fe^{2+} \rightarrow Fe^{3+} + e$ type of situation), then the surface states which exist on the surface of the semiconductor will consist of the first two kinds and may be small in number. In this case the ideal Schottky barrier case probably obtains. However, in such a situation the adsorbed $H_2O$ may also give rise to surface states.

In many reactions involving a semiconductor–solution interface, the interfacial situation has to take into account that a high concentration ($10^{13}$ cm$^{-2}$) of surface states will exist at the surface.

Insofar as this concentration of surface states is high (and this means more than 1% of the total sites of the surface), the Schottky part of the barrier—the amount of potential within the space charge region—tends to decrease, and the potential drop at the Helmholtz part of the double layer will increase.

In an extreme case, when a semiconductor has a large number of surface states, it can be regarded as "metallized," i.e., the change of potential difference within the semiconductor due to applied potential is quite small, say less than $\pm 0.1$ V, whereas the change of potential difference in the Helmholtz double layer is quite large, say, $\pm 1$ V (as in the case of metal) (Figure 15).

The majority of semiconductors will exist in some intermediate situation between these two extremes.

## 5.4. Fermi Levels in the Semiconductor and in the Solution

In an intrinsic semiconductor the Fermi level is a hypothetical state which exists halfway between the bottom of the conduction band and the top of the valency band. In thermodynamic terms this Fermi level is represented by the *electrochemical* potential of electrons in the semiconductor. The fact that the Fermi level exists halfway inside the energy gap, and where ideally no electrons or holes can exist, is of small consequence. The Fermi level represents the energy state at which the probability of existing electron and hole are equal and half each. The Fermi level within the semiconductor represents an ideal situation which is calculable and is in fact equivalent to the *electrochemical* potential inside the semiconductor.

The electrochemical potential $\bar{\mu}_e^{sc}$ is defined as follows:

$$\bar{\mu}_e^{sc} = \mu_e^{sc} - e_0 \phi^{sc} \tag{36}$$

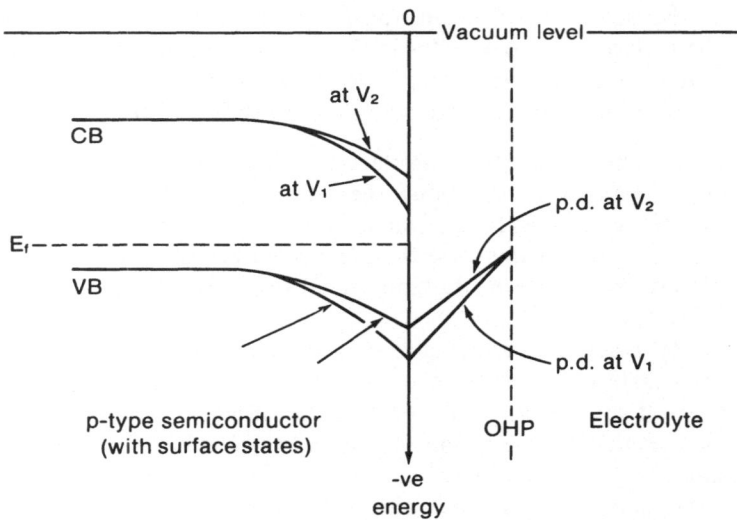

FIGURE 15. Schematic diagram of a *p*-type semiconductor–solution interface at two applied potentials, $V_1$ and $V_2$, in the presence of surface states. The diagram shows the potential drop, p.d., in the solution Helmholtz layer and exhibits a variation in p.d. with applied potentials in this case of having surface states. The Fermi level is pinned and the semiconductor is metallized.

where $\mu_e^{sc}$ is the chemical potential of electron in the semiconductor, and $\phi$ is the inner potential of the semiconductor, representing the work done taking a test $+ve$ unit charge from infinity to a position inside the bulk of the semiconductor.

However, there has been an attempt by some authors to discuss a Fermi level in the solution. The reason why this used to be regarded as important in semiconductor–solution theory was that there must be an equality between the electrochemical potential of the electrons inside the semiconductor and that in the solution. This may not be too good a reference state owing to the tenuous nature of the reality of these two states in respect to the presence of any electrons. However, it was thought to be a good reference state, something like a reference to the reversible hydrogen electrode, and therefore it was interesting to inquire what this value of the Fermi level of electrons in the solution might be. For many years, it was thought that this Fermi level in the solution was equal to the reversible potential of the redox couple in the solution on the vacuum scale potential.[19,20] It was shown[21] in 1983 that this was not true and that in fact the determination of the Fermi level energy of an electron in solution is not practicable. That is, to find the Fermi level in solution one must know the inner potential of solution $\phi^s$ to add to the reversible potential of the redox system on the vacuum scale. $\phi^s$ is not known.

Hence, the Fermi level in the solution is a quantity that is not directly applicable to the semiconductor–solution interface at this time, though it may be possible in future research to evaluate it by means of measurements of individual inner potential.

Hence, the use of the determination of solution Fermi level from standard redox potential (cf. Figure 16) is not applicable in the absence of the value of $\phi^s$. However, the question is whether the Fermi level in solution is

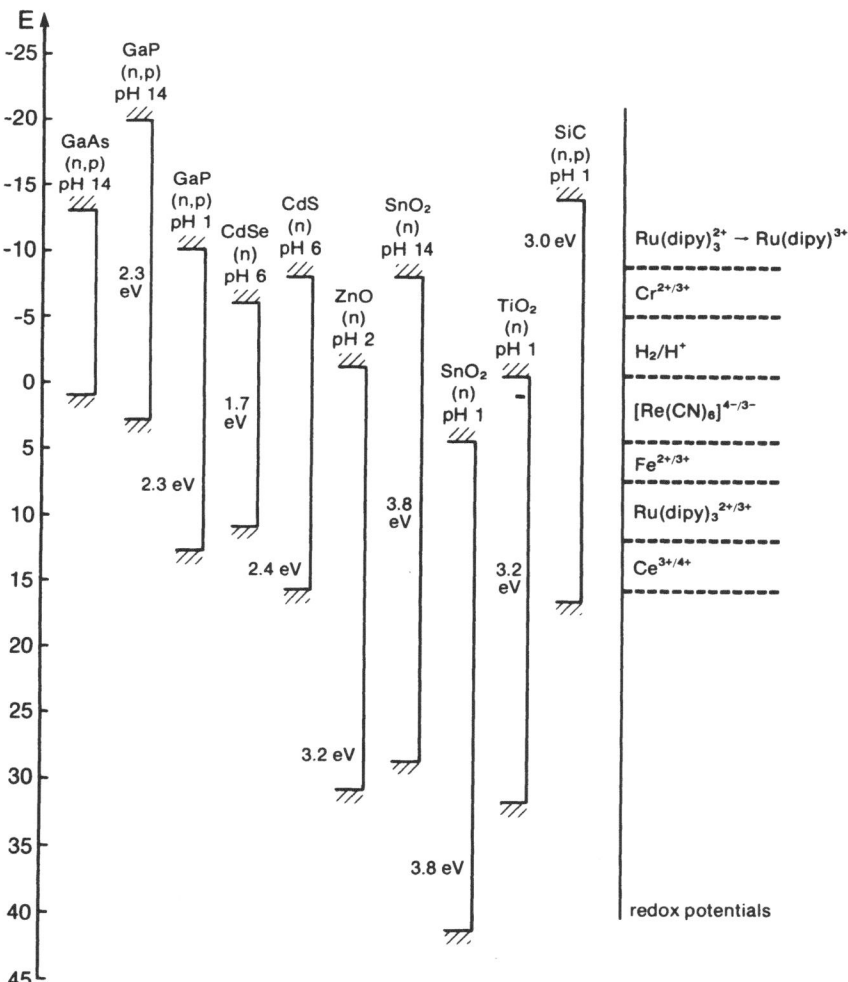

**FIGURE 16.** Relative position of energy bands at the surface of various semiconductor electrodes [values vs. NHE (vacuum scale)] (from Ref. 27).

a desirable quantity for the electrode kinetic calculation. Fortunately, it is not a relevant quantity for the study of kinetic processes of electrochemical and photoelectrochemical reactions at the electronic conductor (metal or semiconductor)–solution interface. The most desirable quantity to know is the distributed electronic donor and acceptor levels of the ions in solution. To determine this distribution, one needs to know the difference of energy, $\Delta E_0$, between the activated and the ground state of acceptor or donor ions in solution. For example, when $Fe^{3+}$ (aq) ion is in solution, one can determine the difference in energy $\Delta E_0$ by using the following cycle:

$$
\begin{array}{ccc}
Fe^{3+}(aq) + M(e)(sol^n) - - -\xrightarrow{\Delta E_0} - -\to Fe^{2+\bullet} \ \ldots\ldots \text{ aq} \\
\uparrow \qquad\qquad\qquad\qquad\qquad\qquad \downarrow \\
\Phi^{ec} - \chi^s \qquad\qquad\qquad\qquad -\Delta F^{\#}(2) \\
\\
Fe^{3+}(aq) + e_{Vac} \qquad\qquad\qquad Fe^{2+}(aq) \qquad\qquad (37) \\
\uparrow \qquad\qquad\qquad\qquad\qquad\qquad \downarrow \\
\Delta F_s(3) + 3\chi^s \qquad\qquad\qquad -\{\Delta F_s(2) + 2\chi^s\} \\
\\
Fe^{3+}(Vac) + e_{Vac} \leftarrow - - -\xleftarrow{I} - - - Fe^{2+}(Vac)
\end{array}
$$

Hence,

$$\Delta E_0 = \Phi^{ec} + \Delta F^{\#}(2) + \Delta F_s(2) - \Delta F_s(3) - I \qquad (38)$$

where $\Phi^{ec}$ is the work function of an electronic conductor (metal or semiconductor), $\Delta F^{\#}(2)$ is the free energy of activation of $(Fe^{2+\bullet}\cdots aq)$ $\Delta F_s(2)$ and $\Delta F_s(3)$ are the free energies of solvation of $Fe^{2+}$ and $Fe^{3+}$ ions, respectively, $I$ is the ionization energy of $Fe^{2+}$ ion, and $\chi^s$ is the surface potential of electron in solution.

## 6. The Current–Potential Relation at the Semiconductor–Solution Interface

### 6.1. General

There are mainly two kinds of current possible at a semiconductor–solution interface, i.e., one net current at the conduction band of the semiconductor and another at the valence band of the semiconductor. However, in a *p*-type semiconductor one will expect dominant anodic current at the valence band due to electron transfer from a donor ion in the solution to holes in the valence band in the semiconductor. On the other hand, in an *n*-type semiconductor one will expect dominant cathodic current across the conduction band due to electron transfer to an acceptor ion in solution. If the concentration of electrons and holes at the surface of the semiconductor is assumed to correspond to the equilibrium values $n_{s,0}$ and $P_{s,0}$ one can write the rate equations as[5]

$$i_c = i_{c,0} \left( 1 - \frac{n_s}{n_{s,0}} \right) \tag{39}$$

and similarly for the valence band in a *p*-type semiconductor

$$i_v = i_{v,0} \left( \frac{P_s}{P_{s,0}} - 1 \right) \tag{40}$$

when $i_{c,0}$ and $i_{v,0}$ are the exchange current densities across the conduction band in an *n*-type semiconductor and valence band a *p*-type semiconductor, respectively. $i_{c,0}$ can be expressed as

$$i_{c,0} = \frac{KT}{h} \delta C_A^0 \, f(E_c) \, P_T(E_c) \, D_a(E_c) \tag{41}$$

when $\delta$ is the width of the double layer at the semiconductor solution interface, $C_A^0$ is the bulk concentration of the acceptor ion in solution, $f(E_c)$ is the Fermi distribution of electron in the semiconductor, $P_T(E_c)$ is the tunneling probability of electrons across the interfacial barrier at the interface (Figure 17), and $D_a(E_c)$ is the distribution of acceptor states in the ions in solution. Since most electrons pass across the bottom of the conduction

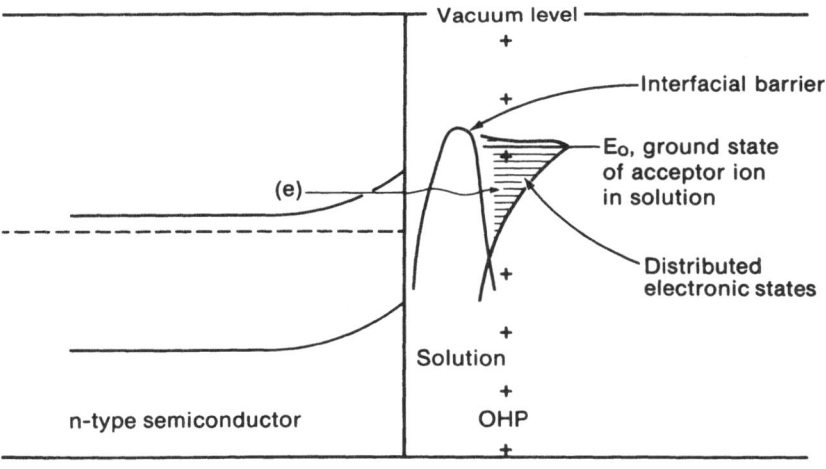

FIGURE 17. Interfacial barrier at the *n*-type semiconductor–solution interface. The distribution of electronic states in ions in solutions is also given. OHP, outer Helmholtz plane.

band edge we have considered $E = E_c$, the band edge energy of the conduction band at the surface.

Similarly,

$$i_{v,0} = \frac{kT}{h} \delta C_D^0 f(E_v) P_T(E_v) D_d(E_v) \tag{42}$$

where $D_d(E_v)$ is the distribution of a donor state and $C_D^0$ is the bulk concentration donor ions in solution and $E_v$ is the energy corresponding to the top of the valence band edge at the surface. With the assumption (in the absence of high concentration of surface states) that the variation of the electrode potential at varying polarizing voltage occurs fully over the space charge layer inside the semiconductor (i.e., Helmholtz layer potential $\phi_H$ remaining constant), one can express

$$\frac{n_s}{n_{s,0}} = \exp\left(-\frac{e_0 \eta_c}{kT}\right) \tag{43}$$

and

$$\frac{P_s}{P_{s,0}} = \exp\left(\frac{e_0 \eta_a}{kT}\right) \tag{44}$$

The effect of overpotential inside the semiconductor is shown in Figure 18. Now using the above two equations (43) and (44), one gets

$$i_c = i_{c,0}[1 - e^{-e_0 \eta_c/kT}] \tag{45}$$

$$i_v = i_{v,0}[e^{e_0 \eta_a/kT}] \tag{46}$$

where $i_{c,0}$ and $i_{v,0}$ should be used from Eqs. (41) and (42), respectively.

These equations (45) and (46) are comparable to the Butler–Volmer equation [Eqs. (11) and (12) for the current density at the metal–solution interface].

For large departure from equilibrium Eqs. (45) and (46) become

$$i_c = i_{c,0} e^{-e_0 \eta_c/kT} \tag{47}$$

and

$$i_v = i_{v,0} e^{-e_0 \eta_a/kT} \tag{48}$$

These equations (47) and (48) are now closer in form to the similar Butler–Volmer equation at large departure from equilibrium, i.e., Eqs. (15) and (16).

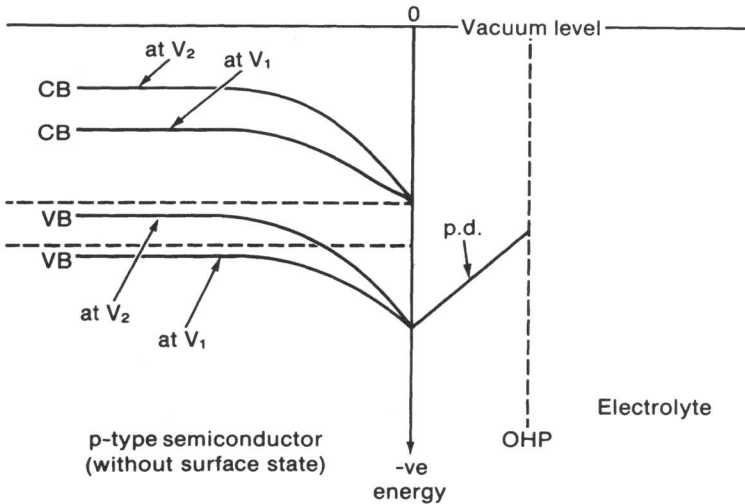

*FIGURE 18.* The schematic diagram of *p*-type semiconductor–solution interface at two applied potentials, $V_1$ and $V_2$, in the absence of surface states. The diagram shows the potential drop in the solution Helmholtz layer and exhibits no variation in Helmholtz layer p.d. with applied potentials. The Fermi level is not pinned. $(V_2 - V_1) = \eta$ when $V_1$ is the equilibrium potential.

The main difference appears in the exponential term. The expressions at the metal–solution interface contains the factor $\alpha$ in the exponential term, the value of which is less than one, but the similar expression at a semiconductor solution interface does not have an $\alpha$ term. This is because the change in the potential drop in the semiconductor–solution interface has been considered to occur inside the semiconductor (Figure 18). However, in the presence of a large number of surface states ($10^{13}$ cm$^{-2}$) the surface of the metal becomes metallized, and in such a situation the change in potential drop mainly occurs in the Helmholtz layer of the double layer. Hence, for such a metallized semiconductor–solution interface the exponential term of the current potential relation involves the transfer coefficient, $\alpha$.

Thus, one gets the slope, $\eta/\log i = RT/F$ for nonmetallized semiconductor and $\eta/\log i = 2RT/F$ for metal and also metallized semiconductor.

## 6.2. Rate-Determining Step at the Semiconductor–Solution Interface

It has been pointed out that the general course of the current–potential curve at metals is given by a S-shaped curve of the type that is given in Figure 9. In the beginning the interfacial reaction is always rate determining, but by the time we are some way up the S curve, when the current is no longer increasing exponentially with overpotential, transport either in the solution or in the electronic conductor begins to play an important part, and finally there is a limiting current above which it is not possible to force the current to increase further.

Qualitatively, the same type of reasoning applies to the rate-determining steps in the semiconductor–solution interface. There is one important difference, and that is that when the semiconductor does not have a preponderance of surface states the $i/V$ curve has a different gradient from that which it would have as a metal, or at a semiconductor with a significant concentration for surface states.

Another difference that may come into the semiconductor–solution interface is the importance of recombination at the surface of the semiconductor. The carrier prior exit to an ion in solution may be trapped and thereby annihilated with surface combination centers on the surface. Some surface states behave as traps. It may be that the final rate of a reaction to the semiconductor–solution interface depends upon recombination of holes and electrons within the semiconductor. Such a situation occurs under illumination condition.

## 7. Insulator–Solution Interfaces

### 7.1. General

So far in our presentation of the fundamentals of charge transfer reactions at interfaces between electron and ion conductors, we have been dealing with the basic case of a metal in contact with a solution (about which more than 90% of all work in electrochemistry has been done), and the semiconductor–solution interface, mainly work of the last 20 years.

The presentation we have made hitherto is then *background* for the real thing in respect to this book because here we are interested virtually in the insulator–solution interface.

Thus, biological substances are generally regarded as conducting poorly. The question of conduction in biological materials will be dealt with in the next chapter, but at the present time it is unusual to find a protein that conducts significantly (less than $10^{-8}$ mho cm$^{-1}$).

The conduction is thus around $10^6$ times less conducting than many semiconductor materials, and for many years it was the major blockage to the application of electrodic models to biological phenomena. Indeed, this blockage is not entirely removed, and there are still doubts concerning the degree to which the electrode–solution interface (as known for metals and semiconductors) can be applied to the interface between what are effectively insulators.

Work by Kallman and Pope in 1960,[23] however, made a difference to the situation. They showed that it was possible to inject charges into insulators from solutions. Correspondingly, Rosenberg[24] showed that the ambiance of a so-called insulating material had great effect upon the degree to which it conducted.

Another aspect of the insulator–solution interface depends upon the fact that the layers of material through which electrons may have to pass is thin. For example, in membranes it is often of the order of 50 Å. The passage of current through a solid is dependent on the length of the passage; this small path length will mean that small potential differences are associated with the passage of current through such thin layers. This to some extent helps the understanding that electrode processes occur at all biological–solution interfaces. Although the assumption of electron transfer in biology is almost universal, it must be clear that insofar as it concerns the protein–solution interface, it rests upon the assumption that the conductivity is sufficient.

This indeed is not an easy task to examine because it may well be that it is only at *certain sites* in a membrane that there is sufficient electronic conductivity for the "electrode" to function. It may well be that our model of a biological electrode (say a membrane) is a model of an insulating layer in which are insulated a number of "wires," and this would mean that the proteins which are part of biomembranes, and which "stick through them," may be the source of the transport between the two sides of the membrane and an origin of an electron and proton transfer site at the protein–solution interface (Figure 19).

## 7.2. Double Layer at the Insulator–Solution Interface

The double layer at the insulator–solution interface is related to the semiconductor solution interface. The difference between a semiconductor and insulator is rather conditional.[26] Like the semiconductor, the insulator has a forbidden energy gap between the valence band and the conduction band, but the gap is much larger compared to that in the semiconductor. But unlike the semiconductor, the insulator is characterized by a small conductivity in comparison with semiconductors such as Si, Ge, etc.

A drop of potential $\phi$ at the insulator–solution interface is similar to that at semiconductor–solution interface. The potential drop takes place in three regions: (1) in the region confined by the space charge in the insulator; (2) in a dense part of a double layer containing no free charges; and (3) in the diffuse part of the double layer in the electrolyte. On the basis of these considerations one can develop a model of the double layer shown in Figures 20a–20c.

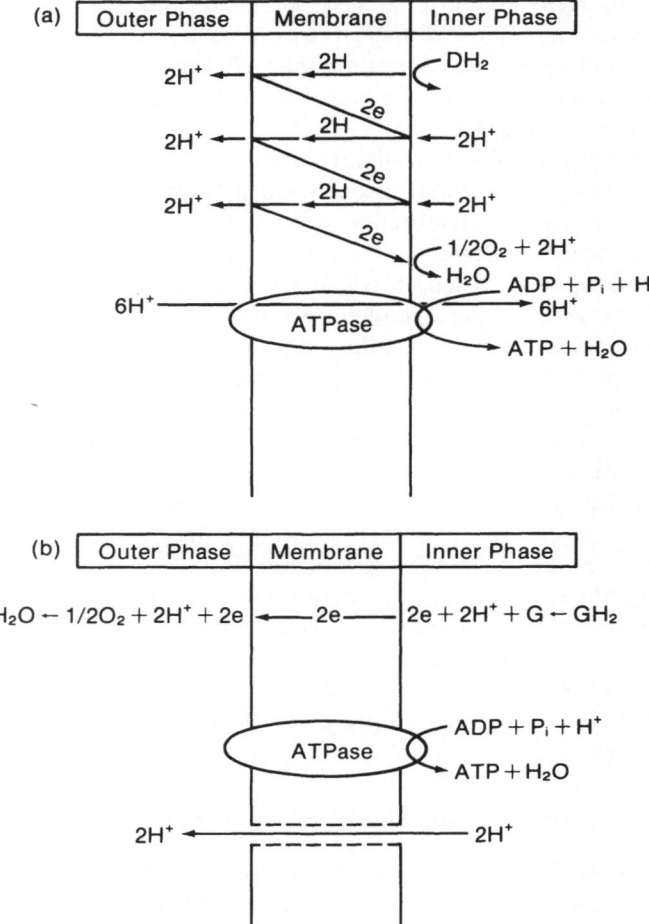

FIGURE 19. (a) Chemiosmotic model of phosphorylation. (b) Electrodic model of phosphorylation.

By using an external power supply, one can change the value of space charge in the insulator. A positive polarization of the insulating electrode makes a positive space charge which increases the electron energy in the surface layer, and then the bands should curve upwards. At negative polarization, on the contrary, the space charge layer accumulates an excess negative charge and the band curves downward.

Hence, a feature of the insulating electrode is the deep electric field penetration into the electrode. By solving the Poisson equation for the

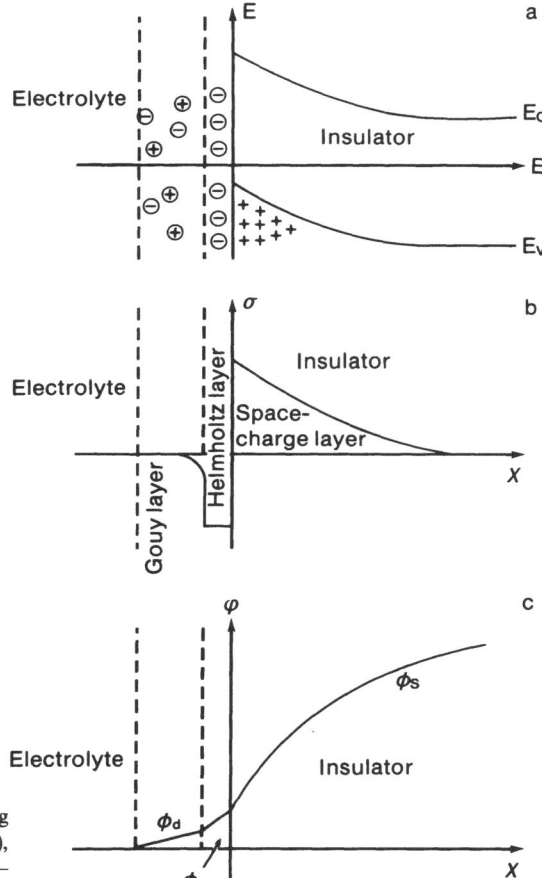

FIGURE 20. The band bending (a), distribution of the charge (b), and potential (c) at the insulator–electrolyte interface.

insulator–solution interface, one can determine the potential distribution in the insulator over a layer thickness $d$, where the potential $V$ ceases to vary due to the screening of the surface charge field. The thickness of the layer after which the surface charge is screened is determined by the Debye length, $d_D$, i.e.,

$$d_D = \left(\frac{\varepsilon_i \varepsilon_0 kT}{4e_0^2 N}\right)^{1/2} \tag{49}$$

where $N$ is the number of electrons in the insulator, $\varepsilon_i$ is the dielectric constant of the insulator, and $\varepsilon_0$ is the permitivity of vacuum.

By using the above equation for $N = 10^5 - 10^6 \text{ cm}^{-3}$ in the insulator the

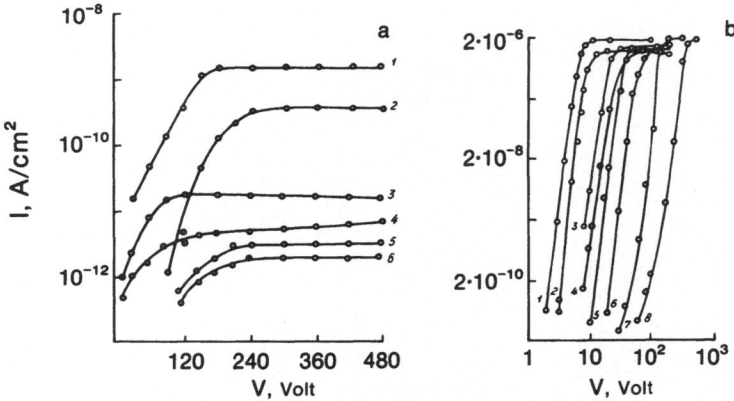

FIGURE 21. Dependence of the cathode current density on the applied voltage on the electrode. (a) From Teflon at various compositions of the contacting solution: 1, $1.5 \times 10^{-2} M$ $Ce^{+4}$ in $1 M H_2SO_4$; 2, $10^{-1} M K MnO_4$ in $1 M H_2SO_4$; 3, $1 M H_2So_4 + O_2$; 4, $0.5 M H_2SO_4 + O_2$; 5, $1 M NaCl + O_2$; 6, $1 M NaOH + O_2$. (b) From anthracene in the system of $10^{-2} M Ce(SO_4)_2$ in $0.5 M H_2SO_4$ (Fig. 6b from Ref. 5, reprinted by permission of Electrochemical Society): 1, 23 $\mu$m; 2, 28 $\mu$m; 3, 42 $\mu$m; 4, 51 $\mu$m; 5, 60 $\mu$m; 6, 130 $\mu$m; 7, 150 $\mu$m. (From Ref. 26.)

Debye length may reach several centimeters. But, in reality, the traps in the insulator reduce the depth of penetration of the field. One distinction of the insulating electrode from the semiconductor is the fact that the thickness of the insulating electrode may be less than the Debye length in it. It is then possible to obtain a mutual influence of the potential distribution at opposite sides of the insulating membrane.

### 7.3. The Current–Potential Relation at the Insulator–Solution Interface

The mathematical formulation of the current–potential relation at the insulator–solution interface is similar to that at the semiconductor–solution interface. The main difference is that the range of potential is much higher in the case of the insulator electrode compared to that at the semiconductor electrode. The applied potential usually ranges from $\pm 10$ to $\pm 10^3$ V at the insulator, but at the semiconductor electrode it ranges from $+0$ to $\pm 2$ V.

One can utilize Eqs. (45) and (46) given for the semiconductor electrode to find the current potential relation at the insulator electrode.

Current–potential dependence at some typical insulating electrode is given in Figure 21.

# References

1. K. J. Vetter, *Electrode Kinetics*, Academic, New York (1967).
2. J. O'M. Bockris and A. K. N. Reddy, *Modern Electrochemistry*, Plenum Press, New York (1970).
3. Gileadi, *Interfacial Electrochemistry*, An Experimental Approach, Addison-Wesley, Reading, Massachusetts (1975).
4. V. A. Myamlin and Y. V. Pleskov, *Electrochemistry of Semiconductors*, Plenum Press, New York (1967).
5. H. Gerischer, *Physical Chemistry: An Advanced Treatise* (H. Eyring ed.), Academic, New York (1970), Vol. 9A, Chap. 5.
6. S. R. Morrison, *Electrochemistry at Semiconductor and Oxidized Metal*, Plenum Press, New York (1980).
7. J. Tafel, *Z. Phys. Chem.* **50**, 641 (1905).
8. J. O'M. Bockris, "Over Potentials," *J. Chem. Ed.* **48**, 352 (1971).
9. R. W. Gurney, "Quantum Mechanics of Electrolysis," *Proc. R. Soc. (London)* **A134**, 137 (1931).
10. S. Arrhenius, *Z. Phys. Chem.* **7**, 226 (1889).
11. G. Gamow, "Quantum Theory and Radioactive Disintegration," *Z. Phys.* **51**, 204 (1928).
12. J. O'M. Bockris and S. U. M. Khan, *Quantum Electrochemistry*, Plenum Press, New York (1979).
13. E. Marzbacher, *Quantum Mechanics*, Wiley, New York (1970).
14. A. M. North, *The Collisional Theory of Chemical Reactions in Liquids*, Barnes and Noble, New York (1964).
15. P. E. Clark and A. B. Garrett, "Photovoltaic Cells: The Spectral Sensitivities of Copper, Silver and Gold Electrodes in Solution of Electrolytes," *J. Am. Chem. Soc.* **61**, 1805 (1939).
16. M. Green, "Electrochemistry of the Semiconductor–Electrolyte Electrode. 1. The Electrical Double Layer," *J. Chem. Phys.* **31**, 200 (1959).
17. M. Green, in *Modern Aspects of Electrochemistry* (J. O'M. Bockris, ed.), Vol. 2, Butterworths, London (1959).
18. C. G. B. Garrett and W. H. Brattain, "Physical Theory of Semiconductor Surfaces," *Phys. Rev.* **99**, 376 (1955).
19. H. Gerischer, "Über den Ablauf von Redox Reaktionen an Metallen und an Halbleiter," *Z. Phys. Chem. N. F.* **26**, 223 (1960).
20. H. Gerischer, "Charge Transfer Processes at Semiconductor–Electrolyte Interfaces in Connection with Problems of Catalysis," *Surf. Sci.* **18**, 97 (1969).
21. J. O'M. Bockris and S. U. M. Khan, "Fermi Levels in Solution," *Appl. Phys. Lett.* **42**, 124 (1983).
22. S. U. M. Khan and J. O'M. Bockris, "The Open Circuit Potential in Solar Cells," *J. Appl. Phys.* **52**, 7270 (1981).
23. H. Kallman and M. Pope, "Positive Hole Injection into Organic Crystals," *J. Chem. Phys.* **32**, 300 (1960).
24. B. Rosenberg, "Electrical Conductivity of Proteins II Semiconduction in Crystalline Bovine Hemoglobin," *J. Chem. Phys.* **36**, 816 (1962); B. Rosenberg and H. C. Pant, "The Semiconducting Rectifier Behavior of a Bimolecular Lipid Membrane," *Chem. Phys. Lipids* **4**, 203 (1970); "Electrochemistry on a Bimolecular Lipid Membrane," *ibid.* **6**, 39 (1971).
25. W. Mehl, J. M. Hale, and F. Lohmann, "Charge Transfer Processes at Organic Insulator Electrode," *J. Electrochem. Soc.* **113**, 1166 (1966).

26. L. I. Boguslavsky, in *Comprehensive Treatise of Electrochemistry* (J. O'M. Bockris, B. E. Conway, and E. Yeager, eds.), Plenum Press, New York (1980), Vol. 1, Chap. 7.
27. R. Memming, in *Electroanalytical Chemistry* (A. J. Bard, ed.), Marcel Dekker, New York (1979), Vol. 11.
28. J. O'M. Bockris and S. D. Argade, "Work Function of Metals and the Potential at Which They Have Zero Charge in Contact with Solution," *J. Chem. Phys.* **49**, 5133 (1968).
29. S. Trasatti, "The Concept of Absolute Electrode Potential, An Attempt at a Calculation," *J. Electroanal. Chem.* **52**, 313 (1974).

# 2

# The Origin of Cellular Electrical Potentials

## Gilbert N. Ling

*ABSTRACT:* A brief history of colloid chemistry and membrane theory is given, with particular attention to the work by Bernstein, Hodgkin, Huxley, and Katz. Available energy is discussed in terms of the energy required to operate a $Na^+$ pump. The association–induction (AI) hypothesis is explored in the context of cell $K^+$ and cell water, *in vitro* and *in vivo* testing. A subsidiary of the AI hypothesis, the surface adsorption theory of cell potential, is discussed with respect to model studies, living cells, and cellular resting potentials involving cooperative interaction among surface anionic sites.

## 1. Early History

The ancient Greeks discovered electrification in about 600 B.C., when they noticed that pieces of amber, when rubbed, attracted light objects. Not until the early seventeenth century was electrical repulsion recognized, and not until the eighteenth century did du Fay (1698–1739) clearly recognize the existence of two kinds of electricity.[1] du Fay was also one of the first to suggest that the nervous activity of living organisms might be electrical in nature.[2] The dual role of du Fay as physicist and physiologist clearly portrayed the common origins of the science of physics and the science of cell physiology. Indeed, if physicists had not discovered ways of storing electricity in Leyden jars, it would have been difficult for Leopoldo Caldani (1725–1813) and Aloisius Galvani (1737–1798) to conduct their famous experiments in which they caused skeletal muscle to contract in response to

*Gilbert N. Ling* • Department of Molecular Biology, Pennsylvania Hospital, Eighth and Spruce Streets, Philadelphia, Pennsylvania 19107.

electrical sparks delivered from the Leyden jars. These experiments of Caldani and Galvani marked the beginning of physiological research on cell functions.

Close linkage between research in physical and biological sciences continued. Johann W. Ritter (1776–1810), often considered as the founder of electrochemistry, likened the living phenomenon to a Galvanic process—a process he demonstrated with the aid of the "Voltaic pile."[2]

History shows that physiologists tended to make fast progress when they were able to make use of knowledge derived from the studies of simpler inanimate systems. On the other hand, when physiologists began to seek answers in religious doctrines or metaphysical concepts, progress as a rule slowed down. Thus it took the crusading efforts of four of the most gifted physiologists, Carl Ludwig, Ernst Brücke, Hermann von Helmholtz, and Emil DuBois Reymond to shake loose from "vitalism" that had become widely accepted in the early nineteenth century.

When Helmholtz was barely 20 years old and a junior military surgeon, he submitted a paper to the "Annalen" and soon learnt that this paper was rejected.[2] Of course, this would have been a trivial event except that this paper happened to be an extremely important one. It bore the title, "Über die Erhaltung der Kraft" (or "On the Conservation of Energy"). In later years Helmholtz reminisced on the comments he received: "This has already been well known to us. What does this young medical man imagine when he thinks it necessary to explain so minutely all this to us?"[3,4] However, Helmholtz did receive much praise and laudation from his military superiors, who mistook Helmholtz' "Kraft" for military power.[2]

### 1.1. The Founding of Colloid Chemistry and the Membrane Theory

Thomas Graham (1805–1869) was a great scientist though his work was not as well remembered as deserved. Graham introduced the name and basic concept of "colloidal chemistry" as the chemistry of gelatinlike substances ($\kappa o\lambda\lambda\alpha$, glue or gelatin). He also correctly interpreted semipermeability, i.e., the selective permeability to water but not to sugar, in these words, "The water of the gelatinous starch (in paper sized by starch) is not available as a medium for diffusion of ... the sugar ..."[5] (see below). Among the many colloidal materials he studied was copper-ferrocyanide, which is formed as a gelatinous precipitate when copper sulfate is mixed with K ferrocyanide. The semipermeable properties of a membrane made of copper-ferrocyanide gel was discovered by Moritz Traube, who proposed the "atomic sieve theory" to explain semipermeability.[6] Osmotic studies of Wilhelm Pfeffer provided the foundation for van 't Hoff's law of osmosis[7] as well as Pfeffer's membrane theory of the living cells.[8] According to this membrane theory the contents of a living cell are essentially that of a dilute salt solution kept apart from the external aqueous medium by a sub-microscopic "precipitation membrane" with properties resembling one made from copper-ferrocyanide gel.

## 1.2. The Founding of the Membrane Theory of Cell Potentials by Bernstein

In a paper on the electrical potential across a copper-ferrocyanide gel membrane, Wilhelm Ostwald suggested that a similar potential represents the electrical potentials of living tissues.[9] Julius Bernstein, a student of Helmholtz, took this suggestion and developed it into the "membrane theory" of cellular electrical potentials.[10] In this theory, he assumed that the cell membrane is permeable to $K^+$ but not to anions and $Na^+$. He also postulated that during an action potential there is a transient local increase of membrane permeability.[11]

At about the same time, Overton carried out an extensive study of the living cells to various water-soluble substances.[12] Based on these studies, he proposed the "lipoidal membrane theory," in which the cell membrane postulated by Pfeffer was considered as lipoid in nature. A model of this theoretical cell membrane widely investigated was a layer of olive oil.[13] The olive oil model demonstrated many similar permeability characteristics seen in living cells. However, it also has serious defects. Thus it exhibits a low relative permeability for water while as a rule living cells are highly permeable to water. To overcome this objection, the lipoid membrane was seen as being perforated by small water-filled pores, just wide enough to allow water and other small molecules to go through, but not the larger solutes.[14] This mosaic membrane was further elaborated by Mond and Netter[15] and by Boyle and Conway.[16] In Boyle and Conway's theory the size of the pores in the cell membrane was seen as so rigid and so perfectly uniform that they would admit smaller hydrated ions like $K^+$ but bar permanently the passage of the larger hydrated $Na^+$.

Advent of radioactive isotopes soon showed the incorrectness of concepts of the cell membrane as atomic sieves (see also Ref. 7) and the concept that the cell membrane is absolutely impermeable to solute found at low levels in the resting cells such as $Na^+$. The Na pump hypothesis, long extant, was gradually accepted but at the beginning with less than total enthusiasm (see Lillie[17]). According to this hypothesis the low level of cell $Na^+$ is due to the continual operation of pumps located in the cell membrane which steadily pump $Na^+$ out at the expense of continual energy expenditure.

Bernstein's original postulation of a transient permeability increase during an action potential was confirmed by the impedance measurements of Cole and Curtis.[18] However, if the permeability increases were entirely nonspecific, as implied in Bernstein's original theory, it would be difficult to understand why the potential change during an action potential goes beyond a transient annulment of the resting potential. That is, the potential actually reverses its sign momentarily producing an "overshoot."[19,20]

## 1.3. The Ionic Theory of Cell Potential by Hodgkin, Huxley, and Katz

It was then that Hodgkin and Katz[21] made the important discovery that the magnitude of this overshoot is quantitatively related to the logarithm of the external $Na^+$ concentration. This discovery led to the formulation of the "ionic theory of cell potentials" described by the Hodgkin–Katz equation, following a prior formulation of Goldman.[22]

$$\psi = \frac{RT}{F} \ln \frac{P_K[K^+]_{in} + P_{Na}[Na^+]_{in} + P_{Cl}[Cl^-]_{ex}}{P_K[K^+]_{ex} + P_{Na}[Na^+]_{ex} + P_{Cl}[Cl^-]_{in}} \tag{1}$$

where $\psi$ is the electrical potential, $RT$ and $F$ are the gas constant, absolute temperature, and Faraday's constant respectively. $[K^+]_{in}$, $[K^+]_{ex}$, etc., are the intracellular and extracellular $K^+$ concentration as indicated, etc. Extensive testing of this equation (and that of Bernstein) led to mixed conclusions (for review, see Refs. 23–25).

(1) Positive: The relation of $\psi$ to the absolute temperature, to the logarithm of external $K^+$ (with some reservation, see below), and external $Na^+$ have been unanimously confirmed.

(2) Negative: External $Cl^-$ does not affect the steady potential.

(3) Mixed: The relations predicted by Eq. (1) between intracellular $K^+$ and $\psi$ were observed in three laboratories and were not observed in six others. To the best of my knowledge, the authors of Eq. (1) did not offer explanations for most of these conflicting findings. However, Hodgkin and Katz did modify their Eq. (1), presumably in response to the discovery of the independence of $\psi$ to $[Cl^-]_{ex}$. In their modified equation[26]

$$\psi = \frac{RT}{F} \ln \frac{P_K[K^+]_{in} + P_{Na}[Na^+]_{in}}{P_K[K^+]_{ex} + P_K[Na^+]_{ex}} \tag{2}$$

the $Cl^-$ terms were eliminated. The legitimacy of such partial deletion from an equation coherently derived has been challenged.[23-25]

In the meanwhile, the temporal sequence of events underlying an action potential was investigated by Hodgkin and Huxley[27-30] with the aid of the voltage clamp technique. These studies led to an extension of the ionic theory to include a theory of the action potential based on the "independence principle," i.e., ionic movements in the cell membrane occur in an homogeneous isotropic medium comprising the cell membrane and are independent of the presence of other ions. They presented a formal treatment of the changes of the potential, including the postulation of specific "gates" for $Na^+$ which open transiently in the early rising phase of the action potential and specific gate for $K^+$ which opens transiently in the falling phase of the action potential. A great deal of research based on this theory has followed. Since the Hodgkin–Huxley theory is widely taught and known, no attempt will be made to describe it in detail. Instead I shall spend the remainder of this chapter on aspects of the theories and facts not widely taught, beginning with the testing of the membrane-pump theory on which the ionic theory of cellular potential is built.

## 2. The Energy Available vs. Energy Required to Operate the Na$^+$ Pump

In the membrane-pump model, the steady inward diffusion of $Na^+$ into the cell is offset by a continual outward active transport of $Na^+$ to maintain the steady low level of $Na^+$ in the cell. Diffusion has a low temperature coefficient; active transport has a high temperature coefficient. This difference in temperature coefficient predicts a rise of the level of cell $Na^+$ and a fall of the level of cell $K^+$ at low temperature. Indeed, there were repeated reports supporting this expectation.[31-34] One may expect similar responses following exposure of the cells to metabolic poisons like iodoacetate (IAA) (which blocks glycolysis) and cyanide or nitrogen (which blocks respiration). Again there were many confirmatory reports.[35-37] Based on these findings, one would have expected that a prompt fall of $K^+$ and rise of $Na^+$ concentration would occur in all living cells, when they are exposed simultaneously to low temperature, IAA, and pure $N_2$. Yet results of trials on frog muscles were quite different from this expectation. Table 1 from Ling[38] revealed that exposure of frog muscle and nerves for five hours to low temperature in addition to 0.5 mM IAA and pure nitrogen produced no significant fall of $K^+$. Table 2 revealed similar lack of change of $K^+$ as well as $Na^+$ concentration following 7.7 hr exposure to low temperature and 5 mM IAA plus pure nitrogen.[39] In the same paper referred to above[38] Ling also mentioned results of two other sets of then still preliminary work: During a period of time when IAA and $N_2$ poisoned muscles showed no rise of cell $K^+$, there was no detectible change in the concentration of creatine phosphate and ATP, which were, under the circumstances, the only remaining energy sources of the poisoned muscles.[40] It was also pointed out that the combined action of low tem-

*TABLE 1.*  Effect of IAA and $N_2$ at Low Temperature upon $K^+$ Loss from Frog Muscles and Nerves[a]

| Frog No. | Type of tissue | Muscle No. | | Weight (g.) | mM $K^+$ per liter of intracellular water |
|---|---|---|---|---|---|
| 1 | Sartorius | 1 | Control | 0.0870 | 60.7 |
|   |   | 2 | Expt. | 0.0750 | 69.8 |
| 1 | Semitendinosus | 1 | Control | 0.0710 | 72.6 |
|   |   | 2 | Expt. | 0.0795 | 81.8 |
| 1 | Tibialis anticus longus | 1 | Control | 0.0938 | 71.1 |
|   |   | 2 | Expt. | 0.0900 | 79.2 |
| 2 | N. ischiadicus + N. tibialis + N. peroneus | 1 | Control | 0.0300 | 38.1 |
|   |   | 2 | Expt. | 0.0260 | 39.5 |
| 3 | Sartorius | 1 | Control | 0.0730 | 73.4 |
|   |   | 2 | Expt. | 0.0700 | 78.0 |
| 3 | Semitendinosus | 1 | Control | 0.0660 | 83.0 |
|   |   | 2 | Expt. | 0.0730 | 77.4 |
| 3 | N. ischiadicus + N. tibialis + N. peroneus | 1 | Control | 0.0260 | 42.8 |
|   |   | 2 | Expt. | 0.0242 | 40.0 |

|  |  | Muscles | Nerves |
|---|---|---|---|
| Average: | Control | 100.0% | 100.0% |
|  | Expt. | 105.2% | 98.5% |

[a] All muscles were first kept at 3°C overnight, to enable $K^+$ in both experimental and control muscles to attain a constant value. For the experiment all tissues were kept for 5 hr at 0°C, the experimental series in Ringer's solution +0.5 mM IAA, and in an atmosphere of pure $N_2$, the controls in Ringer's solution and air. (From Ling[38] by permission of Johns Hopkins Press.)

*TABLE 2.*  $K^+$ and $Na^+$ Contents of Frog Muscle after Prolonged Exposure to Nitrogen and Iodoacetate at 0°C[a]

|  | $K^+$ ($\mu$moles/g fresh tissue) | $Na^+$ ($\mu$moles/g fresh tissue) |
|---|---|---|
| Control | 74.9 ± 1.31 | 28.4 ± 1.21 |
| Pairs after 7.74 hr (0°C) in 5 mM IAA and pure nitrogen | 76.3 ± 1.64 | 29.2 ± 1.72 |
| P | >0.5 | >0.7 |

[a] Reference 39.

perature, IAA, and $N_2$, created no change in the $Na^+$ efflux rate, measured with radioactive $^{22}Na$.[38]

Two years later, Keynes and Maisel[41] confirmed the above-mentioned preliminary work, i.e., metabolic poisons did not alter the Na efflux rate of frog muscles, and so did Conway and co-workers another eight years later (Conway et al.[42]). Neither group regarded their finding as confirmatory of Ling's earlier report, which they were appararently unaware of.

While Ling studied the Na efflux of frog muscle at 0°C, Keynes, Maisel, and Conway et al. studied Na efflux at room temperature. In Ling's work, the absence of change in the $Na^+$ efflux rate was shown to accompany a maintained low level of $Na^+$ and high level of $K^+$ in the muscle cells. The level of $Na^+$ and $K^+$ in frog muscle poisoned at room temperature with IAA and other metabolic poisons was not reported by either Keynes and Maisel or Conway et al. If they did, they would have noticed

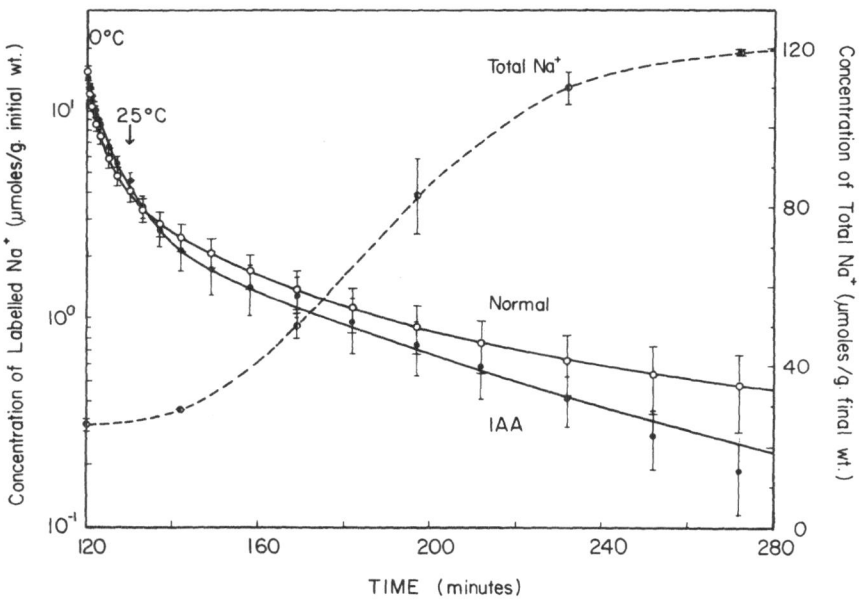

FIGURE 1. Effect of iodoacetamide and nitrogen on the $^{22}Na$-ion efflux of the frog sartorius muscles. Incubation solution contained $^{22}Na$. No iodoacetamide (IAA) was added to the control muscle group. The washing solution used to produce the IAA curve (●) contained 1.18 mM iodoacetamide and was bubbled with purified nitrogen; the normal curve (○) was obtained with normal Ringer phosphate bubbled with air. Each point is the average of four determinations ± standard error. The dashed line represents the level of the $Na^+$ ion in the muscle at equivalent time periods. (Ling and Ochsenfeld[46] by permission of *Physiological Chemistry and Physics*.)

another alarming behavior of the frog muscle from the viewpoint of the membrane-pump theory: the Na$^+$ efflux rate does not appear to be directly related to the level of Na$^+$ in the cell.

Many years later Ling and Ochsenfeld[43] reexamined the effect of room temperature on the Na efflux of poisoned muscles. Their figure, reproduced here as Figure 1, showed that while the Na$^+$ efflux remains unchanging in IAA and N$_2$ at 25°C as Keynes, Maisel, Conway, and co-workers had reported, the muscle cells were actually rapidly gaining more and more Na$^+$. This dissociability of the level of Na$^+$ and the rate of Na$^+$ efflux measured by the widely accepted way first described by Levi and Ussing[44] has serious implications. It threw doubt on the membrane-pump theory in general and the assignment of the slow fraction of Na efflux as one rate-limited by membrane permeability in particular.[45,46]

In the years immediately following 1952, Ling spent much time improving the method of assaying ATP, Na$^+$ efflux rates, etc., so that the definitive work on the calculation of the energy need of the Na pump was not published until 1962.[39] The results of the last three sets of experiments are summarized in Table 3. They show that under the specified conditions (i.e., 0°C, IAA, N$_2$ plus NaCN) when K$^+$ and Na$^+$ concentration were maintained at normal levels, *the minimal energy need of the Na pump is from 15 to 30 times that of the maximally available energy.*

TABLE 3. A Comparison of the Maximum Energy Available for the Minimally Needed Energy for the Na Pump in Frog Muscles at 0°C in the Presence of IAA, Pure Nitrogen and NaCN[a]

| Date | Duration (hr) | Rate of Na exchange, integrated average (M/kg/hr) | $\psi + E_{Na}/F$ integrated average (mV) | Minimum rate of energy required for Na pump (cal/kg/hr) | Maximum rate of energy delivery, (cal/kg/hr) | $\left[ \dfrac{\text{Min. required energy}}{\text{Max. available energy}} \right]$ |
|------|------|------|------|------|------|------|
| 9-12-56 | 10 | 0.138 | 111 | 353 | 11.57 (highest value, 22.19) | 3060% |
| 9-20-56 | 4 | 0.121 | 123 | 343 | 22.25 (highest value, 33.71) | 1542% |
| 9-26-56 | 4.5 | 0.131 | 122 | 368 | 20.47 (highest value, 26.10) | 1800% |

[a] Reference 39.

The data given in Table 3, when viewed in the light of the law of conservation of energy first enunciated by physicist–physiologist Hermann von Helmholtz, refute the $Na^+$ pump theory. It may be added that in the more than 20 years following, I have not known any serious challenge to the results or the conclusions I have drawn. However, three remedial postulations were introduced in attempts to keep the Na pump concept afloat. All were experimentally disproven: the exchange diffusion mechanism; the hypothesis of $Na^+$ sequestration in the sarcoplasmic reticulum; and the non-energy-consuming $Na^+$ pump (for review see Refs. 25, 47).

The significance of the data shown in Table 3 can be fully grasped if one realizes the fact that they represent only the "tip of an iceberg" of the total energy need. A by no means thorough search of the literature made in 1968 revealed that more than 20 pumps besides the Na pump have been postulated (Table 4), and more have been added.[40] All of these postulated pumps are plasma membrane pumps. Now, as a rule, subcellullar structures like the mitochondria, contain ions and other solutes at levels different from those in the cytosol. Thus more pumps must be postulated at

*TABLE 4.* Postulated Membrane Pumps[a]

| Solute | Direction | System |
|---|---|---|
| Na, K[b] | Coupled | Many cells |
| $Ca^{2+}$ | Outward | RBC, striated muscle |
| $Mg^{2+}$ | Outward | Frog sartorius |
| Choline$^+$ | Inward | RBC |
| Amino acids | Inward | RBC, muscle, tumor |
| D-xylose | Inward | Rat diaphragm |
| D-xylose | Outward | Rat diaphragm |
| $Na^+$ | Inward | Frog sartorius |
| Noradrenaline | Inward | Vascular smooth muscle |
| Prostaglandins | Inward | Mammalian liver |
| Curarine | Inward | Mouse diaphragm |
| $Br^-, I^-, ReO_4^-, WO_4^-$ | Outward | Ascites |
| $Cu^{2+}$ | Inward | Ascites |
| Aminopterin | Inward | Yoshida sarcoma |
| $Cl^-$ | Inward | Squid axon, motor neurons |
| $Mn^{2+}$ | Inward | E. coli |
| $Cl^-$ | Outward | E. coli |
| Sugars | Inward | E. coli |
| Amino acids | Inward | E. coli |
| Tetracycline | Inward | E. coli |

[a] Data collection was more or less arbitrary and not intended to be comprehensive (Ling *et al.*[40] by permission of *Annals of the New York Academy of Sciences.*)
[b] See Ling *et al.*[39]

the mitochondrial membrane, membrane of sarcoplasmic reticulum, nuclear membrane, etc. In liver cells the total mitochondrial membrane has been estimated and shown to be 20 times larger than the plasma membrane.[48] Each of the subcellular particle membrane pumps would require that many times more energy to cope with a similar ion gradient across the plasma membrane.

## 3. The Association–Induction (AI) Hypothesis

The association–induction hypothesis was in its early days known as Ling's fixed-charge hypothesis.[38] In years following, the theory has evolved. The newer version plus results of more than 30 years of experimental testing are presented in a book called *In Search of the Physical Basis of Life* (Gilbert N. Ling, Plenum Press, 1984). The following is a brief sketch of the key features of the AI hypothesis.

In the AI hypothesis, a resting living cell exists in a high-energy, metastable equilibrium state. In this state, the three major components of the living cell—water, proteins, and $K^+$—are in close association. The bulk of cell $K^+$ is preferentially adsorbed on $\beta$- and $\gamma$-carboxyl groups of certain cell proteins. The bulk of cell water is adsorbed, in polarized multilayers, on the NH and CO groups of the extended polypeptide chains of "matrix proteins" existing throughout all living cells. The maintenance of this associated, metastable equilibrium state depends on the complexing of the proteins involved with certain minor key components, including ATP and $Ca^{2+}$.

Water in the state of polarized multilayers exhibits a host of properties of the living cells, including size-dependent solute exclusion property[49]; swelling and shrinkage properties[50]; osmotic properties[51]; freezing and thawing properties[52]; NMR relaxation rates[53]; dielectric relaxation (see below); and quasielastic neutron scattering (see below). Thus, the solubility of large and complex solutes like (hydrated) $Na^+$ in this water is reduced; hence the low levels of $Na^+$ found in normal resting cells as well as in model systems.

In the AI hypothesis, the high $K^+$ and low $Na^+$ concentrations are manifestations of an equilbrium state and as such require no continual energy expenditure.

Many pieces of experimental evidence exist in support of the AI hypothesis and against the membrane-pump theory. The reader must consult the aforementioned monograph[25] for a full discussion. Here I shall limit our discussion to two issues: the adsorbed state of $K^+$ and the bulk phase water in living cells.

## 3.1. State of Cell $K^+$

In 1952 when the AI hypothesis was still in its infancy, it was suggested that the selective accumulation of $K^+$ in living cells results from specific preferential adsorption of $K^+$ on the $\beta$- and $\gamma$-carboxyl groups of certain cell proteins.[38] In voluntary muscle, more than 50% of the $\beta$- and $\gamma$-carboxyl groups reside in myosin.[54] In the years following, it was clearly established that myosin constitutes the major protein of the dark or A band in voluntary muscle cells.[54] Incorporating this idea, the AI hypothesis predicts that the bulk of $K^+$ in voluntary muscle should also be localized in the A band. Electron microscopic studies of collagen protofibrils led Hodge and Schmidt[55] to suggest that a positively charged stain, uranyl iron, combines primarily with the $\beta$- and $\gamma$-carboxyl groups of the proteins. Incorporating this second idea into the AI hypothesis, one could make a more refined prediction: $K^+$ and $Cs^+$ and $Tl^+$, which can stoichiometrically and reversibly replace $K^+$ in living cells, should not be localized evenly in the A band in voluntary muscle but more specifically at all the cytological structures that are stained dark with uranium in a conventional EM preparation. These structures comprise the two edges of the A band as well as the Z line in the middle of the light or I band (see Figure 2A).

From 1977 to 1980, both sets of predictions have been confirmed repeatedly in three different laboratories, using four different and independent techniques:

1. Autoradiographical studies of air-dried single muscle cells[56] and of frozen single muscle cells[57];
2. Transmission electron microscopic studies of frozen-dried but unfixed muscle section[58];
3. Dispersive X-ray microanalysis of frozen dried muscle sections[59,60];
4. Laser microprobe mass-spectrometer microanalysis.[61]

All unanimously showed that $K^+$ or other univalent cations like $Cs^+$, $Rb^+$, and $Tl^+$, all of which can stoichiometrically and reversibly replace $K^+$ in muscle, are localized primarily on the two edges of the A bands and at the Z lines.

However, Somlyo et al.,[62] (see also Ref. 63) from their electron dispersion microprobe analysis on cryosections of muscle cells, reached an opposite conclusion: $K^+$ was found more concentrated in the I bands than the A bands. In response, Edelmann[64] made further studies of muscle cryosections in cooperation with Dr. K. Zierold at the Max Planck Institut at Dortmund, FRG. They made cryosections on a FC4 Reichert

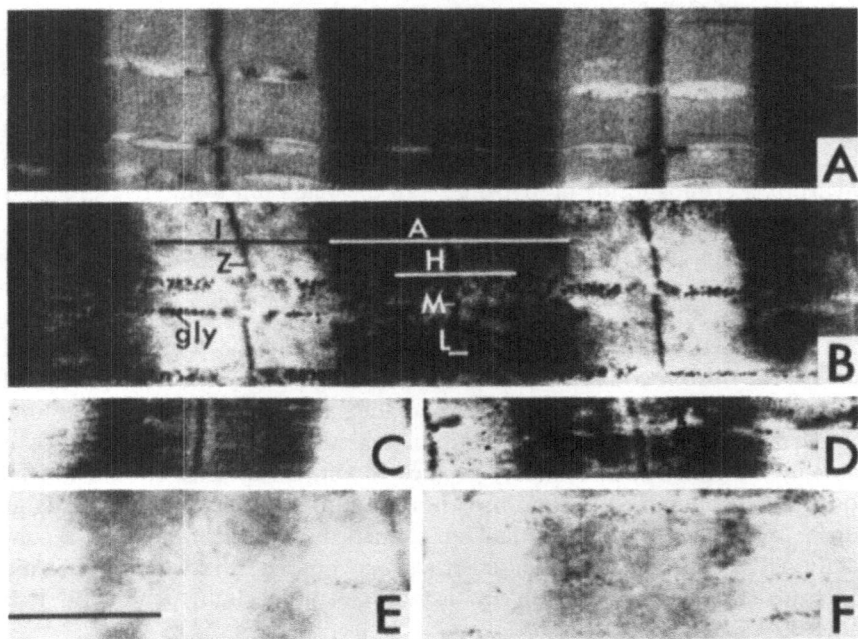

FIGURE 2. Electron micrographs of dry cut, unstained section of freeze-dried frog sartorius muscle. (A) Muscle fixed and stained with uranium–lead by conventional procedure. (B) EM of $Cs^+$-loaded muscle without chemical fixation or staining. (C) $Tl^+$-loaded muscle without chemical fixation or staining. (D) Same as C after exposure of section to moist air, which causes the hitherto even distribution of thallium to form granular deposits in the A-band. (E) Section of central portion of B after loading in distilled water. (F) Normal "$K^+$-loaded" muscle. [A is partial reproduction of EM from Edelmann (unpublished). B–F from Edelmann[58] by permission of *Physiological Chemistry and Physics*.]

cryoultramicrotome and visualized the section on a Siemens ST 100F scanning-transmission electron microscope. First they showed that they could, at will, reproduce Somlyo *et al.*'s results if they followed Somlyo *et al.*'s procedure: for I-band studies, the electron beam was focused at the center of the I band (where the Z line is located) and for A-band studies the beam was focused at the center of the A band where earlier work mentioned above showed less $K^+$ and its substituents, $Cs^+$ and $Tl^+$. However, Edelmann (and Zierold) then showed that if smaller beams were focused at smaller areas, then all that was previously reported by Edelmann, Trombitas, Tigyi-Sebes, and Ling was fully confirmed.

The establishment of localized $K^+$ adsorption in frog muscle by itself represents a disproof of one basic tenet of the membrane-pump theory. There are other important secondary implications:

(1) If the bulk of cell $K^+$ is adsorbed, the osmotic pressure in the cell would be much too low to sustain its normal volume. (The answer to this problem in terms of the AI hypothesis is discussed in Ling.[25,41])

(2) If the bulk of cell $K^+$ is adsorbed, the membrane theory of cellular electrical potential is no longer tenable. The same finding is in full harmony with the theory of cellular potentials according to the AI hypothesis, to be discussed below.

## 3.2. State of Cell Water

According to the polarized multilayer theory of cell water, certain proteins to be referred to as "matrix proteins" exist throughout all living cells. These matrix proteins exist in an *extended* conformation with the polypeptide NHCO groups directly exposed to the bulk phase water. A matrix of chains carrying the alternating positive (P) (e.g., NH groups) and negative (N) sites (e.g., CO groups) is called an NP–NP–NP system. In an NP–NP–NP system, water is polarized in multilayers wherein a reduction of the translational as well as rotational motional freedom of the (bulk phase) water molecules occurs. In such a system the solvency of the water for solutes like $Na^+$ salts, sucrose, etc., is reduced for enthalpic and/or entropic reasons.[25] Variants of an NP–NP–NP system include the NO–NO–NO system and PO–PO–PO system where one type of site is replaced by neutral or vacant (0) sites.

The experimental testing of the polarized multilayer theory of cell water involved two steps. In the first step, one tests if the theory has general validity. This requires verifying predictions of the theory in test tubes. If the theory passes this test, then one can proceed to the next step, i.e., to answer the question, "Does this theory apply to the living cell?" The following is a summary of the results of the testing carried out thus far.

## 3.2.1. Step 1: In Vitro Testing of the Theory

The tests were carried out directed at the following specific predictions of the theory:

(1) Bulk phase will exist in the state of polarized multilayers if there is a high enough concentration of protein molecules which assume an extended conformation with their NHCO groups directly exposed to the bulk-phase water.

(2) Multilayer polarization of the bulk-phase water does not occur if the protein backbone NHCO groups are locked in intermolecular or intramolecular H bonds (e.g., $\beta$-structure, $\alpha$-helix) and are thus not exposed to the bulk-phase water to act as the key components of an NP–NP–NP system.

To test these predictions, we employed a technique that is based on yet another prediction of the theory already mentioned above, namely, that water existing in the state of polarized multilayers has reduced solubility for large molecules and hydrated ions including $Na^+$ salts, sugars, and free amino acids. This criterion of testing has the following merits: experimental simplicity (equilibrium dialysis) and above all, its unambiguity. Thus, if water in a certain model solutions has a $\rho$-value for a $Na^+$ salt ($\rho$-value: the apparent equilibrium distribution coefficient of a probe substance like $Na^+$ between the water under study and the reference normal dilute solution outside the dialysis bag) equal to 0.5, one can then state unequivocally that at least 50% of the water has been affected by the proteins or polymer present.

Results of these experimental tests carried out thus far include the following:

(a) Water in solutions of 13 native globular proteins known to have their NHCO groups locked in $\alpha$-helical or other intramacromolecular H-bonds show no or little solvency reduction for Na salt, sucrose, and the free amino acid glycine.

(b) Gelatin (denatured collagen), known not to form $\alpha$-helical conformation due to its possession of glycine, proline, and hydroxyproline, all well-known helix-breakers,[65] does reduce water solvency for Na salt, sucrose, and glycine.

(c) Globular proteins, ineffective as they are in their native state as mentioned above, behave quite differently when they have been denatured with 10 $M$ urea, which breaks secondary structures exposing NH and CO groups directly to bulk-phase solvent causing solvency reduction for the probe molecules, Na salts, sucrose, and glycine.

(d) SDS and $n$-propanol, which are known to unravel tertiary structure but do not break secondary structure and thus do not expose the NHCO groups, have no effect on water solvency in solutions of native globular proteins.

(e) A solution of synthetic neutral polymers like poly(ethylene oxide) (PEO) $(CH_2-CH_2-O-)_n$, polyvinylmethylether (PVME) $-(CH(OCH_3)CH_2)_n$, polyvinylpyrrolidone (PVP) $(-CH(NC_4H_7O)CH_2-)_n$ satisfy the criteria of NO–NO–NO systems (where N stands for negatively charged oxygen atoms due to its lone pair of electrons and O stands for vacant sites) but do not form intra- or intermolecular H bonds, also exclude $Na^+$ salts, sucrose, and glycine.

An important by-product of these investigations is a redefinition of Thomas Graham's "colloidal state," whose true meaning has been lost in the undue emphasis on large molecule size. The experimental behavior of gelatin suggests that the colloidal state is the one in which the macromolecules exist as an NP–NP–NP system (or its equivalents), and in such state, they polarize in multilayers the solvent molecules in the system. In this new definition much of the observed unusual behavior of gelatin and gelatinlike or colloidal substances reflect properties of the modified water surrounding them.

### 3.2.2. Step 2: In Vivo Testing

(1) Ninety-five percent of the water in frog muscle follows quantitatively the Bradley polarized multilayer adsorption isotherm,[66] as do a number of models of living cell water, including ion exchange resins[67] and gelatin.

(2) Frog muscle excludes sugars and alcohols according to their size and complexity in general agreement with the "size rule," as do a number of cell-water models including ion exchange resin and solution of gelatin and PVME.

(3) Living cells adsorb (in multilayers) large amounts of water vapor at a vapor pressure corresponding to that of plasma or Ringer solution. So do solutions of gelatin, PVP, and DNA, but not those of native globular proteins.

(4) Water in living cells does not freeze in the typical hexagonal pattern seen in normal water, in salt solutions, and in 35% native globular protein such as bovine serum albumin (see Ling and Zhang[52]). Rather, it freezes in living cells in the form of "irregular dendrites" seen also in solutions of gelatin and PVP.

(5) Masszi et al.[68] measured the dielectric relaxation time of frog muscle water and found it to be longer than that of normal liquid water. Clegg et al.[69] also found in brine shrimp cysts a significantly lower dielectric constant in the frequency range of 0.8–70 GHz. Both sets of data resemble the increased dielectric relaxation time of water in model systems of PVP, PEO, and PVME of Kaatze et al.[70]

(6) Very recently Trantham and co-workers[71] studied the quasielastic neutron scattering (QENS) of the living cells of the brine shrimp cysts. The translational diffusion coefficient of the entire population of the water protons in the cell water was reduced by 70% and the rotational diffusion coefficient reduced by 90% of that of normal liquid water. These findings are in full harmony with the polarized multilayer theory of cell water. Then in a highly significant parallel study, Rohrschach[72] showed that a 35% solution of poly(ethylene oxide) (PEO) demonstrates QENS behaviors almost indistinguishable from that of brine shrimp cyst cells, providing another linkage between living cells on one hand and a variety of model systems exhibiting properties of living cells mentioned above.

Earlier Ling[73] had presented reasons for the belief that the cell sur-

face barrier to solute permeation is not due to a continuous lipid layer.
Rather, the weight of evidence favors the view that the cell surface barrier
is primarily proteins and water which they polarize in multilayers. This
model at once answers two so far unanswered questions: (1) How can copper-ferrocyanide gel membranes and cellulose acetate membranes with pore
diameters many times bigger than that of sucrose, be nevertheless virtually
impermeant to sucrose? (2) How can the cell change its surface permeability in response to a small change in the concentration of, for example, external $Ca^{2+}$?

The answer to the first question is as follows: the basis for the impermeability to a solute like sucrose is the size-dependent reduction of
solubility and of diffusion coefficient in the polarized water and not that of
a mechanical sieve as postulated originally by Traube in his atomic sieve
theory, which was disproved and then repeatedly resurrected.[25]

The answer to the second question is, that $Ca^{2+}$ interacts with cardinal sites of the cell surface proteins allosterically controlling the physical
state of the cell surface water and hence its permeability.

## 4. The Surface Adsorption Theory of Cell Potential— A Subsidiary Theory of the AI Hypothesis

### 4.1. The Earlier Model

In 1955 and years following, the suggestion was first made that the
resting potential ($\psi$) of the living cells is not a membrane potential but a
surface adsorption potential[74-76]

$$\psi = \text{const} - \frac{RT}{F} \ln \left( \sum_{i=1}^{n} \tilde{K}_i [p_i^+]_{\text{ex}} \right) \tag{3}$$

where $R$, $T$, and $F$ have the usual meanings. $[p_i^+]_{\text{ex}}$ is the concentration of
the $i$th monovalent cation (among a total of $n$ types) in the bathing
medium. $\tilde{K}_i$ is the adsorption constant of the $i$th ion on the surface anionic
sites (i.e., isolated $\beta$- and $\gamma$-carboxyl groups).

### 4.1.1. Model Studies

Model studies offer confirmatory evidence. Coating a non-$K^+$ sensitive (Corning 015) glass electrode with a thin layer of oxidized collodion
endows the glass electrode with $K^+$ sensitivity indistinguishable from that
of a simple oxidized collodion electrode. Exposure of such a collodion
coated glass electrode to poly-lysine with many fixed $\varepsilon$-amino cationic sites

imparts an anion sensitivity which was not present originally in either the glass electrode or the collodion-coated glass electrode.[77]

In reviewing the history of the search for the origin of the cellular electrical potential, I discovered the following: each of the three models originally chosen for study as models of the cell membranes (oil layer, glass membrane, collodion membrane) eventually was discovered to generate the potential not by virtue of their ionic permeabilities. Rather, they are all surface adsorption potentials.[25]

### 4.1.2. Living Cells

Equation (3), when applied to living cells in a Ringer solution, can be expressed more specifically as

$$\psi = \text{const} - \frac{RT}{F} \ln[\tilde{K}_K[K^+]_{ex} + \tilde{K}_{Na}[Na^+]_{ex}] \tag{4}$$

The data reviewed above showed that, of the relationships predicted between $\psi$ and the variables in the Hodgkin–Katz equation [Eq. (1)], only the relation between $\psi$ and $T$, between $\psi$ and $\ln[K^+]_{ex}$, and between $\psi$ and $\ln[Na^+]_{ex}$ have been unequivocally established. These three relationships are in fact the only ones predicted by Eq. (4). In other words, each of the predictions based on the surface adsorption theory has already been verified.

Since total intracellular $K^+$ and $Na^+$ concentration changed slowly,[78] for experimental measurements of $\psi$ shortly after changes in $[K^+]_{ex}$, etc., the intracellular concentrations $[K^+]_{in}$ and $[Na^+]_{in}$ may be regarded as constant. In that case, Eq. (2) can be written as

$$\psi = \text{const} - \frac{RT}{F} \ln(P_K[K^+]_{ex} + P_{Na}[Na^+]_{ex}) \tag{5}$$

or in more generalized form

$$\psi = \text{const} - \frac{RT}{F} \ln\left(\sum_{i+1}^{n} P_i[p_i^+]_{ex}\right) \tag{6}$$

Obviously Eqs. (5) and (6) are formally analogous to Eqs. (4) and (3), respectively. The two sets of equations differ profoundly in regard to the physical significance of the coefficients, $P_i$ vs. $\tilde{K}_i$.

Edelmann[79] conducted experiments in guinea pig heart muscle to test the alternative Eq. (6) vs. Eq. (3), concluding that it is the surface adsorption constant, $\tilde{K}_i$, and not the membrane permeability constant, $P_i$, that determines $\psi$.

### 4.2. An Improved Theory of Cellular Resting Potential Incorporating Cooperative Interaction among Surface Anionic Sites

Equation (3) was derived on the assumption that the surface anionic sites show no cooperative interaction. In 1979 I published a short article describing a new equation for the cellular resting potential,[80] incorporating the concept of site-to-site cooperative interaction of the cell surface sites as we have argued for and provided experimental evidence in support for the bulk phase (adsorbed) $K^+$[78]:

$$\psi = \text{const} - \frac{RT}{F} \ln \frac{1}{[K^+]_{ex}} \left\{ 1 + \frac{(\xi - 1)}{[(\xi - 1)^2 + 4\xi\theta]^{1/2}} \right\} \tag{7}$$

where

$$\xi = \frac{[K^+]_{ex}}{[Na^+]_{ex}} \cdot K^{00}_{Na \to K} \tag{8}$$

and

$$\theta = \exp(\gamma/RT) \tag{9}$$

$K^{00}_{Na \to K}$ is the intrinsic equilibrium constant for the $Na \to K$ exchange on surface anionic sites giving rise to the potential $\psi$. $-\gamma/2$ is the nearest neighbor interaction energy. When $-\gamma/2 = 0$ ($\theta = 1$), Eq. (3) reduces to Eq. (1); when $-\gamma/2 > 0$ ($\theta < 1$), the adsorption is autocooperative, exhibiting sigmoid type of behavior as in oxygen binding on hemoglobin. Figure 3 is a theoretical plot of $\psi$ against $[K^+]_{ex}/[Na^+]_{ex}$ with varying value of $\theta$ according to Eq. (7).

We have been able to verify Eq. (7) in different ways (see Ling[25]). The cause of electrical potential measured across the inner membrane of liver mitochondria by Maloff, Scordillis, Reynolds, and Tedeschi[81] may be cited. Here $\psi$ behaves almost like most excitable cells. Maloff *et al.* measured but found no effect whatsoever of valinomycin on $K^+$ conductance in the presence of varying $[K^+]_{ex}$. The present theory showed that valinomycin increases the $K^+$ affinity on the surface anionic sites of the mitochondrial inner membrane, by a factor of 3. With this assumption the entire sets of their data can be qualitatively explained by Eq. (7).

### 4.3. The Action Potential

The action potential consists primarily of two sequential events: an inward surge of positive charge largely due to the inward movement of $Na^+$ into the cell followed by an outward surge of positive charges largely

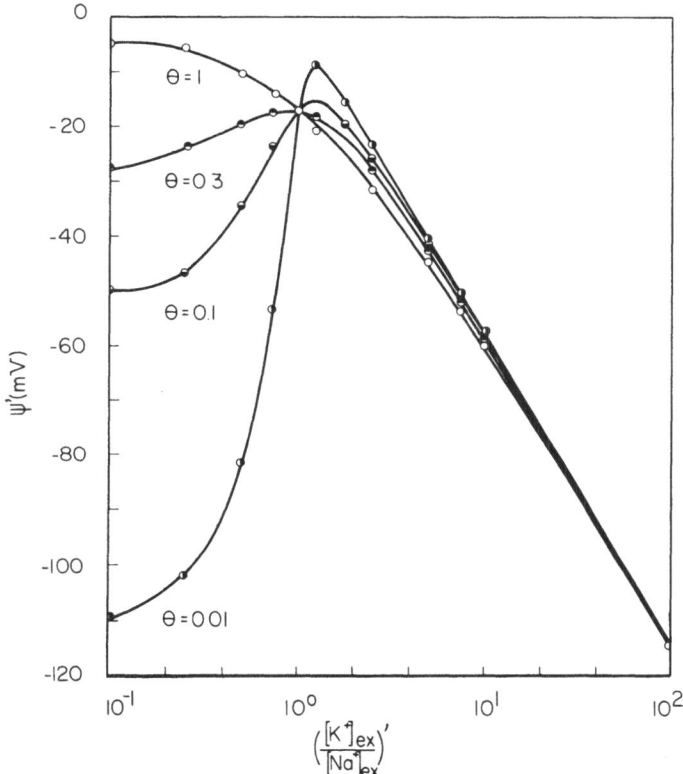

**FIGURE 3.** Plot of resting potential against external $K^+$ and $Na^+$ concentration ratio at different values of $\theta$. Ordinate represents $\psi'$, which is equal to $\psi$-constant. Abscissa represents $([K^+]_{ex}/[Na^+]_{ex})'$, which is $([K^+]_{ex}/[Na^+]_{ex} \cdot K^{00}_{Na \rightarrow K})$. For experiments carried out in the presence of a constant concentration of $Na^+$ (e.g., 100 mM) the abscissa is then $[K]_{ex} \cdot (K^{00}_{Na \rightarrow K}/0.1)$. (Ling[23] by permission of *Physiological Chemistry and Physics*.)

due to $K^+$ leaving the cell. The interpretation offered by Hodgkin and Huxley is that these movements reflect the transient opening of a specific $Na^+$ gate followed by a transient opening of a $K^+$ gate. The basic difficulty with the membrane-pump theory demands an alternative interpretation. The following is one based on the AI hypothesis.

In 1953, Ling extended his theoretical model of the living cell to include a two-dimensional replica of the three-dimensional cell body at the cell surface.[76,82] The $\beta$- and $\gamma$-carboxyl groups at the cell surface like many other $\beta$- and $\gamma$-carboxyl groups existing throughout the cell substance in resting cells, prefer $K^+$ over $Na^+$. It is the selective preference of these surface anionic sites and the high percentage of countercations associated with these sites that has made it possible to explain the competition and saturability shown in the $K^+$ entry into living cells. The experiment of Edelmann[79] described above confirmed yet another postulation of the AI hypothesis, viz., that the same surface anionic sites that determine selective ionic permeability also determined the cellular resting potential.

Theoretical calculations presented briefly in 1960[76] and in full in 1962[39] provided a basis for the postulation of the AI hypothesis that the $\beta$- and $\gamma$-carboxyl groups at the cell surface and elsewhere may alter their electron charge density by allosteric interaction at a distant site, e.g., by the detatchment or adsorption of a cardinal adsorbent, $Ca^{2+}$. As a result, the electron density of the anionic carboxyl groups increases with a consequent rise of the $c$ value. The $c$ value* is the underlying parameter that determines the $pK_a$ of acidic groups.

A molecular interpretation of the action potential in terms of the AI hypothesis is as follows:

(1) In the resting state the cell surface water exists as polarized multilayers (possibly more strongly polarized than the bulk-phase water in the cell). This polarized water then provides size-dependent selective permeability to solutes and ions by the "saltatory route." The $c$ value of the surface fixed anionic site is such that $K^+$ is preferred. As a result, these anionic sites offer additional routes for facile entry of ions like $K^+$ by the adsorption–desorption route than, say, by $Na^+$. Nevertheless both the saltatory and the adsorption–desorption routes are open to $K^+$ as well as $Na^+$, only their quantitative aspects differ. In other words there are no specific $K^+$ routes (or gates) or $Na^+$ routes (or gates).

(2) Vacant surface anionic sites, due to desorption and outward migration of a minute quantity of countercation, $K^+$ from the cell surface, leaves excess of negative charge at the cell surface, giving rise to the normal resting potential.

(3) Activation leads to a cooperative change of the cell surface

---

* The $c$ value is a parameter representing the electron density of a negatively charged atom. It may be described as a way to quantitatively simulate the aggregate effects of the remaining atoms of an oxyacid on the interaction of a hypothetical, prototype, singly charged oxygen atom with a cation, as a displacement (in angstrom units) of the unit electric charge on the oxygen atom from its prototype location at the center of the oxygen atom. Thus, if the aggregate effect produces an overall displacement of electrons in the system toward the oxyacid oxygen, it can be exactly matched by a specific displacement of the unit charge toward the cation, represented as a positive $c$ value, e.g., $+1.0$ Å. On the other hand, if the aggregate effect is to produce the opposite effect, it would be represented as a negative $c$ value, e.g., $-1.0$ Å. Thus, a high $c$ value corresponds to a high $pK_a$ as in acetic acid, and a low $c$ value corresponds to a low $pK_a$ as in trichloracetic acid.

protein–water system, resulting in a transient increase of the $c$ value of the anionic sites, and a rise in the relative preference for $Na^+$ when compared to $K^+$. A large increase of inward migration of external $Na^+$ via the adsorption–desorption route follows. Concomitantly, there is a local depolarization of the cell surface water and an increase of permeability which permits additional $Na^+$ entry via the saltatory route. Together the entrant $Na^+$ gives rise to an excess of positive charges to the cell surface more than enough to neutralize the excess negative charges present in the resting state. An electric polarity reversal represented by the overshoot happens as a result.

The $K^+$ displaced by the entrant $Na^+$ leaves the cells via the adsorption–desorption route and saltatory route, giving rise to the delayed outward current.

For a full comparison of the theoretical prediction of this model and existing data on the properties of the action potential, the reader should consult other more detailed accounts.[23,25] Three outstanding experimental findings that have bearing on this theory will be mentioned.

(1) Villegas et al.[83] have long ago shown that during an action potential, not only was there a transient gain of $Na^+$ permeability in agreement with the $Na^+$ gate concept of the Hodgkin–Huxley theory as well as with the AI hypothesis. In addition they showed that there was also concomitantly an increase of permeability to sucrose and to erythritol which agree only with the AI hypothesis as due to the depolarization of cell surface water giving rise to a nonspecific, size-dependent rise of cell permeability.

(2) Ion entry contains both a competitive, saturable component (in contradiction to the "independence principle") and a nonsaturable component. The nonobedience to the independence principle has been repeatedly reported and needs no further reiteration.

(3) The rank order of selectivity exhibited by the surface anionic sites of resting muscle cells is $Rb^+ > Cs^+ > K^+ > Na^+$;[79,84] and the $pK_a$ is 4.6.[84] Both indicate surface anionic sites of fairly low $c$ value. During the action potential the rank order of ion selectivity of the surface anionic site becomes $Li^+ = Na^+ > K^+ > Rb^+ > Cs^+$ (Hille[85]). Hille's finding thus not only confirms the adsorption–desorption nature of (part of) the inward $Na^+$ movement; it also confirms that the anionic sites have a higher $c$ value, a fact further confirmed by the generally higher $pK_a$ value of these anionic sites measured than those measured at rest (at rest, 4.6; during activation, 4.8–6.5.[86,87]

Since it is generally known now that the "Na channel" is proteinaceous,[88] the only known anionic groups on protein with $pK_a$ value in this range are the $\beta$- and $\gamma$-carboxyl groups, as long ago proposed by the AI hypothesis.

*Acknowledgments*

This investigation was supported by Office of Naval Research Contract N00014-79-0126 and NIH Grants 2-RO1-CA16301 and 2-RO1-GM11422-13.

*References*

1. L. Page and N. I. Adams, *Principles of Electricity*, van Nostrand, New York (1931).
2. K. E. Rothschuh, *History of Physiology* (G. Risse, transl.), p. 150, Robert E. Kreiger Publishing, Huntington, New York (1973).
3. R. H. Murray, *Science and Scientists in the Nineteenth Century*, Shelden, London (1925).
4. B. Barber, *Science* **134**, 596 (1961).
5. T. Graham, *Phil. Trans. R. Soc. London* **151**, 183 (1861).
6. M. Traube, *Arch. Anat. Physiol. Wiss. Med.* **87**, 128 (1867).
7. A. Findlay, *Osmotic Pressure*, 2nd ed., Longmans Green, London (1919).
8. W. Pfeffer, *Osmotishe Untersuchungen: Studien zur Zell-Mechanik*, 2nd ed., Engelmann, Leipzig (1921).
9. W. Ostwald, *Z. Phys. Chem.* **6**, 71 (1890).
10. J. Bernstein, *Pflügers Arch. ges. Physiol.* **92**, 521 (1902).
11. J. Bernstein, *Elektrobiologie*, F. Vieweg und Sohn, Braunschweig (1912).
12. E. Overton, *Vierteiljahrschr. Naturforsch. ges. Zürich* **44**, 88 (1899).
13. R. Collander, in *Plant Physiology* (F. C. Steward, ed.), Academic Press, New York (1959), Vol. 2, p. 3.
14. R. Collander and H. Bärlund, *Acta. Bot. Fenn.* **11**, 1 (1933).
15. R. Mond and H. Netter, *Pflügers Arch.* **224**, 702 (1930).
16. P. J. Boyle and E. J. Conway, *J. Physiol.* **100**, 1 (1941).
17. R. S. Lillie, *Protoplasmic Action and Nervous Action*, University of Chicago Press, Chicago (1923).
18. K. S. Cole and H. J. Curtis, *J. Gen. Physiol.* **22**, 649 (1939).
19. J. S. Burden-Sanderson and F. Gotch, *J. Physiol.* **12**, 5 (1891).
20. A. L. Hodgkin and A. F. Huxley, *J. Physiol. (London)* **104**, 76 (1945).
21. A. L. Hodgkin and B. Katz, *J. Physiol. (London)* **108**, 37 (1949).
22. D. E. J. Goldman, *J. Gen. Physiol.* **27**, 37 (1943).
23. G. N. Ling, *Physiol. Chem. Phys.* **14**, 47 (1982).
24. G. N. Ling, in *Structure and Function in Excitable Cells* (D. C. Chang, I. Tasaki, W. J. Adelman, Jr., and H. R. Leuchtag, eds.), Plenum Press, New York (1983), p. 365.
25. G. N. Ling, *In Search of the Physical Basis of Life*, Plenum Press, New York (1984).
26. B. Katz, *Nerve, Muscle, and Synapse*, McGraw-Hill, New York (1966).
27. A. L. Hodgkin and A. F. Huxley, *J. Physiol.* **116**, 449 (1952a).
28. A. L. Hodgkin and A. F. Huxley, *J. Physiol.* **116**, 473 (1952b).
29. A. L. Hodgkin and A. F. Huxley, *J. Physiol.* **116**, 497 (1952c).
30. A. L. Hodgkin and A. F. Huxley, *J. Physiol.* **116**, 500 (1952c).
31. D. R. Hoagland, P. L. Hibbard, and A. R. Davis, *J. Gen. Physiol.* **10**, 121 (1926).
32. F. C. Steward, *Protoplasma* **15**, 29 (1932).
33. A. M. Brues, L. G. Wesson, and W. E. Cohn, *Anat. Rec.* **94**, 451 (1946).
34. W. Negendank and C. Shaller, *J. Cell Physiol.* **103**, 87 (1980).
35. H. Lundegårdh and H. Burström, *Biochem. Z.* **277**, 223 (1935).

36. J. Harris, *J. Biol. Chem.* **141**, 570 (1941).
37. G. T. Scott and H. R. Hayward, *Biochim. Biophys. Acta* **12**, 401 (1953).
38. G. N. Ling, in *Phosphorous Metabolism* (W. D. McElroy and B. Glass, eds.), Vol. II, p. 748, Johns Hopkins U. P., Baltimore (1952).
39. G. N. Ling, *A Physical Theory of the Living State: The Association–Induction Hypothesis*, Blaisdell, Waltham, Massachusetts (1962).
40. G. N. Ling, C. Miller, and M. M. Ochsenfeld, *Ann. N. Y. Acad. Sci.* **204**, 6 (1973).
41. R. D. Keynes and G. W. Maisel, *Proc. R. Soc. London Ser. B* **142**, 383 (1954).
42. E. J. Conway, R. P. Kernan, and J. A. Zadunaisky, *J. Physiol.* **155**, 263 (1961).
43. G. N. Ling and M. M. Ochsenfeld, *Physiol. Chem. Phys.* **8**, 389 (1976).
44. H. Levi and H. H. Ussing, *Acta Physiol. Scand.* **16**, 232 (1948).
45. G. N. Ling, *Physiol. Chem. Phys.* **12**, 215 (1980).
46. G. N. Ling, C. L. Walton, and M. M. Ochsenfeld, *J. Cell. Physiol.* **106**, 385 (1981).
47. G. N. Ling, C. Walton, and M. R. Ling, *J. Cell. Physiol.* **101**, 261 (1979).
48. A. L. Lehninger, *The Mitochondria*, Benjamin, Menlo Park (1964).
49. G. N. Ling and M. M. Ochsenfeld, *Physiol. Chem. Phys. Med. NMR* **15**, 127 (1983).
50. G. N. Ling, *Physiol. Chem. Phys.* **12**, 383 (1980).
51. G. N. Ling, *Physiol. Chem. Phys. Med. NMR* **15**, 155 (1983).
52. G. N. Ling and Z. L. Zhang, *Physiol. Chem. Phys. Med. NMR* **15**, 391 (1983).
53. G. N. Ling and R. C. Murphy, *Physiol. Chem. Phys. Med. NMR* **15**, 137 (1983).
54. G. N. Ling and M. M. Ochsenfeld, *J. Gen. Physiol.* **49**, 819 (1966).
55. A. J. Hodge and F. O. Schmidt, *Proc. Natl. Acad. Sci. USA* **46**, 186 (1960).
56. G. N. Ling, *Physiol. Chem. Phys.* **9**, 319 (1977).
57. L. Edelmann, *Histochem.* **67**, 233 (1980).
58. L. Edelmann, *Physiol. Chem. Phys.* **9**, 313 (1977).
59. L. Edelmann, *Microsc. Acta Suppl.* **2**, 166 (1978).
60. C. Trombitas and A. Tigyi-Sebes, *Acta Physiol. Acad. Sci. Hung.* **14**, 271 (1979).
61. L. Edelmann, *Physiol. Chem. and Phys.* **12**, 509 (1980).
62. A. V. Somlyo, H. Gonzales-Serratos, H. Shuman, G. McClellan, and A. P. Somlyo, *J. Cell. Biol.* **90**, 577 (1981).
63. M. Sjöström and L. E. Thornell, *J. Microsc.* **103**, 101 (1975).
64. L. Edelmann, *Physiol. Chem. Phys.* **15**, 337 (1983).
65. P. Y. Chou and G. D. Fasman, *Biochem.* **13**, 211 (1974).
66. G. N. Ling and W. Negendank, *Physiol. Chem. Phys.* **2**, 15 (1970).
67. G. N. Ling, *Intern. J. Neuroscience* **1**, 129 (1970).
68. G. Masszi, Z. Szijarto, and P. Grof, *Acta Biochim. Biophys. Acad. Sci. Hung.* **11**, 129 (1976).
69. J. S. Clegg, S. Szwarnowski, Z. E. R. McClean, P. J. Scheppard, and E. H. Grant, *Biochim. Biophys. Acta* **721**, 458 (1982).
70. U. Kaatze, O. Gottman, R. Podbleiski, R. Pottel, and U. Terveer, *J. Phys. Chem.* **82**, 112 (1978).
71. E. C. Trantham, H. E. Rorschach, J. S. Clegg, C. F. Hazlewood, R. M. Nicklow, and N. Wakabayashi, *Biophys. J.* **45**, 927 (1984).
72. H. E. Rorschach, in *Water and Ions in Biological Systems* (V. Vasilescu, ed.), Plenum Press, New York (in preparation, 1985).
73. G. N. Ling, *Biophys. J.* **13**, 807 (1973).
74. G. N. Ling, *Fed. Proc.* **14**, 93 (1955).
75. G. N. Ling, *Fed. Proc.* **18**, 371 (1959).
76. G. N. Ling, *J. Gen. Physiol.* **43**, 149 (1960).
77. G. N. Ling, in *Glass Electrodes for Hydrogen and Other Cations* (G. Eisenman, ed.), Marcel Dekker, New York (1967), Chap. 10, p. 284.

78. G. N. Ling and G. Bohr, *Biophys. J.* **10**, 519 (1970).
79. L. Edelmann, *Ann. N. Y. Acad. Sci.* **204**, 534 (1973).
80. G. N. Ling, *Physiol. Chem. Phys.* **11**, 59 (1979).
81. B. L. Maloff, S. P. Scordillis, C. Reynolds, and H. Tedeschi, *J. Cell. Biol.* **78**, 199 (1978).
82. G. N. Ling, *Proc. 19th Internat. Physio. Congr.*, p. 566, Montreal, Canada (1953).
83. R. Villegas, M. Blei, and G. M. Villegas, *J. Gen. Physiol.* **48**, 41 (1965).
84. G. N. Ling and M. M. Ochsenfeld, *Biophys. J.* **5**, 777 (1965).
85. B. Hille, *Fed. Proc.* **34**, 1318 (1975).
86. I. M. Stillman, D. L. Gilbert, and R. L. Lipicky, *Biophys. J.* **11**, 55a (1971).
87. H. Drouin and R. The, *Pflugers Arch. Ges. Physiol.* **313**, 80 (1969).
88. F. Hucho and W. Scheibler, *Mol. Cell. Biochem.* **18**, 151 (1977).

# 3

# Electrodic Chemistry in Biology

## M. A. Habib and John O'M. Bockris

ABSTRACT: Historical developments of electrochemical concepts in biology are reviewed. Four classical treatments of membrane potential, namely, the treatment of original Nernstian equilibrium, of Donnan equilibrium, the treatments based on Planck's liquid junction potential (Hodgkin, Huxley, and Katz; Goldmann), and the association–induction theory of Ling, respectively, are discussed. The assumptions underlying the classical theories of membrane potential and their implications are pointed out. Developments of the electrodic suggestions in biology are analyzed. It is shown that the heat generation of the body is consistent with electrochemical functioning of the cell. An electrodic theory of membrane potential is outlined. The extent of agreement with experiment is found to be reasonable. It is concluded that biological reactions are largely surface reactions and only to a smaller extent redox reactions in solution.

## 1. Introduction

The application of electrochemistry in biology dates back into the eighteenth century, when Luigi Galvani[1] conducted his first experiment with frog legs. It is well known that electrical potentials are found in biological systems. The thermodynamic approach of Nernst,[2] the originator of the classical theory of the thermodynamics of galvanic cells, was applied by Bernstein[3] as early as 1902 to interpret the potential differences which exist across membranes in biology in general. A number of famous names stand out in the field, of which the most renowned are Donnan,[4,5] Hodgkin, Huxley, and Katz.[6–10] These workers[4–10] are associated with the classical electrochemistry of the conduction of elec-

M. A. Habib and John O'M. Bockris • Department of Chemistry, Texas A&M University, College Station, Texas 77843.

tricity through nerves, and the part played by alkali and alkaline earth ions in determining electrochemical potentials. Most recently, Mitchell[11-13] has entered the field (although, in a different terminology, he has contributed to it since 1961,[12,13] and his work (in terms of protons) descends from the earlier thermodynamic approaches.

## 2. The Membrane Potential

It is universally found that electrical potentials are associated with surfaces in biology. In pre-electrodic days, this called for some special explanation. Many measurements of the so-called "membrane potentials" showed values in the region of 50–100 mV; but occasionally measurements were much lower, or stretched to values of 250 mV[14,15] (Table 1). The large edifice of electrophysiology is intimately associated with these potentials.

The membrane potential in biology came to prominence in the days in which electrode phenomena were treated exclusively in terms of equilibrium thermodynamics. Between 1892 (Nernst[2]) and 1911 (Donnan[4,5]), three treatments were given of membrane potentials. They form such a durable part of electrochemistry, not because of their importance per se, or even of their direct relevance to biological phenomena, but because one of them was the origin of the best-known of bioelectrochemical theories, the Hodgkin–Huxley–Katz[6-9] mechanism for the passage of electricity through nerves.

### 2.1. The Four Classical Membrane Potential Treatments

#### 2.1.1. The Original Nernst Treatment

It is assumed that only one ionic entity can diffuse through the membrane. A concentration gradient pulls the diffusing ion, and a potential gradient holds it back. Equilibrium is thus attained. Then,

$$\Delta E_{\text{membrane}} = \frac{RT}{F} \ln \frac{A_i^\alpha}{A_i^\beta} \tag{2.1}$$

where $A_i$ is the activity of the permeating ion, and $\alpha$ and $\beta$ are the phases.

#### 2.1.2. The Donnan Membrane Equilibrium[4,5]

This is the best-known of the three classical treatments. It differs from the treatment of Nernst, in that it assumes that both cations and anions

*TABLE 1.* Membrane Potentials for Some Biological Membranes

| Cell | Membrane potential (mV) | Reference |
|------|-------------------------|-----------|
| Neurospora | 200 | 14 |
| Chara australis | 170 | 14 |
| Ascaris lumbricoides | 45 | 84 |

permeate the membrane, but that there are some large organic ions on one side which do not pass. Electroneutrality now causes there to be a difference of cation and anion concentration on each side at equilibrium.

### 2.1.3. Planck's Liquid Junction Potential

If two ionic solutions are brought into contact via a permeable membrane, difference in concentration or type of ion will give rise to a liquid junction potential. This concept differs from the earlier ones in that it involves the Nernst–Planck equation which embraces a theory of the influence of both electrical and concentration gradients on the ionic distribution.[16]

By making a number of simplifications of these ideas, Goldmann[17] showed that

$$\Delta E = \frac{RT}{F} \ln \frac{\sum D_i C_i^\beta + \sum D_k C_k^\alpha}{\sum D_i C_i^\alpha + \sum D_k C_k^\beta} \qquad (2.2)$$

where $D_i$ is the diffusion coefficient of the ions, of species $i$.

If this equation is applied to biological systems and the $D$'s are taken as those in aqueous solution, potentials an order of magnitude less than those usually observed across biological membranes are obtained. It is argued that the diffusion coefficients which must be taken are selectively different and account for the discrepancy. This approach involves a steady state rather than an equilibrium.

Hodgkin, Huxley, and Katz[6-9] modified the Goldmann equation by incorporating the differences in the permeabilities of the ions in the Nernst–Planck equation.

### 2.1.4. Association–Induction Theory of Gilbert Ling[18-20]

Gilbert Ling[18-20] suggested that the membrane potential is due to adsorption on fixed charge sites on the cellular membrane, which is, thus, asymmetrical, with the double layers on either side being charged differently. It is this difference in charge, along with differing ion permeabilities caused by the asymmetry, that gives rise to the membrane potential.

Ling has expressed his theory of the generation of cellular membrane potential as follows:

$$\Delta E = k_A - \left(\frac{RT}{F}\right) \ln \left[k_1(K^+) + k_2(Na^+)\right] \qquad (2.3)$$

where $k_1$ and $k_2$ are the adsorption constants of $K^+$ and $Na^+$, respectively. However, the origin of the constant $k_A$, and why only the extracellular fluid affects the surface charge, is not clear.

### 2.1.5. The Assumption Underlying the Classical Theory of Membrane Potentials

The basic equation at the base of the classical theory of membrane potentials is the Nernst–Planck relationship. This is a relationship which does not refer to thermodynamic equilibrium (membrane potentials represent dynamic metabolic processes) but refers to steady states established when both concentration gradients and potential gradients inhabit interface situations. But one basic assumption is made in considerations connected with the application of the Nernst–Planck[16] equation to interfaces, and that is that there is no electron transfer taking place between the solution and the membrane, in parallel to the processes which are under consideration in the Nernst–Planck equation. What the latter does is to assume there is no charge transfer between solid and liquid, no transfer of charges within the solid of the membrane, and that the only forces that exist are those recognized as an electric field gradient which will pull ions in one direction and a concentration gradient which may pull them in the same direction or in the opposite direction. It is the net of these forces which come to balance in the steady state and allows an electrolyte to pass from one side to the other at a certain rate.

It is noteworthy that, although systems certainly exist in laboratories to which the Nernst–Planck equation applies, the corresponding Goldmann[17] equation or the Hodgkin–Huxley–Katz[6–9] equations are not consistent with much of the data available, as shown by Jahn.[21,22]

The widespread evidence that Planck's steady state liquid junction potential is not a sufficient basis to the interpretation of membrane potentials in living systems raises as a question the real origin of potentials in living systems. Correspondingly, it makes questionable the basis of the Hodgkin–Huxley theory.

## 3. The Possibilities of an Electrodic Aspect in Biology

In the classical theory of electrical potentials in biology, these potentials arise from concentration gradients: they are, in fact, work terms associated with the change of chemical potential which arises when one ion passes from the situation of one concentration to a situation of another; and this difference of work is then equal to the potential difference set up, the picture being that ions of one charge have a larger mobility than ions of another, and when the forces set up by the potential gradient are unequal to the forces set up by the concentration gradient, there is a steady state. No charge transfer occurs.

In this case, the electrochemical aspects of biological situations are useful, but to some extent minor in the fact that there is no electrodic reaction actually taking place in any of the reactions concerned.

On the other hand, if electron transfer reactions occur at interfaces, many things are possible. Thus, for example, and in particular, upon leading reactants (e.g., glucose and oxygen) to the surface of some cells, it may be that the glucose reacts at a series of enzymatic sites which may be called anodic (meaning that electrons transfer from the solution to solid), and in another series of enzymatic sites, electrons are emitted to reduce oxygen with the overall reaction probably being the formation of $CO_2$. Such a fuel cell reaction (in fact, very large numbers of them in pairs) should be associated then with the possibility that such a cell (between the electrodes of which there is an electrical driving force) may drive or electrolyze another process within the body.

One of the outstanding ideas of classical biology is the idea of a pump. In these concepts, the direction or passage of ions that cross a membrane is not rational, in the sense that it contradicts the concentration or potential gradients. For these situations in the series, an entity entitled "a pump" is brought in.[23] It is said that "active entities" take the ions and chemically propel them from one site to another. Although such actions are certainly possible, it may be that they become unnecessary if the processes concerned are connected with a process, the overall $\Delta G$ of which is negative. Thus, in an actual macro fuel cell, in which oxygen dissolves at one electrode to form water, and hydrogen dissolves at another to form protons, the ionic current between the two electrodes is carried, not by these entities, but by others, for example, sodium or potassium ions in alkaline solution and also hydroxyl ions. The electrode reactions occur first, and as a result of this, the movement of particles takes place between the two sides of the membrane. Alternatively, are there set up "by means of pumps" differences of concentration between the two sides of the membrane, and does this then in turn feed back to produce a potential difference? It is the former idea which is more associated with modern electrochemistry, and the latter seems to be associated with a concept of Nernst, and potential differences as arising in reversible cells largely from concentration gradients.

## 4. Electrodic Electrochemistry in Biology

There are two principal difficulties in applying electrodic concepts in biology:

1. If an electrodic hypothesis is to be applied to biological situations, there has to be charge transfer within the solid. It may be electronic or protonic, but there must be a way of taking charges within a solid body from the cathodic sites to the anodic ones. However, the situation in respect to conduction and solid bodies in biology is as yet unclear (see below), though it is certainly more positive than it was before, say, 1970.

2. There is a basic experimental difficulty which can be overcome by a thought experiment but has not yet been overcome in actual experiments, and that is to devise a biological system which can be treated as an electrode, not only conceptually, but also in an experiment. One wants to connect an electronic conductor, in some fashion, to a biological entity which is being treated as an electrode, and make its potential vary by means of application of differences of potential from an outside potentiostat, as if the biological membrane were, in fact, a semiconductor or even a metal conductor.

### 4.1. Earlier Electrodic Suggestions

Historically, Lund[24] in 1928 correlated the process of ionic transport, measured as an electrical current, to metabolic reactions, specifically to redox reactions, such as those occurring in mitochondria. Lund[24] measured the amount of $O_2$ consumed by a frog skin, as a function of the current flowing through the skin. Lund's idea of the possible relation of $O_2$ consumption to some bioelectric phenomena was not well accepted at that time.

In 1945, Lundegardh[25] put forward an explanation of ion transport in terms of redox reactions. The redox reactions occurring in respiration were considered as the source of bioelectric phenomena. Describing the oxidation of $Fe^{2+}$ ion to $Fe^{3+}$ in enzymes, Lundegardh[25] proposed that since $Fe^{3+}$ ion attracts one more anion than the $Fe^{2+}$ ion, the process of $Fe^{2+}/Fe^{3+}$ redox reaction causes the movement of anions in the opposite direction to that of the electrons. Since the principal postulate of this theory was regarded as charge separation in connection with ionic transport, the resulting approaches emphasized the latter aspect, rather than the interfacial one.

Freeman Cope[26,27] (1963) was the first person to put forward a model (Figure 1), in which a particular site on an enzyme particle acts like an electrode immersed in a solution containing a redox system functioning cathodically, so that this site develops an electrode potential, which he erroneously thought of as being in reversible equilibrium. Another site, at the other end of the enzyme, acts like a second electrode, at which a second redox process occurs reversibly, it being *assumed* that there is sufficient electronic conductivity along the same path in the enzyme, so that the electrons from and to each process supply each other. It was implied that each enzyme would contain many such $X$ and $Y$ site pairs. Together, they would function without net electron transfer to give an overall chemical reaction.

What Cope suggested was not literally possible because of the reversibility which he assumed, but it *suggests* the application of the Wagner–Traud hypothesis[83] in biology, i.e., the idea of an *overall* reaction with no net electron transfer, but electron transfer controlled. What it needs is the application of the theory of interfacial charge transfer, in which the current density is related to the deviation of the potential drop at the interface, from the reversible value.

The difference of this view, in explaining the potential difference across biological interfaces, compared with that of the earlier workers, is profound. Thus, ion transport through the membrane is a consequence of the electrode potential differences referred to above. Electron transfer at the solid/liquid interface sites is rate determining (one of the two kinds of sites will dominate). The concentration of alkali metal ions on the two sides of the membranes may be the result, rather than the cause, of the potential

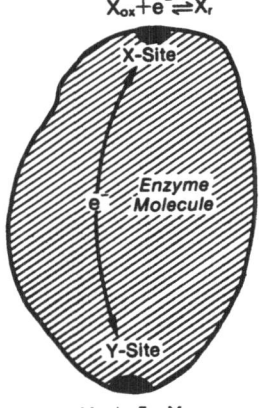

$$X_{ox} + e^- \rightleftharpoons X_r$$

$$Y_{ox} + e^- \rightleftharpoons Y_r$$

*FIGURE 1.* Enzyme particle showing electron transfer between $x$ site and $y$ site, giving rise to oxidation and reduction reactions, the rates of which are governed by the overvoltage phenomena at the sites and the electrical resistance across the enzyme particles.[27]

difference across the membrane. The potential is determined by the thermodynamics and kinetics of the cathodic and anodic processes occurring at the two electrodes. The cathodic process may sometimes be the reduction of oxygen. The anodic reaction will be an organic oxidation reaction (different for membranes of varying function). Those membranes connected with metabolism will be involved in an anodic reaction (the oxidation of, e.g., glucose), which will lead to the formation of $CO_2$.

Del Ducca and Fuscoe[28] then published a paper in 1965 which, for the first time, came out with the suggestion that biological cells could be regarded as fuel cells, and attempted to show the concepts involved.

In 1967, Bockris and Srinivasan[29] published a note in *Nature*, pointing out that the high efficiency of energy conversion in biological metabolism indicated a mechanism which could not be based upon any kind of heat enegine. They suggested that one explanation of the high efficiencies would be to accept a fuel cell mechanism as converting the energy of the chemical reactions occurring in biological organisms to electricity and work.

This hypothesis was discussed by Bockris in 1969,[30] in a comprehensive article in *Nature* entitled, "Are Interfacial Electron Transfer Reactions an Important Step in Biological Reactions?" These ideas were treated in the textbook, *Modern Electrochemistry* (1977).[31]

Such contributions were made with the hypothesis that somehow the difficulty of the lack of sufficient electronic conductivity in biological organisms could be overcome. It was, therefore, with excitement that the work of Rosenberg[32-34] was met, for Rosenberg seemed to have found that the very low conductivity of proteins, which had always been a stumbling block to electron transfer concepts in biology, was, in fact, misleading, because in the experimental arrangements generally used, the proteins were examined in the dry state; whereas, as Rosenberg[32-34] showed experimentally, if the protein were hydrated, then an electronic conductivity could be found, which was several orders of magnitude higher than the conductivity in the dry.

Eley[35] interprets these measurements in terms of semiconductor doping. When water is added, it dissociates to act as a dopant and, hence, an increase of several orders of magnitude of conductivity arises.

These ideas of Rosenberg and Eley[32-34] have, however, recently been met by the work of Pethig, Gascoyne, and Szent-Gyorgyi.[36] These workers have shown that deuteration of the protein gives rise to a change in conductivity, which corresponds to that which would be expected were the proton to be the entity which is carrying charge in the protein.

One of the more famous suggestions Albert Szent-Gyorgyi has made since 1941[37-41] was that electronic conductivity could take place in proteins. The suggestion was initially repudiated, because it was pointed out that the energy gap in proteins is at least 5 eV, so that the conductivity would be negligible. Of course, if there are doping processes in the normal sense, a great increase of conductivity could ensue, and Szent-Gyorgyi's suggestions would appear in principle acceptable.

The recent conclusions of Pethig *et al.*[36] are not simple. According to them, there is a dual transport, partly by protons and partly by electrons, and the two movements are coupled. The present situation, therefore, is that the conductivity of proteins is an ambiguous area, but it seems established that some kind of nonionic conductance in the wet protein does occur and that the order of magnitude of it in specific conductance terms may be as high as $10^{-6}$ mho cm$^{-1}$.

However, the needed degree of conductance is perhaps not a large one. In order for electrodic interfacial charge transfer reactions to be important, it is necessary for the ohmic potential drop across the membrane to be negligible. In electrochemical terms, this may be taken as $< 10$ mV. It is trivial to show[31] that the ohmic drop across a cell is given by $iL/\kappa$, where $L$ is the length of the current path, $i$ the current density, and $\kappa$ the specific conductivity of the cell as a whole (but excluding interface regions outside the cell).* For $iL/\kappa < 0.01$ V and $L \simeq 50 \times 10^{-8}$ cm, $i/\kappa$ is $2 \times 10^4$ V/cm. Data for wet proteins[42] suggest $10^{-10} < \kappa < 10^{-6}$ mho cm$^{-1}$. Taking $\kappa \simeq 10^{-8}$ mho cm$^{-1}$, no meaningful $IR$ drop across a membrane having a conductance in the middle range of that of wet protein would occur, for $i$ up to 0.2 mA cm$^{-2}$. Experience suggests this to be a high current density for an electron transfer reaction: the biological current densities[53] are probably lower than $10^{-6}$ A cm$^{-2}$. It seems reasonable to conclude that the model of electron transfer control of interfacial reactions at membranes will not be impeded by the resistance in the solid phase.

## 4.2. Is Heat Generation of the Body Consistent with the Electrochemical Functioning of Cells?

Two general observations can be thought to give backing to the general electrodic view of biological phenomena.

Classical thermodynamic reasoning identifies ATP-splitting and generating reactions as the major source of cellular heat production. This concept is difficult to reconcile with the empirical observation that heat reduction reflects the "surface law," i.e., for any given temperature gradient

---

* The reasoning given assumes that the charge carriers flow across the entire cross section of the membrane. In protein-containing membranes, they may flow across specific proteins. It is probable that here the cross section for current passage is less, but that the conductivity is greater than that assumed above.

in excess of environmental conditions, heat loss is proportional to the body surface area.[43] If animal heat production represented the sum of the heat generated by a multiplicity of biochemical reactions taking place in the bulk phase, it would be proportional to the volume of the animal concerned, rather than to the surface area. Thus, the result indicates internal surfaces as the locale of the activity.

Corresponding to this, the following argument may be of use. The average current density for the functioning of electrochemical redox processes, which takes place at the interfaces of bilayer lipid membranes, is $10^{-7}$ A cm$^{-2}$. Considering a biological organism with a height of 1.8 m, breadth of 0.5 m, and thickness of 0.3 m, the volume of the organism is $0.27$ m$^3$. Considering it to be "full" of cells, each cell to be a micron $(10^{-6}$ m) long and broad and 0.1 $\mu$m thick, the individual cell volume is $10^{-19}$ m$^3$, so that the number of cells in the organism is around $(0.27/10^{-19}) = 2.7 \times 10^{18}$. Each of these cells has an area, according to the above dimensions, of $10^{-6} \times 10^{-6}$ m$^2 = 10^{-8}$ cm$^2$, so that the total area available is $2.7 \times 10^{18} \times 10^{-8}$ cm$^2 = 2.7 \times 10^{10}$ cm$^2$.

On the other hand, a human body works at a rate of around 20 W, and assuming that the cell potential is of the order of 0.1 V, the effective total current is 200 A. It transpires, therefore, that the current density is $200/(2.7 \times 10^{10}) = 0.74 \times 10^{-8}$ A cm$^{-2}$, consistent with the order of magnitude of current density which can be developed at bilayer lipid membranes.[48]

## 5. Redox Electrode Kinetics at Membrane Bielectrodes

Testing theories of charge transfer at interfaces between biological materials and solutions involves an experimental difficulty: the connecting of an electron source (wire) to the membrane. To avoid this, one may employ low area (0.01 cm$^2$) membranes (BLM) within small holes made in thin Teflon sheets with a potential difference applied across the membrane by means of two platinum electrodes[44,45,48,49] (Figure 2). Reference electrodes on either side of the membrane register the total potential difference across it.

The first workers to carry out such experiments (Crane and Davies[44]) found current–potential curves that were not linear. They interpreted it as a result of membrane breakdown[44] (electrostriction[45]). Mandel,[46,47] developing a suggestion due to Bockris, took data for currents caused to pass across pig's bladder membranes by means of the bielectrode technique and found that the logarithm of current varies linearly with the potential applied across the membrane.

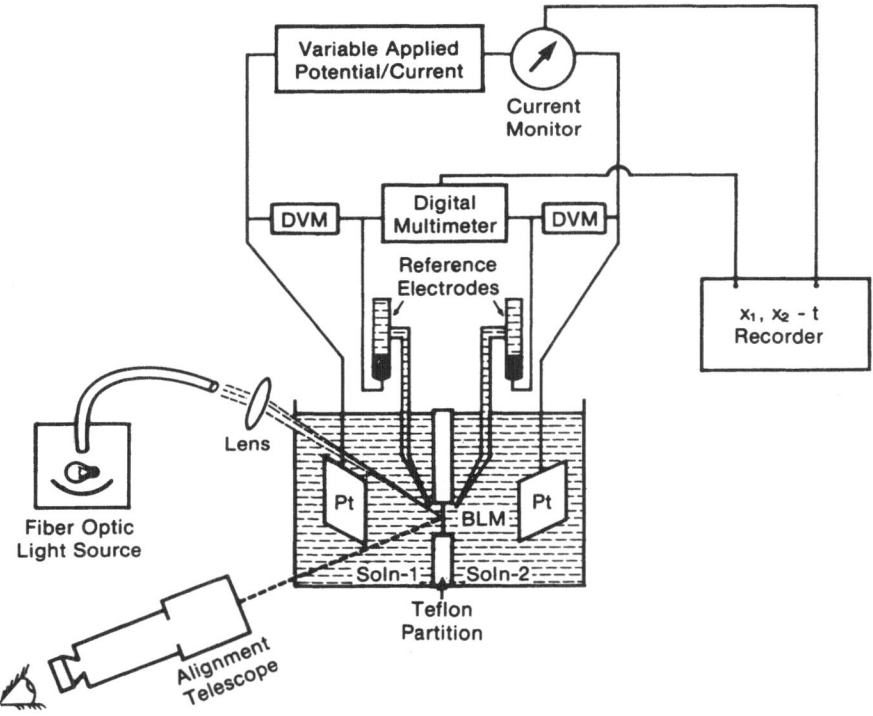

*FIGURE 2.* Experimental setup for the measurement of potentials across a membrane.[48]

Habib and Bockris[48] examined systems in which the redox couple on each side of a BLM was systematically varied. They[48,49] showed that the relation between the current passing across the membrane and the total potential difference across it is Tafelian,[50] and this means that the condition for electron transfer as the rate-determining step prevails at the membrane/solution interfaces.

For a system of membranes with solution on either side, one has two the membrane/solution interfaces, one side being anodic and the other cathodic. If $E_1$ and $E_2$ are the potentials of the corresponding surfaces on the same potential scale, then the p.d. across the membrane is[49]

$$\Delta E = \Delta E_r - \frac{RT}{\alpha_1 F} \ln i_{0,1} - \frac{RT}{\alpha_2 F} \ln i_{0,2} + \frac{RT}{\alpha_1 F} \ln i_1 + \frac{RT}{\alpha_2 F} \ln i_2 \qquad (5.1)$$

For equal areas of side 1 and side 2,

$$i_1 = i_2 = I \qquad (5.2)$$

Thus, Eq. (5.1) can be written as

$$\Delta E = a + B \ln I \tag{5.3}$$

where

$$a = \Delta E_r - \frac{RT}{\alpha_1 F} \ln i_{0,1} - \frac{RT}{\alpha_2 F} \ln i_{0,2} \tag{5.4}$$

and

$$b = \frac{RT}{F} \left( \frac{1}{\alpha_1} + \frac{1}{\alpha_2} \right) \tag{5.5}$$

The above Eq. (5.3) is a pseudo-Tafel equation* for the system of membrane–solution interfaces and is diagnostic for electron transfer to occur across membrane–solution interfaces. The logarithm of the current should vary linearly with p.d. across the membrane, provided the p.d. is not too small ($\geq 20$ mV).

The results found by Habib and Bockris[48] are indeed Tafelian (Figure 3), but attempts to extend the measurements to current density ranges of more than three orders of magnitude caused breakdown of the membranes used. The transfer coefficients were consistent with the electron transfer step as rate determining (Table 2).

More direct experiments to prove the existence of electron transfer through biological membranes involve the deposition of a metal on the membrane surface (Digby,[51,52] Tien[53,54]). Correspondingly, Pohl and Sauer[56,57] found changes in the color of Nile Blue A, indicating its reduction on the outside of an isolated tick salivary gland when current was forced across the gland, so that the negative surface of the gland was in contact with the dye.

Although these experiments are strongly indicative of interfacial electron transfer at the membrane–solution interface, they do not prove it unambiguously, because of the lack of clear values of the potential across the membrane under no applied potential or current condition. Obviously, this should give the difference of the reversible potentials of two different

---

* A normal Tafel equation refers to a single electrode–solution interface, and is based upon a measurement of the potential of the electrode concerned, with a wire attaching to the electrode, then going to a reference electrode. In this case, the prefix "pseudo" is used because the potential being examined is not that of a single interface, with respect to a reference electrode, but that of the difference of potential across two interfaces, together with any significant potential difference across the membrane itself.

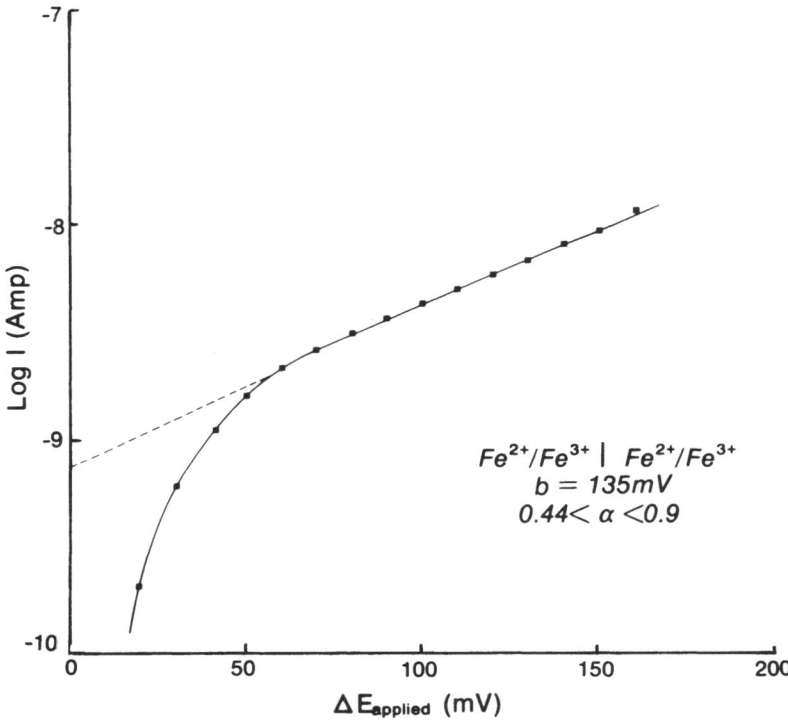

**FIGURE 3.** Variation of applied potential difference, $\Delta E$, across a membrane with logarithm of current flowing across the membrane. The redox couple $Fe^{2+}/Fe^{3+}$ was present in both sides of the membrane in 0.1 $M$ KCl solution.

**TABLE 2.** Kinetic Parameters for the Electron Transfer Processes across the Membrane[a]

| Redox systems | | Tafel slope (mV) | Transfer coefficient $\alpha$ |
|---|---|---|---|
| Side 1 | Side 2 | | |
| $I^-/I_3^-$ | $I^-/I_3^-$ | 165 | $0.36 < \alpha < 0.7$ |
| $Fe^{2+}/Fe^{3+}$ | $Fe^{2+}/Fe^{3+}$ $10^{-2}$ | 135 | $0.44 < \alpha < 0.9$ |
| $Sn^{2+}/Sn^{4+}$ | $Sn^{2+}/Sn^{4+}$ | 300 | $0.2\ \ < \alpha < 0.9$ |
| $Hg^+/Hg^{2+}$ | $Hg^+/Hg^{2+}$ | 400 | $0.15 < \alpha < 0.3$ |
| $I^-/I_3^-$ | $Fe^{2+}/Fe^{3+}$ | 250 | $0.24 < \alpha < 0.5$ |
| $I^-/I_3^-$ | $Sn^{2+}/Sn^{4-}$ | 350 | $0.17 < \alpha < 0.34$ |

[a] Reference 48.

redox systems, one on each side, respectively. However, such reversible potential differences have been recently observed by Feldberg *et al.*[57] and Habib and Bockris.[48] The corresponding electrodic theory based on the difference of redox potentials has recently been developed by Habib and Bockris,[48] as discussed below.

## 6. An Electrodic Model for the Membrane Potential

If some membranes transport electrons, they should exhibit a measurable open-circuit potential when the redox potentials of a redox couple on side 1 of the membrane and another couple on side 2 are different, or when the relative concentrations of the redox couples are different on sides 1 and 2. The potential difference between two sides of the membrane will be then (Figure 4)

$$\Delta E = E_{rev,2} - E_{rev,1} = \Delta E_r \tag{6.1}$$

Thus, under no applied potential condition, within the present hypothesis, the potential difference across the membrane should now refer to the revesible situation, i.e., to the difference of the reversible redox

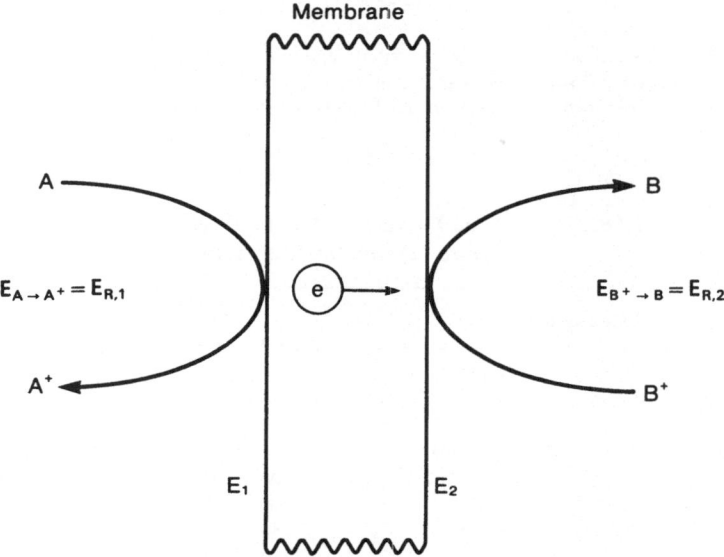

*FIGURE 4.* Schematic diagram showing the oxidation and reduction reactions occurring at membrane–solution interface at side 1 and side 2 of a membrane.

potentials occurring on each side of the membrane, respectively. Thus, $\Delta E$ is measured directly by taking the values of the potential between the reference electrodes. $E_{ra}$ and $E_{rc}$ are the anodic and cathodic redox potentials in solutions 1 and 2, respectively, which are measured by taking the potential of a Pt electrode in each solution with respect to a reference electrode in the same solution.

Habib and Bockris[48] measured the potential differences across bilayer lipid membranes separating two solutions containing redox systems of different redox potentials (Figure 2) under no applied potential condition. They also measured the potential differences when the bilayer membrane was replaced with a Pt membrane, which represents the ideal situation in the presence of electronic condition when the electron transfer hypothesis for the membrane potential is undoubtedly applicable. Correspondingly, they found that the measured membrane potential is, indeed, the difference of the redox potentials on the two sides of the Pt membrane (line I, Figure 5).

Line II in Figure 5 represents similar measurements to that of line I, but the membrane is now a BLM. It is seen that there is a strong correlation between the membrane potential and the difference of the redox potentials, until $\Delta E_r \sim 150$ mV, after which the membrane potential is no longer a function of the redox potential, mainly because of the instability of the BLM under the experimental condition.[48]

Deviations of the membrane potential from the difference in redox potentials are interpreted[48] as the potential observed not being one at equilibrium but one at a steady state. If a redox reaction occurs at a steady state (not at equilibrium) at side 1, say (Figure 4),

$$A \rightarrow A^+ + e \tag{6.2}$$

the electrons released from this reaction would travel across the membrane and initiate another redox reaction at side 2, say,

$$B^+ + e \rightarrow B \tag{6.3}$$

and, thus, overpotentials will be developed at interfaces 1 and 2. The overpotential is the potential of an electrode, set up when a net current passes across it at a value, $i$, and is measured with respect to the potential of a reference electrode at which the reaction concerned is occurring in the thermodynamically reversible condition, i.e., at $i = 0$. The overpotentials at side 1 (anodic) and side 2 (cathodic) are given by

$$E_c - E_{rc} = \eta_c = -|\eta_c| \tag{6.4}$$

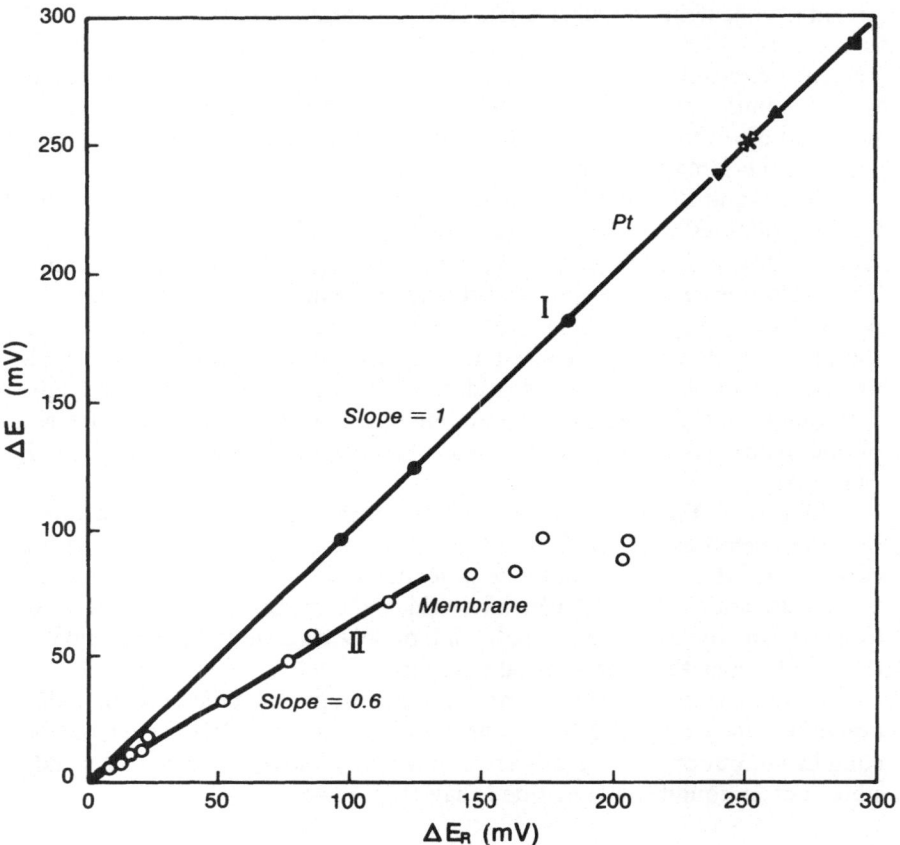

FIGURE 5. Curve I: $\Delta E$ versus $\Delta E_{\text{redox}}$ with a platinum plate in place of a membrane with ▼, $I^{-/3-}/Sn^{2+/4+}$; ★, $Sn^{2+/4+}/Fe^{2+/3+}$; △, $Fe(CN)_6^{3-/4-}/I^{-/3-}$; ■, $Fe(CN)_6^{3-/4-}/Fe^{2+/4+}/Fe^{2+/3+}$; and ●, $Fe(CN)_6^{3-/4-}$, with different concentration in each side. Curve II: $\Delta E$ versus $\Delta E_{\text{redox}}$ for BLM with $Fe(CN)_6^{3-/4-}$ on each side with different concentrations.[48]

and

$$E_a - E_{ra} = \eta_a \tag{6.5}$$

where $E_a$ and $E_c$ are the anodic and cathodic potentials of the membrane surface 1 and 2, respectively, and $E_{ra}$ and $E_{rc}$ are the reversible potentials of the redox systems in side 1 and 2, respectively. On subtraction of (6.5) from (6.4), one gets

$$\Delta E = \Delta E_r - \eta_a - \eta_c \tag{6.6}$$

where

$$\Delta E = E_c - E_a \quad \text{and} \quad \Delta E_r = E_{rc} - E_{ra} \tag{6.7}$$

The slope of the variation of $E$ with $E_r$ will then be given by

$$\frac{\partial \Delta E}{\partial \Delta E_r} = 1 - \frac{\partial \eta_a}{\partial \Delta E_r} - \frac{\partial \eta_c}{\partial \Delta E_r} \tag{6.8}$$

Since

$$\eta_a = \frac{RT}{F} \frac{i_a}{i_{0a}} \tag{6.9}$$

and

$$\eta_c = \frac{RT}{F} \frac{i_c}{i_{0c}} \tag{6.10}$$

Eq. (6.6) becomes

$$\Delta E = \Delta E_r - \frac{RT}{F} i \left( \frac{1}{i_{0a}} + \frac{1}{i_{0c}} \right) \tag{6.11}$$

where, according to the present model, $i_a = -i_c = i$. The Eq. (6.8) may then be represented by the differentiation of Eq. (6.11) with respect to $\Delta E_r$:

$$\frac{\partial \Delta E}{\partial \Delta E_r} = 1 - \frac{RT}{F} i \frac{\partial}{\partial \Delta E_r} \left( \frac{1}{i_{0a}} + \frac{1}{i_{0c}} \right) - \frac{RT}{F} \left( \frac{1}{i_{0a}} + \frac{1}{i_{0c}} \right) \frac{\partial i}{\partial \Delta E_r} \tag{6.12}$$

Now, the exchange current densities are given by[31]

$$i_{0a} = i'_{0a} \exp \left( \alpha E_{ra} F/RT \right) \tag{6.13}$$

and

$$i_{0c} = i'_{0c} \exp \left( -\alpha E_{rc} F/RT \right) \tag{6.14}$$

where $i_0$'s are the constant components of the exchange current densities.

$$\frac{\partial}{\partial \Delta E_r} \left( \frac{1}{i_{0a}} + \frac{1}{i_{0c}} \right) = \frac{1}{i'_{0a}} \exp \left( -\alpha E_{ra} \frac{F}{RT} \right) \left( -\frac{\alpha F}{RT} \frac{\partial E_{ra}}{\partial \Delta E_r} \right)$$
$$+ \frac{1}{i'_{0c}} \exp \left( \alpha E_{rc} \frac{F}{RT} \right) \frac{\alpha F}{RT} \frac{\partial E_{rc}}{\partial \Delta E_r} \tag{6.15}$$

Experimentally, the relative concentrations of the redox couple on the cathodic side were held constant, so that

$$\frac{\partial E_{rc}}{\partial \Delta E_r} = 0 \tag{6.16}$$

with (6.16), Eq. (6.15) may be represented by

$$\frac{\partial \left(\dfrac{1}{i_{0a}} + \dfrac{1}{i_{0c}}\right)}{\partial \Delta E_r} = \frac{1}{i_{0a}}\left(-\frac{\alpha F}{RT}\right)\frac{\partial E_{ra}}{\partial \Delta E_r} \tag{6.17}$$

where $i_{0a}$ is substituted in (6.17) from (6.13).

Since, by definition,

$$\frac{\partial (E_{rc} - E_{ra})}{\partial \Delta E_r} = 1 \tag{6.18}$$

from (6.16) and (6.18), one may have

$$\frac{\partial E_{ra}}{\partial \Delta E_r} = -1 \tag{6.19}$$

which, when substituted in (6.17), yields

$$\frac{\partial}{\partial \Delta E_r}\left(\frac{1}{i_{0a}} + \frac{1}{i_{0c}}\right) = \frac{1}{i_{0a}}\frac{\alpha F}{RT} \tag{6.20}$$

With (6.20), the slope of $\Delta E$ versus $\Delta E_r$, as given by Eq. (6.12), is the given by

$$\frac{\partial \Delta E}{\partial \Delta E_r} = 1 - \frac{\alpha i}{i_{0a}} - \frac{RT}{F}\left(\frac{1}{i_{0a}} + \frac{1}{i_{0c}}\right)\frac{\partial i}{\partial \Delta E_r} \tag{6.21}$$

If it is now assumed that the current, $i$, at the steady state is flowing across the membrane at a constant rate, the $(\partial i/\partial \Delta E_r) = 0$, and hence, from (6.21),

$$\frac{\partial \Delta E}{\partial \Delta E_r} = 1 - \frac{\alpha i}{i_0} \tag{6.22}$$

For metals, it is reasonable to assume that $i_0 \gg i$, and hence the slope $\partial \Delta E/\partial \Delta E_r$ is equal to 1, as found experimentally. When $i \simeq i_0$, which seems more probable for membranes, one obtains

$$\frac{\partial \Delta E}{\partial \Delta E_r} = 0.5 \tag{6.23}$$

when $\alpha$ is assumed to be 0.5. Thus, the electrodic model can explain the variation of the slope of $\Delta E$ versus $\Delta E_r$ plot from 1 to 0.5. The experimental value of $\partial \Delta E/\partial \Delta E_r = 0.6$ is thus explainable in terms of this electrodic model. However, it is desirable to find out the implications of the assumption of $\partial i/\partial \Delta E_r = 0$. From (6.9)

$$i_a = \frac{F}{RT} i_{0a} \eta_a = \frac{F}{RT} i_{0a}(E_a - E_{ra}) \tag{6.24}$$

Therefore,

$$\frac{\partial i_a}{\partial \Delta E_r} = \frac{F}{RT}(E_a - E_{ra})\frac{\partial i_{0a}}{\partial \Delta E_r} + \frac{F}{RT}i_{0a}\frac{\partial(E_a - E_{ra})}{\partial \Delta E_r} \tag{6.25}$$

Differentiation of $i_{0a}$ from Eq. (6.13) with respect to $\Delta E_r$ and substitution in Eq. (6.25) gives

$$\frac{\partial i_a}{\partial \Delta E_r} = -\frac{\alpha F}{RT} i_a + \frac{F}{RT} i_{0a} \frac{\partial \eta_a}{\partial \Delta E_r} \tag{6.26}$$

Using Eq. (6.8) for $\partial \eta_a/\partial \Delta E_r$ and with $\partial E_{rc}/\partial \Delta E_r = 0$ from (6.16), one may obtain

$$\frac{\partial i_a}{\partial \Delta E_r} = -i_a \frac{\alpha F}{RT} + \frac{F}{RT} i_{0a} - \frac{F}{RT} i_{0a} \frac{\partial \Delta E}{\partial \Delta E_r} - \frac{F}{RT} i_{0a} \frac{\partial E_c}{\partial \Delta E_r} \tag{6.27}$$

Substitution of (6.27) in (6.21) and rearrangement gives

$$\frac{\partial \Delta E}{\partial \Delta E_r} = 1 - \frac{\alpha i}{i_{0a}} - \left(1 + \frac{i_{0c}}{i_{0a}}\right) \frac{\partial E_c}{\partial \Delta E_r} \tag{6.28}$$

Use of Eq. (6.10) for $i_c$ and with $-i_c = i_a$ and differentiation with respect to $\Delta E_r$, gives

$$\frac{\partial E_c}{\partial \Delta E_r} = -\frac{RT}{F}\frac{1}{i_{0c}}\frac{\partial i_a}{\partial \Delta E_r} \tag{6.29}$$

Substituting for $\partial i_a/\partial \Delta E_r$ from (6.27) and rearranging gives rise to

$$\frac{\partial E_c}{\partial \Delta E_r} = \frac{\alpha \dfrac{i_a}{i_{0c}} - \dfrac{i_{0a}}{i_{0c}} + \dfrac{i_{0a}}{i_{0c}}\dfrac{\partial \Delta E}{\partial \Delta E_r}}{1 - \dfrac{i_{0a}}{i_{0c}}} \tag{6.30}$$

or

$$\frac{\partial E_c}{\partial \Delta E_r} = \frac{1 - \dfrac{\alpha i}{i_{0a}} - \dfrac{\partial \Delta E}{\partial \Delta E_r}}{1 - \dfrac{i_{0c}}{i_{0a}}} \tag{6.31}$$

Eq. (6.31) may now be used to obtain $\partial E_c/\partial \Delta E_r$ for $i_{0a} = i_{0c}$, $i_{0a} \ll i_{0c}$ or $i_{0a} \gg i_{0c}$, and Eq. (6.28) then gives the expression for $\partial \Delta E/\partial \Delta E_r$, which is the same as Eq. (6.22).[48]

## 7. Application of the Goldmann Equation

The classical theory of membrane potentials is summarized by the Goldmann Eq. (2.2). Utilizing the concentrations used in the present experiments, the fact that KCl was present on both sides at the same concentration, and that this concentration was much higher than that of either of the redox ions, one finds that the Goldmann expression gives only <0.2 mV for a membrane potential (cf. experimental value of 75 mV).

Correspondingly,

$$\Delta E = \frac{RT}{F} \ln \frac{D_1 C_1^{Cl^-} + D_1 C_1^{Fe(CN)_6^{4-}} + D_2 C_2^{K^+} + D_2 C_2^{Fe(NC)_6^{3-}}}{D_1 C_1^{K^+} + D_1 C_1^{Fe(CN)_6^{3-}} + D_2 C_2^{Cl^-} + D_2 C_2^{Fe(CN)_6^{4-}}} \tag{7.1}$$

During the experiment, only $C_1^{Fe(CN)_6^{4-}}$ was varied, while all the other concentrations were kept constant. Thus, the differential coefficient, $\partial \Delta E/\partial \Delta E_r$, is given by

$$\frac{\partial \Delta E}{\partial \Delta E_r} = \frac{\partial \Delta E}{\dfrac{RT}{F} \partial \ln C_1^{Fe(CN)_6^{4-}}}$$

$$= \frac{C_1^{Fe(CN)_6^{4-}}}{C_1^{Fe(CN)_6^{4-}} + C_1^{Cl^-} + C_2^{K^+} + C_2^{Fe(CN)_6^{3-}}} \tag{7.2}$$

The diffusion coefficients are assumed to be equal, as a first approximation. With $C_1^{Fe(CN)_6^{4-}} = 10^{-2}$ mol dm$^{-3}$, the maximum concentration used, and $C_2^{K^+} = C_1^{Cl^-} = 0.1$ mol dm$^{-3} > C_2^{Fe(CN)_6^{4-}} = 10^{-4}$, $(\partial \Delta E/\partial \Delta E_r) < 0.001$.

Thus, using values of the concentrations used in the experiments here, this Eq. (7.2) gives 0.001–0.1 for $\partial \Delta E/\partial \Delta E_r$, whereas values of 0.6–1 were obtained. The Goldmann equation, therefore, appears to be inapplicable to situations such as those described here. Its applicability could be reinstated only if arbitrary, i.e., imaginative changes in $D$ values are introduced.

## 8. Electrochemistry in Biomedical Sciences

Application of electrochemical principles in biomedical siences is widespread. These applications include bone growth and stimulation,[57-59,71] bacterial growth control and detection,[58] blood coagulation,[58,60-64] psychotherapy,[58] control of antibody–antigen reaction,[65] inactivation of pathogens,[66] destruction of tumors,[67,68] hormone and nerve stimulation,[58,69] the generation of electrocadiogram potentials of the heart,[58,70] electroencephalogram potential of the brain,[58] electromyogram potentials of the muscular system,[58] electroanalgesia,[58,72] and a myriad of others. Some of the above applications are described by Eugene Findl,[58] Arthur Pilla,[57,73] Srinivasan *et al.*,[74] Turner *et al.*,[75] and Higgins *et al.*[76]

Some of the above applications suggest that fundamental cellular activity can, indeed, be influenced by an electric field. In all these cases, it is most probable that the living cell senses changes in its microenvironment via its cytoplasmic membrane/extracellular fluid interface, i.e., it is electrified, and as such, will have very analogous properties to the electrode–solution interface. This means that the membrane may respond to changes in local electrical field in much the same manner as an electrode via potential-dependent interfacial processes, such as electrostatic and specific adsorption, charge acceptance and charge transfer. Any of these processes will modify the structure of the interfacial region, resulting in a different interaction of the cell with its environment, thereby eliciting a possible trigger for an appropriate modification in cell function.

## 9. Electrochemical Models for Biological Energy Conversion

Many particles pass through the membrane of a cell with the electrochemical energy gradient, but some against it. Adenosin triphosphate (ATP) is thought to provide the energy for this latter occurrence. An important biochemical reaction is, then, the endothermic formation of ATP by the phosphorylation process. The enzymes which are associated with this are found in the subcellular unit known as the mitochondrion. Electrochemical reactions may function here.

Mitchell[11-13] used Nernstian ideas of electrochemistry to support a concept of the functioning of biological membrane which he called chemiosmotic, as descriptive of a reversible potential difference arising across the membrane as a result of osmotic forces. Mitchell does not consider the transfer of protons as in conduction (or electronic conduction associated with it). He assumes that a carrier system transports $H^+$ from site to site and describes this proton translocation system as consisting of many oxidation–reduction loops forming a chain that transfer protons step by step (Figure 6). Thus, ATP synthesis being a hydrolysis, the pH gradient (or protomotive force) generated by the oxidation–reduction chain will drive the ATP synthesis (Figure 6).

Bockris and Tunuli[77] pointed out that there is no need for special carriers of protons through membranes. It is possible to regard the reduction of $O_2$ as taking place on one part of one side of the membrane, and the oxidation of the organic molecule on the other. The latter will produce an excess of protons and, therefore, drive ADP reaction to ATP. As a result of an organoelectrochemical oxidation producing protons on one side of the membrane, the net reaction is coupling of the latter with the electrochemical oxygen reduction occurring on the other side, with no net electron exchange.

Two substantial reviews and discussions of biological and cellular processes in terms of electrodics have been published by Kell[78] and Berry.[79] Berry emphasized that an exergonic oxidation reaction can initiate an electrochemical process without direct chemical coupling, and, thus, the biological cell is a complex electrochemical device, rather than a chemically powered heat engine, and metabolic reactions are electrochemical processes, giving rise to proton currents.

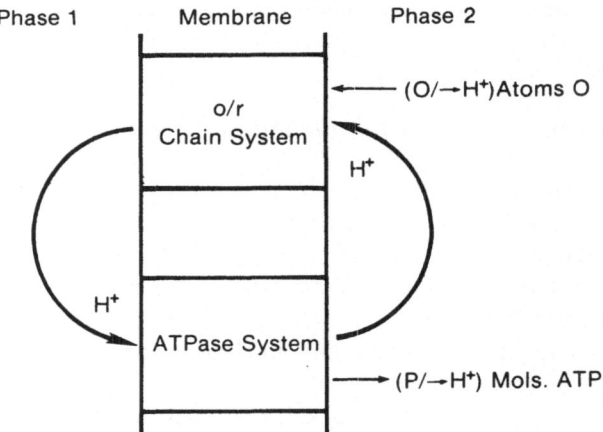

FIGURE 6. Stoichiometry of chemiosmotic coupling. The circulation of one proton is caused by the utilization of a certain number $(O/ \rightarrow H^+)$ of oxygen atoms, and causes the synthesis of a certain number $(P/ \rightarrow H^+)$ of ATP molecules. The P/O quotient is the product $(P/ \rightarrow H^+) \times (\rightarrow H^+/O)$.[11]

## 10. General Considerations

Two views have been contrasted here. First there is the view that the driving force in bioelectrochemical reactions is a concentration gradient arising by means that are not always clearly connected with the metabolic process. This classical view is then contrasted with the fuel cell view, according to which the driving force is the $\Delta G$ of the net electrochemical reaction in what is effectively a fuel cell (typically, perhaps the oxidation on glucose as the anodic reaction and the reduction of oxygen as the cathodic). The driving force of the free energy change gives rise to passage of ions across the membrane in a way that is well established.

Do both mechanisms play a part? At the present time several distinguishing experiments could be devised. One has been done by Bockris and Tunuli,[77] who showed that when photo system I is absorbed upon platinum, and irradiated with light, the ensuing photoelectrochemical reaction is cathodic, but when photo system II replaced photo system I and I was irradiated, the corresponding reaction was anodic. This seems to fit well the theory of photosynthesis, in which photo system I and photo system II are analogous to cathodic and anodic enzyme electrodes, respectively.

A second experiment which is in the future, and is necessary, is to attempt to prove directly the electrodic hypothesis by taking membranes and giving them a background of a conductor. This could be done by using a Langmuir–Blodgett trough, the BLMs being adsorbed on the surface of the solution, plus a conducting arrangement (say a gold-plated glass slide lifted beneath them), the action being repeated several times, producing a sufficiently thick covering of the gold by adsorbed membrane layers. The gold-plated base of the membrane then acts as one electrode contact and the other side of the membrane is placed in the solution and is in effect one side of a bielectrode. This should enable electrical potentials to be changed across the bielectrode and an electrochemical reaction under control of a potentiostat to be carried out at the BLM.

A third approach to experimentation with the electrodic hypothesis in biology might be by means of magnetic fields. As the major difficulty is the attachment of electronic conducting wires to the biological systems, it may be possible to replace this by induction of potential differences across the membranes as is done in electrochemical bone growth.[80]

## 11. Future Applications to Cancer and Arteriosclerosis

Present ideas of electrochemistry in biology are still in their infancy, and biological systems are by their nature complex. However, it does not seem too much to make some extremely crude suggestions about the application of these ideas to our two main degenerative diseases.[81]

Thus, it has been proved by Gingell and Fornes[82] by the variation of the potential of a lead electrode that cells can be made to fall off or stick to the lead, and that falling off or sticking is connected with the potential of zero charge of the lead electrode. Is it possible that such aspects play a role in cancer? One of the major aspects of cancerous cells is that they do not "stick together" but rather act as robber cells breaking through and pushing aside—but not fitting into—the natural cells. It is a feasible hypothesis that the major difference between cancer cells and normal cells is that the electrostatic or double layer interaction between these cells and the neighboring cells is no longer attractive, perhaps owing to the degree of presence of some dielectric agent (e.g., cholesterol or fibronectin). One aspect of cancer would then be interpretable in electrochemical terms.[81]

Alternatively, if the ideas suggested here and earlier[29] are valid then electron movement in biological cells is an essential to their existence. However,[85] this would involve acceleration and deceleration of electronic charges in biological cells, and, according to well-known principles, this implies electromagnetic radiation. Such radiation may be involved in the communication mode by means of which cells know the extent and direction of their growth with respect to other cells. Diminution of the rate-determining electron-transfer reaction at the surface (presumably that due to $O_2$) would diminish the strength of the signal and allow undirected or cancerous growth.

The formation of clots, the origin of arteriosclerosis may also have an electrochemical aspect. The mechanism for clotting is, of course, very complex and involves many steps. However, it is proven[60] that clotting can be turned on and turned off by a change in the potential of metal electrodes. It seems not unreasonable to suggest that whether clots of blood form or not on the arterial walls depends on the potential difference between these walls and their surrounding solution.

This concept would correspond to some work carried out by Sawyer and Srinivasan,[60] in which it was shown that so long as the potential of a metal in contact with both blood suspensions in solution was negative, no clots occurred. Such a principle may well be applicable in the design of compatible prosthetics for use in the body.

## 12. Conclusion

The present electrodic theory of biological reactions, outlined here, would, if confirmed, bring a change in thinking in biology. Thus, biological reactions would be largely surface reactions and only to a smaller extent redox reactions in solution. Electronic conductance would be a prime feature of active biological cells and enzymes, and energy exchange in biological (mitochondrial) cells would bear a relation to the action of fuel cells.

## Acknowledgments

Thanks are due to Dr. Rejou-Michel, Dr. S. G. Canagaratna, and Mr. F. Diniz for several discussions. Financial support from the National Cancer Foundation is gratefully acknowledged.

## References

1. Luigi Galvani, *De Viribus Electricitatus in Motu Musculari Commentarius*, Bonon, Bologna (1791).
2. W. Nernst, *Z. Phys. Chem.* **9**, 137 (1892).
3. J. Bernstein, *Arch. Ges. Physiol. Pfluggers* **92**, 521 (1902).
4. F. G. Donnan, "The Theory of Membrane Equilibria," *Chem. Rev.* **1**, 73 (1924).
5. F. G. Donnan, "Theorie der Membrangleichgewichte und Membranepotentiale bei Vorhandensein von nicht Dialysierenden Elektrolyten; Ein Beitrag zur Physikalisch-Chemischen Physiologie," *Z. Elektrochem.* **17**, 572 (1911).
6. A. L. Hodgkin and R. D. Katz, "The Effect of Sodium Ions on the Electrical Activity of the Giant Axion of the Squid," *J. Physiol.* **108**, 37 (1949).
7. A. L. Hodgkin and A. F. Huxley, "The Dual Effect of Membrane Potential on Sodium Conductance in the Giant Axon of Loligo," *J. Physiol.* **116**, 497 (1952).
8. B. Katz, *Nerve, Muscle and Synapse*, McGraw-Hill, New York (1966).
9. A. L. Hodgkin and A. F. Huxley, "A Quantitative Description of Membrane Current and its Application to Conduction and Excitation in Nerve," *J. Physiol.* **117**, 500 (1952).
10. A. L. Hodgkin, "The Ionic Basis of Electrical Activity in Nerve and Muscle," *Biol. Rev.* **26**, 339 (1951).
11. P. Mitchell, "Chemiosmotic Coupling in Oxidative and Photosynthetic Phosphorylation," *Biol. Rev. Cambridge Philos. Soc.* **41**, 445 (1966).
12. P. Mitchell, "Coupling of Phosphorylation to Electron and Hydrogen Transfer by a Chemiosmotic Type of Mechanism," *Nature* (London) **191**, 144 (1961).
13. P. Mitchell, "General Theory of Membrane Transport from Studies of Bacteria," *Nature* **180**, 134 (1957).
14. C. L. Slayman, "Electrical Properties of Neurospora crassa Respiration and the Intracellular Potential," *J. Gen. Physiol.* **49**, 93 (1965).
15. G. P. Findlay and A. B. Hope, "Ionic Relations of Cells of Chara Australis. VII. Separate Electrical Characteristics of the Plasmalemma and Tonoplast," *Australian J. Biol. Sci.* **17**, 62 (1964).
16. M. Planck, *Ann. Phys. Chem.* **39**, 161 (1890).
17. D. E. Goldmann, "Potential, Impedance and Rectification in Membranes," *J. Gen. Physiol.* **27**, 37 (1943).

18. G. N. Ling, "A New Model for the Living Cell: A Summary of the Theory and Recent Experimental Evidence in its Support," *Int. Rev. Cyt.* **26**, 1 (1969).
19. G. N. Ling, "Two Opposing Theories of the Cellular Electrical Potential: A Quarter of a Century of Experimental Testing," *Bioelectrochem. Bioenerg.* **5**, 411 (1978).
20. G. N. Ling, "The Cellular Resting and Action Potentials Interpretation Based on the Association-Induction Hypothesis," *Physiol. Chem. Phys.* **14**, 47 (1982).
21. T. L. Jahn, "Fixed Charge Theory of the Action Potential and Related Ionic Phenomena," *Bioelectrochemistry and Bioenergetics* **1**, 217 (1974).
22. T. L. Jahn, in: *Proceedings of a Workshop in Bioelectrochemistry*, Princeton, New Jersey, October 10–14 (1971) (A. A. Pilla, ed.), pp. 225–267. National Science Foundation, Washington, D. C., and ESB, Inc., Technology Center, Yardley, Pennsylvania.
23. B. Sarkadi and D. C. Tusteson, in *Membrane Transport in Biology* (G. Giebisch, D. C. Tusteson, and H. H. Ussing, eds.), Springer Verlag, Berlin (1979).
24. E. J. Lund, "Relation Between Continuous Bio-Electric Currents and Cell Respiration," *J. Exp. Zool.* **51**, 265 (1928).
25. H. Lundegardh, "An Electrochemical Theory of Salt Absorption and Respiration," *Nature* **143**, 203 (1939).
26. F. W. Cope, "A Generalized Theory of Particulate Electron Conduction Enzymes Applied to Cytochrome Oxydase. A Theory of Coupled Electron and/or Ion Transport Applied to Pyruvate Carboxylase," *Bull. Math. Biophys.* **27**, 237 (1966).
27. F. W. Cope, "Overvoltage and Solid State Kinetics of Reactions at Biological Interfaces. Cytochrome Oxidase, Photobiology and Cation Transport. Therapy of Heart Disease and Cancer," in *Bioelectrochemistry* (H. Keyzer and F. Gutmann, eds.), Plenum Press, New York (1980).
28. F. G. Del Ducca and J. M. Fuscoe, "Electrons, Enzymes and Energy," *Int. Sci. Tech.* **3**, 56 (1956).
29. J. O'M. Bockris and S. Srinivasan, "Predominantly Electrochemical Nature of Biological Power-Producing Reactions," *Nature* **215**, 197 (1967).
30. J. O'M. Bockris, "A Basic Biological Step?" *Nature* **224**, 775 (1969).
31. J. O'M. Bockris and A. K. N. Reddy, *Modern Electrochemistry*, Vol. 2, Plenum Rosetta Ed., Plenum Press, New York (1977).
32. B. Rosenberg, "Electrical Conductivity of Proteins. II. Semiconduction in Crystalline Bovine Hemoglobin," *J. Chem. Phys.* **36**, 816 (1962).
33. B. Rosenberg and H. C. Pant, "The Semiconducting Rectifier Behavior of a Bimolecular Lipid Membrane," *Chem. Phys. Lipids* **4**, 203 (1970).
34. H. C. Pant and B. Rosenberg, "Electrochemistry on a Bimolecular Lipid Membrane," *Chem. Phys. Lipids* **6**, 39 (1971).
35. D. D. Eley, in *Horizons in Biochemistry* (M. Kasha and B. Pullman, eds.), Academic, New York (1962).
36. P. R. C. Gascoyne, R. Pething, and A. Szent-Gyorgyi, "Water Structure Dependent Charge Transport in Proteins," *Proc. Natl. Acad. Sci. USA* **78**, 261 (1981).
37. A. Szent-Gyorgyi, *Nature* **148**, 157 (1941).
38. A. Szent-Gyorgyi, *Introduction to Submolecular Biology*, Academic, New York (1969).
39. A. Szent-Gyorgyi, *The Living State*, Academic, New York (1972).
40. A. Szent-Gyorgyi, *Electronic Biology and Cancer*, Marcel Dekker, New York (1976).
41. A. Szent-Gyorgyi, *The Living State and Cancer*, Marcel Dekker, New York (1978).
42. R. Pething and A. Szent-Gyorgyi, "Bioelectrochemistry, the Living State and Electronic Conduction in Proteins," in *Bioelectrochemistry* (H. Keyzer and F. Gutmann, eds.), Plenum Press, New York (1980).
43. M. Berry, School of Medical Sciences, Flinders University of South Australia, Bedford Park, South Australia, Private Communication, 1983.

44. E. E. Crane and R. E. Davies, "Electrical Potential Difference and Resistance of Isolated Frog Gastric Mucosa and Other Secretary Membranes," *Trans. Faraday Soc.* **46**, 598 (1950).
45. M. Thompson, R. B. Lenox, and R. A. McClelland, "Structure and Electrochemical Properties of Microfiltration Filter–Lipid Membrane System," *Anal. Chem.* **54**, 76 (1982).
46. L. J. Mandel, "Electrodic Process in Biological Cells," Ph.D. dissertation, University of Pennsylvania, Philadelphia (1969).
47. L. J. Mandel, "Electrochemical Processes at Biological Interfaces," in *Modern Aspects of Electrochemistry*, Vol. 8 (J. O'M. Bockris and B. E. Conway, eds.), Plenum Press, New York (1972).
48. M. A. Habib and J. O'M. Bockris, "Charge Transfer Across Biological Membrane/ Solution Interface: Test of an Electrodic Model," *J. Bioelectricity* **3**, 247 (1984).
49. M. A. Habib and J. O'M. Bockris, "Interpretation of Current Potential Relationships Across Biological Membranes," *J. Bioelectricity* **1**(2), 289 (1982).
50. J. Tafel, *Z. Phys. Chemie* (Leipzig) **50**, 641 (1905).
51. P. S. B. Digby, "Semiconduction and Electrode Processes in Biological Material. I. Crustacea and Certain Soft-Bodied Forms," *Proc. R. Soc. London* **161B**, 504 (1965).
52. P. S. B. Digby, *Symp. Soc. Expt. Biology* **26**, 445 (1972).
53. H. T. Tien, *Bilayer Lipid Membrane*, Marcel Dekker, New York (1974).
54. H. T. Tien, B. Karvaly, and P. K. Shiela, "Electrostenolysis in Bilayer Lipid Membranes," *J. Coll. Int. Sci.* **52**, 185 (1977).
55. H. A. Pohl and J. R. Sauer, "Electron Transport Through the Salivary Gland of the Lone Star Tock," *J. Biol. Phys.* **6**, 118 (1978).
56. H. A. Pohl, in *Bioelectrochemistry* (H. Keyzer and F. Gutmann, eds.), Plenum Press, New York (1980).
57. R. O. Becker and A. A. Pilla, "Electrochemical Mechanisms and the Control of Biological Growth Processes," in *Modern Aspects of Electrochemistry*, Vol. 10 (J. O'M. Bockris and B. E. Conway, eds.), Plenum Press, New York (1972).
58. E. Findl, "Bioelectrochemistry—Electrophysiology—Electrobiology," in *Modern Aspects of Electrochemistry*, Vol. 14 (J. O'M. Bockris and B. E. Conway and R. White, eds.), Plenum Press, New York (1983).
59. F. Burney, E. Herbst, and M. Hinsenkamp, eds., *Electric Stimulation of Bone Growth and Repair*, Springer Verlag, Berlin (1978).
60. S. Srinivasan and P. N. Sawyer, in *Electrochemical Bioscience and Bioengineering* (H. T. Silverman, I. F. Miller, and A. J. Salkind, eds.), Electrochem. Soc., Princeton, New Jersey (1973), pp. 17–36.
61. S. Srinivasan and B. R. Weiss, in *Colloid Dispersion and Mocellar Behavior* (K. L. Mittal, ed.), ACS Symposium Series No. 9, Washington, D.C. (1975), Chap. 24.
62. P. N. Sawyer and S. Srinivasan, "Studies on the Biophysics of Intravascular Thrombosis," *Am. J. Surg.* **114**, 42 (1967).
63. P. N. Sawyer and S. Srinivasan, "New Approaches in the Selection of Materials Compatible with Blood," Annual Reports on Contract PH-68-75 for 1968-1970 (submitted to Medical Devices and Application Program), National Heart Institute, National Institute of Health.
64. G. E. Stoner, J. R. Wilkins, and J. M. Lomeland, "Automated Microbial Detection and Qualification," *J. Electrochem. Soc.* **122**, 109C (1975).
65. C. Mathod, A. Rothen, and J. Casals, "A New Sensitive Method for Detecting Immunological Reactions," *Nature* **202**, 1181 (1964).
66. G. E. Stoner, G. L. Cahen, Jr., and J. Parceltes, "Electrochemical Inactivation of Pathogens," *J. Electrochem. Soc.* **122**, 106C (1975).
67. *Medical World News*, 40U (May 7, 1971).

68. *Medical Tribune*, **16** (May 15, 1971).
69. S. B. Brummer and T. J. Turner, "Electrical Stimulation of the Nervous Systems: The Principle of Safe Charge Injection with Noble Metal Electrodes," *Bioelectrochem. Bioenerg.* **2**, 13 (1975).
70. R. J. Kurtz, E. Findl, A. B. Kurtz, and L. C. Stormo, "Turbulent Flow Streaming Potentials in Large Bore Tubing," *J. Colloid Interface Sci.* **57**, 28 (1976).
71. S. M. Rose, "Regeneration in Denervated Limbs of Salamanders after Induction by Applied Direct Current," *Bioelectrochem. Bioenergetics* **5**, 88 (1978).
72. H. G. L. Coster, D. R. Laver, and J. R. Smith, "On a Molecular Basis of Anaesthesia," in *Bioelectrochemistry* (H. Keyzer and F. Gutmann, eds.), Plenum Press, New York (1980), p. 331.
73. A. A. Pilla, "Electrochemical Information Transfer at Cell Surfaces and Junctions—Applications to the Study and Manipulation of Cell Regulation," in *Bioelectrochemistry* (H. Keyzer and F. Gutmann, eds.), Plenum Press, New York (1980).
74. S. Srinivasan, G. L. Cahen, Jr., and G. E. Stoner, "Electrochemistry in Biomedical Sciences," in *Electrochemistry: The Past Thirty and the Next Thirty Years* (H. Bloom and F. Gutmann, eds.), Plenum Press, New York (1977).
75. A. P. F. Turner, W. J. Aston, I. J. Higgins, G. Davis, and H. A. O. Hill, *Biotechnology and Bioengineering Symposium*, No. 12, Wiley, New York (1982), pp. 413–439.
76. I. J. Higgins, R. C. Hammond, E. Plotkin, H. A. O. Hill, K. Uosaki, M. J. Eddows, and A. E. G. Cass, in: *Hydrocarbons in Biotechnology* (D. E. F. Hansen, I. J. Higgins, and R. Watson, eds.), Heyden, London (1980).
77. J. O'M. Bockris and M. S. Tunuli, "An Electrochemical Model of Biological Energy Storage," *J. Electroanal. Chem.* **100**, 7 (1979).
78. D. B. Kell, "On the Functional Proton Current Pathway of Electron Transport Phosphorylation—An Electrodic View," *Biochim. Biophys. Acta* **549**, 55 (1979).
79. M. N. Berry, "The Function of Energy-Dependent Redox Reactions in Cell Metabolism," *FEBS Lett.* **117**, K106 (1980).
80. S. D. Smith and J. M. Feola, "Pulsed Magnetic Field Modulation of LSA Tumors in Mice," *J. Bioelectricity* **1**, 207 (1982).
81. J. O'M. Bockris and M. A. Habib, "Are There Electrochemical Aspects of Cancer?" *J. Biol. Phys.* **10**, 227 (1982).
82. D. Gingell and J. A. Fornes, "Demonstration of Intermolecular Forces in Cell Adhesion Using a New Electrochemical Technique," *Nature* **256**, 210 (1975).
83. C. Wagner and W. Traud, "Über die Deutung von Korrosionvergangen durch Überlagerung von Electrochemischen Teilvorgangen und über die Potentialbildung an Mischelktroden," *Z. Elektrochem.* **44**, 391 (1938).
84. A. F. Brading and P. C. Caldwell, "The Resting Membrane Potential of the Somatic Muscle Cells of Ascaris umbricoides," *J. Physiol.* **217**, 605 (1971).
85. J. O'M. Bockris, F. Gutmann, and M. A. Habib, *J. Biol. Phys.* (in press, 1985).

# 4

# Elementary Analysis of Chemical Electric Field Effects in Biological Macromolecules

## I. Thermodynamic Foundations

### Eberhard Neumann

*ABSTRACT:* The analysis of bioelectric phenomena requires knowledge of the thermodynamics and kinetics of electric field effects on chemical reactions. Chemical relaxation kinetics in high electric fields is the method of choice in order to imitate the high electric fields operative in living entities like membranes or close to fixed charges like those in proteins and nucleic acids. The present account covers elementary aspects of chemical electric field effects. Part I deals with the thermodynamic foundations of the analytical formalism required for a rigorous treatment of chemical field effects. Part II utilizes this frame of concepts and provides kinetic information as to how to investigate chemical and orientational contributions to structural changes in macromolecules and membrane organizations. The basic formalism established so far for isolated macromolecular systems may be extended to treat more complex bioelectric phenomena on the level of membranes and of cells.

## 1. Introduction

Electric field effects play an important role in many biological cell processes. Phenomena as different as nerve excitation,[1–4] electrogenic ion transport, neurostimulated secretion of hormones and transmitter substances, or the photosynthesis of ATP[5,6] involve cell functions in which biochemical reactions are inseparably coupled to electric field forces.[7,8]

*Eberhard Neumann* • Department of Physical and Biophysical Chemistry, University of Bielefeld, P.O. Box 8640, D-4800 Bielefeld 1, Federal Republic of Germany.

It has been recognized that macromolecules and macromolecular organizations such as biological membranes are particularly effective media for the coupling of high electric fields with biochemical reactivity. Indeed, all biomembranes appear to be associated with electric membrane potentials.

Electrical chemical membrane processes are most evident in the rapid electric communication system of living entities. For example, the generation and rapid transmission of electric signals such as nerve impulses are based on interactions between electric fields and macromolecular membrane organizations. The acquisition and processing of external information, short-term storage, and retrieval of learned experience in the central nervous system are also believed to involve electric field changes coupled to structural transformations in the neuronal membranes.[9]

In order to understand the functional role of electric fields in the usually very complicated biological systems, basic knowledge of electric field effects on simple molecules and on (bio)chemical reactions is an essential prerequisite.

In experimental physics and physical chemistry, external electric fields have traditionally been applied in order to probe the electric–ionic properties of atoms and molecules and to study the electronic and optical details of matter. In particular, the combination of electrical and optical techniques represents a powerful tool for the investigation of overall shape and structure and of the dynamic properties of molecules and molecular interactions.

A particularly instructive example for the power of electro-optic analysis is the membrane-bound bacteriorhodopsin. In this macromolecular system electric fields cause major structural transitions which involve orientation changes of the chromophores retinal, tyrosine, and/or tryptophan residues and $pK$ changes of at least two types of $H^+$-binding sites. The conformational changes are based on a saturable induced dipole mechanism associated with an extremely large anisotropic (electric) polarizability. On a molecular level the induced polarization appears to involve a restricted electric displacement of ionic groups (ion pairs) within the protein in a highly cooperative manner. The electric field effects observed in bacteriorhodopsin membrane fragments are of functional importance for this light-driven $H^+$-pumping system. The results are also suggestive of a possibly quite general polarization mechanism for a very effective interaction of macromolecular organizations with electric fields.[10,11]

Recent electro-optic data on linear polyelectrolytes like the $K^+$-salt of polyriboadenylate, poly$(rA, K^+)$ demonstrate that the dissociation of counterions from the inner atmosphere and, coupled to it, the destacking of the adenine bases in high electric fields is highly anisotropic.[12] The anisotropy of the counterion movement along polyionic surfaces suggests that counterion exchange as well as influx and efflux of counterionic substrates or hormones occur preferably at the border lines of the ionic atmospheres which cover polyionic regions on macromolecular enzyme and receptor proteins and membranes. Once part of the ion cloud, such substrates and activator substances may reach the active sites via surface diffusion. In this model the border regions of the counterion atmosphere serve as a preferable cross section for trapping counterionic substrates.

Chemical thermodynamics and kinetics provide the formalism to describe the observed dependencies of chemical-conformational reactions on the external physical state variables: temperature, pressure, electric and magnetic fields. In the present account the theoretical foundations for the analysis of electrical–chemical processes are developed on an elementary level. It should be remarked that in most treatments of electric field effects on chemical processes the theoretical expressions are based on the "homogeneous-field approximation" of the continuum relationship between the total polarization and the electric field strength (Maxwell field). When, however, conversion factors that account for the molecular (inhomogeneous) nature of real systems are given, they are usually only applicable for nonpolar solvents and thus exclude aqueous solutions. Therefore, in the present study, particular emphasis is placed on expressions which relate experimentally observable system properties (such as optical or electrical quantities) with the applied (measured) electric field, and which include applications to aqueous solutions.

Since molecular-dynamical details of chemical-conformational transitions are derivable from relaxation kinetic measurements, kinetic analysis is therefore included in some theoretical and practical detail.

## 2. Primary Aspects of Matter in Electric Fields

An externally applied electric field is a vectorial perturbation for chemical or orientational distributions involving interacting molecules or molecular organizations. Unlike the isotropic temperature and pressure effects on chemical-conformational transformations, direct sensitivity to electric field forces is bound to certain electrical properties of the chemical structures involved. Major structural-chemical changes in electric fields require the presence of ions, or ionized groups, or permanent or induced dipolar charge configurations, preferably in macromolecular structures.

The primary molecular-mechanical effects of electric fields involve (a) the orientation of permanent dipoles or of dipolar parts in a more complex structure, in the direction of the applied field; (b) the deformation of polarizable systems (and also, but not necessarily, the subsequent orientation of the induced dipoles in electrically anisotropic particles) including changes in the distance between the charge centers of an ion pair in a macromolecular structure.

## 2.1. Electrical–Chemical Coupling

Chemically, molecular conformations with large electric moments increase in concentration at the expense of those configurations with smaller moments. Secondly, the presence of electric fields increases the dissociation of weak acids and bases and promotes the separation of ion pairs into the corresponding free ions (dissociation field effect, second Wien effect). The free ions or ionized structures then may move in the direction of the electric field (electrophoresis) and a field-dependent stationary state in the ion distribution may be established.

Thus, basically two types of electric–chemical coupling may be differentiated, (a) permanent or induced *dipolar equilibria,* and (b) *ionic* (dissociation and association) *processes* involving (macro-)ions and low molecular weight ions (of preferably opposite charge sign). Whereas dipolar equilibria in electric fields are accessible to thermodynamic analysis, ionic processes involving free ions require a kinetic approach.*[13–16]

## 2.2. Elementary Chemical Processes

Changes in the concentration of chemically interacting reaction partners may arise from two types of elementary chemical reactions: *intramolecular* (or monomolecular) and *bimolecular* elementary steps.

If the molecules B of a system equilibrate between two alternative structures or conformations, $B_1$ and $B_2$, according to

$$B_1(\uparrow) \rightleftharpoons B_2(\uparrow) \qquad (2.1)$$

where $B_2$ has the higher electric dipole moment (indicated by the longer arrow), an external electric field will shift this intramolecular equilibrium to the side of higher moments.

---

* It should be mentioned that even in the absence of dipolar, polarizable, or ionic reaction partners, high electric fields may cause shifts in chemical distributions.[13–15] Such a field effect requires, however, that the solvent phase has a finite temperature coefficient of the dielectric permittivity or a finite coefficient of electrostriction; an additional condition is that the chemical reactions proceed with a finite reaction enthalpy ($\Delta H$) or a finite partial volume change ($\Delta V$). Electric field induced temperature and pressure effects of this type are usually very small; they may, however, gain importance for isochoric reactions in the membrane phase.

A bimolecular reaction step is involved in all dimerization processes like

$$B(\uparrow) \cdot B(\downarrow) \rightleftharpoons 2B(\uparrow) \tag{2.2}$$

When, for instance, the dipole moments compensate each other upon complex formation, the reaction is associated with a dipole moment change which is of the order of the monomer dipole moment. This is a particularly favorable condition for the electric field-induced shift to the right-hand side.

The bimolecular process in the reaction

$$(L^+ \cdot B^-) \rightleftharpoons L^+ + B^- \tag{2.3}$$

involves ion-pair formation to $(L^+ \cdot B^-)$ and may lead to neutralization via $(L^+ \cdot B^-) \rightleftharpoons LB$. Such equilibria are always shifted by electric fields to the side of the freely mobile ions as far as the overall change is concerned; see, however, Ref. 12.

The exchange reaction according to

$$L + CB \rightleftharpoons LB + C \tag{2.4}$$

may proceed through an LBC intermediate or may involve the bimolecular elementary steps $BC \rightleftharpoons C + B$ and $LB \rightleftharpoons L + B$; in any case, on an elementary scale the reaction equation (2.4) only involves bimolecular steps.

In general, the equilibrium state of a chemical process between several interaction partners $B_j$,

$$0 \rightleftharpoons \sum_j v_j B_j$$

or

$$|v_1| B_1 + |v_2| B_2 + \cdots \rightleftharpoons \cdots v_n B_n \tag{2.5}$$

where the $v_j$ are the stoichiometric coefficients, may be characterized by an *apparent equilibrium constant K* (concentration ratio) according to

$$K = \prod \bar{c}_j^{v_j} \tag{2.6}$$

In this form, $\prod$ stands for product over species $B_j$; $\bar{c}_j$ is the equilibrium concentration (mol dm$^{-3}$) of $B_j$. The $v_j$ are negative for the reactants and positive for the products; the reaction equation (2.5) is read from left to right. [For Eq. (2.1) we have $K = \bar{c}_2/\bar{c}_1$ and for Eq. (2.3), $K = \bar{c}_L \cdot \bar{c}_B/\bar{c}_{LB}$.]

Whereas the $K$ values are usually concentration dependent, the actual thermodynamic equilibrium constants, $K^{\ominus}$, defined as thermodynamic activity ratios $[K^{\ominus} = K \cdot \bar{Y} = \prod (\bar{c}_j \cdot \bar{y}_j)^{\nu_j}]$, are independent of concentration.

From a practical point of view it is frequently the concentration of a species which may be directly determined, for instance, by optical or electro-optic monitoring techniques. On the other hand the general theoretical analysis of electric field-induced concentration shifts or conformational shifts is, however, intrinsically bound to a formalism which describes the dependence of equilibrium and stationary-state distribution constants on the electric field intensity.

## 2.3. Biological and Experimental Electric Fields

In living organisms electric fields of sufficiently high intensity and of variations large enough to affect chemical processes are encountered not only within membrane phases, but also near the surfaces of membranes and protein organizations, for instance at the active sites of enzymes and receptors.[17]

The observed membrane potential differences, $\Delta\psi_m$, of up to 100 mV may correspond to average values of the electric field strength, $\bar{E} = \Delta\psi_m/d$, of about 100 kV cm$^{-1}$ when the thickness, $d$, of the dielectric membrane part is about 10 nm.

### 2.3.1. Polyionic Field Effects

Besides the powerful field changes occurring within membranes, there are inhomogeneous electric fields originating from the surface of polyionic macromolecules and membranes. The electric potential $\psi(\mathbf{r})$ in the environment of these structures decays with increasing distance $\mathbf{r}$ from the surface of fixed ionized groups (or absorbed ions). The corresponding electric field forces $\mathbf{E} = -\text{grad}\,\psi(\mathbf{r})$, however, are largely screened by counterion atmospheres at physiological ionic strengths (0.1–0.15 mol dm$^{-3}$). An effective direct interaction of these inhomogeneous fields with chemical reactions is limited to a short range of about 1 nm at 0.1 mol dm$^{-3}$ ionic strength and can involve only low molecular weight species. The electric fields of polyionic surfaces may, however, indirectly affect chemical reactions by accumulating small ionic species in their immediate environment. In these regions of higher ionic strengths, rate and extent of chemical reactions between ionic reaction partners will be different from the behavior in the bulk solution. This catalytic effect will be very pronounced for polyelectrolyte structures. Theoretical approaches aimed at understanding polyionic electric field effects are being advanced; for instance, partial dehydration of ionic reaction partners in the high local electric field close to a polyionic surface appears to be one of the important factors.[18] A point not considered so far is that a kind of orientational "fixation" of a reactant in the locally high electric field of the polyion may either favour or disfavour a reaction. Practically, the ionic strength dependence of rate and equilibrium constants may be used in order to establish the mechanism of polyionic field effects.[17,19]

### 2.3.2. Experimental Limitations

The field intensities which are experimentally accessible are limited by dielectric breakdown. In aqueous solutions, fields up to $150 \, \text{kV cm}^{-1}$ may be controlled over distances in the millimeter and centimeter range. It is an additional limitation that in ionic solutions electric fields cannot be maintained for a long time. Owing to ionic currents the field will decrease and Joule heating may cause appreciable temperature increases. These problems can be minimized by applying field pulses of limited duration to ionic solutions and suspensions. In any case, the maximum homogeneous fields that can be experimentally achieved are comparable to the maximum values of electric fields encountered in biomembranes.

### 2.4. Biopolymers

Among the early examples of the successful use of electric fields to probe ionic structures and electrical and optical anisotropies are the linear polyelectrolytes. Basic information about macromolecular dimensions, size, and shape have been derived from the relaxation of field-induced changes in optical properties[20–22] and in electrical parameters of the electrically and optically anisotropic systems.[21,22] The analysis of electric conductivity measurements has demonstrated that linear polyelectrolytes are electrically anisotropic.[23–25] It was established that the extremely large dipole moments, which the electric field produces by displacement of the counterion atmosphere parallel to the long axis of the polyions, are responsible for their orientations in the direction of the external field.

Interest in electric field effects on macromolecules was appreciably revived when it was found that electric fields are capable of producing structural-conformational changes in biopolymers and membranes. Here, too, optical properties are a convenient indicator of field-induced processes. Initial hints of presumably chemical contributions to field-induced changes in birefringence were reported for DNA solutions of low ionic strength.[26] Dielectric measurements have shown that polypeptides in viscous organic solvents may undergo intramolecular helix-coil transitions in the presence of electric fields.[27] In the meantime there are many reports on field-induced conformation changes in multistranded as well as in single-stranded polyelectrolytes.[28-35]

Of particular interest in the discussion of electric field effects in biological structures is the observation of threshold phenomena. It has been found that electric impulses above a certain threshold intensity are capable of triggering conformational transitions in metastable polynucleotide structures. A similar threshold effect is associated with electric field-induced permeability changes in vesicle membranes[29,36] as well as in cellular systems.[37,38] Recently, nonlinear field dependencies of base stacking in single-stranded polynucleotides have been discussed as a threshold effect.[34] See also Part II, Chapter 5 of this volume.

## 3. General Thermodynamic Foundations

Electric field-induced chemical transformations in macromolecules and macromolecular organizations such as membranes cannot be analyzed satisfactorily in all cases because adequate theoretical approaches are lacking. The observed dependence of biopolymer reactions on the electric field intensity seems, however, to be very similar to that of small molecules. Therefore, it appears pertinent to introduce the analysis of field-induced macromolecular changes with relationships which are derived to describe field effects in reactions of small molecules.

### 3.1. General Reaction Parameters

It is well known that *chemical* processes are dependent on the intensive *physical* variables $(z)$, e.g., temperature $(T)$, pressure $(P)$, or external electric field $(E)$. This observation may be generally described by the $z$ dependence of the thermodynamic and apparent equilibrium constants, $K^{\ominus}(z)$ and $K(z)$, and in terms of DeDonder's reaction variable, $\xi$(mol), or of a degree of transition, $\Theta$. According to DeDonder, the differential change $dn_j$ in the amount of substance $n_j$(mol) of the reaction partner $j$ in a chemical process may be related to the stoichiometric coefficient $v_j$ (with the appropriate sign):

$$d\xi = dn_j/v_j = V dc_j/v_j \tag{3.1}$$

where $V$ is the volume $(c_j = n_j/V)$, or in integral form:

$$n_j = n_j(\text{ref}) + v_j\xi, \qquad c_j = c_j(\text{ref}) + v_j\xi/V \tag{3.2}$$

where $n_j(\text{ref})$ and $c_j(\text{ref})$ are the reference values (for instance, at given values of $P$, $T$, and $E$). As $K(z) = \prod \bar{c}_j{}^{v_j}$ is a function of $z$, $\xi$ is also dependent on $z$. It is now important to realize that not only are the equilibrium values $\bar{c}_j$ and thus $\xi$ a function of $z$ but additionally the extent of a $z$-induced change, $\Delta c_j$ or $\Delta \xi$, in $c_j$ and thus in $\xi$ depends on the actual "position" of the equilibrium. Indeed, there are optimum conditions of $\bar{c}_j$ (or $\xi$ and $\bar{\Theta}$) for major $z$-induced chemical transformations. In order to describe this experimental experience, it is useful to express the $z$-induced changes in $\xi$ or $\Theta$, for instance by

$$d \ln \xi = (\partial \ln K/\partial \ln \xi)_z^{-1} d \ln K \tag{3.3}$$

$$d \ln \Theta = (\partial \ln K/\partial \ln \Theta)_z^{-1} d \ln K \tag{3.4}$$

where the subscript $z$ refers to constant values of the $z$ parameters.

For the intramolecular two-state transition represented by Eq. (2.1), we have $\Theta = \bar{c}_2/(\bar{c}_2 + \bar{c}_1)$ and $K = \bar{c}_2/\bar{c}_1 = \Theta/(1 - \Theta)$. In this case the term $(\partial \ln K/\partial \ln \Theta)_z = \Theta(\partial \ln K/\partial \Theta)_z$ is equal to $1/(1 - \Theta)$, and Eq. (3.4) takes the simple form

$$d \ln \Theta = (1 - \Theta) d \ln K \tag{3.5}$$

The corresponding expressions for the bimolecular reaction in Eq. (2.3) are $\Theta = \bar{c}_B/c^0 = \bar{c}_L/c^0$ and $K/c^0 = \Theta^2/(1 - \Theta)$, where the total concentration $c^0$ of a $1:1$ component ratio is given by $c^0 = \bar{c}_L + \bar{c}_{LB} = \bar{c}_B + \bar{c}_{LB}$. Differentiation according to Eq. (3.4) yields

$$d \ln \Theta = [(1 - \Theta)/(2 - \Theta)] d \ln K \tag{3.6}$$

It is readily seen that in both these elementary cases of the intramolecular and bimolecular reactions, a $z$-induced *relative* shift in $\Theta$ and thus in $\xi$ and $c_j$ is maximal at $\beta \to 0$. The *absolute* displacements, however, have maximum values around $\Theta = 0.5$ as outlined below.

When the changes in $K$ (and thus in $\Theta$, $\xi$, or $c_j$) produced by the external perturbation steps $\delta z_i$ are small, we may use linear approximations. For instance, Eq. (3.4) then reads

$$\delta \Theta/\Theta_{ref} = \{\Theta[\partial \ln K/\partial \Theta]_z\}^{-1} \delta K/K_{ref} \tag{3.7}$$

where

$$\delta K = K(z_i) - K_{ref} \ll K_{ref} \tag{3.7a}$$

and

$$\delta \Theta = \Theta(z_i) - \Theta_{ref} \ll \Theta_{ref} \tag{3.7b}$$

holds. The sign $\delta$ is used for small differences. $K_{ref}$ and $\Theta_{ref}$ are appropriate reference values; for instance, these values may refer to $E = 0$. With the help of Eq. (3.7) estimates for the relative shifts of dipolar and ionic equilibria by external electric fields were calculated.[7,8]

### 3.2. General van't Hoff Relations

The dependence of the apparent equilibrium constant $K(z_i)$ on the intensive variable $z_i(P, T, E)$ may be expressed by a generalized van't Hoff relationship according to[13]

$$(\partial \ln K / \partial z_i)_{z \neq z_i} = \Delta Z_i / RT \tag{3.8}$$

where $R$ is the gas constant, $T$ the Kelvin temperature, and

$$\Delta Z_i = RT(\partial \ln K / \partial z_i)_{z \neq z_i} \tag{3.9}$$

is the extensive reaction quantity complementary to $z_i$. (The subscript $z \neq z_i$ means all $z$ constant except $z_i$.)

It is recalled that when $z_i = T$, $\Delta Z_i = \Delta H / T$, where $\Delta H$ is the reaction enthalpy representing the enthalpy difference of *one* stoichiometric transition. When $z_i = P$, $\Delta Z_i = -\Delta \bar{V}$, where $\Delta \bar{V}$ is the molar partial volume change for one stoichiometric transformation. Finally, when $z_i = E$, the measured electric field, then $\Delta Z_i = \Delta M$, where $\Delta M$ is the molar reaction dipole moment.[13-15] It will be shown below that $\Delta M$ of dipolar equilibria refers to the components parallel to $E$, of the dipole moments of the interacting dipolar molecules or macromolecular substructures.

The $\Delta$-sign used in the context of Eqs. (3.8) and (3.9) is defined by the partial differential $\partial / \partial \xi$. The extensive quantity $Z_i$ is given by

$$Z_i = \sum n_j Z_{i,j} \tag{3.10}$$

where

$$Z_{i,j} = (\partial Z_i / \partial n_j)_{n \neq n_j} \tag{3.10a}$$

is the average value of the partial molar quantity $Z_i$ of species $j$, By differentiation of Eq. (3.10) with respect to the reaction variable $\xi$ and using Eq. (3.1) we obtain

$$\Delta Z_i \equiv (\partial Z / \partial \xi)_z = \sum \nu_j Z_{i,j} \tag{3.11}$$

The reaction quantity $\Delta Z_i$ may, in general, be dependent on the actual value of $z_i$ and on $\xi$ because of the dependence of the activity coefficient ratio $Y$ on $z_i$ and $\xi$; see below.

Formally we may introduce a van't Hoff relationship for the thermodynamic equilibrium constant $K^{\ominus}$:

$$(\partial \ln K^{\ominus}/\partial z_i)_{z \neq z_i} = \Delta Z_i^{\ominus}/RT \tag{3.12}$$

where at given values of $z \neq z_i$ the reaction parameter $\Delta Z_i^{\ominus}$ is a constant, independent of $\xi$ and of $Y$.

## 3.3. Transition Curves

The $z$ dependence of the general reaction variables, extent of reaction $\xi(z)$, degree of transition or degree of dissociation $\Theta(z)$, can be formulated in terms of thermodynamic quantities. By applying the chain rule of differentiation and using Eq. (3.8) the $z_i$-induced change in $\xi(z_i)$ is given by

$$(\partial \xi/\partial z_i)_{z \neq z_i} = (\partial \xi/\partial \ln K)_z \, (\partial \ln K/\partial z_i)_{z \neq z_i}$$

$$= (\partial \xi/\partial \ln K)_z \, \Delta Z_i/RT \tag{3.13}$$

Since from Eq. (3.1), $dc_j = V^{-1}v_j \, d\xi$, the $z$ dependence of the reaction can be expressed in terms of the concentration change of one of the reaction partners $j$ by

$$dc_j = v_j \Gamma(\Delta Z_i/RT) \, dz_i \tag{3.14}$$

where the definition (3.15) is introduced[13]:

$$\Gamma \equiv (\partial c_j/\partial \ln K)_z = 1 \bigg/ \sum_j v_j^2/c_j \tag{3.15}$$

By analogy to Eq. (3.13) the $z_i$ dependence of the fractional transformation variable $\Theta$ is given by

$$(\partial \Theta/\partial z_i)_{z \neq z_i} = (\partial \ln K/\partial \Theta)_z^{-1} \, \Delta Z_i/RT = \Gamma_\Theta \, \Delta Z_i/RT \tag{3.16}$$

As discussed in the context of Eq. (3.5), the quantity $\Gamma_\Theta$ of an intramolecular transition step according to Eq. (2.1) is given by

$$\Gamma_\Theta = (\partial \ln K/\partial \Theta)_z^{-1} = \Theta(1 - \Theta) \tag{3.17}$$

Hence, in this case, the maximum change in $\Theta$ can be achieved by a change in $z_i$ at $\Theta = \Theta_m = 0.5$. For conditions where Eqs. (3.7) hold the maximum effect produced by a change $\delta z_i$ is generally expressed as

$$(\delta \Theta/\Theta_m)_{z \neq z_i} = (1 - \Theta_m)(\delta K/K)_{z \neq z_i} \tag{3.18}$$

Similarly, the respective expressions for bimolecular steps like that in Eq. (2.3) with $1:1$ component ratio, are

$$\Gamma_\Theta = \Theta(1 - \Theta)/(2 - \Theta) \tag{3.19}$$

$$(\delta\Theta/\Theta_m)_{z \neq z_i} = [(1 - \Theta_m)/(2 - \Theta_m)](\delta K/K)_{z \neq z_i} \tag{3.20}$$

In this case, $\Gamma_\Theta$ has a maximum at $\Theta_m = 2 - \sqrt{2} = 0.586$, where $\delta z_i$ produces a maximum concentration shift.

### 3.3.1. Inflection Point

Structural transitions in macromolecules and membranes are frequently cooperative, resulting in rather steep transition curves $\Theta(z_i)$, starting sigmoidal at low values of $z_i$, going through an inflection point, and finally reaching a saturation value. Very often the transition curves are symmetric having the maximum slope at $\Theta = 0.5$. The $z_i$ value at $\Theta = 0.5$, $z_i(0.5)$, is commonly called the midpoint of transition.

In the vicinity of the inflection point the slope of the transition curve can be graphically determined with some reliability. In the case of a structural two-state transition step like that in Eq. (2.1), the Eqs. (3.16) and (3.17) can be used to derive an expression for the slope of the transition curve at $\Theta = 0.5$, where $z_i = z_i(0.5)$:

$$(\partial\Theta/\partial z_i)_{z \neq z_i, 0.5} = (\Delta Z_i)_{0.5}/4RT \tag{3.21}$$

Thus, the slope value $(\partial\Theta/\partial z_i)_{0.5}$ yields an estimate of the reaction quantity $\Delta Z_i$ at $\Theta = 0.5$.

### 3.3.2. Integrated van't Hoff Relations

When $\Delta Z_i$ is independent of $z_i$ and $\xi$ this van't Hoff reaction quantity is a constant describing the respective transition at constant $z \neq z_i$. If $\Delta Z_i$ is only a function of $z_i$, integration of Eq.(3.16) in the limits $\Theta(z_i)$ and $\Theta = \Theta^0$ at $z_i^{(0)}$ yields $\int \Gamma_\Theta^{-1} d\Theta = \int \Delta Z_i \, dz_i/RT$. From Eq. (3.17) we have $d \ln K = \Gamma_\Theta^{-1} d\Theta$, and in terms of the apparent equilibrium constant we obtain the integrated general van't Hoff relationship

$$\ln K(z_i) = \ln K(z_i^{(0)}) + \int \Delta Z_i \, dz_i/RT \tag{3.22}$$

For computational analysis it is useful to apply Eq. (3.22) in the form of

$$K(z_i) = K(z_i^{(0)}) \cdot e^x \tag{3.23}$$

where the $x$ quantity is defined as

$$x \equiv \int \Delta Z_i \, dz_i / RT \tag{3.24}$$

### 3.4. Chemical Affinity

Before starting the thermodynamic analysis of electric–chemical field effects it is necessary to recall some relations familiar from processes in the absence of external electric fields. The most general equilibrium condition for processes where the temperature and the pressure are under experimental control is that the characteristic Gibbs free energy is at its minimum.[39] The Gibbs free energy is defined by

$$G \equiv U + PV - TS \tag{3.25}$$

where $V$ is the volume, $S$ is the entropy, and $U$ is the inner energy of the system. The general Gibbs equation

$$dU = T \, dS - P \, dV + \sum_j \mu_j \, dn_j \tag{3.26}$$

combines the thermodynamic state functions with the reversible work term $dW$, which comprises all types of differential work $dW$.

The work term of chemical systems includes the chemical work $dW^{(\text{ch})}$ of changing the amount of substance $n_j$ at given chemical potentials $\mu_j$ of (neutral) molecules $j$. Thus, in addition to the volume work $dW^{(v)} = -P \, dV$, we have the (reversible) chemical work

$$dW^{(\text{ch})} = \sum_j \mu_j \, dn_j \tag{3.27}$$

From Eq. (3.25) we obtain $dG = dU + P \, dV + V \, dP - T \, dS - S \, dT$. Substitution of equations (3.26) and (3.27) leads to the classical Gibbs equation:

$$dG = -S \, dT + V \, dP + \sum_j \mu_j \, dn_j \tag{3.28}$$

In the absence of an external electric field, the work function of an isobaric reaction is

$$dG_P(T) = -(H/T) \, dT + \sum_j \mu_j \, dn_j \leqslant 0 \tag{3.29}$$

where the identity $(H/T) = S$ was introduced. The corresponding expression of an isothermal reaction is

$$dG_T(P) = V \, dP + \sum_j \mu_j \, dn_j \tag{3.30}$$

At constant $P$ and $T$, the work function characterizing chemical systems (in the absence of electric fields) is given by

$$(dG)_{P,T} = \sum_j \mu_j \, dn_j \leqslant 0 \tag{3.31}$$

Generally, for equilibrium processes $dG = 0$. For nonequilibrium states and irreversible processes we have $dG < 0$. Under isothermal–isobaric conditions, any dissipation of Gibbs free energy

$$(dG^{\mathrm{irr}})_{P,T} \equiv -T d_i S = -\sum \mu_j \, dn_j \tag{3.32}$$

where $d_i S$ is the internal entropy production, can only arise from irreversibly running chemical processes.[29]

From Eq. (3.29) we recall that the chemical potential of the (neutral) molecule $j$ is defined by

$$\mu_j = (\partial G / \partial n_j)_{n \neq n_j} \tag{3.33}$$

where all $n$ except $n_j$ are held constant. For practical purposes a standard chemical potential $\mu_j^{\ominus}$ is introduced such that

$$\mu_j = \mu_j^{\ominus} + RT \ln a_j \tag{3.34}$$

where $a_j = c_j y_j$ is the thermodynamic activity and $y_j$ the thermodynamic activity coefficient of species $j$. At unit activity $(a_j = 1)$ we obtain $\mu_j = \mu_j^{\ominus}$.)

In a closed chemical system the $n_j$ change when a chemical reaction is occurring. The chemical work term may then be rewritten as

$$dW^{(\mathrm{ch})} = \sum_j \nu_j \mu_j \, dn_j / \nu_j \tag{3.35}$$

DeDonder's chemical affinity $A$ is defined by

$$A = -\sum_j \nu_j \mu_j \tag{3.36}$$

Recalling Eq. (3.1) and substituting Eq. (3.36) into Eq. (3.35) we obtain

$$dW^{(\mathrm{ch})} = \sum_j \mu_j \, dn_j = -A \, d\xi \tag{3.37}$$

Combination of Eqs. (3.27), (3.31), and (3.37) leads to a general expression for chemical processes in terms of the chemical affinity:

$$A = -(\partial G/\partial \xi)_{P,T} \geqslant 0 \tag{3.38}$$

In line with Eq. (3.30), the equilibrium condition is $A = 0$ and, corresponding to $dG < 0$, a nonequilibrium state is associated with $A > 0$.

Using Eqs. (3.34) and (3.36) we see that

$$A = -\sum_j v_j \mu_j^\ominus - RT \sum_j v_j \ln a_j \tag{3.39}$$

At equilibrium, $A = 0$ and all activities assume their equilibrium values $\bar{a}_j$. Since the thermodynamic equilibrium constant $K^\ominus$ is defined by

$$K^\ominus = \prod \bar{a}_j^{\,v_j} = \prod (\bar{c}_j \, \bar{y}_j)^{v_j} = K \cdot \bar{Y} \tag{3.40}$$

where

$$K = \prod \bar{c}_j^{\,v_j} \quad \text{and} \quad \bar{Y} = \prod \bar{y}_j^{v_j} \tag{3.40a}$$

we immediately see from Eq. (3.39) that (at $A = 0$)

$$\sum_j v_j \mu_j^\ominus = -RT \ln K^\ominus \tag{3.41}$$

Substitution of this expression into Eq. (3.39) yields

$$A = RT(\ln K^\ominus - \ln Q^\ominus) \tag{3.42}$$

where the notation of a nonequilibrium distribution is introduced according to

$$Q^\ominus = \prod a_j^{\,v_j} = \prod (c_j \, y_j)^{v_j} = Q \cdot Y \tag{3.43}$$

Clearly, analogous to Eq. (3.40), the definitions

$$Q = \prod c_j^{\,v_j} \quad \text{and} \quad Y = \prod y_j^{\,v_j} \tag{3.43a}$$

apply.

From Eq. (3.42) it may be seen that the chemical affinity represents a kind of "thermodynamic distance" of a nonequilibrium distribution from its equilibrium distribution. On the other hand Eq. (3.42) may be used to specify the conditions of applying the van't Hoff relations and their integrated forms to the analysis of experimental data, for instance, to relaxation kinetic amplitudes.

*Step Perturbations.* Suppose that an external parameter $z_i$ ($P$, $T$, or $E$) can be "suddenly" changed in a practically rectangular fashion from an initial value $z_i^{(0)}$ to $z_i$. Immediately after this change, the previous activity ratio $K^{\ominus}(z_i^{(0)})$ at $z_i^{(0)}$ becomes a nonequilibrium ratio $Q^{\ominus}(z_i)$ at $z_i$. Thus, in systems with "inertia," initially we have $Q^{\ominus}(z_i) = K^{\ominus}(z_i^{(0)})$; then the non-equilibrium will relax until the new equilibrium characterized by $K^{\ominus}(z_i)$ is attained. Hence the sequence

$$K^{\ominus}(z_i^{(0)}) \to Q^{\ominus}(z_i) \to K^{\ominus}(z_i)$$

is a general thermodynamic representation of a chemical relaxation initiated by a "jump" in a physical state variable $z_i$.

### 3.5. Application Limits

There are some limitations for a straightforward application of the general van't Hoff relations in the form of Eq. (3.8) and all other equations based on it. Explicitly, the reaction quantity $\Delta Z_i$ may be dependent on $z_i$ and on $\xi$.

We recall that the Gibbs Eq. (3.29) for the $z_i(= T, P)$ dependence of a closed chemically interacting system, to which Eq. (3.37) applies at constant $z \neq z_i$, may be expressed as

$$dG(z_i \xi)_{z \neq z_i} = Z_i \, dz_i - A \, d\xi \tag{3.44}$$

where $Z_i = -S$, $V$ and $z_i = T$, $P$, respectively. For $Z_i = M$ and $z_i = E$, see below. Because $dG$ is a total differential the second cross differentials are equal. Hence from Eq. (3.44) we derive

$$(\partial Z_i/\partial \xi)_z = (\partial A/\partial z_i)_{\xi, z \neq z_i} \tag{3.45}$$

With the definition of $\Delta Z_i \equiv (\partial Z_i/\partial \xi)_z$, Eq. (3.45) may be rewritten

$$(\partial [A/RT]/\partial z_i)_{\xi, z \neq z_i} = \Delta Z_i/RT \tag{3.46}$$

As seen in Eqs. (3.8) and (3.9), the reaction quantity $\Delta Z_i$ refers to the $z_i$ dependence of the apparent equilibrium constant $K = K^{\ominus}/\bar{Y}$. Therefore

$$(\partial \ln K/\partial z_i)_{z \neq z_i} = (\partial \ln [K^{\ominus}/\bar{Y}]/\partial z_i)_{z \neq z_i} \tag{3.47}$$

Since $\bar{Y}$ is the value of $Y$ at equilibrium, i.e., at $A = 0$, we may use Eq. (3.42) in the form of

$$A/RT = \ln (K^{\ominus}/Y) - \ln Q \tag{3.48}$$

and specify Eq. (3.47) as

$$(\partial \ln K/\partial z_i) = (\partial \ln [K^{\ominus}/Y]/\partial z_i)_{z \neq z_i, A = 0} \tag{3.49}$$

The $z_i$ dependence of $K^{\ominus}/Y$ may now be expressed in terms of $\xi(z_i)$ at constant $A$ by

$$(\partial \ln [K^{\ominus}/Y]/\partial z_i)_{z \neq z_i} = (\partial \ln [K^{\ominus}/Y]/\partial z_i)_{\xi, z \neq z_i}$$
$$+ (\partial \ln [K^{\ominus}/Y]/\partial \xi)_z \cdot (\partial \xi/\partial z_i)_{A, z \neq z_i} \tag{3.50}$$

On the other hand Eq. (3.84) implies that the affinity $A(\xi, z_i)_z$ is a function of both $\xi$ and $z_i$ when all other $z$ parameters are held constant. Hence

$$dA(\xi, z_i)_z = (\partial A/\partial \xi)_z \, d\xi + (\partial A/\partial z_i)_{\xi, z \neq z_i} \, dz_i \tag{3.51}$$

When $A$ is constant, $dA = 0$; for $A = 0$, too. Equations (3.51) and (3.46) thus yield

$$(\partial \xi/\partial z_i)_{A, z \neq z_i} = -(\Delta Z_i/RT)[\partial(A/RT)/\partial \xi]_z^{-1} \tag{3.52}$$

The denominator of the right-hand side of Eq. (3.52) may be expressed in terms of Eq. (3.84) by

$$(\partial[A/RT]/\partial \xi)_z = (\partial \ln[K^{\ominus}/Y]/\partial \xi)_z + (\partial \ln Q/\partial \xi)_z \tag{3.53}$$

Since $K^{\ominus}$ only depends on the state variables $z$, we have at constant $z$

$$(\partial \ln K^{\ominus}/\partial \xi)_z = 0 \tag{3.54}$$

Furthermore, from Eqs. (3.1) and (3.43a) it is readily seen that

$$(\partial \ln Q/\partial \xi)_z = \sum (v_j/c_j)(\partial c_j/\partial \xi)_z$$

Using now Eq. (3.54) and the definition of $\Gamma = (\sum v_j^2/c_j)^{-1}$ in Eq. (3.15), we may rewrite Eq. (3.53) as

$$(\partial[A/RT]/\partial \xi)_z = -(\partial \ln Y/\partial \xi)_z - (V\Gamma)^{-1} = -(V\Gamma^*)^{-1} \tag{3.55}$$

where

$$(\Gamma^*)^{-1} \equiv \Gamma^{-1}[1 + V\Gamma(\partial \ln Y/\partial \xi)_z] \tag{3.56}$$

We now substitute Eq. (3.55) into Eq. (3.52):

$$(\partial \xi/\partial z_i)_{A, z \neq z_i} = V\Gamma^* \Delta Z_i/RT \tag{3.57}$$

Finally, the $z_i$ dependence of $(K^\ominus/Y)_\xi$ is obtained by differentiation of Eq. (3.48) at constant $\xi$:

$$(\partial[A/RT]/\partial z_i)_{\xi,z \neq z_i} = (\partial \ln [K^\ominus/Y]/\partial z_i)_\xi + (\partial \ln Q/\partial z_i)_{\xi,z \neq z_i} \quad (3.58)$$

At constant $\xi$, $d\xi = 0$; hence, from Eq. (3.1), all $dc_j$ are zero. Therefore, recalling Eq. (3.43a) we see that

$$(\partial \ln Q/\partial z_i)_{\xi,z \neq z_i} = 0 \quad (3.59)$$

Substitution of Eqs. (3.59) and (3.46) into Eq. (3.58) leads to

$$(\partial \ln[K^\ominus/Y]/\partial z_i)_{\xi,z \neq z_i} = \Delta Z_i/RT \quad (3.60)$$

We now introduce this expression together with Eqs. (3.57) and (3.54) into Eq. (3.50) and obtain

$$(\partial \ln [K^\ominus/Y]/\partial z_i)_{z \neq z_i} = (\Delta Z_i/RT)[1 - V\Gamma^*(\partial \ln Y/\partial\xi)_z] \quad (3.61)$$

At equilibrium, where $A = 0$, the quantities $\Gamma^*$ (and $\Gamma$) and $Y$ have their equilibrium values $\bar\Gamma^*$ (and $\bar\Gamma$) and $\bar Y$. Combining now the Eqs. (3.61) and (3.49) we obtain the final expression

$$(\partial \ln K/\partial z_i)_{z \neq z_i} = (\Delta Z_i/RT)[1 - V\bar\Gamma^*(\partial \ln \bar Y/\partial\xi)_z] \quad (3.62)$$

This relationship permits rigorous evaluation of the $z_i$ dependence of equilibrium concentrations $\bar c_j(z_i)$ or $\Theta(z_i)$.

It is readily seen that Eq. (3.62) reduces to the commonly used Eq. (3.8) provided that the dependence of the activity coefficient ratio $Y$ on $\xi$ is negligibly small. This condition is usually fulfilled if (a) the change $\delta z_i$ only produces a small shift in $K$ and thus in $Y$, or (b) the value of $Y$ is determined by an excess of components other than the reaction partners $j$, a condition commonly met when ionic reactions occur in the presence of an excess of inert electrolyte.

When $z_i = T$ or $P$ the dependence of $\Delta Z_i = \Delta H/T$ or $-\Delta\bar V$, respectively, on $T$ or $P$ may be solely expressed in terms of $Y$.

Owing to Eq. (3.54) we have $(\partial \ln K^\ominus/\partial z_i) = (\partial \ln K^\ominus/\partial z_i)_\xi$. Using now Eq. (3.60) at $A = 0$, where $Y = \bar Y$, and Eq. (3.12) we obtain

$$\Delta Z_i/RT = (\partial \ln K^\ominus/\partial z_i)_{\xi,z \neq z_i} - (\partial \ln \bar Y/\partial z_i)_{\xi,z \neq z_i} \quad (3.63)$$

Therefore Eq. (3.64) holds[13]

$$\Delta Z_i = \Delta Z_i^\ominus - (\partial \ln \bar Y/\partial z_i)_{\xi,z \neq z_i} \quad (3.64)$$

It is instructive to recall Eqs. (3.10) and (3.11) and compare with

Eq. (3.64). Obviously, $\Delta Z_i = \sum_j v_j Z_{i,j}$ refers to the general nonideal behavior. Ideal additivity refers to $Y = 1$; for this limiting case Eq. (3.11) reads

$$\Delta Z_i^\ominus = \sum_j v_j Z_{i,j}^\ominus \tag{3.65}$$

Thus, as usual, nonidealities are covered by the introduction of activity coefficients. Finally, it is shown below that in external electric fields ($z_i = \mathbf{E}$, $\Delta Z_i = \Delta M$) the specific expression for $\Delta M$ as a function of $\mathbf{E}$ depends on the mechanism of the field–dipole interactions.

## 3.6. Electrochemical Potential

The analytical treatment of electric field effects on chemical distributions may be started by recalling Guggenheim's original concept of the electrochemical potential $\tilde{\mu}_k$.[39] For a single (*isolated*) ion $B_k$, $\tilde{\mu}_k$ is written in the form

$$\tilde{\mu}_k = \mu_k + F\tilde{z}_k \psi_k^\ominus \tag{3.66}$$

where $\mu_k$ is the ordinary chemical potential, $F$ the Faraday constant, $\tilde{z}_k$ the charge number (with sign), and $\psi_k^\ominus$ the ideal electrostatic Coulomb potential of the *isolated* ion $B_k$.

In the presence of other ions it is necessary to account for the screening effect of the ionic atmosphere. It is then useful to introduce a more general form of the electrostatic potential term of Eq. (3.66) by a charging integral

$$\tilde{\mu}_k = \mu_k + F \int \bar{\psi} \, d\tilde{z}_k \tag{3.67}$$

where $\bar{\psi}$ is the mean electric potential. Note that, at a given distance $r$ from the charge center, usually $|\psi(r)| < |\psi^\ominus(r)|$.

In a collection of species $B_k$ the total electric work, $dW^{(\mathrm{el})}$, of charging $B_k$ from $\tilde{z}_k = 0$ to $\tilde{z}_k$ is the sum over the charging integrals of all species:

$$dW^{(\mathrm{el})} = F \cdot \sum_k \left( \int \bar{\psi} \, d\tilde{z}_k \right) dn_k \tag{3.68}$$

Recalling the (neutral) chemical work term from Eq. (3.27):

$$dW^{(\mathrm{ch})} = \sum \mu_k \, dn_k \tag{3.69}$$

we may define an electrochemical work term by summation according to $dW^{(ch)} + dW^{(el)} = dW^{(ech)}$. Thus, with Eqs. (3.67)–(3.69) we obtain

$$\sum_k \tilde{\mu}_k \, dn_k = \sum_k \left( \mu_k + F\left[ \int \bar{\psi} \, d\tilde{z}_k \right] \right) dn_k \tag{3.70}$$

The fundamental Gibbs Eq. (3.26) for ionic species may then be expressed as

$$dU = T \, dS - P \, dV + \sum_k \tilde{\mu}_k \, dn_k \tag{3.71}$$

and the Gibbs function $dG$ for ionic systems is given by

$$dG = -S \, dT + V \, dP + \sum_k \tilde{\mu}_k \, dn_k \tag{3.72}$$

Analogous to Eq. (3.33) we have the familiar expression for the electrochemical potential:

$$\tilde{\mu}_k = (\partial \tilde{G}/\partial n_k)_{P,T,n \neq n_k} \tag{3.73}$$

### 3.6.1. Electrochemical Affinity

In line with Eq. (3.36), the electrochemical affinity of chemically reacting ions is defined by

$$\tilde{A} = -\sum_k \nu_k \tilde{\mu}_k \tag{3.74}$$

Introducing Eq. (3.70) into Eq. (3.74) we obtain

$$\tilde{A} = -\sum_k \nu_k \mu_k - F \sum_k \nu_k \int \bar{\psi} \, d\tilde{z}_k \tag{3.75}$$

Since in a chemical reaction in a closed system the total charge is conserved, i.e., $\sum \nu_k \tilde{z}_k = 0$, we see with Eq. (3.70) that

$$\tilde{A} = A \tag{3.76}$$

Recalling Eqs. (3.37) and (3.38) it is readily seen that

$$\tilde{A} = -(\partial G/\partial \xi)_{P,T} \geq 0 \tag{3.77}$$

Thus the electrochemical affinity is equal to the ordinary chemical affinity (in the absence of electric fields). Nevertheless, it is useful to introduce a standard value of the electrochemical potential according to

$$\tilde{\mu}_k^{\ominus} = \mu_k^{\ominus} + F \int \psi_k^{\ominus} \, d\tilde{z}_k \tag{3.78}$$

where $\mu_k^{\ominus}$ is the ordinary standard value for the case when $B_k$ is neutral. Hence[19]

$$\tilde{\mu}_k = \tilde{\mu}_k^{\ominus} + RT \ln a_k \tag{3.79}$$

### 3.6.2. Electrochemical Activity Coefficient

The formalism of the Eqs. (3.67), (3.78), and (3.79) is suited to explicitly showing that the (electric) activity coefficient of ionic species accounts for deviations from the ideal (unscreened) Coulomb behavior. It will be demonstrated below that an analogous formalism describes non-idealities in the interactions between dipolar species.

If the ionic species $B_k$ is uncharged, the chemical potential is written in the familiar form of Eq. (3.34) as

$$\mu_k = \mu_k^{\ominus} + RT \ln a_k^{(0)} \tag{3.80}$$

where the superscript (0) is used to indicate the neutral form of the species $B_k$. Owing to the ionic character of $B_k$ the activity, $a_k$, of the ion is different from the activity, $a_k^{(0)}$, in the neutral form.

By this formalism the quantity $\tilde{\mu}_k$ is once expressed in terms of $a_k^{(0)}$ and $\bar{\psi}$ and, alternatively, as a function of $\psi_k^{\ominus}$ and $a_k$:

$$\tilde{\mu}_k = \mu_k^{\ominus} + RT \ln a_k^{(0)} + F \int \bar{\psi} \, d\tilde{z}_k$$

$$= \mu_k^{\ominus} + F \int \psi_k^{\ominus} \, d\tilde{z}_k + RT \ln a_k \tag{3.81}$$

Rearrangement leads to

$$F \int (\bar{\psi} - \psi_k^{\ominus}) \, d\tilde{z}_k = RT \ln(a_k/a_k^{(0)}) \tag{3.82}$$

Since we refer to the same amount of species $B_k$ in the uncharged form and in the ionic form, the concentrations are equal, i.e., $c_k = c_k^{(0)}$. Hence

$$(a_k/a_k^{(0)}) = y_k/y_k^{(0)} \tag{3.83}$$

Remembering that solely electrostatic interactions are covered by the terms $\psi$ and $\psi_k^\ominus$,

$$\psi_k^\ominus(\mathbf{r}) = \tilde{z}_k e_0/(4\pi\varepsilon_0\varepsilon \cdot r) \tag{3.84}$$

where $e_0 = 1.6 \times 10^{-19}$C is the (positive) elementary charge, $\varepsilon_0 = 8.854 \times 10^{-14}$ CV$^{-1}$ cm$^{-1}$ the permittivity of the vacuum, and $\varepsilon$ the dielectric permittivity of the medium. Therefore, $y_k^{(0)} = 1$ and thus $y_k/y_k^{(0)} = y_k$. When we now rewrite Eq. (3.82) for this case as

$$F \cdot \int (\bar{\psi} - \psi_k^\ominus) \, d\tilde{z}_k = RT \ln y_k \tag{3.85}$$

it is readily seen that the (electrostatic) activity coefficient of ions indeed covers the difference between the actual and the ideal Coulomb potential of the formal charge $z_k e_0$.

Introducing Eq. (3.79) into Eq. (3.74), we obtain

$$\tilde{A} = -\sum_k \nu_k \tilde{\mu}_k = -\sum_k \nu_k \tilde{\mu}_k^\ominus - RT \sum_k \nu_k \ln (c_k \cdot y_k) \tag{3.86}$$

At equilibrium we have $\tilde{A} = 0$ and $c_k \cdot y_k = \bar{c}_k \cdot \bar{y}_k$. Since, by analogy to Eq. (3.41), the relation

$$\sum \nu_k \tilde{\mu}_k^\ominus = -RT \ln K^\ominus \tag{3.87}$$

holds, the application of Eq. (3.40a) leads to

$$K^\ominus = K \cdot \bar{Y}$$

and, by using Eq. (3.85), we obtain

$$RT \ln \bar{Y} = F \sum_k \nu_k \int (\bar{\psi} - \psi_k^\ominus) \, d\tilde{z}_k \tag{3.88}$$

As shown elsewhere,[17,19] relationship (3.88) may be used to estimate charge numbers of ionic binding sites on macromolecules in the framework of the Debye–Hückel approximations.

## 3.7. "Dielectrochemical" Potential*

For the description of interactions between ionic reaction partners which may associate to ion pairs, it is useful to extend Guggenheim's con-

---

* The treatment of this section is formalistic, but the content is pictorial and instructive. The rigorous thermodynamic definition of the dielectrochemical potential is given in Section 4.

cept of an *electrochemical potential* of a single ion and to define a "*dielectrochemical* potential" of a dipolar ion pair.

Suppose that two ions $B_q$ and $B_k$ are in equilibrium with an ion pair $B_q^{(z_q)} \cdot B_k^{(z_k)}$ according to

$$B_q^{z_q} \cdot B_k^{z_k} \rightleftharpoons B_q^{z_q} + B_k^{z_k} \tag{3.89}$$

For simplicity we assume that $B_q$ and $B_k$ are equally but oppositely charged, i.e., $\tilde{z}_k = -\tilde{z}_q = \tilde{z}_j$. The position vectors of the charge centers in the pairing process are $\mathbf{r}_k$ and $\mathbf{r}_q$.

The pairing process may be quite formally viewed as a superposition of the "individual" electrochemical potentials $\tilde{\mu}_k(\mathbf{r}_k)$ and $\tilde{\mu}_q(\mathbf{r}_q)$. The sum is then a function of $\mathbf{r} = \mathbf{r}_k - \mathbf{r}_q$, which is the (average) vectorial distance between the charge centers. This distance dependence of the sum

$$\tilde{\mu}_{kq}(\mathbf{r}) = \tilde{\mu}_k(\mathbf{r}_k) + \tilde{\mu}_q(\mathbf{r}_q)$$

may be expressed in differential form as

$$d\tilde{\mu}_{kq}(\mathbf{r}_k) = d\tilde{\mu}_k(\mathbf{r}_k) + d\tilde{\mu}_q(\mathbf{r}_q) \tag{3.90}$$

At constant charge numbers $\tilde{z}_k$ and $\tilde{z}_q$, differentiation of Eq. (3.67) yields, respectively,

$$d\tilde{\mu}_k(\mathbf{r}_k) = d\mu_k(\mathbf{r}_k) + N_A e_0 \tilde{z}_k \, d\bar{\psi}(\mathbf{r}_k) \tag{3.91}$$

and

$$d\tilde{\mu}_q(\mathbf{r}_q) = d\mu_q(\mathbf{r}_k) + N_A e_0 \tilde{z}_q \, d\bar{\psi}(\mathbf{r}_q)$$

Substituting these expressions into Eq. (3.90) and using $\tilde{z}_k = -\tilde{z}_q = \tilde{z}_j$ and $d\mathbf{r} = d\mathbf{r}_k - d\mathbf{r}_q$, we may write

$$d\tilde{\mu}_{kq}(\mathbf{r}) = d\mu_{kq}(\mathbf{r}) + N_A e_0 \tilde{z}_j [d\bar{\psi}(\mathbf{r}_k) - d\bar{\psi}(\mathbf{r}_q)] \tag{3.92}$$

The electrostatic potential resulting from the superposition of the potentials of the individual charges defines an average potential $\bar{\psi}(\mathbf{r})$; thus $d\bar{\psi}(\mathbf{r}_k) - d\bar{\psi}(\mathbf{r}_q) = d\bar{\psi}(\mathbf{r})$. Introducing this definition into Eq. (3.85), and using the equation $d\bar{\psi} = (d\bar{\psi}/d\mathbf{r}) \, d\mathbf{r}$, we obtain

$$d\tilde{\mu}_{kq}(\mathbf{r}) = d\mu_{kq}(\mathbf{r}) + N_A e_0 \tilde{z}_j (d\bar{\psi}/d\mathbf{r}) \, d\mathbf{r} \tag{3.93}$$

The differential $(d\bar{\psi}/dr)$ defines the electric field $\mathbf{E}$ arising from the ion pair. According to Maxwell's definition

$$d\bar{\psi}/d\mathbf{r} = -\mathbf{E} \tag{3.94}$$

On the other hand the product $\tilde{z}_j \cdot e_0 \cdot d\mathbf{r}$ is the increment of the electric dipole moment

$$d\mathbf{m}_j = \tilde{z}_j e_0 \, d\mathbf{r} \tag{3.95}$$

Thus the ion pair $B_q \cdot B_k$ represents a dipole $B_j$ which is associated with a dipole moment $\mathbf{m}_j$.

We now substitute Eqs. (3.94) and (3.95) into Eq. (3.93) and apply the definitions

$$\tilde{\mu}_{kq} = \tilde{\mu}_j, \qquad \mu_{kq} = \mu_j$$

In this way the "individual" electrochemical potentials of the ions of an ion pair are expressed in terms of the electric dipole work as

$$d\tilde{\mu}_j(\mathbf{r}) = d\mu_j(r) - N_A \mathbf{E} \, d\mathbf{m}_j(\mathbf{r}) \tag{3.96}$$

Since the electric moment of an *isolated* dipole is given by the product of charge $\tilde{z}_j e_0$ and distance, the integration of Eq. (3.96) in the limits $\mathbf{r} = 0$ and $\mathbf{r}$ yields

$$\tilde{\mu}_j = \mu_j - N_A \mathbf{m}_j^{\ominus} \mathbf{E}(\mathbf{r}) \tag{3.97}$$

In this form, $\tilde{\mu}_j$ may be called the *dielectrochemical potential* of one mole of *isolated* dipolar species $B_j$ with the individual moment $\mathbf{m}_j^{\ominus}$.

For analytical purposes it is convenient to use a parallel-plate capacitor as a measuring cell. In this geometry only the component $m_j$ (of $\mathbf{m}_j$) that is *parallel* to the electric field lines between the plates (or to an eventually applied external field creating the Maxwell field $\mathbf{E}$ in the dielectric) contributes to the measurable polarization. For capacitor geometry the scalar product of $\mathbf{m}_j \mathbf{E}(r)$ in Eq. (3.97) is given by the parallel component $(\mathbf{m}_j)_{\parallel} = \mathbf{m}_j \cos \vartheta_j$ and by $\mathbf{E}$, that is, the Maxwell field vector perpendicular to the capacitor plates; $\vartheta_j$ is the angle between $\mathbf{E}$ and $\mathbf{m}_j$. Therefore, Eq. (3.97) applies in the form

$$\tilde{\mu}_j = \mu_j - N_A (\mathbf{m}_j)_{\parallel} E \tag{3.98}$$

In a collection of many species $B_j$ with different orientations of their dipole moments $\mathbf{m}_j$ relative to $\mathbf{E}$, an average contribution may be defined by

$$m_j = \langle \mathbf{m}_j \cos \vartheta_j \rangle = \mathbf{m}_j \langle \cos \vartheta_j \rangle \tag{3.99}$$

This average value will depend on the average field $\mathbf{E}$ resulting from the different orientations. Therefore Eq. (3.98) may be more generally written as

$$\tilde{\mu}_j = \mu_j - N_A \int m_j \, dE \tag{3.100}$$

When there is no preferential orientation like in a random distribution of dipoles, the contributions of all dipole moments parallel to $\mathbf{E}$ cancel each other, because $\langle \cos \vartheta_j \rangle$ in Eq. (3.99) for a random distribution of permanent dipoles is zero. In this case $\tilde{\mu}_j = \mu_j$ and no macroscopic polarization of the medium occurs.

Macroscopic organizations possessing a permanent electric dipole moment like electrets have a finite Maxwell field $\mathbf{E}$ which may be externally measured, for instance, as a plate capacitor field.

The measurable polarization of the dielectric electret within the capacitor plates is then given by

$$M = (\mathbf{M})_\| = N_A \sum_j n_j \langle \mathbf{m}_j \cos \vartheta_j \rangle$$

$$= N_A \sum_j n_j m_j = \sum_j n_j M_j \tag{3.101}$$

where the parallel component, $M_j$, of the average partial *molar* dipole moment is defined by

$$M_j = (\partial M / \partial n_j)_{n \neq n_j} = N_A \mathbf{m}_j \langle \cos \vartheta_j \rangle \tag{3.102}$$

Hence, Eq. (3.100) may be used in the form

$$\tilde{\mu}_j = \mu_j - \int M_j \, dE \tag{3.103}$$

If on average there is no macroscopic electric field across the dielectric, we have $\tilde{\mu}_j = \mu_j$.

Recalling Eqs. (3.95), (3.96), and (3.101) and noting that $F = N_A e_0$, we may express Eq. (3.103) as

$$\tilde{\mu}_j(\mathbf{r}) = \mu_j - F\tilde{z}_j \int \mathbf{r} \, d\mathbf{E} \tag{3.104}$$

In this form the dielectrochemical potential introduced here has a similar formal structure as Guggenheim's electrochemical potential as expressed in Eq. (3.67).

### 3.7.1. Polyionic Macromolecules

In macromolecular biological structures ion-pair formation frequently occurs. For instance, the inner counterions surrounding the polynucleotide macroanions as well as in all other linear polyelectrolytes appear to form ion pairs with the fixed polyionic matrix. Externally applied electric fields can compete with the inner fields and shift these counterions relative to the polyion, thus producing large dipole moments.[23–25] It is known that proteins may contain inner salt bridges, i.e., ion pairs between fixed ionized side chains of the amino acid residues of the polypeptide chains and/or other ionic groups. In particular membrane proteins like the bacteriorhodopsin of the purple membranes of halobacteria appear to contain an unusually large number of charged groups within the protein structure. In such a case we may group together oppositely charged groups into ion pairs. The total moment $\mathbf{M}$ of the macromolecule is then the vector sum over all individual contributions of the single ion pairs according to Eq. (3.101). When we now inspect Eq. (3.93) and rewrite as

$$d\tilde{\mu}_j(r) = d\mu_j(\mathbf{r}) - N_A \tilde{z}_j e_0 \mathbf{E}(d\mathbf{r}_k - d\mathbf{r}_q) \tag{3.105}$$

we see that changes in the distance between two charged groups with $\tilde{z}_k(\mathbf{r}_k)$ and $\tilde{z}_q(\mathbf{r}_q)$, respectively, will change the dielectrochemical potential and thus the contribution of this ion pair to the total polarization. On the other hand an externally applied electric field $\mathbf{E}$ of a sufficiently high field strength may change the distance between the charge centers of an ion pair. This type of distance variation by external fields appears to be responsible for the large induced dipole moments in bacteriorhodopsin of purple membranes.[10,41]

## 4. Thermodynamics in Electric Fields

### 4.1. The Characteristic Gibbs Function

The concept of a dielectrochemical potential introduced in the previous section is already implicit in Guggenheim's treatment of dielectrics in the presence of external fields. In order to apply the familiar criteria for reversible (equilibrium) processes and irreversible (nonequilibrium) processes in terms of an appropriate Gibbs function, Guggenheim introduced the characteristic Gibbs function in the presence of electric fields by a transformation.[39] We may express the transformed Gibbs free energy as

$$\tilde{G} = G - W^{(\mathrm{el})} \tag{4.1}$$

where $G$ is the ordinary Gibbs free energy at $\mathbf{E}$ and $W^{(el)}$ is the (reversible) electric work.

The differential work term of the field-dipole interaction is

$$dW^{(el)} = \mathbf{E}d\langle\mathbf{M}\rangle = E\,dM \tag{4.2}$$

consistent with the work terms in Eq. (3.28). From Eq. (3.101) we recall that the total moment parallel to $\mathbf{E}$ is given by $M = \sum_j n_j M_j$, representing a specific case of the general expression (3.10). The partial molar dipole moment of species $B_j$ is

$$M_j = (\partial M/\partial n_j)_{n \neq n_j, z} \tag{4.3}$$

By $M_j = N_A \mathbf{m}_j \langle\cos\vartheta_j\rangle$, $M_j$ refers to the average of the field-parallel contributions of all individual moments $\mathbf{m}_j$.

Substitution of the integrated Eq. (4.2) into Eq. (4.1) yields

$$\tilde{G} = G - ME \tag{4.4}$$

In this context, Eigen and DeMaeyer[13] used the relationship $G^* = G - D^*E$, with $D^* = V \cdot P$. where $P$ is the polarization per unit volume and $V$ is the volume.

It is now useful to denote the chemical potential of the (dipolar) species $B_j$ in the presence of an electric field by a special symbol:

$$\mu(E) = \tilde{\mu} \tag{4.5}$$

The (reversible) chemical work term, analoguous to Eq. (3.27), is then of a dielectrochemical nature:

$$dW^{(ch)} = \sum_j \tilde{\mu}_j\,dn_j \tag{4.6}$$

The total differential work term of a chemically open system derives from an extension of Eq. (3.28). Using Eqs. (4.2) and (4.6) we find

$$\sum dW = -PdV + \sum_j \tilde{\mu}_j\,dn_j + EdM \tag{4.7}$$

The general Gibbs equation for the inner energy in the presence of $\mathbf{E}$ is obtained by substitution of Eq. (4.7) into (3.26):

$$dU = TdS - PdV + \sum_j \tilde{\mu}_j\,dn_j + EdM \tag{4.8}$$

From Eq. (3.25) we have

$$dG = dU + d(PV - TS) \tag{4.9}$$

Substitution of Eq. (4.8) into (4.9) results in

$$dG = -SdT + VdP + \sum_j \tilde{\mu}_j \, dn_j + EdM \tag{4.10}$$

We now clearly see that the ordinary Gibbs free energy increases in the electric field compared to $E = 0$.

In order to have the electric field as the independent variable instead of $M$, a Legendre transformation is required. By Eq. (4.4) the transformed Gibbs free energy is

$$d\tilde{G} = dG - d(EM) \tag{4.11}$$

Substituting Eq. (4.10) into (4.11) finally leads to the characteristic Gibbs function for chemically open systems in electric fields:

$$d\tilde{G} = -SdT + VdP + \sum_j \tilde{\mu}_j \, dn_j - MdE \tag{4.12}$$

The transformed Gibbs free energy clearly decreases in the presence of electric fields. This property is required for a consistent thermodynamic treatment of electric-chemical field effects.

For isobaric–isothermal conditions the characteristic Gibbs function reduces to

$$d\tilde{G}_{T,P} = \sum_j \tilde{\mu}_j \, dn_j - MdE \tag{4.13}$$

The fundamental relationship may now be used to rigorously derive an expression for the chemical potential of species $B_j$ in the presence of field-dipole interactions. Consistent with general thermodynamic formalism we obtain from Eq. (4.13)

$$\tilde{\mu}_j = (\partial \tilde{G}/\partial n_j)_{T,P,E,n \neq n_j} \tag{4.14}$$

Following the rules of cross differentiation, Eq. (4.13) leads to

$$(\partial M/\partial n_j)_{T,P,E,n \neq n_j} = -(\partial \tilde{\mu}_j/\partial E)_{T,P,n} \tag{4.15}$$

Inspection of Eq. (4.3) results in

$$(\partial \tilde{\mu}_j/\partial E)_{T,P,n} = -M_j \tag{4.16}$$

According to Kirkwood and Oppenheim,[40] integration of Eq. (4.16) between $E$ and $E = 0$ provides the relationship

$$\tilde{\mu}_j(E) = \tilde{\mu}_j(0) - \int M_j \, dE \tag{4.17}$$

Since $\tilde{\mu}_j(0)$ is the ordinary chemical potential $\mu_j$ at $\mathbf{E} = 0$, Eq. (4.17) is rewritten as

$$\tilde{\mu}_j = \mu_j - \int M_j \, dE \tag{4.18}$$

and $\tilde{\mu}_j$ may be called the *dielectrochemical potential* of species $B_j$ in external electric fields. By Eq. (3.100) it is evident that for completely random distributions of dipoles or at $\mathbf{E} = 0$, the relation $\tilde{\mu}_j = \mu_j$ holds. In these cases the dielectrochemical potential equals the chemical potential.

### 4.2. Dielectrochemical Affinity

By analogy to Eq. (3.74), we may now define a "dielectrochemical affinity" for chemically interacting dipolar species $B_j$ by

$$\tilde{A} = -\sum_j v_j \tilde{\mu}_j \tag{4.19}$$

Substitution of Eq. (4.18) into Eq. (4.19) yields

$$\tilde{A} = -\sum_j v_j \tilde{\mu}_j = -\sum_j v_j \mu_j + \sum_j v_j \int M_j \, dE \tag{4.20}$$

Introducing Eq. (4.19) with $dn_j = v_j \, d\xi$ into Eq. (4.13) we obtain the Gibbs function of chemically reacting systems in external electric fields:

$$(d\tilde{G})_{P,T} = -\tilde{A} d\xi - M dE \tag{4.21}$$

From this expression it is readily seen that the dielectrochemical affinity is consistently defined in terms of $\tilde{A}$ and $\xi$:

$$\tilde{A} = -(\partial \tilde{G}/\partial \xi)_{P,T,E} \tag{4.22}$$

It is remarked that reversible and irreversible processes in external electric fields are characterized by

$$d\tilde{G} \leqslant 0 \quad \text{and} \quad \tilde{A} \geqslant 0 \tag{4.23}$$

Thus the definitions of $\tilde{G}$ and $\tilde{A}$ permit the thermodynamic treatment of electric field effects in the framework of concepts which are familiar from ordinary chemical thermodynamics in the absence of electric fields.

## 4.3. Activity Coefficients

As in the case of ionic reactions it is also useful to define a standard value, $\tilde{\mu}_j^{\ominus}$, of the dielectrochemical potential for the dipolar species $B_j$:

$$\tilde{\mu}_j = \tilde{\mu}_j^{\ominus} + RT \ln \tilde{a}_j \qquad (4.24)$$

where $\tilde{a}_j$ is the activity of $B_j$ in the presence of $\mathbf{E}$. Analogous to Eq. (4.18) the standard dielectrochemical potential is given by

$$\tilde{\mu}_j^{\ominus} = \mu_j^{\ominus} - \int M_j^{\ominus} \, dE \qquad (4.25)$$

where $\mu_j^{\ominus}$ is the ordinary standard potential used in

$$\mu_j = \mu_j^{\ominus} + RT \ln a_j \qquad (4.26)$$

Note that $a_j$ is the activity of $B_j$ at $\mathbf{E} = 0$. $M_j^{\ominus}$ is the standard value of $M_j$ of the isolated dipoles behaving ideally. There is a formal similarity between Eq. (4.25) for dipoles and Eq. (3.78) for ionic interaction partners. As in the case of the electrochemical potential, the dielectrochemical potential may also be expressed in two ways.

From Eqs. (4.18), (4.26) and (4.24), (4.25) we obtain

$$\tilde{\mu}_j = \mu_j^{\ominus} + RT \ln a_j - \int M_j \, dE$$

$$= \mu_j^{\ominus} + RT \ln \tilde{a}_j - \int M_j^{\ominus} \, dE \qquad (4.27)$$

Hence, analogous to Eq. (3.82) the relation

$$-\int (M_j - M_j^{\ominus}) \, dE = RT \ln (\tilde{a}_j/a_j) \qquad (4.28)$$

holds. Since we refer to the same mount of components, the concentrations are equal, i.e., $\tilde{c}_j = c_j$. Again, the activity ratio is thus given by the activity coefficients:

$$\tilde{a}_j/a_j = \tilde{y}_j/y_j \qquad (4.29)$$

If only dipolar interactions are considered we set $y = 1$. In this case Eq. (4.28) reads

$$-\int (M_j - M_j^{\ominus}) \, dE = RT \ln \tilde{y}_j \qquad (4.30)$$

In the form of this equation it is obvious that for dipolar species, too, the activity coefficient covers deviations from simple ideal additivities.

For the *ideal* case of pure additive superposition of the formal charge-distance products $\tilde{z}_j e_0 \mathbf{r}_j = m_j^\ominus$, where $\tilde{z}_j$ is the integer charge number, the total (standard) polarization is given by

$$M^\ominus = \sum_j n_j M_j^\ominus = N_A \sum_j n_j m_j^\ominus \tag{4.31}$$

where

$$m_j^\ominus = \langle \mathbf{m}_j \cos \vartheta_j \rangle = m_j^\ominus \langle \cos \vartheta_j \rangle$$

and

$$M_j^\ominus = N_A m_j^\ominus \tag{4.32}$$

### 4.4. Van't Hoff Relationship

It is now pertinent to derive a rigorous expression for the dependence of the thermodynamic equilibrium constant $K^\ominus$ on the externally applied electric field. For this purpose we recall Eqs. (4.19) and (4.24):

$$\tilde{A} = -\sum_j v_j \tilde{\mu}_j^\ominus - RT \sum_j v_j \ln \tilde{a}_j \tag{4.33}$$

At equilibrium we have $\tilde{A} = 0$ and all activities are equilibrium values. For this case $K^\ominus(E) = \prod \tilde{a}_j^{v_j}$ and Eq. (4.33) yields

$$RT \ln K^\ominus(E) = -\sum_j v_j \tilde{\mu}_j^\ominus \tag{4.34}$$

Further, from Eq. (4.26) we have

$$\sum_j v_j \tilde{\mu}_j^\ominus = \sum_j v_j \mu_j^\ominus - \sum_j v_j \int M_j^\ominus \, dE \tag{4.35}$$

When we now differentiate $M^\ominus$, defined by Eq. (4.31) as $M^\ominus = \sum_j n_j M_j^\ominus$, with respect to $\xi$ and use Eq. (3.1) in the form $dn_j/d\xi = v_j$, we obtain

$$(\partial M^\ominus / \partial \xi)_{E,T,P} = \sum_j v_j M_j^\ominus = \Delta M^\ominus \tag{4.36}$$

The introduction of Eqs. (4.35) and (3.41) in the form of $RT \ln K^\ominus(0) = -\sum v_j \tilde{\mu}_j^\ominus$ at $\mathbf{E} = 0$, into Eq. (4.34) finally leads to

$$\ln K^\ominus(E) = \ln K^\ominus(0) + \int \Delta M^\ominus \, dE/(RT) \tag{4.37}$$

which is the integrated van't Hoff relationship for a dipolar equilibrium in an external electric field.

From Eq. (4.21) we obtain by cross differentiation

$$(\partial \tilde{A}/\partial E)_{\xi,T,P} = (\partial M/\partial \xi)_{E,T,P} = \Delta M = \sum_j v_j M_j \qquad (4.38)$$

The relationship between the affinity and the $M_j$ terms refers to constant $\xi$ and thus to given values $c_j$. For chemically interacting species Eq. (4.30) is rewritten as

$$-\sum_j v_j \int (M_j - M_j^{\ominus}) \, dE = RT \sum v_j \ln \tilde{y}_j \qquad (4.39)$$

Substituting Eqs. (4.36) and (4.37) and recalling the definition $\tilde{Y} = \prod \tilde{y}^{v_j}$ yields the expression

$$\Delta M = \Delta M^{\ominus} - RT(\partial \ln \tilde{Y}/\partial E)_{\xi} \qquad (4.40)$$

In this equation the quantities $\Delta M$, $\Delta M^{\ominus}$, and $\tilde{Y}$ represent terms at a given field intensity **E**.

Since the thermodynamic and the apparent equilibrium constants are connected by $K^{\ominus} = K \cdot \tilde{Y}$, a comparison of the van't Hoff relations

$$(\partial \ln K^{\ominus}/\partial E)_{P,T} = \Delta M^{\ominus}/RT \qquad (4.41)$$

and

$$(\partial \ln K/\partial E)_{P,T} = \Delta M/RT = (\Delta M^{\ominus}/RT) - (\partial \ln \tilde{Y}/\partial E)_{\xi,P,T} \qquad (4.42)$$

with Eq. (4.30) shows that the activity coefficient product is given by

$$RT \ln [\tilde{Y}(E)/\tilde{Y}(0)]_{\xi} = -\int (\Delta M - \Delta M^{\ominus}) \, dE \qquad (4.43)$$

If $\tilde{Y}$ is independent of $E$, by $\Delta M = \Delta M^{\ominus}$ the reaction moment is independent of $\xi$ (or of the concentrations of the reaction partners). In any case, the relationship (4.42) provides the basis for the analysis of electric field-induced concentration shift in dipolar equilibria. Whereas this part of the account dealt with thermodynamic foundations of the analysis of chemical electric field effects, the second part (Chapter 5) covers some kinetic and mechanistic aspects of macromolecular bioelectric processes.

## Acknowledgments

I thank Miss B. Wilkenloh for the careful processing of the manuscript. The financial support of the Stiftung Volkswagenwerk, Grant No. I 34-706, and of the Deutsche Forschungsgemeinschaft, Grant No. NE 227, is gratefully acknowledged.

## List of Symbols

| | |
|---|---|
| $A$ | Chemical affinity (DeDonder) |
| $\tilde{A}$ | Dielectrochemical affinity in the presence of electric fields |
| $A_\lambda$ | Absorbance (per centimeter) at wavelength $\lambda$ |
| $\delta A$ | Electric-field-induced absorbance change |
| $\delta A_\sigma$ | Absorbance change at the light polarization angle $\sigma$ |
| $\Delta A$ | Linear dichroism |
| $A_q$ | Shape factor of the ellipsoid, $q$ axis |
| $a_j$ | Thermodynamic activity of species $B_j$; $\bar{a}_j$, equilibrium value; $a_j = c_j y_j$, dimensionless by division by $c^\ominus = 1$ mol dm$^{-3}$ |
| $\tilde{a}_j$ | Value of $a_j$ in the presence of an electric field |
| $c_j$ | Concentration of species $B_j$ (mol dm$^{-3}$); $\bar{c}_j$, equilibrium value |
| $e_0$ | Elementary charge ($1.6 \times 10^{-19}$ C) |
| $\mathbf{E}(E)$ | Electric field strength vector (absolute value), Maxwell field |
| $\mathbf{E}_F(E_F)$ | External electric field vector (absolute value), Fröhlich field |
| $\mathbf{E}_{int}$ | Internal field vector |
| $\mathbf{E}_{dir}$ | Directing field vector |
| $f$ | Reaction field factor |
| $F$ | Faraday constant ($9.65 \times 10^4$ C mol$^{-1}$) |
| $G$ | Gibbs free energy (free enthalpy) (in J) |
| $\tilde{G}$ | Guggenheim's characteristic free energy in electric fields, transformed Gibbs free energy |
| $g, \tilde{g}$ | Conversion factors |
| $g_K$ | Kirkwood correlation factor |
| $I_c$ | Ionic strength (mol dm$^{-3}$) |
| $K$ | Apparent equilibrium constant (concentration ratio) |
| $K^\ominus$ | Thermodynamic equilibrium constant (activity ratio) |
| $k, k^\ominus$ | Rate constants corresponding to $K$, $K^\ominus$ |
| $k$ | Boltzmann constant ($1.38 \times 10^{-23}$ J K$^{-1}$), thermal energy $kT$ (J) |
| $L[r]$ | Langevin function of $r$ |
| $\mathbf{M}$ | Total polarization vector, macroscopic dipole moment, thermal average $\langle \mathbf{M} \rangle = \langle \mathbf{P} \rangle V$ |
| $M$ | Component of $\langle \mathbf{M} \rangle$ parallel to $\mathbf{E}$ |
| $M_j$ | Partial molar dipole moment (contribution to $\mathbf{M}$ of $B_j$) |

| | |
|---|---|
| $\Delta M$ | Reaction dipole moment, $\Delta M = (\partial M / \partial \xi)_z = \sum_j v_j M_j$ |
| $\Delta M^{\ominus}$ | Standard value of $\Delta M$ |
| $\mathbf{m}_j$ | Individual dipole moment of species $B_j$, or charge configuration $\tilde{z}_j e_0 \mathbf{r}_j$ |
| $m_j$ | Average value of field-parallel component of $\mathbf{m}_j$ |
| $\mathbf{m}_{(\alpha)}$ | Induced dipole moment contribution to $\mathbf{m}$ |
| $\mathbf{m}_{(p)}$ | Permanent dipole moment contribution to $\mathbf{m}$ |
| $N_A$ | Avogadro constant ($6.02 \times 10^{23}$ mol$^{-1}$) |
| $n_j$ | Amount of substance of species $B_j$ (mol) |
| $N_j$ | Number of species $B_j$ molecules |
| $\mathbf{P}$ | Electric polarization per unit volume |
| $P$ | Pressure |
| $\mathbf{p}$ | Permanent dipole moment |
| $q$ | Half-axis of ellipsoid polarization |
| $Q, Q^{\ominus}$ | Concentration ratio, activity ratio |
| $R$ | Gas constant ($R = kN_A = 8.31$ J K$^{-1}$ mol$^{-1}$) |
| $\mathbf{r}$ | Radius vector (position vector) |
| $T$ | Kelvin temperature (K) |
| $t$ | Time |
| $V$ | Volume |
| $v$ | Reaction rate |
| $v_p$ | Rate of product formation |
| $v_r$ | Rate of reactant formation |
| $W$ | Work (J) |
| $x$ | Field-dependent exponent in $K(E) = K(0)e^x$ |
| $y_j$ | Thermodynamic activity coefficient of $B_j$ |
| $\tilde{y}_j$ | Activity coefficient in the present of $\mathbf{E}$ |
| $Y, \tilde{Y}$ | Activity coefficient ratio at $\mathbf{E} = 0$, at $\mathbf{E}$ |
| $Z_i$ | Extensive state variable ($V, S, M, ...$) conjugate to the intensive property $z_i$ |
| $\Delta Z_i$ | Reaction quantity conjugate to $z_i$ |
| $\Delta Z_i^{\ominus}$ | Standard value of $\Delta Z_i$ |
| $z_i$ | Intensive property ($P, T, E$) |
| $\tilde{z}_k$ | Formal charge number (with sign) of ion $B_k$ |
| $\tilde{z}_j$ | "Formal (positive) charge" of the dipole $B_j$ with $\mathbf{m}_j = \lvert \tilde{z}_j \rvert e_0 \mathbf{r}_j$ |
| $\alpha$ | Polarizability tensor |
| $\alpha_q$ | Polarizability component of the ellipsoidal axis $q$ |
| $\Gamma^*$ | Amplitude factor containing $\Gamma$ and concentration dependence of $Y$ |
| $\Gamma$ | Amplitude factor [$\Gamma = (\sum_j v_j^2 / c_j)^{-1}$] |
| $\gamma$ | Field factor of the field dissociation effect |
| $\delta$ | Small change |
| $\vartheta(\delta)$ | Angle between dipole axis of $\mathbf{m}_j$ and the electric field vector $\mathbf{E}$ |

| | |
|---|---|
| $\Delta$ | Differential operator, $\Delta = \partial/\partial\xi$ |
| $\varepsilon$ | Permittivity tensor (dielectric constant) |
| $\varepsilon_0$ | Vacuum permittivity ($8.85 \times 10^{-14}$ F Cm$^{-1}$) |
| $\varepsilon_\infty$ | Permittivity characteristic for the induced polarization |
| $\varepsilon_j$ | Extinction coefficient of species $B_j$; $\bar{\varepsilon}_j$ random average value of $\varepsilon_j$ |
| $\Theta$ | Degree of transition |
| $\kappa$ | Electric conductance (S) |
| $\mu$ | Chemical potential (J mol$^{-1}$) |
| $\tilde{\mu}_k$ | Electrochemical potential of ion $B_k$ |
| $\tilde{\mu}_k^{\ominus}$ | Standard value of $\tilde{\mu}_k$ |
| $\tilde{\mu}_j$ | Dielectrochemical potential of dipolar species $B_j$ in an electric field |
| $\tilde{\mu}_j^{\ominus}$ | Standard value of $\tilde{\mu}_j$ |
| $\nu_j$ | Stoichiometric coefficient (with sign) |
| $\xi$ | Extent of reaction (mol) |
| $\sigma$ | Light polarization angle between plane of polarization and electric field vector |
| $\tau$ | Relaxation time |
| $\phi$ | Orientation factor |
| $\phi^{(\text{ch})}$ | Chemical transformation factor |
| $\Psi_k^{\ominus}(\mathbf{r}_k)$ | Electric potential of the isolated charge |
| $\Psi(\mathbf{r})$ | Mean electric potential at position $\mathbf{r}$ |

## References

1. K. S. Cole, *Membranes, Ions and Impulses*, University of California Press (1968).
2. I. Tasaki, *Physiology and Electrochemistry of Nerve Fibers*, Academic, New York (1982).
3. D. Nachmansohn and E. Neumann, *Chemical and Molecular Basis of Nerve Activity*, Rev., Academic, New York (1975).
4. P. L. Dorogi and E. Neumann, *Proc. Natl. Acad. Sci. USA* **77**, 6582–6586 (1980).
5. H. T. Witt, E. Schlodder, and P. Gräber, *FEBS Lett.* **69**, 272–276 (1976).
6. E. Schlodder, M. Rögner, and H. T. Witt, *FEBS Lett.* **138**, 13–18 (1982).
7. E. Neumann, in *Topics of Bioelectrochemistry and Bioenergetics* (G. Milazzo, ed.), Wiley, New York (1981), Vol. 4, pp. 113–160.
8. E. Neumann, in *Ions in Macromolecular and Biological Systems* (D. H. Everett and B. Vincent, eds.), Scientechnica, Bristol (1978), pp. 170–191.
9. A. Katchalsky and E. Neumann, *Int. J. Neurosci.* **3**, 175–182 (1972).
10. K. Tsuji and E. Neumann, *Int. J. Biol. Macromol.* **3**, 231–242 (1981).
11. K. Tsuji and E. Neumann, *FEBS Lett.* **128**, 265–268 (1981).
12. D. Schallreuter, Ph.D. thesis, Konstanz and Martinsried (1982); E. Neumann, K. Tsuji, and D. Schallreuter in *Biological Structures and Coupled Flows* (A. Oplatka and M. Balaban, eds.), Academic, New York (1983), pp. 135–138.
13. M. Eigen and L. C. M. DeMaeyer, *Tech. Org. Chem.* **8**(2), 895–1054 (1963).

14. K. Bergmann, M. Eigen, and L. C. M. DeMaeyer, *Ber. Bunsenges. Phys. Chem.* **67**, 819–826 (1963).
15. G. Schwarz, *J. Phys. Chem.* **71**, 4021–4030 (1967).
16. L. Onsager, *J. Chem. Phys.* **2**, 599–615 (1934).
17. H.-J. Nolte, T. L. Rosenberry, and E. Neumann, *Biochemistry* **19**, 3705–3711 (1980).
18. A. Enokida, T. Okubo, and N. Ise, *Macromolecules* **13**, 49–53 (1980).
19. E. Neumann and H. J. Nolte, *Bioelectrochem. Bioenerg.* **8**, 89–101 (1981).
20. C. T. O'Konski and A. J. Haltner, *J. Am. Chem. Soc.* **79**, 5634–5649 (1957).
21. M. Tricot and C. Houssier, in *Polyelectrolytes* (K. C. Frisch, D. Klempner, and A. V. Patsis, eds.), pp. 43–90, Technomic, Westport (1976).
22. E. Fredericq and C. Houssier, *Electric Dichroism and Electric Birefringence*, Clarendon, Oxford (1973).
23. M. Eigen and G. Schwarz, *Z. Phys. Chem. N.F.* **4**, 380–385 (1955).
24. M. Eigen and G. Schwarz, *J. Coll. Sci.* **12**, 181–194 (1957).
25. M. Eigen and G. Schwarz, in *Electrolytes* (B. Pesce, ed.), Pergamon, Oxford (1962), pp. 309–335.
26. C. T. O'Konski and N. C. Stellwagen, *Biophys. J.* **5**, 607–613 (1965).
27. G. Schwarz and J. Seelig, *Biopolymers* **6**, 1263–1277 (1968).
28. E. Neumann and A. Katchalsky, *Proc. Natl. Acad. Sci. USA* **69**, 993–997 (1972).
29. E. Neumann, *Angew. Chem. Intern. Ed.* **12**, 356–369 (1973).
30. K. Kikuchi and K. Yoshioka, *Biopolymers* **12**, 2667–2679 (1973).
31. T. Yasunaga, T. Sano, K. Takahashi, H. Takenaka, and S. Ito, *Chem. Lett. (Japan)*, 405–408 (1973).
32. A. Revzin and E. Neumann, *Biophys. Chem.* **2**, 144–150 (1974).
33. D. Pörschke, *Nucl. Acid. Res.* **1**, 1601–1618 (1974).
34. D. Pörschke, *Biopolymers* **15**, 1917–1928 (1976).
35. K. Kikuchi and K. Yoshioka, *Biopolymers* **15**, 583–587 (1976).
36. E. Neumann and K. Rosenheck, *J. Membrane Biol.* **10**, 279–290 (1972).
37. U. Zimmermann, P. Scheurich, G. Pilwat, and R. Benz, *Angew. Chem.* **93**, 332–351 (1981).
38. J. Teissie and T. Y. Tsong, *Biochemistry* **20**, 1548–1554 (1981).
39. E. A. Guggenheim, *Thermodynamics*, 5th rev. ed., North-Holland, Amsterdam (1967); *J. Phys. Chem.* **33**, 842 (1929).
40. J. G. Kirkwood and I. Oppenheim, *Chemical Thermodynamics*, McGraw-Hill, New York (1961).
41. K. Tsuji and E. Neumann, *Biophys. Chem.* **17**, 153–163 (1983).

# 5

# Elementary Analysis of Chemical Electric Field Effects in Biological Macromolecules

## II. Kinetic Aspects of Electro-Optic and Conductometric Relaxations

*Eberhard Neumann*

*ABSTRACT:* Electric field effects in macromolecular organizations such as proteins, nucleic acids, and membranes frequently involve both chemical-conformational changes and physical-orientational displacements of molecular subgroups. Electro-optic techniques in conjunction with relaxation kinetics in high electric fields provide a tool for the investigation of the complex processes encountered in bioelectric phenomena on the level of macromolecules, membrane fragments, and other cellular units. Whereas Part I covers the thermodynamics of electric field effects, this part deals with practical and theoretical aspects of kinetics and mechanisms in aqueous solutions of macromolecules and membranes.

## 1. Introduction

The electrophysiological voltage clamp technique is a widely used method to approach mechanisms of ion transport across cell membranes. Basically, the voltage clamp is the application of a rectangular electric field and the measurement of relaxations of electric currents which are frequently rate-controlled by structural changes in the ion transport gating proteins. In a similar manner chemical relaxtion kinetics appears to be the method of

*Eberhard Neumann* • Department of Physical and Biophysical Chemistry, University of Bielefeld, P.O. Box 8640, D-4800 Bielefeld 1, Federal Republic of Germany.

choice to investigate electric field effects of biological systems in solution or suspension in the presence of electric fields which are as high as those in natural cell membranes. Since, in principle, dynamic details of chemical-conformational transitions as well as rotational processes are accessible from relaxation kinetic measurements, the kinetic analysis of chemical electric field effects and electrical chemical mechanisms are outlined in some detail.

## 2. Rate Constants in Electric Fields

### 2.1. Dipolar Equilibria

The equilibrium constant of an elementary chemical reaction is a function of the physical state variables $z_i (= P, T, E)$. Therefore the rate constants must depend differently on the $z_i$ values. Consider a chemical reaction

$$\sum_r v_r B_r \underset{k_r}{\overset{k_p}{\rightleftharpoons}} \sum_p v_p B_p \tag{2.1}$$

where the subscripts $r$ and $p$ refer to reactants and products, respectively. In this notation the stoichiometric coefficients $v_r$ and $v_p$ are positive definite and the apparent equilibrium constant $K$ is expressed as

$$K = \prod \bar{c}_p{}^{v_p} \Big/ \prod \bar{c}_r{}^{v_r} = k_p / k_r \tag{2.2}$$

where $k_p$ and $k_r$ are the rate constants of the forward and reverse direction, respectively. The thermodynamic equilibrium constant $K$ and the rate constants for unit activity $k_p^{\ominus}$ and $k_r^{\ominus}$ are given by

$$K^{\ominus} = \prod \bar{a}_p{}^{v_p} \Big/ \prod \bar{a}_r{}^{v_r} = k_p^{\ominus} / k_r^{\ominus} \tag{2.3}$$

At equilibrium we have

$$k_p \prod \bar{c}_r{}^{v_r} = k_r \prod \bar{c}_p{}^{v_p} \tag{2.4}$$

$$k_p^{\ominus} \prod \bar{y}_r{}^{v_r} \cdot \prod \bar{c}_r{}^{v_r} = k_r^{\ominus} \prod \bar{y}_p{}^{v_p} \cdot \prod \bar{c}_p{}^{v_p} \tag{2.5}$$

Recalling Eq. (3.24) of part I,[1] the $z_i$ dependence of $K$ is

$$\ln K(z_i) = \ln K(z_i^{(0)}) + \int \Delta Z_i \, dz_i / (RT) \tag{2.6}$$

where $z_i^{(0)}$ is a reference value. In line with Eq. (2.2), the relation (3.10) of part I is rewritten as

$$\Delta Z_i = \sum_j v_j Z_{i,j} = \sum_p v_p Z_{i,p} - \sum_r v_r Z_{i,r} \tag{2.7}$$

where

$$\Delta Z_{i,p} = \sum_p v_p Z_{i,p} \tag{2.8}$$

and

$$\Delta Z_{i,r} = \sum_r v_r Z_{i,r} \tag{2.9}$$

the contributions of the products and those of the reactants are explicitly separated. With these expressions and with Eq. (2.2) we may express Eq. (2.6) in terms of the rate coefficients as

$$\ln[k_p(z_i)/k_r(z_i)] = \ln[k_p(z_i^{(0)})/k_r(z_i^{(0)})]$$
$$+ \int (\Delta Z_{i,p} - \Delta Z_{i,r}) \, dz_i/RT \tag{2.10}$$

After term separation we obtain

$$k_p(z_i) = k_p(z_i^{(0)}) \exp\left[\int \Delta Z_{i,p} \, dz_i/RT\right]$$
$$k_r(z_i) = k_r(z_i^{(0)}) \exp\left[\int \Delta Z_{i,r} \, dz_i/RT\right] \tag{2.11}$$

The relations between the unit-activity quantities $k^{\ominus}(z_i)$ and $\Delta Z_i^{\ominus}$ are analogous to Eqs. (2.11).

The general formalism developed here is particularly useful for the description of electric field effects on the rate constants of dipolar equilibria. We may choose $z_i^{(0)} = E = 0$ as a suitable reference and specify Eqs. (2.11) as

$$k_p(E) = k_p(0) \exp\left[\int \Delta M_p \, dE/RT\right]$$
$$k_r(E) = k_r(0) \exp\left[\int \Delta M_r \, dE/RT\right] \tag{2.12}$$

Thus the rate constants of the product formation are dependent on the dipole moment contributions of the products and the rate constants of the reactant formation are a function of the dipole moment contributions of the reactants. Whereas the field dependence of the equilibrium constant only yields the difference $\Delta M$ of the reaction partners, the rate constants provide a means to determine the dipole moments of the reactants and, separately, those of the products. Equations (2.12) were used to discuss the rate aspects of electric-field-induced permeability changes in membranes[2] in the context of electric membrane fusion[3] and electric gene transfer[4] by electroporation.[5]

## 2.2. Ionic Equilibria

According to Onsager it is the dissociation rate constant, $k_d$, of the separation of an ion pair that is mainly affected by the electric field; the association rate constant, $k_a$, remaining practically unchanged.[6] Consider an ion-pairing equilibrium:

$$\mathbf{L}^+ \cdot \mathbf{B}^- \underset{k_a}{\overset{k_d}{\rightleftharpoons}} \mathbf{L}^+ + \mathbf{B}^- \tag{2.13}$$

The Onsager treatment provides an expression for the electric field dependence of $k_d^{\ominus}$. Note that $k_d^{\ominus} = k_d Y_{L \cdot B}$. The electric-field-induced increase in the conductivity of electrolytes usually starts nonlinear, followed by a range where the relative conductivity change, $\Delta \kappa / \kappa(0)$, is linearly dependent on the electric field strength and finally approaches a field-independent saturation value. In the linear range Onsager's theory of diluted weak electrolytes yields

$$\left( \frac{\partial \ln k_d^{\ominus}}{\partial |\mathbf{E}|} \right)_{P,T} = \frac{(\tilde{z}_L u_L - \tilde{z}_B u_B)|\tilde{z}_L \tilde{z}_B| e_0^3}{(u_L + u_B) \, 8\pi\varepsilon_0\varepsilon(kT)^2} \tag{2.14}$$

where $u$ is the ionic mobility of the free ions ($u \approx 10^{-4} \, \mathrm{cm^2 \, V^{-1} \, s^{-1}}$).

For symmetric electrolytes where $\tilde{z}_L = -\tilde{z}_B = |\tilde{z}|$, Eq. (2.14) is reduced to

$$\left( \frac{\partial \ln k_d^{\ominus}}{\partial |\mathbf{E}|} \right)_{P,T} = |z^3| e_0^3 / [8\pi\varepsilon_0\varepsilon(kT)^2] = \gamma \tag{2.15}$$

Note the strong dependence on the charge number. As shown elsewhere, the equilibrium constant of a 1:1 weak electrolyte like acetic acid is increased by an electric field of $100 \, \mathrm{kV \, cm^{-1}}$ to about 14%, that for a 2:2 electrolyte like $MgSO_4$ to about 110%.[9,10] Compared to simple dipolar equilibria of small molecules where electric-field-induced changes in $K$ are very small, we see that the dissociation step of simple ion pairs is associated with a relatively large electric field effect. If the activity coefficient of the ion pair can be taken equal to unity, we have $k^{\ominus} = k$, and Eqs. (2.14) and (2.15) apply to $k$.

Following the analysis in the previous section the association step in Eq. (2.13) must also be field dependent. According to Eqs. (2.12) we readily see that (with $|\mathbf{E}| = E$)

$$\left(\frac{\partial \ln k_a^{\ominus}}{\partial E}\right)_{P,T} = \Delta M_a^{\ominus}/RT \tag{2.16}$$

where $\Delta M_a$ is given by

$$\Delta M_a^{\ominus} = M_{L \cdot B}^{\ominus} = N_A m_{L \cdot B}^{\ominus} \tag{2.17}$$

Note that in line with Eq. (4.31) of part I[1] we have

$$m_{LB}^{\ominus} = \mathbf{m}_{L \cdot B}^{\ominus} \langle \cos \vartheta_{LB} \rangle \tag{2.18}$$

Therefore the actual field dissociation effect (or second Wien effect) as a whole is also determined by the dipole moment of the ion pair which can dissociate into the free ions.

At finite electrolyte concentration the activity coefficients have to be considered. Since for Eq. (2.13), $k_a^{\ominus} = k_a \cdot y_L \cdot y_B$,

$$(\partial \ln k_a^{\ominus}/\partial E)_{P,T} = (\partial \ln k_a/\partial E)_{P,T} + (\partial \ln[y_L \cdot y_B]/\partial E)_{P,T} \tag{2.19}$$

and from Eq. (4.40) of part I:

$$\Delta M_a^{\ominus} = \Delta M_a + RT(\partial \ln[y_L y_B]/\partial E)_{\xi,P,T} \tag{2.20}$$

Usually the effect of an electric field on the activity coefficient of free ions is apparent from the first Wien effect, i.e., from the perturbation (and finally the destruction) of the ionic atmosphere by an external electric field.

As shown for conductivity data on $MgSO_4$, the contribution of the first Wien effect can be quantitatively covered in terms of the Wilson theory.[8]

We recall that the (practical) equilibrium constant (concentration ratio) is given by $K = k_d/k_a$. Thus we may formally express $K$ of ionic (-dipolar) equilibria of the type (2.13) by

$$(\partial \ln K/\partial E)_{P,T} = \gamma - (\Delta M_a^{\ominus}/RT) - (\partial \ln \tilde{Y}/\partial E)_{\xi,P,T} \tag{2.21}$$

where $\tilde{Y}$ is given by $\tilde{Y} = \tilde{y}_L \tilde{y}_B/\tilde{y}_{LB}$ at the field strength $\mathbf{E}$. At very high field strengths the ionic atmosphere screening may be reduced to a large extent such that $\tilde{Y} = 1$; also, at very diluted solutions we may set $\tilde{Y} = 1$. In both cases the approximation $(\partial \ln \tilde{Y}/\partial E)_{\xi,P,T} = 0$ may be used.

## 2.2.1. Polyelectrolytes

It is well established that the second Wien effect is particularly large in linear polyelectrolytes.[11,12] Compared to simple electrolytes the linear range of the conductivity increase with increasing electric fields starts already at relatively low field strengths ($5$–$10$ kV cm$^{-1}$). Because of the rather extended linear region Onsager's equation for the dissociation field effect has been used as the basis for a qualitative discussion of the second Wien effect in polyelectrolytes.

Denoting by $B_n$ the polyion and by L the counterion, the large mass difference justifies the approximation $u_B \ll u_L$ for small observation times, $n$ being the degree of polymerization. The counterions which interact in multiple ion pair equilibria with the fixed charges of the polyion experience a larger attraction potential than that arising from a single charged residue. Owing to the neighboring fixed charges the effective charge, $z_B^{\text{eff}} e_0$ per residue, B, is larger than $z_B e_0$ itself. On the other hand, the accumulation of the counterions lowers the overall attraction potential because repulsive contributions are superimposed. It should also be mentioned that counterion accumulation creates a lower (local) dielectric constant as compared to that of the bulk aqueous solution.

If now the diffuse counterion binding is viewed in terms of one residue of the polyion, we may apply Eq. (2.21) in the suggestive form

$$(\partial \ln k_d/\partial E)_{P,T} = (\tilde{z}_L e_0)^2 |\tilde{z}_B^{\text{eff}}| e_0/[8\pi\varepsilon_0\varepsilon(kT)^2] \tag{2.22}$$

and thus quantify the influences of counterion valency and effective polyionic charge per residue.

The analysis of electric conductivity relaxations of the linear polyelectrolyte poly(riboadenylate, K$^+$) according to Eq. (2.22) at 293 K yields a formal effective charge of about $-6$ for the interaction of a K$^+$ ion with the inner counterion atmosphere of this polyanion.[7,8]

It has been found that the dissociation field effect in polyelectrolytes is generally larger by a factor of about $10$–$100$ as compared to simple electrolytes. Thus the values of the relative displacements of the distribution constant, $\delta K/K(0) = \partial \ln K/\partial E$, in polyelectrolytes for the same conditions that have been used for simple $1:1$ electrolytes (see above), are in the range of $1.4$ for $E = 10^4$ V cm$^{-1}$ and $14$ for $E = 10^5$ V cm$^{-1}$. We conclude from these estimates that already for moderately large changes in the field intensity the degree of counterion binding may change to a large extent. Since usually the conformation and the degree of stretching of flexible polyelectrolytes depends on counterion binding, external electric field changes may readily affect structural changes. Owing to the efficiency of dissociation field effects in polyelectrolytes, it can be expected that appreciably large structural changes are induced by already moderately high electric fields.

As an additional remark, the decrease in the number of counterions near the polyion in the presence of high electric fields will also decrease the counterion polarization and thus the magnitude of the induced dipole moment. This, in turn, will change the reaction moment of chemical transformations involving induced dipoles. In any case, macromolecular complexes in which polyelectrolytic subunits are associated decrease in stability with increasing electric fields.

A theoretical approach for the second Wien effect of polyelectrolytes has been initiated in terms of the counterion condensation model by Manning.[13] According to this theory the degree of counterion dissociation from the condensed layer is linearly dependent on the field strength.

## 3. Reaction Moment and Electric–Chemical Mechanism

One of the main aims of investigating electric field effects on chemical transformations is to determine the reaction mechanism and, in simple cases, the dipole moments of the reaction partners. As outlined above, the rate constants provide the key information.

The reaction dipole moment $\Delta M$ of a dipolar equilibrium may be obtained from the measurement of continuum properties such as the dielectric permittivity as well as from direct monitoring of concentration shifts produced by an externally applied electric field. In both approaches to reaction properties it is primarily the chemical part of the total polarization that is aimed at. However, the *chemical* processes are intimately connected with the *physical* processes of polarization and dipole rotation. In the case of small molecules the orientational relaxations are usually rapid compared to the diffusion limited chemical reactions. When, however, macromolecular structures are involved, the rotational processes of the macromolecular dipoles may control a major part of the chemical relaxations.[7,8] Two types of processes may be involved if a vectorial perturbation like an external electric field is applied: a chemical concentration change and a change in the orientation of the reaction partners.

It is known that in a random distribution of permanent dipolar or induced dipolar reaction partners the (local) extent of the electric field effect depends on the orientation of the individual dipoles relative to the field direction.[14–16] Therefore the measured bulk effects always represent orientational averages. In this context it is stressed that the total macroscopic polarization, M, caused by an electric field in a random distribution of particles, is a statistical average that results from the polarizing and orienting action of the field vector against the randomizing thermal agitations.[17]

For plane-plate capacitor geometry, which is experimentally most adequate, the field-parallel component $M$ of $\mathbf{M}$ is the sum over all field-parallel components $m_j$ of the individual moments $\mathbf{m}_j$. We recall that[1]

$$M = \langle \mathbf{M} \rangle = N_A \sum_j n_j \langle \mathbf{m}_j \cos \vartheta_j \rangle = N_A \sum_j n_j m_j \tag{3.1}$$

Whereas Eq. (3.1) expresses $\mathbf{M}$ in terms of the average contributions of the individual molecular moments $\mathbf{m}_j$, the continuum approach to $\mathbf{M}$ represents the total moment in terms of an overall macroscopic dielectric permittivity $\varepsilon$.

The fundamental relationship between the total moment $\mathbf{M}$ and the measured electric field $\mathbf{E}$ may be written in terms of the absolute amounts ($M$ and $E$) as

$$M = \varepsilon_0(\varepsilon - 1) \, VE \tag{3.2}$$

where the vector $\mathbf{M}$ is in the same direction as the field vector $\mathbf{E}$.

In classical electrostatics the polarization is given by the macroscopic dipole moment per unit volume, $\mathbf{P}$. Hence $\mathbf{M} = V \cdot \mathbf{P}$. Further on, $\mathbf{P}$ is represented in terms of linear and nonlinear contributions:

$$\mathbf{P} = \chi\mathbf{E} + \chi'E^2\mathbf{E} + \cdots \tag{3.3}$$

where the susceptibility tensor is given by $\chi = \varepsilon_0(\varepsilon - 1)$. In this form the dielectric permittivity tensor $\varepsilon$ is considered as a constant and is independent of $\mathbf{E}$. Chemical contributions to $\mathbf{P}$ only appear as odd powers of $\mathbf{E}$ and are qualified as nonlinear terms.

When Eq. (3.2) is applied as a general expression for homogeneous dielectrics in electric fields, the permittivity tensor is an overall quantity that depends on the intensive variables $z(= T, P, E)$. If, in addition, chemical transformations are caused by changes in $z$, then $\varepsilon$ also depends on the extent of reaction $\xi$. In this manner "nonlinearities" are hidden in $\varepsilon$ $(T, P, E, \xi)$.

## 3.1. Reaction Moments from Dielectric Data

The chemical reaction moment $\Delta M$ refers to that part of the total moment $\mathbf{M}$ which changes in the course of a chemical-conformational transition. Differentiation of Eq. (3.2) with respect to the reaction variable $\xi$ results in

$$\Delta M = (\partial M/\partial \xi)_{V,E} = \varepsilon_0 VE(\partial \varepsilon/\partial \xi)_{V,E} \tag{3.4}$$

Equation (3.4) may be called the continuum expression for the chemical reaction moment at constant total volume between the capacitor plates, usually realized at low field intensities and diluted solutions of the reaction partners. When volume changes occur electrostriction terms must be explicitly considered.[18,19]

Under isothermal–isobaric–isochoric conditions, $M$ is solely a function of $\xi$ and $E$. The dependence on $E$ at constant $\xi$ defines a (normal) physical part whereas the dependence of $M$ on $\xi$ at constant $E$ may be referred to as the chemical contribution to a change in $M$. The field dependence of the total moment may then be expressed as

$$(\partial M/\partial E)_V = (\partial M/\partial E)_{V,\xi} + (\partial M/\partial \xi)_{V,E}(\partial \xi/\partial E)_V \tag{3.5}$$

At equilibrium in the presence of $E$ the characteristic (dielectro-)chemical affinity $\tilde{A}$ of dipole systems is zero and $d\tilde{A} = 0$. Hence the term $(\partial \xi/\partial E)_{z,\tilde{A}=0}$ can be calculated; from Eq. (3.57) of part I[1] we obtain

$$(\partial \xi/\partial E)_{V,\tilde{A}=0} = V\bar{\Gamma}^* \Delta M/RT \tag{3.6}$$

At equilibrium there is no further change of $M$ in the field $E$. Therefore $(\partial M/\partial E)_{V,\tilde{A}=0}$ is a constant at a given field strength.

We may now call the term $(\partial M/\partial E)_{V,\xi}$ the (normal) physical term, because it refers to a fixed value of $\xi$. From Eq. (3.2) we obtain

$$\Delta \varepsilon^{(ph)} = (\partial M/\partial E)_{V,\xi}/(\varepsilon_0 V) = (\partial \varepsilon/\partial E)_{V,\xi} \tag{3.7}$$

In a similar way we may define the chemical contribution of changes in $M$ by

$$\Delta \varepsilon^{(ch)} = (\partial \varepsilon/\partial \xi)_{V,E} = (\partial M/\partial \xi)_{V,E}(\partial \xi/\partial E)_{V,\tilde{A}=0}/(\varepsilon_0 V) \tag{3.8}$$

Substitution of Eqs. (3.4) and (3.6) into (3.8) leads to the well-known relation

$$\Delta \varepsilon^{(ch)} = \Gamma^*(\Delta M)^2/(\varepsilon_0 RT) \tag{3.9}$$

Thus the chemical part of a change in $M$ by $E$ may be derived from the electric field dependence of dielectric relaxation curves.[14,19]

### 3.2. Permanent and Induced Dipole Moments

Two types of polarization processes may contribute to the total macroscopic polarization:

$$\mathbf{M} = \mathbf{M}_{(\alpha)}(\mathbf{E}_{int}) + \mathbf{M}_{(p)}(\mathbf{E}_{dir}) \tag{3.10}$$

According to Onsager the induced moment term $M_{(\alpha)}$ is determined by the internal or local field $E_{int}$, whereas the permanent dipole term $M_{(p)}$ is related to the directing field $E_{dir}$ orienting the permanent dipoles $p$. The combination of Eqs. (3.10) and (3.2) requires that the two different field vectors must be expressed in terms of the measured Maxwell field. The calculations of the terms $M_{(\alpha)}$ and $M_{(p)}$ as functions of $E_{int}$ and $E_{dir}$ usually are approximations. The final expressions may be written in terms of conversion factors ($g$ factors[20]) which are a function of particle anisotropies as well as of the properties of the medium in which the particles are embedded (polar, nonpolar, gas phase, or fluid phase).

### 3.2.1. Individual Dipole Moments

In line with Eq. (3.10) the molecular dipole moments $m$ may generally be expressed as

$$m = m_{(\alpha)} + p \tag{3.11}$$

where $m_{(\alpha)}$ represents the induced moment and $p$ is the permanent dipole moment.

In anisotropic molecules $m$ represents the vector sum of all dipolar contributions. The calculations are readily performed for ellipsoidal molecules and for simple geometries like a sphere, long cylinders, or flat disks.[17] The total moment of an ellipsoid where the main polarization axes are the (half-)axes $q = a, b, c$, is given by the vector sum $m = m_a + m_b + m_c$.

The dipole moment component along the axis $q$ is expressed analogously to Eq. (3.11) as

$$m_q = m_{(\alpha)q} + p_q \tag{3.12}$$

The induced dipole moment is given by

$$m_{(\alpha)} = \alpha_q (E_{int})_q \tag{3.13}$$

where $\alpha_q$ is the component of the polarizability tensor $\alpha$ in the direction of the $q$ axis and $(E_{int})_q$ the internal field in the $q$ direction.

When $\alpha$ is independent of $E$, then $m_{(\alpha)}$ is obviously linearly dependent on $E$. Generally, however, the polarizability tensor reflecting charge displaceability may depend on $E$. Thus a more general definition of the polarizability is given by

$$\alpha = (\partial m_{(\alpha)}/\partial E_{int})_{E \to 0} \tag{3.14}$$

As outlined previously the parallel component $(\mathbf{m})_{\parallel}$ of $\mathbf{m}$ contributes to the total polarization. Thus for the $q$ axis we have

$$(\mathbf{m}_q)_{\parallel} = \mathbf{m}_q \cos \vartheta_q \qquad (3.15)$$

where $\vartheta_q$ is the angle between the $q$ axis and the electric field.

The actual field-parallel contribution of $\mathbf{m}$ can be calculated from Onsager's concepts of the cavity field and of the reaction field.[17,21] At first, Eq. (3.15) is specified as

$$(\mathbf{m}_q)_{\parallel} = \alpha \cos \vartheta_q (\mathbf{E}_{int})_q + p_q \cos \vartheta_q \qquad (3.16)$$

where $\alpha_q = \alpha \cos \vartheta_q$. The internal field in the $q$ direction is given by

$$(\mathbf{E}_{int})_q = \tilde{g}_q g_q \mathbf{E} \cos \vartheta_q \qquad (3.17)$$

It is obvious that in isotropic particles where the polarizability is equal in all directions the internal field is simply

$$\mathbf{E}_{int} = \tilde{g} g \mathbf{E} \qquad (3.18)$$

The factor $\tilde{g}_q$ represents the reaction field contribution of permanent dipoles according to

$$\tilde{g}_q = [1 + f_q(1 - \alpha_q f_q)^{-1} \langle \mathbf{p}_q \rangle] \qquad (3.19)$$

where $f_q$ is the reaction field factor[17,21] and $\langle \mathbf{p}_q \rangle$ is the average contribution of the permanent dipoles. For nonpolar particles (where $p_q = 0$), $\tilde{g}_q = 1$ and $(\mathbf{E}_{int}) = g_q \mathbf{E} \cos \vartheta_q$.

Obviously, the total value of the field-parallel components is $(\mathbf{m})_{\parallel} = \sum_q (\mathbf{m}_q)_{\parallel}$.

In a collection of statistically distributed, mobile dipolar species, the total field-parallel contribution to the polarization is the statistical average over the $\cos \vartheta$ projections on the field vector. For the sake of transparentness we shall confine the further analysis to *uniaxial anisotropic particles* $B_j$, i.e., to uniaxial dipole moments $\mathbf{m}_j$. In this case Eqs. (3.12) and (3.16) read, respectively,

$$\mathbf{m}_j = \alpha_j (\mathbf{E}_{int})_j + \mathbf{p}_j \qquad (3.20)$$

$$(\mathbf{m}_j)_{\parallel} = \alpha_j \tilde{g}_j g_j \mathbf{E} \cos^2 \vartheta_j + \mathbf{p}_j \cos \vartheta_j \qquad (3.21)$$

The effective average contribution to $\mathbf{M}$ is finally given by

$$m_j = \langle (\mathbf{m}_j)_{\parallel} \rangle = \alpha_j \tilde{g}_j g_j E \langle \cos^2 \vartheta_j \rangle + p_j \langle \cos \vartheta_j \rangle \qquad (3.22)$$

### 3.2.2. Total Polarization

In a mixture of $N_j$ molecules to type $B_j$, the total polarization moment can be generally expressed as

$$M = \sum_j N_j m_j = N_A \sum_j n_j m_j$$

$$= \sum_j N_j \langle \alpha_j (E_{\text{int}})_j \rangle + \sum_j N_j \langle \mathbf{p}_j \rangle \tag{3.23}$$

In order to reduce the complexity the two contributions to $M$, the induced moment $M_{(\alpha)}$ and the permanent moment $M_{(p)}$ will be treated separately. Applying Eq. (3.22) to (3.23) we obtain

$$M_{(\alpha)} = \sum_j N_j \alpha_j \tilde{g}_j g_j E \langle \cos^2 \vartheta_j \rangle \tag{3.24}$$

$$M_{(p)} = \sum_j N_j p_j \langle \cos \vartheta_j \rangle \tag{3.25}$$

where the summation is over all particle types. The average values $\langle \cos^2 \vartheta_j \rangle$ and $\langle \mathbf{p}_j \cos \vartheta_j \rangle$ are thermal averages under the polarizing and orienting action of the electric field; they are therefore dependent on $\mathbf{E}$, on the molecular shape and size, and on temperature.

### 3.2.3. Induced Moment

In the absence of permanent dipoles (i.e., $\tilde{g} = 1$), the absolute value $M_{(\alpha)}$ of the total induced moment $\mathbf{M}_{(\alpha)}$ is derived from Eq. (3.24):

$$M_{(\alpha)} = \sum_j N_j \alpha_j g_j \langle \cos^2 \vartheta_j \rangle E \tag{3.26}$$

In isotropic particles the total induced moment is given by

$$M_{(\alpha)} = \sum_j N_j \alpha_j g_j E \tag{3.27}$$

In the case of uniaxial anisotropic molecules the thermal average $\langle \cos^2 \vartheta_j \rangle$ may be expressed as a function of the orientation factor

$$\phi_j = [3 \langle \cos^2 \vartheta_j \rangle - 1]/2 \tag{3.28}$$

This factor can be directly obtained from electro-optic data, for instance, linear dichroism[22,24] and birefringence.[20,23,24] The orientation can also be expressed in terms of the dipole moments involved.[22–24]

Since from Eq. (3.28), $\langle \cos^2 \vartheta_j \rangle = \frac{1}{3}(1 + 2\phi_j)$, substitution into Eq. (3.26) yields

$$M_{(\alpha)} = \left[ \sum_j N_j \alpha_j g_j (1 + 2\phi_j) E \right] \Big/ 3 \qquad (3.29)$$

Two limiting cases are of practical importance. The low-field condition $m_j E \ll kT$ means negligible orientation in the field direction, i.e., $\phi \ll 1$. Hence Eq. (3.29) reduces to

$$M_{(\alpha)} = \left( \sum_j N_j \alpha_j g_j E \right) \Big/ 3 \qquad (3.30)$$

At high fields when $\phi \to 1$, corresponding to total alignment of the induced moments in the field direction, we obtain

$$M_{(\alpha)} = \sum_j N_j \alpha_j g_j E \qquad (3.31)$$

Should the induced moment be saturated at high fields, i.e., $m_{(\alpha)} = m_s$, then for $\phi \to 1$,

$$M_{(\alpha),s} = \sum_j N_j (m_s)_j \qquad (3.32)$$

In this case the induced moment is independent of the field strength (dielectric saturation).

### 3.2.4. Permanent Moment

When the molecules $B_j$ have a permanent dipole moment $\mathbf{p}_j$ and are freely mobile, the thermal average of $\cos \vartheta_j$ in the total permanent moment

$$M_{(p)} = \sum_j N_j p_j \langle \cos \vartheta_j \rangle \qquad (3.33)$$

is given by the Langevin function $L[r_j]$ of the directing field $(\mathbf{E}_{\text{dir}})_j$, where

$$(\mathbf{E}_{\text{dir}})_j = g_j \mathbf{E} \qquad (3.34)$$

Since $r_j = p_j q_j E/(kT)$, we have

$$\langle \cos \vartheta_j \rangle = L[p_j g_j E/(kT)] \qquad (3.35)$$

Note that $L[r_j] = \coth r_j - r_j^{-1}$. Introducing Eq. (3.35) into (3.25) leads to

$$M_{(p)} = \sum_j N_j p_j L[p_j g_j E/(kT)] \tag{3.36}$$

At low field strengths $(r_j \ll 1)$, $L[r_j] = g_j p_j E/(3kT)$ and

$$M_p = \sum_j N_j p_j^2 g_j E/(3kT) \tag{3.37}$$

At high fields $(r_j \gg 1)$, $L[r_j] \to 1$ and Eq. (3.36) reduces to

$$M_{(p)} = \sum_j N_j p_j \tag{3.38}$$

From Eqs. (3.30) and (3.37) it is seen that at low field strengths both the contributions of the induced moment $M_{(\alpha)}$ and of the permanent dipole moment $M_{(p)}$ are linear in $E$. This correlates well with the continuum expression for $M$ as discussed in the context of Eq. (3.2).

### 3.2.5. Form Factors and g Factors

The conversion factors $g$ and $\tilde{g}$ contain the so-called form factors which account for shape anisotropies. In an ellipsoidal molecule the form factors (sometimes called depolarizing factors, which are the components of the depolarizing tensor) of the main polarization axes are $A_q = A_a$, $A_b$, $A_c$. In line with the vectorial character of the internal and directing fields the $g$ factors of anisotropic molecules are tensors.[17,20] If the environment of the molecules (which are characterized by the polarizability tensor $\alpha$ and the permanent dipole moment **p**), can be considered as *nonpolar* and the overall dielectric permittivity is $\varepsilon$, the $g$ factor of the $q$ axis is given by

$$g_q = \varepsilon\{1 + [(\varepsilon_\infty)_q - 1]A_q\}/\{\varepsilon + [(\varepsilon_\infty)_q - \varepsilon]A_q\} \tag{3.39}$$

The polarizability along the $q$ axis is

$$\alpha_q = \varepsilon_0 V_j[(\varepsilon_\infty)_q - 1]/\{1 + [(\varepsilon_\infty)_q - 1]A_q\} \tag{3.40}$$

where $V_j = (4/3)\pi abc$ is the molecular volume of the ellipsoidal molecules $B_j$ with the half-axes $q = a, b, c$. It is stressed that the value of $\varepsilon$ in Eqs. (3.39) and (3.40) is the effective dielectric permittivity of the total system. The quantity $\varepsilon_\infty$ may be considered as the "molecular permittivity" at frequencies of the polarizing field, where the permanent dipoles do not contribute any more to the total polarization; $(\varepsilon_\infty)_q$ is the $q$ component of the permittivity tensor $\varepsilon_\infty$.

To facilitate comparison with familiar representations of dielectrics[17] note that the reaction field factor of the $q$ axis is given by

$$f_q = A_q(1 - A_q)(\varepsilon - 1)/\{\varepsilon_0 V_j[\varepsilon + (1 - \varepsilon)A_q]\} \tag{3.41}$$

At particular geometries the form factors are analytically expressed in a simple form. For very long cylinders the depolarizing factors along the long axis is zero; thus $g = 1$. Therefore the local field which affects counterion polarization in linear polyelectrolytes is equal to the externally applied electric field. Another type of shape which is relevant for biological systems is the flat disk; flat patches of biological membranes may be described in terms of the disk geometry. The depolarizing factor for the polarization direction along the disk-normal, i.e., perpendicular to the disk plane, is $A_\perp = 1$; therefore, $g = 1/\varepsilon$. The form factor of spherical isotropic systems is $A_q = 1/3$. In the case of polarizable dipolar spheres $B_j$ [with $(\varepsilon_\infty)_j$ and $\mathbf{p}_j$] immersed in (a large excess of) a nonpolar medium of the effective bulk permittivity $\varepsilon$, Eqs. (3.39) and (3.40) yield

$$g_j = \varepsilon[(\varepsilon_\infty)_j + 2]/[2\varepsilon + (\varepsilon_\infty)_j] \tag{3.42}$$

Substituting $A_q = 1/3$ into Eq. (3.40), we obtain the familiar Clausius–Mosotti equation:

$$\alpha_j = 3\varepsilon_0 V_j[(\varepsilon_\infty)_j - 1]/[(\varepsilon_\infty)_j + 2] \tag{3.43}$$

generally valid at high frequencies of the polarizing electric field where permanent dipoles do not contribute to the polarization; $V_j$ is the "volume of particle $B_j$." The reaction field factor of spherical molecules for the same conditions is given by

$$f_j = 2(\varepsilon - 1)/[3\varepsilon_0 V_j(2\varepsilon + 1)] \tag{3.44}$$

In a pure condensed medium of nonpolar molecules we have $\varepsilon_\infty = \varepsilon$. For isotropic spheres we have

$$g = (\varepsilon + 2)/3$$
$$\alpha = 3\varepsilon_0 V_j(\varepsilon - 1)/(\varepsilon + 2) \tag{3.45}$$

In the case of a pure liquid of nonpolarizable dipoles ($\mathbf{p}$) we have $\varepsilon_\infty = 1$; hence the conversion factor of the spherical permanent point dipoles is

$$g = 3\varepsilon/(2\varepsilon + 1) \tag{3.46}$$

Finally, the $g$ factor of spherical molecules in the gas phase is

$$g = (\varepsilon + 2)/3 \tag{3.47}$$

and the polarizability is given by Eq. (3.45). The $g$ factors for the various specific cases are compiled in Table 1.

### 3.3. Reaction Moment and Equilibrium Constant

When changes in the concentration of the reaction partners can be measured directly, say by an optical method, the analysis of equilibrium properties is based on Eqs. (3.23) and (4.42) of part I. The electric field

*TABLE 1.* Conversion Factors ($g$, $g_j$) for Spherical Molecules, Relating Internal Field ($\mathbf{E}_{int}$), Directing Field ($\mathbf{E}_{dir}$), and Fröhlich Field ($\mathbf{E}_F = \mathbf{E}_{dir}$) to the (measured) Maxwell Field ($\mathbf{E}$)[a]

| I. Pure liquids ($\varepsilon$) | $g$ | $\mathbf{E}_{int}$ | $\mathbf{E}_{dir}$ |
|---|---|---|---|
| (a) Polarizable polar spheres (Onsager) $(p, \varepsilon_\infty, \varepsilon)$ | $\dfrac{\varepsilon(\varepsilon_\infty + 2)}{2\varepsilon + \varepsilon_\infty}$ | $\tilde{g}g\mathbf{E}_{\parallel}$ | $g\mathbf{E}$ |
| (b) Polarizable nonpolar spheres $(p = 0, \varepsilon_\infty = \varepsilon)$ | $\dfrac{\varepsilon + 2}{3}$ | $g\mathbf{E}_{\parallel}$ | |
| (c) Polar nonpolarizable spheres $(p, \varepsilon_\infty = 1, \varepsilon)$ | $\dfrac{3\varepsilon}{2\varepsilon + 1}$ | | $g\mathbf{E}$ |
| (d) Polarizable permanent dipoles [Fröhlich, $p = p_G(\varepsilon_\infty + 2)/3; \varepsilon$] | $\dfrac{3\varepsilon}{2\varepsilon + \varepsilon_\infty}$ | | $g\mathbf{E}$ |
| II. Molecules [$p_j$, $(\varepsilon_\infty)_j$] diluted | $g_j$ | $(\mathbf{E}_{int})_j$ | $(\mathbf{E}_{dir})_j$ |
| (1) In nonpolar fluid medium ($\varepsilon$) | | | |
|   (a) Polarizable polar spheres $[p_j, (\varepsilon_\infty)_j]$ | | $\tilde{g}_j g_j \mathbf{E}_{\parallel}$ | $g_j\mathbf{E}$ |
|   (b) Polarizable nonpolar spheres $[p_j, (\varepsilon_\infty)_j = 1; \varepsilon]$ | $\dfrac{\varepsilon[(\varepsilon_\infty)_j + 2]}{2\varepsilon + (\varepsilon_\infty)_j}$ | $g_j\mathbf{E}_{\parallel}$ | |
|   (c) Polar nonpolarizable spheres $[p_j, (\varepsilon_\infty)_j = 1; \varepsilon]$ | $\dfrac{3\varepsilon}{2\varepsilon + 1}$ | | $g_j\mathbf{E}$ |
| (2) In polar fluid medium ($\varepsilon$) (Fröhlich, Kirkwood, $g_K$ factor), $p_j = (p_j)_G[(\varepsilon_\infty)_j + 2]/3$ | $\dfrac{3\varepsilon}{2\varepsilon + (\varepsilon_\infty)_j}$ | | $g_j\mathbf{E}$ |
| (3) In gas phase ($\varepsilon$), polar polarizable spheres $[(p_j)_G, (\varepsilon_\infty)_j, \varepsilon]$ | $\dfrac{(\varepsilon_\infty)_j + 2}{3}$ | $g_j\mathbf{E}_{\parallel}$ | $g_j\mathbf{E}$ |

[a] In the case of isotropic polarization, $\mathbf{E}_{\parallel} = E$; for uniaxial anisotropic polarizability $\mathbf{E}_{\parallel} = E \cos \vartheta_j$, where $\vartheta_j$ is the angle between the dipole axis and $\mathbf{E}$. $\varepsilon$, total dielectric permittivity; $\varepsilon_\infty$, dielectric permittivity of the induced (high-frequency) polarization. The factor $\tilde{g}$ refers to the permanent dipoles' contribution of $\mathbf{E}_{int}$; $\tilde{g}_j = 1 + f_j(1 - \alpha_j f_j)^{-1}\langle p_j \rangle$, where $f_j$ is the reaction field factor and $\alpha_j$ the polarizability tensor of the molecule $j$ (if $p_j = 0$ or $\langle p_j \rangle = 0$, $\tilde{g}_j = 1$).

dependence of the apparent equilibrium constant (concentration ratio) is given by

$$K(E) = K(0)e^x \tag{3.48}$$

where the following definition holds:

$$x = \int_0^E \Delta M \, dE/RT \tag{3.49}$$

It is frequently observed that the reaction moment $\Delta M$ is dominated either by the induced dipole term $M_{(\alpha)}$ or by the permanent dipole term $M_{(p)}$. It is therefore useful to follow Eq. (3.10) and to write $\Delta M$ in two terms:

$$\Delta M = \Delta M_{(\alpha)} + \Delta M_{(p)} \tag{3.50}$$

corresponding to a separation of the quantity $x$:

$$x = x_{(\alpha)} + x_{(p)} \tag{3.51}$$

The induced polarization is thus characterized by

$$x_{(\alpha)} = \int \Delta M_{(\alpha)} \, dE/RT \tag{3.52}$$

and the permanent dipole contribution is given by

$$x_{(p)} = \int \Delta M_{(p)} \, dE/RT \tag{3.53}$$

The investigation of chemical processes in solutions is preferably performed under the condition of higher dilution such that the individual reaction partners can be considered independent and the thermodynamic activity coefficients are either constant or equal to one. As to chemical and physical processes in the presence of applied electric fields, the solvent may be treated as an infinite fluid dielectric in which the molecules or particles are immersed. On a microscopic scale, however, the solvent molecules are more or less densely packed, probably forming dynamic fluctuating clusters. The space in between the molecules and clusters certainly is vacuum. Even if the internal and directing fields which actually work on the molecules are homogeneous, the local Maxwell field in the vicinity of the molecules is inhomogeneous. The calculation of the internal and the directing field in terms of an inhomogeneous Maxwell field is extremely intricate.[17] Therefore the classical relations between $\mathbf{E}_{int}$ and $\mathbf{E}_{dir}$ and the macroscopic average field (Maxwell field) of real molecular dielectrics are only approximations, related to the homogeneous part of the Maxwell field, $\mathbf{E}$. The approximations involve the assumption that the bulk of the dielectric can be represented by an effective average permittivity $(\varepsilon)$, being a constant over the dielectric. In the framework of this assumption the total polarization of a fluid dielectric and the chemical contributions to polarization changes may be treated in a similar way as the homogeneous approximation. In any case, the specific expressions of the reaction moment and the field dependence of the equilibrium constant reflect not only the different dielectric properties of the interacting molecules but also the reaction mechanism.

### 3.3.1. Nonpolar Polarizable Spheres

In the case of pure induced polarization the solute molecules $B_j$ may be considered as polarizable spheres associated with a "molecular" permittivity $(\varepsilon_\infty)_j$ and $\mathbf{p}_j = 0$. When these molecules are immersed in a large quantity of nonpolar solvent the total permittivity of the solution $\varepsilon$ is practically that of the solvent.

For particles of uniaxial anisotropic polarizability $\alpha_j$, Eqs. (3.2) and (3.23) with $M = M_{(\alpha)}$ are combined with Eq. (3.29). Since $N_j = N_A n_j$, the resulting expression reads

$$\varepsilon_0(\varepsilon - 1)\,V = N_A \sum_j n_j \alpha_j (1 + 2\phi_j)\, g_j/3 \tag{3.54}$$

The appropriate relations of $g_j$ and $\alpha_j$ are obtained from Eqs. (3.44) and (3.45), respectively. Assuming now that all $(\varepsilon_\infty)_j$ are equal such that all $B_j$ have $(\varepsilon_\infty)_j = \varepsilon_\infty$, Eq. (3.54) may be formulated for the low-field limiting case $\phi_j \to 0$ as

$$\varepsilon_0(\varepsilon - 1)(2\varepsilon + \varepsilon_\infty)/\varepsilon = N_A \sum_j n_j \alpha_j (\varepsilon_\infty + 2)/3V \tag{3.55}$$

In this way the quantities which change upon a chemical transformation, the total permittivity $\varepsilon$ and the mole quantities $n_j$, are separated. Hence the differentiation with respect to the reaction variable $\xi$ is readily performed separately at both sides of the equation. Using Eq. (3.1) of part I in the form $dn_j/d\xi = v_j$ we finally obtain

$$(\partial \varepsilon/\partial \xi)_{E,V} = N_A \varepsilon^2 \sum_j v_j \alpha_j (\varepsilon_\infty + 2)/[3V\varepsilon_0(2\varepsilon^2 + \varepsilon_\infty)] \tag{3.56}$$

Substitution of Eq. (3.56) into (3.4) leads to

$$\Delta M_{(\alpha)} = [N_A \varepsilon^2/(2\varepsilon^2 + \varepsilon_\infty)] \sum v_j \alpha_j [(\varepsilon_\infty + 2)/3] E \tag{3.57}$$

Insertion of Eq. (3.57) into (3.52) and integration yields the $x$ quantity of the low-field range:

$$x_{(\alpha)} = [3\varepsilon^2/(2\varepsilon^2 + \varepsilon_\infty)] \left\{ \sum_j \nu_j \alpha [(\varepsilon_\infty + 2)/3]/(6kT) \right\} E^2 \qquad (3.58)$$

For isotropic polarizable spheres Eq. (3.27) applies and

$$x_{(\alpha)} = [3\varepsilon^2/(2\varepsilon^2 + \varepsilon_\infty)] \left\{ \sum_j \nu_j \alpha [(\varepsilon_\infty + 2)/3]/2kT \right\} E^2 \qquad (3.59)$$

In the limiting case of saturated induced dipole moments, Eq. (3.32) is used to obtain the reaction moment $\Delta M$ according to Eq. (3.4):

$$\Delta M_s = N_A \sum_j \nu_j (m_s)_j \qquad (3.60)$$

The $x$ quantity of this case is

$$x_s = \sum_j \nu_j (m_s)_j E/(kT) \qquad (3.61)$$

Inspecting the Eqs. (3.58), (3.59), and (3.61) we realize that an induced dipole mechanism is associated with a quadratic dependence of the $x$ quantity on the field strength. The temperature dependence is linear in $1/T$. At high field strengths saturation may occur and a transition to a linear dependence on $E/T$ will be observed.[9,10]

### 3.3.2. Polar Nonpolarizable Spheres

The special case of uniaxial nonpolarizable point dipoles refers to $(\varepsilon_\infty)_j = 1$ and a finite value of the permanent dipole moment $\mathbf{p}_j$ in the general expressions. The solvent is nonpolar and the total permittivity of the solution is $\varepsilon$. The combination of Eqs. (3.32) and (3.4) only leads to general analytical forms of $(\partial \varepsilon/\partial \xi)_{V,E}$ and of the $x_{(p)}$ factor if the $g$ factors were independent of $\xi$. Simple analytical expressions can only be derived for the limiting cases of small and large field strengths, respectively.

At low field strengths where Eq. (3.37) applies we derive from the general Eq. (3.2) that

$$\varepsilon_0(\varepsilon - 1)V = N_A \sum_j n_j p_j^2 g_j/(3kT) \qquad (3.62)$$

The $g$ factor of nonpolarizable spherical point dipoles is given by Eq. (3.46):

$$g_j = g = 3\varepsilon/(2\varepsilon + 1)$$

Term separation in Eq. (3.62) leads to

$$\varepsilon_0(\varepsilon - 1)(2\varepsilon + 1)/3\varepsilon = (N_A/V) \sum n_j p_j^2/(3kT) \tag{3.63}$$

and

$$(\partial\varepsilon/\partial\xi)_{V,E} = (N_A/\varepsilon_0 V)[3\varepsilon^2/(2\varepsilon^2 + 1)] \sum v_j p_j^2/(3kT) \tag{3.64}$$

Introducing now Eq. (3.64) into (3.4) we obtain for spherical point dipoles

$$\Delta M_{(p)} = N_A[3\varepsilon^2/(2\varepsilon^2 + 1)] \left\{ \sum v_j p_j^2/(3kT) \right\} E \tag{3.65}$$

Substitution into Eq. (3.53) and integration yield the $x$ quantity for the low-field range of chemically interacting point dipoles

$$x_{(p)} = [(3\varepsilon^2/(2\varepsilon^2 + 1)] \left\{ \sum v_j p_j^2/[6(kT)^2] \right\} E^2 \tag{3.66}$$

In the high field strength range Eq. (3.38) applies and the orientational saturation is given by

$$\Delta M_s = N_A \sum_j v_j p_j \tag{3.67}$$

and

$$x_s = \sum_j v_j p_j E/kT \tag{3.68}$$

Thus, also a permanent dipole mechanism may be characterized by a transition of the $x$ quantity from a quadratic field strength dependence to a linear one; concomitant with the E dependence the temperature variation changes from $T^{-2}$ to $T^{-1}$. Therefore the temperature dependence of equilibrium and rate constants may be used to differentiate between permanent and induced moments.

The combination of the continuum expression (3.2) with the molecular representation in terms of the Langevin function $L[r_j]$ yields

$$\varepsilon_0(\varepsilon - 1)V = N_A \sum_j n_j p_j L[p_j g_j E/kT] \tag{3.69}$$

The differentiation with respect to $\xi$ involves the term $\partial L[p_j g_j E/kT]/\partial \xi$ that cannot be treated in a closed analytical form. Even if the approximation $\Delta M_{(p)} = \sum_j v_j p_j L(r_j)$ could be applied, the integration of the function $L[r_j]$ to obtain the $x$ quantity, according to

$$x_{(p)} = \frac{\sum_j v_j p_j \int L[r_j]\, dE}{kT} = \sum_j (v_j/g_j) \ln \frac{\sinh r_j}{r_j} \qquad (3.70)$$

is valid only if the $g_j$ factors can be considered independent of $E$, i.e., $(\partial \varepsilon / \partial E) = 0$.

The assumption of constant $\varepsilon$ is inherent in all integrations according to Eqs. (3.52) and (3.53). The $\varepsilon$ values in the specific expressions for the $x$ quantities refer to the actual value of $\varepsilon$ in the presence of **E**.

### 3.3.3. Polar Polarizable Spheres (Onsager)

Real molecules are always polarizable. When these particles have a permanent dipole moment ($\mathbf{p}_j$) they are characterized by the set ($p_j, \alpha_j$); the polarizability may be expressed in terms of the dielectric permittivity ($\varepsilon_\infty$)$_j$ of the induced high-frequency polarization (by the Clausius–Mosotti equation).

In the case of isotropic polarizabilities and uniaxial permanent dipoles the electric fields which actually work on the molecules are given by Eqs. (3.18) and (3.34). The internal field of polar polarizable spheres is given by

$$(\mathbf{E}_{\text{int}})_j = \tilde{g}_j g_j \mathbf{E}$$
$$= [1 + f_j(1 - \alpha_j f_j)^{-1}\langle \mathbf{p}_j\rangle] g_j \mathbf{E} \qquad (3.71)$$

For spherical particle geometry the factors $g_j$, $\alpha_j$, and $f_j$ are obtained from Eqs. (3.42)–(3.44), respectively. The low-field approximation of $\langle \mathbf{p}_j\rangle$ is given by

$$\langle \mathbf{p}_j\rangle = p_j\langle\cos\vartheta_j\rangle = p_j^2 g_j E/(3kT) \qquad (3.72)$$

The total moment in Eq. (3.23) may be rewritten as

$$M = N_A \left\{ \sum_j n_j \alpha_j (\mathbf{E}_{\text{int}})_j + \sum_j n_j \langle \mathbf{p}_j\rangle \right\} \qquad (3.73)$$

In the low-field range we substitute Eqs. (3.71) and (3.72) into (3.73). Using now Eqs. (3.42)–(3.44) for $g_j$, $\alpha_j$, and $f_j$, respectively, together with the approximation that all ($\varepsilon_\infty$)$_j$ are equal, i.e., ($\varepsilon_\infty$)$_j = \varepsilon_\infty$, we obtain the familiar Onsager equation for pure dipole liquids in the suggestive form

$$\varepsilon_0(\varepsilon - \varepsilon_\infty)(2\varepsilon + \varepsilon_\infty)/3\varepsilon = (N_A/V)\sum_j n_j p_j^2[(\varepsilon_\infty + 2)/3]^2/(3kT) \qquad (3.74)$$

Differentiation with respect to $\zeta$ yields

$$(\partial\varepsilon/\partial\zeta)_{V,E} = (N_A/\varepsilon_0 V)[3\varepsilon^2/(2\varepsilon^2 + \varepsilon_\infty)] \sum v_j p_j^2 [(\varepsilon_\infty + 2)/3]^2/(3kT) \qquad (3.75)$$

and applying Eq. (3.4) results in

$$\Delta M = [N_A 3\varepsilon^2/(2\varepsilon^2 + \varepsilon_\infty^2)] \left\{ \sum v_j p_j^2 [(\varepsilon_\infty + 2)/3]^2 \right\} E \qquad (3.76)$$

With Eq. (3.49) the equilibrium shift by small fields is described by

$$x = [3\varepsilon^2/(2\varepsilon^2 + \varepsilon_\infty^2)] \left\{ \sum v_j p_j^2 [(\varepsilon_\infty + 2)/3]^2/[6(kT)^2] \right\} E \qquad (3.77)$$

In a similar manner we may derive the expressions of $\Delta M$ and $x$ for anisotropic polarizable permanent dipoles, interacting in nonpolar media. Here again, general expressions in terms of the Langevin function cannot be derived in closed analytical form.

The Onsager approach appears quite adequate for the analysis of chemical reactions in nonpolar media.[14,19,25,26] For polar liquids like water, which is of particular interest for biochemical processes, a modified model developed by Fröhlich appears to be more adequate.[27,28]

### 3.4. Reactions in Polar Media

According to Fröhlich, a pure condensed dielectric consisting of polarizable molecules with a permanent dipole moment **p** may be formally represented by a continuum permittivity $\varepsilon_\infty$ accounting for the "molecular" polarizability, embedded in the bulk continuum with the effective permittivity $\varepsilon$. The fundamental polarization equation for such a polar dielectrics is

$$M = \varepsilon_0(\varepsilon_\infty - 1) VE + N_A \sum_j n_j \langle \mathbf{p}_j \rangle \qquad (3.78)$$

Note that the induced part is formally separated from the permanent dipole moment contribution in a particular manner. In Fröhlich's version of the Onsager model the spherical dipoles have effective dipole moments

$$\mathbf{p}_j = (\mathbf{p}_j)_G [(\varepsilon_\infty)_j + 2]/3 \qquad (3.79)$$

where $(\mathbf{p}_j)_G$ is the dipole moment of $B_j$ in the gas phase.

The actual field working as the directing field is the Fröhlich field $\mathbf{E}_F$, which is given by

$$(\mathbf{E}_F)_j = g_j \mathbf{E} \qquad (3.80)$$

where

$$g_j = 3\varepsilon/[2\varepsilon + (\varepsilon_\infty)_j] \qquad (3.81)$$

In a medium of polar molecules specific intermolecular interactions such as, for instance, H-bridges in water may occur. The effect of this property is accounted for by the Kirkwood correlation factor $g_K$.

The average contribution of the "Fröhlich dipoles" to the total moment is, analogous to Eq. (3.72), given by

$$\langle p_j \rangle = p_j(g_K)_j L[p_j g_j E/kT] \qquad (3.82)$$

The low-field approximation reads

$$\langle p_j \rangle = p_j^2(g_K)_j g_j E/(3kT) \qquad (3.83)$$

and the limiting case of orientational saturation has an average moment contribution of

$$\langle p \rangle_s = p_j(g_K)_j \qquad (3.84)$$

### 3.4.1. Kirkwood–Fröhlich Equation

For pure liquids where all $(\varepsilon_\infty)_j = \varepsilon_\infty$, the combination of Eqs. (3.2), (3.78), (3.81), and (3.83) leads to the familiar Kirkwood–Fröhlich equation, written here in the suggestive form of the separated variables $\varepsilon$ and $n_j$:

$$\varepsilon_0(\varepsilon - \varepsilon_\infty)(2\varepsilon + \varepsilon_\infty)/3\varepsilon = (N_A/V)\sum n_j(g_K)_j(p_j)_G^2[(\varepsilon_\infty + 2)/3]^2/(3kT) \qquad (3.85)$$

Since in a pure liquid $\sum_j n_j(g_K)_j(p_j)_G^2 = n g_K p_G^2$, Eq. (3.85) is useful for the determination of dipole moments of polar liquids by dielectric measurements. The $g_K$ factor must be calculated.[17]

### 3.4.2. Ion-pair Equilibria in Aqueous Solution

Chemical equilibria such as ion-pair formation of electrolytes in aqueous solution where the hydrated ion pairs behave as polarizable dipoles, may be treated in terms of the Fröhlich formalism.

If the $(\varepsilon_\infty)_j$ values of the individual ions and ion pairs are basically determined by the hydration spheres we may use the approximation $(\varepsilon_\infty)_j = \varepsilon_\infty$ and $\varepsilon_\infty = 5(\pm 1)$ at 20°C.[28]

Since the H-bond coordination number of a hydrated ion or ion pair is not known, plausibility arguments of a symmetric interaction pattern with the bulk water may justify the approximation $(g_K)_j = 1$.

The low-field limiting case of ion-pair equilibria in water may be derived from Eq. (3.85):

$$\varepsilon_0(\varepsilon - \varepsilon_\infty)(2\varepsilon + \varepsilon_\infty)/3\varepsilon = (N_A/V) \sum_j n_j p_j^2/(3kT) \tag{3.86}$$

where $p_j$ is given by Eq. (3.79). Substitution of

$$(\partial\varepsilon/\partial\xi)_{V,E} = (N_A/\varepsilon_0 V)[3\varepsilon^2/(2\varepsilon^2 + \varepsilon_\infty^2) \sum_j v_j p_j^2/(3kT) \tag{3.87}$$

into (3.4) yields the reaction moment

$$\Delta M = N_A[3\varepsilon^2/(2\varepsilon^2 + \varepsilon_\infty^2)]\left[\sum_j v_j p_j^2/(3kT)\right] E \tag{3.88}$$

The use of Eq. (3.49) and integration result in the $x$ factor of ion-pair equilibrium displacements at low field strengths:

$$x = [3\varepsilon^2/(2\varepsilon^2 + \varepsilon_\infty^2)]\left\{\sum_j v_j p_j^2/[6(kT)^2]\right\} E^2 \tag{3.89}$$

At high field strengths leading to orientational saturation $(\langle p_j \rangle = p_j)$, combination of Eq. (3.2) and (3.78) yields

$$\varepsilon_0(\varepsilon - \varepsilon_\infty) V \cdot E = N_A \sum_j n_j p_j \tag{3.90}$$

$$\Delta M = \varepsilon_0 VE(\partial\varepsilon/\partial\xi)_{V,E} = N_A \sum_j v_j p_j \tag{3.91}$$

$$x = \sum_j v_j p_j/(kT) \tag{3.92}$$

Note that in the Fröhlich version of the Onsager model due to Eq. (3.79), Eqs. (3.85) and (3.86) are identical to (3.74) and Eqs. (3.87) and (3.88) are identical to Eqs. (3.75)–(3.77), respectively. On the basis of of Eq. (3.87) the dipole moments of ion-pairs formed by $Mg^{2+}$ and $SO_4^{2-}$ in aqueous solution have been estimated.[7,8]

### 3.5. Induced Dipole Moments in Polyionic Macromolecules

As already mentioned, large induced dipole moments may result from atomic polarization. When the polarization is caused by displacements of ionic groups within macromolecules the conformational folding of the

(a)                                          (b)

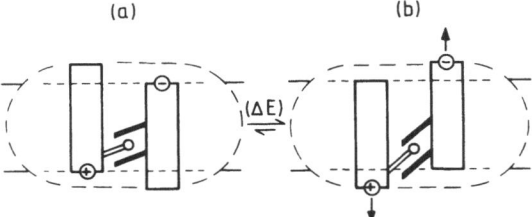

FIGURE 1. Principle of the saturatable induced dipole mechanism causing positional changes of side chains in helical membrane proteins. In bacteriorhodopsin helical parts with different net charge may move transversal to the membrane plane in opposite directions when the electric membrane field is increased, (a) → (b). The geometrically limited increase in the distance of the charge centers is equivalent to a saturatable induced dipole moment. The transversal displacement of at least one of the two helical parts can thereby cause a concerted rotational shift of the retinal ($=$o) and of aromatic amino acid side chains which may sandwich (T. H. Haines) the retinal chromophore.

polymer structure may restrict the local mobility of the charged groups and dipolar ion pairs. Restricted conformational displaceability may then lead to a saturable induced dipole moment[30]; see Figure 1.

Consider a dipolar ion pair of a positively charged (lysine) group and a negatively charged (glutamic acid) group in a protein. An external electric field may induce an increase, $\delta r$, of the distance vector $r$ between the charge centers of this ion pair. When this protein is part of a membrane structure, then the field-induced distance change $\delta r$ may not only lead to an increase in the scalar amount of the dipole moment, but may also be accompanied by a rotation of the dipole vector (Figure 2). According to

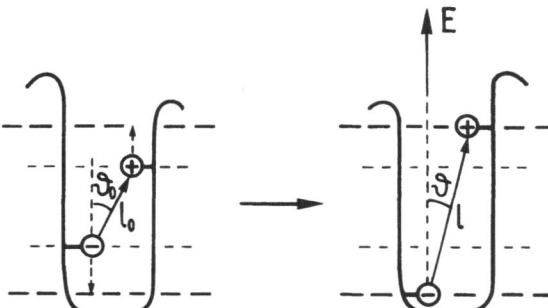

FIGURE 2. Scheme for the anisotropic mobility of a section in a membrane protein. Field-induced displacement of two oppositely charged groups (of an ion pair) along the membrane normal, leading to an increase of the apparently permanent dipole moment $m_0 = e_0 \cdot l_0$ to $m = e_0 \cdot l$, equivalent to an *induced* dipole moment of $m - m_0 = p_{(\alpha)} = e_0 \delta l$, where $\delta l = l - l_0$. The dipole moment increase is accompanied by a rotation of the dipole axis $l$ toward the membrane normal; the electric field vector is parallel to the normal.

this model an electric field $\mathbf{E}$ increases the dipole moment of an ion pair from a value $\mathbf{m}(0) = |\tilde{z}|\, e_0 r_0$ at $E = 0$ to a value $\mathbf{m}(E) = |\tilde{z}|\, e_0 \mathbf{r}$, with $\mathbf{r} > \mathbf{r}_0$. The distance increase corresponds to an induced dipole moment:

$$\mathbf{m}_{(\alpha)} = \mathbf{m}(r) - \mathbf{m}(r_0) = |\tilde{z}|\, e_0 \delta \mathbf{r} = \alpha \mathbf{E}_{int} \tag{3.93}$$

The limited conformational flexibility of (membrane) proteins will only permit motions of restricted extent. The membrane structure may further limit the motions in certain directions. It is likely that a major unidirectional charge displacement may only occur along the membrane normal.

Due to these limits the atomic polarizability $\alpha$ refers to the displaceability of charged protein groups in directions closely along the membrane normal. Furthermore, due to conformational restrictions, $\alpha$ will decrease from an initial value $\alpha$ at $\mathbf{E} = 0$, with increasing field strength until, at the saturation of the charge displacement, the limit $\alpha \to 0$ is reached.

Thus, in general, the induced dipole moment $\mathbf{m}_{(\alpha)}$ will reach a saturation value $\mathbf{m}_s$. As pictured in Figure 2, a restricted charge displacement may be accompanied by an orientational charge of the dipole axis. The electric field dependence of the total moment may be described in terms of coth functions as in the case of the counterion polarization in linear polyelectrolytes.[31] See also Yoshioka *et al.*[29]

The model outlined in Figure 1 for bacteriorhodopsin is suggestive not only for a possible control function of the electric field of the bacterial membrane during the photocycle of bacteriorhodopsin, but also for a possibly general, induced-dipole mechanism for electric-field-dependent

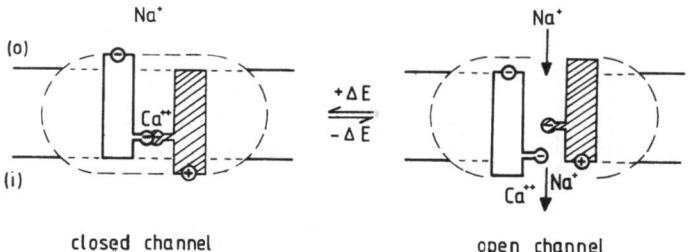

FIGURE 3. Model for a gating element of the axonal $Na^+$ channel. At high electric field (resting membrane potential) between the outside (o) and the inside (i) of the excitable membrane, the gating protein is in the closed conformation. $Ca^{2+}$ ions are bound to anionic side chains of amino acids of helical parts, kept by the membrane field in a high dipole moment configuration (large distance between the cationic and anionic groups). Depolarization reduces the electric membrane field, which, in turn, reduces the electric polarization by transversal movements of the helical parts, decreasing the distance between the fixed charges. The $Ca^{2+}$ ion of the former bridge is exchanged by $Na^+$ ions which flow through the open, "field-relaxed" conformation into the cell interior.

structural changes in membrane transport proteins such as the gating proteins in the excitable membranes of nerve and muscle cells. Analogous to Figure 1, we may suggest an induced-dipole gating mechanism for the axonal $Na^+$ channel in a schematic form (Figure 3), also incorporating ion exchange $Ca^{2+}/Na^+$.[32,33]

## 4. Measurement of Electric Field Effects

The majority of biological processes involves ionic species in aqueous environments. The stability of many biopolymer systems requires a finite ionic strength. Adequately, electric field effects in these systems have to be studied in conducting solutions and suspensions. Traditionally, dielectric measurements, conductivity relaxations, and electro-optic techniques were used to study electric field effects on chemical reactions.[18,19] In order to minimize Joule heating, high electric field strengths are applied in the form of short field pulses. The analysis of field-induced changes is particularly straightforward for rectangular pulses.

However, if the system is simple and only one process occurs or if the field-induced changes are longlived and relax with time constants large compared to the field duration, Joule heating temperature jump spectrometers may be used.[2,31,33]

Relaxation kinetic methods provide the most powerful approach to chemical field effects. The quantitative analysis of relaxation kinetic data is appreciably simplified when the $z$-induced changes are small. A general expression for chemical relaxation conditions may be found in terms of the appropriate chemical reaction affinity according to

$$A \ll RT \tag{4.1}$$

In terms of Eq. (2.1) the reaction rate $v$ is

$$v = v_p - v_r \tag{4.2}$$

the rates of product and reactant formation, respectively, are given by

$$v_p = k_p^\ominus \prod a_r^{v_r}, \qquad v_r = k_r^\ominus \prod a_p^{v_p} \tag{4.3}$$

At equilibrium, $\bar{v} = 0$, thus $\bar{v}_p = \bar{v}_r$ and all $a_j$ are given by the equilibrium values $\bar{a}_j$.

We now rewrite Eq. (4.2) as

$$v = v_p(1 - v_r/v_p) = v_r(v_p/v_r - 1) \tag{4.4}$$

and recall that $Q^{\ominus} = \prod a_p{}^{v_p}/\prod a_r{}^{v_r}$, $K^{\ominus} = \prod \bar{a}_p{}^{v_p}/\prod \bar{a}_r{}^{v_r}$. Since $A = RT \ln(K^{\ominus}/Q^{\ominus})$, substitution into Eq. (4.4) leads to

$$v = v_p(1 - e^{-(A/RT)}) = v_r(e^{A/RT} - 1) \tag{4.5}$$

Near equilibrium where $A \ll RT$, we may set $v_p = \bar{v}_p$ and $v = \bar{v}_r$. Series expansion of the exponentials in Eq. (4.5) and neglect of higher-order terms result in an expression for the near-equilibrium chemical rate:

$$v_{eq} = \bar{v}_r A/RT = \bar{v}_p A/RT \tag{4.6}$$

### 4.1. Chemical Relaxations

In Eq. (4.6) the chemical reaction rate close to equilibrium is linearly dependent on the affinity. Close to equilibrium the $z$-induced changes in the extent of the reaction, $\xi$, also depend linearly on $z$. Thus, for small perturbations we may generally specify

$$\delta\xi = (\partial\xi/\partial z_i)_{z \neq z_i} \partial z_i = \xi(z_i) - \xi_{ref} \ll \xi_{ref} \tag{4.7}$$

We now recall Eq. (3.14) of Part I and express the chemical relaxation condition in electric fields in terms of the concentration of the reaction partner $B_j$:

$$\delta c_j = v_j \Gamma \Delta M E/(RT) \tag{4.8}$$

The condition $\delta c_j \ll c_j(0)$ must hold and the definition (3.15) applied to equilibrium at $E$ is

$$\Gamma = \left(\sum_j v_j^2/\bar{c}_j\right)^{-1} \tag{4.9}$$

If the thermodynamic activity coefficients depend on the extent of reaction, then the quantity $\Gamma$ must be replaced by the term $\Gamma^*$; see Eq. (3.56) of part I. Therefore the amplitude of the $E$-induced concentration shift is

$$(\delta c_j)_{\infty} = v_j \bar{\Gamma}^* \Delta M E/(RT) \tag{4.10}$$

If an external perturbation is applied faster than the chemical equilibration time, the response to a step perturbation is a relaxation spectrum containing exponentials of time, $t$. The time course of the chemical relaxation with $q$ (normal mode) processes is given by

$$\delta\xi(t) = \sum_q (\delta\xi)_{\infty,q} e^{-t/\tau_q} \tag{4.11}$$

where $(\delta\xi)_{\infty,q}$ is the amplitude and $\tau_q$ the relaxation time, respectively, of the relaxation mode $q$.

An elementary step is always described by

$$\delta c_j(t) = (\delta c_j)_\infty e^{-t/\tau} \tag{4.12}$$

This expression is generally applicable for intramolecular elementary steps. It is applicable for bimolecular steps only if the perturbation is small such that $\delta\xi \ll \xi_{\text{ref}}$ or $\delta c_j \ll (c_j)_{\text{ref}}$.

Recent theoretical developments offer a simple formalism to evaluate time constants and amplitudes in terms of total concentrations, $c_j^0$, rather than equilibrium concentrations, $\bar{c}_j$.[9,34–36]

In a kinetic titration according to $L + B = LB$, where for instance $c_B^0$ is kept constant and $c_L^0$ is varied, the amplitude factor $\bar{\Gamma}$ is zero at $c_L^0 = 0$, passes then through a maximum at $(c_L^0)_{\Gamma\text{max}} = c_B^0 + K$, and finally approaches zero again. Provided the conditions can be chosen such that $K < c_B^0$, also the relaxation time passes through a maximum, at $(c_L^0)_{\tau\text{max}} = c_B^0 - K$. We may therefore use the two maxima and determine the value of $K$ as well as that of the total number of binding sites participating in the relaxation process by

$$K = [(c_L^0)_{\Gamma\text{max}} - (c_L^0)_{\tau\text{max}}]/2 \tag{4.13}$$

$$c_B^0 = [(c_L^0)_{\Gamma\text{max}} + (c_L^0)_{\tau\text{max}}]/2 \tag{4.14}$$

In Table 2 the key relations for the elementary chemical reactions are summarized.

It is recalled that the aim of chemical relaxation kinetics is the elucidation of reaction mechanisms: number and nature of elementary steps; identity and properties of reactants, intermediates, and products; kinetic and thermodynamic parameters (rate constants, reaction enthalpy, and entropy) characterizing individual equilibria and reaction pathways.[37,38]

### 4.2. Indication of Concentration Changes

For the measurement of concentration changes and for the recording of orientational changes in solutions of optically anisotropic molecules, optical techniques have proven to be widely applicable. If ionic species are involved, conductivity measurements are suitable to monitor concentration as well as orientation changes in electrically anisotropic molecules. The Wien effects are directly accessible from the conductivity change $\delta\kappa/\kappa(0)$ relative to the $\kappa$ value at $E = 0$. For a 1:1 ionic equilibrium like that in

**TABLE 2.** Relaxation Parameters of Elementary Chemical Reactions
(Kinetic Titration)[a]

| Reaction | Relaxation time | Amplitude factor |
|---|---|---|
| $L + B \underset{k_{-1}}{\overset{k_1}{\rightleftharpoons}} LB$ | $\tau = \dfrac{1}{k_1[c_L^0 + c_B^0 + K)^2 - 4c_L^0 c_B^0]^{1/2}}$ | $\Gamma = \dfrac{K}{2}\left\{\left[1 - \dfrac{4c_L^0 c_B^0}{(c_L^0 + c_B^0 + K)^2}\right]^{1/2} - 1\right\}$ |
| $c_B^0 = \text{const}, c_L^0 = 0$ | $\tau_0 = \dfrac{1}{k_1(c_B^0 + K)}$ | $\Gamma_0 = 0$ |
| $c_B^0 = \text{const}, (c_L^0)_m$ | $\tau_m = \dfrac{1}{2k_1(Kc_B^0)^{1/2}}, c_B^0 > K$ | $\Gamma_m = \dfrac{K}{2}\left\{\left[\dfrac{c_B^0 + K}{K}\right]^{1/2} - 1\right\}$ |
| | at $(c_L^0)_{\tau_m} = c_B^0 - K$ | at $(c_L^0)_{\Gamma_m} = c_B^0 + K$ |
| $2B \underset{k_{-1}}{\overset{k_1}{\rightleftharpoons}} BB$ | $\tau = \dfrac{1}{k_1[K(K + 8c_B^0)]^{1/2}}$ | $\Gamma = \dfrac{K}{8}\left\{\dfrac{K + 4c_B^0}{[K(K + 8c_B^0)]^{1/2}} - 1\right\}$ |
| $c_B^0 = 0$ | $\tau_0 = (k_1 K^2)^{-1} = (k_{-1}K)^{-1}$ | $\Gamma_0 = 0$ |
| $B \underset{k_{-1}}{\overset{k_1}{\rightleftharpoons}} B$ | $\tau = \dfrac{1}{k_1 + k_{-1}} = \dfrac{1}{k_1(1 + K)}$ | $\Gamma = \dfrac{c_B^0 \cdot K}{(1 + K)^2}$ |

[a] Superscript zero refers to total (analytical) concentration; the amplitude factor $\Gamma$ is defined by Eq. (4.9) of the text.

Eq. (2.13), where the degree of dissociation may be written as $\Theta = c_L/c_B^0$, we have

$$\delta c_L/c_L(0) = \delta\kappa/\kappa(0) = \delta\Theta/\Theta^0 \tag{4.15}$$

With the help of Eqs. (4.15), (3.16), and (3.19) of part I[(1)] we may express the relaxation amplitude by

$$(\delta\kappa)_\infty = \kappa(0)[(1 - \Theta^0)/(2 - \Theta^0)](\partial \ln K/\partial E)_{P,T} E \tag{4.16}$$

In the linear range substitution of Eq. (2.21) into (4.16) leads to

$$(\delta\kappa)_\infty/\kappa(0) = [(1 - \Theta^0)/(2 - \Theta^0)][\gamma - \Delta M_a^\ominus/RT - (\partial \ln \tilde{Y}/\partial E)_\xi] E \tag{4.17}$$

In a similar way the relaxation time of an ionic process may be derived from the conductivity relaxation according to

$$\delta\kappa(t) = (\delta\kappa)_\infty e^{-t/\tau} \tag{4.18}$$

It is, however, remarked that in electrically anisotropic systems like the linear polyelectrolytes the measured conductivity relaxation may not be determined by the rate of the chemical reaction ($\tau = \tau^{(ch)}$), but may rather be rate controlled by orientational processes, i.e., $\tau = \tau^{(rot)}$.[(7,8)]

When optical changes are induced by the electric fields, light transmission and fluorescence emission appear to cover, in general, both concentration changes and rotational contributions in anisotropic systems. The linear dichroism seems to yield maximum information on molecular shape or chromophore position relative to rotation axis.[39,40]

The absorbance $A$ of polarized light is directly correlated to concentration and absorption anisotropy of molecules through the Lambert–Beer law:

$$A_\lambda = \sum_j (A_\lambda)_j = l \sum_j \varepsilon_j c_j \qquad (4.19)$$

where $\varepsilon_j$ is the (decadic) absorption coefficient of component $j$ in a composite system, $l$ is the optical pathway, and $\lambda$ is the wavelength of the light. When the absorbance is measured with normal, unpolarized light, then from Eq. (4.19) we obtain the absorbance change per centimeter:

$$\delta A_\lambda^1 = \sum_j \varepsilon_j \delta c_j \qquad (4.20)$$

at constant values of $\varepsilon_j$ as a function of the concentration changes $\delta c_j$. On the same line, absorbance relaxations directly reflect concentration relaxations [Eq. (4.12)].

Experimentally, recent progress in instrumentation has opened the way for measuring field-induced, rotational and chemical relaxations in parallel, both optically and electrically in the nanosecond time range.[7,8]

### 4.3. Component Contributions to Absorbance

In general, the total optical signal change, $\delta S$, produced by a perturbation will contain several contributions as, for instance, demonstrated for electric-field-induced absorbance changes in ribosomal RNA.[41] The concentration shifts of the components $B_j$ in a composite interacting system are summarized in a chemical term $\delta S^{(ch)}$. Density (volume) changes and changes in the intrinsic optical and electrical properties of the system constitute a physical term $\delta S^{(ph)}$.

In isotropic systems and at isotropic perturbations such as temperature and pressure changes, the term $\delta S^{(ph)}$ only reflects density (volume) changes.

In solutions and suspensions of anisotropic molecules directing external forces such as a hydrodynamic flow or electric field forces may cause signal changes $\delta S^{(rot)}$ originating from molecule rotations. In particular optically and electrically anisotropic (dipolar or polarizable) macromolecules exhibit major electric dichroism and electric birefringence.

In general, the total signal change is given by[9]

$$\delta S = \delta S^{(ch)} + \delta S^{(rot)} + \delta S^{(ph)}$$

There are numerous reviews on dichroism and birefringence as well as on the use of optical signals to indicate concentration changes. Less frequent, however, are accounts where it is emphasized that both chemical and orientational changes may contribute to the measured optical signals.[41,42]

We now recall the basic absorbance equation for a multicomponent system. Per centimeter light path

$$A_\lambda^1 = \sum_j \varepsilon_j c_j \tag{4.21}$$

As outlined below it is of great practical relevance to use linearly polarized light. Furthermore, in electric field experiments it is customary to choose the direction (of the electric field of a parallel-plate capacitor measuring cell) as a reference for the light polarization plane.

When an electric field is applied to a chemical system which exhibits both electrical and optical anisotropy, both the $\varepsilon_j$ and the $c_j$ terms in the fundamental Eq. (4.21) may be field dependent. Note that the usual extinction coefficients of optically anisotropic molecules reflect random average values $\bar{\varepsilon}_j$ of all chromophore orientations of the system when measured with polarized light.

In order to cover field effects on $\varepsilon_j$ and $c_j$ the field induced absorbance (per centimeter) has two types of components:

$$dA_\lambda^1(E) = \left[ \sum_j c_j \left( \frac{\partial \varepsilon_j}{\partial E} \right)_{c_j} + \sum_j \varepsilon_j \left( \frac{\partial c_j}{\partial E} \right)_{\varepsilon_j} \right] dE \tag{4.22}$$

one at constant $c_j$ (orientational) and one at constant $\varepsilon_j$ (chemical concentration shifts). Therefore the field-induced absorbance change,

$$\delta A_\sigma = A_\sigma(E) - A(0) \tag{4.23}$$

measured with light polarized at the angle $\sigma$ relative to the field vector, where $A_\sigma(E)$ is the absorbance in the presence of the field and $A(0)$ that at $E = 0$, may not only involve orientational changes $\delta(\varepsilon_j)_\sigma$ but also concentration changes $\delta c_j$.

Whereas in the absence of $E$ the absorbance is independent of $\sigma$,

$$A^1(0) = \sum_j \bar{\varepsilon}_j c_j(0) \tag{4.24}$$

the absorbance in the field,

$$A_\sigma^1(E) = \sum_j (\varepsilon_j)_\sigma \, c_j(E) \tag{4.25}$$

is dependent on $\sigma$ because of $(\varepsilon_j)_\sigma$.[43]

Denoting the field-induced changes in $\varepsilon_j$ and $c_j$ relative to the zero-field values by

$$\delta(\varepsilon_j)_\sigma = (\varepsilon_j)_\sigma - \bar{\varepsilon}_j \tag{4.26}$$

and

$$\delta c_j = c_j(E) - c_j(0) \tag{4.27}$$

respectively, Eq. (4.23) is rewritten in terms of Eqs. (4.24) and (4.25):

$$\delta A_\sigma^1 = \sum_j \left[ (\varepsilon_j)_\sigma \, c_j(E) - \bar{\varepsilon}_j c_j(0) \right]$$

$$= \sum_j \left\{ \delta(\varepsilon_j)_\sigma [c_j(0) + \delta c_j] + \bar{\varepsilon}_j \, \delta c_j \right\} \tag{4.28}$$

where the separation of the terms depending on $\sigma$ from those independent of $\sigma$ is evident.

Introducing the definitions

$$\delta A_\sigma^{(\mathrm{rot})} = \sum_j \delta(\varepsilon_j)_\sigma [c_j(0) + \delta c_j] \tag{4.29}$$

$$\delta A^{(\mathrm{ch})} = \sum_j \bar{\varepsilon}_j \, \delta c_j \tag{4.30}$$

Eq. (4.28) may be generally written in terms of a rotational and a chemical contribution[41,43]:

$$\delta A_\sigma^1 = \delta A_\sigma^{(\mathrm{rot})} + \delta A^{(\mathrm{ch})} \tag{4.31}$$

For axially symmetric measuring geometry like that of a parallel-plate capacitor cell, $\delta A^{(\mathrm{ch})}$ can be experimentally obtained in two independent ways, using the three light polarization modes $\sigma = 0$, $\sigma = \pi/2$, and $\sigma = \sigma^*$. Provided that $\delta A \ll A(0)$ we have $\sigma^* = 0.955$ (54.7°). Axial symmetry provides the relationship

$$\delta A_\parallel^{(\mathrm{rot})} + 2\delta A_\perp^{(\mathrm{rot})} = 0 \tag{4.32}$$

where the subscript $\parallel$ refers to $\sigma = 0$ (parallel mode) and $\perp$ denotes $\sigma = \pi/2$ (perpendicular mode). At $\sigma = \sigma^*$, $\delta A_{\sigma^*}^{(\mathrm{rot})} = 0$. If $\sigma^* = 0.955$, then

$$\delta A^{(\mathrm{ch})} = \delta A_{0.955}^1 \tag{4.33}$$

On the other hand Eqs. (4.31) and (4.32) can be combined to

$$\delta A^{(ch)} = \tfrac{1}{3}(\delta A_{\parallel}^1 + 2\delta A_{\perp}^1)$$    (4.34)

In the framework of this formalism the nature of the absorbance change, either chemical or purely rotational, may be derived from the amplitudes. More detailed information can, of course, only be obtained from the analysis of the total time course of the relaxations. Owing to coupling between chemical-conformational transitions to the orientations of the molecules relative to the electric field vector, orientational and chemical relaxations can be coupled, and the rate-limiting process may determine both chemical and rotational contributions. The analysis is straightforward when the time scales of chemical and rotational processes are different. In any case, from a practical point of view, the analysis of the time course of the measured signal is indispensable to determining the various components and their amplitudes.[9,41]

### 4.4. Linear Dichroism

The time course of orientational changes induced by electric fields contains information on the orientation mechanism, and on the electrical and geometrical properties (main dipole axis, length) of the aligning and deorienting molecules. For instance, permanent dipole orientation of a given particle type in the presence of a constant electric field builds up with zero slope and has two modes, whereas the build-up of induced dipole orientation starts with maximum slope and is characterized by only one time constant. The deorientation relaxation of a system of identical particles, after termination of the step pulse, is monophasic, independently of the presence of permanent or induced dipoles. Table 3 summarizes the characteristic features of the rotational kinetics indicated by electric dichroism and birefringence for small perturbations.[39,40] We see that there are a number of specific relationships to differentiate between permanent and induced dipole mechanism. In particular, the technique of field-reversal is a sensitive indicator for the relative contributions of permanent or induced dipoles.[44,45]

The analysis of orientational changes faces problems when nonrigid molecules or molecules of nonhomogeneous length distribution are present. The quantitative treatment of field-induced changes in molecule shape is still very difficult. Chain bending or stretching, structural changes, dimer formation, or multimeric aggregation will change the anisotropy components arising from long-range optical interactions as well as short-range interactions with the solvent having, in general, a refractive index different from that of the absorbing molecules considered (form anisotropy). Such

*TABLE 3.* Rotational Relaxation (Dichroism)$^a$

| $\delta A(t)/(\delta A)_\infty$ | Signal build-up $(t_0 \leqslant t \leqslant t_e)$ | $\left[ \dfrac{d\delta A(\mathrm{rot})}{dt} \right]_{t_0}$ | Decay $(t_e \leqslant t \leqslant \infty)$ | $\dfrac{\int_0^{t_e} \delta A(t)\, dt}{\int_{t_e}^\infty \delta A(t)\, dt}$ |
|---|---|---|---|---|
| (a) General, $r \neq 0$ | $\dfrac{3r\exp(-2D_r t) - (r-2)\exp(-6D_r t)}{2(r+1)}$ | $0$ | $\exp(-6D_r t)$ | $\dfrac{4r+1}{r+1}$ |
| (b) Permanent and saturated induced dipoles, $r = \infty$ | $\exp(-2D_r t) + \exp(-6D_r t)$ | $0$ | $\exp(-6D_r t)$ | $4$ |
| (c) Induced dipoles, $r = 0$ | $\exp(-6D_r t)$ | $(6D_r)^{-1}$ | $\exp(-6D_r t)$ | $1$ |

$^a$ Rectangular field pulse starting at $t_0$ and ending at $t_e$ in a solution of anisotropic molecules ($B_j$); $\delta A(t)/(\delta A)_\infty$, relative deviation from steady-state or equilibrium value, $(\delta A)_\infty$ being the maximum deviation at $t_0$ for the build-up of $\delta A$, and at $t_e$ for the field-free relaxation, respectively; $(d\delta A/dt)_{t_0}$ represents the initial slope at $t_0$ of the signal build-up; $D_r$ is the rotational diffusion coefficient, $r$ is the ratio between permanent and induced dipole terms: $r = \beta/2\gamma$ where $\beta = g_j p_j E/(kT)$ ($g_j = 1$ for elongated particles) and $\gamma = \frac{1}{2} \Delta \alpha_j g_j E^2/(kT)$, $\Delta \alpha_j$ being the excess polarizability; the integral ratio represents the area above the rise curve ($t_0 \leqslant t \leqslant t_e$) and that below the zero-field relaxation curve ($t_e \leqslant t \leqslant \infty$).$^{(50,51)}$ The $g$ factors are listed in Table 1.

contributions, however, are small for solutions of low turbidity, i.e., if the size of the aggregates remains small compared to the wavelength of the monochromatic light used.[39]

As already mentioned, dipolar or polarizable molecules aligning in the direction of an external field show linear dichroism at wavelengths corresponding to optical transitions the moments of which, in many examples, may be considered fixed with respect to the main dipole axis. The optical effects of rigid macromolecules of cylindrical symmetry such as linear (rodlike) polyelectrolytes in dilute solutions of low ionic strength are quantitatively analyzable. Owing to anisotropic counterion polarization, parallel to the long axis, the dipole axis of polyelectrolytes coincides with the long molecule axis. If, in addition, the polyions contain planar chromophores (ring structures such as the purine or pyrimidine bases in polynucleotides) with absorption bands resulting from $\pi \to \pi^*$ transitions with transition moments in the plane of the chromophore and eventually $n \to \pi^*$ transitions with moments perpendicular to this plane, these two components are associated with different absorption coefficients.[39]

The linear dichroism $\Delta A$ has been originally defined for absorbance changes of purely rotational origin. In the notation used here, we have

$$\Delta A = A_{\parallel}^{(\text{rot})} - A_{\perp}^{(\text{rot})} = \delta A_{\parallel}^{(\text{rot})} - \delta A_{\perp}^{(\text{rot})} \tag{4.35}$$

It can, however, be shown that a more general definition of $\Delta A$ (as the difference between the absorbances at $\sigma = 0$ and at $\sigma = \pi/2$), holds, independentley of whether there are chemical contributions present or not. Because $\delta A^{(\text{ch})}$ is independent of $\sigma$, i.e., $\delta A_{\parallel}^{(\text{ch})} = \delta A_{\perp}^{(\text{ch})}$, the application of Eq. (4.31) leads to the general form

$$\Delta A = A_{\parallel} - A_{\perp} = \delta A_{\parallel} - \delta A_{\perp} \tag{4.36}$$

Note that the linear dichroism is given by the *measured* absorbance changes in the electric field.

At high field strengths orientational changes may reach saturation. For $E \to \infty$, $\Delta A = \Delta A_s$. The degree of orientation may then be defined by the orientation factor [Eq. (3.28)] as

$$\Phi = \Delta A / \Delta A_s = (\delta A^{(\text{rot})} / \delta A_s^{(\text{rot})})_\sigma \tag{4.37}$$

relative to the saturation values.[46]

The dependence of $\Phi$ on the electric field strength contains information on the electrical properties of the molecules: permanent dipole moment **p** and/or polarizability tensor $\alpha$. General equations for the field dependence of $\phi$ are given by O'Konski *et al.*[24]

It should be remarked that Eq. (4.37) provides the basis for the rigorous analysis of chemical contributions of the induced or permanent dipole moments of the reaction partners, according to Eqs. (3.29) and (3.36). Note that the Langevin function can also be expressed in terms of $\Phi$: $L[r] = (r/3)(1 - \Phi)$, valid for $\Phi \ll 1$.

### 4.5. Chemical Transition Factor

The mechanism of electric field effects on a chemical equilibrium is reflected in the dependence of rate and equilibrium constants on the electric field strength. We recall Eq. (3.48): $K(E) = K(0)e^x$, $x = \int \Delta M \, dE/RT$. If the field-induced concentration shifts can be measured by absorbance changes, then Eq. (4.30) can be used. For example, the intramolecular transition $B_1 \rightleftharpoons B_2$ is described by

$$\delta A^{(ch)} = \bar{\varepsilon}_2 \, \delta c_2 + \bar{\varepsilon}_1 \, \delta c_1 \tag{4.38}$$

where $\bar{\varepsilon}_2$ and $\bar{\varepsilon}_1$ are the (random) average values of the extinction coefficients of the conformations $B_2$ and $B_1$, respectively. Mass conservation dictates that $\delta c_2 + \delta c_1 = 0$. By definition, $\Theta = c_2/(c_1 + c_2) = c_2/c^0$. With $\delta c_2 = c^0 \delta \Theta$ Eq. (4.38) is rewritten as

$$\delta A^{(ch)} = (\bar{\varepsilon}_2 - \bar{\varepsilon}_1)c^0 \delta \Theta \tag{4.39}$$

where

$$\delta \Theta = \Theta(E) - \Theta^0 \tag{4.40}$$

$\Theta^0$ being the $\Theta$ value at $E = 0$.

Because in a two-state transition $K = \Theta/(1 - \Theta)$ holds we have $K(E) = \Theta(E)/[1 - \Theta(E)]$ and at $E = 0$, $K(0) = \Theta^0/(1 - \Theta^0)$. Hence,

$$\Theta(E) = K(0) \, e^x/[1 + K(0)e^x] \tag{4.41}$$

Substitution of Eqs. (4.40) and (4.41) into (4.39) leads to

$$\delta A^{(ch)} = (\bar{\varepsilon}_2 - \bar{\varepsilon}_1)c^0 \frac{(e^x - 1)(1 - \Theta^0)}{e^x + (1 - \Theta^0/\Theta^0)} \tag{4.42}$$

providing a relationship between the field induced absorbance change (per centimeter) and the electric field factor $x$ and $\Theta^0$.[30]

Analogous to the orientation factor $\Phi$ of the linear dichroism and the birefringence, we may define a chemical transition factor according to

$$\Phi^{(ch)} = \delta A^{(ch)}/\delta A_s^{(ch)} \tag{4.43}$$

where $\delta A_s^{(ch)}$ is the saturation value of the chemical absorbance amplitude at high field strengths. Equations (4.39) and (4.40) yield

$$\Phi^{(ch)} = [\Theta(E_0) - \Theta^0]/(1 - \Theta^0) \tag{4.44}$$

It is evident that in the limiting case $\Theta^0 = 0$ at $\mathbf{E} = 0$, $\Phi^{(ch)} = \Theta$ holds. Finally, combination of Eqs. (4.41) and (4.44) yields

$$\Phi^{(ch)} = (e^x - 1)/[e^x + (1 - \Theta^0)/\Theta^0] \tag{4.45}$$

For calculational purposes, this is a very useful expression to describe chemical transitions as a function of the externally applied electric field.[30]

### 4.6. Differentiation between Component Contributions

Among the extreme cases, the analysis of chemical electric field effects is simplest when the rotational equilibria are established faster than the diffusion-limited chemical processes. The other extreme is the complete control of the chemical processes by the rate of the orientational relaxations. As seen in Table 3, bimolecular chemical reactions exhibit a characteristic dependence of time constant and amplitude on concentration.

Independently of the time course of the absorbance changes there are a number of ways to differentiate between chemical and orientational contributions of anisotropic molecules and particles.

If chemical transformations are associated with isobestic or isochromic wavelengths, then at these wavelengths chemical contributions are zero, i.e., $\delta A^{(ch)} = 0$. When plane-polarized light is used at the light polarization angle $\sigma^*$, the rotational contributions cancel, i.e., $\delta A_{\sigma^*} = 0$. A purely chemical concentration shift in randomly distributed reaction partners is associated with an absorbance change which is independent of $\sigma$. The pure rotational contributions always obey Eq. (4.32). It is obvious that the time constants of the chemical and rotational parts of the relaxations must also be independent of $\sigma$.

It should be remarked that chemical contributions of interacting anisotropic molecules are usually negligibly small if simple dipolar equilibria are concerned. Appreciable field effects are encountered only in macromolecular dipolar systems at high field intensities. On the other hand, the second Wien effect and structural changes coupled to ionic dissociation–association processes may occur at already low field intensities.

## 5. *Macromolecular Cooperativity and Hysteresis*

Experimental experience demonstrates that in any case large reaction dipole moments $(\Delta M)$ are required to produce major displacements of dipolar equilibria; high ionic valencies are necessary for larger dissociation field effects in ionic association–dissociation reactions.[9]

These conditions generally require that the reaction partners themselves have either large dipole moments or large polarizabilities or a high density of fixed ionic groups. The structures which fulfill these conditions are macromolecules and macromolecular organizations such as polyionic biopolymers, biopolymer complexes, or biomembranes.

There is, however, another important feature of macromolecules and of organizations involving biopolymers, which mark them as attractive candidates for very efficient field effects. It is well known that macromolecular systems are very often capable of undergoing structural changes which are highly cooperative in nature. Among the immediate consequences of this cooperativity are far-reaching conformational changes produced by only small changes in the environmental conditions.

In a cooperative process, a larger sequence of residues in a polymer chain or an entire subunit (domain) of a macromolecular system are the reaction units which transform "as a whole." Returning to dipolar systems, even if the dipole moment of a single residue in such a sequence is small, the total cooperative unit may have a very large dipole moment. In this manner, cooperativity sums small reaction moments of elementary steps into concerted action; it thus represents a powerful amplification mechanism.

If a cooperative chemical transformation is coupled to an electric field effect, a relatively small change of the field intensity may suffice to cause a practically complete transition. Thus, electrical chemical coupling amplified by cooperativity is probably also a powerful mechanism for a direct and very efficient electrical control of biochemical reactivity. This principle is certainly very suggestive for the exploration of bioelectric mechanisms in general.

There is, however, an interesting alternative to strong equilibrium cooperativity for producing large structural changes by only small changes in external parameters. This alternative comprises thermodynamically metastable states and nonequilibrium transitions in cooperatively stabilized systems. The dissipative element of metastability and nonequilibrium processes endows the structures involved with threshold and trigger properties.[9,33] Electric field changes going beyond the stability point (threshold) of a metastable configuration will trigger abrupt nonequilibrium transitions to more stable (equilibrium) states.

It is well known that in certain cases the occurrence of structural metastability leads to pronounced hysteresis phenomena. Besides threshold and trigger features, hysteresis in structural transitions is a mechanism for chemical oscillations[47] and molecular memory.[33,48] In particular, the memory principle expressed in hysteresis is of appreciable cell-cybernetic interest in biology.[33]

In an electrical–chemical hysteresis, the nonequilibrium transitions underlying memory imprint on the one hand and erasure on the other hand, are triggered by electric field changes which go beyond the thresholds of the respective metastable states.

In an attempt to estimate the energetics and kinetics of nonequilibrium transformations in metastable biopolymers the physical chemical behavior of a model system exhibiting pronounced hysteresis loops was investigated.[33] The model hysteresis to be briefly discussed results from the acid–base titration of the polyelectrolyte complex poly(A) · 2poly(U). The overall process underlying the hysteresis loop is the cyclic transition between two helical structures: the triple helix poly(A) · 2poly(U) and the protonated double helix poly(A) · poly(A).

The overall reaction may be written in terms of the reactive residues:

$$2(U \cdot A \cdot U) \rightleftharpoons (A \cdot A) + 4(U) \qquad (5.1)$$

It has been found that in the course of the acid titration, the $(U \cdot A \cdot U)$ complex does not directly transform to the protonated $(A \cdot A)$ double helix, but is at first protonated (metastable protonation equilibrium) and then converts irreversibly to $(A \cdot A)$ and free $(U)$ along the lower branch of the hysteresis. Reversing the direction of the pH change by adding alkali, a new curve is traced as the base branch of the loop.

Here, it is appropriate to note that electric impulses exceeding a threshold of about $20 \, kV \, cm^{-1}$ directly induce the conformational transition to $(A \cdot A)$ sequences in Eq. (5.1). The process that is primarily affected by the electric field is the helix-coil transition of base-paired $(A \cdot U)$ regions to the separated base residues according to

$$(U \cdot A \cdot U)_n = (A)_n + 2(U)_n + \Delta n \, C^+ \qquad (5.2)$$

The field effect on this reaction includes liberation of $\Delta n$ counterions, $C^+$, from the ionic atmosphere of the triple helix. Under the experimental conditions of acid pH values the field-induced reaction is coupled to the protonation of the (A) residues

$$(A) \rightleftharpoons (AH^+) \qquad (5.3)$$

and thus to the irreversible formation of the protonated poly(A) · poly(A)

double helix. In this example the electric impulse acts as a trigger transiently opening the triple helix. The coupling with an irreversible process
prevents relaxation to closed $(U \cdot A \cdot U)$ base pairs after the impulse is ter-
minated. It is thus possible to gradually cross the hysteresis from the acid
toward the base branch. The fraction of triple helix converted per impulse
is constant, suggesting that only a fraction of the randomly distributed
complexes is favorably oriented to the external field pulse.[31] This key
observation points to an end effect in this anisotropic system: at terminal
regions counterion association is reduced and electric field effects decreasing the local ion pairing are most efficient. A particular end effect is caused
by counterion displacement along the polyion.

A counterion polarization mechanism has been proposed to explain
the electric induction of conformational changes in polyelectrolyte complexes such as the $(U \cdot A \cdot U)$ triple helix. In accordance with this idea, the
external electric field shifts the ionic atmosphere of the $(U \cdot A \cdot U)$ complex
and thereby induces a dipole moment. At the negative pole of the induced
macrodipole, the screening by the ion cloud of the negative phosphate
residues is reduced. This, in turn, causes repulsion between the ends of the
polyanions and leads finally to the unwinding of the triple helix.[31] It later
came to our attention that a polarization mechanism had already been
proposed for strand separation of DNA by Pollak and Rein.[49]

It is now worth mentioning that the magnitudes of impulse intensity
and duration, used in our investigations of the polynucleotides and membrane proteins, is well within the range of biological interest. This has
revived the discussion on electrically induced conformational changes in
macromolecules and membranes as a possible mechanism for electrically
controlled regulatory processes in general and for a recording of electric
signals in particular. With this in mind, we may consider directed structural
transitions induced by electric impulses in biopolymers as model reactions
for the process of imprinting the information of the nerve impulses in the
neuronal network of the brain.

After these more speculative remarks, it finally appears appropriate to
mention that the challenging field of bioelectric–chemical research requires
a basic knowledge of the fundamental principles of electric field effects in
elementary (bio)chemical reactions and molecular-rotational processes.
The present elementary account on analytical aspects of chemical and
orientational effects induced by electric fields in macromolecules summarizes some useful information as to how to investigate mechanisms of
bioelectric phenomena on the macromolecular level.

## Note Added in Proof

After the preparation of this review further progress in the physical chemical analysis of electric field effects in biological macromolecules[52-58] and in the membranes of isolated cells and organelles[59] has been documented. A few additional references are selected.[60-63]

## Acknowledgments

I thank Miss B. Wilkenloh for the careful processing of the manuscript. The financial support of the Stiftung Volkswagenwerk, Grant No. I 34-706, and of the Deutsche Forschungsgemeinschaft, Grant No. NE 227, is gratefully acknowledged.

## References

1. E. Neumann, Chapter 4 in this volume, *Modern Bioelectrochemistry* (F. Gutmann and H. Keyzer, eds.) Plenum Press, New York (1985).
2. E. Neumann and K. Rosenheck, *J. Membrane Biol.* 10, 279–290 (1972).
3. E. Neumann, G. Gerisch, and K. Opatz, *Naturwissenschaften* 67, 414–415 (1980).
4. E. Neumann, M. Schaefer-Ridder, Y. Wang, and H. P. Hofschneider, *The EMBO J.* 1, 841–845 (1982).
5. I. Sugar and E. Neumann, *Biophys. Chem.* 19, 211–225 (1984).
6. L. Onsager, *J. Chem. Phys.* 2, 599–615 (1934).
7. E. Neumann, K. Tsuji, and D. Schallreuter, in *Biological Structures and Coupled Flows* (A. Oplatka and M. Balaban, eds.), pp. 135–138, Academic, New York (1983).
8. D. Schallreuter, Ph.D. thesis, Konstanz and Martinsried (1982).
9. E. Neumann, in *Topics of Bioelectrochemistry and Bioenergetics* (G. Milazzo, ed.), Vol. 4, pp. 113–160, Wiley, New York (1981).
10. E. Neumann, in *Ions in Macromolecular and Biological Systems* (D. H. Everett and B. Vincent, eds.), pp. 170–191, Scientechnica, Bristol (1978).
11. U. Schödel, R. Schögl, and M. Eigen, *Z. Phys. Chem. N.F.* 15, 350–362 (1958).
12. K. F. Wissbrun and A. Patterson, Jr., *J. Polymer Sci.* 33, 235–249 (1958).
13. G. S. Manning, *Biophys. Chem.* 9, 189–192 (1977).
14. G. Schwarz, *J. Phys. Chem.* 71, 4021–4030 (1967).
15. L. C. M. DeMaeyer, *Methods Enzymol.* 16, 80–118 (1969).
16. L. C. M. DeMaeyer and A. Persoons, *Tech. Chem. (N.Y.)* 6(2), 211–235 (1973).
17. C. J. F. Böttcher, O. C. Van Belle, P. Bordewijk, and A. Rip, *Theory of Electric Polarization*, Elsevier, Amsterdam (1973), Vol. 1.
18. M. Eigen and L. C. M. DeMaeyer, in *Techniques of Organic Chemistry* (S. L. Friess, E. S. Lewis, and A. Weissberger, eds.), Wiley, New York (1963), Vol. 8(2), pp. 895–1054.
19. K. Bergmann, M. Eigen, and L. C. M. DeMaeyer, *Ber. Bunsenges. Phys. Chem.* 67, 819–826 (1963).
20. A. Peterlin and H. A. Stuart, *Z. Phys.* 112, 1–19 (1938).
21. L. Onsager, *J. Am. Chem. Soc.* 58. 1486–1493 (1936).
22. W. Kuhn, H. Dürkop, and H. Martin, *Z. Phys. Chem. B* 45, 121–155 (1939).

23. A. Peterlin and H. A. Stuart, *Z. Phys.* **112**, 129–147 (1939).
24. C. T. O'Konski, K. Yoshioka, and W. H. Orttung, *J. Phys. Chem.* **63**, 1558–1565 (1959).
25. L. C. M. DeMaeyer, M. Eigen, and J. Suarez, *J. Am. Chem. Soc.* **90**, 3157–3161 (1968).
26. L. Hellemans and L. C. M. DeMaeyer, *J. Chem. Phys.* **63**, 3490–3498 (1975).
27. H. Fröhlich, *Theory of Dielectrics*, Clarendon, Oxford (1958).
28. E. H. Grant, R. J. Sheppard, and G. P. South, *Dielectric Behaviour of Biological Molecules in Solution*, Clarendon, Oxford (1978).
29. K. Yoshioka and K. Takakusaki, *Sci. Pap. Coll. Gen. Educ.*, University of Tokyo **31**, 111–124 (1981).
30. K. Tsuji and E. Neumann, *Biophys. Chem.* **17**, 153–163 (1983).
31. E. Neumann and A. Katchalsky, *Proc. Natl. Acad. Sci. USA* **69**, 993–997 (1972).
32. D. Nachmansohn and E. Neumann, *Chemical and Molecular Basis of Nerve Activity*, Rev., Academic, New York (1975).
33. E. Neumann, *Angew. Chem. Intern. Ed.* **12**, 356–369 (1973).
34. R. Winkler-Oswatitsch and M. Eigen, *Angew. Chem. Int. Ed.* **18**, 20–49 (1979).
35. E. Neumann and H. W. Chang, *Proc. Natl. Acad. Sci. USA* **73**, 3994–3998 (1976).
36. H. J. Nolte and E. Neumann, *Biophys. Chem.* **10**, 253–260 (1979).
37. M. Eigen and L. C. M. DeMaeyer, *Tech. Chem. (N.Y.)* **6**(2), 63–146 (1973).
38. T. M. Jovin, in *Biochemical Fluorescence: Concepts* (R. F. Chen and H. Edelhoch, eds.), Marcel Dekker, New York (1976), pp. 305–374.
39. M. Tricot and C. Houssier, in *Polyelectrolytes* (K. C. Frisch, D. Klempner, and A. V. Patsis, eds.), pp. 43–90, Technomic, Westport (1976).
40. E. Fredericq and C. Houssier, *Electric Dichroism and Electric Birefringence*, Clarendon, Oxford (1973).
41. A. Revzin and E. Neumann, *Biophys. Chem.* **2**, 144–150 (1974).
42. M. Dourlent, J. F. Hogrel, and C. Hélène, *J. Am. Chem. Soc.* **96**, 3398–3406 (1974).
43. K. Tsuji and E. Neumann, *Int. J. Biol. Macromol.* **3**, 231–242 (1981).
44. C. T. O'Konski and A. J. Haltner, *J. Am. Chem. Soc.* **79**, 5634–5649 (1957).
45. I. Tinoco and K. Yamaoka, *J. Phys. Chem.* **63**, 423–427 (1959).
46. S. P. Stoylov, *Adv. Coll. Interf. Sci.* **3**, 45–110 (1971).
47. A. Katchalsky and R. Spangler, *Quart. Rev. Biophys.* **1**, 127–175 (1968).
48. A. Katchalsky and E. Neumann, *Int. J. Neurosci.* **3**, 175–182 (1972).
49. M. Pollak and R. Rein, *J. theoret. Biol.* **19**, 333–336 (1968).
50. A. Peterlin, *Z. Phys.* **111**, 232–236 (1938).
51. K. Nishinari and K. Yoshioka, *Kolloid Z. Z. Polymere* **240**, 831–836 (1970).
52. S. Diekmann, W. Hillen, M. Jung, R. D. Wells, and D. Pörschke, *Biophys. Chem.* **15**, 157–167 (1982).
53. S. Diekmann, W. Hillen, B. Morgeneyer, R. D. Wells, and D. Pörschke, *Biophys. Chem.* **15**, 263–270 (1982).
54. S. Diekmann and D. Pörschke. *Biophys. Chem.* **16**, 261–267 (1982).
55. D. C. Rau and E. Charney, *Biophys. Chem.* **17**, 35–50 (1983).
56. D. Pörschke, H.-J. Meier, and J. Ronnenberg, *Biophys. Chem.* **20**, 225–235 (1984).
57. D. Pörschke, W. Hillen, and M. Takahashi, *EMBO J.* **3**(12), 2873–2878 (1984).
58. Z. A. Schelly and R. D. Astumian, *J. Phys. Chem.* **88**(6), 1152–1156 (1984).
59. P. Gräber, U. Junesch, and G. H. Schatz, *Ber. Bunsenges. Phys. Chem.* **88**, 599–608 (1984).
60. E. H. Serpersu and T. Y. Tsong, *J. Biol. Chem.* **259**(11), 7155–7162 (1984).
61. R. Korenstein, D. Somjen, H. Fischler, and I. Binderman, *Biochim. Biophys. Acta* **803**, 302–307 (1984).
62. H. Berg, E. Bauer, D. Berg, W. Förster, M. Hamann, H.-E. Jacob, A. Kurischko, P. Mühlig, and H. Weber, *Stud. Biophys.* **94**, 93–96 (1983).
63. E. Neumann, *Bioelectrochem. Bioenerg.* **13**, 219–223 (1984).

# 6

# Some Aspects of Charge Transfer in Biological Systems

## Felix Gutmann

*ABSTRACT:* The following modes of biological charge transfer are discussed: via solitons, as electronic conductance as hydrated proteins, via surface charge transfer complexes, and by protonic charge transfer complexes. Electron transfer involving the cytochromes is treated in the light of electrochemical measurements including voltammetry. A hypothesis for the lateral, surface conductance of biological membranes is proposed, based on electron donor–acceptor surface interactions giving rise to polaritons.

## 1. Introduction

In a way, all biological charge transfer is electrochemical in nature. Thus, this field is so huge that the following discussion will have to be confined to a few selected aspects. Many relevant topics are treated by the other contributors to this volume.

Regular and periodic organization of molecules is essential for the living state, and fundamental to its processes are structured instabilities which are reflected in the solid state by, e.g., charge transfer complexes. Many *in vivo* properties are difficult to explain by classical chemical processes in solution but appear to fit solid state physical processes in cells. Solid state events involving conduction are evident in animate aggregations and may well be an essential characteristic of life, which may be an electromagnetic phenomenon. A growing body of reviews and texts[1] is

*Felix Gutmann* • School of Chemistry, Macquarie University, North Ryde, N.S.W. 2113, Australia.

available to support these views. Chelate metal complexes of purines, pyrimidines, and their nucleosides have been reviewed in detail.[2]

For charge transfer to occur, free charges must either be initially, inherently present, or injected from outside, or else be generated within the system by means of an injection of energy.

When a molecule is excited by a sufficiently energetic event, an electron can be ejected. The resulting molecular cation polarizes electrons from its immediate environment in a time of about $10^{-16}$ s, and this polarization energy stabilizes the final state of the system.

When two opposite charges in a medium of relative permittivity $\varepsilon_r$ approach one another, the Coulombic energy of attraction between them equals $kT$ at a critical distance $R_c$ so that[3]

$$R_c = e^2/4\varepsilon_r\varepsilon_0 kT$$

or                                                                                                                        (1)

$$R_c = 55.9/\varepsilon_r \text{ nm}$$

A limit therefore exists for the quantity of charge of one kind which can be imposed without resulting in recombination.[3] This limits the amounts to about $10^{17}$ electronic charges/cm$^3$, while there are about $10^{21}$ molecules/cm$^3$. Electric neutrality need not be preserved in organic media. In an organic layer of about 5 nm the total of such uncompensated space charge is approximately $10^{-10}$ C/cm$^2$, which, if it were uniformly distributed, would yield a rather small electric field of about $10^3$ V cm$^{-1}$. Potential differences of about 5 mV across 5 nm, i.e., $10^4$ V cm$^{-1}$, do not represent a serious problem for space charge limitation of current flow through biological systems typically several nanometers thick. It also follows that only one molecule in several thousand can be charged if compensating charges are absent. Thus, modeling a biological system, a localized representation based on individual charged species is preferable to a representation of a charge density distributed over all the molecules.[3]

If the distance between $M^+$ and $M^-$ is less than $R_c$, they exist as an ion pair, i.e., in a charge transfer state. If $M^+$ and $M^-$ are locked into original lattice states then the distance can have only certain values. From the energies for linear polyacenes, determined accurately by Silinsh,[4] it appears that, as the number of rings increased,* charge separation in the

---

* Silinsh estimated[4] the energy ($E_{CT}$) of the charge transfer state of pentacene to be about one order of magnitude smaller than that of naphthalene. As the number of conjugated rings increases it becomes easier to separate the charges in a CT state; the photoinduced $M^+M^-$ thermalized CT state of pentacene has, under some conditions, charge separation distances as large as 12.5 nm.[5]

charge transfer state is facilitated. This could be significant for biological systems containing highly conjugated molecules such as chlorophyll.[3]

Electrochemical methods have been applied to the study of a variety of biological charge transfer problems. Thus, to give just a few examples, voltammetry has been used to study tissue,[6] zeta potentials have been related to cystic fibrosis,[7] an electrophoretic mobility test of macrophage for malignancy has been suggested[8] and polarography has been applied, *inter alia*, to the study of carcinogenicity of several chemical compounds by the Czech school.[9]

## 2. Solitons[73]

The important role played by solitons* in biological transfer remains to be clarified: solitons are said[11] to arise from localized disordered regions on, say, a membrane surface in the presence of the transmembrane electric field; the local perturbation then tends to spread and to move leading to changes in the orientation of the lipid membrane molecules. The soliton energy is considerably below the "energy band gap" of a polypeptide chain but may initiate proton transfer[12] in a hydrogen bonded chain in the presence of an electrostatic field.

Solitons may result in energy transporting along biological macromolecules. Their genesis and propagation as phonons in macromolecules establish important system properties in these molecules. Electromagnetic energy may be converted to soliton conducted energy as a transductive step. These solitons propagate quite slowly along the molecule, exhibiting threshold and possibly "windowed" characteristics in relation to the exciting energy. In transductive coupling of exciting events on cell membranes, they thus offer a vehicle for highly selective relationships in the electrochemical environment. In the transmembrane coupling of cell surface events to specific intracellular enzyme systems or organelles, either at internal membrane sites or via the cytoskeletal system to submembrane regions, solitons may also play a major role.

Binding of drug molecules to helical protein molecules[67] may inhibit soliton propagation. Further studies along these lines may allow "tailoring" of new drugs with optimal therapeutic capabilities. Their actions may be modulated by concurrent exposure to selected electromagnetic fields. Similar selective modulations of tissue functions by modification of the

---

* A soliton[10] is a localized collective excited state likely to arise in nonlinear dispersive media. The soliton is capable of migrating over relatively long distances with very little energy dissipation. It may be electrically charged or may be neutral such as a kink in a polymeric chain, or a domain wall.

electromagnetic environment may be seen in wound healing, tissue regeneration, and cancer therapy. These studies in biological interactions with the electromagnetic environment may yield important new knowledge about biophysical bases of circadian and other biological rhythms, with important therapeutic implications in endrocrine and immune disorders.

The solitons are thought[68] to induce simultaneously a displacement of the vibrational energy levels of the macromolecular chain and a rearrangement of the aqueous environment, until some sort of resonance is attained between these two processes, and a transition between charge and discharge regimes occurs. The result is a region of highly structured water and what often is called a Fröhlich regime, characterized by the existence of coherent electrical oscillations having frequencies of the order of 10–100 gHz, and discussed by Fröhlich himself in Chapter 8 of this volume. In other words, the metabolic energy reaches definite, localized sites within the macromolecular chain, e.g., DNA or protein. The energy causes a local deformation which travels along the chain as a soliton, causing the two effects mentioned above. Eventually, the soliton energy is transferred into the aqueous phase as a Fröhlich polarization wave. Maintained by the continuing supply of metabolic energy, this wave then plays a major part in affecting the cell organization.

One main problem still remains, viz., the isolation and characterization of the solitons as appearing in biological systems.

## 3. Conduction and Biological Structure

Much of what follows is based on discussions by Pethig and co-workers.[13] (See his contribution to this volume, Chapter 7.)

Proteins have the general structure of polypeptide chains with –C–C–N– as a repeat unit; see Figure 1.

Each peptide unit lies in a plane because it consists of a delocalized system of $\pi$ electrons associated with $\pi$ orbitals of the C and O atoms together with the lone electron-pair orbital of the N atom. Such an electron resonance structure is sufficient to produce significant diamagnetic anisotropy in the protein.[14] Some two dozen amino-acid residues make up polypeptides, but only glycine and proline have a first atom in the side chain R which is not a carbon atom. Thus, many features of regularity are present which may cooperate in forming energy bands, particularly in the extended arrays of $\alpha$ helices and $\beta$-pleated sheets. In fact, spectra of DNA

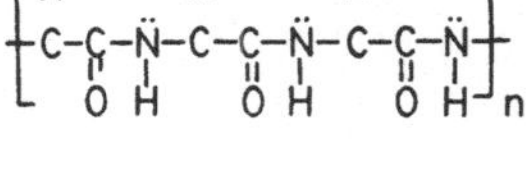

FIGURE 1. General structure of proteins.

and RNA are not basically different from those of their constituent groups because randomization of the constituents washes out all features of the electronic spectrum. Tong[15] suggests that this is due to the presence of geometry-independent lines and allowed energy bands. Altieri and Krizan,[16] using a self-consistent method, show term energies of reasonable magnitude to exist if DNA and related models are characterized as intrinsic semiconductors with a band energy gap of approximately 2 eV. Many proteins have a semiconduction activation energy of about this value.[17]

Whether charge carriers can migrate freely through proteins depends greatly on these energy bands. Coherent electronic motion can occur if broad bands of electronic states are available for the ground states (valency bands) and the excited states (conduction bands). If the bands of the extended states are less than about 2 kT, or if the energy states are localized, charge transport takes place by activated hopping or by tunneling processes.[13]

It appears that all proteins in the natural "pure" state are insulators. The valence band of the extended states is completely occupied by electrons, and the band gap is so large that at physiological temperatures the possibility of promoting electrons across the gap is negligible. Then, calculations[13] indicate that the energy band widths are greater for atomic interactions along the polypeptide backbone chains (–C–C–N–C–C–N–) than those arising from the hydrogen-bonded network interactions which stabilize the protein's tertiary structure. Thus, water molecules in a gramicidine A channel form a linear structure along the channel axis and containing about seven molecules.[96] The majority of semiconduction experiments have involved soluble proteins, probably because these are easily purified and crystallized. However, the structural proteins, those which support the main functional elements, are usually discarded in purification processes, as residues. Experimental evidence has existed for some time for electronic conduction in sea-animal integuments, e.g., crab shells.[12] In order to function, such structural proteins will generally be complexed with other molecules in a manner analogous to dopants and thereby facilitate charge transport.

## 4. Hydration and Charge Transfer

### 4.1. Proteins

In their natural state many proteins are bound into hydrophobic lipid matrices. The energetics of charge separation and mobility are governed largely by the dielectric constant of the surrounding medium. Calculations[13] have shown that the internal structures have an effective high-frequency relative permittivity of about 2.6, which value does not increase much upon hydration. Even in the presence of a fully extended hydrogen-bonded network, conduction along the backbone is more likely, but the particular properties of a protein will depend on its precise conformation, environment, and interactions with other molecules. Thus, it has been shown[19] that the measured conductances and permittivities of bovine serum albumin and lysozyme are determined by the degree of hydration as well as of the NaCl concentration in the salt solution used as the medium. The electrical behavior is governed by protein–water interactions; the conductance is due to a proton site-to-site hopping process.[19]

Considerable differences in the resistivity between that of ordinary proteins and that of electron transfer proteins have been reported[20,21] and also between ferrocytochrome-$c$ and ferricytochrome-$c$. This is attributed to the electronic state of the central metal atom.[20] The anhydrous cytochromes exhibit resistivities many orders of magnitude higher; thus anhydrous ferricytochrome $c_3$ has a room temperature of $4.1 \times 10^{12} \, \Omega$ cm and the ferrocytochrome $c_3$ a value of $1.6 \times 10^{10}$; these data are even more remarkable because the ferri compound is said[24] not to follow an Arrhenius type relationship in its temperature dependence while the ferro compound is reported to yield a negative value for its thermal activation energy. Cytochromes will be discussed in more detail in the next section.

Another interesting example for severe hydration effects in relation to charge transfer is the purple membrane of *Halobacterium halobium*: it acts as a photoelectret, producing at least semipermanent electric charges upon illumination; this activity has been shown to be greatly affected by hydration.[95]

Water in proteins has been determined from hydration isotherms obtained gravimetrically and the more recent quartz crystal resonator techniques.[39] The steady state conductivity of protein samples increases rapidly with water absorbed. The weight percentage, $m$, relates to the conductivity according to the Spivey equation,[40]

$$\lambda = \lambda_D \exp(\alpha m) \tag{2}$$

where $\lambda_D$ is the dry-state conductivity and $\alpha$ a constant. For $m$ of about 5 wt %, a change occurs such that $\alpha = 1.3$ for $m < 5$ wt % and $\alpha = 0.9$ for $m > 5$ wt %. This may be explained by assuming that at 5 wt % nearly all the protein primary sorption sites are occupied by water molecules and the population of secondary hydration sites begins. The frequency $f_m$ of maximum loss of dielectric dispersion is also found[13] to increase with increasing hydration following the relationship

$$\tau = \tau_0 \exp(-\beta m) \tag{3}$$

where $\tau$ is the relaxation time and $\beta$ a constant. Again at 5 wt % the value of the constant changes.[13] Although $f_m$ changes, the area under the dielectric loss factors $\varepsilon''$ vs. $\log(f)$ curve does not change with hydration, suggesting[13] that the observed dielectric dispersions are not directly related to the relaxation of water dipoles.

## 4.2. Cytochromes

Cytochrome-$c_3$ is specially interesting because this gobular hemoprotein (molecular weight about 12,500) is a defined chemical entity and known[22] to be the electron carrier in the respiratory chain of, *inter alia, Desulfovibrio vulgaris*. The electron transfer rate in its reversible electrochemical electrode[98] redox reaction has the enormous value of 0.1 cm sec$^{-1}$. Its resistivity in the ferro- as well as the ferri-form is shown in Figure 2. The conductivity was ohmic and appears to be electronic in nature[20]; values are said[20] to be well reproducible. The low-temperature branch of ferrocytochrome-$c$ exhibits the usual Arrhenius form, but the high-temperature region shows a most unusual behavior, as seen from Figure 2. The resistivities of these two forms are reported[20] to differ by 11 orders of magnitude, with the reduced form having the extraordinarily low value of 57 $\Omega$ cm at the transition temperature $T_M$, a value comparable to that of germanium. It is remarkable that even anhydrous films of tetrahemoprotein cytochrome $C_3$ exhibit[23] good conductivity at 292 K.

The polypeptide chain within the cytochrome molecule has virtually no conductivity at all, while the complete hemoprotein exhibits a conductivity of the order of $10^{-9}$–$10^{-11}(\Omega$ cm$)^{-1}$, suggesting a considerable contribution of the heme units to the overall electron transfer. Since the porphyrin subunit is about 13 Å diameter, sufficient delocalization of $\pi$ electrons should be expected[20] to impart conductivity. It has been

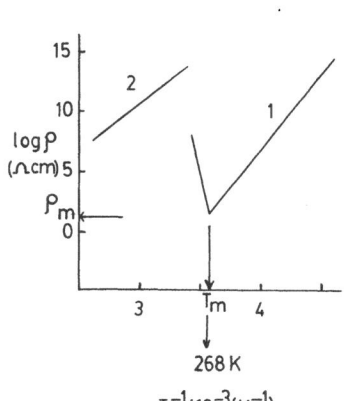

FIGURE 2. Temperature dependence of the resistivity of (1) ferrocytochrome-$c_3$ and (2) ferricytochrome-$c_3$.[20]

hypothesized that some part of the hemes in cytochrome-$c_3$ are exposed to the protein surface so that electrons may tunnel through a peptide residue.[19] The heme–heme interaction in the cytochrome $c_3$ molecule was inferred from NMR and Mössbauer spectroscopy. Cytochrome $c_3$ undergoes a reversible oxidoreduction by catalytic action of hydrogenase.[26]

Table 1 collects the electrical conductivity data of several protein solids.[27] The ferri-forms of both cytochromes have a similar resistivity in magnitude at room temperature. On the other hand, very drastic differences appeared in the resistivity of the ferro-form.

The contribution of heme to the electrical conduction of biological materials is clear because the polypeptide chain itself has a resistivity $10^{14}$ $\Omega$ cm. On the other hand, the resistivity of one hemoprotein is about $10^9$–$10^{11}$ $\Omega$ cm.

Taking into account the size of porphyrin, 1.3 nm in diameter, the distance is small enough to cause the overlapping of $\pi$ electrons. Electrochemical measurement of the reduction of cytochrome $c_3$ on a mercury electrode also suggests[31] that some parts of hemes are exposed to the protein surface and/or the electrons tunnel through a peptide residue.

Rapid direct electron transfer between cytochrome $c$ and a gold electrode takes place in the presence of 4,4′-bipyridyl and 1,2-bis(4-pyridyl)ethylene, which form an adsorbed layer on the electrode surface thus facilitating the electron-transfer reaction.[32] There are some strik-

TABLE 1.  Electrical Conductivity of Several Protein Solids[a]

| Substance | Temperature range (°C) | $E$ in $E/2kT$ (eV) | Resistivity ($\Omega$ cm) at temperature indicated (°C) |
|---|---|---|---|
| Cytochrome $c$ oxidized[b] | — | 2.7 | $6.1 \times 10^{16}$(30) |
| Cytochrome $c$ oxidized[b] | 50 to 85 | 1.2 | $3.1 \times 10^{11}$(30) |
| Cytochrome $c$ reduced[b] | 10 to 60 | 1.2 | $3.1 \times 10^9$(30) |
| Cytochrome $c_3$ oxidized[b] | 20 to 70 | 3.3 | $2.3 \times 10^{12}$(30) |
| Cytochrome $c_3$ reduced[b] | −40 to −5 | 7.7(?) | $5.7 \times 10^1$(−5) |
| Ferricytochrome $c$ | — | 1.2 | $6.5 \times 10^{10}$(55) |
| Ferrocytochrome $c$ | — | 1.2 | $6.5 \times 10^8$(55) |
| Lysozyme | — | — | $> 10^{14}$(30) |
| Myoglobin | — | 0.31 | $3.6 \times 10^{10}$(30) |
| Ribonuclease | — | — | $> 5 \times 10^{14}$(30) |
| Trypsin | — | — | $> 10^{14}$(30) |

[a] After Nakahara, Kimura, and Inokuchi.[30]
[b] All samples in the condensed form measured in surface cells with gold electrodes.

ing analogies between the reaction of cytochrome *c* at this electrode and its reaction with cytochrome oxidase with respect to the effect of chemical modification of the cytochrome *c* lysine residues and the effect of poly-L-lysine on the electrode reaction. The results suggest that cytochrome *c* binds to the 4,4'-bipyridyl-modified gold electrode surface prior to electron transfer in a manner similar to its interaction with the oxidase.

The electron transfer reaction between cytochrome *c* and cytochrome oxidase proceeds via a protein complex in which the ε-amino groups of the cytochrome *c* lysine residues are believed[33] to play an imporant role. Chemical modification of these lysines is well known[34] to affect the cytochrome *c*–oxidase electron-transfer reaction. The importance of lysine residues in the protein–protein interaction is further illustrated by the effect[35] of poly-L-lysine, a competitive inhibitor of the cytochrome *c*–oxidase reaction which acts by binding to the oxidase.[36] Both *N*-acetimidyllysyl and *N*-guanidinyllysyl horse heart cytochrome *c* are electroactive at a gold electrode in the presence of 4,4'-bipyridyl or 1,2-bis(4-pyridyl)ethylene, giving rise to quasireversible diffusion-controlled dc and ac voltammograms indistinguishable from those (Figure 3) of the native protein,[37] with a half-wave potential, $E_{1/2}r = 0.25$ V vs. NHE. Similarly[36] both are enzymatically active[37] in the cytochrome–oxidase system. The enzymatically inactive *N*-trifluoroacetyl and *N*-maleyl derivatives are both electroinactive.[32]

Poly-L-lysine inhibits the electrode reaction of native horse heart cytochrome *c*, as shown by dc voltammetry (Figure 3), again analogous to its inhibiting effect on the cytochrome *c*–oxidase reaction. The effect on the ac cyclic voltammetry peak current, $i_P(ac)$, is more marked. The varation with poly-L-lysine concentration is consistent with adsorption of poly-L-lysine onto the electrode surface, decreasing the effective free electrode area.

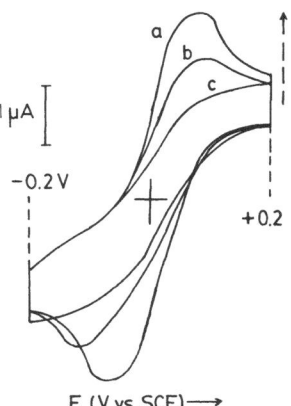

FIGURE 3. dc Voltammogram of horse heart ferri-cytochrome *c*, 0.4 nM in 0.1 *M* NaClO$_4$, 0.02 *M* phosphate buffer, pH 7, saturated solution of 1,2-bis(4-pyridyl)ethylene in the potential range from +0.2 to −0.2 V vs. SCE with poly-L-lysine: (a) 0 mg ml$^{-1}$, (b) 1 mg ml$^{-1}$, (c) 1.5 mg ml$^{-1}$, dc potential scan rate, 100 mV s$^{-1}$. After Eddowes *et al.*[32]

The electron-transfer reaction of the native protein has been studied[32] by ac impedance measurements. The heterogeneous electron-transfer rate was found to be as fast as that determined[38] for the ferri/ferrocyanide couple $(0.3-3.0 \times 10^{-4} \text{ msec}^{-1})$.

Double-potential-step chronocoulometry, rotating disk, and ac rotating ring disk indicate[32] that the native cytochrome binds to the 4,4'-bipyridyl-modified gold electrode surface prior to electron transfer. Thus it appears that 4,4'-bipyridyl acts by forming a suitable surface at the electrode–solution interface to which the cytochrome can bind. The binding of the protein to the electrode may therefore be similar in kind to that observed in the cytochrome $c$–oxidase reaction, involving the $\varepsilon$-amino groups of lysine residues in the vicinity of the cytochrome $c$ heme crevice. The similarities to the cytochrome $c$–oxidase reaction may result from a preferred orientation of the cytochrome when bound at the electrode surface or to cytochrome oxidase such that the exposed heme edge is adjacent to the electrode surface or the cytochrome oxidase, thereby enabling rapid electron transfer to occur.

Hydrated melanin exhibits a reversible switching behavior[41]; upon application of an electric field above a certain threshold value this substance changes reversibly and reproducibly from a state of high resistivity to one of low resistivity. The value of the critical switching field depends on the source, nature, and mode of preparation of the pigment; only hydrated specimens show the effect. Cope proposes[42] that the high-resistivity form may be considered as an emulsion of electrons in an electron-poor matrix; copious electron injection by virtue of the switching field then inverts the emulsion yielding electron-poor islands in an electron-rich matrix. This should affect the redox potential, and this has indeed been observed.[43] The model is further supported by the negative photoconductivity reported[44] for melanin: this effect, viz., a drop in conductivity upon illumination, is rare in organic materials; it was first reported[45] in 1965 for chlorpromazine. It is thought to be associated with the presence of centers capable of existing in two different valency, or electronic, states.

## 5. Charge Transfer and Adsorption. Surface Effects

### 5.1. Introduction

Surface effects in aqueous media always involve electric double layers characterized by more or less—depending on location—immobilized charges and by a firmly held layer of structured water exhibiting a permittivity considerably below that of its bulk value. There is an extensive literature on this subject[46]; see also the discussion in Pethig's article in this volume.

An adsorbate behaves as an electron donor if the energy of its lowest unoccupied level evaluated in its adsorbed state is situated at or above the Fermi level; the adsorbate, conversely, acts as an acceptor if the highest occupied level of the system substrate plus adsorbate is at or below the Fermi level. Surface complexes are often dimerlike with their energy states well localized within the energy gap, and resulting in a depletion of states in the vicinity of the Fermi levels; any excess charge tends to accumulate on the surface complex,[47] raising the carrier hopping probability. Some surface complexes consist of an adatom plus its nearest neighbor on the surface, again a dimerlike structure.[49] The first stage of an adsorption process involves a charge transfer, donor–acceptor, interaction between the adsorbent and the adsorbate,[48] resulting in an increased effective ionization potential, or electron affinity, of the adsorbed donor or acceptor, and a rise in the local energy levels of the substrate. This stage is followed[48] by electron, or hole, transfer into stabilized vacant energy levels of the adsorbate. Surface excitons and polaritons are likely to arise on surfaces,[50] their binding energy is reported to be of the order of $0.1$ eV.[51]

The topic of surface CTC is basic to the theories of catalysis and is usually treated in that context.[52]

Surface CTC of chlorpromazine (CPZ) has been shown to be involved[53] in hypothermia and sedation of mice; *in vitro* surface adducts between chlorpromazine and collagen have been demonstrated, as well as with lignocain and with phenytoin.[54] The existence of such CPZ surface adducts has been directly confirmed by electroreflectance measurements.[94]

If an acceptor is incorporated into a cationic micelle, then the rate of reactions involving hydrated electrons is increased by a factor of about 60 times that prevailing at outside negatively charged anionic micelles.[55]

Since counterions interact with, and are attracted by the, say, hydrophillic heads of the colloidal particles, micellar assemblages of such particles should be catalytically active. This, in general, has been found to be the case,[56] and should be a rewarding field for the study of micellar CTC as well as electrocatalysis. Formation of micellar CTC is also reported to result in considerably shifted ion exchange equilibria.[57]

## 5.2. Membranes

### 5.2.1. The Transmembrane Conductance

This topic has been extensively studied and many reviews and treatises dealing with it are available.[59] In this volume, it is discussed in contributions by Pethig (Chapter 7) and by Habib and Bockris (Chapter 3).

## 5.2.2. Membrane Surfaces

The surface of a biological membrane exhibits electron donating as well as electron accepting centers, though the surface as a whole is not, in general, electrically neutral, exhibiting the well-known membrane potential. We now wish to suggest that interactions with an external molecular electron acceptor or donor produces a conformational change of the complementary region on the membrane surface, perhaps initiated by Poole–Frenkel, thermally assisted field emission of carriers. The resulting transient voltage pulse may then be reversed by an opposing membrane potential which transfers the carrier back and thus restores the initial steric configuration on the membrane surface.

This hypothesis requires a finite, nonzero, conductance in the surface plane of the membrane because the electron, or hole, initially transferred forms a movable excited surface state, i.e., a polariton.[60] This may well give rise to a soliton, given a supply of energy at the proper locale.

No direct contact between donor and acceptor is envisaged; charge transfer is expected to occur once the donor approaches the acceptor, or vice versa, within tunneling distance, say 20 Å. The donor–acceptor charge transfer thus yields a pattern of surface charges and of sterically oriented proteins. Moreover, the highly nonuniform, inhomogeneous local electrical fields in the vicinity of the charges produce large, local, mechanical forces which are liable to lead to the opening or closing of ionically conducting transmembrane channels. One could then still further hypothesize that it is this surface charge pattern which is involved in the recognition of "self" and the disturbance of which plays a part in uncontrolled growth, viz., carcinogenesis. It may be significant in this context that of 289 carcinogens listed in the latest, Tenth, Collective Index of Chemical Abstracts, all but one—benzene—are electron acceptors or have pronounced electron accepting sites in their molecular structure.

The transmembrane, transcellular transient mentioned has indeed been observed,[61] e.g., in the hyphae of the water mould *Achyla bisexualis* as well as in other fungi.

The electron involved in the donor–acceptor interaction need not originate from the actual surface layer but may well come from the interior of the bilipid membrane, for example, from intramembrane proteins. The electron escape depth in organic materials is comparable to intermolecular distances.[62]

Alternatively, it may arise from an "activating ion" which enters into a surface complex with an adsorbed enzyme; it has been shown[97] that the stability of such complexes is greatly affected by the electric double layer structure of the membrane surface. As the electrostatic screening is increased, the complex tends to become less stable and the enzyme activity drops.[97]

It is notable that all condensed polycyclic aromatics with a planar molecular structure are carcinogenic—and all have the same and relatively high value of polarization energy, viz., 1.7 eV and thus are capable of acting as efficient electron donors as well as acceptors.[63] Whether a given protein is capable of insertion into the membrane is chiefly determined by hydrophobic forces, but the final orientation of the protein depends on electrostatic interactions such as the forces arising from surface charge asymmetries—which are a feature of the hypothesis proposed above—and from local electrochemical events.[64] Any changes in the electrical surface pattern are likely to propagate to adjacent cell membranes by a process akin to epitaxy, as has been demonstrated for inorganic layers by Distler[65] for distances of up to about 230 Å. A photoelectret process is said[65] to be involved.

## 6. Proton Transfer Complexes

The electron transfer involved in the formation of a charge transfer complex may in suitable systems be coupled to a proton transfer,[69] resulting in proton transfer complexes. This holds, especially for surface reactions, e.g., at electrodes, and a fortiori in nonaqueous media, because the proton affinity of water is so very high, viz., about 8.9 eV.[70]

The proton affinity $A_p$ is defined as the enthalpy of the reaction

$$X + H^+ \rightarrow XH^+ \tag{4}$$

In the gas phase, where there are energy contributions from solvation or polarization, $A_p$ may be measured by means of ion cyclotron spectroscopy.[80] Proton affinities are often given in kJ/mole; 1 kJ/mole equals 0.0104 eV; 1 kcal/mole equals 0.0436 eV. Exchange repulsion energies do not enter into proton affinities.[72] The "relative protonicity" of solvents, akin to the donicity concept for electrons, is discussed by Bayless *et al.*[74]

Hydroxy-dinitro pyridines, e.g., act as electron acceptors and/or as proton donors to, e.g., napthalene derivatives.[71]

The field of protonic CTC has been opened up by Matsunaga and his school,[69] and has been reviewed by Morokuma.[72] This author points to the close linkage existing in such adducts between proton and concurrent electron transfer; one does not occur without the other. The proton complex is said[72] to involve an electron transfer from, e.g., an amine to the proton. There is evidence that the electron transfer in some cases may be catalyzed by the presence of a proton and thus enhanced above its thermally controlled rate.[75]

Proton transfer may, sometimes, be really nothing more than a case of conventional hydrogen bonding, but in many cases the simultaneous transfer of an electron and a proton produces a new different type of adduct; then the difference between mere hydrogen bonding; and complex formation may indeed be dramatic, as pointed out by Arnett and Mitchell.[76] There is no correlation between the heats of protonation and of hydrogen bonding.[76]

Ion pair and hydrogen bonded complexes[77] are beyond the scope of this paper. Here, only a few representative examples of protonic complexes can be discussed.

The 1:1 and 1:2 complexes between oxalic acid and $\varepsilon$ amino acids also involve a proton transfer from the carboxyl group of the oxalix acid to the carboxyl ion of the amino acid; there is a similar interaction in the 1:1 complex between malonic acid and glycine.[78] These adducts may well involve a great deal of proton delocalization resulting in the formation of proton energy bands rather than energy levels. Proton tunneling is a well-established fact,[79] and even halogen ions have been shown to be able to tunnel between two suitable energy levels. Ionic charge transfer complexes between benzyl radicals and halide ions have also been reported.[80]

The average time of residence of the proton at a temporary equilibrium site is of the order of one vibrational period of the OH group and of the average lifetime of $H_3O^+$, viz., about $2.5 \times 10^{-13}$ s; since the dielectric relaxation time of water is about $10^{-11}$ s, it appears that dielectric relaxation and frictional processes are not involved in proton transfer.[81] However, in what appears to be protonic charge transfer complexes between, e.g., aliphatic amines and phenols, a considerable increase in viscosity is reported.[82]

Intramolecular proton transfer in electronically excited molecules has been reviewed[83]; in salicyclic acid esters it is said to lead to deexcitation of the excited electron[89]; the free energy change involved in that proton transfer is reported as about 0.13–0.22 eV.[84]

The effect has been shown to accompany at least some cases of electronic excitation[85] in organic molecules, such as aromatic compounds. These effects have hardly been touched upon as far as they affect biologically important transitions, as they are bound to do. This is an important and rather new field which should be explored.

Electron excitation causes only a very small change in the enthalpy of hydrogen bonding; a value of about 0.0044 eV is reported.[86]

Bacteriorhodopsin is a pigment found as a single protein component of the purple membrane of *Halobacterium halobium* and similar extreme halophiles.[87] The purple membrane converts light energy by translocating protons across the membrane to generate an electrochemical potential. A review[88] of this subject has been published. The naturally recurring crystalline structure has been determined with electron microscopy at 7 Å resolution. The protein field is folded into seven $\alpha$-helical chains, all of which span the hydrophobic core of the membrane.[88] The chromophore has been located in a lysine residue in a sequence of the second helical chain from the amino terminal group[89]; cf. discussions of methyl-glyoxal-protein complexes. The chromophere's double bond chain makes an angle of about 20° to the plane of the membrane,[90] with its ionone ring close to the center of the membrane.[91]

The role of bacterio-rhodopsin in the proton pumping activity of halobacteria has been further investigated by Bagyinka *et al.*,[92] while the light transduction via the pigmented bilipid membranes, of the purple membrane of *H. halobium*, has been studied by Ti Tien.[93]

## Acknowledgments

The author is indebted to Professor Gerald C. Huth of the School of Medicine, University of Southern California, and to Dr. Bevan Reid, Queen Elizabeth II Institute of Medical Research, University of Sydney, for many stimulating discussions. Thanks are also due to Professor John O'M. Bockris and to the Chemistry Department of Texas A & M University for their hospitality.

## References

1. *Advances in Biological and Medical Physics* (J. W. Lawrence, J. W. Gofman and T. L. Hayes, eds.), Academic, New York (1978).
2. T. J. Kistenmacher and L. G. Marzilli, "Chelate Metal Complexes of Purines Pyrimidines and Their Nucleosides: Metal–Ligand and Ligand–Ligand Interactions." *Jerusalem Symp. Quantum Chem. Biochem.* **9**(1), 7–40 (1977).
3. L. E. Lyons, in *Bioelectrochemistry*, Proceedings U.S.–Australia Joint Seminar on Bioelectrochemistry, Pasadena, 1979 (H. Keyzer and F. Gutmann, eds.), Plenum Press, New York (1980).
4. E. Silinsh, A. I. Belkind, D. Balode, A. Biseniece, V. V. Grechov, L. Taure, M. V. Kurik, Ya. Vertsimaka, and I. Bok, "Photoelectric Properties, Energy Level Spectra and Photogeneration Mechanisms of Pentacene," *Phys. Status Solidi (a)* **25**, 339–347 (1974).
5. E. Silinsh, Electronic States of Organic Molecular Crystals, Zinat., *Academy of Science, Latvian SSSR, Riga, 1978.*
6. R. L. McGreery *et al.*, *Brain Res.* **73**, 23 (1974).
7. M. A. Kahn *et al.*, *Tex. Rep. Biol. Med.* **31**, 665 (1973).
8. M. M. Lubran, *Ann. Clinic. Lab. Sci.* **4**, 121 (1974).

9. V. Podany, A. Vachalkova, and L. Bahna, "Electrochemical Properties of Polycyclic Compounds Studied by the Polarographic Method in Anhydrous Systems. III. Polarographic Reduction Potentials of Carcinogenic Nitrogen Compounds in Dimethyl Sulfoxide," *Neoplasma* **23**, 617–622 (1976); cf. also L. Bahna, V. Podany, M. Benesova, A. Godal, and A. Vachalkova, "Carcinogenicity and Polarographic Behavior of dibenz (a, h) anthracene, dibenz (a, h) acridine, and dibenz (a, h) phenazine," *Neoplasma* **25**, 641–645 (1978); and A. Vachalkova, V. Podany, and L. Bahna, "Electrochemical Properties of Polycyclic Compounds Studied by the Polarographic Method in Anhydrous Systems. IV. Polarographic Study of Carcinogenic and Noncarcinogenic Hydrocarbons in Ethylene Glycol Monomethyl Ether," *Neoplasma* **24**, 565–571 (1977); and V. Podany, E. Rezabkova, and L. Bahna, "Electrochemical Properties of Polycyclic Compounds Studied by the Polarographic Method in Anhydrous Systems. V. Oxidation Potentials of Carcinogenic Hydrocarbons in Acetonitrile," *Neoplasma* **25**, 57–65 (1978).

10. A. S. Davydov, "Solitons in Molecular Systems," *Phys. Scr.* **20**, 387–394 (1979); A. S. Davydov, "Solitons in Bioenergetics, and the Mechanism of Muscle Contraction," *Int. J. Quantum Chem.* **16**, 5–17 (1979); cf. also J. M. Hyman, D. W. McLaughlin, and A. C. Scott, *Physica* **3D**, 23 (1981); A. C. Scott, "Dynamics of Davydov Solitons," *Phys. Rev.* **26A**, 578 (1982); A. C. Scott, "The Vibrational Structure of Davydov Solitons," *Phys. Scr.* **25**, 651–658 (1982).

11. Nan-Ming Chao and S. White, "Orientational Waves in Cell Membranes," *Mol. Cryst. Liq. Cryst.* **88**(1–4), 127 (1982).

12. Y. Kashimori, T. Kikuchi, and K. Nishimoto, "The Solitonic Mechanism for Proton Transport in a Hydrogen Bonded Chian," *J. Chem. Phys.* **77**, 1904–1907 (1982); cf. also T. Kikuchi and K. Nishimoto, "Theoretical Studies of Hemoproteins, I. Mathematical Description of the Allosteric Effect," *Int. J. Quantum Chem.* **15**, 379–387 (1979).

13. R. Pethig and A. Szent-Gyorgyi, in Bioelectrochemistry, Proceedings of the U.S.–Australia Joint Seminar on *Bioelectrochemistry*, 1979 (H. Keyzer and F. Gutmann, eds.), Plenum Press, New York (1980), p. 227–252.

14. D. L. Worcester, "Structural Origins of Diamagnetic Anisotropy in Proteins," *Proc. Natl. Acad. Sci., U.S.A.* **75**, 5475 (1978).

15. B. Y. Tong, *J. Non-Cryst. Solids* **4**, 455 (1978).

16. J. Altieri and J. E. Kirzan, "Self-Consistent Band Theoretic Models of DNA," *J. Biol. Phys.* **3**, 103–110 (1975).

17. F. Gutmann and L. E. Lyons, *Organic Semiconductors*, Wiley, New York, 1967; F. Gutmann, H. Keyzer, and L. E. Lyons, *Organic Semiconductors, Part B*, Krieger Publ. Co., Malabar, Florida (1983).

18. P. S. B. Digby, *Proc. R. Soc. London Ser. B* **161**, 502 (1965); *Proc. Linn. Soc. London* **178**, 129 (1967); *Symp. Zool. Soc. London* **19**, 159 (1967).

19. S. Bone, J. Eden, and R. Pethig, "Electrical Properties of Proteins as a Function of Hydration and Sodium Chloride Content," *Int. J. Quantum Chem. Quantum Biol. Symp. 1981*, **8**, 307–316.

20. Y. Nakahara *et al.*, *Chem. Lett.* **1979**, 877.

21. Y. Nakahara *et al.*, *Chem. Phys. Lett.* **47**, 251 (1977).

22. C. M. Dobson, N. J. Hoyle, C. F. Geraldes, P. E. Wright, R. J. P. Williams, M. Bruschi, and J. LeGall, "Outline Structure of Cytochrome $c_3$ and its Properties," *Nature (London)* **249**, 524–429 (1974); K. Ono *et al.*, see Ref. 25; K. Kimura *et al.*, *Biochem. Biophys. Acta* **567**, 96 (1979).

23. U. Ichimura *et al.*, *Chem. Lett.* **1982**(1), 19; cf. also F. I. Adamosov, G. B. Postnikova, V. K. Sadydov, and M. V. A. Volkenstein, "Study of Electron Transport in Hemoproteins. II. Relation to the Rate of Reduction of Ferricytochrome $c$ by Oxmyoglobin to the pH," *Soc. Mol. Biol. (Moscow)*, *Molekylar Biologiya* **11**, 441 (1977).

24. K. Kimura, Y. Nakahara, T. Yagi, and H. Inokuchi, "Electrical Conductivity of

Hemoprotein in the Solid Phase: Anhydrous Cytochrome $c_3$ Film," *J. Chem. Phys.* **70**, 3317–3323.

25. K. Ono, K. Kimura, T. Yagi, and H. Inokuchi, "Mössbauer Study of Cytochrome $c_3$," *J. Chem. Phys.* **63**, 1640–1642 (1975).

26. Y. Nakahara, K. Kimura, and H. Inokuchi, "Electrical Conductivity of an Anhydrous Cytochrome $c_3$ Film as a Function of Temperature and Ambient Pressure," *Chem. Phys. Lett.* **73**(1), 31 (1980).

27. Y. Nakahara, K. Kimura, H. Inokuchi, and T. Yagi, "Electrical Conductivity of Solid State Proteins: Simple Protein Cytochrome $c_3$ as Anhydrous Film," *Chem. Lett.* **1979**, 877–880 (1979); cf. also K. Kimura, Y. Nakahara, T. Yagi, and H. Inokuchi, "Electrical Conduction of Hemoprotein in the Solid Phase: Anhydrous Cytochrome $c_3$ Film," *J. Chem. Phys.* **80**, 3317–3323 (1979).

28. D. D. Eley and D. I. Spivey, *Trans. Faraday Soc.* **56**, 1432 (1960).

29. P. Taylor, *Disc. Faraday Soc.* **27**, 239 (1959).

30. Y. Nakahara, K. Kimura, and H. Inokuchi, "Electrical Conductivity of Cytochrome $c$ Anhydrous Film," *Chem. Phys. Lett.* **47**, 251 (1977).

31. K. Niki, T. Yagi, H. Inokuchi, and K. Kimura, *J. Electrochem. Soc.* **124**, 1889 (1970).

32. M. J. Eddowes, H. A. O. Hill, and K. Uosaki, "Electrochemistry of Cytochrome $c$. Comparison of the Electron Transfer at a Surface-Modified Gold Electrode with that to Cytochrome Oxidase," *J. Am. Chem. Soc.* **101**, 7113–7114 (1979); cf. also A. E. G. Cass, M. J. Eddowes, H. A. O. Hill, K. Uosaki, R. C. Hammond, I. J. Higgins, and E. Plotkin, *Nature (London)* **285**, 673 (1980).

33. S. Ferguson-Miller, D. L. Brautigan, and E. Margoliash, *J. Biol. Chem.* **253**, 149 (1979).

34. S. Takemori, K. Wada, K. Ando, M. Hosokawa, I. Sezuku, and K. Okunuki, *J. Biochem. (Tokyo)* **52**, 28 (1962).

35. B. S. Mochan, W. B. Elliott, and P. Nicholls, "Patterns of Cytochrome Oxidase Inhibition by Polycations," *J. Bioenerg.* **4**, 329–345 (1973).

36. C. H. A. Seiter, R. Margalit, and R. A. Perreault, *Biochem. Biophys. Res. Commun.* **68**, 807 (1976).

37. M. J. Eddowes and H. A. O. Hill, "Electrochemistry of Horse Heart Cytochrome $c$," *J. Am. Chem. Soc.* **101**, 4461–4464 (1979).

38. V. Marecek, Z. Samec, and J. Weber, "The Dependence of the Electrochemical Charge-Transfer Coefficient on the Electrode Potential. Study of the Hexacyanoferrate (III) Hexacyanferrate (IV) Redox Reaction on Polycrystalline Gold Electrode in Potassium Fluoride Solutions," *J. Electroanal. Chem. Interfacial Electrochem.* **94**, 169–185 (1978); cf. also J. Weber, Z. Samec, and V. Marecek, "The Effect of Anion Adsorption on the Kinetics of the Iron $(3_+)$/Iron $(2_+)$ Reaction on Platinum and Gold Electrodes in Perchloric Acid," *J. Electroanal. Chem. Interfacial Electrochem.* **89**, 271–288 (1978).

39. P. R. C. Gascoyne and R. Pethig, "Experimental and Theoretical Aspects of Hydration Isotherms for Biomolecules," *J. Chem. Soc. Faraday Trans. I*, **73**, 171–180 (1977); S. Bone, P. R. C. Gascoyne, and R. Pethig, "Dielectric Properties of Hydrated Proteins at 9.9 GHz," *J. Chem. Soc. Faraday Trans. I*, **73**, 1605–1611 (1977).

40. D. Spivey, *Discuss. Faraday Soc.* **27**, 239 (1959).

41. P. M. Conny, J. E. McGinness, and E. Armour, Proc. Int. Pigm. Cell Conf. 9th, *Pigm. Cell* **1976**(2), 321.

42. F. W. Cope, "Inversion of Emulsions of Aggregated Electrons as a Possible Mechanism for Electrical Switching in Wet Melanin and in Amorphous Inorganic Semiconductors. A Manifestation of Cooperative Electron Interactions," *Physiol. Chem. Phys.* **9**, 543–546 (1977).

43. E. V. Gan, H. F. Haberman, and I. A. Menon, "Electron Transfer Properties of Melanin," *Arch. Biochem. Biophys.* **173**, 666–672 (1976).

44. P. Crippa, V. Christofoletti, and N. Romeo, "A Band Model for Melanin Deduced from

Optical Absorption and Photoconductivity Experiments," *Biochim. Biophys. Acta.* **538**(1), 164–170 (1978); cf. also P. Baraldi, R. Capelletti, P. R. Crippa, and N. Romeo, "Electrical Characteristics and Electret Behavior of Melanin," *J. Electrochem. Soc.* **126**, 1207–1212 (1979).

45. F. Gutmann and H. Keyzer, "Electrical Conduction in Chlorpromazine," *Nature (London)* **205**, 1102 (1965).

46. *Comprehensive Treatise of Electrochemistry*, Vol. 1, *The Double Layers* (J. O'M. Bockris, B. E. Conway, and E. Yeager, eds.), Plenum Press, New York (1980); Sh. U. M. Khan, *Mod. Aspects Electrochem.* **15**, 305 (1983); M. D. Levi, B. B. Damaskin, and I. A. Bagotskaya, *Itogi Nauki Tekh. Ser. Elektrokhim.* **19**, 47 (1983); A. Hamelin, T. Vitanov, E. S. Sevastyanov, and A. Popov, *J. Electroanal. Chem. Interfacial Electrochem.* **145**(2), 225 (1983); B. E. Conway, "The Solid–Electrolyte Interface," *Nato Conf. Ser., Ser.* **6**(5), 497 (1983); G. A. Martynov and R. R. Salem, "Electronic Capacitor at a Metal/Electrolyte Interface," *Elektrokhimiya* **19**, 1060–1070 (1983); and G. A. Martynov and R. R. Salem, *Lecture Notes in Chemistry*, Vol. 33: "Electrical Double Layer at a Metal–Dilute Electrolyte Solution Interface," Springer-Verlag, Berlin (1983); also B. W. Ninham, "Surface Forces—The Last 30 Ångstrom," *Pure Appl. Chem.* **53**, 2135–2147 (1981).

47. T. L. Einstein, "Changes in Density of States Caused by Chemisorption," *Phys. Rev. B* **12**, 1262–1274 (1975).

48. S. G. Gagarin and Yu. A. Kolbanovskii, *Kinet. Katal.* **19**, 1463 (1980).

49. T. B. Grimley, "Electronic Structure of Adsorbed Atoms and Molecules," *J. Vac. Sci. Technol.* **8**, 31–38 (1971).

50. J. M. Turlet, J. Bernard, and P. Kottis, "Fluorescence from (001) Surface and Subsurface Excitons in Anthracene Crystal: Some Experimental Evidences," *J. Lumin.* **18–19**, 47–50 (1979); cf. also R. T. Holm and E. D. Palik, "Internal Reflection Spectroscopy Studies of Thin Films and Surfaces," *Opt. Eng.* **17**, 512–524 (1978); see also *Phys. Rev. B* **17**, 2173 (1978).

51. R. Del Sole and E. Tossatti, *Solid State Commun.* **22**, 307 (1970).

52. There is a huge body of literature on this topic, e.g., *Comprehensive Treatise of Electrochemistry*, Vols. 2 and 7 (J. O'M. Bockris, B. E. Conway, E. Yeager, R. E. White, and S. U. M. Khan, eds.), Plenum Press, New York (1981 and 1983); *Proc. Symp. on Electrocatalysis* (W. E. O. Grady, P. N. Ross and F. G. Will, eds.), Electrochem. Soc., Princeton, New Jersey (1982); K. Tamara and M. Ichinawa, *Catalysis by Electron Donor–Acceptor Complexes*, Kodansha, Tokyo (1975); J. O'M. Bockris and S. U. M. Khan, *Quantum Electrochemistry*, Plenum Press, New York (1979).

53. H. Keyzer *et al.*, in *"4th Internatl. Symp. on Phenothiazines and Related Drugs"* (H. Eckert, I. S. Forrest and E. Usdin, eds.), Elsevier, Amsterdam (1980).

54. J. P. Farges and F. Gutmann, unpublished.

55. H. J. Frank *et al.*, *Ber. Bunsen Ges. Phys. Chem.* **80**, 547 (1970); M. Gratzel, A. Henglein, and E. Janata, "Mechanism of Electron Transfer from $\rho_{aq-}$ to Acceptors in Micelles," *Ber. Bunsen Ges. Phys. Chem.* **79**, 475–480 (1975); cf. also M. Graetzel, J. J. Kozak, and J. K. Thomas, "Electron Reactions and Electron Transfer Reactions Catalyzed by Micellar Systems," *J. Chem. Phys.* **62**, 1632–1640 (1975).

56. C. A. Bunton, *Progr. Solid State Chem.* **8**, 167 (1973).

57. V. T. Gorshkov *et al.*, *Zh. Fiz. Khim.* **51**, 2680 (1977).

58. A. M. Kolber, "Mono- and Divalent Competitive Adsorption to a Charged Membrane in a Closed System: A Comparative Study," *J. Theor. Biol.* **94**(3), 633–649 (1982).

59. For example, R. A. Klein, *NATO Adv. Sci. Inst. Ser.; Ser. A* **59**, 301–317 (1983); cf. also "The Movement of Molecules Across Membranes," *Q. Rev. Biophys.*, **15**, 667 ff. (1982); H. T. Tien, *Bilayer Lipid Membranes*, Marcel Dekker, New York (1974); C. F. Fox, "The Structure of Cell Membranes," *Sci. Am.* **226**(2), 30 (1972); R. A. Capaldi, "A Dynamical

Model of Cell Membranes," *Sci. Am.* **230**(3), 26 (1974); see also the specialized journal, *J. Membrane Biol.*; B. Lutenberg and L. Van Alphen, "Molecular Architecture and Functioning of the Outer Membrane of Escherichia Coli and Other Gram-Negative Bacteria," *Biochim. Biophys. Acta.* **737**, 51–115 (1983); G. Cevc and D. Marsh, "Properties of the Electrical Double Layer near the Interface Between a Charged Bilayer Membrane and Electrolyte Solution: Experiment vs. Theory," *J. Phys. Chem.* **87**, 376–379 (1983).

60. See Ref. 18, p. 79 ff.

61. D. L. Knopf, M. D. Lupa, J. H. Caldwell, and F. M. Harold, *Science* **220**, 1385 (1983); I. F. Jaffe and R. Nucnelli, *J. Cell Biol.* **68**, 614 (1974).

62. S. Hino and H. Inokuchi, "Electron Escape Depths of Organic Solids. II. The Energy Dependence of Naphthacene and Perylene Films," *J. Chem. Phys.* **70**, 1142–1146 (1979).

63. N. Sato, H. Inokuchi, K. Seki, J. Aoki, and S. Iwashima, "Ultraviolet Photoemission Spectroscopic Studies of Six Nanocyclic Aromatic Hydrocarbons in the Gaseous and Solid States," *J. Chem. Soc. Faraday Trans. 2* **78**, 1929–1936 (1982); N. Sato, K. Seki, and H. Inokuchi, "Polarization Energies of Organic Solids Determined by Ultraviolet Photoelectron Spectroscopy," *J. Chem. Soc. Faraday Trans., 2* **77**, 1621–1633 (1981); N. Sato, K. Seki, and H. Inokuchi, "Ultraviolet Photoelectron Spectra of Tetrahalo-*P*-Benzoquinones and Hexahalobenzenes in the Solid State," *J. Chem. Soc. Faraday Trans. 2* **77**, 47–54 (1981); I. Ikemoto, Y. Sato, T. Sugano, N. Kosugi, H. Kuroda, K. Ishii, N. Sato, K. Seki, and H. Inokuchi, "Photoelectron Spectroscopy of the Molecule and Solid of 11,11,12,12-Tetracyanonaphthoquinodimethane (TNAP)," *Chem. Phys. Lett.* **61**, 50–53 (1979); K. Seki, S. Hashimoto, N. Sato, Y. Harada, K. Ishii, H. Inokuchi, and J. Kanbe, "Vacuum-Ultraviolet Photoelectron Spectroscopy of Hexatricontane (N-C36-H74) Polycrystal: A Model Compound of Polyethylene," *J. Chem. Phys.* **66**, 3644–3649 (1977).

64. J. N. Weinstein, R. Blumental, J. Van Renswoude, C. Kempf, and R. D. Klausner, "Charge Clusters and the Orientation of Membrane Proteins," *J. Membr. Biol.* **66**, 203–212 (1982); C. Kempf, R. D. Klausner, J. N. Weinstein, J. Van Renswoude, M. Pincus, and R. Blumenthal, *J. Biol. Chem.* **257**(5), 2469 (1982).

65. G. I. Distler and V. G. Obronov, "Photoelectret Mechanism of Long Range Transmission of Structural Information," *Nature (London)* **224**, 261–262 (1969); G. I. Distler and V. G. Obronov, "Induced Polarization Structure in Interfacial Epitaxial Layers," *Naturwissenschaften* **57**, 495 (1970); cf. also G. I. Distler, *J. Crystal Growth* **9**, 76 (1971).

66. D. C. Reynolds and T. C. Collins, *Excitations: Their Properties and Uses*, Academic, New York (1981); G. Nicklus and I. Prigogine, *Self Organization in Non-Equilibrium Systems*, Wiley, New York (1977); A. S. Davydov, *Solid State Theory*, Nauka, Moscow (1976); A. S. Davydov, *Biology and Quantum Mechanics*, Pergamon Press, Oxford (1982); A. S. Davydov, *Phys. Scr.* **20**, 387 (1979); H. Matsumoto, M. Tachiki, and H. Umezava, *Thermo-field Dynamics and Condensed States*, Elsevier, North Holland, Amsterdam (1982).

67. A. S. Davydov, A. A. Eremko, and A. A. Zegeenko, *Ukr. Fiz. Zhurn.* **23**, 983 (1978); A. S. Davydov and A. D. Suprun, *Configurational Changes and Optical Properties of Alpha-Spiral Protein Molecules*, Inst. of Theor. Phys. Kiev, Publ. *ITF-73-1-P*, (1973); see also A. S. Davydov, Ref. 73.

68. E. Del Guidice, S. Doglia, M. Milani, and G. Vitiello, *A Quantum Field Theoretical Approach to the Collective Behavior of Biological Systems*, preprint of paper read at the Conference on Non-Linear Electrodynamics in Biological Systems, Loma Linda, California 1983.

69. S. G. Christov, *J. Res. Inst. Catal., Hokkaido Univ.* **24**, 27 (1976); G. Saito and Y. Matsunaga, *Bull. Chem. Soc. Jpn* **46**, 1609 (1973); **45**, 963 (1972); **47**, 2873 (1974); **47**, 1020 (1974); **46**, 714 (1973); Y. Matsunaga, *ibid.* **48**, 37 (1975); Y. Matsunaga and R. Osawa, *ibid.* **47**, 1589 (1974); J. M. Dumai *et al.*, *J. Chem. Phys., Phys-Chem. Biol.* **72**, 1185 (1975);

H. Ratajcak *et al.*, *Chem. Phys.* **17**, 197 (1976); A. Kofler, *A. Electrochem.* **50**, 200 (1974); G. I. Krishtalik *et al.*, *J. Res. Inst. Catal.*, *Hokkaido Univ.* **22**, 101 (1974); R. R. Dogonadze and A. M. Kuznetsov, *ibid.* **26**, 15 (1978); I. Yu. Martynov *et al.*, *Russ. Chem. Rev.* **46**, 1 (1977); W. Klopffer, *Adv. Photochem.* **10**, 311 (1977).

70. M. J. Rice and W. L. Roth, *J. Solid State Chem.* **4**, 294 (1972); L. J. Gagliardi, see Ref. 81.
71. J. Koziol and P. Tomasik, *Bull. Acad. Pol. Sci. Ser. Sci. Chim.* **25**, 689 (1977).
72. K. Morokuma, *Acc. Chem. Res.* **10**, 294 (1977).
73. J. L. Beauchamp, *Ann. Rev. Phys. Chem.* **22**, 527 (1971); also in *Interactions between Ions and Molecules* (P. Ausloos, ed.), Plenum Press, New York (1975), p. 413; N. Hartmann *et al.*, *Top. Curr. Chem.* **43**, 57 (1973).
74. J. H. Bayless, L. Friedmann, F. B. Cook, and H. Shechter, "The Effect of Solvent of the Course of the Bamford-Stevens Reaction," *J. Am. Chem. Soc.* **90**, 531 (1968).
75. K. Kalnins *et al.*, *Dokl. Akad. Nauk. SSSR* **244**, 400 (1979).
76. E. M. Arnett and E. J. Mitchell, "Hydrogen Bonding. VI. A Dramatic Difference Between Proton Transfer and Hydrogen Bonding," *J. Am. Chem. Soc.* **93**, 4052–4053 (1971).
77. T. Erdey-Gruz and S. Lengyel, *Mod. Aspects of Electrochem.* **12**, 1 (1977); J. E. Crooks and B. H. Robinson, in *Proton Transfer, Faraday Symp. Chem. Soc. London*, No. 10 (1975), p. 29; C. E. Bannister *et al.*, *ibid.*, p. 78.
78. J. Nishijo, *Bull. Chem. Soc. Jpn* **47**, 1539 (1974).
79. P. W. Anderson *et al.*, *Phil. Mag.* **25**, 1 (1972); J. Tauc, *Phys. Today*, **Oct. 1976**, 27; S. G. Christov, *Contemp. Phys.* **13**, 199 (1972); *Phys. Status Solidi* **7**, 371 (1971); *Croatica Chim. Acta.* **44**, 67 (1972); G. Gusman and R. Deltowi, *Solid State Commun.* **15**, 1587 (1974); R. R. Dogonadze and A. M. Kuznetsov, *J. Res. Inst. Catalysis, Hokkaido Univ.* **22**, 93 (1974); J. H. Bush and J. R. De la Vega, "Symmetry and Tunneling in Proton Transfer Reactions, Proton Exchange between Methyloxonium Ion and Methylalcohol, Methylalcohol and Methoxide Ion, Hydronium Ion and Water, and Water and Hydroxyl Ion," *J. Am. Chem. Soc.* **90-99**, 2397–2406 (1977).
80. T. Izumida, T. Ichikawa, and H. Yoshida, "Effect of Matrix Polarity on the Charge Transfer Band of the Benzyl Radical–Halide Complex," *J. Chem. Phys.* **83**, 373–375 (1979).
81. L. J. Gagliardi, "Dielectric Friction and Protonic Mobility," *J. Chem. Phys.* **58**, 2193–2194 (1973).
82. G. N. Felix and P. C. Huyskens, "Influence of the Formation of Ions on the Viscosity of Phenol–Amine Mixtures," *J. Phys. Chem.* **79**, 2316–2322 (1975).
83. W. Klopffer, *Adv. Photochem.* **10**, 311 (1977); I. Yu. Martynov *et al.*, *Russ. Chem. Revs.* **46**, 1 (1977).
84. Yu. I. Martinov *et al.*, *Khim. Vys. Energ.* **11**, 443 (1977).
85. I. Deperasinska and J. Prochorov, *Adv. Molec. Relax. Interact. Process* **11**, 51 (1977).
86. V. P. Klindukhov and T. G. Meister, "Determination of the Intermolecular Hydrogen Bond Energy of Some Systems in Ground and Primary Excited Electron States," *Molekulyar Spektroskopya (1977)*, 12–41; cf. also *Adv. Molec. Relax. Interact. Processes* **13**, 107 (1978).
87. F. Babler and A. Von Zelewski, *Helv. Chim. Acta* **60**, 2723 (1977); F. Gutmann and H. Keyzer, *Electrochim. Acta* **13**, 693 (1968); M. A. Slifkin, *Charge Transfer Interactions of Biomolecules*, Academic Press, London (1971), p. 251.
88. E. A. Chandross and J. Ferguson, *J. Chem. Phys.* **47**, 2557 (1967); H. A. Staab and V. Schwendemann, "Charge Transfer Interactions 17: Cyclophane Quinhydrone—A Donor–Acceptor Cyclophane with Extremely Short Transannular Distance," *Angew. Chem.* **90**, 805–807 (1978); H. A. Staab and V. Zapf, "Charge Transfer Interactions 18: Indirect Donor–Acceptor Interactions in Quinhydrones of a 4-layered 2,2 Parachyclphane," *Angew. Chem.* **90**, 807–808 (1978).
89. W. W. Robertson, A. D. King, Jr., and O. E. Weigand, Jr., "Determination of Excited

State Dipole Moments of Anthracene," *J. Chem. Phys.* **35**, 464–466 (1961); N. Tyutyuckov *et al.*, *Theor. Chim. Acta (Berlin)* **20**, 385 (1971).

90. J. K. Roy and D. G. Whitten, *J. Am. Chem. Soc.* **95**, 7162 (1972); D. V. O'Connor and W. R. Ware, *ibid.* **101**, 121 (1979).

91. M. Gordon and W. R. Ware, *The Exciplex*, Academic, New York (1975); *Molecular Association* (R. Foster, ed.), Academic, New York (1979), p. 2; P. Fröhlich and E. L. Wehry, "The Study of Excited State Complexes (Exciplexes)," in *Modern Fluorescence Spectroscopy* (E. L. Wehry, ed.), Plenum Press, New York (1976), Vol. 2, p. 319.

92. C. Bagyimka *et al.*, *Acta Biol. Acad. Sci. Hung.* **32**(3-4), 311 (1981).

93. H. T. Tien, "Photoelectric Bilayer Lipid Membrane: A Model for the Thylakoid Membrane,"Brookhaven Natl. Lab. Rep. BNL-50530 (1977); cf. also H. T. Tien, *Bioelectrochem. Bioenerget.* **5**, 318 (1978).

94. A. B. Ershler, A. M. Funtikov, and I. M. Levinson, *Elektrokhimiya*, **18**(11), 1577 (1982).

95. G. Varo and L. Keszthely, "Photoelectric Signals from Dried Membranes of Halobacterium Halobium," *Biophys. J.* **43**(1), 47–51 (1983).

96. V. E. Khutorskii, "Water Structure in the Transmembrane Gramicidin A Channel," *Bio-org. Khim.* **9**(6), 846–848 (1983).

97. M. L. Ahrens, "Electrostatic Control by Lipids upon the Membrane-Bound Sodium–Potassium ATPase. II. The Influence of Surface Potential upon the Activating Ion Equilibria," *Biochim. Biophys. Acta* **732**(1), 1–10 (1983).

98. K. Niki, T. Yagi, H. Inokuchi, and K. Kimura, "Electrode Reactions of Cytochrome $c_3$ of *Desulfovibrio vulgaris, Miyazaki*," *J. Electrochem. Soc.* **124**, 1889–1891 (1977).

# Ion, Electron, and Proton Transport in Membranes: A Review of the Physical Processes Involved

## Ronald Pethig

*ABSTRACT:* The physical concepts and problems associated with the understanding of ionic, electronic, and protonic transport in membrane structures are discussed. Ion transport is intimately associated with the existence of a transmembrane potential, and the contributions that ion concentration gradients, membrane surface charges, and surface redox reactions may give to this are described, together with the physical features required of ion channels and pores. An understanding of the coupling of electron transport processes to proton motive forces is a central task for modern bioenergetics, and some of the factors involved are discussed, as well as the physical mechanisms that control electron and proton transport processes in membrane structures. Other topics included are the dielectric properties of biological electrolytes, electronic induction, and dipole interactions in proteins, and proton transport in water, ice, and model systems.

## 1. Introduction

This article will explore some of the physical concepts and problems of relevance to charge transfer and charge interactions in biological membranes. An increased understanding of such processes, some of which are depicted in Figures 1 and 2, represents a central task for modern bioenergetics and bioelectrochemistry. The object will not be to delve into the biochemical data that is of relevance to this area of study, but rather to discuss the physical aspects of charge transport and translocation processes

*Ronald Pethig* • School of Electronic Engineering Science, University College of North Wales, Dean Street, Bangor, Gwynedd LL57 1UT, United Kingdom.

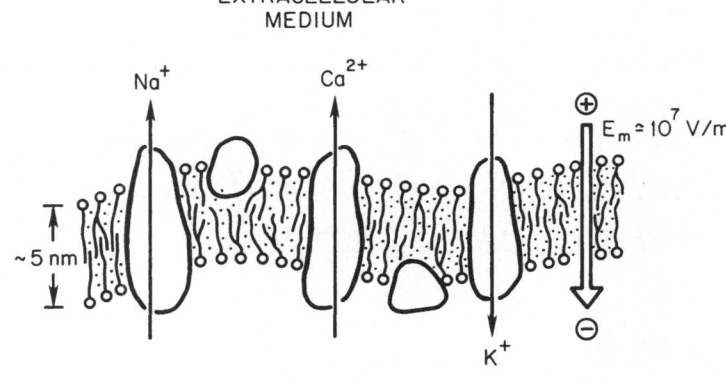

EXTRACELLULAR
MEDIUM

$Na^+$          $Ca^{2+}$

$E_m \simeq 10^7$ V/m

~5 nm

$K^+$

CYTOPLASM

*FIGURE 1.* Schematic representation of the fluid-mosaic model of a cell membrane showing protein molecules incorporated into a lipid bilayer structure. Lateral diffusion of the proteins and lipid molecules occurs, but the lipids very rarely migrate from one side of the membrane to the other. A transmembrane electrical field $E_m$ arises from the action of vectorial ion pumps (ATPase proteins) in producing ionic concentration differences across the membrane, and from the presence of membrane surface charges. Surface redox reactions may affect this membrane field.

in general, keeping where possible within the bounds of biological relevance. Hopefully, such an approach will be of assistance to biochemists in the interpretation of some aspects of their studies and in the formulation of models and experiments. At the same time this article aims to provide a review of some of the molecular and submolecular physical concepts currently being investigated and developed in the study of the electrical properties of biological materials.

## 1.1. Ions, Electrons, and Protons in Cell Membranes

The basic features of a cell membrane are given in Figure 1 to show how it consists essentially of protein molecules incorporated into a semifluid liquid bilayer structure. As a rough guide the plasma membrane of most cells is composed, on a dry weight basis, of nearly equal components of protein and lipid. Because of its nonpolar nature the lipid membrane structure is intrinsically impermeable to polar and electrically charged molecules. For example, turbidity measurements on sarcoplasmic reticulum membranes provide membrane resistance values of $2.6 \times 10^7$ and $2.5 \times 10^6$ $\Omega$ cm$^2$ for the permeability of calcium ions and protons, respectively, while for sodium and potassium ions the corresponding values are

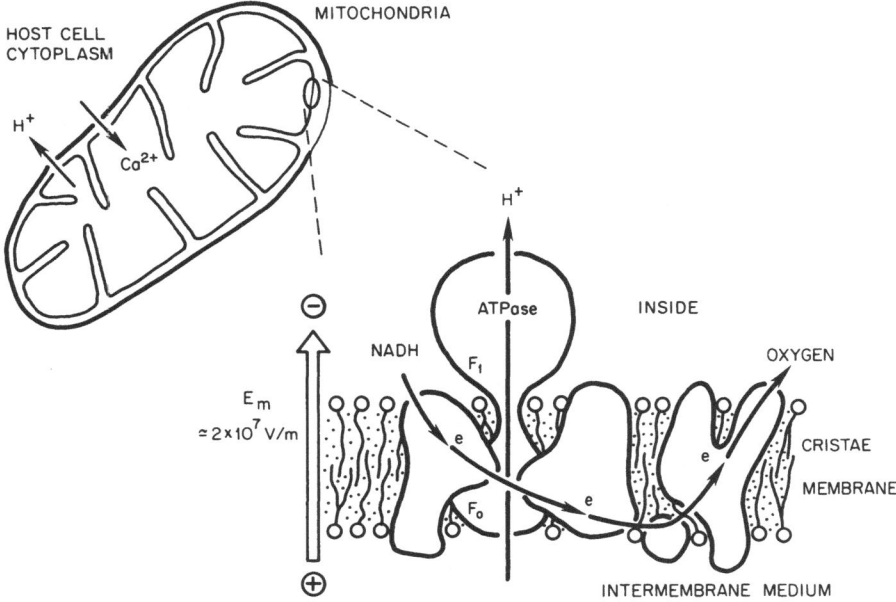

*FIGURE 2.* The vectorial pumping of calcium ions and protons across the mitochondrion membranes. A schematic enlargement of the inner (cristae) membrane is shown to indicate the existence of protein-based electron ($e$) and proton ($H^+$) conduction pathways. On average, the cristae membrane is composed of 75% protein and 25% lipid.

around $5 \times 10^4 \, \Omega \, cm^2$.[1] Such large resistance values are essential not only for preventing charged metabolites within the cell from leaking out, but also for maintaining the large electric fields that are generated across these membranes. However, as indicated in Figure 1, there are systems located within membranes which transport ions across the membrane to produce inside the cell a large concentration of potassium and low concentrations of sodium and calcium ions. A number of membrane proteins have been isolated which facilitate this active transport of ions against their own gradients of concentration and induced electrical potential, and the free energy $\Delta G$ that needs to be supplied to achieve this is given by

$$\Delta G = RT \ln C + ZF\Delta\psi \qquad (1)$$

where $C$ is the transmembrane ion concentration ratio, $\Delta\psi$ is the electric potential difference created across the membrane, and $R$, $T$, $Z$, and $F$ have their usual meanings of the gas constant, absolute temperature, charge valency of the ion, and the Faraday, respectively. In general, $\Delta\psi$ is found to be of value around 60–80 mV, equivalent to a field $E_m$ of the order

$10^7 \, \text{Vm}^{-1}$ acting across the membrane, with the outside of the cell at the more positive potential. It should be noted that Eq. (1) does not take account of the loss of free energy due to leakages of ions back across the membrane. In excitable tissues such as nerves, electrical impulses are produced as a result of the sudden increase in back-leakage of sodium and potassium ions, which causes a transient collapse of the transmembrane potential gradient. Some of the physical aspects associated with the membrane potential and associated diffuse electrical double layer, together with the active and passive transport of ions, will be discussed in this article.

Charge transport processes also occur in the membranes of intracellular organelles such as mitochondria, chloroplasts, and in the energy-transducing membranes of bacteria. A simple outline is given in Figure 2 for the case of a mitochondrion, within whose inner (cristae) membrane a complex system of enzymes controls and mediates the transport of electrons, protons, and calcium ions. The flow of electrons along an electron transport chain of proteins is an elementary step preceding the transport of ions and protons, the formation of adenosine-triphosphate (ATP), and other physiological processes. The electron transport chain will be discussed later in this chapter, and mention will now be made of the proton-ATPase family of enzymes which catalyze the hydrolysis and synthesis of ATP according to the general scheme

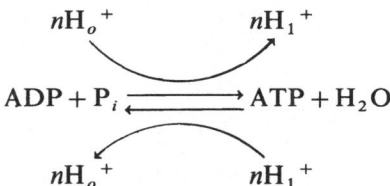

where ADP and $P_i$ represent adenosine-diphosphate and inorganic orthophosphate, respectively. The free energy required for the active transport of ions is provided by the free energy released on the hydrolysis of ATP, and a review of the current concepts concerning this process has been given by Tanford.[2] The ATPases appear to consist of two parts, a membrane enclosed hydrophobic moiety $(F_o)$ which contains a proton-conducting channel and a hydrophilic moiety $(F_1)$ which catalyzes the reversible reaction scheme depicted above and protrudes into the aqueous medium.[3] The designations $nH_o^+$ and $nH_1^+$ thus refer to protons (or hydronium ions) on the $F_o$ and $F_1$ sides of the cristae membrane, respectively, with suffix "$o$" meaning "oligomycin sensitivity."

Since the introduction of the so-called chemiosmotic hypothesis by Mitchell,[4-6] extensive experimental data obtained from mitochondria, chloroplasts, and bacteria have confirmed the general concept that the electron transport proteins and the proton ATPase complex are coupled to each other by an electrochemical gradient of protons. However, the basic process by which this coupling is attained is still under active debate. In essence, Mitchell has proposed the following:

(i) The electron transport chain is topologically organized so that as electrons are passed from the succinate substrate to oxygen, protons are transferred from the inside of the mitochondrial cristae membrane to the outside.

(ii) As a result of the relative impermeability of the cristae membrane to ions and protons, this electrogenic pumping of protons creates a proton-motive force $\Delta p$, composed of a membrane potential and a proton concentration (pH) gradient, of magnitude

$$\Delta p = \Delta \psi - (2.3RT/F)\,\Delta \text{pH}$$

(For mitochondria that are respiring but not synthesizing ATP, an average value for $\Delta p$ is 200 mV, with $\Delta \psi \cong 170$ mV.[7])

(iii) The macroscopic thermodynamic potential so established drives the synthesis of ATP as protons flow down their own electrochemical gradient along a reversible proton conduction pathway in the ATPase.

On the other hand, concurrent with Mitchell's scheme has been that initiated by Williams,[8–10] in which the following is envisaged:

(i) The protons initially generated by the redox chain of electron carriers are confined to local regions in the membrane and are not in thermodynamic equilibrium with protons in the aqueous phases on either side of the membrane.

(ii) The protons so generated are translocated within the membrane region to a lower free energy level, with the released energy being used to drive the synthesis of ATP by ATPase.

This type of approach has been actively developed and extended by Kell,[11–13] and the essential difference between these two main lines of reasoning is outlined in Figure 3. At the physical level the difference between the two models rests in the kinetics of the processes involved, and to a major extent depends on whether pathways exist which facilitate larger proton mobilities than can occur in bulk water. This aspect is given particular attention in this chapter, and it is concluded that membrane associated pathways for the rapid translocation of protons is a feasible concept. If local proton coupling does occur, rather than via the bulk aqueous phases on either side of the membrane, then the experimentally determined (bulk phase) values for the membrane potential and proton-motive force could significantly underestimate the true driving force for ATP synthesis.

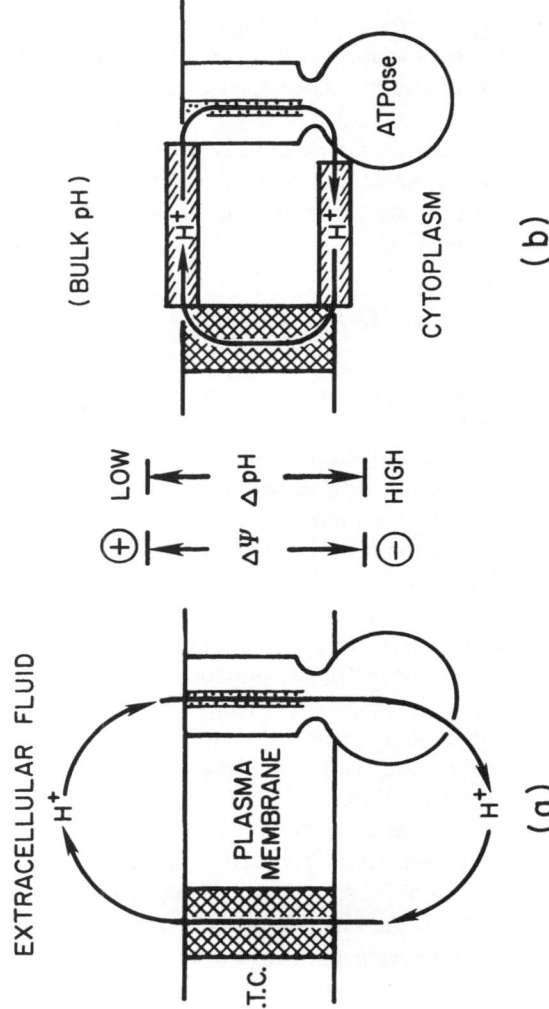

*FIGURE 3.* Schematic representations of a bacterial cytoplasmic membrane to show the essential difference between (a) the bulk chemiosmotic, and (b) the local "electrodic" mechanism currently under active discussion as possible mechanisms by which the electron transport chain (e.t.c.) and ATP synthesis are coupled to vectorial proton translocation. (Based on Kell and Hitchens.[13])

The physical concepts that are relevant to the conduction or translocation of ions, electrons, and protons in biological membranes will now be outlined. In considering protons separately from the other ions, the intention is to emphasize the unique character of these elementary particles.

## 2. Ion Transport

### 2.1. Membrane Potential

Considerations of the physical features that could control the transport and permeability of ions in membranes need to take into account the transmembrane potential. This potential, as determined experimentally, corresponds to the difference in the electrical potentials between the two aqueous phases away from the inner and outer membrane surfaces, and its steady-state value is related to such factors as the difference in ionic concentrations of the two aqueous phases, ionic permeability of the membrane, and the presence of any fixed charges or dipolar species in the membrane structure. Excitable cells such as those of nerve or muscle tissue exhibit resting membrane potentials of up to 90 mV and greater, whereas many of the small and nonexcitable cells (e.g., erythrocytes) exhibit smaller values of around 20 mV or so.

It has already been mentioned that the outside of the cell is at the more positive potential, but of course this does not necessarily imply that the outside surface carries a net positive charge. In fact, various experiments, the most common being electrophoretic measurements, have shown that at neutral pH the external surfaces of most cells are negatively charged and that this arises from the predominance of ionizable carboxyl groups ($pK \cong 4.6$) over that of ionizable basic species. An average value for the external surface charge density of several nonexcitable animal cells is around one electronic charge per 20 nm$^2$ ($-0.8 \ \mu C/cm^2$) of membrane surface, (e.g., Ref. 14). Such studies are not without problems, however. For example, varying results have been reported for the membrane of the giant squid axon, with Aono and Ohki[15] giving estimates of the negative charge densities on the outer and inner membrane surfaces of $-27$ and $-17 \ \mu C/cm^2$, respectively, as compared to other workers, who give values of $-13$[16] and $-1 \ \mu C/cm^2$,[17] respectively. Segal[18] gives a much lower estimate of $-0.02 \ \mu C/cm^2$ for the outside charge density, while the recent work of Ohki[19] indicates the most appropriate value to be around $-8 \ \mu C/cm^2$. Although there are large differences reported for the magnitudes of the surface charge densities, there does appear to be a trend that suggests that the inside membrane surface is less charged than the outside. This difference could be thought to result mainly from the fact that for animal cells surrounded by physiological fluids the intracellular pH is typically 0.1–0.2 pH units lower than that of the external medium,[20] so that the inner membrane surface is closer to its isoelectric point than the external surface. However, this presupposes that the internal and external membrane surfaces of cells have effectively similar physical and chemical properties, and there is sufficient evidence to indicate that this is not the

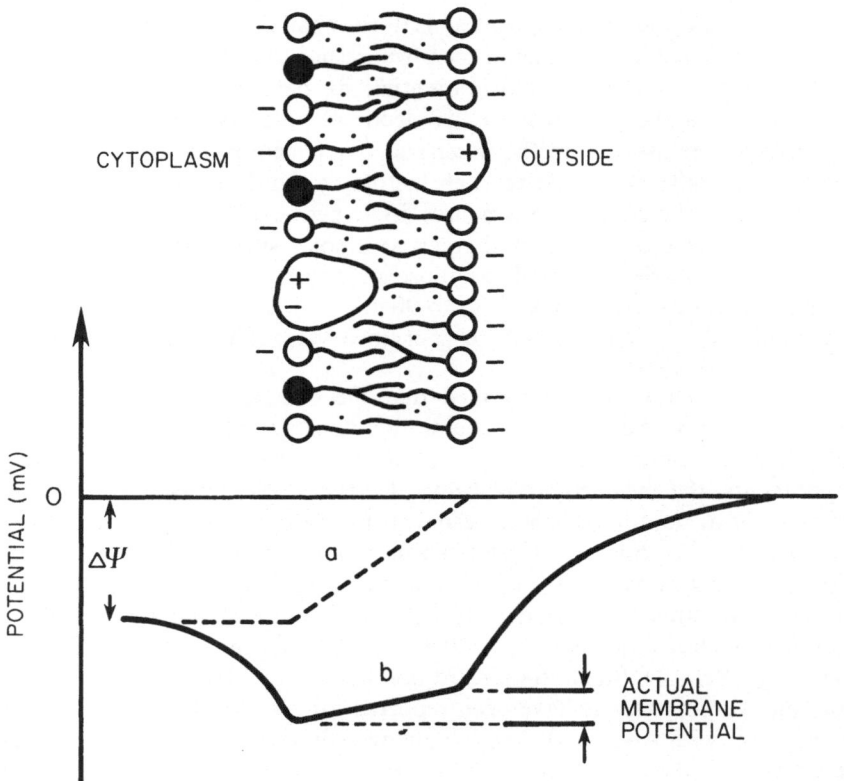

FIGURE 4. Schematic representation of a cell membrane to show the asymmetrical nature of the lipid bilayer structure and the presence of fixed negative charges on the membrane surfaces. Ion pumps generate intracellular concentrations of ions (typically $Na^+ \cong 20$ mM, $K^+ \cong 150$ mM) that differs appreciably from the extracellular concentrations ($Na^+ \cong 150$ mM, $K^+ \cong 5$ mM). The back diffusion of sodium ions is the predominant factor controlling the membrane potential $\Delta\psi$, and the model of Goldman–Hodgkin–Katz predicts a potential profile of the form of (a). The presence of surface charges gives rise to the profiles shown in (b). (See text for details.)

case.[21] The lipid composition on each side of the membrane may consist of differing classes, and the ionization constants of the acidic and basic groups of proteins can vary significantly with changes in the structural and ionic nature of their local environments. For example, the oligosaccharide-containing lipid molecules called glycolipids are found only on the outer surface of animal cell membranes, and some of these (called gangliosides) contain one or more ionizable sialic acid residues. In human red blood cells, negatively charged phosphatidylserines are mainly located in the inner membrane surface and most of the negative charge of the outer cell surface is carried by the sialic acid groups attached to the carbohydrate coated tails of glycophorin transmembrane proteins. In fact, most of the negative charge on the outer surface of animal cells is likely to be associated with their having a carbohydrate-rich coat of glycoproteins and proteoglycans. Another membrane to have been extensively studied is that of the thylakoid membrane of chloroplasts where surface charge densities ranging from $-0.46$ to $-7.4$ $\mu C/cm^2$ have been reported,[22] and which are considered to arise from ionized carboxyl groups of the glutamic and aspartic acid residues of exposed membrane proteins.

The purpose here of drawing attention to the existence of fixed electrical charges on the membrane surfaces is that they may have a significant influence on the magnitude of the potential difference acting across the membrane interior. This is illustrated in Figure 4 to show the essential differences between the potential profiles deduced from the ion diffusion model of Goldman–Hodgkin–Katz, and those associated with fixed membrane charges.

### 2.1.1. Ion Diffusion Model

The diffusion potential theory has its origins in the Nernst theory formulated to describe the liquid junction potential developed at the interface of two electrolytes with different ionic concentrations and ionic mobilities. For two univalent ionic solutions the liquid junction potential is given by

$$\Delta\psi = \psi_1 - \psi_2 = \frac{RT}{F}\frac{u^+ - u^-}{u^+ + u^-}\ln\frac{[C_2]}{[C_1]} \tag{2}$$

where $u^+$ and $u^-$ are the mobilities of the positive and negative ions, respectively, and $\psi_1$, $\psi_2$ and $C_1$, $C_2$ are the electrical potentials and ionic concentrations phases 1 and 2, respectively. The potential $\Delta\psi$ thus arises from a difference in ion mobilities. The refinements of the theory by Goldman, Hodgkin, and Katz[23] have led to the following expression for the transmembrane potential arising from the extracellular and intracellular $K^+$, $Na^+$, and $Cl^-$ concentration and permeability differences:

$$\Delta\psi = \frac{RT}{F} \ln \frac{P_K[K^+]_o + P_{Na}[Na^+]_o + P_{Cl}[Cl^-]_i}{P_K[K^+]_i + P_{Na}[Na^+]_i + P_{Cl}[Cl^-]_o} \tag{3}$$

where $P$ is the permeability constant and subscripts $o$ and $i$ refer to the outside and inside phases, respectively. Typically, for animal cells, $[Na_o^+] \cong [K_i^+] \cong 150$ mM; $[Cl_o^-] \cong 110$ mM; $[Cl_i^-] \cong 80$ mM; $[Na_i^+] \cong 20$ mM; and $[K_o^+] \cong 5$ mM. The excitable membranes of nerve and muscle exhibit a high permeability for potassium and a low one for sodium, so that it is widely accepted that Eq. (3) can be simplified to give the membrane potential at a value close to the potassium equilibrium potential at 298 K of

$$\Delta\psi = \frac{RT}{F} \ln \frac{[K_o^+]}{[K_i^+]} \cong 26 \ln \frac{5}{150} \cong -90 \text{ mV}$$

It is also commonly considered that the low membrane potential of non-excitable cells arises from a near equality of the sodium and potassium permeabilities setting its value near the midpoint of the potassium and sodium equilibrium potentials. The membranes of many animal cells are highly permeable to $Cl^-$, and so according to the diffusion model this must also make an important contribution to the membrane potential.

Another approach, also based on thermodynamic equilibrium concepts, invokes the Donnan equilibrium potential, which for univalent ionic solutions separated by a membrane gives

$$\Delta\psi = \psi_i - \psi_0 = \frac{RT}{F} \ln \frac{a_0^+}{a_i^+} = \frac{RT}{F} \ln \frac{a_i^-}{a_o^-} \tag{4}$$

where $a$ is the activity of the particular ionic species, and the sub- and superscripts have the meanings employed in Eq. (3). Equation (4) is derived from the requirement that at thermodynamic equilibrium there must be a balance of the electrochemical potentials $\mu_i$ and $\mu_o$ on each side of the membrane. The electrochemical potential is given by the general expression

$$\mu = \mu^0 + RT \ln a + ZF\psi$$

where $\mu^0$ is the standard reference chemical potential. By taking into account the diffusion of ions across the membrane and combining Eqs. (3) and (4), a fixed charge membrane model was developed by Teorell.[24] In this case the total membrane potential is composed of the two boundary potentials given by the Donnan equilibrium potentials plus a diffusion potential, and a uniformly distributed fixed charge is assumed to be distributed across the membrane. Ohki[25] provides a detailed discussion of the various aspects and applicability of these two models. Later in this section, when discussing ion pore models, the point will be made that the presence of bare charges in the membrane interior is an energetically unfavorable situation.

### 2.1.2. Surface Charge Model

The existence of a net electrical charge on a membrane surface will give rise to a surface potential $\psi_s$. Since at neutral pH the membranes are negatively charged, diffusible cations will be attracted to the surface and anions will be repelled, and an electrical double layer will be created. Analyses of these effects have largely been based on the theorems of Gauss, Poisson, and Boltzmann as first formulated by Gouy and Chapman and later extended by Grahame.[26] A detailed account of the theory has been given by Brown.[27]

If the membrane surface is assumed to be planar and the surface charges smeared uniformly over the surface, then the relationship between $\psi_s$ and the surface charge density ($\sigma$) for an electrolyte composed of mixed valency ions is given by

$$\sigma = \left\{ 2\varepsilon_0\varepsilon_r kT \sum_j n_{jb}[\exp(-Z_j F\psi_s/RT) - 1] \right\}^{1/2} \tag{5}$$

where $k$ is Bolzmann's constant, $\varepsilon_0$ and $\varepsilon_r$ are the permittivity of free space and of the bulk electrolyte, respectively, and $n_{jb}$ is the ionic concentration (ions dm$^{-3}$) in the bulk electrolyte away from the influence of the membrane surface. The electrolyte permittivity is assumed to be constant up to the membrane surface, and the potential of the bulk electrolyte is taken as reference zero. By assuming that $\psi_s$ is only a little more negative than the zeta potential at the hydrodynamic plane of shear, measurements of electrophoretic mobility have commonly been used to determine membrane surface charge densities. Although it is, therefore, a circular argument to show that Eq. (5) gives acceptable values for $\psi_s$ in terms of $\sigma$, it is of value to see how the potential profiles vary with surface charge density. For simplification we shall assume that the electrolyte is not of mixed valency, in which case Eq. (5) can be simplified to give

$$\sigma = (8n_b\varepsilon_r\varepsilon_0 kT)^{1/2} \sinh(ZF\psi_s/2RT)$$

Since $n_b = NC_b$, where $N$ is the Avogadro constant, then numerical substitution into this equation ($\varepsilon_r = 78$, $T = 298$ K) gives

$$\sigma = 0.117\sqrt{C_b}\,\sinh\,(Z\psi_s/51.4) \qquad (6)$$

In Eq. (6), $\sigma$ is in C/m$^2$, $C_b$ is in moles per liter, and $\psi_s$ is in millivolts. In Figure 5a the potential profiles are shown for three values of the surface charge density with a monovalent electrolyte such as KCl of concentration

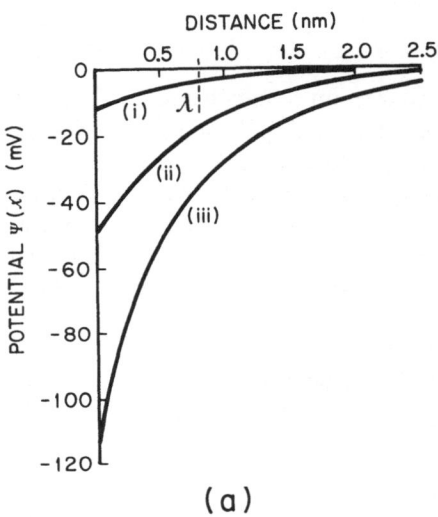

(a)

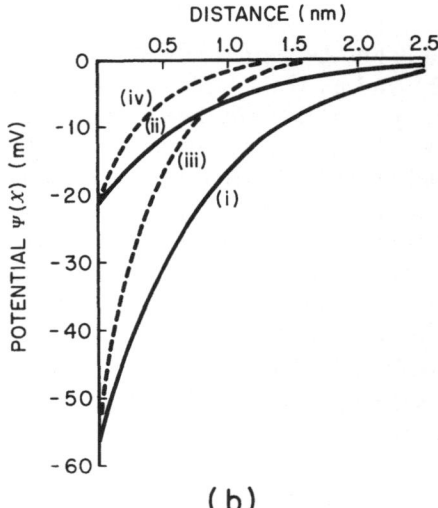

(b)

FIGURE 5. (a) The variation of electric potential near a charged membrane surface according to the Gouy–Chapman theory, at 25°C and for a 0.15 M uni-univalent electrolyte such as NaCl. Three values for the surface charge density were chosen: (i) = 1 μC/cm$^2$; (ii) = 5 μC/cm$^2$; (iii) = 20 μC/cm$^2$. The Debye length $\lambda$ is indicated. (b) The potential profiles for a membrane surface charge of 5 μC/cm$^2$ and for a uni-univalent electrolyte of concentration (i) 100 mM and (ii) 1 M. Profiles are also given for a divalent electrolyte such as MgSO$_4$ of concentrations (iii) 7.2 mM and (iv) 210 mM to show the greater effect of divalent counterions in changing the surface potential as compared with monovalent ions.

0.15 $M$. A typical phospholipid head group occupies an area of the order 0.5–0.6 $nm^2$, so that a surface change density of 5 $\mu C/cm^2$, equivalent to one charge per 3.2 $nm^2$, would correspond to about 17% of the lipids being charged. This would appear to be a reasonable situation, and the corresponding surface potential of 49 mV calculated from Eq. (6) is also of the correct magnitude. The effects of changing the strengths of a monovalent and divalent electrolyte are shown in Figure 5b.

The way in which the potential decreases with distance $(x)$ from the surface is given by

$$\psi(x) = \frac{2RT}{ZF} \ln \left[ \frac{1 + \alpha \exp(-x/\lambda)}{1 - \alpha \exp(-x/\lambda)} \right] \qquad (7)$$

where

$$\alpha = \frac{\exp(ZF\psi_s/2RT) - 1}{\exp(ZF\psi_s/2RT) + 1}$$

and

$$\lambda = \left( \frac{\varepsilon_0 \varepsilon_r kT}{2q^2 n_b Z^2} \right)^{1/2}$$

where $q$ is the charge on an electron. The factor $\lambda$ is known as the Debye–Hückel screening length. For monovalent electrolytes, then for surface potentials less then around 20 mV, Eq. (7) can be simplified to the form

$$\psi(x) = \psi_s \exp(-x/\lambda) \qquad (8)$$

and $\psi(x)$ falls to a value of $0.37\psi_s$ at a distance $\lambda$ from the membrane surface. At 298 K and assuming $\varepsilon_r$ has the bulk water value of 78, then for a 0.15 $M$ KCl electrolyte $\lambda$ has a value of 0.8 nm. In Figure 4b the fall-off in potential is shown on both sides of the membrane, and the potential difference across the membrane interior is given by the difference in the surface potentials of the outer and inner membrane surface. The magnitude and polarity of the bulk membrane potential gradient may therefore differ appreciably from that indicated by $\Delta\psi$ in Figure 4.

The ionic concentration $C_j(x)$ will be governed by the Boltzmann equation

$$C_j(x) = C_{jb} \exp[-Z_j F\psi(x)/RT] \qquad (9)$$

which indicates that divalent cations $(Z = 2)$ rather than monovalent cations will be preferentially attracted to a negatively charged surface. Likewise, anions $(Z = -1$ or $-2)$ will be repelled and diffuse away from a negatively charged surface. [Physicists will more readily recognize Eq. (9) by noting the equality $kT/q = RT/F \cong 26$ mV at 300 K.] Equation (9) also predicts that at any distance $x$ from the membrane surface

$$n^+ n^- = n_b^2$$

so that, for example, for a 0.15 $M$ KCl electrolyte and with $\psi_s = -49$ mV there will be about 1 $M$ K$^+$ and 22.5 mM Cl$^-$ adjacent to the membrane surface. Moving away from the surface the K$^+$ concentration will decrease and the Cl$^-$ concentration will increase, until at about 4 nm from the surface the cation and anion concentrations will be equal at 0.15 $M$. Electrical engineers will have a better feel for the form of the electrical double layer produced by noting that for values of $\psi_s$ less than 20 mV, Eq. (5) reduces to

$$\sigma = \varepsilon_0 \varepsilon_r \psi_s / \lambda$$

This is identical to the standard equation $Q = VC$ applicable to a parallel-plate capacitor, with $\lambda$ being the distance between the plates.

The Gouy–Chapman theory of the electrical double layer at charged surfaces makes assumptions such as that the surface charge is uniformly smeared over the (planar) membrane surface, that the ions can be represented as point charges, and that the image charges they induce in the membrane are negligible. The relative permittivity of the bulk electrolyte is also assumed to have a constant value right up to the membrane surface. Improvements of the theory have been developed to take into account the sizes of the ions and the interactions that occur between water dipoles and the ions and membrane charges. The main effects of these improvements are summarized schematically in Figure 6 for the case of a negatively charged membrane, to show a hydration layer adjacent to the membrane in which the potential falls linearly before the Gouy–Chapman potential profile commences. A detailed treatment of the various features of Figure 6, such as the Stern layer, have been given by Bockris and Reddy.[28]

### 2.1.3. Discrete Surface Charges

The assumption that the electrical charges on the membrane surface are evenly smeared out may mask important effects occurring at the molecular scale depicted in Figure 6. For a point charge on the surface, the variation of the potential with distance can be given in terms of Coulomb's law and in the same form of Eq. (8) as

$$\psi(x) = \frac{q}{4\pi\varepsilon_0\varepsilon_r} \frac{1}{x} \exp(-x/\lambda) \tag{10}$$

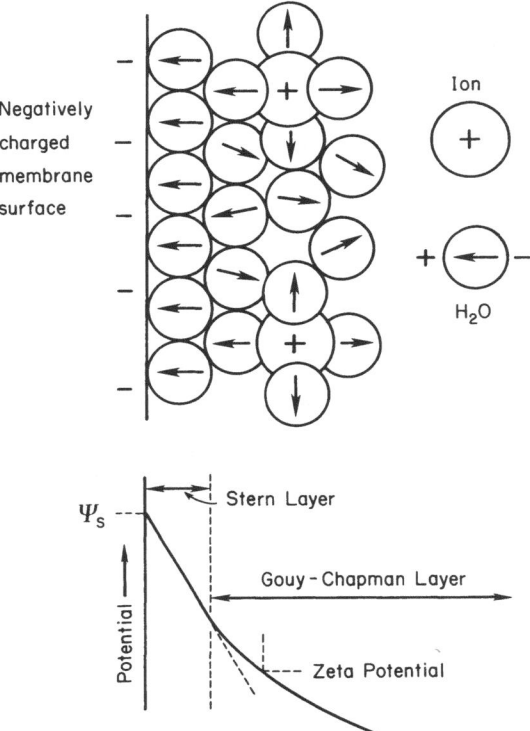

*FIGURE 6.* The arrangement of water molecules and counterions near to a negatively charged membrane surface according to the Stern model. Within the Stern layer of polarized water molecules the electric potential falls linearly, and for distances further than this the potential profile follows that predicted by the Gouy–Chapman theory of electrical double layers. For ascites cells the potential drop between the surface potential $\psi_s$ and the zeta potential has been determined to be around 5 mV.[31]

If the effective radius ($b$) of the charge needs to be taken into account, then Eq. (10) should be multiplied by the factor[29] $\exp(-b/\lambda)/(1 + b/\lambda)$.

To describe the potential generated by charges at a membrane–solution interface, the effect of induced charges in the membrane should also be taken into account. Brown[27] outlines how this may be accomplished by the method of images, and he shows how membrane thicknesses of 7 nm or more may be taken as being of semi-infinite extent for such calculations. By way of illustrating the effect of treating the surfaces as discrete electronic charges, Figure 7 has been constructed using the formulas given by Brown[27] to show the potential contours around a "channel opening" in a membrane surface containing an average charge

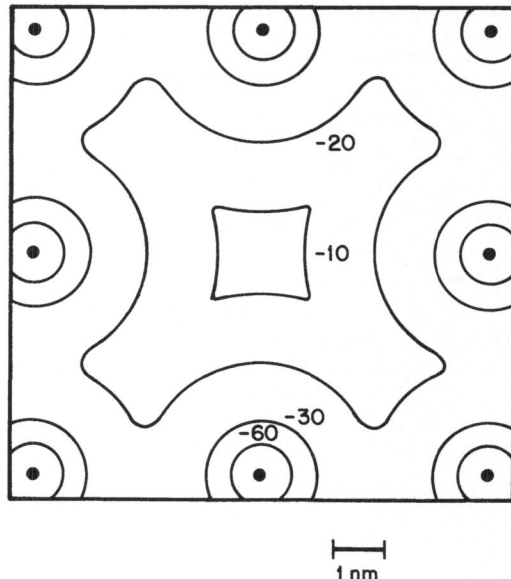

FIGURE 7. The potential contours (mV) around a channel opening in a membrane with a uniformly distributed and fixed charge density of $-0.8\ \mu C/cm^2$. The plane shown is situated 0.1 nm above the planar surface of the membrane.

density of $-0.8\ \mu C/cm^2$. The plane of interest is taken as 0.1 nm above the surface and is thus in the Stern layer region. A value of $\varepsilon_r = 5$ has been taken for the permittivity of the polarized water in the Stern layer (more on this later), and a minimum screening by counterions has been assumed.

It should be stressed that Figure 7 represents a static picture. The lipid and protein molecules in the membrane are capable of rapid lateral movement, with the lipids having diffusion constants around $10^{-8}\ cm^2/s$. This indicates that in a time period of 1 $\mu s$ a lipid molecule could diffuse over a distance of 2 nm, which would certainly have the effect of smearing out the detail presented in Figure 7, but *only* if the time period of interest is of 1 $\mu s$ or greater. The time for diffusion of charged species across the membrane is less than 1 $\mu s$ for electrons and protons, and approaches this value for ions. Considerations of the discreteness of the surface charges are therefore of relevance for such processes.

### 2.1.4. Surface Charge Effects on Charge Transfer Reaction Rates

The Boltzmann equation (9) results from the equilibrium condition that the electrochemical potential of charged species must be the same at any distance from the membrane, provided the membrane surface charge is evenly smeared out. If the activity coefficient $f_j$ does not vary with distance, then everywhere the activity $a_j$ of the species is given by

$$a_j = C_j f_j$$

and Eq. (9) can be written in the form

$$a_{js} = a_{jb} \exp\left(-Z_j F \psi_s / RT\right) \tag{11}$$

If a charge transfer or redox reaction, involving a reagent A in the electrolyte and a membrane-bound component B, of the form

$$A^- + B \underset{k}{\overset{k}{\rightleftharpoons}} A + B^-$$

takes place at the membrane surface, then the rate of oxidation of $A^-$ will depend on its activity at the membrane surface. This in turn will depend on the surface potential as expressed by Eq. (11). The apparent rate constant ($k$) will therefore be related to the true rate constant ($k_0$) by the expression

$$\ln k = \ln k_0 - ZF\psi_s / RT \tag{12}$$

The slope of a plot of $\log k$ against $C_b^{-1/2}$ should therefore provide a value of the surface density in the vicinity of the reaction site in the membrane surface. This approach has been adopted to determine the internal and external membrane charges of chloroplasts by monitoring redox reactions,[30] and the external membrane charge of cells by measurements of the rate of quenching of semiquinone and ascorbate-free radicals.[31]

### 2.1.5. Redox Potential Model

There have been proposals that the membrane potential is produced mainly as a result of oxidation–reduction potentials at the membrane surfaces. The consequences of such proposals, if proven, would completely revolutionize present concepts concerning membrane biology. Thus, for example, ion diffusion and transport across membranes would be conceived to result as a *consequence* of the transmembrane potential and *not* determine the potential as described by the Nernstian concepts already outlined (Figure 8). One side of the membrane is conceived as supporting an anodic reaction, the other side a cathodic reaction. In the concepts outlined by Habib and Bockris[32] the principal anodic reaction is envisaged to be of the type

$$RN_n \rightarrow R + nH^+ + ne^-$$

FIGURE 8. The Redox and Nernstian Question.

where RH is an organic molecule such as glucose. The cathodic reaction would involve the reduction of oxygen:

$$O_2 + 4H^+ + 4e^- \rightarrow 2H_2O$$

The coupling of these two reactions requires that charge is capable of being transferred across the membrane, most probably through the membrane-bound protein structures. The experimental and theoretical studies of charge transport in protein structures has been extensively reviewed elsewhere.[33–35] The potential difference $\Delta V$ across the membrane would then have the value

$$\Delta V = \Delta V_R + V_0 + iR_m$$

where $\Delta V_R$ is the difference between the equilibrium redox potentials of the two surface reactions, $\Delta V_0$ is the difference between the small overpotentials developed at the two membrane surfaces, and $iR_m$ is the ohmic voltage drop across the membrane resistance. These concepts are described in more detail elsewhere in this volume.[36]

## 2.2. Dielectric Properties of Biological Water

The relative permittivity $\varepsilon_r$ was introduced into Eq. (5) and values for the electrolyte of 78 and then 5 were used for this factor in deriving the potential profiles of Figures 5 and 7, respectively. The dielectric properties of water associated with biological structures are of considerable relevance to many of the concepts considered in this chapter, and so a brief discussion will now be given of some of the factors that determine these properties.

The Onsager expression (Ref. 33, Chap. 1) for calculating the relative permittivity of polar liquids, and which takes into account the local electric "reaction" field acting on the individual polar molecules, has the form

$$\frac{(\varepsilon_s - \varepsilon_\infty)(2\varepsilon_s + \varepsilon_\infty)}{\varepsilon_s(\varepsilon_\infty + 2)^2} = \frac{Nm^2}{9\varepsilon_0 kT} \tag{13}$$

where $\varepsilon_s$ is the static or limiting low-frequency relative permittivity and $\varepsilon_\infty$ is the value at a frequency high enough that dipole orientation effects are unable to contribute to the overall polarizability of the liquid. Each molecular dipole has a dipole moment $m$ and there are assumed to be $N$ dipoles per unit volume. From vapor phase measurements, the dipole moment for an individual water molecule is known to be 1.84 Debye units. Since it is generally considered that intermolecular motions of the water molecule should not be included in Eq. (13) to calculate the effect of its self-induced reaction field, the value for $\varepsilon_\infty$ is taken as 1.85 (the square of the refractive index) rather than as 4.5, the relative permittivity measured at infrared frequencies. Insertion of $\varepsilon_\infty = 1.85$ into Eq. (13) for $T = 20°C$ yields a value for $\varepsilon_s$ of 27, a value much lower than the experimental value of 80.4. The reason for the discrepancy is of considerable importance and is mainly associated with the structural correlations that exist through hydrogen bonding between a water molecule and its immediate neighbors.

The tetrahedral hydrogen-bonding structure for a central water molecule linked to its four nearest neighbors is shown in Figure 9. (Strictly speaking, this refers to the ice I structure since for bulk water at 20°C there are on average 4.5 water molecules coordinated to any one central water molecule.) As a result of these structural links, any one water molecule can-

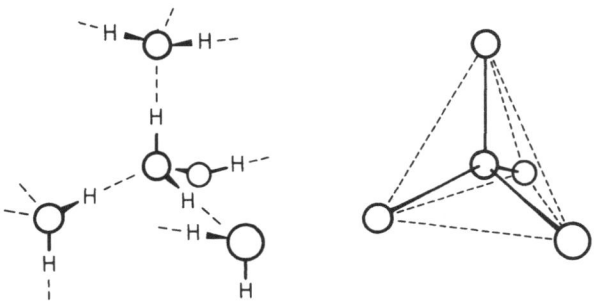

*FIGURE 9.* The tetradral hydrogen-bonding structure of ice I.

not respond to the orienting influence of an applied electric field without also influencing the orientation of its neighbors. To account for this correlated ordering of neighboring molecules the effective dipole moment factor $m^2$ in Eq. (13) should be replaced by $gm^2$, with $g$ being the correlation parameter given by

$$g = 1 + N\langle \cos \theta \rangle$$

where $\langle \cos \theta \rangle$ is the mean of the cosines of the angles between the dipole moments of the $N$ nearest neighbors and of the central water molecule. For a tetrahedral group $N = 4$ and $\langle \cos \theta \rangle = 1/3$, so that ignoring other possible influences $g = 2.33$. The generally accepted value for $g$ for water at 20°C is 2.82, corresponding to a value for $\varepsilon_s$ in Eq. (13) of 80.4. With increasing temperature the structure correlation can be expected to decrease, and this is reflected in the value for $g$ decreasing to a value of 1.64 at 300°C. The theories that have been developed to account for the various correlation effects have been discussed by Hasted.[37] The main point to be stressed is that the value for $\varepsilon_s$ of around 80 applies *only* for normal bulk water having the type of hydrogen bond associations shown in Figure 9, and where each water molecule has orientational *freedom*.

The relative permittivity for water varies with temperature, and to a good approximation between 0 and 40°C

$$\varepsilon_s = 87.74 - 0.4T$$

with $T$ being the temperature in °C. The frequency variation of the relative permittivity for water and ice is shown in Figure 10, and although in pas-

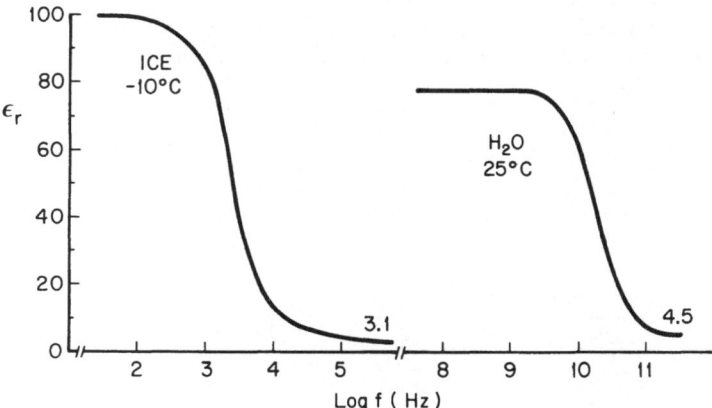

FIGURE 10. The variation of the relative permittivity with frequency for water and ice.

sing between the ice and liquid phase at 0°C there are only small changes in the dynamic nature of the hydrogen-bond networks, the difference in dielectric behavior is very significant. It might reasonably be assumed that water molecules closely associated with biological structures will exhibit dielectric properties somewhere between those shown for water and ice in Figure 10. Dielectric measurements[38] on lysozyme have shown that 36 water molecules are so bound to the protein structure that they are prevented from rotating under the influence of applied electric fields, and another 70 water molecules are so bound that they exhibit a dielectric dispersion at frequencies around 100 MHz and not at around 10 GHz as expected for normal bulk water. Measurements[39] at microwave frequencies for DNA at various stages of hydration have shown that around 280 water molecules per helix turn are so influenced by the electrostatic forces associated with the DNA structure that they do not "freeze" until temperatures of $-100$°C and lower are attained. Bearing in mind these effects, and the temperature and frequency variation, it can be understood why physicists prefer to use the term "permittivity" and not "dielectric *constant.*"

What value of the relative permittivity should be assigned to the water molecules depicted in Figure 6, where the normal tetrahedral bonding structure has been disrupted by the presence of the charged membrane surface and solvated ions? The mono-layer of water molecules in the Stern layer adjacent to the charged membrane surface is perhaps the simplest to consider, especially if we neglect for the present the ions in the electrical double layer. The dipole moments of these water molecules will be strongly oriented by the strong electric field associated with the membrane charges, and so will not be able to respond to any reorientation forces associated with externally applied electric fields. Intramolecular vibrations should be relatively unrestricted, however, and so an appropriate value for the relative permittivity of the water molecules in the Stern layer would be around 5, corresponding to the value for water at infrared frequencies. This type of discussion is in fact not permissible, for the reason that the concept of permittivity is a macroscopic one and we have attempted to describe the dielectric properties of a single layer of water molecules. At this microscopic level, the concept of permittivity is more properly considered in terms of volume polarizability, but because we need not consider orientational contributions to the total polarizability of water in the Stern layer, our final conclusion is appropriate. The second layer of water molecules will also be partly polarized by the charged membrane surface (now partially screened by the dipoles of the Stern layer water molecules) and tetrahedral hydrogen-bond associations will be restricted. A value of around 30–40 would be appropriate for the relative permittivity of this layer of water. By the fourth hydration layer the influence of the charged membrane surface should be considerably reduced and the normal bulk dielectric property of water can reasonably be assumed to have been attained.

What will be the effect of the ions that are present in the electrical double layer? As already described, Eq. (9) predicts that for a KCl electrolyte of bulk concentration 0.15 $M$ and a membrane surface of potential $-50$ mV, the concentration of $K^+$ ions at the membrane surface will be around 1 molar. For a cubic lattice of ions of concentration $C$ moles per liter, the average interionic distance $r$ may be calculated from the equation

$$r = 0.95C^{-1/3} \text{ nm}$$

so that for the above example the average distance between $K^+$ ions near the membrane surface will be around 1 nm, giving room for only 3–4 water molecules between individual ions. The opportunity for extensive networks of the normal tetrahedral structure for water is very limited. The four or so water molecules that form the primary hydration sheath around an ion will be rotationally immobilized by the ion's strong electric field and can be expected to have dielectric properties similar to the water molecules in the Stern layer. A significant disruption of the normal bulk water structure can be expected to have occurred for the remaining water molecules. This disruption results in a lowering of the overall relative permittivity and for electrolytes of ionic strength up to 1 or 2 molar, the relative permittivity $\varepsilon$ decreases linearly with ionic strength according to a relationship of the form

$$\varepsilon = \varepsilon_r C(\delta^+ + \delta^-)$$

where $C$ is the ionic concentration and $\delta^+$ and $\delta^-$ are the so-called dielectric decrement values for the cations and anions, respectively. Since cell membranes are negatively charged, we need only consider the effect of cations in the region close to the membrane. For $Na^+$ and $K^+$ ions the value for $\delta^+$ is 8, and for $Mg^{2+}$ ions it is 24 (Ref. 33, p. 140). Accordingly, for a membrane with a surface potential of $-50$ mV the bulk relative permittivity of a 0.15 $M$ NaCl or KCl electrolyte will have a value of around 72 in the region near to the membrane surface. Considerations of this effect could usefully be incorporated into Eqs. (5) and (10).

## 2.3. Ion Pores

We can now consider the passage of ions through cell membranes. Firstly, we note that since cell membranes possess a negative potential at physiological pH, arising from one or a combination of the various processes already described in this section, then as predicted by the Boltzmann relationship of Eq. (9) only cations will tend to diffuse towards the membrane. This is the reason why only cations are shown being transported across the membranes in Figures 1 and 2, anions being repelled away from cell membranes.

One's first thought might be to assume that the transmembrane potential presents the greatest barrier or aid (depending on transport direction) for an ion crossing the membrane. In fact, the greatest barrier is associated with the fact that in order for an ion to pass between the aqueous regions on each side of the membrane it must travel through a region of low permittivity. As depicted in Figure 6, because of its electrical charge an ion exerts a strong local electrical field which strongly polarizes nearby water molecules. The dipoles of the neighboring water molecules are aligned so as to reduce the surface electrostatic potential of the ion. This potential $V$ is given by

$$V = \frac{q}{4\pi\varepsilon_0\varepsilon_r r_i}$$

and its electrostatic self-energy $W$ is given by

$$W = \frac{q^2}{8\pi\varepsilon_0\varepsilon_r r_i}$$

where $q$ is the ion's charge and $r_i$ its radius. For sodium and potassium the ionic radius is 0.095 and 0.133 nm, respectively. The electrostatic potential at the surface of such ions in water is then around 0.19 and 0.14 V for $Na^+$ and $K^+$, respectively, and their electrostatic self-energies are 9 and 6.7 kJ/mole, respectively. The relative permittivity of the inner membrane region, and also of protein interiors, can be estimated to have a value of about 2.5 (Ref. 33, pp. 65 and 211). This means that an ion located within a membrane, either in the lipid phase or within a transmembrane protein, will have an electrostatic self-energy some 30 times greater than when it was solvated in the bulk electrolyte. The potential energy barrier confronting a sodium ion is shown in Figure 11, and as an aid in visualizing the energy scale this has been given in units of $kT$, corresponding to the mean thermodynamic energy available to an ion when it is in equilibrium with its environment at temperature $T$. Since a potassium ion is of a larger radius, the barrier it will need to penetrate is smaller than that for the sodium ion. The total barrier shown in Figure 11 includes the potential energy associated with the membrane potential of the order of 80 mV, which corresponds to an energy difference across the membrane of around $3kT$ (7.7 kJ/mole) and so is negligible compared with the total barrier height of around $100kT$ (257 kJ/mole). The concepts dealt with here are exactly those of the Born theory for the free energy of ion–solvent interac-

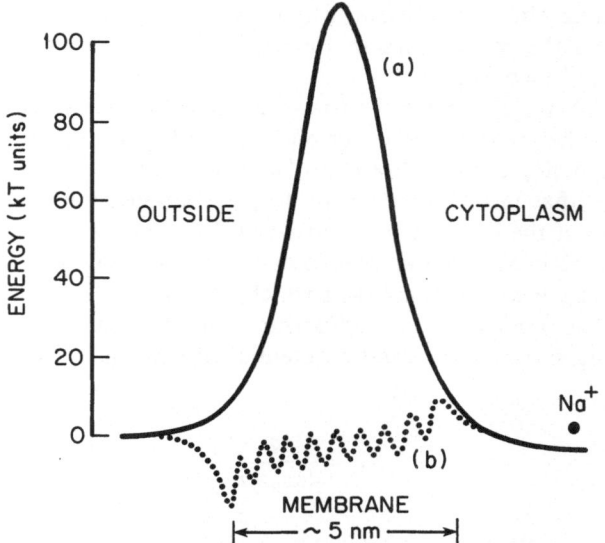

FIGURE 11.   (a) The electrostatic potential energy barrier confronting a sodium ion passing through a cell membrane. The energy scale is in kT units, and at physiological temperatures $100 \, kT \cong 260 \, kJ/mole$. (b) The potential energy barrier associated with an ion passing through a water-lined channel according to the model of Edmonds.[41]

tions, and an excellent treatment of this theory has been given by Bockris and Reddy (Ref. 28, Chap. 2).

The barrier shown in Figure 11 is too high for an ion to be thermally activated over it, and as will be appreciated from the forthcoming section on tunneling phenomena there is a negligible probability (because of its mass and the barrier width) of an ion being able to quantum mechanically tunnel its way across the membrane. Since the existence of the barrier results from the ion possessing a net electrical charge, it might be thought that minimizing the effect of this charge, either by pairing the ion with another one of opposite charge or wrapping it in a "coat" of neutral polarizable molecules to reduce the surface potential, or having the ion pass through a channel whose walls are lined with counter charges, would significantly lower the membrane barrier. These possibilities have been examined in detail[40,41] and have been found to offer no real solution as to how the barrier opposing ion transport across the membrane can be significantly reduced. As can be appreciated from the discussion of the electrostatic self-energy of charges, a channel lined with countercharges would represent an energetically unfavorable situation. This argument also places the concept proposed by Teorell[24] concerning a fixed charge model for the membrane potential as an unrealistic one.

The ideal situation would be for the ion to pass from one aqueous phase to the other through water-filled channels in the membrane so that during transit the electrostatic energy of the ion would be comparable to that when fully hydrated. Such a model has often been proposed and a recent refinement has been suggested by Edmonds.[41] The suggestion is that the channel wall is defined by protein molecules that span the membrane. More specifically he proposes that the channel is formed by five helical polypeptide chains which provide hydrogen-bond links to stabilize a clathrate-like ordered structure of water molecules within the channel. The magnitude and polarity of the transmembrane potential is such that the dipoles of these water molecules will tend to be directed towards the cell interior. The full energy of solvation of an ion such as $Na^+$ is achieved only if it is in close contact with at least four of five water molecules and, for this reason, the water structure in the channel must be continuous and cagelike. As a result of the polarization of the water molecules along the direction of the membrane field, the negative ends (oxygen end) of the water dipoles will be directed towards the extracellular medium, so that the channel entrance at the cell interior will have the more positive potential. It is envisaged[41] that as a result of this an ion within the channel will tend to be propelled in a direction opposite to that which would result from the membrane potential acting alone, as shown in Figure 11. This model proposed by Edmonds has some interesting and potentially valuable features. It appears to provide a basis by which membrane structures can select ions on the basis of their size and charge, and a possible understanding of such phenomena as channel gating, switching and nerve action potentials is also offered. The difference in potential energy between a water molecule oriented along the membrane field and one against it will only be of the order of $4kT$, and so any small external influence could readily cause a channel water molecule to "flip over" and so significantly alter the rate of ion transport through the channel. For example, a negatively charged or hydrophobic molecule coming into contact with the outer channel opening could easily cause a reversal of all the water dipoles in the channel and lead to a reversal of the charge polarity at the other end of the membrane channel. Such effects could also be coupled with, or mediated by, allosteric transitions occurring in the transmembrane proteins that form the channel walls. The local electric field in the membrane pore can be expected to be sensitive to changes in the surface potentials of the two membrane surfaces, and Jordan[42] has presented calculations of the electrical potential within a hydrated membrane pore and shows how this influences the effective ionic conductance of ion pores.

## 3. Electron Transport

Electron transport processes are known to occur in photosynthetic systems, in the respiratory chain in mitochondria, and in redox reactions involved in the functioning of microsomes and enzymes such as catalase and the peroxidases. Witt[43] and Konings and Boonstra,[44] have provided excellent reviews of the electronic steps involved in green plant and bacterial photosynthesis, respectively. In green plants photosynthesis takes place in the membranes of thylakoids that are contained within subcellular organelles called chloroplasts. About $10^5$ pigment molecules (chlorophylls-*a* and -*b*, and carotenoids) are embedded into each thylakoid membrane, and light energy absorbed by the pigments is used to drive electrons derived from the oxidation of water via various electron carriers to NADP$^+$. The electron carriers include quinones, copper- and iron-containing proteins (plastocyanin and ferredoxin) and cytochrome molecules, and there are about 200 such electron transport chains in each thylakoid membrane. The separation of electronic charge from the primary electron donor ($H_2O$) to NADP, and the resulting electric field produced, is the primary act of photosynthesis. The electric field is used to drive protons for the synthesis of ATP (Figure 3), and its vectoral nature emphasizes the fact that the electron transport must follow specific pathways provided by the membrane protein structures. Mitochondria are organelles found in the cells of all oxygen-utilizing plants and animals, and they must have developed after the photosynthetic organisms had provided an appreciable supply of oxygen to the environment. As shown in Figure 2, the electron transport chain in the inner mitochondrial membrane is also highly vectorial, and involves cytochrome molecules as the main pathway for electron transport. The cytochrome proteins have incorporated into their structure a porphyrin group containing a central Fe ion. This Fe ion is directly involved in the electron transfer process, and repeatedly switches between the two valency states, $Fe^{2+}$ and $Fe^{3+}$, as it acts as an electron donor or acceptor, respectively.

We shall restrict ourselves here to some aspcts of the processes by which the electrons can be vectorially tansferred through membrane protein structures. The two basic processes to be considered, namely, tunneling or electronic delocalization, can be summarized in the form of Figure 12. A brief discussion will also be given of electron induction effects.

### 3.1. Tunneling

In the subject of quantum mechanics, particles such as electrons and protons can be described in terms of their wave nature. The lighter the particle the larger is its wavelength, and the wavelength of an electron can be

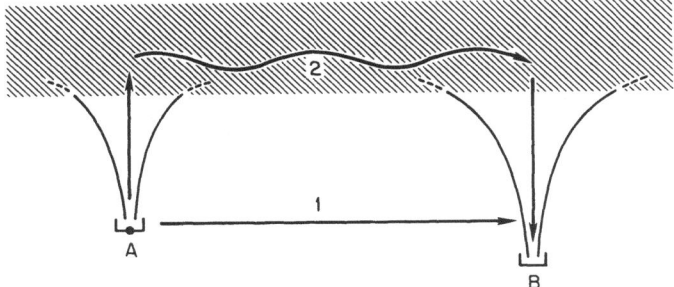

FIGURE 12. The two paths an electron can take in transferring between two localized sites A and B in a protein structure. Route 1 is achieved by tunneling through the intermediate medium, while route 2 involves activation into extended and delocalized electronic states of the protein structure.

much larger than the size of an atom. Associated with the wavelength of the particle is its wave function, and the only physical meaning that can be given to this function is that is specifies the probability of finding that particle at any finite position in space. In Figure 13a an electron, with its associated wave function $\psi(r)$ is shown occupying an electronic energy level in a potential energy well $A$. The electron is bound to the vicinity of

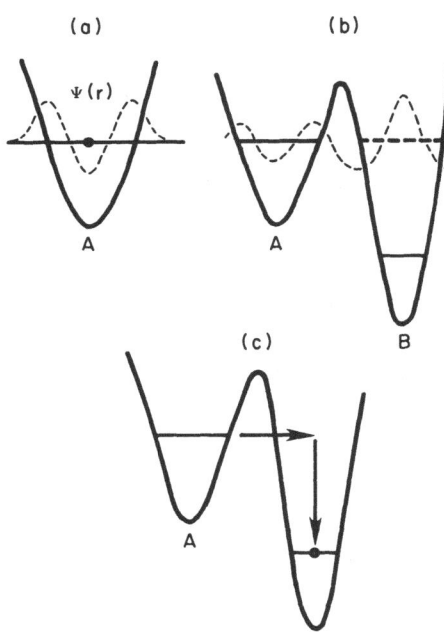

FIGURE 13. (a) An electron in site A with its associated wave function $\psi(r)$. (b) An electron oscillating between two sites A and B that are isolated from any external energy sinks or sources. (c) Sites A and B are coupled to an energy sink so that the electron in A can tunnel into site B and fall down into its electronic ground state.

site $A$ by the electrostatic potential barrier defining the energy well (site $A$ could be a positively charged Fe ion, for example), but as indicated by the spatial extent of the wave function, there is a finite probability of the electron appearing at locations outside the electrostatic "domain" of site $A$. In Figure 13b another electronic site $B$, unoccupied by an electron, is shown situated within the spatial extent of the wave function of the electron in site $A$. For the present we assume that sites $A$ and $B$ are isolated from any surrounding medium which could couple them into an energy sink or source by means of light interactions (photons) or lattice vibrations (photons). In this case the electron at site $A$ can appear in the electrostatic well of site $B$, with the amplitude of the corresponding wave function in site $B$ dependent on the height and width of the potential energy barrier between sites $A$ and $B$. In this condition the electron will oscillate between $A$ and $B$, spending most of its time (according to the wave function form shown in Figure 13b) in the vicinity of site $B$. If the two sites are now coupled to an energy sink and source, then, once the electron has penetrated into the vicinity of $B$, it can fall to the energy ground state of site $B$ (as shown in Figure 13c) by emitting photons or by giving up its excess energy to the surrounding medium in the form of lattic vibrations (heat). The process by which the electron has passed from $A$ to $B$ is termed "tunneling" and it is a process that is not possible in classical mechanics. In a tunneling process the particle need not be a well-defined entity such as an electron or proton, and the barrier could be, for example, a classically forbidden molecular rearrangement. The more massive the particle, the smaller is the extent of its wave function and its tunneling capability.

This qualitative description of the tunneling process has been given to aid an understanding of the essential physical concepts involved. Quantitative descriptions applicable to biological systems have been given by Redi and Hopfield[45] and DeVault.[46] These treatments show that the rate of electron tunneling between two sites is critically dependent on the distance between the sites, on the form of the potential energy barrier between them, and on the thermodynamic coupling of the sites in their local environment. These factors are used by biological structures to define specific electron pathways.

For the case of the cytochrome molecules, sites $A$ and $B$ of Figure 13 would represent the heme groups located within either the same or adjacent protein molecules, where the tunneling distance would be of the order 1–2 nm. The protein structure acts as the supporting matrix to fix the heme sites in the positions which maximize the tunneling of electrons between them, and at the same time it provides the necessary energy source and sink through vibrational coupling of the heme to the local polypeptide chains. Biochemists often characterize the specifice heme groups in the electron transport chains in terms of their reversible redox potentials, and redox energy is used as a measure of the energy of an electron at one of these sites. The zero of energy for the standard redox potential scheme is the energy released by one electron at a hydrogen electrode, when causing molecular hydrogen to be evolved in aqueous solution, whereas the absolute redox energy can be defined as the energy of the electron at its site in question relative to that when it is at rest at an infinite distance away. Thus, although the standard redox energy differences between sites are identical to the absolute redox energy differences, the redox energy of a state is the negative of the redox potential. In considering electron transfer processes, then an electron will tend to be transferred from a high redox energy state (low redox potential) to a lower redox energy state (higher redox potential). This concept can sometimes confuse physical scientists who are used to defining both the absolute zero of energy and potential for an electron as being that when it is brought to rest at infinity.

Analyses[45,46] of the electron transfer rates that occur in the electron transport chains of mitochondria and some photosynthesis systems indicate that electron tunneling between heme groups is the dominant process. This may not always be the case, however. For example, in modeling the cytochrome-*c* peroxidase: cytochrome-*c* complex it was concluded[47] that the iron-to-iron distance at 2.56 nm was too great to rely on a straightforward tunnelling process alone. A system of inter- and intramolecular pi-orbital, ionic, and hydrogen-bonding interactions was found to form a bridge between the hemes, suggesting that this might be the pathway for electron transfer. This would correspond to route 2 in Figure 12. Distances of around 2.5 nm between oxidation–reduction centers also seem to occur in other cytochrome systems.[47] Also, since these active redox centers are protected by surrounding polypeptide structures to shield against nonspecific interactions via random collisions with cellular oxidants and reductants, it would be reasonable to assume that the protein structures provide, in part, specific pathways for the electron transfers. Theoretical considerations of the role that protein structures can make in facilitating electron transfer between redox centers have been presented by Petrov.[48]

## 3.2. Electron Delocalization

The packing density of the constituent atoms in the interior regions of protein molecules is, on average, equivalent to that of most organic compounds in the crystalline state.[49] As such, it is possible to consider that membrane proteins may exhibit some of the solid-state electronic properties which have been extensively studied in elemental amorphous materials and organic polymers. Such properties include electronic conduction via localized and delocalized electronic states. When localized states are involved, the electron transport takes place by a series of tunneling or hopping processes, such that the average time taken by the electron to travel between two sites is much less than the time spent immobilized at these sites. The concept of electron delocalization involves an electron being excited into a band of energy levels that extend over ten or more of the atoms in the material, and the electron is envisaged to travel in the form of a wave rather than as a discrete particle. Most of the electron energy levels in a solid are not in fact randomly distributed in their energy levels, but instead tend to group together into bands, according to the nature and extent of the solid's interatomic interactions. Glass, an amorphous material with no long-range regular structure, is transparent because of this very fact. The width of such energy bands in a material, their spatial extent, and the electron occupancy of the bands determine the electronic conductivity of the material. Even for the case of relatively small protein structures there will be a large number of atomic interactions. For example, a protein molecule of molecular weight 100,000 will contain in round figures about 7000 hydrogen, 4400 carbon, 1400 oxygen, 1200 nitrogen, and several tens of sulfur atoms. Confronting the theoretician is the problem of calculating the form the energy bands will take in typical protein structures, and the experimentalist has the task of demonstrating the existence or otherwise of these bands. Extensive reviews of the theoretical and experimental progress that has been made in this field of study have been given elsewhere. [33–35]

At present, it may be said that the extent to which protein structures can support delocalized electron transport is not clearly understood, although there is sufficient evidence to show that proteins, other than the cytochromes, can sustain electronic conduction processes under certain conditions. Microwave Hall effect measurements [50] for proteins of low water content have indicated that the electron mobilities have values that can be considered on the border line between a band-type conduction and one involving a hopping-type process. It is not yet known whether increasing the hydration level of the proteins will increase the mobility values. Other studies [51,52] have been directed towards investigating whether electron-accepting molecules such as methylglyoxal can react with protein structures to form charge-transfer complexes. Theoretical and experimental studies of electronic conduction effects in protein structures is a rather neglected field of study, and more attention could usefully be given to it.

## 3.3. Electronic Induction and Displacement Effects

Apart from effects associated with the translocation of electrons over macroscopic dimensions, a number of effects can also occur which involve subtle and small changes in the local distribution of electronic charge. One example of this is electronic rearrangement within some molecules that gives rise to a permanent electric dipole moment. Partly as a result of the difference in electronegativity of the oxygen and nitrogen atoms, amino acids possess a net dipole moment. In a peptide unit this dipole moment has a value of about 3.7 Debye units, and because in an alpha helix the peptide moments lie nearly parallel to the helix axis, the total moment for a helix of $N$ residues will have the value $3.6N$ Debye units (Ref. 33, pp. 42–53). This is equivalent to there being an excess of half an electronic charge at the helix C-terminus and an equal deficit at the N-terminus. The three-dimensional structure of many globular proteins is stabilized by interactions between such helix moments,[53] and the presence of an effective half-charge at the ends of helices can influence the binding of charged substrates in the active sites of enzymes,[54] and also modify the ionization constants of nearby acidic or basic side chains in the polypeptide structure.

The effect of electronegative atoms in causing electronic charge displacement in a molecule is known by chemists as an inductive effect and is, for example, particularly evident in the ways substituents can alter the strength of an acid. Consider acetic acid ($CH_3COOH$) having a p$K$ of 4.8. If electronegative chlorine is attached to form $ClCH_2COOH$, electronic charge is withdrawn from the carboxyl and has the effect of stabilizing the anionic form of the acid and thus reducing the p$K$ to 2.9, increasing the acidity by almost 100-fold. Adding another chlorine to form $Cl_2CHCOOH$ strengthens the acidity even more to given the p$K$ as 1.3. This inductive effect can act through several atoms, so that $ClCH_2CH_2CH_2COOH$ has a p$K$ of 4.5, for example. If unsaturated, hyperconjugated pi-bonds separated the chlorine and carbonyl groups rather than saturated sigma-bonds, then the inductive effect would be transmitted over a greater distance. Inductive effects can be caused by the presence of electrical charges, and a good example of this is the way lysine side groups in proteins can have their p$K$ values lowered from 10.7 to as low as 6.0 due to nearby positive charges. The basis of Ling's association–induction hypothesis[55] is that cooperative effects in proteins can arise from inductive interactions between adjacent sites in the protein structure. For example, by comparing the p$K$ values of the amino group in glycine, diglycine, triglycine, etc., Ling found that the effect of substituting a glycyl residue for an OH group extended to glycine residues further down the polypeptide chain. This was considered to demonstrate that electronic induction effects can be transmitted over long distances along the partially resonating-bond structure of a polypeptide. One effect of this would be that adjacent ionizable side chains would "sense" the state of ionization of its neighbor and also the nature of any counterions, and it was proposed that such effects could account for the specificity of alkali ion binding on cell membranes.

## 4. Proton Transport

Protons are ions—hydrogen atom ions. Why then treat protons apart from other ions? The reason is that protons are apart in almost all their physical and chemical characteristics. Consisting solely of the hydrogen atom nucleus with no orbiting electrons, the proton is smaller than the other ions by an effective factor of around $10^5$-fold. It is an elementary particle of Fermi dimension and can be considered to exhibit properties that lie on the borderline between classical physics and quantum mechanics. Such properties lead to protons having interesting chemical and transport properties.

As a consequence of their small size, the electrostatic potential at the surface of a proton is very large and so it strongly attracts electrons and molecular dipoles. Its smallness also endows it which a high chemical reactivity limited by few steric hindrances. These properties lead to such features as a proton being able to share itself between two lone electron-pair orbitals (a total of four electrons) to form hydrogen bonds, and its having an enormous heat of hydration of around 1150 kJ/mole, an abnormally high translocational mobility, and through rapid transfer reactions its providing the basis for acidic and basic reactions. Such effects imply that a free isolated proton rarely exists (Ref. 28, Chap. 5).

With its large heat of hydration and strong attraction for dipoles we can expect protons to exist in aqueous solution as an $H_3O^+$ ion, also termed an hydronium ion. An hydronium ion has an effective radius equivalent to that of a water molecular (0.14 nm) and so is roughly midway between the size of a potassium and sodium ion, and as such can be expected to have a heat of hydration lying between the values of 340 and 426 kJ/mole for $K^+$ and $Na^+$, respectively. The total heat of hydration of a proton is about three times larger than this, suggesting that a proton interacts with more than just one water molecule. The accepted concept is that the hydronium ion interacts with three water molecules to form an $H_9O_4^+$ cluster, as shown in Figure 14.

The mobility of ions in aqueous solution is well described by Stokes' law where the limiting mobility is determined by the ion's radius and the viscosity of the solution. Since an hydronium ion has a radius midway between that of a potassium and sodium ion, we might expect its mobility to lie somewhere between $5.8 \times 10^{-8}$ and $7.6 \times 10^{-8} \, m^2 \, V^{-1} \, s^{-1}$ (Ref. 33, p. 104). In fact, the mobility of protons and hydronium ions in aqueous solution are similar and of a value $(36.3 \times 10^{-8} \, m^2 \, V^{-1} \, s^{-1})$ considerably greater than that predicted by Stokes' law. The key to an understanding of this lies in the observation that proton mobilities in nonpolar liquids such as hexane or propanol are not abnormally larger. Hydrogen-bond networks therefore seem important, and in fact it is the translocation of

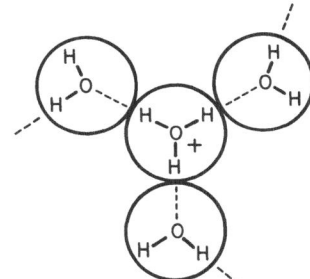

FIGURE 14. The hydration model of the proton, to show the formation of the $H_9O_4{}^+$ cluster.

protons via the hydrogen bonds between water molecules that gives the $H_3O^+$ ion such an *apparently* large mobility. An individual $H_3O^+$ ion as such does not in fact move at all, and this can be appreciated by a "gedanken experiment" using Figure 14. If the lower protons in the $H_3O^+$ ion of Figure 14 is imagined to hop (tunnel) to the bottom water molecule, this is equivalent to the $H_3O^+$ ion being translated downward with the mobility of the much smaller proton. The classical literature treatments of the atomistic mechanisms involved in the transport of protons via hydrogen-bond networks in water and ice are those presented by Conway et al.[56] and by Eigen and DeMayer.[57]

## 4.1. Water and Ice Models

Because of its biological importance and the extent to which water is involved in the functioning of living systems, the tendency has been to consider that proton transport across cell membranes occurs in essentially the same manner as proton transport in water or ice. The mechanism for this has often been interpreted in terms of the diffusion of Bjerrum defects which can arise when, as a result of the rotation of a water molecule in its crystal lattic site, an oxygen–oxygen link is left without a proton while a neighboring oxygen–oxygen link acquires an extra proton. A site that is deficient in a proton is termed and $L$ defect, and one that has acquired an extra proton is called a $D$ defect. The formation and translocation of these defects are outlined in Figure 15. The translocation of a $D$ defect represents the movement of an excess proton, while the $D$ defect could act as an acceptor site for a proton injected into the network from the surrounding medium. Proton transport in such water structures can, therefore, be considered in terms of the formation and diffusion of defects in the hydrogen-bond network.

The proton transfer processes depicted in Figure 15 could also occur if some or all of the water molecules were replaced by the hydrophilic sidechains of proteins, provided that a continuous hydrogen-bond network

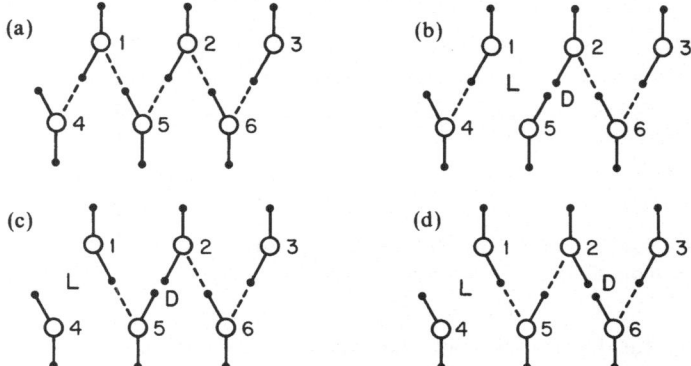

FIGURE 15. (a) Lattice of six hydrogen-bonded water molecules. (b) Rotation of water molecule 5 produces an L and a D defect. (c) Rotation of water molecule 1 results in migration of the L defect. (d) Rotation of water molecule 2 results in migration of the D defect.

were formed. For example, the O–H groups could be provided by car-boxyls of the side-chains of aspartic or glutamic acid, or by the hydroxyl groups of serine, threonine, and tyrosine. Proton donors for the hydrogen bonds could also be provided by the amino groups of arginine, asparagine, glutamine, histidine, lysine, and tryptophan, and the complete hydrogen-bond network could also include H bonds formed between adjacent polypeptide chains by the main-chain carbonyl and amino groups. In such a system the creation of mobile $L$ and $D$ defects could occur not only as a result of the rotation of O–H groups but also through the translocation of protons between the ionizable acidic and basic side-chain groups. Several such molecular approaches to the transport of protons in biological systems have been described in the literature, and one model[58] has led to interesting conclusions regarding the effects of electrochemical potentials created by pH and voltage gradients. For example, under acid conditions the rate of proton transport was found to be limited by the injection of $L$ defects into the hydrogen-bonded network and the effect of an electric field was equivalent to that expected from the application of Ohm's law for con-ventional electrical conductions. These kinds of proton-transport models have been specifically directed towards an understanding of the functioning of bacteriorhodopsin membranes, where the alpha-helical conformations of the transmembrane proteins are considered to be ideal for such conduction processes. Such concepts could, of course, also be applied to the water-filled membrane pores envisaged to control ion transport. The concept that protein structures are able to support significant proton conductivity is also supported by electrical and dielectric measurements[59,60] where the transport mechanisms are found to be sensitive to the water structure associated with the proteins. Also, the various studies that have been made of the rates at which hydrogen atoms in protein structures can exchange with surrounding aqueous media indicate that the more labile H atoms should not be thought of as fixed components of the protein structure, but rather as forming a loose "coat" of mobile positive charges around the protein. Protons have unique properties.

## 4.2. Other Model Systems

Based on the concept that protons exhibit behavior lying between that expected for classical corpuscles and quantum particles such as electrons, Freund[61] has made the interesting proposal that proton transport in some biological systems may occur as a wavelike motion in bands of delocalized and extended energy states, called proton bands. This is equivalent to transport process 2 described for electrons in Figure 12. The effective mobility of a proton in such a proton band could be greater than that arising from the defect diffusion model of Figure 15 by a factor of at least $10^6$. Evidence cited[61] for such a mechanism was obtained from electrical measurements on solid ionic hydroxides such as $Al(OH)_3$ and $Ca(OH)_2$, and although such materials may appear to have little relevance to biological structures, the point may be made that ice may also not be an appropriate model for proton transport in membrane structures.

Polysaccharides are known to be involved as structural elements in cell walls and intercellular spaces, so studies of their protonic conductivity and of how water interacts with such materials may prove of value. A step in this direction has been taken with studies of hydration-dependent proton transport in cyclodextrins. The cyclodextrins consist of six or more $\alpha(1-4)$-linked glucopyranose $(C_6H_{10}O_5)$ molecules which are closed to form roughly circular structures with a cavity in the center. They have been used as artificial enzymes. In their hydrates, water molecules form an extensive hydrogen-bond network between themselves as well as with the hydroxyl oxygen atoms of the parent cyclodextrin. Neutron diffraction studies[62] on $\beta$-cyclodextrin (cycloheptaamylose) have shown that for each molecular hydrate 19 of the possible total of 45 hydrogen-bonds are of the type O–H··H–O in which the oxygen atoms are the normal distance apart for hydrogen bonds, but the apparent H–H contacts are much shorter than the sum of the van der Waals' radii for two hydrogen atoms. The hydrogen-atom positions are found to be statistically half-filled, so that if one of the two H atoms is present in the hydrogen bond, the other one is forced to form a bond with a neighboring oxygen atom, and vice versa. The oxygen atoms act as fixed centers around which the hydrogen atoms arrange to form optimum bonding patterns. Long chains of such networks are formed and the cooperative and concerted rearrangements of the H-atom locations, together with the domino-type effect that occurs in this process, has led to the term "flip-flop hydrogen bond" being adopted.[62] Interconnected chains of such flip-flop hydrogen bonds should be able to provide pathways for rapid, and vectorial, long-range movement of excess protons.

Electrical and dielectric measurements on $\alpha$- and $\beta$-cyclodextrin[63,64] using palladium black proton-injecting electrodes have shown that when fully hydrated they exhibit protonic conductivities larger than that of ice by a factor of $10^3$, with $\beta$-cyclodextrin having a conductivity of $10^{-4}$ mho m$^{-1}$ at 294 K. (The protonic conductivity of water and ice is around $5.7 \times 10^{-6}$ and $1 \times 10^{-7}$ mho m$^{-1}$, respectively). However, this represents the macroscopic conductivity and so will be dominated by the resistive pathways between the conducting regions of extended flip-flop hydrogen bonds. Such a heterogeneous model is suggested by the presence of a dielectric dispersion whose characteristics are dependent on the degree of hydration and hence also of the extent of the hydrogen-bond networks.[64] The physical origin of such a dispersion can be understood in terms of the percolation of protons between regions of high and low conductivity. The electrical properties of heterogeneous materials can involve complicated relationships between the geometry and dielectric properties of the constituent phases (Ref. 33, Chap. 5). More work is required on the cyclodextrins before these factors can be quantified, but as a first estimate the results so far obtained[64] would be consistent with there being conducting "islands" of average conductivity $10^{-3}$ mho m$^{-1}$ and spatial extent 3 nm separated some 110 nm from each other by resistive regions where the extent of flip-flop hydrogen bonding is relatively limited. The possible existence in membrane structures of proton conduction pathways of such extent and conductivity would be of considerable importance concerning the two schemes shown in Figure 3.

## 5. Concluding Remarks

We have considered here some of the physical factors involved in ion, electron, and proton transport in membrane structures, as well as the electric fields and electrical double layers that are intimately associated with these transport processes. In considering the membrane potential, which more accurately should be termed the transmembrane potential drop, it can be seen that the presence of membrane surface charges may be significant and could produce effects not expected from a consideration of Nernstian diffussion processes alone. For example, from Eq. (3) and the values quoted in the text for the ion concentration differences across a typical cell membrane, the potassium equilibrium potential $V_K$ is around $-90$ mV, for sodium $V_{Na} \cong +52$ mV, and for chlorine $V_{Cl} \cong -8$ mV. The net electric field tending to drive these ions from the cell, according to the Nernstian concepts, would be $\varDelta\psi - V_K$, $\varDelta\psi - V_{Na}$, $\varDelta\psi - V_{Cl}$, and the actual ion fluxes would be equal to the product of these voltage differences with the particular ionic conductance value. This assumes that the trans-membrane potential is $\varDelta\psi$, and from Figure 4 it can be seen that this is only true if there are zero, or equal, surface charges on both sides of the membrane. Figure 5a shows that for a surface charge of $5 \, \mu C/cm^2$, for example, the surface potential for a $0.15 \, M$ univalent electrolyte is $-49$ mV. Relatively small differences or changes of the surface potentials of the two membrane surfaces could significantly modify the actual transmembrane potential from the value $\varDelta\psi$ given in Eq. (3) and as shown in Figure 4a. Very little attention appears to have been given to this possible complication. It may be that the newly described method[31] of probing cell surface potentials by observing the quenching of anionic free radicals at specific charge transfer sites on the cell membrane will have a useful role to play in this respect. The possibility that charge transfer or redox reactions at the membrane surfaces may influence the transmembrane ionic fluxes has also received little attention. There will be those biologists for whom the cartoon of Figure 8 will appear provocative and stupid—in no way amusing. Science should be fun, and sometimes in asking stupid questions one can be surprised by the answers.

In considering the theoretical modeling of the electrical double layer at the cell surface, it may be useful to consider the way in which the value of the relative permittivity of the electrolyte varies with distance from the membrane surface. Usually the permittivity is assumed to be constant up to the surface and to have a value equivalent to that of normal bulk water. By comparing the value obtained for the zeta potential from electrophoresis measurements with the surface potential obtained from free radical quenching studies, more detailed information concerning the structure of the electrical double layer may be forthcoming.

The properties of water associated with biological structures are also of great relevance to an increased understanding of ion transport through membrane pores, and to the processes that might influence proton transport and the magnitude of proton-motive forces across membranes. Present considerations of how electron transport and the proton ATPase complex are coupled via proton gradients pay particular attention to whether protons can be more mobile in membrane structures than in aqueous solution. It may be concluded from the details given here that the existence of pathways for the rapid transport of protons in membranes is a physical possibility. Bulk water may not be the best matrix for efficient proton transfer, and more studies of this would be useful. Also, considerably more attention could be directed towards understanding electron delocalization processes in protein structures.

The path that the biological and medical sciences has taken in reaching the stage of considering submolecular processes involving ions, electrons, and protons has been a long one. Hippocrates and Galen, and much later the Renaissance anatomists, recognized that the human body is composed of organs. In the hands of Leeuwenhoek the microscope revealed the existence of cells, and this century has witnessed amazing discoveries at the subcellular level of molecules. Slowly but surely, in the form of advances in bioelectrochemistry and bioelectronics, progress is being made at the submolecular level. Perhaps this will be the lowest level that biology will need to descend, for to probe to the subatomic level will be to tamper with energies way beyond those normally associated with living systems. But perhaps there may be surprises here too! The task will then be to work backwards, to mold all levels of understanding into one unit and in doing so approach a real understanding of the living state.[65]

## References

1. W. Hasselbach and H. Oetliker, "Energetics and Electrogenicity of the Sarcoplasmic Reticulum Calcium Pump," *Ann. Rev. Physiol.* **45**, 325–339 (1983).
2. C. Tanford, "Mechanism of Free Energy Coupling in Active Transport," *Ann. Rev. Biochem.* **52**, 379–409 (1983).
3. L. M. Amzel and P. O. Pedersen, "Proton ATPases: Structure and Mechanism," *Ann. Rev. Biochem.* **52**, 801–824 (1983).
4. P. Mitchell, "Coupling of Phosphorylation to Electron and Hydrogen Transfer by a Chemi-osmotic Type of Mechanism," *Nature* **191**, 144–148 (1961).
5. P. Mitchell, "Chemiosmotic Coupling in Oxidative and Photosynthetic Phosphorylation," *Biol. Rev.* **41**, 445–502 (1966).
6. P. Mitchell, "Compartmentation and Communication in Living Systems. Ligand Conduction: A General Catalytic Principle in Chemical, Osmotic and Chemiosmotic Reaction Systems," *Eur. J. Biochem.* **95**, 1–20 (1979).
7. S. J. Ferguson and M. C. Sorgato, "Proton Electrochemical Gradients and Energy Transduction Processes," *Ann. Rev. Biochem.* **51**, 185–217 (1982).
8. R. J. P. Williams, "Possible Functions of Chains of Catalysts," *J. Theor. Biol.* **1**, 1–13 (1961).
9. R. J. P. Williams, "Proton-Driven Phosphorylation Reactions in Mitochondrial and Chloroplast Membranes," *FEBS Lett.* **53**, 123–125 (1975).
10. R. J. P. Williams, "The Multifarious Couplings of Energy Transduction," *Biochim. Biophys. Acta.* **505**, 1–44 (1978).
11. D. B. Kell, "On the Functional Proton Current Pathway of Electron Transport Phosphorylation. An Electrodic View," *Biochim. Biophys. Acta.* **549**, 55–99 (1979).
12. D. B. Kell, D. J. Clarke, and J. G. Morris, "On Proton-Coupled Information Transfer

along the Surface of Biological Membranes and the Mode of Action of Certain Colicins," *FEMS Microbiol. Lett.* **11**, 1–11 (1981).

13. D. B. Kell and G. D. Hitchens, "Proton-Coupled Energy Transduction by Biological Membranes. Principles, Pathways and Praxis," *Faraday Discuss. Chem. Soc.* **74**, 377–388 (1982).

14. E. Elul, "Fixed Charge in the Cell Membrane," *J. Physiol.* **189**, 351–365 (1967).

15. O. Aono and S. Ohki, "Origin of Resting Potential of Axon Membrane," *J. Theor. Biol.* **37**, 273–282 (1972).

16. D. L. Gilbert and G. Ehrenstein, "Effect of Divalent Cations on Potassium Conductance of Squid Axons: Determination of Surface Charge," *Biophys. J.* **9**, 447–463 (1969).

17. E. Rojas and I. Atwater, "An Experimental Approach to Determine Membrane Charges in Squid Giant Axon," *J. Gen. Physiol.* **51**, 131s–145s (1968).

18. J. R. Segal, "Surface Charge of Giant Axons of Squid and Lobster," *Biophys. J.* **8**, 470–489 (1968).

19. S. Ohki, "Membrane Potential of Squid Axons: Comparison between the Goldman–Hodgkin–Katz Equation and the Diffusion/Surface Potential Equation," in *Charge and Field Effects in Biosystems* (M. J. Allen and P. N. R. Usherwood, eds.), pp. 147–156, Abacus Press, Tunbridge Wells (1984).

20. A. Roos and W. F. Boran, "Intracellular pH," *Physiol. Rev.* **61**, 296–434 (1981).

21. J. Rothman and J. Lenard, "Membrane Asymmetry," *Science* **195**, 743–753 (1977).

22. J. Barber, "Influence of Surface Charges on Thylakoid Structure and Function," *Ann. Rev. Plant Physiol.* **33**, 261–295 (1982).

23. A. L. Hodgkin and B. Katz, "The Effect of Sodium Ions on the Electrical Activity of the Giant Axon of the Squid," *J. Physiol.* **108**, 37–77 (1949).

24. T. Teorell, "Transport Processes and Electrical Phenomena in Ionic Membranes," *Progr. Biophys.* **3**, 305–369 (1953).

25. S. Ohki, "Membrane Potential of Phospholipid Bilayer and Biological Membranes," *Progr. Surf. Membrane Sci.* **10**, 117–252 (1976).

26. D. C. Grahame, "The Electrical Double Layer and the Theory of Electrocapillarity," *Chem. Rev.* **41**, 441–501 (1947).

27. R. H. Brown, "Membrane Surface Charge: Discrete and Uniform Modelling," *Progr. Biophys. Molec. Biol.* **28**, 343–370 (1974).

28. J. O'M. Bockris and A. K. N. Reddy, *Modern Electrochemistry*, Chap. 7, Plenum Press, New York (1970).

29. R. A. Robinson and R. M. Stokes, *Electrolyte Solutions*, Butterworth, London (1970).

30. C. T. Yerkes and G. T. Babcock, "Surface Charge Asymmetry and a Specific Calcium Ions Effect in Chloroplast Photosystem II," *Biochim. Biophys. Acta* **634**, 19–29 (1981).

31. R. Pethig, P. R. C. Gascoyne, J. A. McLaughlin, and A. Szent-Györgyi, "Interaction of the 2,6-dimethoxysemiquinone and Ascorbyl Free Radicals with Ehrlich Ascites Cells: A Probe of Cell Surface Charge," *Proc. Natl. Acad. Sci. USA* **81**, 2088–2091 (1984).

32. M. A. Habib and J. O'M. Bockris, "Interpretation of Current–Potential Relationships across Biological Membranes," *J. Bioelectricity* **1**, 289–294 (1982).

33. R. Pethig, *Dielectric and Electronic Properties of Biological Materials*, Wiley, Chichester (1979).

34. R. Pethig, "Electronic Conduction in Biopolymers," Electronic Conduction and Mechanoelectrical Transduction in Biological Materials (B. Lipinski, ed.), Marcel Dekker, New York (1982), pp. 1–98.

35. R. Pethig, "Biological Polymers," in *Noncrystalline Semiconductors* (M. Pollak, ed.), CRC Press, Boca Raton (in press).

36. M. A. Habib and J. O'M. Bockris, Chapter 3, this volume.

37. J. B. Hasted, "Liquid Water: Dielectric Properties," in *Water: A Comprehensive Treatise* (F. Franks, ed.), Vol. 1, pp. 255–309, Plenum, New York (1972).
38. S. Bone and R. Pethig, "Dielectric Studies of the Binding of Water to Lysozyme," *J. Mol. Biol.* **157**, 571–575 (1982).
39. T. E. Cross and R. Pethig, "Microwave Studies of the Interaction of DNA and Water in the Temperature Range 90–300 K," *Int. J. Quantum Chem: Quantum Biol. Symp.* **10**, 143–152 (1983).
40. V. A. Parsegian, "Energy of an Ion Crossing a Low Dielectric Membrane: Solutions to Four Relevant Electrostatic Problems," *Nature (London)* **221**, 844–846 (1969).
41. D. T. Edmonds, "Membrane Ion Channels and Ionic Hydration Energies," *Proc. R. Soc. London Ser. B* **211**, 51–62 (1980).
42. P. C. Jordan, "Electrostatic Modeling of Ion Pores," *Biophys. J.* **39**, 157–164 (1982); **41**, 189–195 (1983).
43. H. T. Witt, "Energy Conversion in the Functional Membrane of Photosynthesis. Analysis by Light Pulse and Electric Pulse Methods. The Central Role of the Electric Field," *Biochim. Biophys. Acta* **505**, 355–427 (1979).
44. W. N. Konigs and J. Boonstra, "Anaerobic Electron Transfer and Active Transport in Bacteria," *Curr. Topics Membr. Transp.* **9**, 177–231 (1977).
45. M. Redi and J. J. Hopfield, "Theory of Thermal and Photoassisted Electron Tunneling," *J. Chem. Phys.* **72**, 6651–6660 (1980).
46. D. DeVault, "Quantum Mechanical Tunnelling in Biological Systems," *Quart. Rev. Biophys.* **13**, 387–564 (1980).
47. T. L. Poulos and J. Kraut, "A Hypothetical Model of the Cytochrome *c* Peroxidase. Cytochrome *c* Electron Transfer Complex," *J. Biol. Chem.* **255**, 10322–10330 (1980).
48. E. G. Petrov, "Role of Polypeptide Chain Structure in Donor–Acceptor Electron Transfer through Proteins," *Studia Biophysica* **93**, 237–240 (1983).
49. F. M. Richards, "Areas, Volume, Packing, and Protein Structure," *Ann. Rev. Biophys. Bioeng.* **6**, 151–176 (1977).
50. T. E. Cross and R. Pethig, "Microwave Hall Effect Measurements on Biopolymers," *Int. J. Quantum Chem.: Quantum Biol. Symp.* **7**, 389–395 (1980).
51. R. Pethig and A. Szent-Györgyi, "Electronic Properties of the Casein–Methylglyoxal Complex," *Proc. Natl. Acad. Sci. USA* **74**, 226–228 (1977)
52. S. Bone and R. Pethig, "Dielectric Properties of Protein–Methylglyoxal Adducts: Interfacial and Bulk Effects ," *J. Chem. Soc., Faraday Trans. I* **78**, 1785–1794 (1982).
53. H. A. Scheraga, K. C. Chou, and G. Nemethy, "Interactions between the Fundamental Structures of Polypeptide Chains," in *Conformation in Biology* (R. Srinivasan and R. H. Sarma, eds.), pp. 1–10, Adenine Press, New York (1982).
54. J. Warwicker and H. C. Watson, "Calculations of the Electrical Potential in the Active Site Cleft Due to α-Helix Dipoles," *J. Mol. Biol.* **157**, 671–679 (1982).
55. G. N. Ling, *A Physical Theory of the Living State: The Association–Induction Hypothesis*, Blaisdell, Waltham, Massachusets (1962).
56. B. E. Conway, J. O'M. Bockris, and H. Linton, "Proton Conductance and the Existence of the $H_3O^{\cdot}$ Ion," *J. Chem. Phys.* **24**, 834–850 (1956).
57. M. Eigen and L. De Maeyer, "Self-dissociation and Protonic Charge Transport in Water and Ice," *Proc. R. Soc. London Ser. A* **247**, 505–533 (1958).
58. E. W. Knapp, K. Schulten, and Z. Schulten, "Proton Conduction in Linear Hydrogen-bonded Systems," *Chem. Phys.* **46**, 215–229 (1980).
59. P. R. C. Gascoyne, R. Pethig, and A. Szent-Györgyi, "Water Structure Dependent Charge Transport in Proteins," *Proc. Natl. Acad. Sci. USA* **78**, 261–265 (1981).
60. J. Behi, S. Bone, H. Morgan, and R. Pethig, "Effect of Deuterium–Hydrogen Exchange on

the Electrical Conduction in Lysozyme," *Int. J. Quantum Chem: Quantum Biol. Symp.* **9**, 367–374 (1982).

61. F. Freund, "Proton Highlife and Midway Tunneling," *Trends Biochem. Sci.* **6**, 142–145 (1981).

62. W. Saenger, Ch. Betzel, B. Hingerty, and G. M. Brown, "Flip-Flop Hydrogen Bonding in a Partially Disordered System," *Nature (London)* **296**, 581–583 (1982).

63. S. Bone and R. Pethig, "Cyclodextrins as Model Systems for the Study of Proton Transport," *Int. J. Quantum Chem: Quantum Biol. Symp.* **10**, 133–141 (1983).

64. J. Behi, S. Bone, H. Morgan, and R. Pethig, "Protonic Charge Transport Studies in Cyclodextrins," in *Charge and Field Effects in Biosystems* (M. J. Allen and P. N. R. Usherwood, eds.), pp. 139–146, Abacus Press, Tunbridge Wells (1984).

65. A. Szent-Györgyi, *Introduction to a Submolecular Biology*, pp. 135, Academic Press, New York (1960).

# 8

# Coherent Excitation in Active Biological Systems

## H. Fröhlich

*ABSTRACT:* The principal types of coherent excitations: metastable state with high electric dipole moment, and coherent high-frequency electric vibrations are derived from simple models. Far-reaching consequences for biological systems are discussed. The conjecture is supported by a great variety of experiments.

## 1. General

From the point of view of atomic and molecular physics (and chemistry), biological materials are extremely complex and complicated systems. Yet they function in a most systematic way. Molecular biology succeeded in establishing the structure of the relevant giant molecules. Thus, e.g., it was found that enzymes possess an active site where the relevant processes can take place according to the laws of chemistry. Yet, this fails to explain their enormous catalytic power. Clearly, function must be treated in terms of dynamic, rather than static, properties. As a next step one may be tempted to introduce the known interaction between the atoms and ions of the system and, based on the known structure, calculate their dynamical properties. Such a procedure would be relevant in the range of linear displacements from a rigid structure like a solid. Enzymes, however, do not form such a rigid structure. Attempts to carry out the appropriate calculation have been discussed recently.[1] Even for small enzymes they

*H. Fröhlich* • Department of Physics, Oliver Lodge Laboratory, Oxford Street, P.O. Box 147, Liverpool L69 3BX, England.

require quite considerable computer time. Such endeavors, thus, while useful as a demonstration of the importance of dynamic properties, will not result in a general theory of the relevant processes, if such a theory exists.

One might then try an approach that emphasizes the common features of various systems—enzymes in the above example—and treat them in a general way thus creating a framework into which the individual cases will fit. To take an example from pure physics, consider the dynamic properties of fluids. Here we do not proceed by first considering the structure of a particular fluid, a most formidable task, and then, based on this knowledge, investigate the dynamic properties. If this were possible then we would obtain the motion of individual atoms or molecules which constitute the fluid, a result which is not very interesting. What is interesting is the space-time dependence of the mean density, $\rho(x, t)$ and of the mean current density, $j(x, t)$. Both are connected by a set of differential equations, the Navier–Stokes equations, from which their space-time dependence can be derived in terms of certain parameters like compressibility, frictional constants, which define the particular fluid. These Navier–Stokes equations can be derived exactly from the basic multiparticle equation, the Schrödinger equation,[2] not by solving it in terms of the multiparticle coordinates, but rather by first expressing $\rho$ and $j$ in terms of individual particle properties, and then finding the differential equations satisfied by them, which essentially arise as symmetry properties of the multiparticle system.

The success of this procedure rests in the definition of the two multiparticle quantities $\rho$ and $j$, which describe the essential properties of hydrodynamic flow. Derivation of the actual values of the various parameters entering the equations of flow would require, of course, deeper investigation of the basic multiparticle equation. Could we not apply a corresponding method in the treatment of the dynamic properties of biological systems? The answer is no, not for the time being, for we do not yet know the quantities that would take a role corresponding to $\rho$ and $j$ in the previous case, though later in this chapter we shall make proposals in this respect and discuss their experimental verification. The difficulty arises from the enormous number of states of which excited multiparticle systems, like the biological ones, are capable and which make a systematic investigation and inspection impossible. Thus the lowest energy state of a system of $N$ particles, the ground state, usually—though not always—is a crystal. Small displacements of the particles from their equilibrium positions can be treated in a systematic way as linear excitations. In biological systems, however, we are very far from these states, in the realm of nonlinear excitation. The number of states then rises exponentially with the number $N$ of particles—in some particular case say as $10^N$. Quite clearly it is impossible, even with a supercomputer, to evaluate and investigate all these states. Hence we must find a way of deciding on "interesting" properties, usually connected with some kind of order. This order need not, however, be of a spatial nature. In a machine, for instance, this order consists in its way of functioning. Normally we treat this with the laws of macrophysics, which, in turn, follow from the laws of microphysics.

Early microbiologists were tempted to consider spatial order as the only order existing in physics. Thus, after no spatial order had been found in the arrangement of amino acids in an enzyme, Monod[3] says that knowing the arrangement of 199 amino acids would not permit us to predict the 200th. He concludes that the arrangements are random. Actually, however, the whole chain must function as an enzyme; replacing the "correct" 200th amino acid by a "wrong one" disturbs this function.

In looking to physics for guidance to find a concept capable of describing biological "order" we are led to the concept of coherence, which applies to cases as different as superconductors, superfluids, lasers, and masers. Roughly speaking it implies that if a certain property is known at a space-time region near $(x, t)$ it is also determined at another, $(x', t')$. Thus in a coherent wave, for instance, amplitude and phase at $(x, t)$ determine their values at another, $(x', t')$. Considering now biological systems, periodic processes are very common among them with periods ranging from that of life cycles to very high-frequency electric vibrations. In contrast to most physical periodic processes, periods of biological processes often vary from case to case. Thus the periods of the life cycles of cells grown under exactly the same conditions vary within certain limits, i.e., each cell exhibits a certain amount of personality. Also, if an apparent cause for a certain biological process is removed, another takes over because biological processes are frequently multicausal.

We shall presently describe the derivation of certain coherent excitations on the basis of definite physical models. Such developments show that random energy supplied to a certain system need not lead to heating but may result in the excitation of ordered (coherent) states. This need not imply that other models could not lead to similar excitations, i.e., that a multicausal situation exists—in other words that experimental verification of the excitation need not necessarily be considered as proof of a particular model. Clearly a situation then arises which requires close collaboration between theory and experiment. Thus, e.g., theory has predicted certain coherent excitations and experiment verifies their existance. This presents a development from the point of view of physics. From the point of view of biology, however, we may ask a question that is prohibited in physics: what is the purpose of these excitations? Evidence will be presented later in this chapter for the first (physical) stage, but the second, biological question has hardly been touched yet. Its solution will, of course, lead back again to physics, i.e., a certain process has certain consequences.

If these consequences are vital biologically then multicausal possibilities must be considered again, i.e., what other processes could have the same consequences.

## 2. Coherent Excitations

In looking for a physical characteristic common to most biological systems, one is at once led to their dielectric properties.[4] Most startling among these is the high electric field of about $10^5$ V/cm present in biological membranes. This field acts in the range of nonlinear response. Materials inside membranes are, therefore, strongly, nonlinearly polarized. Furthermore, the many ions contained in biomolecules make an essential contribution to their stabilization through their interaction[5] as well as through their interaction with the surrounding medium, e.g., water.[6] On the other end biological systems are capable of reacting to extremely weak external impulses as reported in Section II of Ref. 7. Hence, by appropriate arrangements and excitations, biological systems may form most sensitive instruments that in normal technology would be devised in completely different ways with very different materials.

As a first case of coherent excitation, a metastable state with very high electric polarization (ferroelectric state) will be shown to exist in materials with high polarizability, combined with elastic softness, as discussed in Section III B of Ref. 7, and Ref. 8. Such excitation is coherent as the polarization at a point $x$ determines the value at another, $x'$. The metastability of this excitation arises from the shape dependence of the electrostatic energy of a material. Thus, e.g., the ratio of the energies of a slab and that of a sphere both having equal volume and polarization is $3\varepsilon/(\varepsilon + 2)$, where $\varepsilon$ is the dielectric constant (cf. p. 168 of Ref. 9). To the electrostatic energy must be added the elastic energy arising from a deformation of the material.

Thus consider a sphere with polarization $P$ and electrostatic energy $\frac{1}{2}\gamma P^2$. If deformed into an ellipsoid with excentricity $\eta$, this energy is changed by terms proportional to $\eta$, and higher powers, as the electrostatic energy does not have an *extremum* for the sphere, $\eta = 0$. The elastic energy arising from the displacement will, however, have no linear term in $\eta$ but be proportional to $\eta^2$, as elastically the sphere is the equilibrium state. The total energy $U$, therefore, for small $\eta$ will have the form

$$U = \tfrac{1}{2}\gamma P^2(1 + c_1\eta + \tfrac{1}{2}c_2\eta^2) + \tfrac{1}{4}C\eta^2 \tag{1}$$

where $C > 0$ and $\gamma > 0$, and the $C_2$ term has been inserted as it might influence the $C$ term. Minimizing with regard to $\eta$ yields

$$\eta = -c_1\gamma P^2/(C + c_2\gamma P^2) \tag{2}$$

and hence

$$U = \tfrac{1}{2}\gamma P^2 - (c_1 \tfrac{1}{2}\gamma P^2)^2/(C + c_2\gamma P^2) \tag{3}$$

Thus $U$ rises more slowly than it would without the deformation ($\eta = 0$), and it reaches a maximum beyond which it decreases with increasing $P^2$, provided $c_2 > 0$, and $2c_2 < c_1^2$. For still larger $P^2$, higher-order $P^2$ terms will be relevant, so that $U$ goes through a minimum. Should that be at negative $U$, then this represents the ground state and the material is ferroelectric. Otherwise this minimum represents a metastable state with very high polarization, a quasi-ferro-electric state.

On a more formal basis we may consider a homogeneous polarization field $P$, energy density $\tfrac{1}{2}\omega^2 P^2$, interaction with an elastic field with potential energy density $\tfrac{1}{2}s^2\rho^2$ for longitudinal elastic displacements yielding a density $\rho$. Their interaction must be proportional to $\rho P^2$, to provide the effect discussed above.

The total potential energy has then the form

$$W = \tfrac{1}{2}\omega^2 P^2 + (\tfrac{1}{2}s^2\rho^2 + cP^2\rho)(1 - d^2 P^2) \tag{4}$$

where $c$ and $d$ are constants and the $d^2 P^2$ term has been included to take care of nonlinearity required to stabilize the material. Minimizing with regard to $\rho$ yields

$$\rho = -\frac{c}{s^2} P^2 \tag{5}$$

and hence

$$W = \frac{1}{2}\omega^2 P^2 - \frac{1}{2}\frac{c^2}{s^2}P^4 + \frac{1}{2}\frac{c^2 d^2}{s^2}P^6 \tag{6}$$

Thus this potential energy increases proportional to $P^2$ for small $P^2$, and proportional to $P^6$ for large $P^2$. In between it has a maximum, followed by a minimum provided a parameter $\lambda$ satisfies

$$\lambda = 3s^2 d^2 \omega^2/c^2 < 1 \tag{6a}$$

and this minimum represents a metastable state when $\tfrac{3}{4} < \lambda < 1$, while for $\lambda < \tfrac{3}{4}$ this highly polar state is the ground state. This confirms our previous results, and (6a), moreover, indicates the conditions the material must satisfy. They are, in particular, elastic softness (small $s$) and high polarizability (small $\omega$) in conjunction with large coupling between the fields (large $c^2$).

Existence of highly polar metastable states should thus be a general property of biological materials. One would expect this to hold in less simplified models, introducing particular structures together with anisotropy. One would expect then to find more than one metastable state. A small generalization of the above is possible, in fact, if the various parameters $s$, $\omega$, $c$ slowly depend on space coordinates, provided derivatives of $P$ and $s$ need not be introduced into the expression for the potential energy.

Most generally, of course, small excitations in materials with definite molecular structure are described through frequency–wave number bands, which can be classified as polar or nonpolar. Introduction of nonlinear features, as required here, must then be investigated for various particular cases, though our above treatment should always be correct as a limiting case. From the structural point of view, metastable excited states would then be classified as arising from conformational changes.

A candidate frequently mentioned in this respect is the displacement of $H^+$ in an H bond (e.g., Refs. 10–12 and related papers), for which also evidence from an analysis of dielectric properties in the millimeter region is available.[13,14] Displacement of various $H^+$ ions does, of course, yield electric polarization, though this need not be homogeneous. It is important, however, to realize that this polarization necessarily is connected with electric polarization of the rest of these molecules, which in turn leads to elastic displacements in it, as follows from our general model. Within this frame, polarization of the H bond represents a subsystem which enhances the stability of the metastable state. Detailed calculations have not been carried out yet.

So far we have considered the potential energy $W$ of our system only, and found it to have a minimum at $P = 0$ and at $P = \pm P_0$ if $P_0^2$ is the value of $P^2$ at which the metastable state occurs. The system can of course oscillate harmonically if the displacement from $P = 0$ or $P = \pm P_0$ is small. More generally, however, the system will carry out anharmonic oscillations. In fact it is shown in Ref. 8 that to a good approximation $P$ satisfies the equation of motion

$$\ddot{P} + \frac{\partial V(P)}{\partial P} = 0 \tag{7}$$

where

$$V = A + \tfrac{1}{2}BP^2 + CP^4 \tag{8}$$

i.e., as long as $P^4$ can be neglected, the oscillations are harmonic. Yet even then the system possesses highly nonlinear properties as $A$, $B$, $C$ are not constant but depend on the time average $p$ of $P^2$,

$$p = \overline{P^2} \tag{9}$$

by

$$A = \frac{1}{2}\frac{c^2}{s^2}p, \qquad B = \omega^2 - 2\frac{c^2}{s^2}p - \frac{d^2c^2}{s^2}p^2, \qquad C = \frac{c^2d^2}{s^2}p \qquad (10)$$

Thus each amplitude of $P$ defines its own potential $V$. Clearly for small amplitudes when $p$ can be neglected in $B$, and the $P^4$ terms are also negligible, the frequency is $\omega$. For growing amplitude, but when $P^4$ is still negligible, the frequency of oscillations is softened, i.e., $B$ becomes smaller until finally it changes sign; $V$ then has a maximum at $P = 0$ and a minimum further outside.

Near the metastable state, $P$ may, of course carry out small, harmonic oscillations again, but with a frequency different from $\omega$.

The coherent excitation considered so far is essentially static; its establishment requires a supply of energy larger than the energy of the metastable state above the round state $P = 0$. A dynamic, very different type of coherent excitation may exist whose maintenance requires a continuous supply of energy at a rate larger than a critical $s_0$. The model is again based on polar oscillations of a large biological system, forming a band of modes with frequencies $\omega_1$, $\omega_2$,..., $\omega_z$. Metabolic energy is considered to be supplied at a rate $s_j$ to mode $\omega_j$. This systems of polar modes is considered to interact with a heat bath which is kept at a constant temperature $T$. In cells, this will largely consist of cell water with its ions. Linear interaction will then impose a Planck distribution for the excitation of the polar modes. This interaction is described in terms of emission and absorption of single quanta $h\omega_j$ between the modes and the heat bath. When nonlinear interactions are taken into account, however, then energy exchange in terms of two quanta will occur. Most important among these are simultaneous emissions and absorption of quanta $h\omega_j$ and $h\omega_l$. If $n_j$ represents the number of quanta by which mode $\omega_j$ is excited, then the following equation arises for the rate of change, $n_j$ (cf. Section III C of Ref. 7):

$$\dot{n}_j = s_j - \phi_j(n_j e^{\beta\omega_j} - (1 + n_j))$$

$$- \sum_l \chi_{jl}[n_j(1 + n_j) e^{\beta\omega_j} - n_l(1 + nj) e^{\beta\omega_l}] \qquad (11)$$

where $\beta = h/kT$, and the $\phi$ and $\chi$ are transition probabilities. The ratio of absorption and emission terms, proportional to $n$ and $1 + n$, respectively, follows from the principle of detailed balance. Note that after further summation over $j$, the $\chi$ terms vanish so that in the stationary case when $\dot{n}_j$ vanishes, the total number $\sum n_j$ of quanta that arises from the total rate of supply $\sum s_j$ is determined by the first-order terms only through the $\phi$. The actual distribution will, however, be largerly determined by the $\chi$ terms, provided that they are relatively large. In the stationary case then they must vanish so that

$$n_j = \frac{1}{\exp[\beta(\omega_j - \mu)] - 1} \tag{12}$$

follows where $\mu$ must be determined from the already known total number of quanta $\sum n_j$. When this number exceeds a critical one, i.e., when $\sum s_j$ exceeds $s_0$, then $\mu$ must approach the lowest $\omega_1$, yielding a large $n_1$, as in Bose condensation. This frequency is then very strongly, i.e., coherently, excited. Detailed considerations[7] show that in certain circumstances other frequencies might be strongly excited.

To understand the possibility of this type of coherent excitation, the strong interaction with the heat bath must be emphasized. It endeavors to impose its temperature on the distribution which is demonstrated by the form (12), which represents a Bose distribution at temperature $T$. But while in a Bose gas the number of particles is fixed so that the Einstein condensation arises only when the temperature is sufficiently lowered, in our case the temperature is fixed, but the number of particles (quanta) increases with increasing rate of energy supply.

It must also be emphasized that this type of coherent excitation arises from the strong nonlinear coupling with the heat bath ($\chi$ terms). Other terms could be introduced, in principle, which would counteract this effect. Thus a very strong nonlinear interaction among the quanta of the polar modes, without the use of the heat bath, would simply tend to increase the temperature. We thus have found a particular example of the two different ways energy supply to a system may act, resulting in either an increase in temperature, or in the creation of a new order, that may be dynamic. Our considerations have emphasized the second possibility. Experiment must decide whether biological systems make use of it.

Finally, in this section, possible frequencies for the excitation of coherent vibrations should be considered. Originally special reference was made to membrane oscillations as their high polarization will then yield coherent electric oscillations (cf. Ref. 7, with earlier literature). Since the thickness of a membrane is about $10^{-6}$ cm and its elastic constant is equivalent to a velocity of sound of about $10^5$ cm/s, a frequency of the order of $10^{11}$ Hz was expected, corresponding to electromagnetic waves in the millimeter region. When based on proteins or DNA, however, both higher and lower frequencies may be expected.

It must be emphasized at this point that most of the polar modes of a system interact only weakly with electromagnetic waves of the same frequency, as usually the wave lengths do not match. Thus, e.g., excitation of sections of the outer membranes of a cell may result in a very high multipole electric oscillation, and hence in small interactions with electromagnetic waves of the same frequency.

## 3. Consequences

The two coherent excitations, metastable highly polar states, and coherent electric vibrations, may have far-reaching consequences arising from the forces exerted by the respective fields which are absent in the non-excited states. The electric field in the metastable state is static; it thus acts inside and near the excited system, while at larger distances it is screened by the ions in the surrounding medium. For molecules, e.g., proteins, surrounded by cell water, the screening distance naturally is much shorter than for proteins dissolved in membranes. In the case of excitation of coherent vibrations, on the other hand, screening is negligible once the frequency exceeds a critical one.

Forces arising from the excitation of the metastable state may then be employed to reduce activation energies in enzyme processes.[15,4] The system enzyme + substrate will change with the motion of ions in it and, hopefully, the direction and magnitude of the internal field will change as required by the process. The whole system, enzyme + substrate, while active, possesses a high electric dipole moment which, in principle, could be measured by standard dielectric methods. Such measurements have not been attempted yet; they face difficulties arising from the presence of unattached ions in the medium in which the enzyme reaction occurs.

For a protein dissolved in a membrane the possibility arises that it may act as a switch element. For if $\mu$ is the dipole moment in the metastable state, and $F$ is the electric field in the membrane, then if the directions of $\mu$ and $F$ are appropriate the interaction lowers the energy by $(\mu - \mu_0)F$, where $\mu_0$ is the dipole moment in the ground state. If this energy equals the energy of excitation of the metastable state, then the system is degenerate with two equal energy states. They then can be switched by the addition of relatively weak fields.

It should also be noticed that in proteins, aligned in a membrane, the excitation energy may be transferred from a particular excited protein to its neighbor, thus providing transport of energy that is nearly loss free.

Solitons, furthermore, represent a particular case of excitation to a metastable state.[16,17] In the Davydov soliton, in particular, stretching of the $C = 0$ bond forms the polar excitation, stabilized by the surroundings which play the part of the elastic field. This soliton, too, provides a loss-free transport of energy.

Excitation of a metastable state requires a certain minimum amount of energy. Excitation of the coherent polar mode, on the other hand, requires supply of a minimum rate of energy, $s_0$; its maintenance thus requires biological pumping. Its excitation results in a long-range, frequency-selective interaction that may lead to attraction between two, or many molecules, provided they oscillate with equal frequency. In particular cases this may also result in selective repulsion. In the case of the enzyme–substrate system, in the first stage the (frequency) correct substrate is attracted to the enzyme (long-range selectivity); in the second stage the usual short-range chemical processes take place—an example of a multicausal process.

As indicated in connection with Eq. (11), the energy supply $s_j$ may occur at any, or all, frequencies of the polar modes. The subsequent excitation of a particular mode, e.g., $\omega_1$, then requires some time even when the total rate of supply exceeds $s_0$. The system thus possesses storing ability. One would expect, however, that this time is particularly short if the energy is supplied at the frequency of the mode that will be excited coherently. Furthermore, arising from the considerations of mode softening, this mode, when excited, may be detached from the band and hence provide a very frequency-sensitive target for further energy supply. Detailed calculations on this question have not been carried out yet.

Consider now biological processes that make use of coherent excitation of a particular polar mode, requiring pumping at a rate $\sum s_j > s_0$. Let (cf. Section IV C of Ref. 7) $t_1$ be the time required for this process; then, we expect it to be the faster the larger $\sum s_j$; but it will never occur ($t_1 = \infty$) as long as $\sum s_j < s_0$. Thus if $a$ and $n$ are positive numbers we may assume

$$t_1 = a \Big/ \left( \sum s_j - s_0 \right)^n, \qquad \sum s_j \geqslant s \tag{13}$$

If we now consider a complex biological process consisting of many steps, only one of which depends on the excitation, i.e., on $s_j$, then if $t_2$ is the time required for all these $s_j$-independent processes, the total time for the process is

$$t = t_1 + t_2 \tag{14}$$

i.e., the total rate is

$$\frac{1}{t} = \frac{(\sum s_j - s_0)^n}{\alpha + t_2(\sum s_j - s_0)^n}, \qquad \sum s_j > s_0 \tag{15}$$

It thus depends in a steplike manner on the rate of supply $\sum s_j$.

We must assume now that the biological system is controlled in such a manner that the rate of energy supply is switched on at a period that integrates the particular process considered here with the whole biological development. External supply of energy, e.g., by microwaves at a rate say $s_e$, will then influence the rate $1/t$ and hence interfere with the general development. The effect of even a small outside $s_e$ can be very big at a stage when the biological $\sum s_j$ is close to $s_0$, because replacing $\sum s_j$ by $\sum s_j + s_e$ in Eq. (15) is equivalent to replacing $s_0$ by $s_0 - s_e$, i.e., it shifts the steplike curve Eq. (15). At particular times in the biological development, small $s_e$ hence may cause large effects, in particular when $s_e$ is supplied at the "correct" frequency, which involves little time for the reaching of the required quasistationary distribution of the excitation of modes.

In considering now candidates that may act as carriers for the electric vibrations arising in the coherent excitation of a particular mode, recent results on the structure of cells must be presented. It has been found[18] that an intricate network of protein structures known as the cytoskeleton exists, among which the so-called microtrabecular lattice extends from the cell surface throughout the cytoplasm. It has also been proposed[18] that enzymes do not float in the cytoplasm but are attached to this lattice, which may act as carrier of the coherent electric vibrations that provide frequency-selective interaction between appropriately excited molecules. Supporting this conjecture is the conclusion that "microtubules $\cdots$ form a vectorial framework upon which some other force generating component is organized."[19] This also is supported by theoretical investigations on self-focusing.[20] To avoid misunderstandings it should be repeated again that while the relevant frequencies might be in the range of electromagnetic millimeter waves, interaction with such radiation will be weak as the wavelengths will not be in the millimeter region, but probably will be less than the cross section of the tubules, possibly of the order of molecular distances.

It will now be shown that combination of our two coherent excitations may result in low-frequency periodic enzyme reactions giving rise to corresponding electric vibrations—another coherent electric vibration (Ref. 21 and Section IV E of Ref. 7). Consider an enzyme process consisting of the interaction of a substrate and an excited enzyme and assume the energy thus made available to excite a nonexcited enzyme. Then if $N$, $Z$, $S$ are the respective numbers of excited enzymes, nonexcited enzymes, and substrates, the rate of change of $N$ is proportional to $NZS$ with a rate constant $\alpha$. Also if there exists a spontaneous decay of excitation of an enzyme, rate constant $\beta$, then

$$\frac{dN}{dt} = \alpha NZS - \beta N \tag{16}$$

PROBABILITY

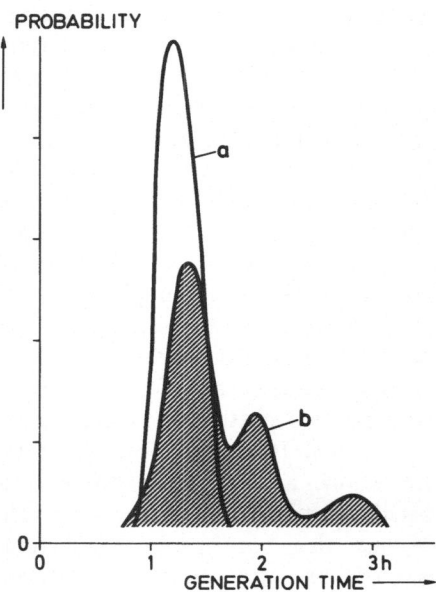

FIGURE 1. Probability of time of cell cycle when irradiated (a) with 83800 MHz or without irradiation, (b) with 83835 MHz (W. Grundler, personal communication).

GENERATION TIME ⟶

and the number of substrates decreases at the rate $\alpha NZS$. Consider now that during this process substrates are coherently excited to a frequency that attracts other substrates to the region (rate constant $\gamma$); then

$$\frac{dS}{dt} = -\alpha NZS + \gamma S \tag{17}$$

Consider the case that $Z$ is so large that it can be treated as time independent; then time-independent solutions $N_0$, $S_0$ exist,

$$N_0 = \gamma/\alpha Z, \qquad S_0 = \beta/\alpha Z \tag{18}$$

Hence if

$$N = N_0 + v, \qquad S = S_0 + \sigma \tag{19}$$

then

$$\dot{v} = \gamma\sigma + \alpha Z\sigma v, \qquad \dot{\sigma} = -\beta v - \alpha Z\sigma v \tag{20}$$

These are the well-known Lotka–Volterra equations that have periodic solutions. In the limit that $\sigma$ and $v$ are small so that the $\sigma v$ term can be neglected, $(\beta\gamma)^{1/2}$ represents the frequency of oscillation, which thus is determined by the rate $\gamma$ of spontaneous decay of enzyme excitation, and by the rate $\beta$ of attraction of substrate molecules. According to our first type of coherent excitation, the excited enzyme carries a large electric dipole moment, which thus also will oscillate with the frequency $(\beta\gamma)^{1/2}$. Outside electric observation will normally integrate the contribution of all enzymes and thus will also depend on their spatial disposition.

This scheme was originally derived for brain waves, and extended by taking into account a possible short-range interaction between excited enzymes, which then tends to organize them into a ferroelectric system. This results in an extension of Eqs. (19) and (20) into a complicated set of equations whose solutions are limit cycles; this means that the system will then tend to oscillate with a fixed amplitude.

This case has been investigated in detail[22] and it has been shown that a weak, appropriate, external perturbation may cause the collapse of the limit cycle and hence lead to the liberation of the considerable energy stored in it. This was intended as a model to demonstrate how weak electric perturbations can cause large effects in the brain (e.g., a nerve impulse).

The considerations given here are of a very general nature, and modifications of the theory can easily be performed without altering the principal result of coherent excitations. One will expect the excitation of a polar mode to be relevant whenever long-range coordination is required, as in cell division, tissue growth, and the cancer problem; cf. Section IV D of Ref. 7 and Ref. 23.

The theoretical derivation of the existence of a coherently excited polar mode following from (11) can also be phrased in a field theoretical model, or in a more exact way be supported by numerical work as quoted in Ref. 7, as well as by more recent work—the present chapter is not meant to provide a complete review. It is of importance, however, that other models involving nonlinear coupling of oscillations can lead to coherent excitation of a single mode.[24,25]

It has also been suggested that the assembly of biological oscillators may be modeled as parameter tunnel diode memory system with couplings. The frequencies of such a system can include values very much lower than the "design" frequency, which thus supports our prediction of mode softening.[26]

## 4. Experiments

A great variety of experiments support the conjecture of coherent excitations in biological systems. The complexity of biological materials might permit different interpretations in a number of these experiments. The possibility of understanding them from a common point of view gives strength, however, to our interpretation. In fact the conclusions of a recent meeting[27] were "that coherent excitation had now been shown to be an intimate feature of biological activity; further progress would rely on detailed studies...." The proceedings of this meeting were published in the "Green Book."[28]

A number of these experiments require subtle preparation of the biological material, which is not always described in detail and hence may lead to failure of reproducibility. As a first example consider the Raman spectrum of life algae. If certain frequencies are nonthermally excited, as required in coherent excitation, then the ratio of anti-Stokes to Stokes lines should be larger than in thermal equilibrium. This was actually found.[29] A repetition of the experiment showed no such effect.[30] The authors forgot, however, to supply $CO_2$, which is required to maintain the biological activity, giving rise to the excitation.

A great number of investigations on the Raman effect in active bacteria have been carried out.[31] To be active, considerable dilation must be maintained, and it has been estimated that an enhancement of excitation by a factor of $10^5$ is required to obtain sufficient intensity to be noticiable.[32] Such enhancement cannot be expected for a single cell. A number of cells must, therefore, be assumed to oscillate coherently because the scattering intensity then is proportional to the square of the number of scattering units. While thus, seeing the lines at all already provides an indication for coherent excitation, measurement of the ratio $R$ anti-Stokes: Stokes intensity, carried out in one case, gives a more direct proof. Thus at about $120 \text{ cm}^{-7}$, in E. coli B bacteria, $R \simeq 1$ was measured while in thermal equilibrium $R \simeq 0.55$. Note that $R = 1$ is the largest possible value.[33] To maintain the required synchronization is a task that so far has not been properly documented. Adverse influences may arise from the laser beam, or from turbulance that may develop in a flow instrumentation, used to avoid destructive action of the laser beam on bacteria.

To avoid some of these difficulties a special method of synchronization has been used, whereby appropriate nutrients are provided, suddenly, after a period of starvation.[34] It was found that a nutrient line at $980 \text{ cm}^{-1}$ is highly excited, the enhancement above thermal being by a factor between 8 and 15; and it showed a slow periodicity of about 10 min. For a tentative interpretation we assume that nutrient molecules in or near an active cell (the measurements integrate excited and nonexcited molecules) are very highly excited and attract outside nutrient molecules to the cell, or to a particular region in it. This would then correspond to the term $\gamma S$ in Eq. (17). The periodicity, then, might follow in a similar way. By further investigation of the activity of this system, it should be possible to separate the diffusion part of the molecules from the region in which they are subjected to forces, and hence, possibly, determine the relevant biochemical process.

While this separation of movement due to forces, from diffusion, remains to be carried out, a very thorough separation of this type has been performed in connection with the formation of so-called rouleaux of red blood cells, i.e., stacks of cells that form in their plasma.[35,36] It had been observed that when such cells approach to a certain distance they seem to rush towards each other. Analysis with the help of the Smoluchowski theory of diffusion revealed, quantitatively, the existence of a long-range attractive force. Its properties were found to agree with the requirements of the theory of long-range interaction arising from coherent excitation of electric vibrations in the membrane. Thus on removal of the membrane potential, the interaction disappears, but is restored when the potential is restored. Similarly it is found that the interaction disappears when the metabolism, required to maintain the excitation [terms $s_j$ in Eq. (11)] is inhibited. Furthermore the interaction is specific in that a mixture of red blood cells from different animals segregate into rouleaux of equal origin.

It was also found that the interaction is carried by long molecules which, inside cells, corresponds to the role attributed[18,20] to the microtrabecular lattice as discussed in Section 3. Furthermore it has been shown[37] that albumin, a globular protein, does not transmit the interaction at physiological concentrations. Albumin can, however, be converted into a transmitter by a treatment that unravels the tertiary structure and allows it to polymerize.

The frequencies for the excitation discussed here have not been measured directly, but the involvement of the membranes suggests a region near $10^{11}$ Hz, though both lower frequencies (mode softening) and higher frequencies (proteins dissolved in the membrane) may be contemplated. In addition, much lower frequencies arise from periodic enzyme reactions, discussed in Eqs. (16)–(20). Among other experiments that support the existence of coherent excitation[38] are the attraction of particles of a dielectric powder to active cells,[39] by the method of "microdielectrophoresis." Investigations have been carried out on a great variety of living cells. They show that an oscillating electric field with minimum frequency of about 10 kHz exists near the surfaces of living but not of dead cells. This can be suppressed by metabolic inhibitors and is maximal (order 100 V/cm) near mitosis (in yeast).

Since the dielectric particles used in these experiments are much smaller than the cells to which they are attracted, they do not give much information about the relative phase on various sections of the cell surface. Thus a dipolar oscillation would interact strongly with electromagnetic waves of the same frequency, and the field will then decrease slowly with the distance from the cell, in contrast to the case of a high multipole whose electric field will be significant near the cell surface only. Nevertheless, within this region it will give rise to a secondary magnetic field, which might result in the attraction of small magnetic particles, as has been observed recently.[40] While thus observation of emission of electromagnetic radiation would present a useful identification, lack of such observation does not imply absence of electric vibrations; for this might require measurement instruments to be placed very close to the cell surface.

An enormous literature has accumulated on the action of microwaves on biological systems, much of it, however, at single frequencies or single intensities only, which makes it difficult to analyze the results. An exception is formed by investigation of the action of coherent millimeter waves on biological systems. These wavelengths are strongly absorbed by cell water, and great care has been taken, therefore, to eliminate thermal effects by using low intensities. In this case direct action of the electromagnetic field is excluded as this would require intensities many millions of times stronger than those used in the relevant experiments.[41] Early experiments are discussed in Section V B of Ref. 7. Their results are best summarized as follows[42]: "(a) the effect of irradiation depends strongly on the frequency of the microwaves; (b) in certain microwave-power ranges, the effect of exposure depends weakly on variation of the power through several orders of magnitude; (c) the effects are observed to depend significantly on the time of irradiation."

These results seem to confirm all theoretical predictions first made in 1968[43]: As shown in previous sections, the excitation of coherent membrane oscillations should be in the $10^{11}$ Hz region at very sharp frequencies; the response should be steplike as a function of the supplied energy [cf. Eq. (15)], and the system has storing capability. However, the supplied energy, in the theory, is largely of metabolic nature—this has to be so as the energy supplied externally alone would be much too small to cause big effects. It has been pointed out, however, in the discussion following Eq. (15) that at an appropriate instant of the development of the biological system, a small external energy supply $s_e$ may shift the position of the step in the response function, i.e., it will act strongly at a very sensitive stage of development. This may be compared to a phase transition in the development of the whole biological system; the externally supplied energy then changes its timing of this process relative to other processes, and the consequences for the whole system may be specific in particular cases.

It seems obvious that such subtle interference might be influenced by the natural fluctuations present in highly excited systems, like biological ones. This is in particular so as it has been shown that noise may cause phase transitions in nonlinear systems.[44] It might be suggested, therefore, that details of the experimentation are relevant for the outcome of the investigation.

Consider now, in particular, the sharp resonances in the rate of growth of yeast under the influence of microwaves in the 42-GHz region.[45,46] A great number of very sharp maxima and minima in the rate of growth versus frequency curve was found. The experiment emphasizes high-frequency stability and provides the use of two very different electrodes to supply the radiation to the yeast cultures that grow in liquid suspension. Cross correlation between the two geometries in the range between 41700 and 41800 MHz reveals a central line of about 10 MHz width with three side bands on each side, separated by about 18 MHz. The effects are relatively small, about $\pm 10\%$, but very well resolved and reproducible. Another feature, apart from the extremely sharp resonances, lies in the absence of a difference in magnitude between the two electrodes, although their surfaces differ by a factor of 10, and the radiation penetrates from the surface only a fraction of a millimeter. The suspension is constantly stirred, and it can be concluded that a single yeast cell is at most affected once during its life cycle, and then only in a particular state of development.

Clearly one would expect such extraordinary effects to arise from important, highly sensitive properties which will not reveal themselves in population effects, though these have the property of high reproducibility.

An investigation of the time of a life cycle of individual cells, and the influence of millimeter waves on it, was therefore undertaken. Frequencies in the 83-GHz region were used—for organizational reasons different from those of the source used in the experiments described above. The cell cycle time was not influenced by radiation of 83800 MHz, but very strongly so by the neighboring 83835 MHz, as shown in Figure 1. Remarkably, however, about half of the cells are not affected, whereas the others are affected very strongly. This leaves the question open whether cells show a "personality" by responding to one or another frequency, or whether a secondary effect, like a particular fluctuation at a critical time, is required to produce the effect, as suggested above.

Further experimentation should result in a more detailed knowledge of the particular cell processes involved. The sharpness of the resonances makes it quite clear, already at the present stage, that coherent excitations must be involved, and the fine structure of 18 MHz might arise from periodic enzyme reactions, Eq. (20).

While the above results are highly reproducible, it is not yet quite clear which particular items in the instrumentation, and the handling of the biological materials, are vital. Thus, at the same laboratory, a similar instrumentation has been produced which shows no effect. Up to the time of writing, the reason for this had not yet been discovered.

Other investigations in the field are published in Ref. 28. It should be expected that multicellular systems might show a larger effect. In fact[47] the growth of cress roots can be completely stopped within 100 seconds after irradiation with low-intensity 56-GHz microwaves, provided the polarization is parallel to the root. From the theoretical discussion it can be concluded that the process, which owing to the effect of external energy supply is started at an inappropriate time, completely disrupts the growth. Further biological and physical investigations are required to find this process. Clearly it must, at its own time, make use of coherent excitation.

Other biological systems exist, however, whose dynamic properties are much better known than those of cress root. Among these, properties of the membranes systems of electron transport phosphorylation have been studied from the point of view of bioenergies.[48] This requires energy transport over relatively long distances, and other concepts that can all be supplied by various coherent excitations. Measurements to confirm this are now planned.

Finally, to return to the metastable highly polar state of individual molecules, it should be possible, in sufficiently strong external fields, to lift the molecules into this state. Transitions into a metastable state were actually discovered when strong fields were applied to so-called Langmuir–Blodget multilayers and the low-frequency dielectric response was then measured.[49] Haemoglobin and some other molecules were investigated. When the applied field reached about $10^5$ V/cm, characteristic changes in the dielectric response were observed. They are maintained even when they are remeasured in low fields, i.e., the layers have been transferred into a metastable state. This state persists for up to six weeks. The gradual reversion to the initial characteristics can be accelerated by heating for short periods to about 40°C. It is to be expected that, owing to the presence of water and ions, the high polarization is screened. The fact, however, that the metastable state is reached through the application of very high fields indicates its basic high polarity.

Another support for the existence of a highly polar metastable state arises in the so-called electret method. Here the material is polarized in strong fields, and the polarization is frozen in by cooling to very low temperatures. Gradual increase in temperature then releases polarization at characteristic temperatures. In our metastable state, the release should begin at temperatures above physiological ones. Both investigated enzymes, trypsin and urease, show this effect,[50] though the polarizing field amounted to only 25000 V/cm.

In a very different kind of experiment it was found that laser irradiation increases the activity of chymotrypsin by a factor of up to 3.[51] This was investigated as arising from lifting the enzyme into the active, metastable state through the Raman effect. Detailed investigations of the dependence on the density of molecules and on laser power support this interpretation. For other possibilities see Ref. 52.

In summary it is seen that a variety of different experiments support the conjecture of two types of coherent excitation: a metastable state with large electric polarization, and existance of coherent electric vibrations. It should be emphasized again that the theoretical models show that such excitations should be possible. They are not unique; other models might achieve similar results. In fact if, as we think, biological activity makes great use of these excitations, the manner of establishing them need not be unique.

## References

1. Roger H. Pain, "Dynamic Proteins," *Nature* **305**, 581–582 (1983).
2. H. Fröhlich, "The Connection Between Macro- and Microphysics," *Riv. Nuovo Cimento* **3**, 490–534 (1973).
3. J. Monod, *Chance and Necessity*, transl., Collins, London (1972), p. 95.
4. H. Fröhlich, "The Extraordinary Dielectric Properties of Biological Materials," *Proc. Natl. Acad. Sci. USA* **72**, 4211–4215 (1975).
5. A. Wada and H. Nakamura, "Nature of the Charges Distribution in Proteins," *Nature* **293**, 757–758 (1981).
6. A. Warshel, "Electrostatic Basis of Structure–Function Correlation in Proteins," *Acc. Chem. Res.* **14**, 284–290 (1981).
7. H. Fröhlich, in *Advances in Electronics and Electron Physics* (L. Marton, ed.), Academic, New York (1980), Vol. 53, pp. 85–152.
8. H. Fröhlich, "Collective Behaviour of Non-Linearly Coupled Oscillating Fields," *J. Collect. Phenom.* **1**, 101–109 (1973).
9. H. Fröhlich, *Theory of Dielectrics*, second edition, Oxford Press, Oxford (1958).
10. G. Zundel, in *The Hydrogen Bond—Recent Developments in Theory and Experiment* (P. Schuster, G. Zundel, and C. Sandorfy, eds.), North-Holland, Amsterdam (1976), Vol. 2, pp. 683–766.
11. W. Kristof and G. Zundel, "Structurally Symmetrical, Easily Polarisable Hydrogen Bonds Between Side Chains in Proteins III," *Biopolymers* **19**, 1753–1769 (1980).
12. P. Rastogi, W. Kristof, and G. Zundel, "Easily Polarisable Proton Transfer Hydrogen Bonds Between the Side Chains of Histidine and the Carboxylic Acid Groups of Glutamic and Aspartic Acid Residues in Proteins," *Int. J. Biol. Macromol.* **3**, 154–158 (1981).
13. L. Genzel, F. Kremer, A. Poglitsch, and G. Bechtold, "Relaxation Processes on a Picosecond Time Scale in Hemoglobin and Poly (L-Alanine) Observed by Millimeter-Wave Spectroscopy," *Biopolymers* **22**, 1715–1729 (1983).
14. L. Genzel, F. Kremer, A. Poglitsch, and G. Bechtold, in *Coherent Excitations in Biological Systems* (H. Fröhlich and F. Kremer, eds.), Springer-Verlag, Berlin (1983), pp. 58–70.
15. H. Fröhlich, "Long Range Coherence and the Action of Enzymes," *Nature* **228**, 1093 (1970).
16. A. S. Davydov, "Solitons, Bioenergetics, and the Mechanism of Muscle Concentration," *Int. J. Quantum Chem.* **16**, 5–17 (1979).
17. H. Bilz, H. Büttner, and H. Fröhlich, "Electret Model for Collective Behaviour of Biological Systems," *Z. Naturforsch.* **36b**, 208–212 (1981).
18. J. Clegg, in *Coherent Excitations in Biological Systems* (H. Fröhlich and F. Kremer, eds.), Springer-Verlag, Berlin (1983), pp. 162–175.
19. J. Hymans, "Hooked on Microtubules," *Nature* **291**, 107–108 (1981).
20. E. Del Guidice, S. Doglia, and M. Milani, in *Coherent Excitations in Biological Systems* (H. Fröhlich and F. Kremer, eds.), Springer-Verlag, Berlin (1983), pp. 123–129.

21. H. Fröhlich, "Possibilities of Long- and Short-Range Electric Interactions of Biological Systems," *Neurosci. Res. Program Bull.* **15**, 67–70 (1977).
22. F. Kaiser, in *Coherent Excitations in Biological Systems* (H. Fröhlich and F. Kremer, eds.), Springer-Verlag, Berlin (1983), pp. 128–133.
23. H. Fröhlich, "Coherent Electric Vibrations in Biological Systems and the Cancer Problem," *IEEE Trans. Microwave Theory Tech.* **26**, 613–617 (1978).
24. H. Fröhlich, "Conditions for Coherent Excitations in Biological Systems," *Phys. Lett.* **93A**, 105–106 (1982).
25. E. Fermi, J. R. Pasta, S. M. Pasta, and S. M. Ulam, in *Collected Works of Fermi*, Vol. 2, University of Chicago Press, Chicago (1965), pp. 978–988.
26. B. S. Thornton, "Solid State Memory Problems Support Mode-Softening for Larger Assemblies of Biological Dipole Oscillations," *Phys. Lett.* **102A**, 77–79 (1984).
27. F. Keilman and D. Kell, "Coherent Excitation in Biology," *Nature* **301**, 656–657 (1983).
28. *Coherent Excitations in Biological Systems* (H. Fröhlich and F. Kremer, eds.), Springer-Verlag, Berlin (1983).
29. F. Drissler and R. MacFarlane, "Enhanced Anti-Stokes Raman Scattering from Living Cells of Chlorella Pyrenoids," *Phys. Lett.* **69A**, 65–67 (1978).
30. S. Kimoshita, K. Hirato, and T. Kushida, "Population of Vibrational State of Carotenoid Molecules in Living Cells of Chlorella," *J. Phys. Soc. Jpn* **49**, 314–321 (1980).
31. S. Webb, "Laser Raman Spectroscopy of Living Cells," *Phys. Rep.* **60**, 201–224 (1980).
32. H. Fröhlich, in *Physics as Natural Philosophy* (A. Shimony and H. Feshbach, eds.), MIT Press, Cambridge, Massachusetts (1982), pp. 287–293.
33. S. Webb, M. Stoneham, and H. Fröhlich, "Evidence for Non-Thermal Excitation of Energy Levels in Active Biological Systems," *Phys. Lett.* **63A**, 407–408 (1977).
34. F. Drissler and L. Santo, in *Coherent Excitations in Biological Systems* (H. Fröhlich and F. Kremer, eds.), Springer-Verlag, Berlin (1983), pp. 6–9.
35. S. Rowlands, in *Coherent Excitations in Biological Systems* (H. Fröhlich and F. Kremer, eds.), Springer-Verlag, Berlin (1983), pp. 154–161.
36. S. Rowlands, C. Eisenberg, and L. Sewchand, "Contractiles: Quantum Mechanical Fibrils," *J. Biol. Phys.* **11**, 1–4 (1983).
37. S. Rowlands, L. Sewchand, and L. Skibo, "Conversion of Albumin into a Transmitter of the Ultra Long-Range Interaction of Human Erythrocytes," *Cell Biophys.* **5**, 197–203 (1983).
38. H. Fröhlich, "Evidence for Coherent Excitation in Biological Systems," *Int. J. Quantum Chem.* **23**, 1589–1595 (1983).
39. H. Pohl, in *Coherent Excitations in Biological Systems* (H. Fröhlich and F. Kremer, eds.), Springer-Verlag, Berlin (1983), pp. 199–210.
40. M. Akhalaya, M. Kakiasvili, and K. Zakaraya, "On Biomagnetism of Cells," *Phys. Lett.* **101A**, 367–370 (1984).
41. H. Fröhlich, "What are Non-Thermal Electric Biological Effects?," *Bioelectromagnetics* **3**, 45–46 (1982).
42. N. Devyatkov, "Influence of Millimeter-Band Electromagnetic Radiation on Biological Objects," *Sov. Phys. Usp.* **10**, 568–569 (1974).
43. H. Fröhlich, "Long Range Coherence and Energy Storage in Biological Systems," *Int. J. Quantum Chem.* **2**, 641–649 (1968).
44. L. Arnold, W. Horsthemke, and R. Lefever, "White and Coloured Noise and Transition Phenomena in Nonlinear Systems," *Z. Physik.* B **29**, 367–373 (1978).
45. W. Grundler, F. Keilmann, V. Putterlik, L. Santo, D. Strube, and I. Zimmerman, in *Coherent Excitations in Biological Systems* (H. Fröhlich and F. Kremer, eds.), Springer-Verlag, Berlin (1983), pp. 21–37.

46. W. Grundler and F. Keilmann, "Sharp Resonances in Yeast Poove Nonthermal Sensitivity to Microwaves," *Phys. Rev. Lett.* **51**, 1214–1216 (1983).
47. F. Kremer, A. Poglitch, L. Santo, D. Sperher, and L. Genzel, "Low Intensity Millimeter Waves Inhibit the Growth of Cress Roots," *Z. S. Naturf. C.* (in print).
48. D. Kell and G. Hitchens, in *Coherent Excitations in Biological Systems* (H. Fröhlich and F. Kremer, eds.), Springer-Verlag, Berlin (1983), pp. 178–198.
49. J. Hasted, S. Husain, A. Ko, D. Rosen, E. Nicol, and J. Birch, in *Coherent Excitations in Biological Systems* (H. Fröhlich and F. Kremer, eds.), Springer-Verlag, Berlin (1983), pp. 71–83.
50. S. Mascarenhas, "Electrets in Biophysics," *J. Electrostat.* **1**, 141–146 (1975).
51. N. Kolias and W. Melander, "Laser-Induced Stimulation of Chymotrypsin Activity," *Phys. Lett.* **57A**, 102–104 (1976).
52. J. Pokorny, *J. Theor. Biol.* **98**, 20–27 (1982).

# 9

# Collective Properties of Biological Systems

## Solitons and Coherent Electric Waves in a Quantum Field Theoretical Approach

*Emilio Del Giudice, Silvia Doglia, Marziale Milani, and Giuseppe Vitiello*

*ABSTRACT:* We present a dynamical scheme for biological systems. We use methods and techniques of quantum field theory since our analysis is at a microscopic molecular level. Davydov solitons on biomolecular chains and coherent electric dipole waves are described as collective dynamical modes. Electric polarization waves predicted by Fröhlich are identified with the Goldstone massless modes of the theory with spontaneous breakdown of the dipole-rotational symmetry. Self-organization, dissipativity, and stability of biological systems appear as observable manifestations of the microscopic quantum dynamics.

## 1. Introduction

Biological systems are collective systems indeed. They seem to be satisfactorily described when the list of their components is supplemented by the prescription of their location inside the system and the timing of their operation. Biologists refer to such a timing as a "cell clock"; what is now

*Emilio Del Giudice* • Dipartimento di Fisica dell'Università, Via Celoria, 16-20133 Milano, Italy, and Istituto Nazionale di Fisica Nucleare, Sez. di Milano, Italy. *Silvia Doglia* • Dipartimento di Fisica dell'Università, Via Celoria, 16-20133 Milano, Italy, and Gruppo Nazionale di Struttura della Materia del C.N.R., Milano, Italy. *Marziale Milani* • Dipartimento di Fisica dell'Università, Via Celoria, 16-20133 Milano, Italy, and Istituto Nazionale di Fisica Nucleare, Sez. di Milano, Italy. *Giuseppe Vitiello* • Dipartimento di Fisica dell'Università, 84100-Salerno, Italy, and Istituto Nazionale di Fisica Nucleare, Sez. di Napoli, Italy.

required is the engine responsible for such a dynamical ordering. Dynamical mechanisms that could give account for collective behavior have been proposed in the frame of different approaches.

H. Fröhlich[1] has introduced, in a framework of far-from-equilibrium processes, coherent electric polarization waves as the physical agent able to control the working of distant and separate parts of the system, making them cooperative. On the other hand, it has been shown that biomolecules are able to host on their own chains, in a conservative framework, highly nonlinear subdynamics which give rise to deep structural and conformational changes.

Very interesting subdynamics have been investigated by A. S. Davydov,[2,3] who has shown that metabolic reactions can result in strictly confined "bumps" traveling along biomolecular chains and carrying energy in an efficient way a long distance away from the point of release.

Our aim is to provide a systematic approach to the dynamics of a biological system[4,5] by using the widely accepted tool for exploring the collective properties of a system made from a large number of elementary microscopic components, namely, quantum field theory (QFT). We choose quantum field theory instead of classical field theory since quantum theory has proved to be the most adequate tool for describing atoms, molecules, and their interactions, and we wish actually to carry out an investigation of the biological system at a molecular level. Moreover, we choose quantum field theory instead of quantum mechanics because we are not interested in singling out just a few degrees of freedom of the system; on the contrary we would like to get a comprehensive description of the dynamics of the whole system (collective behavior).

The QFT approach could be sketched as follows. In a preliminary stage the microscopic components of the system are enumerated; the state of minimum energy is called the ground state or the vacuum. It is possible that more than one ground state exists for the same system, each of them being characterized by different macroscopic properties. In such a case different regimes or phases will correspond to different vacua. Subsequently, the microscopic components are linked to field operators whose dynamical evolution is obtained through equations of motion derived from a Lagrangian. The usual prescription is to choose this Lagrangian as symmetric as possible in order to get the simplest equations of motion while that part of the dynamics which is less symmetric is taken into account via the vacuum.

Where the vacuum does not exhibit the same invariance of the Lagrangian, one says that a spontaneous breakdown of the symmetry occurs. General theorems have been derived which elucidate the dynamical consequences on the whole system of the interplay between a symmetric Lagrangian and a nonsymmetric vacuum. The Goldstone theorem[6] is a very important one; it states that massless modes (usually called Goldstone bosons) appear when the Lagrangian is invariant, while the vacuum is non-invariant under a certain group of transformations. Notice that the invariance or noninvariance of the Lagrangian and of the vacuum under a transformation compels us to take into account the whole of the system degrees of freedom so that the Goldstone modes are really collective modes. A massless mode actually describes a long-range interaction among the system components and such an interaction establishes a correlation (i.e., order) among the components. The noninvariance of the vacuum under a certain transformation thus implies that this vacuum is an ordered state, in which absorption and emission of quanta of the Goldstone field, undergoing the transformation of noninvariance, take place. The mathematical structure of the theorem makes evident that the above-mentioned boson quanta exist in a condensed form and their condensation onto the ground state corresponds to a sort of "printing" of the symmetry pattern on the vacuum.

It is interesting to compare the collective properties arising from a microscopic theory with the macroscopic predictions obtained at a thermodynamic level in the framework of different approaches. The ordering of biological systems by coherent electric waves as proposed by Fröhlich could be understood, for instance, as the phenomenological manifestation of the breakdown of some symmetry at a microscopic level.[4,5,7] In this case the fact that the condensation, i.e. the emission and absorption, of Goldstone boson quanta does not require any energy expense is the microscopic counterpart of the thermodynamic dissipativity condition, which is a necessary condition[1] for the appearance of these waves. Dissipativity is a very general and important feature of living systems as far-from-equilibrium systems.[8] If we consider the Goldstone modes not as particles but as waves, the condensation of quanta producing an ordered state would lead us to take into account coherent waves. Coherent electric polarization waves require then, in the above framework, a vacuum with a nonzero value for the relevant field, namely, for the polarization field. As a matter of fact, water around biomolecules exhibits a nonzero polarization, i.e., an electret.[9,10]* Zero frequency modes have been also observed, which point to the existence of massless modes.[11]

Nonlinear dynamical effects on one-dimensional systems such as biomolecular chains, e.g., Davydov solitons, can also be analyzed in terms of boson condensation.[12-14] In the Fröhlich wave case, the condensation

---

* We define the electret state of a dielectric material, in agreement with the definition of Refs. 9 and 10, as a metastable state characterized by a stored polarization. Its decay time must be assumed long with respect to the characteristic time of experiments performed on the material.

induced by a symmetry breakdown occurs on large regions homogeneously, whereas in one-dimensional molecular systems one has quite localized condensations producing solitons. QFT can provide a unified description of such physically different situations. It is also interesting to note that an analysis by means of boson condensation can be performed in a variety of systems in many-body physics.[12] Thus a QFT approach has the appealing feature of extending methods and mathematics currently used in condensed matter physics to living matter.

In the following sections we present such a unified treatment. First we sketch qualitatively the interplay between Davydov and Fröhlich mechanisms (Section 2). Then we present an account of both of them and we discuss the transition from one regime to the other (Section 3). Finally we discuss the phenomenological implications of the theoretical scheme and the available supporting experimental evidence (Section 4).

## 2. Solitons, Electrets, and Fröhlich Waves

A very appealing schematization features a biological system as a set of electric dipoles, a dielectric with peculiar properties. Both water molecules—which account for a large amount of the global weight—and the macromolecules having a role in the biochemical machinery (DNA, RNA, proteins, and so on) have relevant dipole values or exhibit atomic subgroups with large dipole values. The starting point of the Fröhlich approach, for instance, is just to treat living matter as a dielectric, where the correlations among different parts are established by electric polarization waves. On the other hand the dynamics of a one-dimensional set of dipoles gives rise to vibrational solitons on biomolecular chains as proposed by Davydov.

Let us first discuss these two mechanisms separately. Biological systems are considered by Fröhlich as three-dimensional sets of dipoles. It is well known that the basic dipole–dipole interaction is rotationally invariant, so that we expect that the Lagrangian describing the whole system—whatever its detailed mathematical form is—be rotationally invariant too. On the other hand an electret (not a ferroelectret) has been detected for almost all the biomolecules.[9,10] Furthermore, this electret does not belong to the bare molecular structure but to its "dressing," the surrounding hydration water. Some sort of molecular dynamics produces a nonzero electric polarization in the surroundings, thus breaking the rotational symmetry of the dipole dynamics. It is possible consequently to introduce the formalism and theorems concerning spontaneously broken symmetries and their dynamical rearrangement (see Section 3).[15] In a first rough treatment one could use the Goldstone theorem, which predicts the appearance of massless bosons. A more adequate treatment would consider the electric field as a gauge field; a massive electric quantum excitation would then appear. It is known that the mass of a quantum of a field is inversely proportional to its propagation range; consequently, while a rigorously massless mode would correspond to an infinite propagation range (infinite size of the system), a nonzero mass corresponds to the observed finite range correlations (finite size of the system). A further improvement would concern the introduction of finite temperature effects which will allow thermal excitations. The problem will then arise of the interaction of coherent electric waves with thermal effects.

Let us now consider the one-dimensional dynamics occurring on the macromolecular chains. Davydov has shown[2,3,16] that a vibrational soliton—namely, a localized excitation strictly bound to a localized deformation—occurs on one-dimensional chains made by weakly bound monomers under two conditions:

i. the initial excitation is localized, i.e., energy is supplied to the system at small sized sites;
ii. the coupling between monomers exhibits a positive anharmonicity.

Under the above conditions solitons are created and travel quite slowly on the chain exhibiting a remarkable stability when their lifetime is compared with the one of the original excitation. Solitons have been therefore proposed as the long-range carriers of the energy produced by metabolic reactions and stored on the molecules.[17] For this reason they were used in a model of muscle contraction by Davydov.

Apart from the storage of energy, many additional and partially unexplored consequences arise on the biomolecules because of solitons. Conformational and structural changes may occur as well as changes in the vibrational frequency patterns. The binding of hydration water molecules to biomolecules is also affected. Davydov has recently pointed out[18] that a soliton on the $\alpha$-helix of a protein behaves as a mobile trap for loosely bound electrons, giving rise to an electrosoliton which has the features of a supercurrent. In other words, once a vibrational soliton is produced when a certain amount of energy is supplied, an electric current is obtained without further expense of energy. It is reasonable to assume that such a supercurrent, via its associated fields, might induce a preferred direction (an orientation) among water molecules surrounding the biomolecule. Actually, it is a fact[9] that the activation energy that one must supply to a biomolecule in order that an electret may appear around it is just of the same order of magnitude (0.2–0.4 eV) as the energy required for the production of a soliton on an $\alpha$-helix protein through the ATP hydrolysis.

The above-discussed mechanisms are particular ones. It is possible to integrate them in a provisional, unified scheme giving a partial qualitative picture of cell operation.[19] The metabolic reaction, governed by nutrition and membrane activity, produces localized excitations at selected points of selected molecules (DNA). Such excitations propagate in soliton form along the chains storing energy on them; the rate of accumulation is controlled by the difference between the number of solitons produced per unit time (metabolic reaction rate) and the number of solitons decaying in unit time. This rate is linked to the chain elasticity, a time-dependent parameter. We have described what we call the energy charge phase of the cell cycle. Simultaneously, a water electret is created around the molecule because of the soliton accumulation. When the electret reaches a threshold corresponding to the breakdown of the rotational symmetry of the dipole set (representing a finite-sized cell), then a different dynamical regime is switched on, where electric polarization waves must appear according to the previous discussion on the dynamical rearrangement of symmetry. These waves, propagating in the system, are instrumental in its biochemical organization, as discussed elsewhere.[19,20]

## 3. Boson Condensation and Collective Modes

Quantum field theory provides a two-level description for a system:

a. the dynamical level, where basic microscopic fields, say $\psi(x)$, are introduced, which are called the Heisenberg fields and obey the dynamic equations

$$\Lambda(\partial)\,\psi(x) = J[\psi(x)] \tag{3.1}$$

b. the phenomenological level, where observable quantities are expressed in terms of physical fields, say $\phi(x)$, which are called also asymptotic or quasiparticle fields and satisfy free field equations

$$K(\partial)\,\phi(x) = 0 \tag{3.2}$$

In Eqs. (3.1) and (3.2), $\Lambda(\partial)$ and $K(\partial)$ are kinetic differential operators, $x$ stands for $(\mathbf{x}, t)$, and $J[\psi(x)]$ is a nonlinear functional of $\psi(x)$.

As a next step, the set of the physical states, or Fock space, of the system is introduced. It is denoted by $\{|0\rangle\}$ and contains all the states obtained by repeated application of the "quasiparticle" creation operators on the ground state, or "vacuum," $|0\rangle$. The "quasiparticle" creation operators are functions of the fields $\phi(x)$.

At the first level of description the "forces" acting among the system components appear through the functional $J[\psi(x)]$, while, at the second level, the fields are considered free, because only fairly independent quantities are observable. In a scattering process, for instance, the particles are observed only before and after the collision. In a complex system either the whole compound system or the correlations among the con- stituents—namely, the free particles acting as messengers (phonons and so on)—are observed. These "free" phenomenological fields are of course the output of the inner dynamics described at the first level and such relationship is provided by the "dynamical map"[12,15]:

$$\langle a| \psi(x) |b\rangle = \langle a| \Psi[\phi(x)] |b\rangle \tag{3.3}$$

Equation (3.3) is the bridge between the deep underworld where the "forces" are at work and the apparent world where observations are perfor- med. In Eq. (3.3) $|a\rangle$ and $|b\rangle$ are wave-packet states, which belong to the Fock space $\{|0\rangle\}$ and $\Psi[\phi(x)]$ is a functional of (normally ordered) products of the $\phi(x)$ fields.

Let us now investigate how a symmetry at the dynamical level would affect the phenomenological description. Consider a transformation which leaves Eq. (3.1) invariant, namely, let $g$ be an element of the symmetry group of Eq. (3.1):

$$\psi'(x) = g\psi(x) \tag{3.4}$$

$$\Lambda(\partial) \psi'(x) = J[\psi'(x)] \tag{3.5}$$

The dynamical map (3.3) would then select a field $\phi'(x)$ corresponding to $\psi'(x)$, defining a transformation $h$ which would give $\phi'(x)$ as a function of the field $\phi(x)$ corresponding to the original $\psi(x)$ via the dynamical map

$$\langle a| g\psi(x) |b\rangle = \langle a| \Psi[h\phi(x)] |b\rangle \tag{3.6}$$

Usually the functional $\Psi[\phi(x)]$ is highly nonlinear, so that $h$ is expected to be different from $g$. However, since $g$ leaves Eq. (3.1) invariant (i.e., it corresponds to a physical process which requires no expense of energy), $h$ leaves in turn Eq. (3.2) invariant

$$\phi'(x) = h\phi(x) \tag{3.7}$$

$$K(\partial) \phi'(x) = 0 \tag{3.8}$$

If the transformations $g$ form a group $G$, then the transformations $h$ form a group $H$, too.

We are interested in the situation $H \neq G$, namely, when the dynamical map (3.6) rearranges the symmetry transformations so that the system acquires a symmetry at the phenomenological level, which is quite different from the one at the microscopic dynamical level. In this case we say that a dynamical rearrangement of the symmetry occurs.

We are going to explore the consequences of the dynamical rearrangement of the symmetry in the two cases of the Davydov vibrational solitons on macromolecules and of the Fröhlich electric waves in the whole biological system. In the first case we deal with an explicit form of the equations, so that a complete treatment is possible, while in the second case we limit ourselves to the consideration of the dynamical rearrangement of the symmetry and of the consequent appearance of massless modes.

Let us consider first the soliton dynamics on the biomolecular chains.

### 3.1. Davydov Soliton Regime

A biomolecular chain can be schematized as a one-dimensional chain of weakly bound monomers of mass $M$, whose electric dipole moment is aligned along the chain. Consequently the internal dipole quantum number, giving the spatial orientation of the dipole, can be taken as frozen; the electric dipole, which would be described by a spinor in the usual three dimensions, is described in this one-dimensional case by a scalar.

In a first approximation the molecule length is assumed infinite and corrections due to the finite length are introduced as a second step. The intramolecular excitation brought by an external supply of energy (e.g., the output of a chemical reaction) is described by a complex Heisenberg boson field $\psi(\xi, t)$, solution of a nonlinear Schrödinger equation[2,3,16]:

$$\left(i\hbar \frac{\partial}{\partial t} - \varLambda + J \frac{\partial^2}{\partial \xi^2}\right) \psi(\xi, t) = -G\psi^+(\xi, t)\,\psi(\xi, t)\,\psi(\xi, t) \qquad (3.9)$$

The right-hand side of Eq. (3.9) is understood to be normally ordered. $\varLambda$ is a constant energy term which may depend on the free exciton energy, on the static interaction energy, and on the deformation energy. $J$ is the resonance interaction energy between neighboring monomers. The quasiparticle free equation (3.2) becomes in this case

$$\left(i\hbar \frac{\partial}{\partial t} - \varLambda + J \frac{\partial^2}{\partial \xi^2}\right) \phi(\xi, t) = 0 \qquad (3.10)$$

The operator $\phi(\xi, t)$ is the free exciton field. As a first step we are going to show that the $c$-number Davydov soliton[2,3,16]

$$\chi_c(\xi, t) = \left(\frac{\mu}{2}\right)^{1/2} \frac{\exp i\left[\dfrac{\hbar v}{2J}(\xi - \xi_0) - E_v \dfrac{t}{\hbar}\right]}{\cosh \mu(\xi - \xi_0 - vt)} \qquad (3.11)$$

$$\int |\chi_c(\xi, t)|^2 \, d\xi = 1 \qquad (3.12)$$

is just a convenient vacuum expectation value for $\psi(\xi, t)$ in the "classical" limit. In Eq. (3.11) $\hbar$ is the Planck constant and

$$E_v = \Lambda + \frac{\hbar v^2}{4J} - J\mu^2, \qquad \mu = \frac{G}{4J} \tag{3.13}$$

Let us assume a transformation $B$ such that

$$\psi(\xi, t) \to \psi'(\xi, t) = B\psi(\xi, t) \tag{3.14}$$

$$\chi_c(\xi, t) = \langle 0| \, \psi'(\xi, t)|0\rangle \,|_{\hbar = 0} = \langle 0| \, B\psi(\xi, t) \,|0\rangle \,|_{\hbar = 0} \tag{3.15}$$

which leaves Eq. (3.9) invariant. The dynamical map (3.3) then becomes

$$\langle 0| \, B\psi(\xi, t) \,|0\rangle = \langle 0| \, \psi[\beta\phi(\xi, t)] \,|0\rangle \tag{3.16}$$

which defines $\beta$ as an invariant transformation for Eq. (3.10). Let us require also that $\beta$ be a canonical transformation, namely, $\beta$ preserves the canonical commutation relations

$$[\phi(\xi, t), \dot{\phi}(\xi', t)] = [\phi'(\xi, t), \dot{\phi}'(\xi', t)] = i\delta(\xi - \xi')$$
$$[\phi(\xi, t), \phi(\xi', t)] = [\phi'(\xi, t), \phi'(\xi', t)] = 0 \tag{3.17}$$
$$[\phi(\xi, t), \phi(\xi', t)] = [\phi'(\xi, t), \phi'(\xi', t)] = 0$$

where

$$\phi'(\xi, t) = \beta\phi(\xi, t) \tag{3.18}$$

We consider now the general class of canonical transformations

$$\phi'(\xi, t) = \beta\phi(\xi, t) = \phi(\xi, t) + f(\xi, t) \tag{3.19}$$

where the field $\phi(\xi, t)$ is translated by a $c$-number function $f(\xi, t)$. The transformation (3.19) would leave Eq. (3.10) invariant, provided that $f(\xi, t)$ is a ($c$-number) solution of the same equation, namely,

$$f(\xi, t) = ce^{-\mu(\xi - vt)} \exp\left[i\left(\frac{\hbar v}{2J}\xi - \frac{E_v t}{\hbar}\right)\right] \tag{3.20}$$

Let us show that the transformation (3.19) on the fields $\phi$, which corresponds to the "translation" (3.20), induces through the dynamical map just the transformation (3.14) on the fields $\psi$, so that Eq. (3.15) is satisfied. The generator of (3.19) in the $v = 0$ frame of reference is

$$D = -\int_{-\infty}^{\infty} d\xi [g(\xi, t) \dot{\phi}(\xi, t) - \dot{g}(\xi, t) \phi(\xi, t)] \tag{3.21}$$

where

$$g(\xi, t) \doteq \theta(\xi) f_0(\xi, t)$$
$$f_0(\xi, t) = f(\xi, t)\big|_{v=0} \tag{3.22}$$

$\theta(\xi)$ is the usual step function: $\theta(\xi) = 0$ for $\xi < 0$ and $\theta(\xi) = 1$ for $\xi > 0$. It is possible to read Eq. (3.19) as

$$\phi'(\xi, t) = \beta\phi(\xi, t) = \lim_{g \to f_0} e^{-iD}\phi(\xi, t) e^{iD} \qquad \text{at} \quad v = 0 \tag{3.23}$$

and

$$\langle 0| e^{-iD}\phi(\xi, t) e^{iD} |0\rangle\big|_{g=f_0} = \langle f_0| \phi(\xi, t) |f_0\rangle = f_0(\xi, t) \tag{3.24}$$

The state $|f_0\rangle$ is defined as

$$|f_0\rangle \equiv e^{iD} |0\rangle\big|_{g=f_0}, \qquad \langle f_0 | f_0\rangle = 1 \tag{3.25}$$

We introduce now the annihilation operator $a_{\bar{k}}$ of a quantum of the field $\phi(\xi, t)$ whose momentum is $\bar{k}$; let us also consider the $c$-number function $\alpha(\bar{k})$, the Fourier transform of a function which coincides with $f_0(\xi, t)$ in a suitable $\xi$ domain; then

$$a_{\bar{k}} |f_0\rangle = \alpha(\bar{k}) |f_0\rangle \tag{3.26}$$

i.e., $|f_0\rangle$ is a coherent state for $\phi(\xi, t)$, namely, any functional of $\phi$ acts on $|f_0\rangle$ as a $c$-number.

The dynamical map (3.3) then becomes

$$\langle f_0| \psi(\xi, t) |f_0\rangle = \Psi[f_0(\xi, t)] \tag{3.27}$$

$$\langle f_0| J[\psi(\xi, t)] |f_0\rangle = J[\Psi[f_0(\xi, t)]] \tag{3.28}$$

A boost to the original frame of reference and the substitution of Eqs. (3.27) and (3.28) into Eq. (3.9) produce at last

$$\left( i\hbar \frac{\partial}{\partial t} - A + J\frac{\partial^2}{\partial \xi^2} \right) \psi[f] = -G |\psi[f]|^2 \psi[f] \tag{3.29}$$

Equation (3.29) is a $c$-number equation which coincides with the one proposed by Davydov [see Eq. (2.10) of Ref. 2]. The solution of (3.29) is, then, just the soliton (3.11)

$$\Psi[f(\xi, t)] = \chi_c(\xi, t) \tag{3.30}$$

which allows us to determine the constant $c$ in Eq. (3.20):

$$c = (2\mu)^{1/2} e^{\mu\xi_0} \exp\left(-i\frac{\hbar v}{2J}\xi_0\right) \tag{3.31}$$

In conclusion from the above formalism we can deduce that the Davydov soliton is the macroscopic envelope of the localized boson condensation of the excitation quanta $\phi(\xi, t)$ induced by the $\beta$ transformation (3.19).[12–14] The vanishing of the coefficient $G$ of the nonlinear term of Eq. (3.29), via Eq. (3.13), implies the disappearance of the soliton, which then looks like a necessary consequence of the nonlinearities of the dynamics.

As a further step, we are going now to show how the QFT approach could recover all the relevant Davydov results starting from a restricted set of assumptions.[4]

We started actually from the nonlinear Schrödinger equation (3.9) governing the $\psi(\xi, t)$ field only. A deformation (phonon) field has not, so far, been independently introduced. But the density probability $\rho^f$ of $\psi$ in the soliton state $|f\rangle$

$$\rho^f(\xi, t) = \gamma \langle f| \psi^+(\xi, t) \psi(\xi, t) |f\rangle = \gamma |\chi_c(\xi, t)|^2 \tag{3.32}$$

or [see Eq. (3.11)]

$$\rho^f(\xi, t) = \gamma \frac{\mu}{2} \frac{1}{\cosh^2 \mu(\xi - \xi_0 - vt)} \tag{3.33}$$

is a solution of the (phonon) equation

$$\left(\frac{\partial^2}{\partial t^2} - v^2 \frac{\partial^2}{\partial \xi^2}\right) \rho^f(\xi, t) = 0 \tag{3.34}$$

The phonon equation is then a direct consequence of the self-interacting intramolecular excitation field $\psi$, which is the only independent physical object. As a matter of fact, the $\psi$-excitation field dynamics induces conformational changes in the chain lattice which are described by a phonon field.

It is useful to recast Eq. (3.34) in the more familiar form of the Davydov Eq. (2.56) of Ref. 2. Let us define

$$v_{ac} = \left(\frac{W}{M}\right)^{1/2}$$

$$v^2 = v_{ac}^2 - v_{ac}^2(1 - s^2) \tag{3.35}$$

$$s = \frac{v}{v_{ac}} \ll 1$$

where $w$ is the chain elasticity coefficient and $v_{ac}$ is the sound velocity on the chain in absence of solitons. Equation (3.34) is consequently transformed into

$$\left(\frac{\partial^2}{\partial t^2} - v_{ac}^2 \frac{\partial^2}{\partial \xi^2}\right) \beta^f(\xi, t) - \frac{2K}{M} \frac{\partial}{\partial \xi} |\chi_c(\xi, t)|^2 = 0 \qquad (3.36)$$

where

$$K = 1/2w(1-s^2)\gamma \qquad (3.37)$$

$$-\frac{\partial}{\partial \xi} \beta^f(\xi, t) = \rho^f(\xi, t) \qquad (3.38)$$

Equation (3.36) is just the Davydov equation (2.56) of Ref. 2. Moreover Eq. (3.29) can be transformed as follows:

$$\left[i\hbar\frac{\partial}{\partial t} - \Lambda + J\frac{\partial^2}{\partial \xi^2} - 2K\frac{\partial \beta^f(\xi, t)}{\partial \xi}\right]\chi_c(\xi, t) = 0 \qquad (3.39)$$

where

$$\mu = \frac{K^2}{w(1-s^2)J} \qquad (3.40)$$

Equation (3.39) coincides with Eq. (2.5a) of Ref. 2. We conclude that the present QFT approach is equivalent to the Davydov analysis. All the results of the latter theory can be recovered in the QFT framework, such as, for instance, the soliton stability, the soliton–photon interaction and the attraction of electrons by the self-generating potential wells leading to quasisuperconducting properties of the molecular chains, hosting solitons. The chains considered above were assumed to be infinitely long; let us now drop this assumption. This change affects mainly the stability of the soliton. The soliton is stable when the local deformation $\beta^f(\xi, t)$ [or $\rho^f(\xi, t)$] is stable and vice versa [see Eqs. (3.36) and (3.39)]. The conditions for the stability of $\beta^f$ have been investigated in Ref. 14. For the sake of conciseness, we do not report here such an analysis (see also Ref. 4), but we give the conclusion. In a finite chain the soliton state has a nonzero transition amplitude with any state of $n$ free excitation quanta, so that its lifetime has to be finite. The longer the chain, the longer the lifetime.

Very important features of solitons, such as their rate of creation and decay, their density on the chain, the effects of external perturbations (for instance, changes of the soliton density and decay rate induced by variations of the physical and chemical structure of the chain) are yet a matter of investigation.

All the above analysis shows that an incoherent energy supply can be transformed into a coherent but localized form (the soliton) through the inner nonlinear dynamics occurring on the molecules constituting the biological system. The next step is the dynamical appearance of self-organization in the system without further expense of energy; energy stored in soliton form during the "charge" or Davydov regime is completely released outward during the "discharge" or Fröhlich regine.

## 3.2. Fröhlich Regime

It has been experimentally found that water surrounding biomolecules can be brought in an electret state by low electric fields at low frequency.

We have said in Section 2 that the electret activation energy is of the same order of magnitude as the energy required for the production of a soliton on a biomolecule. The numerical coincidence suggests that the soliton might induce the observed electret while traveling on the chain. How it could happen is at present under investigation; we limit ourselves here to suggesting that the soliton could be a trigger for the water polarization. As a consequence the dipole quantum number carried by the dipole excitation $\psi$ field would no longer be frozen since the water electret has transformed the previously one-dimensional system into a three-dimensional one. Let us describe the dipole quantum number by the spinor:

$$\psi(x) = \begin{pmatrix} \psi\uparrow(x) \\ \psi\downarrow(x) \end{pmatrix} \tag{3.41}$$

where $x \equiv (\tilde{x}, t)$. The Lagrangian $\mathscr{L}$ of the whole biological system is very complicated, but it is safe to assume its invariance under the rotation of the dipole degree of freedom of the doublet field (3.4). More technically, we assume its invariance under the $SU(2)$ group of transformations

$$\psi(x) \to \psi'(x) = e^{i\theta_i \sigma_i} \psi(x), \qquad i = 1, 2, 3 \tag{3.42}$$

$\theta_i$ are real parameters and $\sigma_i$ are the Pauli matrices. Let us now get as much information as possible by using symmetry arguments only.

The path integral formalism of QFT appears to be the most suitable one. We introduce[20,21,22] the generating functional

$$W[J, j, n] = \frac{1}{N} \int [d\psi][d\psi^+] \exp \left\{ i \int d^4x \right.$$

$$\times \; [\mathscr{L}[\psi(x)] + J^+(x)\,\psi(x) + \psi^+(x)\,J(x)$$

$$\left. + \; j^+(x)\,D_\psi^{(-)}(x) + D_\psi^+(x)\,j(x) + D_\psi^{(3)}(x)\,n(x) - i\epsilon\,D_\psi^3(x)] \right\}$$

$$\tag{3.43}$$

$D_\psi^{(i)}(x)(i = +, -, 3)$ is the dipole density operator, whose explicit form is not needed here, but whose transformation law under $SU(2)$ group is

$$D_\psi^{(i)} \to D_\psi^{(i)}(x) - \lambda_j \varepsilon_{ijk} D_\psi^{(k)}(x) \qquad (i, j, k = 1, 2, 3) \qquad (3.44)$$

Notice that $D_\psi^{(\pm)}(x) = D_\psi^{(1)}(x) \pm i D_\psi^{(2)}(x)$ The number $N$ in Eq. (3.43) is

$$N = \int [d\psi][d\psi^+] \exp\{i \, d^4x [\mathscr{L}[\psi(x)] - i\varepsilon D_\psi^{(3)}(x)]\} \qquad (3.45)$$

The fields $\psi$ and $\psi^+$ and the sources $J$ and $J^+$ anticommute, while $j$ and $j^+$ commute. The $\varepsilon$ term in (3.43) takes care, as usual, of the vacuum non-vanishing polarization:

$$\langle 0| D_\psi^{(3)}(x) |0\rangle = P \neq 0 \qquad (3.46)$$

The parameter $\varepsilon$ has to be put to zero only *after* all the mathematical manipulations have been performed. Equation (3.46) is referred to as the spontaneous breakdown of the $SU(2)$ dipole rotational symmetry.

The ground state does not share the Lagrangian full invariance, but it is invariant only under rotations around the third axis (i.e., the direction of the electret). This group of residual invariance transformations is a subgroup of the original $SU(2)$ symmetry group. The nonvanishing electric polarization (3.46) is the trigger of the Fröhlich regime.

In Refs. 20, 21, 22 a detailed mathematical treatment is given which illustrates the consequences of the condition (3.46) on the formalism derived from the generating functional (3.43). Here we sketch only the main points and the conclusions. The first point is the validity, in the absence of long-range fields (gauge fields), of the Goldstone theorem. Two massless boson modes $P(x)$ and $P^+(x)$ must appear under the condition (3.46). The set of "phenomenological" fields [namely, the solution of Eq. (3.2)] in our case is then the field $\phi(x)$ of the free excitation quanta and the Goldstone fields $P(x)$ and $P^+(x)$:

$$\Lambda(\partial) \phi(x) = 0 \qquad (3.47)$$

$$\left[ i\frac{\partial}{\partial t} + \omega(\partial) \right] P(x) = 0 \qquad (3.48)$$

The form of the differential operators $\Lambda(\partial)$ and $\omega(\partial)$ depends on the details of the dynamics and cannot be given here. Of course, more excitation fields can be added, but here we are interested *in* the Goldstone modes only. In particular we leave for further research the effects of the finite temperature, which imply the consideration of thermal excitations.

Let us concentrate on the Goldstone modes. They are the necessary consequence, through the Goldstone theorem, of the nonvanishing vacuum polarization (water electret). The interaction of these modes with the "matter" field $\phi$ is embodied in the scattering matrix $S$:

$$S(\phi, \phi^+, P, P^+) = \langle :\exp[-iA(\phi, \phi^+, P, P^+)]: \rangle \qquad (3.49)$$

while the dynamical map (3.3) is expressed in a path integral formalism as

$$SD^{(i)}(x, \phi, \phi^+, P, P^+) = \langle D_\psi^{(i)}(x): \exp[-iA(\phi, \phi^+, P, P^+)]: \rangle \qquad (3.50)$$

and

$$
\begin{aligned}
A(\phi, \phi^+, P, P^+) = \int d^4x [ & \rho^{-1/2} P(x) \, K(\vec{\partial}) \, D_\psi^{(-)}(x) \\
& + \rho^{-1/2} D_\psi^{(+)}(x) \, K(-\overleftarrow{\partial}) \, P^+(x) \\
& + z^{-1/2} \phi^+(x) \, \Lambda(\vec{\partial}) \, \psi(x) \\
& + z^{-1/2} \psi^+(x) \, \Lambda(-\overleftarrow{\partial}) \, \phi(x) ]
\end{aligned} \qquad (3.51)
$$

$\rho$ and $z$ are the wave function renormalization constants, $\vec{\partial}$ and $\overleftarrow{\partial}$ denote space-time derivatives operating on the right and left side, respectively, and

$$K(\partial) = -\left[ i\frac{\partial}{\partial t} + \omega(\partial) \right] \qquad (3.52)$$

By standard techniques it is possible to get the rules of transformation of the fields $\phi$ and $P(P^+)$ when the field $\psi(x)$ undergoes an $SU(2)$ transformation (3.42):

(a)  case $\theta_2 = \theta_3 = 0$:

$$\phi(x) \to \phi'(x) = \phi(x)$$

$$P(x) \to P'(x) = P(x) + i\theta_1 \left(\frac{P}{2}\right)^{1/2}$$

$$P^+(x) \to P'^+(x) = P^+(x) - i\theta_1 \left(\frac{P}{2}\right)^{1/2} \qquad (3.53)$$

(b)  case $\theta_1 = \theta_3 = 0$:

$$\phi(x) \to \phi'(x) = \phi(x)$$

$$P(x) \to P'(x) = P(x) - \theta_2 \left(\frac{P}{2}\right)^{1/2}$$

$$P^+(x) \to P'^+(x) = P^+(x) - \theta_2 \left(\frac{P}{2}\right)^{1/2} \qquad (3.54)$$

(c)   case $\theta_1 = \theta_2 = 0$:

$$\phi(x) \to \phi'(x) = e^{(1/2)i\theta_3\sigma_3}\phi(x)$$

$$P(x) \to P'(x) = e^{-(1/2)i\theta_3} P(x) \tag{3.55}$$

$$P^+(x) \to P'^+(x) = e^{+(1/2)i\theta_3} P^+(x)$$

The generators of the above three transformations are, respectively, in the space of physical states,

$$D^{(1)} = \left(\frac{P}{2}\right)^{1/2} \int d^3x [P(x) f(x) + P^+(x) f^*(x)]$$

$$D^{(2)} = -i \left(\frac{P}{2}\right)^{1/2} \int d^3x [P(x) f(x) - P^+(x) f^*(x)] \tag{3.56}$$

$$D^{(3)} = \int d^3x \left[ \phi^+(x) \frac{\sigma_3}{2} \phi(x) - P^+(x) P(x) \right]$$

The function $f(x)$ is a square-integrable $c$-number solution of the free field equation (3.48):

$$K(\partial) f(x) = 0 \tag{3.57}$$

and makes the integrals (3.56) convergent. The condition (3.57) makes also the generators (3.56) time independent, so that the corresponding transformations can be physically implemented without any expense of energy.

In other words the transformations (3.53)–(3.55) generated by the operators (3.56) are invariant transformations of the free-field "phenomenological" equations (3.47) and (3.48) for any choice of $f(x)$. Notice that the $P(x)$ field "translations" in Eqs. (3.53) and (3.54) are understood as the limit for $f(x) \to 1$ of the transformation generated by $D^{(1)}$ and $D^{(2)}$, namely,

$$P(x) \to P'(x) = \lim_{f(x) \to 1} \left[ P(x) + i\theta_1 f(x) \left(\frac{P}{2}\right)^{1/2} \right] \tag{3.58}$$

The transformations (3.53)–(3.55) do not belong to the original $SU(2)$ group of symmetry [see Eq. (3.42)] but to the so-called $E(2)$ group—the group of the transformations of the Euclidean two-dimensional space into itself. The $E(2)$ group is the group "contraction" of $SU(2)$. We have shown explicitly that the group of symmetry at the level of observable quantities is different from the group of symmetry at the level of the basic fields. This difference vanishes when the vacuum net polarization $P$ vanishes.

Let us show now that the vacuum electric polarization is induced by a coherent homogeneous condensation of the massless dipole-wave quanta $P(x)$ in the vacuum $|0\rangle$, which is controlled in turn by the field "translation" (3.53) and (3.54). We get indeed, by using Eq. (3.56),

$$\langle 0| e^{-iD^{(2)}}e^{-iD^{(1)}}P^+(x)\, P(x)\, e^{iD^{(1)}}e^{iD^{(2)}} |0\rangle = P\,|f(x)|^2 \xrightarrow[f \to 1]{} P \qquad (3.59)$$

We define

$$|f\rangle = e^{iD^{(1)}}e^{iD^{(2)}} |0\rangle \qquad (3.60)$$

Then Eq. (3.59) becomes

$$\langle f|\, P^+(x)\, P(x)\, |f\rangle = P\,|f(x)|^2 \xrightarrow[f \to 1]{} P \qquad (3.61)$$

By introducing the annihilation operator $P_{\bar{K}}$ of a quantum of the field $P(x)$ with momentum $\bar{K}$ and taking the Fourier amplitude $f_{\bar{K}}$ of $f(x)$, we get

$$P_{\bar{K}} |f\rangle = P^{1/2}f_{\bar{K}} |f\rangle \qquad (3.62)$$

which shows that $|f\rangle$ is a coherent state for the $P(x)$ field. In conclusion, a coherent homogeneous (i.e., not space dependent) condensation of dipole-wave quanta $P(x)$ is responsible for the ordered state of polarization. No expense of energy is required since $D^{(1)}$ and $D^{(2)}$ are time independent. We can then identify the dipole-wave Goldstone fields $P(x)$ and $P^+(x)$ with the coherent electric polarization waves predicted by Fröhlich.

A relevant feature of this theory is the so-called "low-energy theorem," analogous to corresponding theorems which hold in ferromagnetism and in particle physics. The invariance of the "phenomenological" field equations under the transformations (3.53)–(3.55) requires a corresponding invariance of the $S$ matrix. In the case of an infinite volume [namely, $f(x) \equiv 1$], this requirement is shown to imply

$$S(\phi(x), \phi^+(x), P(x), P^+(x)) = S(\phi(x), \phi^+(x), \partial P(x), \partial P^+(x)) \qquad (3.63)$$

The $S$ matrix, then, would be independent of $P(x)$ in the zero-momentum limit; in the infinite volume case, the Fröhlich regime would be stable against external perturbations exciting soft (zero momentum) modes.

In realistic systems, however, the volume is finite and surface effects become important. For a finite volume $V$ it has been shown[4] that

$$\langle 0| D_P^{(3)}(x) |0\rangle = \lim_{\varepsilon \to 0} i\varepsilon \int_{-1/\eta}^{1/\eta} d^3x \int \frac{d^3p}{2\pi} e^{i\vec{p}\cdot\vec{x}} \rho(p)$$

$$\times \left( \frac{1}{\omega(p) - i\varepsilon a} - \frac{1}{\omega(p) + i\varepsilon a} \right)$$

$$= \lim_{\varepsilon \to 0} i\varepsilon \int d^3p \, \delta_\eta(p) \, \rho(p) \left( \frac{1}{\omega(p) - i\varepsilon a} - \frac{1}{\omega(p) + i\varepsilon a} \right)$$

$$= \lim_{\varepsilon \to 0} i\varepsilon\rho(\eta) \left( \frac{1}{\omega(p=\eta) + i\varepsilon a} - \frac{1}{\omega(p=\eta) - i\varepsilon a} \right) \quad \text{at } \eta \sim 0$$

$$(3.64)$$

where $\eta = V^{-1/3}$, $\delta_\eta(p)$ approaches $\delta(p)$ as $\eta \to 0$ and the following relation has been used:

$$\lim_{\eta \to 0} \int d^3p \, \delta_\eta(p) \, f(p) = f(0) = \lim_{\eta \to 0} \int d^3p \, \delta(p - \eta) \, f(p)$$

The function $\omega(p)$ is the energy of a dipole-wave quantum of momentum $p$; for $V \to \infty$ ($\eta \to 0$) $\omega(p=0) = 0$. Then, as $V \to \infty$, or, alternatively "far from the boundaries" of the system, $\omega(p=\eta) \to 0$ and Eq. (3.64) reduce to Eq. (3.46) [notice that $\rho(0) = p/2$]. Near the boundaries ($\eta$ different from zero), $\omega(p=\eta)$ behaves as an "effective mass" of the quantum of the $P(x)$ field.

Consequently the range of the correlation mediated by $P(x)$ and the domain of the coherence will extend over a distance $\eta^{-1}$. Finite (but long) range correlation forces are then expected in realistic (finite volume) systems. A similar conclusion can be drawn from Eq. (3.58) when an appropriate cut-off function $f_\eta(x)$ is introduced. In conclusion a relation between the size of the system and the not exactly zero mass of the $P$-field quantum is obtained. As announced in Section 2 a more correct treatment, which considers the electric field as a gauge field, would lead to a dynamically constructed (nonzero) mass of the quanta, producing thus a finite volume for the biological system.

Let us now look at the form the low-energy theorem assumes for $\eta \neq 0$. The field equation becomes

$$K_\eta(\partial) P(x) = 0 \qquad (3.65)$$

The operator $K_\eta(\partial)$ depends upon the effective mass. Then the transformation rule of $P(x)$ changes to

$$P(x) \rightarrow P(x) + f_\eta(x) \text{ const} \qquad (3.66)$$

where $f_\eta(x)$ is prescribed to be a $c$-number solution of Eq. (3.65). In Eq. (3.63) $\partial P(x)$ must be replaced by $K_\eta(\partial) P(x)$ in order to have an $S$ matrix invariant under the $E(2)$ group transformations. Thus an energy threshold, equal to the effective mass of the quantum, appears, through $K_\eta(\partial)$, in the $S$ matrix. The consequence is that external perturbations whose energy is below threshold cannot perturb the system. Since the effective mass increases as $1/\eta$, the smaller the system the higher the threshold. The existence of an energy threshold matches a corresponding feature of the Fröhlich theory.

## 4. Phenomenological Evidence

In the above dynamical scheme a crucial role is played by solitons, water electrets, and coherent electric waves. In this section we present some experimental findings which provide evidence for the existence of such collective modes in biological systems.

### 4.1. Solitons on Macromolecules

Support for solitons on different kinds of molecules has been accumulated in these last years; $\alpha$-helix proteins have been extensively investigated.[3] It has been recognized that the hydrogen-bond filaments along $\alpha$-helix chains support a pathway for solitons in the nonlinear dynamics of the chains. It has been shown that a simple positive anharmonicity in the interaction among nearest-neighbour monomers is the required nonlinearity in the chain dynamics ($C=O$ bonds in $\alpha$-helix).[27,28]

A measure of the dipole–dipole interaction energy between two adjacent monomers and of the anharmonicity of the chain can be obtained by an inspection of the bandwidth of the absorption band of the monomer excitation, that forms the exciton. For $\alpha$-helix proteins in which inhomogeneous broadening has been eliminated by ordering processes of the samples, the bandwidth of the $C=O$ absorption at $1660 \text{ cm}^{-1}$ [29] gives evidence of the excitonlike collectivization of the vibrational $C=O$ excitation along the chain. This is a prerequisite for the existence of Davydov solitons on the chain.[2]

In recent years, it has been suggested that Davydov solitons could actually occur on a number of one-dimensional chains of biological interest. Raman experiments on green algae *Chlorella P.*[30] suggested the possible formation of Davydov solitons on carotenoid chains.[31] Infrared absorption studies of acetanilide, a model system for one-dimensional chains of H bonds, have been recently interpreted in terms of Davydov-like solitons, involving hydrogen-bonded amide groups.[28,32] Of particular interest is the possible existence of solitons on nucleic acids.[19,33-35] Whether Davydov solitons could actually occur on DNA, has been examined.[19] The stretching vibration of the P–O bond in the phosphodiester backbone of the double helix, being directed along the axis of the chain, satisfies the symmetry and polarity requirements for a collectivization of the excitation along the chain. However, only DNA in B-form displays the proper anharmonicity for the formation of a vibrational P–O exciton. This can be seen from the width of the P–O Raman bond at 830 cm$^{-1}$ [36,37] For DNA in A-form the band of the P–O stretching is shifted at 807 cm$^{-1}$ and narrowed in a sharper line, showing that the P–O excitation remains localized on individual P–O groups and does not spread along the chain in exciton form. A prerequisite for the appearance of Davydov solitons on DNA is then fulfilled only for B helix structures. Different biological functions are associated to B (resting) and A (active) DNA; a soliton dynamics could have a major role in DNA structural and functional transitions. Indeed, B → A,[35] B → Z[34] transitions as well as the opening of DNA helices[33] have been described in terms of soliton dynamics.

### 4.2. Water in Biological Systems

Water in biological systems exhibits peculiar properties. Only a percentage of total water appears in the form of usual liquid water.[38,39] A relevant part is made up of water molecules which are bound to different sites of macromolecules in the form of hydration water, while a sizable amount, although not fixed to any definite molecular site, is strongly affected by macromolecular fields; this kind of water has been termed "vicinal" water.[39] "Vicinal" water has been extensively investigated in the last decade and it exhibits noticeable properties.[40] It does not have a unique freezing temperature but an interval ranging from −70 to −50°C; it is a very bad solvent for electrolytes, but nonelectrolytes have the same solubility properties in it as in usual water. Its viscosity is enhanced and its NMR response is anomalous. The whole of these physical properties point to the existence of some dynamic correlation among its constituents.

It is interesting to notice that most of the "vicinal" water surrounds the elements of the cell cytoskeleton.[39] It is not unlikely that the coherent electric waves discussed in this paper are responsible for such a correlation.[20] Bound water also exhibits some kind of collective behavior;

actually it has been shown that bound water produces electrets around many macromolecules. Activation energies of about 0.3–0.4 eV have been detected and relaxation times up to $10^5$ s have been measured, suggesting the existence of zero-frequency modes of oscillation.[9,10] A zero-frequency mode, i.e., in our theoretical framework a massless mode, appears when dynamical rearrangements of symmetry are at work. Since the electret activation energy is of the same order of magnitude as the energy required for a Davydov soliton, we think that the electret is the result, via a spontaneous symmetry breakdown, of the soliton regime occurring on biomolecules.

It is worthwhile noticing that hysteresis has been observed in all cases[9,10]; in our model then the electret would appear different in subsequent cycles (alternance of Davydov and Fröhlich regimes) during the cell lifetime, giving rise to a nonstationary evolution instead of a periodic one (on a time scale comparable with the lifetime of the biological system under consideration).

## 4.3. Coherent Electric Waves in Living Matter

Up to now, different kinds of evidence can be quoted, which support the Fröhlich proposal of electric waves in living matter. First of all, we mention those resonance experiments where external electromagnetic waves of selected frequencies interact with biological targets inducing macroscopic nonthermal consequences. There is a growing literature[41–43] about microwave interaction with different types of cells. It has been observed that at given frequencies the effects occur only when a threshold is overcome and are independent from different intensities of irradiation.

A similar phenomenon, as far as the biological effects are concerned, has been observed in the visible region for a long time,[44] but only recently has satisfactory phenomenological evidence been achieved.[45] In this case, astonishingly, there is only one frequency in the visible domain (at 632.8 nm, the wavelength of He–Ne lasers) which is the same for different types of cells (for instance *E. coli* and He–La). Moreover, active cells emit electric and electromagnetic radiation on a very large frequency spectrum. Measurements have been performed at frequencies lower than 100 GHz[46,47] and in the optical range.[48] This last type of emission, which could be of coherent type, has been attributed to the dynamics of DNA in the cell.

Further evidence for coherent electric waves comes from Raman experiments on living cells.[49] These experiments show a time-dependent pattern of lines in IR and IR regions. The connections between Raman spectroscopic data and the underlying (spatial and temporal) order in living matter is discussed elsewhere.[50–52]

A different kind of support for Fröhlich ideas has been brought by experiments on rouleaux formation in erythrocyte suspensions.[53] These experiments show that selective long-range interactions appear among metabolically active red blood cells. The connection between the experimental outcome and Fröhlich theory has been recently discussed.[20,54]

In conclusion, phenomenological evidence supports the main elements of our QFT approach. We feel that the task of getting a satisfactory dynamical description of living systems is not out of reach.

## References

1. H. Fröhlich, "The Biological Effects of Microwaves and Related Questions," *Adv. Electron. Electron Phys.* **53**, 85–152 (1980); H. Fröhlich, "Long-Range Coherence in Biological Systems," *Riv. Nuovo Cimento* **7**, 399–418 (1977).
2. A. S. Davydov, "Solitons in Molecular Systems," *Phys. Scr.* **20**, 387–394 (1979).
3. A. S. Davydov, *Biology and Quantum Mechanics*, Pergamon, Oxford (1982).
4. E. Del Giudice, S. Doglia, M. Milani, and G. Vitiello, "Spontaneous Symmetry Breakdown and Boson Condensation in Biology," *Phys. Lett.* A **95**, 508–510 (1983); E. Del Giudice, S. Doglia, M. Milani, and G. Vitiello, "A Quantum Field Theoretical Approach to the Collective Behaviour of Biological Systems," *Nucl. Phys.* B **251** [FS 13], 375–400 (1985).
5. G. Vitiello, E. Del Giudice, S. Doglia, and M. Milani, "Boson Condensation in Biological Systems," *Report to the International Conference on Nonlinear Electrodynamics in Biological Systems*, Loma Linda, California, June 1983; E. Del Giudice, S. Doglia, M. Milani, and G. Vitiello, Solitons and Self-Organization in Biological Systems," in *Proceedings of the Second International Workshop on Nonlinear and Turbulent Processes*, Kiev (USSR), October 1983, Gordon and Breach, in press.
6. J. Goldstone, "Field Theories with Superconductor Solutions," *Nuovo Cimento* **19**, 154–164 (1961); J. Goldstone, A. Salam, and S. Weinberg, "Broken Symmetries," *Phys. Rev.* **127**, 965–970 (1962).
7. R. Paul, "Production of Coherent States in Biological Systems," *Phys. Lett.* A **96**, 263–268 (1983).
8. I. Prigogine and G. Nicolis, *Self-Organization in Non-equilibrium Systems, from Dissipative Structures to Order Through Fluctuations*, Wiley, New York (1977).
9. S. Celaschi and S. Mascarenhas, *Biophys. J.* **20**, 273–278 (1977).
10. J. B. Hasted, H. M. Millany, and D. Rosen, *J. Chem. Soc. Faraday Trans.* **77**, 2289–2297 (1981).
11. S. Mascarenhas, personal communication (August 1982).
12. H. Matsumoto, M. Tachiki, and H. Umezawa, *Thermo-field Dynamics and Condensed States*, North-Holland, Amsterdam (1982).
13. H. Matsumoto, P. Sodano, and H. Umezawa, "Extended Objects in Quantum Field Theory and Soliton Solutions," *Phys. Rev.* D **19**, 511–516 (1979).
14. L. Mercaldo, I. Rabuffo, and G. Vitiello, "Canonical Transformations in Quantum Field Theory and Solitons," *Nucl. Phys.* **188B**, 193–204 (1981).
15. H. Umezawa, "Dynamical Rearrangement of Symmetries," *Nuovo Cimento* **40**, 450–475 (1965); L. Leplae, R. N. Sen, and H. Umezawa, "Asymmetric Ground States in Invariant

Many-Body Theories," *Nuovo Cimento* **49B**, 1–31 (1967). H. Umezawa, "Self-consistent Quantum Field Theory and Symmetry Breaking," in *Renormalization and Invariance in Quantum Field Theory* (E. R. Caianiello, ed.), Plenum Press, New York (1974), pp. 275–328.

16. A. S. Davydov and V. I. Kislukha, "Solitons in One-Dimensional Molecular Chains," *Sov. Phys. JETP* **44**, 571–575 (1976).

17. A. S. Davydov, "Energy Transfer Along α-Helical Proteins," in *Structure and Dynamics: Nucleic Acids and Proteins* (F. Clementi and R. H. Sarma, eds.), New York (1983), pp. 377–387 Adenine.

18. A. S. Davydov, "Solitons in Molecular Systems,"Preprint, Kiev ITP-83-115E (September 1983).

19. E. Del Giudice, S. Doglia, and M. Milani, "A Collective Dynamics in Metabolically Active Cells, *Phys. Scr.* **26**, 232–238 (1982).

20. E. Del Giudice, S. Doglia, and M. Milani, "Self-focusing of Fröhlich Waves and Cytoskeleton Dynamics," *Phys. Lett.* **90A**, 104–106 (1982); E. Del Giudice, S. Doglia, and M. Milani, "Actin Polymerization in Cell Cytoplasm" in *The Application of Laser Light Scattering to the Study of Biological Motion* (J. C. Earnshaw and M. W. Steer, eds.), Plenum Press, New York (1983); pp. 493–497; E. Del Giudice, S. Doglia, and M. Milani, "Self-focusing and Ponderomotive Forces of Coherent Electric Waves: A Mechanism for Cytoskeleton Formation and Dynamics," in *Coherent Excitations of Biological Systems* (H. Fröhlich and F. Kremer, eds.), Springer, Berlin (1983), pp. 124–127; E. Del Giudice, S. Doglia, and M. Milani, "Order and Structures in Living Systems," *Report to the International Conference on Nonlinear Electrodynamics in Biological Systems*, Loma Linda, California (USA) June 1983.

21. H. Matsumoto, H. Umezawa, G. Vitiello, and J. K. Wyly, "Spontaneous Breakdown of a Non-Abelian Symmetry," *Phys. Rev. D* **9**, 2806–2813 (1974).

22. M. N. Shah, H. Umezawa, and G. Vitiello, "Relation Among Spin Operators and Magnons," *Phys. Rev. B* **10**, 4724–4736 (1974).

23. C. De Concini and G. Vitiello, "Spontaneous Breakdown of Symmetry and Group Contractions," *Nucl. Phys.* **116B**, 141–156 (1976). C. De Concini and G. Vitiello, "Relation Between Projective Geometry and Group Contraction in Spontaneously Broken Symmetries," *Phys. Lett.* **70B**, 355–357 (1977).

24. F. J. Dyson, "General Theory of Spin Wave Interactions," *Phys. Rev.* **102**, 1217–1230 (1956).

25. S. L. Adler, "Consistency Conditions on Strong Interaction Implied by a PCA Vector Current. I," *Phys. Rev.* **137B**, 1022–1033 (1965). S. L. Adler, "Consistency Conditions on Strong Interaction Implied by a PCA Vector Current. II," *Phys. Rev.* **139B**, 1638–1643 (1965).

26. M. N. Shah and G. Vitiello, "Self-consistent formulation of itinerant Electron Ferromagnet," *Nuovo Cimento* **30B**, 21–42 (1975).

27. A. S. Davydov and A. V. Zolotariuk, "Electrons and Excitons in Nonlinear Molecular Chains," *Phys. Scr.* **28**, 249–256 (1983).

28. G. Careri, U. Buontempo, F. Carta, E. Gratton, and A. C. Scott, "Infrared Absorption in Acetanilide by Solitons," *Phys. Rev. Lett.* **51**, 304–307 (1983).

29. N. A. Nevskaia and Yu. N. Chirgadze, "Infrared Spectra and Resonance Interactions of amide-I and amide-II vibrations of α-helix," *Biopolymers* **15**, 637–648 (1976).

30. F. Drissler and R. M. MacFarlane, "Enhanced Anti-Stokes Raman Scattering from Living Cells of Chlorella Pyrenoidosa," *Phys. Lett.* **A69**, 65–68 (1978); F. Drissler, "Discovery of Phase Transitions in Photosynthetic Systems," **77A**, 207–210 (1980).

31. E. Del Giudice, S. Doglia, and M. Milani, "Solitons in Biological Systems at Low Temperature," *Phys. Scr.* **23**, 307–312 (1981).

32. A. C. Scott, "Dynamics of Davydov Solitons," *Phys. Rev. A* **26**, 578–593 (1982).

33. S. W. Englander, N. R. Kallenbach, A. J. Heeger, J. A. Krumhanols, and S. Litwin, "Nature of the Open State in Long Polynucleotide Double Helices," *Proc. Natl. Acad. Sci. USA* **77**, 7222–7226 (1980).

34. P. Jensen, M. V. Jaric, and K. H. Bennemann, "Soliton-like Processes during Right–Left Transition," *Phys. Lett.* **95A**, 204–208 (1983).

35. P. Beaconsfield and E. Balanovski, "EM-induced B-DNA to A-DNA Transition: Signal Stimulating Conditions for DNA-Mediated Insulin Production and Cell Replication," *Phys. Lett.* **100A**, 172–174 (1984).

36. S.C. Erfurth, E. J. Kier, and W. L. Peticolas, "Determination of the Backbone Structure of Nucleic Acids and Nucleic Acid Oligomers by Laser Raman Spectroscopy," *Proc. Natl. Acad. Sci. USA* **69**, 938–941 (1972).

37. G. J. Thomas and K. A. Hartman, "Raman Studies of Nucleic Acids VIII. Estimation of RNA Secondary Structure from Raman Scattering by Phosphate-Group Vibrations," *Biochem. Biophys. Acta* **312**, 311–322 (1973).

38. R. Cooke and I. D. Kuntz, "The Properties of Water in Biological Systems," *Ann. Rev. Biophys. Bioeng.* **3**, 95–123 (1974).

39. J. Clegg, "Intracellular Water, Metabolism and Cellular Architecture," *Collect. Phenom.* **3**, 289–312 (1981).

40. F. Franks, *Water, A Comprehensive Treatise*, Pergamon (1975).

41. N. D. Devyatkov, "Influence of Millimeter-Band Electromagnetic Radiation on Biological Objects," *Sov. Phys. Usp.* **16**, 568–569 (1974).

42. S. J. Webb, "Nutrition and in Vivo Rotational Motion: A Microwave Study," *Int. J. Quantum Chem. Quantum Biol. Symp. No. 1*, 245–251 (1974); S. J. Webb, "Genetic Continuity and Metabolic Regulation as Seen by the Effects of Various Microwave and Black Light Frequencies on these Phenomena," *Ann. N. Y. Acad. Sci.* **247**, 327–344 (1975).

43. W. Grundler, F. Keilmann, V. Putterlik, L. Santo, D. Strube, and I. Zimmermann, "Nonthermal Resonant Effects at 42-GH Microwaves on the Growth of Yeast Cultures," in *Coherent Excitations of Biological Systems* (H. Fröhlich and F. Kremer, eds.), Springer, Berlin (1983), pp. 21–37.

44. F. Hillenkamp, "Interaction between Laser Radiation and Biological Systems," in *Lasers in Biology and Medicine* (E. F. Hillenkamp, R. Pratesi, and C. A. Sacchi, eds.), Plenum, New York (1979), pp. 37–68.

45. T. J. Karu, O. A. Tiphlova, V. S. Letokhov, and Y. V. Lobko, "Stimulation of *E. Coli* Growth by Laser and Incoherent Red Light," *Nuovo Cimento* **2D**, 1138–1149 (1983).

46. H. Pohl, "Natural Oscillating Fields of Cells," in *Coherent Excitations in Biological Systems* (H. Fröhlich and F. Kremer, eds.), Springer, Berlin (1983), pp. 199–210.

47. A. H. Japary-Asl and C. W. Smith, "Biological Dielectrics in Electric and Magnetic Fields," in 1983 Annual Report Conference on Electrical Insulation and Dielectric Phenomena.

48. K. H. Li, F. A. Popp, W. Nagl, and H. Klima, "Indications of Optical Coherence in Biological Systems and Its Possible Significance," in *Coherent Excitations in Biological Systems* (H. Fröhlich and F. Kremer, eds.), Springer, Berlin (1983), pp. 117–122.

49. S. J. Webb, "Laser Raman Spectroscopy of Living Cells," *Phys. Rep.* **60**, 201–224 (1980).

50. E. Del Giudice, S. Doglia, M. Milani, and M. P. Fontana, "Raman Spectroscopy and Order in Biological Systems," *Cell Biophys.* **6**, 117–129 (1984).

51. E. Del Giudice, S. Doglia, M. Milani, and S. J. Webb, "A Time Consistent Feature as Seen in the Raman spectra of Metabolically Active Cells," *Phys. Lett.* **91A**, 257–260 (1982).

52. E. Del Giudice, S. Doglia, M. Milani, and S. J. Webb, "In vivo Ordered Structures as Seen by Laser Raman Spectroscopy," in *Proceedings of the 2nd International Seminar on the Living State*, Bhopal (India) (1983).

53. S. Rowlands, L. S. Sewchand, and E. G. Enns, "Further Evidence for a Fröhlich Interaction of Erythrocytes," *Phys. Lett.* **87A**, 256–260 (1982); S. Rowlands, "Coherent Excitations in Blood," in *Coherent Excitations in Biological System* (H. Fröhlich and F. Kremer, eds.), Springer, Berlin (1983), pp. 145–161.

54. R. Paul, R. Chatterjee, J. A. Tuszynski, and O. G. Fritz, "Theory of Long-Range Coherence in Biological Systems. I. The Anomalous Behaviour of Human Erythrocytes," *J. Theor. Biol.* **104**, 169–185 (1983).

# 10

# Electrostatic Modulation of Electromagnetically Induced Nonthermal Responses in Biological Membranes

*James D. Bond and Gerald C. Huth*

*ABSTRACT:* A discussion of possible nonthermal responses of biological membranes to very low intensity electromagnetic fields is presented. The role of membrane surface charge in mediating such responses is examined. In particular, means whereby electrostatic surface properties can be systematically altered in order to determine their role in influencing the membrane's response to weak fields is discussed. The very important role played by the membrane's electrochemical environment in determining the charge state of the membrane per se as well in determining the distribution of various ionic species in this environment is discussed in the context of the system (membrane and ambient electrolyte) responding to weak external perturbations.

## 1. Introduction

The existence of nonthermally induced effects in biological systems by weak external electromagnetic fields is a subject of much controversy. Weak fields refer to the fact that the magnitude of the applied is too small to produce effects attributable to thermal coupling. More and more evidence is being presented which suggests, however, that external electromagnetic fields can induce effects of a nonthermal origin in various kinds of biological tissue.[1] A nonthermal effect is defined here to mean an effect which does not result from the thermalization of the electromagnetic

*James D. Bond* • Science Applications, Inc., 1710 Goodridge Drive, McLean, Virginia 22102. *Gerald C. Huth* • Institute for Physics, University of Southern California, 4676 Admiralty Way, Marina del Rey, California 90291.

energy received by the biological system. In other words it is an effect which cannot be caused by heating the system.

Presently there exists no generally accepted theory of how very weak electromagnetic fields couple nonthermally to a biosystem. This is the basis for part of the controversy. Some of the hypotheses that have been suggested to explain such nonthermal interactions of the electric field component will be reviewed herein with an emphasis on those that focus on interactions with biological membranes. It will then be shown how some of these hypotheses might be tested by electrostatically modulating the response. Electrostatic modulation can be achieved by modifying the electrochemical environment of the system.

It must be kept in mind that this general area of nonthermal effects is far from being a mature, well-developed subdiscipline of electromagnetic field effects on biological systems. The material presented in this chapter was selected with the intent of suggesting new protocols, both experimental and theoretical, that will possibly further the development of this field.

## 2. Background and Review of Theoretical Models to Explain Coupling of Electromagnetic Fields to Membranes

The fact that electric fields play an important role in governing a variety of physiological processes has been well established.[2] In fact, Newton, during the seventeenth century, viewed nerve propagation as a phenomenon whose basis resided in optical vibrations within a solid, transparent medium.[3] The correlation between the nature of light and electromagnetic radiation had not yet been established. It was during the middle of the nineteenth century that duBois-Reymond conclusively demonstrated the existence of electric currents in both nerve and muscle behavior.[4] Not until the twentieth century, however, was an idea advanced that established a connection between biological membranes and electric field effects. This was the membrane hypothesis of Bernstein, and it was significant in that it established a functional correlation between the plasma membrane of a nerve cell and the permeabilities of the ionic species known to contribute to the resting potential across the membrane.

In the early 1950s, the membrane hypothesis of Bernstein was modified by Hodgkin and Huxley in an effort to explain certain experimental observations on nerve cell behavior.[5] One of their most significant contributions during this time was to construct an empirical model that accurately described the time dependence of the nerve action potential and the time dependence of the potassium and sodium currents during excitation. They hypothesized that charged particles, possibly electric dipoles within the membrane, are redistributed when the electric field across the membrane is altered, and that these particles govern the movement of various ionic species across the membrane. The Hodgkin–Huxley theory was phenomenological and did not provide a physicochemical basis of membrane excitation.

Although there now exists considerably more knowledge concerning membrane structure, and correlations between membrane structure and ionic permeability changes under a host of different conditions have been documented, there still does not exist a molecular theory explaining excitable membrane phenomena in the usual sense where the application of statistical mechanics yields empirically observed and derived thermodynamic relationships, e.g., the Hodgkin–Huxley equations.

The foregoing remarks serve as a remainder that electric field effects, and the response of biological membranes to changes in the electric fields to which they are exposed, can have important physiological consequences. Membranes serve as viable substrates to which internal fields can couple to regulate physiological processes. The magnitudes of the fields governing excitable membrane phenomena are, however, many orders of magnitude larger than the external electric field strengths responsible for eliciting nonthermal biological responses. The small field magnitude is the argument often used as the basis to preclude the direct action of an external electromagnetic field, e.g., causing the rotation of a polar molecule within the membrane matrix, as being a viable nonthermal means of membrane-field coupling.[6]

The remaining discussion in this section is devoted to the examination of various models and model systems that have been developed in attempts to characterize, in terms of physical and chemical mechanisms, the response of a biomembrane to weak electromagnetic fields. The role and utility of modeling is often misunderstood, and thus a brief explanation is in order here. Models have proved to be an important component in methods that have been successfully employed as "deductively manipulative constructs essential to the evolution of theory from observation."[4] As such, they should serve as supplements to experimentation by suggesting new experiments. As well, they should possess the obvious capability of explaining existing experimental observations.

In order to determine mechanisms of coupling of an external electric field to a biological system, e.g., a membrane and its electrolytic environment, information regarding the nature and role of intrinsic fields within the system is of interest. Intrinsic fields refer here to chemical bonding, electrostatic interactions, as well those internal electric potential gradients established by the asymmetric distribution of various ionic species, e.g., across a nerve cell membrane. There still exists much uncertainty, however, regarding the nature of many of the interactions within composite macromolecular systems, the nature of lipid–protein interactions in biomembranes being a prime example.

Cognizance of the magnitude of long-range van der Waals' interactions, consisting of electrostatic, induction, and dispersion components, is of primary interest in examining the coupling as well as the structural and functional integrity of systems subjected to external perturbations. Of importance also, especially for amphiphilic macromolecular systems such as membranes, are hydrophobic interactions. The origin of the hydrophobic effect does not reside in van der Waals' interactions; in fact van der Waals' interactions are of minor significance in the overall hydrophobic effect.[7] Hydrogen bonding is another important consideration in understanding the structure and properties of water in biological systems, in accounting for the stability of helices formed by polypeptides, and in determining the physical state of various membrane systems. Intrafacial hydrogen bonds, e.g., along the plane of the membrane surface, can make a significant contribution to the overall surface free energy. Hydrogen bonds between macromolecular structures, e.g., within a cellular membrane, have often been neglected because they have either been deemed insignificant relative to hydrogen bonds to water molecules or they have been difficult to detect. Recently, however, their importance has been demonstrated in their ability to influence the thermal phase transition temperature in certain membrane systems.[8,9]

The magnitude of the electric field encountered in initiating athermal responses apparently dictates that the nature of the coupling with a given system be other than via a direct action of the field.[6] In view of this fact, possible cooperative modes of interactions among the components of a macromolecular composite, such as a membrane, are likely candidates for consideration as coupling mechanisms. Cooperative phenomena refers here to those molecular processes that result in physical changes in the structure of matter which are associated with a definite value of an intensive thermodynamic variable, for example the absolute temperature or electric field strength. The idea of cooperative behavior is related to the fact that transitions between different states of matter cannot be understood in terms of individual molecular components; interactions among the components must be taken into account. The possibility of cooperative effects in membranes has been discussed extensively.[10–12] These effects stem from changes in membrane structure triggered by changes in the concentration of bound ligands as well as cooperative responses triggered by an electric field.

The notion of collective phenomena refers also to the fact that certain properties of an assembly of particles, however large or small an assembly, cannot be explained solely in terms of the properties of the individual component particles, but can only be explained in terms of properties of the system of particles as a whole. In a general sense, collective phenomena is synonymous with cooperative phenomena in that interactions among the constituent parts of a given system must be considered. Further discussion of cooperative and/or collective phenomena will be presented in a later section.

The most concerted effort to date that invokes the concept of collective phenomena as a vehicle whereby electromagnetic radiation can couple nonthermally to a biological system, with specific reference to membranes, is that of H. Fröhlich.[13] The ideas concerning the coupling of external electromagnetic fields to biological systems, membranes in particular, stem from his earlier thoughts on metabolically induced coherent electric vibrations and also the existence of highly polarized metastable states within such systems.[14–16]

Specific importance is given to the role of the large electric fields, of the order of $1 \times 10^7$ V/m, encountered in cell membranes. The presence of such large field strenghs could indeed lead to highly polarized states within the membrane. It should be pointed out here that a variety of cells, not just nerve cells, effectively utilize the plasma membrane potential as a means of controlling their physiological function. Examples are nerve and muscle membranes in regulating ionic permeabilities; the release of neurotransmitters by presynaptic membranes; the control of contractile processes in muscle membrane; the secretion of insulin by pancreatic cells; and possibly the secretion of adrenalin by adrenal medulla cells.[17]

The general approach adopted by Fröhlich in one of his models, namely, the vibrational model, is based upon the assumption of electric dipole oscillations of units or subunits of a biological system; e.g., the system might conceivably be a biomembrane, and the units correspond to specific polar segments of the membrane, capable of oscillating with a frequency, $\omega_0$. Interactions among the oscillators can then produce a branch of longitudinal modes in a narrow frequency range such that

$$0 < \omega_1 \leqslant \omega \leqslant \omega_2 \tag{1}$$

where $\omega$ might be significantly different from $\omega_0$. These modes correspond to the establishment of longitudinal electric waves within the given systems.

The precise nature of origin of such oscillating dipoles in biological membranes is speculative, as Fröhlich has pointed out.[13] However, it is conceivable that membrane segments consisting of integral protein could sustain such oscillations. In view of the dielectric properties and the hydrophobic character of the hydrocarbon core associated with the lipid matrix, it is not likely that such oscillations could be sustained by the lipid moieties, at least not in the sense suggested by Fröhlich.[15]

Starting with an assumed form for a rate equation that governs the time rate of exchange of energy quanta with a heat bath in which the putative dipole oscillators are immersed, and considering only first and second order interactions with the heat bath, Fröhlich is able to show that the stationary state solutions for such a system can result in a phase transition analogous to Bose condensation. Such a transition is manifested when the rate of energy supplied to the longitudinal electrical modes exceeds a critical value, and subsequently all energy is channeled into a single mode having the lowest frequency. Externally applied electromagnetic energy could serve to trigger such an event. Fröhlich proposes that such electric vibrations in the range $10^{11}$–$10^{12}$ s$^{-1}$ could be excited; these estimates are based upon a membrane of thickness $1 \times 10^{-8}$ m capable of accommodating an acoustic velocity on the order of $10^3$–$10^4$ m/sec, all reasonable estimates consistent with membrane mechanical properties.

The prediction of the existence of a phenomenon analogous to a Bose–Einstein condensation, as in Fröhlich's vibrational model, has been confirmed via several other approaches, namely, via a transport theory formalism by Kaiser[18] and a molecular Hamiltonian approach by Bhaumik et al.,[19] and also by Wu and Austin[20,21]; the basic concept has been shown to be on firm theoretical ground. The difficulty in these ideas gaining wide acceptance is the lack of conclusive experimental evidence to provide confirmation that such effects occur in biological membranes. Furthermore, the theory presented thus far is not at the stage where it can be used in an analytical sense to explain those data thus far reported or be used in a predictive manner.

It would be difficult to incorporate detailed microstructural properties of membranes into Fröhlich's vibrational model, as he points out. However, examination of how variations of certain macroscopic characteristics of a given membrane system might modulate the proposed states of coherent excitation could be conducted. This might be accomplished empirically by systematically investigating the effects of changes in ambient pH and ionic strength on metabolically intact membrane systems, since metabolic input is a necessary ingredient in this model. For example, modulation as a consequence of altering the membrane surface charge density could conceivably be manifested in a variety of ways, some of which are discussed in later sections.

The general concept of the excitation of coherent oscillations in biological systems, with limited reference to membranes, has been nicely reviewed by Kaiser.[22–24] A discussion of Fröhlich's model is presented in terms of how it possesses the requisite characteristics to undergo various types of phase transitions subsequent to external perturbations. Reference, however, to the specific involvement of biological membranes in such phase transitions is only briefly mentioned. The emphasis in Kaiser's analysis of Fröhlich's basic hypothesis is on the idea that stable limit cycle behavior serves as an adequate description for coherent oscillations in biosystems, and that such limit cycles can be driven by external fields. This concept provides a basis of explanation for the coupling of weak external fields to biological tissue. For example, the inherent limit cycle behavior can serve as a means of storing metabolic energy provided by the system, and weak external perturbations, e.g., an electromagnetic field, can ultimately result in limit cycle collapse.

Another model proposed by Fröhlich, and extended by Kaiser, is referred to as the high-polarization model.[13,20,23,25,26] This idea is predicated on the establishment of metastable strongly polarized states. The fact that biological membranes can sustain a high degree of electric polarization—for example, protein imbedded within the membrane matrix can conceivably reside in states of high polarization—allows for membranes to serve as a possible biological complex in which such a state could be realized. According to Fröhlich[13] and Bhaumik *et al.*,[27] the strong excitation of polar modes results in the deformation of the given system, and thus the activation of elastic constraints. Such a response can lead to polarization mode softening and the establishment of a ferroelectric state. Bhaumik *et al.*[27] have shown that the energy threshold required to initiate the Bose-condensation-like phenomenon discussed previously is actually lowered when elastic deformations are considered.

The relevance of the models proposed by Fröhlich and extended by Kaiser, Bhaumik *et al.*, and Wu and Austin, to biological membranes as being the coupling substrate for external radiofrequency fields resides in the fact that

1. membranes can be considered as open systems typically found in nonequilibrium steady states;
2. membranes possess the macromolecular components, as well as the appropriate ambient electrolytic environment, compatible and consistent with the basic tenets of the Fröhlich concepts;
3. membranes can accommodate the requirement of both models for energy input via the utilization of metabolic energy produced by the cell;
4. membranes can be viewed as composite macromolecular systems capable of supporting collective or cooperative interactions that can be triggered via various stimuli, low intensity external electromagnetic fields being only one example.

A model that invokes the concept of cooperative behavior and simultaneously incorporates to some extent membrane structure has been formulated by Grodsky.[28–30] This model is based upon a type of lattice statistics formulation in which the polar head groups of membrane lipid moieties occupy one set of lattice sites, and displaced normal to the plane of this lattice is another lattice configuration corresponding to cationic binding sites associated with membrane glycoprotein residues. A Hamiltonian is developed describing the interaction among neighboring sites within a given lattice, interactions between the two lattices, and interactions of both lattices with an external field. The model represents an attempt to incorporate basic membrane physical and chemical characteristics, for example dipole–dipole coupling, charge–dipole coupling, and ligand binding into a framework that is not purely phenomenological. The formalism used by Grodsky is analogous to that developed to describe ferromagnetic and antiferromagnetic systems and phase transitions in such systems.[31]

The assumption of the various configurations of the polar head groups of the phospholipid molecules in Grodsky's model is questionable in view of the stereochemical constraints associated with a given polar head group and the presence of intramembrane hydrogen bonding along the plane of the membrane between adjacent lipids. It is true that the polar head groups of the phospholipid molecules within the membrane are to some degree flexible; however, in view of the aforementioned constraints it is difficult to imagine, for instance, an antiparallel configuration as postulated by Grodsky. Attempts to employ similar lattice statistics arguments based on cooperative dipole–dipole coupling among phospholipid head groups to account for changes in membrane permeability in excitable membranes have not been successful.[32] The general concept of a critical value of an externally applied electric field precipitating a phase-transition-like event has previously been suggested as a possible mechanism to account for various other membrane phenomena and is certainly worthy of further consideration.[10]

One major conceptual difference between Grodsky's model and those proposed by Fröhlich is the role of metabolism; Grodsky's model does not require metabolic input. Grodsky postulates the existence of resonance behavior with the inherent frequencies of the coupling structures, and suggests, too, that the rectification of time-varying electromagnetic perturbations could be accommodated by the cationic lattice in his model behaving as a natural diode. These suggestions are highly speculative, and there is no experimental evidence to support them.

Perhaps it is appropriate to interject here that many of the attempts at modeling the effects of electromagnetic fields on biological systems have been, to put it mildly, highly conjectural. An attempt to justify such a conjectural approach has been discussed by Fröhlich.[13] It is apparent, however, that there should be a more conscious effort to correlate theory and experiment, especially in the area of membrane effects.

Grodsky does suggest many interesting analogies between known processes in various physical systems, for example, the physics of solids, and how similar processes might exist in electromagnetic field–membrane interactions. However, these ideas have not been carefully developed within an analytical framework appropriate to correlate with membrane properties.

Lawrence and Adey have presented a model of membrane–field interactions based upon the possible establishment of solitary wave motion, solitons, within the membrane complex.[33] While there exists no clear-cut evidence of soliton formation in cell membranes, it is conceivable that membranes could support the propagation of such solitary waves via nonlinear coupling among the constitutent molecules. Fröhlich has previously suggested the possibility of soliton excitation in biological systems in connection with his high polarization model.[13] The basis of the Lawrence–Adey model is Davydov's work[34] on soliton formation in biological systems coupled with work by Vaccaro and Green[35] and Triffet and Green[36] in which they model the current and conductance changes in excitable membranes in terms of nonlinear plasma oscillations.*

Cain has presented a phenomenological description[37] of the effects of radiofrequency fields on excitable membranes by examining the effects of a time varying potential, $V(t)$, on the voltage-dependent rate constants, $\alpha$ and $\beta$, as originally employed by Hodgkin and Huxley[5] to account for ionic conductance changes in squid axon membranes. Assuming $V(t)$ of the form

$$V(t) = V_0 + V_m \cos \omega t \qquad (2)$$

where $V_0$ is the membrane resting potential, $V_m$ is the amplitude of applied potential, $\omega$ is the angular frequency, and $t$ is time, a derivation of the corresponding $\alpha$ and $\beta$ is presented, and the sodium and potassium conductance calculated via the Hodgkin–Huxley formalism. Although such an approach does not explicitly yield information regarding molecular mechanisms at the membrane level, it is informative in the sense that it does predict how the ionic conductances of the Hodgkin–Huxley theory should vary according to such a time-varying perturbation, and thus could be tested experimentally. Indirectly such approaches can provide insight into possible mechanisms in the same manner in which numerous modifications of the Hodgkin–Huxley phenomenology have provided insight into possible molecular mechanisms associated with membrane excitation. Cain hypothesizes that the presence of an oscillating field in an excitable biological membrane can lead to conductance changes through an alteration of the state distribution of gating particles within the membrane.[38,39] Although there is no experimental evidence of such, it

---

* See also Chapter 9 in this volume by Del Giudice, Doglia, Milani, and Vitiello.

would be interesting to examine experimentally the effects of time varying fields on gating currents per se.

The suggestion by Cain that the effect occurs through an actual dipolar reorientation would require unusually large field strengths or extremely large dipole moments. The prediction of an inhibitory effect due to membrane hyperpolarization because of increased steady-state potassium conductance concomitant with decreased steady-state sodium conductance certainly should merit enough attention to design experiments to test the existence of such a response.

Barnes and Hu[40,41] examined the effect of a time-varying field on the concentration profile of ionic species adjacent to either surface of the plasma membrane. The concentration profile was described by a Boltzmann distribution and the potential associated with the impressed field was described by an expression of the form of Eq. (2). Estimates of the field strength within the membrane as a function of the dielectric constants and resistivities of the membrane and its aqueous environment and calculated values for the field strength within the aqueous surroundings were obtained via an equivalent circuit analysis. Numerical estimates of the shift in the ionic concentration profile for an incident power density of $20 \text{ mW/cm}^2$, which yields a 9-$\mu$V potential drop across the membrane, are of the order of one part in $10^6$, and an associated transmembrane ionic current density of $6 \times 10^{-11} \text{ A/cm}^2$, according to the model. We do not discuss the possible specific biological implications of such effects here, other than to state that alterations in the distribution of the concentration of ions adjacent to either membrane surface could influence a number of surface-specific membrane processes.

The above model of Barnes and Hu does not address directly the question of how very low intensity electromagnetic fields can couple to a membrane. Their suggestion of the orientation of polar molecules due to the torque they experience in the presence of an external field would not be a viable mechanism for field strengths associated with nonthermal phenomena.

An approach somewhat analogous to that of Barnes and Hu was adopted by Pickard and Rosenbaum.[42] They examined the effects of an impressed radiofrequency field on the resting potential across the plasma membrane and the effects of such fields with regard to ion transit time within an ion channel spanning the membrane. Of interest is their conclusion that the frequency range at which transit time effects are no longer important lies well below the microwave portion of the spectrum. This estimate was based upon data for the sodium ion current in the squid axon membrane.

Pickard and Rosenbaum also presented some rough calculations to determine the frequencies at which electromechanical resonance effects between an impinging field and the displacement of so-called gating particles within excitable membranes might be important. Their estimates fall within the microwave range. However, the models they used to represent the "ion gates" are, as they point out, extremely crude, and their results should at best be considered only as an order of magnitude estimate. The concept of gating particle resonance with an applied field is an interesting one, and in principle could be studied experimentally by looking for alterations in gating currents under various exposure conditions.

While the modeling efforts of Barnes and Hu, Cain, Pickard, and Rosenbaum are not likely to reveal the precise nature of athermal coupling between field and membrane, they are useful in that correlations can conceivably be made, in the case of excitable membranes, between possible changes in membrane permeability to various ionic species, effects on gating currents, displacement of the resting potential, etc., to the frequency and intensity of the applied radiation, as well as to the duration of exposure.

Nazarea[43] has adopted an interesting approach to studying electromagnetic field effects on ionic conductance through biological membranes by examining the noise spectra associated with transmembrane conductance. Analysis of electrical noise spectra in general has provided valuable insight into conductance mechanisms in the excitable membranes of nerve and muscle.[44-46] The fluid-mosaic model serves as the structural basis for his analysis with the assumption that a conformational change in transport proteins alters the conductance properties of these proteins. Furthermore, it is assumed that this conformational change is initiated by a perturbation of the orientational order of the phospholipid matrix in which these proteins are imbedded, and that fluctuations in the conductance has an autocorrelation function proportional to the autocorrelation of the phospholipid order parameter. The time-varying field perturbation is assumed to couple only to first-order fluctuations of the dynamic parameter that characterizes the phospholipid orientational order. Nazarea is able to show that a field perturbation can indeed alter membrane properties such that variations in the conductance noise spectra can be expected. Unfortunately, no specific calculations or comparisons with experimental data were presented. This technique, however, could in principle be a very powerful tool for elucidating the coupling mechanisms between field and membrane.

Having surveyed the literature concerned with the possible mechanisms of nonthermal interaction of nonionizing electromagnetic fields with biological membranes, it becomes readily apparent that this is an area in which there is a significant need for additional theory and/or modeling. As evidenced by many of the papers referenced above, there exists a great deal of speculation in a number of modeling efforts thus far developed. This is not necessarily bad as long as it can provide insight for the development of new experimental protocols or serve as a basis for the development of more rigorous theoretical treatments that can be realistically compared to experiment. Conjectural treatments that provide no insight for new experimental and/or theoretical approaches and that are inconsistent with existing data quite obviously serve no useful purpose.

The ideas that have been advanced by Fröhlich are based on sound theoretical principles even though they, too, as Fröhlich himself states, are somewhat speculative. It would be of considerable value if his ideas were examined more closely in an effort to devise additional experiments that would serve to test these concepts. There exists a tremendous literature on cooperative or collective phenomena in physics, chemistry, biophysics, and biochemistry. As discussed in the text of this review, this appears to be the vehicle by which many of the observations that implicate an athermal membrane response will be explained, and therefore more theoretical effort in this area would be desirable.

## 3. Cooperative Coupling between Membranes and External Fields

In view of the improbability of direct field effects accounting for nonthermal coupling, dipole reorientation being an example, it has been suggested that the underlying mechanism must be explained in terms of a cooperative or collective response triggered by the electric field component of the perturbing influence.[18,47] There exists an extensive literature discussing cooperative phenomena in biological membranes per se as well as certain macromolecular components that reside within the membrane complex.[10–12,48–51] Many of the arguments previously used that invoke cooperativity are based on equilibrium thermodynamics and statistical mechanics. Some of these ideas will be explored in more detail later in this section. While hopefully much insight can be gained by cautiously using such approaches, techniques and concepts in nonequilibrium thermodynamics and statistical mechanics are maturing rapidly, and it is likely that major breakthroughs in theoretical molecular biophysics will result from the application of such. In reality, many biological systems are far from equilibrium and comprise what are called dissipative structures.[12,52] Such structures by the way are consistent with the ideas proposed by Fröhlich[13] in the sense that they can be "switched" to a coherent mode. A nonequilibrium phase transition that manifest itself as a limit cycle is a familiar example of such a phenomenon.[24,25] The possible role of these types of nonequilibrium phase transitions in governing the interaction of external electromagnetic fields with biological systems has been discussed by Kaiser.[24]

One of the first attempts to examine the consequences of an electric-field-induced cooperative response was by Hill.[10] This effort was the basis for the later work of Blumenthal, Changeux, and Lefever[12] and subsequently the work of Hill and Chen.[53-55] Most of these efforts, however, were directed toward understanding electric-field-induced excitability in those classes of membranes commonly referred to as excitable membranes, for example as in nerve and muscle. They were not attempts to model the interaction of external time-varying electric fields with biological membranes. Most all of these formulations were based on some type of mean-field theory and the use of lattice statistics. More recently Grodsky[28-30] and Denner and Kaiser[56] performed somewhat analogous calculations with reference to a specific dipole model of an excitable membrane. Both analyses used a type of mean field theory to generate the thermodynamic expressions used to describe the behavior of the systems.

In the remainder of this section, certain notions suggesting the possibility of electric field triggered cooperative responses, introduced originally by Hill,[10] shall be examined to explore their possible relevance in explaining nonthermal coupling. Recall that this is an equilibrium model, the rationale being that even though the membrane system under study resides in a nonequilibrium state, the degrees of freedom of interest are in thermal equilibrium.[56] An alternative hypothesis is the one adopted by Blumenthal, Changeux, and Lefever[12] in which they assumed the membrane is a stable, dissipative structure in a nonequilibrium state. An external perturbation in the environment could cause such a system to make the transition from an initial nonequilibrium stable configuration to another nonequilibrium, but stable, configuration. An interesting and perhaps significant point made in their analysis was that cooperative molecular interactions of membrane subunits can be enhanced by the dissipation of energy in the membrane system. This is especially important in view of the fact that certain experimental data, for example alterations in $Ca^{+2}$ efflux from the membrane surface,[57] suggest that cooperative structural transitions are involved in membrane–field interactions. The specific coupling, for example, might occur via some collective, coherent excitation as suggested by Fröhlich,[13] which ultimately is manifested in a cooperative structural transition, perhaps a structural phase transition in certain cases.

Consider the membrane to be a two-dimensional lattice. Each lattice site is assumed to exist in one of two states, state $A$ or state $B$. Also each site may be occupied by a bound ligand or it may be "empty," meaning no bound ligand. For a membrane of $M$ lattice sites, Hill[10] has shown that the grand partition, $\Xi$, is of the form

$$\Xi(T, E, \lambda, M) = \xi^M \tag{3}$$

where $\xi$ is the grand partition function for a subunit, $T$ is the temperature,

$\lambda$ is the absolute activity of the ligand, and $E$ is the electric field strength at the lattice site; the characteristic thermodynamic equation takes the form

$$-d(\mu'M) = d(kT \ln \Xi) = S \, dT + P \, dE - \mu' \, dM + N \, d\mu \qquad (4)$$

where $\mu'$ is the chemical potential of a subunit, $k$ is Boltzmann's constant, $S$ is the entropy, $P$ is the total electric polarization, $N$ is the number of bound ligands, and $\mu$ is the chemical potential of the ligand. The product $(\mu'M)$ is analogous to what is sometimes referred to as the Landau potential, $-(pV)$, in thermodynamics where $p$ is pressure and $V$ is volume. The explicit representation for $\xi$ takes the form

$$\xi = j_A(\exp(-u_A/kT))[\exp(\alpha_A E^2/2kT)](1 + q_A\lambda)$$
$$+ j_B(\exp(-u_B/kT))[\exp(\alpha_B E^2/2kT)](1 + q_B\lambda) \qquad (5)$$

where $j_A$ is a partition function for a unit in state $A$, $q_A$ is a partition function of a bound ligand at a site in state $A$, $\alpha_A$ is the polarizability of subunit in state $A$, and $u_A$ is the interaction free energy between a unit in state $A$ and its nearest neighbor. Similar definitions hold for symbols with the subscript $B$. Hill used the Bragg–Williams approximation in determining the interaction free energies. The result obtained by Hill based on the above formalism that is of interest to us is the condition for a phase transition to occur. That condition is

$$\frac{j_A \exp(-cw_{AA}/2kT)}{j_B \exp(-cw_{BB}/2kT)} \left(\frac{1 + q_A\lambda}{1 + q_B\lambda}\right) \exp[(\alpha_A - \alpha_B)E^2/2kT] = y \qquad (6)$$

where the $w_{AA}$ and $w_{BB}$ are the nearest neighbor interaction free energies between $AA$ and $BB$ sites, respectively, and $y = 1$ at the phase transition. From Eq. (6), it is obvious that $\lambda$ and/or $E$ can initiate or "trigger" a phase transition. The field, $E$, that was the original focus of Hill's analysis was the field associated with the transmembrane potential. The point is often made that this field is exceptionally large, on the order of $1 \times 10^7$ V/m. It is important to recognize that according to Eq. (6) $E$ serves only to trigger the transition, so all that might be necessary for the transition to occur is to bring about a small change in $E$. It is not necessarily required that the stimulus be of the same magnitude as $E$. The possibility of an external field coupling specifically to a membrane bound enzyme, and how such coupling can lead to variations in membrane potential, will be explored in more detail in a later section.

Equation (6) suggests another means of learning more about the role of a membrane electric-field-induced cooperative response, which may be a phase transition if $y = 1$. That means is through $\lambda$, which is proportional to the concentration of ligand molecules, and which, from Eq. (6), can also serve to trigger a response. An electric-field-induced cooperative response at some given value, $E$, can be modulated by varying $\lambda$. Examining the resulting values of $E$ at which the putative phase transition occurs, treating $\lambda$ as a variable parameter, might yield additional information concerning the nature of membrane–field coupling, assuming we know how varying $\lambda$ alters the physiochemical state of the membrane. As Hill has pointed out $\lambda$ or $E$ or both can serve to initiate a phase transition. One experimental approach to take in investigating such effects would be to examine the effect on $E$ of a systematic variation in the concentration of a known ligand in a membrane system that is well characterized in terms of membrane–ligand interactions. Experiments of this kind could provide additional guidance for future theoretical investigations. The ligand here does not necessarily have to be charged. The case of charged ligands and the role of membrane surface charge will be discussed in a later section. It must be remembered, too, that there still exist many uncertainties about membrane structure and membrane structure–function relationships in both excitable and nonexcitable membrane systems, so it would be highly desirable to select a system that is reasonably well understood.

Another possibility, of course, is that the cooperative response, whether it is manifested as a structural phase transition or some higher-order collective interaction, e.g., a Fröhlich type transition, or some combination of both, does not necessarily have to directly involve the membrane potential or membrane field, $E$, as defined by Hill.

One way in which to incorporate such an effect is to add an additional term to the Hamiltonian assumed by Hill. Although he did not explicitly write out the Hamiltonian, $H$, describing his system, he tacitly assumed it to be separable and of the form

$$H = H_j + H_E + H_u + H_q \tag{7}$$

where the terms $H_j$, $H_E$, $H_u$, and $H_q$ correspond to the contributions from the individual sites $(H_j)$, the interaction of a site with $E$, i.e., polarization effects $(H_E)$, the interaction among or between sites $(H_u)$, and the absorption of ligands $(H_q)$. We can add an additional term to the Hamiltonian, $H$, which takes into account the interaction with a time-varying field. It is assumed that any polarization effects due to this contribution are negligible compared to those in $H_E$. Equation (6) now becomes

$$\frac{j_A \exp(-cw_{AA}/2kT)}{j_B \exp(-cw_{BB}/2kT)} \left(\frac{1+q_A}{1+q_B}\right)$$
$$\times \exp[(\alpha_A - \alpha_B)E^2/2kT][\exp(w_A - w_B)/kT] = y \tag{8}$$

where $w_A$ and $w_B$ are the interaction energies of sites in state $A$ and state $B$, respectively, with the impressed external field. No attempt is made here to write down the explicit functional form of $w_A$ or $w_B$. Both are dependent on the magnitude and frequency of the applied field, $E'$. That is $w_A = w_A(E')$ and $w_B = w_B(E')$. Thus from Eq. (8) for a given $E$ and $\lambda$, $E'$ can serve to trigger the transition. It has been reported in a variety of experiments that the obsrved effects are very frequency and field intensity specific. Equation (8) suggests, as did Eq. (7) for $\lambda$ and $E$, that $\lambda$, $E$, and $E'$ collectively might control the onset of cooperative response. It would be of interest to experimentally examine the effects of systematic variations in $\lambda$ and $E$. Since $E'$ cannot be measured directly because it is the local field seen by the tissue, this could be accomplished indirectly by looking for variations in the power density and/or frequency windows reported to date.[57]

If cooperative structural transitions or collective interactions, or some combination thereof, are involved in the electromagnetic nonthermal coupling of fields to biological systems, membranes in particular, it appears that such effects could be readily modulated by altering the electrochemical environment in a systematic manner as suggested above in the case of ligand concentration. The sensitivity of the observables to such changes in the various kinds of experiments in which this could be done is difficult to predict *a priori*. It can be done, however, by performing specific model calculations. Computer simulations of such experiments are currently in progress for the specific case of the reported alterations in $Ca^{+2}$ efflux.[58] In view of all the uncertainties ranging from those associated with membrane structure to membrane–field coupling mechanisms, it would be useful in the development of theoretical models to have access to a range of experimental data reflecting the role of the electrochemical environment of the system. Aspects other than ligand binding will be discussed in subsequent sections.

## 4. Role of Membrane Surface Charge in Modulating Cooperative Responses to External Stimuli

Phospholipid molecules can be viewed as forming the membrane structural matrix within which various protein molecules are imbedded. Those specific proteins are generically referred to as integral proteins (they comprise approximately 70% of the total membrane protein) to distinguish them from the more loosely bound peripheral protein. The lipid moieties are either acidic or zwitterionic at physiological pH. The proteins, both integral and peripheral, normally are characterized by acidic groups and can be charged or strongly polarized. Electrochemical characteristics of the electrolytic environment in opposition with the membrane strongly influence the electrostatic state of the membrane surface, and while under physiological conditions only small variations in pH or ionic strength might occur, effecting changes in these variables externally under controlled experimental conditions permits them to be used as effective probes to examine membrane structure–function relationships.

It is interesting to note that, historically, protein played the favorite role in attempts to understand the functional dynamics of cellular metabolism and membrane physiology. Once the functional importance of the lipid component of membranes was recognized, most of the initial experimental effort was aimed at hydrophobic effects. However, it is now recognized that surface charge effects associated with membrane phospholipids have been shown to profoundly affect enzymatic activity and influence the action of various hormones, vitamins, and drugs.[59]

There exists considerable evidence for the occurrence of thermal phase transitions in pure phospholipid membranes and intact biological membranes.[60] At the transition temperature the following changes have been observed for a transition from the ordered to fluid state:

a. the formation of rotational isomers within the hydrocarbon chains;
b. the onset of rapid lateral diffusion of the lipid molecules;
c. an expansion of the bilayer structure.

Biological systems, however, are for the most part reasonably constant in temperature. As described in the last section another possible "trigger" to initiate a phase transition might reside in the form of an external electromagnetic field provided that the field can couple to the membrane via some direct or indirect mechanism. Whatever the coupling mechanism might be, if indeed there is a phase-transition-like event analogous to a thermal phase transition with concomitant structural changes, then the electrostatic interactions at the membrane surface are likely to be important in mediating the response. Although electrostatic interactions between surface polar groups do not alter the basic bilayer structure, they can influence the conformation, stability, and binding affinity of the component macromolecules. The role of surface electrostatic interactions in affecting the onset of thermal phase transitions in certain phospholipid membranes has been clearly demonstrated by Trauble *et al.*[61] They examined the effects of surface electrostatic interactions on membrane structure by studying the pH and ionic strength dependency of the phase transition temperature. Assuming the phase transition to be a reversible two-state transition they showed that a shift in transition temperature could be accounted for in terms of a change in the electrostatic contribution to the molar Gibbs free energy. That is, using their notation

$$\Delta T_t = \frac{\Delta G^{\text{el}}}{\Delta S^*} \tag{9}$$

where $T_t$ is the phase transition temperature, $G^{\text{el}}$ are the electrostatic contributions to the molar Gibbs free energy, and $S^*$ is the entropy of an identical membrane bearing zero net charge. An explicit functional form for $\Delta G^{\text{el}}$ was derived, that is

$$\Delta G^{\text{el}} = -\left(\frac{\varepsilon}{\pi}\right)\left(\frac{kT}{e}\right)^2 \kappa [\cosh(e\psi_0/2kT) - 1]\, \Delta f \tag{10}$$

where $\varepsilon$ is the dielectric constant (permittivity), $k$ is Boltzmann's constant, $e$ is the protonic charge, $\kappa$ is the inverse Debye length, $\psi_0$ is the membrane surface potential, and $f$ is the area occupied per charged lipid molecule. Consider now the following notion. Let $W$ be the energy absorbed by a membrane exposed to an external electric field. Define an effective or equivalent temperature, $T_{\text{eff}}$, such that

$$T_t \rightarrow T_{\text{eff}} = \frac{W}{k} \tag{11}$$

Then,

$$\Delta T_t \rightarrow \Delta T_{\text{eff}} = \frac{\Delta W}{k} = \frac{\Delta G^{\text{el}}}{\Delta S^*} \tag{12}$$

where $G^{\text{el}}$ is governed by Eq. (10) and values for $\Delta S^*$ are readily obtained from the literature.[61] $\psi_0$ in Eq. (10) is a function of membrane surface charge density and ionic strength. Trauble *et al.* used a well-known approximation, namely the Gouy–Chapman approximation,[62] to relate $\psi_0$ to the surface charge density and ionic strength. The various other assumptions, approximations, etc. are discussed in detail in Ref. 61. The point is that Eq. (12) tells us that shifts in $W$ could be brought about by changing the electrostatic molar free energy. This can be accomplished by systematically varying the ambient pH and ionic strength. Admittedly Eqs. (11) and (12) were introduced in an *ad hoc* fashion, and they do not provide us with explicit information on how the membrane–field coupling takes place. However, if the external field does trigger a phase transition, or merely a cooperative response that is a function of membrane surface charge density, manifested via $\psi_0$, then it is likely that the $T_{\text{eff}}$ can be influenced by systematically altering $G^{\text{el}}$ as already mentioned. Again it is difficult to predict *a priori* the details of any shift in $T_{\text{eff}}$. Experimentally, however, a search for such shifts could be readily conducted as a function of pH and ionic strength, and examination of the resulting data for trends in shifts, etc. might provide information into the nature of the specific coupling. Forgetting the distinction between cooperative structural phase transitions and collective interactions that do not necessarily manifest themselves through structural charges, it is possible that by systematically altering $G^{el}$ and examining the consequences, i.e., any differences in previously reported nonthermal effects involving membranes, useful information might be obtained about the nature of the membrane–field interaction mechanism. For example, a means of discriminating between membrane–field coupling at the level of integral membrane proteins functioning as enzymes and coupling at a higher level of organization involving the composite membrane structure could be determined by looking at the system's response over a range of pH values and a range of ionic strengths. Modulation of the coupling to specific enzymes is not likely to be the same as for phospholipid domains, for example. Further discussion of the direct alteration of enzymic activity is presented in the next section.

## 5. Alterations in Electrical Double Layer Structure by an External Field Coupling to the Membrane

Another important consideration in attempts to understand the coupling of electromagnetic fields to biomembranes is the subsequent or concomitant effect on the structure of the electrical double layer adjacent to the membrane surface upon exposure. Alterations in double layer structure could, for example, occur if the net effect of the membrane–field interaction results in a redistribution of membrane surface change. There exist other possible means of effecting changes in the structure of the double layer, some of which will also be discussed below.

Consider a planar membrane bearing a uniform negative surface charge density. A field-induced change in the surface charge density would alter the profiles of cationic build-up and anionic depletion in the "diffuse" region adjacent to the membrane surface.[61,62] The thickness of the diffuse double layer, however, would not be altered. This can be seen by examining the functional form of the inverse Debye length, $\kappa$, whose reciprocal is a measure of the extension from the membrane surface of the double layer. For example for a 1:1 electrolyte

$$\kappa = \left[ \left( \frac{8\pi}{\varepsilon} \right) \frac{ne^2}{kT} \right]^{1/2} \tag{13}$$

where $n$ is the bulk salt concentration. The other symbols have been previously defined. As seen from Eq. (13) $\kappa$ does not depend on surface charge density. The potential profile in the diffuse region is given by[62]

$$\psi(x) = \psi_0 \exp(-\kappa x) \tag{14}$$

where $x$ is the distance from the membrane surface located at $x = 0$, and $\psi_0$ is the membrane surface potential, i.e., at $x = 0$. However, $\psi_0$ is a function of the membrane surface charge density, $\sigma$.[62] That is

$$\psi_0 = \left(\frac{kT}{e}\right) \sinh^{-1} \left(\frac{2\pi\sigma e}{\kappa\varepsilon kT}\right) \tag{15}$$

Changes in $\sigma$ yield changes in $\psi(x)$. Further discussion of specific changes in the ion concentration profiles, which depend on $\psi(x)$, will be presented later in this section.

There exists another likely possibility whereby membrane–field coupling results in an alteration of the double layer structure. Assume that the external field interacts specifically with a membrane bound enzyme such that the reaction rate, $J_R$, between this enzyme and a charged substrate in the electrolytic environment is altered. That is, the effect of the field is to alter the enzymic activity which is manifested as a change in $J_R$. Most biological membranes contain active enzymes, and it has been previously postulated that external time-varying fields can alter the activity of membrane-bound enzymes.[63] The specific influence of enzymic surface reactions on the structure of the electrical double layer has also been considered.[64,65] Therefore, alterations in enzyme–substrate reaction rate by an external perturbation will lead to alterations in double layer structure. DeSimone[64] and Pennline et al.[65] have called attention to the physiological significance of structural modifications in the double layer due to the enzymic activity of membrane-bound enzymes. They demonstrated that the potential profile originating at the membrane surface and extending into the ambient extracellular electrolyte can differ markedly from that derived from Gouy–Chapman theory.[62] The deviations in the calculated potential profile are the result of a nonequilibrium contribution to the potential. The functional form of the parameter, $\mu$, which characterizes the nonequilibrium contribution to the potential is

$$\mu = \frac{J_R}{2\kappa c} \left(\frac{1}{D_2} - \frac{1}{D_1}\right) \tag{16}$$

where $c$ is the total anion or cation concentration, $D_1$ is the diffusion coefficient for the substrate reacting with the bound enzyme, $D_2$ is the diffusion coefficient for the product of the enzyme–substrate reaction, and the other parameters have been previously defined. An external perturbation in the form of an electromagnetic field could modify $\mu$ through $J_R$ as suggested above. For certain physically realizable values of the parameters governing the nonequilibrium contribution to the double layer potential profile, DeSimone[64] showed that surface reactions between enzyme and substrate could result in an almost discontinuous change from a case of low surface concentration to a case of high surface concentration. The imposition of an external electromagnetic field that could affect the physiological integrity of a natural concentration switching mechanism could obviously be significant. The possibility of such a mechanism governing the reported alterations in divalent cationic calcium efflux from such cerebral tissue is presently under study.[58] A case where interference caused by electromagnetic effects with enzyme activity might play a role in mediating the membrane response is found in those instances where either capacitive or inductive coupling techniques have been employed to enhance the union of recalcitrant bone fractures.[66]

Any modification in the double layer potential affects the concentration profile of ions that reside within the double layer structure. In the case of an enzyme–substrate reaction, in which a nonequilibrium contribution to the potential arises due to differences in substrate and product diffusion coefficient, as seen from Eq. (16), variation in the equilibrium concentration profiles of not only charged reactant and charged product but also variations in the concentration profiles of nonreacting ionic species present is possible.[64,65] The concentration profiles of all ionic species adjust according to changes in the reaction rate, $J_R$.[64] In order to calculate $J_R$ a specific kinetic formalism must be adopted. This has been done for the case of Michael–Menten kinetics.[64,65] In a more general sense, however, the relationship between reaction rate, $J_R$, potential profile, $\psi(x)$, and ion concentration profile, $c(x)$, is seen by starting with an examination of the Nernst–Planck electrodiffusion equations,

$$J_i = -D_i \left[ \left( \frac{dc_i}{dx} \right) + \left( \frac{z_i e}{kT} \right) c_i \frac{d\psi}{dx} \right] \tag{17}$$

where the subscript $i$ refers to the $i$th ionic species and $z_i$ is the valence of species $i$. The steady state solution to Eq. (17) for the $c_i$ as a function of the potential, $\psi$, and of the $J_i$ can be readily obtained for a given system.[65] The equation governing the potential is Poisson's equation,

$$\frac{d^2\psi}{dx^2} = -\frac{4\pi\rho(x)}{\varepsilon} \tag{18}$$

$$\rho(x) = \sum_{i=1}^{n} z_i e c_i(x) \tag{19}$$

the $c_i(x)$ being obtained from the solution of Eq. (17) and $n$ is the number of different ionic species present. Combining Eqs. (18), (19) and the solutions of the $c_i(x)$ from Eq. (17) results generally in a set of nonlinear ordinary integrodifferential equations. The solution to this set of equations has been found by Pennline et al.[65] for a given boundary value problem. The primary point to be made is that the potential profile, $\psi(x)$, and the concentration profiles, $c_i(x)$, are closely coupled and explicitly dependent on the $J_i$. Influencing $J_i$ by an external field would be manifested in both $c_i(x)$ and $\psi(x)$.

## 6. Conclusion

As indicated throughout the text of this chapter there exists much uncertainty about the nonthermal interaction of weak electromagnetic fields with biological membranes. It is likely that a number of different mechanisms are involved depending upon the particular frequency and field intensity of the applied field. It is suggested that a means of learning more about the specific nature of field–membrane coupling is to attempt to modulate the response by systematically altering the electrostatic state of the membrane and its environment.

## Acknowledgments

We gratefully acknowledge the assistance of Carol A. Jordan in the preparation of parts of this chapter. This work was partially supported by Office of Naval Research contract No. N00014-83-C-0008.

## References

1. H. Fröhlich and F. Kremer, *Coherent Excitations in Biological Systems*, Springer-Verlag, New York (1983).
2. B. Katz, *Nerve, Muscle, and Synapse*, McGraw-Hill, New York (1966).
3. I. Newton, *Opticks*, Dover, New York (1952).
4. L. D. Harmon and E. R. Lewis, "Neural Modeling," *Physiol. Rev.* **46**, 513–591 (1966).
5. A. L. Hodgkin, *The Conduction of the Nervous Impulse*, Liverpool University Press, Liverpool (1964).
6. H. Fröhlich, "What are Nonthermal Electric Biological Effects?" *Bioelectromagnetics* **3**, 45–46 (1982).
7. C. Tanford, *The Hydrophobic Effect: Formation of Micelles and Biological Membranes*, Wiley, New York (1973).
8. H. Eibl and P. Woolley, "Electrostatic Interactions at Charged Lipid Membranes. Hydrogen Bonds in Lipid Membrane Surfaces," *Biophys. Chem.* **10**, 261–271 (1979).

9. K. Toko and K. Yamafuji, "Stabilization Effect of Protons and Divalent Cations on Membrane Structures of Lipids," *Biophys. Chem.* **14**, 11–23 (1981).

10. T. Hill, "Electric Fields and the Cooperativity of Biological Membranes," *Proc. Natl. Acad. Sci. USA* **58**, 111–114 (1967).

11. J. Changeux, J. Thiery, Y. Tung, and C. Kittel, "On the Cooperativity of Biological Membranes," *Proc. Natl. Acad. Sci. USA* **57**, 335–344 (1967).

12. R. Blumenthal, J. Changeux, and R. Lefever, "Membrane Excitability and Dissipative Instabilities," *J. Membr. Biol.* **2**, 351–374 (1970).

13. H. Fröhlich, "The Biological Effects of Microwaves and Related Quetions," *Adv. Electron. Electron Phys.* **53**, 85–152 (1980).

14. H. Fröhlich, "Bose Condensation of Strongly Excited Longitudinal Electric Modes," *Phys. Lett.* **26A**(9), 402–403 (1968).

15. H. Fröhlich, "Long-Range Coherence and Energy Storage in Biological Systems," *Int. J. Quant. Chem.* **2**, 641–649 (1968).

16. H. Fröhlich, in *Theoretical Physics and Biology* (M. Marois, ed.), Wiley, New York (1969), pp. 13–22.

17. W. Almers, "Gating Currents and Charge Movements in Excitable Membranes," *Rev. Physiol. Biochem. Pharmacol.* **82**, 96–190 (1978).

18. F. Kaiser, "Boltzmann Equation Approach to Fröhlich's Vibrational Model of Bose Condensation-Like Excitations of Coherent Modes in Biological Systems," *Z. Naturforsch.* **34a**, 134–146.

19. D. Bhaumik, K. Bhaumik, B. Dutta-Roy, and M. Engineer, "A Microscopic Approach to the Fröhlich Model of Bose Condensation of Phonons in Biological Systems," *Phys. Lett.* **59A**, 77–80 (1976).

20. T. M. Wu and S. Austin, "Bose Condensation in Biosystems," *Phys. Lett.* **64A**, 151–152 (1977).

21. T. M. Wu and S. Austin, "Cooperative Behavior in Biological Systems," *Phys. Lett.* **65A**, 74–76 (1978).

22. F. Kaiser, "Coherent Oscillations in Biological Systems. I. Bifurcation Phenomena and Phase Transitions in an Enzyme–Substrate Reaction with Ferroelectric Behavior," *Z. Naturforsch.* **33a**, 294–304 (1978).

23. F. Kaiser, "Coherent Oscillations in Biological Systems. II. Limit Cycle Collapse and the Onset of Travelling Waves in Fröhlich's Brain Wave Model," *Z. Naturforsch.* **33a**, 418–431 (1978).

24. F. Kaiser, in *Biological Effects of Nonionizing Radiation* (K. Illinger, ed.), American Chemical Society Symposium Series 157, Washington, D.C. (1981), pp. 213–241.

25. H. Fröhlich, "Long-Range Coherence in Biological Systems," *Riv. Nuovo Cimento* **7**(3), 399–418 (1977).

26. F. Kaiser, "Limit Cycle Model for Brain Waves," *Biol. Cybernetics* **27**, 155–163 (1977).

27. D. Bhaumik, K. Bhaumik, B. Dutta-Roy, and M. Engineer, "Polar Modes with Elastic Restoring Forces, Bose Condensation, and the Possibility of a Metastable Ferroelectric State," *Phys. Lett.* **62A**, 197–200 (1977).

28. I. Grodsky, "Possible Physical Substrates for the Interaction of Electromagnetic Fields with Biological Membranes," *Ann. N. Y. Acad. Sci.* **247**, 117–124 (1975).

29. I. Grodsky, "Neuronal Membrane: A Physical Synthesis," *Math. Biosci.* **28**, 191–219 (1976).

30. I. Grodsky, "Biophysical Bases of Tissue Interactions," *Neurosci. Res. Program Bull.* **15**, 72–80 (1977).

31. K. Huang, *Statistical Mechanics*, Wiley, New York (1963).

32. D. Van Lamsweerde-Gallez and A. Meessen, "The Role of Proteins in a Dipole Model for Steady-State Ionic Transport Through Biological Membranes," *J. Membr. Biol.* **23**, 103–137 (1975).

33. A. Lawrence and W. Adey, "Nonlinear Wave Mechanisms in Interactions Between Excitable Tissue and Electromagnetic Fields," *Neurol. Res.* **4**(1/2), 115–152 (1982).
34. A. Davydov, "Solitons in Molecular Systems," *Phys. Scr.* **20**, 387–294 (1979).
35. S. Vaccaro and H. Green, "Ionic Processes in Excitable Membranes," *J. Theor. Biol.* **81**, 771–802 (1979).
36. T. Triffet and H. Green, "Information and Energy Flow in a Simple System," *J. Theor. Biol.* **86**, 3–44 (1980).
37. C. Cain, "A Theoretical Basis for Microwave and RF Field Effects on Excitable Cellular Membranes," *IEEE Trans. Microwave Theory Tech.* **28**(2), 142–147 (1980).
38. C. Cain, "Biological Effects of Oscillating Electric Fields: Role of Voltage Sensitive Ion Channels," *Bioelectromagnetics* **2**, 23–32 (1981).
39. C. Cain, in *Biological Effects of Nonionizing Radiation* (K. Illinger, ed.), American Chemical Society Symposium Series 157, Washington, D.C. (1981), pp. 147–160.
40. F. Barnes and C. Hu, "Model for Some Nonthermal Effects of Radio and Microwave Fields on Biological Membranes," *IEEE Trans. Microwave Theory Tech.* **25**, 742–746 (1977).
41. F. Barnes and C. Hu, in *Nonlinear Electromagnetics* (P. Uslenghi, ed.), Academic, New York (1980), pp. 391–426.
42. W. Pickard and F. Rosenbaum, "Biological Effects of Microwaves at the Membrane Level: Two Possible Athermal Electrophysical Mechanisms and a Proposed Experimental Test," *Math. Biosci.* **39**, 235–253 (1978).
43. A. Nazarea, in *Aeromedical Review: USAF Radiofrequency Radiation Bioeffects Research Program—A Review* (J. C. Mitchell, ed.), Review 4-81, Brooks Air Force Base, San Antonio, Texas (1981).
44. C. Stevens, "Inferences About Membrane Properties from Electrical Noise Measurements," *Biophys. J.* **12**, 1028–1047 (1972).
45. Y. Chen, "Differentiation of Channel Models by Noise Analysis," *Biophys. J.* **16**, 965–971 (1976).
46. A. Verveen and L. De Felice, "Membrane Noise," *Prog. Biophys. Molec. Biol.* **28**, 189–265 (1974).
47. T. Tenforde, "Thermal Aspects of Electromagnetic Field Interactions with Bound Calcium Ions at the Nerve Cell Surface," *J. Theor. Biol.* **83**, 517–521 (1980).
48. T. Hill, "Some Possible Biological Effects of Electric Fields Acting on Nucleic Acids or Proteins," *J. Am. Chem. Soc.* **80**, 2142–2147 (1958).
49. Y. Chemitskii, Y. Lin, and S. Konev, "Cooperative Transitions in the Supermolecular Structure of Nerve Protein," *Biofizika* **14**, 1023–1026 (1969).
50. D. Engelman, "X-Ray Diffraction Studies of Phase Transitions in the Membrane of Mycoplasma Laidlawii," *J. Mol. Biol.* **47**, 115–117 (1970).
51. H. Kijima and S. Kijima, "Cooperative Response of Chemically Excitable Membranes," *J. Theor. Biol.* **71**, 567–585 (1978).
52. M. Malek-Mansour, G. Nicolis, and I. Prigogine, in *Thermodynamics and Kinetics of Biological Processes* (I. Lamprecht and A. Zotin, ed.), Walter de Gruyter & Co., New York (1982), pp. 75–103.
53. T. Hill and Y. Chen, "Cooperative Effects in Models of Steady-State Transport Across Membranes," *Proc. Natl. Acad. Sci. USA* **65**, 1069–1076 (1970).
54. T. Hill and Y. Chen, "Cooperative Effects in Models of Steady-State Transport Across Membranes; Oscillating Phase Transition," *Proc. Natl. Acad. Sci. USA* **66**(1), 189–196 (1970).
55. T. Hill and Y. Chen, "Cooperative Effects in Models of Steady-State Transport Across Membranes; Simulation of Potassium Ion Transport in Nerve," *Proc. Natl. Acad. Sci. USA* **66**(3), 607–614 (1970).

56. V. Denner and F. Kaiser, "Phase Transition Behavior of a Greater Membrane Model," *Int. J. Quantum Chem.: Quantum Biology Symp.* 9, 41–57 (1982).
57. W. Adey, "Tissue Interactions with Nonionizing Electromagnetic Fields," *Physiol. Rev.* 61(2), 435–514 (1981).
58. J. Bond, D. Mikulecky, and C. Jordan, "A Model for Alterations in RF Induced Ca$^{+2}$ Efflux Based on Changes in the Structure of the Electrical Double Layer at an Enzymatic Surface," *Bioelectromagnetics Abstracts, Sixth Annual Meeting of the Biolelectromagnetics Society*, Atlanta (1984).
59. S. Hubbard and S. Brody, "Glycerophospholipid Variation in Choline and Inositol Autotrophs of Neurospora Crassa," *J. Biol. Chem.* 250, 7173–7179 (1975).
60. A. Trauble, *Biomembranes* 3, 197–227 (1972).
61. H. Trauble, M. Teubner, P. Woolley, and E. Eibl, "Electrostatic Interactions at Charged Lipid Membranes." I. Effects of pH and Univalent Cations on Membrane Structure, *Biophys. Chem.* 4, 319–342 (1976).
62. J. Bockris and A. Reddy, *Modern Electrochemistry*, Vol. 2, Plenum Press, New York (1977).
63. H. Fröhlich, "The Extraordinary Dielectric Properties of Biological Materials and the Action of Enzymes," *Proc. Natl. Acad. Sci. USA* 72(11), 4211–4215 (1975).
64. J. DeSimone, "Perturbations in the Structures of the Double Layer at an Enzymic Surface," *J. Theor. Biol.* 68, 225–240 (1977).
65. J. Pennline, J. Rosenbaum, J. DeSimone, and D. Mikulecky, "A Nonlinear Boundary Value Problem Arising in the Structure of the Double Layer at an Enzymatic Surface," *Math. Biosci.* 37, 1–17 (1977).
66. A. Pilla, in *Bioelectrochemistry* (H. Keyzer and F. Gutmann, eds.), Plenum Press, New York (1980), pp. 353–396.

# 11

# Differential Energy Control and the Dielectric Structure of Cells

## B. S. Thornton, B. L. Reid, and S. J. Webb

*ABSTRACT:* The transformation of normal cells towards tumor is discussed in terms of dielectric phenomena. The laser Raman spectroscopic results obtained by Webb are analyzed by inversion of the differences in spectral response, considering the cell as dielectrically divisible into five distinct layers. Maxima of energy transfer to the cell membrane are obtained at 45, 90, 140, and 180 cm$^{-1}$ for tumorous cells, while normal cells exhibit lower levels of such transfer. The altered dielectric cell structure indicates a change in the pathway for energy transfer in tumors. These pathways appear to involve the fibronectin layer as well as the microtrabecular lattice which, in the carcinoma cell, appears to become disrupted, so as to prevent or inhibit energy migration in certain directions. The cell thus loses control over its metabolic rate. The magnetic permeability appears to be larger than unity in tumor cells which exhibit a higher degree of ordering, of crystallinity in their nuclei, than do normal cells. For some oscillatory modes and frequencies, the tumor cell appears to be electrically isolated and thus unresponsive to contact inhibition.

## 1. Introduction

This paper deals with the phenomenon of the transformation of cells from the normal to the tumorous state in terms of dielectric alterations. It therefore seems pertinent to discuss briefly what is known about transformation and why electrodynamics might aid in its understanding.

The sciences of chemistry and physics advanced first by their isolated

---

*B. S. Thornton* • Faculty of Mathematical and Computing Sciences, The New South Wales Institute of Technology, P.O. Box 123, Broadway, N.S.W. 2007, Australia. *B. L. Reid* • Queen Elizabeth II Research Institute for Mothers & Children, University of Sydney, Sydney, Australia. *S. J. Webb* • Visiting Professor, The New South Wales Institute of Technology, Sydney, Australia.

independent progress, but later progress came only when these sciences were integrated into physical chemistry, quantum mechanics, and solid state science. Today the two are almost inseparable, especially with regard to the theories that surround them. Biology, on the other hand, has spent much more time in the descriptive stages of naming species and the development of evolutionary theories which cannot be tested in any conclusive way in the laboratory. Only in the past four decades, after it embraced chemistry in the form of biochemistry, did progress become rapid, but in its enthusiasm for the value of this chemical approach it has seemingly failed to keep up with and utilize the equally rapid progress in the other areas of natural science. Thus, the early successful biological concepts, based on random collision phenomena of *in vitro* chemistry, can no longer explain modern knowledge concerning the kinetics of *in vivo* biochemistry, and so on, although attempts to do so, by vast complications of the original ideas, are still current. Life clearly can be explained only in terms of both chemistry and physics and so the modern interpretation of physics and chemistry must be brought to bear on biological phenomena if they are to be understood. One such area of scientific integration is electrodynamics, as this area plays a large role in biological behavior.

## 2. Cell Transformation

Often, a particular change in the structure of a molecule or metabolic process, thought at first to be involved in this transition, is found in normal cells, and this, plus the myriad of agents with seemingly unrelated chemical and physical properties able to initiate the oncogenic process, has led to the idea that many different types of changes *in vivo* may have the same consequence: the production of an abnormal cell which, through its proliferation, may become a tumor. In the body of a particular animal with an inefficient immune system, therefore, such a cell may run rampant infiltrating other tissues and thus set up the disease known today as cancer.

It is generally agreed that abnomal cells are produced *in vivo*, almost every day, via mistakes in cell metabolism leading to genetic alterations, but these are generally destroyed by the immune system and the "actions" of normal cells which surround them. Indeed, it has been suggested and demonstrated in tissue culture that a single abnormal cell which is completely surrounded by normal ones is unable to proliferate. Only when a single abnormal cell is in direct contact with another are they able to proliferate. The implications from such knowledge are that in order to produce a tumor either the division of a cell and its transformation must be concommitant phenomena or the same transformation must occur simultaneously in two or more adjacent cells. The above considerations, however, may apply only to certain types of tumors while other types may be able to inhibit the growth or even destroy neighboring normal cells in order to affect their own proliferation. Such a phenomenon is not uncommon among microbial cells where a single cell or spore in isolation from others of its kind will not divide or germinate.

Another factor that appears evident from the wealth of data available is that many carcinogenic chemicals, viruses, and radiations are able to transform only those cells which are in the process of differentiation to produce new tissues. Fully differentiated ones, with their limited lifetime, seldom, if ever, are transformed. In fact, some sarcoma-inducing agents are effective only if injected into the new born and have no action on adult animals. This has led some researchers to believe that a link exists between differentiation and the control of cell proliferation, a link which becomes uncoupled when oncogenic transformation occurs during cell differentiation. The latter is a process that is ongoing throughout the lifetime of any animal or plant, first for its embryonic development and later for the renewal of its tissues and organs.

At the root of the disease we call cancer is the change in the cell which endows it with the ability to resist a phenomenon exhibited by normal cells coined as "contact inhibition." Giving the phenomenon a name, however, in no way implies that its mechanism is understood. At the moment, no one is clear on just what changes produce this ability, since different insults from chemicals, radiations, and viruses can induce it *via* a variety of changes in macromolecular structure and altered metabolism. Many of these factors now have been observed and studied from their structural and biochemical aspects but no common parameter along these lines has emerged. One of the problems that face investigators is that the agent of the insult, with the exception of a few viruses, is never present in the formed tumor cells, and indeed, the original alteration induced in molecular structures may no longer exist, with only some manifestation of it remaining.

The result of these findings has been an emphasis on the genetic nature of the problem, since whatever change is induced it must clearly be passed on to successive generations of tumor cells. The approach has been a traditional one in the sense that researchers have looked for a particular area of DNA, a gene, involved in the oncogenic process and have coined the name *Oncogene* to describe it. Most viral oncogenes have been found to have the structural organization of genes found in normal cells and seem to be copies of them, and this has added to the idea that the tumor virus induces the synthesis of too much of a given protein after its genome has entered a cell. Are these genes of normal cells which clearly have been handed down through the years to successive generations of animal, just dormant ones waiting to be activated by some agent in order to induce oncogenic transformation? The answer clearly is "no"; there are too many agents able to switch them on for any species to have survived the evolutionary process. These genes and their products must be, therefore, vital parts of the normal cell's anatomy and its metabolic activities. Now, investigation into the activities of these normal genes has demonstrated that alterations in their positions, in a cell genome, relative to one another can result in transformation of the cell. Since it is well known today *in vivo* a cell time clock is in operation by which each gene is read and a metabolic event occurs at a specific time in the lifetime of a cell, it is apparent that a change in the cell clock causing an event to occur too soon or too late can also be oncogenic. As has been pointed out,[1] the anatomy, or architecture, of a normal cell is a reflection of its time clock and vice-versa, that is, they are one and the same thing. Any change in the clock alters the anatomy and any alteration in the anatomy alters the clock.

If, therefore, all modern knowledge, emanating from biochemistry, cell biology, genetics, and microbiology, concerning the working of the normal and tumor cell is put together, it becomes clear that the difference between these two types of cell rests not on the presence or absence of a given gene or molecule, but on the quantity and hence, relative positions of these entities one to another. Thus any change in the normal architecture of a cell can be oncogenic, particularly if the change alters the vital sequences, in time, by which DNA is synthesized. It is the linear thinking, DNA → RNA → protein, which seems to have clouded the issue as to what is normal and what is abnormal.

A life form is a self-perpetuating cyclic system, with no master molecule governing its beginning and end; the entire system controls itself by the laws of nature (i.e., the laws of chemistry, mathematics, and physics), which seem to govern energy and entropy relationships.

As knowledge has grown regarding the phenomenon of life, so it has become abundantly clear that its basic unit, the cell, is a specific arrangement of molecules in which the whole has properties that are much more than the sum total of those of its individual parts. These unique properties reside in the manner in which the energy, present in any individual component, is redistributed among the whole, as macromolecular complexes and new daughter cells are formed. Thus energy, which resides in the vibrational and oscillatory modes of individual molecules, is redistributed to bring them "in phase" with one another, setting up motions of electrically charged entities unique to the organized system.[2] The formed cell, therefore, being a unique assembly of molecules, has properties related to crystals; thus one area of science which could aid in the understanding of what is a normal or an abnormal life form is solid state physics and another is electrodynamics.

In attempts to understand how vibrational, and other oscillatory, energy is distributed within normal and abnormal cells, Webb[3] has brought millimeter microwave and Raman spectroscopy into the whole cell biology. His results have provided a new and unique form of data from the "whole cell" regarding life processes which lends itself to a variety of physical methods of analysis. One such form of analysis is ascertaining whether the data supply information regarding the dielectric properties of normal and abnormal cells, since these clearly must reflect the *in vivo* arrangement of molecules and distribution of electrical energy.

Dipole oscillations in an assembly of molecules in the membrane of cells can be modeled as phase-locked solid state oscillators by a basic circuit[4,5] as in Figure 1. Loose coupling between such circuits imposes an eigenvalue problem from which significant mode softening can be shown to result and this has been suggested to be an important requirement in the energetics associated with the reproduction and mutation of cells.[6] As each individual unit oscillator can operate at subharmonics as well as harmonics, the above model is consistent with the idea that *in vivo* a number of discrete frequencies exist in the cell.

The control of energy transactions in specific frequency bands thus becomes dependent on biological parameters which are the analogs of those of an electromagnetic oscillator model with solid state components. The analogies do of course include the dielectric constant between charges. Present work by Thornton[21] applying a new type of analysis to Webb's results indicates that the dielectric spatial structure and surface morphology of a cell are important in the control of such energy transactions.

## 3. Determination of Dielectric Structure of Cells by Inversion of Their Raman Spectra

Preliminary studies suggest that the general form (not the line splitting) of the changes observed by Webb in the Raman scattering spectral response of cells virologically in transition from normal to tumor ones can be explained in terms of dielectric alterations in the cells.

The existence of dielectric layers may be expected if one considers a model similar to the Simone model of double layers at enzymic surfaces. Consider a uniformly charged cell surface in contact with a diffusion zone which is in contact with a stirred reservoir containing a number of types of ions, as in the cell's membrane.

The equations of electrodiffusion at the boundaries reduce to a differential equation closely related to a class studied recently by Dobrokhotov and Maslov[7] from which layers can be shown to form

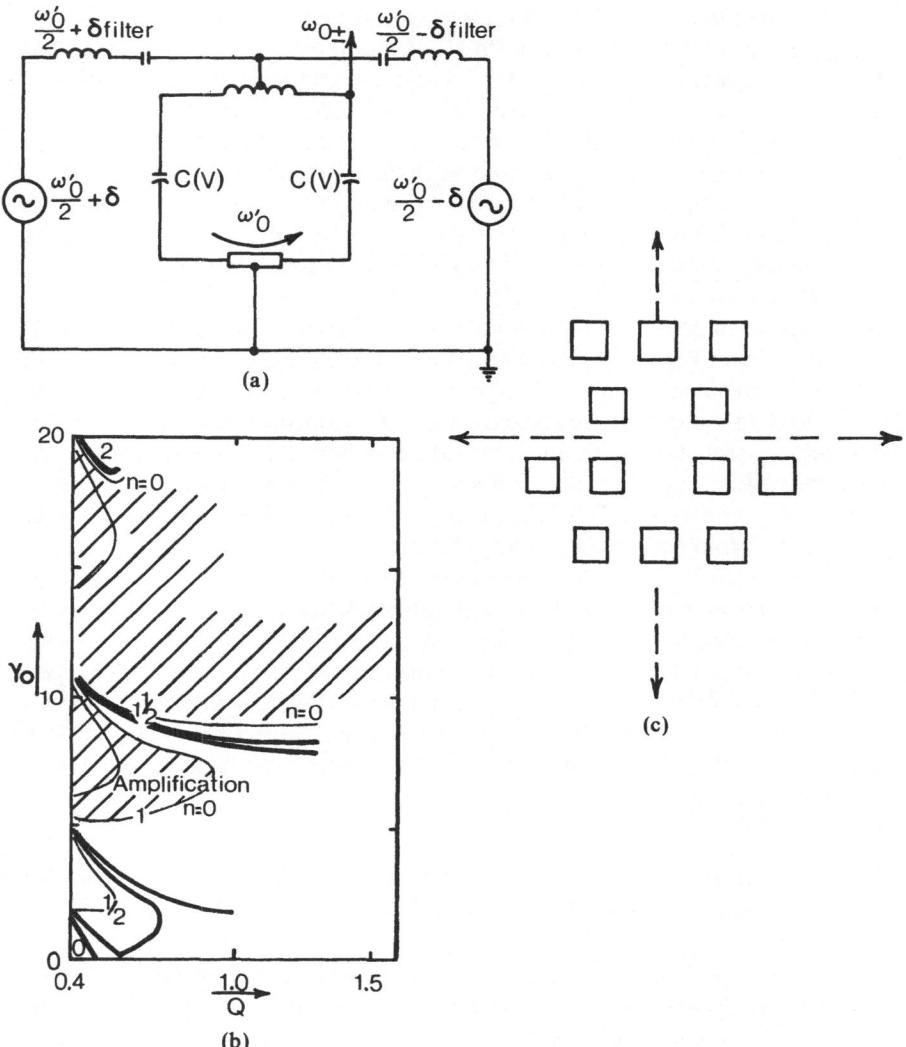

(a)

(b)

(c)

FIGURE 1. (a) Parametron semiconductor circuit with a one-to-one correspondence analogue of a dipole double layer. Inductance $L$, resistance $R$, and capacitors $C$, $C(V)$. (b) Stability diagram of behavior of the analog model (a); $Q = \omega L/R$ is a performance parameter of the circuit and $\gamma_0$ is a function of $Q$, $\omega_0$, $\Phi$ (the junction potential of the semiconductor diode), and the amplitude of sinusoidal voltage. Numbers $0$, $\frac{1}{2}$, $1$, $1\frac{1}{2}$, $2$,... are dominant frequency in units of $\omega_0$. The characteristic curves $(n = 0)$ are common boundaries of the divergent (amplification) and stable regions. (c) Model of a collective group of individual units of type (a) extending in two dimensions and with inductive and/or capacitive couplings between themselves. The eigenvalues for the frequencies of the system show that considerable mode softening can occur as proposed by Fröhlich.

about stable geodesics. This, in combination with the invagination of the membrane throughout much of the cell including the endoplasmic reticulum, could give rise to layers of differing electrical characteristics and their status would change with differing boundary conditions.

In a more traditional sense, if metabolism is considered governed by the diffusion of metabolites the same kind of layering would occur due to the mutual inhibiting and stimulating effects of "inward" and "outward" diffusion processes. Thus, although it is generally recognized today that metabolic rates are too rapid and too precise in terms of time of occurrence and place in the cell to be governed by diffusion, the concept of boundaries of different electrical characteristics realized from the diffusion hypothesis is still valid. If this is taken as a start and the extracellular material outside the cell membrane is included, five dielectrically different zones from the nucleus outwards can be visualized (Figure 2). With the concept of layering, from whatever theory is more applicable and the knowledge of the compartmentalization of metabolic events, there seems some justification in considering the dielectric layering in cells, in transition from normal to tumorous. The laser Raman spectroscopic experiments of whole cells conducted by Webb[8] have been analyzed by the inversion of the differences in spectral response to try to determine if layering changes may have caused these spectral differences. Webb's data are in the form of spectral responses of the whole cell, (see Figure 3). Changes within the cell give rise to changes in the spectral response from the complex processes and possible dielectric restructuring within the cell. The latter would then change the observed response of the initial Raman lines after their passage through the layering which acts as an optical filter. The problem of determining the layered dielectric structure from the cell's spectral response is an "inverse problem" in contrast to the direct problem of evaluating responses from a specified set of dielectric layers in the cell.

The cell was considered first as dielectrically divided into five (as yet unknown) layers and a new data inversion technique[9,10] then used to determine the specific dielectric structure that would account for the responses observed. The inversion method reduces the nonuniqueness inherent in conventional formulations of such inverse problems (such as determination of strata impedances in goephysics and determination of potential wells in scatterings from atomic nuclei). The theory reduces this class of nonlinear problem to a linear one without making approximations, thereby giving greater accuracy to the results obtained. Also any sensitivity to changes or errors in the layer thickness is small [a 10% change in one layer changes the permittivity $(\varepsilon = \varepsilon' - i\varepsilon'')$ values for that layer by approximately 10% and causes almost no change in the results for the other layers]. The method is based on the concept of "reflection loss" and energy transfer through terminal interfaces of the model. The associated

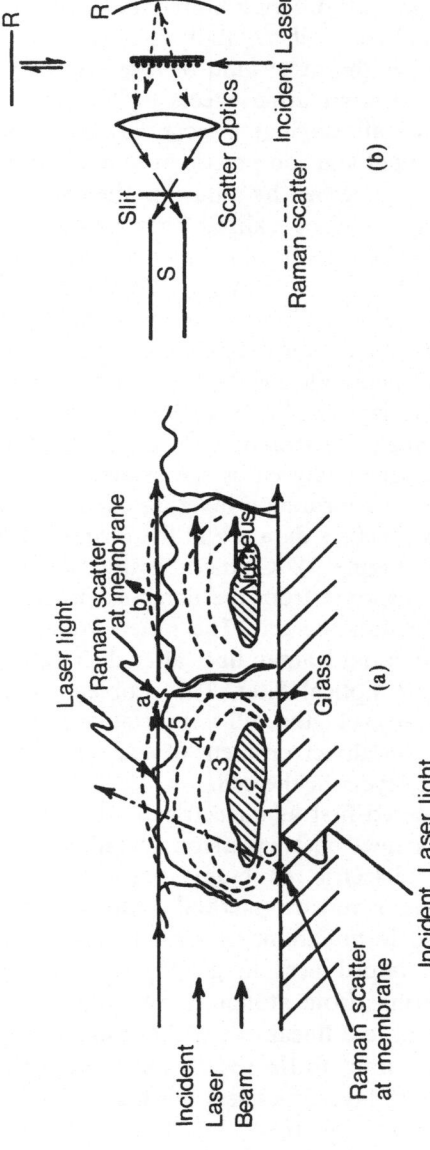

FIGURE 2. (a) Monolayer of cells on a glass slide illuminated by laser light as in Raman spectrometer (b). Each cell was considered as dielectrically layered (see text), and the spectral response was inverted in order to determine what the initially unknown dielectric layers would be in order to produce the spectral responses observed.

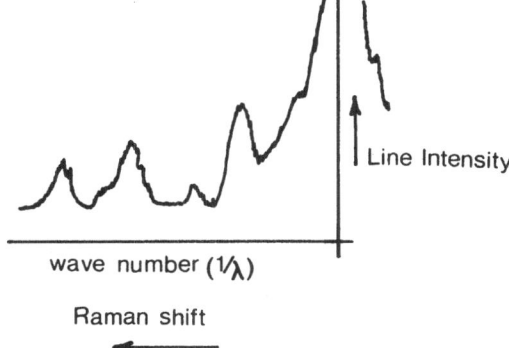

*FIGURE 3.* Illustrative type of response of Raman spectra of cells analyzed by inversion to obtain values for dielectric layers of cell. Specific responses are shown in Ref. 8.

eigenvalue formulation gives a solution for the maximum, minimum, and intermediate levels of electromagnetic energy absorption in, and transfer out of, the cell.

## 4. Results

Since the Raman spectra and its modification of line shape and frequency is present only in metabolically active cells, we conclude that the dielectric layering is an effect of metabolic processes. These processes would also be responsible for the change in boundary conditions in producing the layering referred to previously in the differential equation of electrodiffusion. In other words, the layering requires energy for its ordering. Once established, the dielectric layering acts as a filter to the Raman scattered lines, broadening them and displacing them in frequency. To determine the layering structure from these observed effects is, as mentioned previously, the objective of the inversion method.

An assumption was made which appeared justified from other biophysical studies that most of the Raman scattering originated from sites in the cell's membrane zone. Also the Raman spectral response observed from the normal, active cell was assumed to have come from transmissions through a dielectric structure which gives a flat uniform response, i.e., the filter function $F(\lambda)$ was a straight line and the Raman shifts for the normal cell were assumed to be the true shape and true shifts in frequency. (The assumption was subsequently verified from the model of the layers.) The model requires specification of cell diameter mean value(s). The cells were assumed to be oblate spheroids in a monocellular layer and with the general orientation of the nucleus as in Figure 2. The responses at 48 and at 72 hr after infection were compared with the true response of the normal

TABLE 1. Normal Cell, Metabolically Active, Two Average Sizes in Population, 40 $\mu$ and 80 $\mu$ Diameter[a,b]

| Layer $d/$ cell diameter $D$ | $\varepsilon = \varepsilon' - i\varepsilon''$ | | Magnetic permeability |
|---|---|---|---|
| | $\varepsilon'$ | $\varepsilon''$ | $\mu$ |
| Air | 1.0 | 0 | 1 |
| (a) Time $t = 0$ | | | |
| 0.06 | 3.92 | 1.02 | 1 |
| 0.11 | 4.61 | 1.20 | 1 |
| 0.31 | 3.91 | 1.02 | 1 |
| 0.20 | 3.32 | 0.86 | 1 |
| 0.31 | 3.91 | 1.02 | 1 |
| (b) Time $t = 48$ hr after infection | | | |
| 0.06 | 3.84[c] | 0.50[c] | 1 |
| 0.11 | 4.71 | 0.61 | 1 |
| Zone of 0.31 | 4.00 | 0.52 | 1 |
| nucleus → 0.20 | 3.42 | 0.44 | 1[d] |
| 0.31 | 4.00 | 0.52 | 1 |
| (c) Time $t = 72$ hr after infection | | | |
| 0.07 | 3.92 | 0.05 | 1 |
| 0.12 | 4.61 | 0.46 | 1 |
| Zone of 0.33 | 3.91 | 0 | 1 |
| nucleus → 0.16 | 2.04 | 0.37 | 2.04[e] |
| 0.32 | 3.91 | 0.43 | 1 |
| Glass | 6 | 2 | 1 |

[a] Tables 1a–1c show the transition, in a model used to analyze Webb's results, of dielectric layering of cells from normal to tumorous over 72 hr. Values shown are representative of the trend observed in the frequency band rather than definitive since the computed $\varepsilon'$, $\varepsilon''$ values vary with frequency in the range 50–200 cm$^{-1}$.
[b] At some frequencies the eigenvalue solutions for the 48-hr and 72-hr cases are equal except for one value in the set, indicating a bifurcation process operating in the transformation.[14,15]
[c] $\varepsilon' \to 2.14$, $\varepsilon'' \to 0.43$ at some frequencies.
[d] $\mu > 1$ indicated at some frequencies.
[e] $\mu > 1$ indicated at most frequencies in range.

cell at time $t = 0$. This yielded the filter function $F_{48}(\lambda)$ and $F_{72}(\lambda)$ involved in each of the two cases. The inversion program was then invoked in order to determine what layering would yield the salient features relevant to these observed spectral changes at 48 and 72 hr. The dielectric layering thus obtained was checked by using these results in a direct evaluation program in order to verify consistency. The results obtained in initial trials are given in Tables 1a, 1b, and 1c for the three time periods $t = 0$, 48, and 72 hr after infection.

## 5. Discussion of Results

The results show that for cases consistent with the Raman spectral response for cells 48 hr after infection in Webb's experiments, there is a maximum energy transfer $\int$ (line intensity) $d(1/\lambda)$ to the cell membrane at approximately 45, 90, 140, and 180 cm$^{-1}$. This does not appear to be so for the normal cell, where lower levels of energy transfer seem associated with "normal" behavior. The dielectric structure of layers calculated for normal cells does not affect the shape or frequency of the Raman lines observed for the normal cell. (The dielectric layer model used cannot account for the splitting of lines seen in the spectra of tumor cells.[8] Such splitting is of great importance in terms of the possible degeneracy in the oscillatory modes of molecules—perhaps from breaks in the fibronectin layer.)

The work presented here indicates that the dielectric change of the overall system can alter the energy filtering in the cell. Some specific deductions can be made as follows:

(i) The different dielectric structure of a cell, when it changes from normal to tumorous, indicates a change in the pathways for energy transfer. Energy can diffuse more readily in the normal cell, with a greater number of pathways possible, than in the tumor cell which has a disrupted dielectric architecture (but the energy flow may be more "ordered"). These pathways appear to include the fibronectin layer and the microtrabecular lattice of the cell, where changes in the dielectric structure could indicate a disruption of the lattice and thus in the energy transfer routes of solitons.[11,12]

All biopolymers have locked into them considerable internal energy which can be described as $\sum N_r E_r$ (where $N_r$ is the number of oscillators in energy level $E_r$). By Boltzman's statistics it can be shown that, in such a system, any change in the number of oscillators in a given energy level will result in a redistribution of the total energy, causing changes in bond distances and the making and breaking of others, resulting in a conformational change.

Dielectric disruptions of the type indicated in tumor cells would prevent or inhibit migration of energy in certain directions leading to loss of metabolic rate control. The rates of synthesis of DNA, RNA, and protein would be different when energy pathways are disrupted or inhibited. This is possibly related to Sach's[13] findings on growth control whereby normal control by proteins is uncoupled from differentiation in the tumor cell.

(ii) An observation with important implications is that the transition from one state to another represented by the three cases of Tables 1a, 1b, and 1c, was by bifurcation of eigenvalues of the energy transfer (reflection loss) formulation for the inversion problem. Common eigenvalues remain for each transition corresponding to the different spectral types for normal, 48-hr transformed, and 72-hr transformed cells. The bifurcation and corresponding jump from one eigenstate to another with the possibility of a completely new type of behavior is related to concepts in nonlinear thermodynamics presented by Glansdorff and Prigogine.[14,15]

(iii) The overall refractive index of the nucleus, or sometimes of the overall cell, is indicated as a little higher for tumor cells in most of the models used. Experimental observations at optical frequencies also show this effect.

(iv) The values of the imaginary part $\varepsilon''$ of the dielectric permittivity $\varepsilon = \varepsilon' - i\varepsilon''$ within the normal and 48-hr infected cell, obtained from the inversion results, are much lower ($\times \frac{1}{2}, \frac{1}{4}$) than the value for normal water in the frequency range concerned. This indicates that the cellular water is different and more "bound" than normal water—a subject discussed by previous researchers.[16,17]

(v) In one of the models used, the dielectric changes within tumor cells in Webb's experiments showed that a much lower conductivity layer can exist, meaning that a cavity effect is possible. Also, there are changes at the surface zone of the tumor cell where the fibronectin is altered. According to Hynes,[18] properties of fibronectin may explain the observation that different areas of the surface of many cells are not uniform but differ markedly in their properties. Although a domain of altered fibronectin in a tumor cell can be considered as a perforated aperture in the cavity, through which electromagnetic energy might leak, analysis indicates that a Zenneck surface wave could not, in effect, exist.

Obviously the electromagnetic energy does escape, as observed in the Raman spectra, and can actually be propagated through conductive (mesh) regions of such an aperture which survive the fibronectin disruption instead of through the perforations. Also, in the more general case in vivo, it would seem that a different mode of propagation along or within the membrane can also occur by processes discussed by Pohl[19] (e.g., dipole rotational waves or ionic waves) and their effects observed by him, using dielectrophoresis and cellular spin resonance. Increased time spent by electromagnetic waves within the tumor cells in vivo could account for increased energy loss by heat if the thermal conductivity of the outer surface zone were higher (arising from a more crystalline structure) for tumor cells than for normal cells.

(vi) The calculated value of magnetic permeability $\mu$ is greater than unity in the zone which includes the nucleus in transformed cells, 72 hr after infection. This occurs over a range of frequencies but is present at only some frequencies for the 48-hr case. It can be interpreted as an increase in structural ordering within the nucleus of tumor cells, but further study is needed.

(vii) The results given in (v) above also indicate that tumor cells can, for some modes and frequencies, be rendered electrically isolated, and this may be the reason why tumor cells do not respond to contact inhibition.

(viii) Further study based on the work presented herein may be a fruitful avenue for further investigations in the several such areas[20,14,15] in which the results have implications.

## References

1. S. J. Webb, *I.R.C.S.J. (Med. Sci.) U.K.* **2**, 18–22 (1974).
2. H. Fröhlich, *Nature* **228**, 1093 (1970).
3. S. J. Webb, *Phys. Rep.* **60**, 201, (1980).
4. G. C. Huth, J. D. Bond, and P. A. Tove, *Intl. Conf. on Non-Linear Electro-Chemistry in Biological Systems*, Jerry L. Pettis Memorial Veteran's Hospital, San Bernadino, California, June 5–9, 1983.
5. B. S. Thornton, *Phys. Lett.* **102A**, 77 (1984).
6. H. Fröhlich, *Phys. Lett. A* **39**, 153 (1972).
7. S. Yu Dobrokhotov and V. P. Maslov, *Sov. Phys. Dokl.* **23**, 894–896 (1978).
8. S. J. Webb, R. Lee, and M. E. Stoneham, *Int. J. Quant. Chem. Quant. Biol. Symp.* **4**, 277 (1977).
9. B. S. Thornton, I.S. Air Force Avionics Laboratory Report No. TR-74-181, Wright-Patterson AFB, Ohio (1974).
10. B. S. Thornton, *Geophysics* **44**, 801–819 (1979).
11. A. S. Davydov, *Sov. Phys. JETP* **51**, 397 (1980).
12. A. C. Scott, *Phys. Rev. A* **26**, 578–595 (1982).
13. L. Sachs, *J. Cell Physiol. Suppl.* **1**, 151–164 (1982).
14. P. Glansdorff and I. Prigogine, *Thermodynamic Theory of Structure, Stability and Fluctuations*, Wiley, London (1971).
15. P. Glansdorff, "Evolution of Complex Networks of Reaction," in *Living Systems as Energy Converters* (R. Buvet, M. J. Allen, and J. P. Massué, eds.), North-Holland, Amsterdam (1977), pp. 41–54.
16. W. Drost-Hansen and J. Clegg, *Cell Associated Water*, Academic, New York (1979).
17. S. J. Webb, *Bound Water in Biological Integrity*, C. C. Thomas (Academic Lecture Series), Springfield, Massachusetts (1965).
18. R. O. Hynes, "Relationship Between Fibronectin and the Cytoskeleton," in *Cytoskeleton Elements and Plasma Membrane Organisation*, Elsevier/North-Holland, Amsterdam (1981), p. 100.
19. H. A. Pohl, "Micro-dielectrophoresis of Dividing Cells," in *Bioelectrochemistry* (E. H. Keyzer and F. Gutmann, eds.), Plenum Press, New York (1981), pp. 273–295.
20. A. Szent-Györgi, *The Living State and Some Observations on Cancer* Dekker, New York (1978).
21. B. S. Thornton, *Phys. Lett.* **106A.**, 198 (1984).

# 12

# Biological Dielectrophoresis

## The Behavior of Biologically Significant Materials in Nonuniform Electric Fields

### Herbert A. Pohl and J. Kent Pollock

*ABSTRACT:* A brief history of the behavior of materials in nonuniform electrical fields is presented, followed by a theory of dielectrophoretic force and the derivation of the general force equation. Attention is paid to the several classes of polarization which lead to the experimental considerations of induced cellular dielectrophoresis. A distinction between batch and continuous methods is discussed, with a focus on a new microtechnique. While dielectrophoresis can induce aggregation of materials, i.e., cells, other orientational applications exist. Cell division, cellular spin resonance, and pulse-fusion of cells form topics appropriate to the realm of high-frequency electrical oscillations and are discussed in the context of living material.

## 1. Introduction

When living cells or their parts are in a nonuniform electric field, a force upon them usually arises. This can be used to help analyze, sort, or study such materials. This is not to say that such nonuniform field effects are unique to living matter, but rather to call attention to the fact that such forces can be uniquely useful in the biological milieu even though they can be do occur in the case of inanimate matter. As we shall see, such nonuniform field forces (*dielectrophoretic* forces) resemble (but are not the same as) those exerted by a magnet upon a piece of iron, or by the earth

*Herbert A. Pohl* • National Magnet Laboratory, Massachusetts Institute of Technology, Building NW-14, Cambridge, Massachusetts 02139. *J. Kent Pollock* • Pohl Cancer Research Laboratory, 515 Harned Avenue, Stillwater, Oklahoma 74075.

gravitationally pulling down a stone. All are the result of an inhomogeneous field pulling something towards the region of greatest field intensity. The events using the action of nonuniform electric fields are much less familiar, but are nevertheless quite useful—as in mineral separations, image production (xerography), evoking a feathery snowflake shape, aiding field ion microscopy, in crop dusting, and in cell-sorting.[1] Modern applications of nonuniform field effects to biologically significant materials such as cells, organelles, virus particles, DNA, proteins, etc., include wound and bone healing, disease analysis, water quality control, and aid in the understanding of the very makeup and behavior of cells[1,2] and their parts.

The ever-widening fronts upon which nonuniform field effects such as dielectrophoresis open are already broad and growing. Without claiming completeness one might note that dielectrophoresis and related nonuniform field effects play a role in bone and wound healing[1]; in the separation of live and dead cells (using, incidentally, purely physical means)[2,3]; in the recognition of abnormal factor VIII in hemophilia[4]; in the continuous separation of normal and abnormal cells[5]; in the levitation of single cells at various frequencies to obtain useful and diagnostic spectra[1,6–8]; in the concentrating of biological products such as soybean phosphatides or bacterial spores[1]; in the elucidation of distinctions between the thrombocytes of hemophilic, transmitter, and normal blood[4]; in the sorting of cells which differ only in diet[10]; in the formation of tissue models from normally separate organisms[11]; in the determination of diffusion coefficients of proteins in solution[12]; in the concentrating of cellular suspension as for water quality control analysis, or other bioassays[1]; and in the study of radiofrequency electric fields emitted by living cells.[13–16]

## 2. Dielectrophoresis

Nonuniform electric fields produce a force upon all real bodies, even those which are electrically neutral. This somewhat surprising fact has been long overlooked, although it is now evident that the earliest recorded observations on electricity were just such effects. Thales of Miletus in what is now Turkey recorded in about 600 B.C. that when amber was rubbed, it attracted small bits of lint and fluff. We now know that the amber was charged by frictional action to produce triboelectrically generated charge of one kind on the amber and the opposite charge on the rubbing cloth. The charged amber was then able to attract the small bit of (neutral) matter by the processes of what we now call polarization in which the free charges left on the amber push away like charges on the lint particle and pull on the unlike charges. Since the pulled charges in the lint are nearer the amber and hence in a stronger field than the repelled ones, a net force towards the amber arose, as noted by Miletus. We nowadays call it the "dielectrophoretic force."

The motion that results from the action of a nonuniform electric field upon a neutral is called *dielectrophoresis* (DEP) from the Greek word *phoresis* meaning "motion." It is to be distinguished from that motion caused by the response of a *free charge* to an external electric field. The latter is known as *electrophoresis*.

The cause of dielectrophoresis is rather easily described. Any electric field, whether uniform or nonuniform, will exert a force upon a charged object. What is important for us is that divergent, inhomogeneous, or nonuniform electric fields act upon *neutral* objects, producing forces and events which are both interesting and useful.

The performance of neutral matter in a nonuniform electric field can perhaps best be understood by comparing the responses of charged and of neutral bodies in both uniform and in nonuniform fields. Referring to Figure 1a, we first consider the two particles in a *uniform field*. A charged object, such as an ion or charged pingpong ball, will be pulled along the field lines towards the electrode carrying charge of opposite sign. This is the behavior typical of electrophoresis. It is familiar in the electrolysis of ions, for example. The *neutral* object, on the other hand can be thought of as, first, being polarized by the presence of the external field. As a result of this

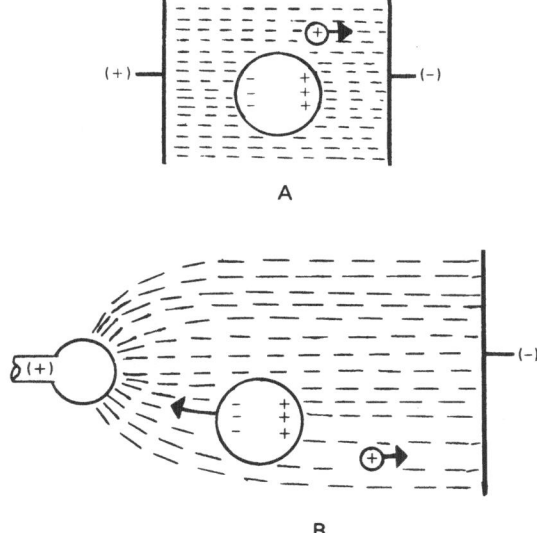

*FIGURE 1.* (A) Two particles in a uniform dielectric field. (B) Two particles in a nonuniform dielectric field.

polarization and partial separation of its internal charges it may experience a twist or torque in the field, but, and this is crucial, it is not impelled to move as a whole translationally towards either electrode. The occurrence of the torque arises only if the material being polarized is elongated or anisotropic. The dipole evoked by the external field or even a preexisting one would then tend to align itself in the field to achieve minimum energy.

In the case of a *nonuniform* field, and referring to Figure 1b, where again both charged and a neutral objects are shown being present, we observe a still different comparative behavior. The charged object moves much as before. It is impelled towards the electrode of opposite charge. The neutral object will now, however, attempt to move translationally. This occurs because the external field in effect produces a positive charge upon it on the side nearest the negative electrode, and a negative charge upon that side nearest the positive electrode. Because the object is neutral overall, the two effective charges, positive and negative, are in fact equal in magnitude although of opposite sign. Since, as here, the externally applied field is non-uniform, it spreads and weakens as it crosses the particle and produces *unequal* forces on the two effective charges. After all, the force $F$, upon a charge $q$, is dependent upon the electric field strength, $E_{local}$, in the region of the charge: $F = qE_{local}$. The net result will be an overall force on the neutral body, and normally results in the neutral body being impelled or pushed into the region of stronger field. To sum up, if a neutral object (i.e., one having *exactly* equal numbers of positive and negative charges in its composition) has, for some reason, had average centers of negative and positive charge even slightly displaced from coincidence, then a nonuniform electric field will act differently upon each centrum, evoking a net overall force on the object. The usual result is for the object to tend to go to the region of highest field strength. This can, of course, be modified by the presence of a surrounding medium, if the latter is more polarizable than the body under consideration. The analogy with gravitational effects is both helpful and suggestive. Just as in the case of gravitation where we see the stone (denser) sink in water, water cork (less dense) floats, so we can observe that while in a nonuniform electric field, a body submerged in a fluid medium will seek the region of higher field intensity if it is more polarizable than the medium, and vice versa. If it is less polarizable than the surrounding medium, it is repelled from the region of higher field intensity. The attractive case is called *positive dielectrophoresis* on the body, while the latter case, in which the body is impelled out to regions of weaker fields, is termed *negative*[1] *dielectrophoresis*.

Yet another thing is remarkable about dielectrophoresis, which the reader has perhaps already noticed. The force exerted by the nonuniform field upon the neutral, polarizable body is in the *same* direction no matter which electrode is positive or negative. The applied field could therefore well be that due to an alternating potential. This is so because the polarization induced in the object is reversed when the field direction reverses, at least in the usual circumstances, putting the attracted end of the dipole always nearest the strongest field. The contrast between the behavior of neutral and charged objects when in an alternating field (called ac) is very noticeable. The charged object, if it is only under the influence of electrophoreses, will merely tend to "shudder" about its original position. A neutral object in an ac field, on the other hand, and under the influence of positive dielectrophoresis, will move rather steadily toward the region of higher field intensity.

Whereas we have used the picture of the nonuniform field acting upon a "body" for clarity of exposition, it is obvious that the DEP force also acts in a similar manner on each volume element of the media present, whether it be a homogeneous medium or an inhomogeneous one such as a suspension of particles in another phase of matter. DEP force can thus be considered as acting throughout the volume of the field and media, and varying perhaps from point to point.

One further point might be made for clarity. As we have seen, *dielectrophoresis* is the translational motion evoked by a nonuniform electric field. In the case of some solid materials and in certain semisolid ones (e.g., liquid crystals) there is seen still another mechanical response of a neutral body to an electric field, that of a distortion. This is *electrostriction*,[17-21] and refers to the distortional response or strain resulting from an imposed electrical stress. Electrostrictive strains are used in sound transducers, for example. Historically speaking the two effects, translational (dielectrophoresis) and distortional (electrostriction), where both at times referred to as "electrostriction" with resultant confusion. Modern usage has tended to restrict the term electrostriction[21] to the discussion of distortional strain that has been induced electrically. For the sake of brevity, we shall frequently use the abbreviation DEP response as that referred to properly as dielectrophoresis. One can, of course, couple a moment arm to the dielectrophoretic force (e.g., DEP force) producing a torsion, and possibly a realignment of the body in the field.

## 3. Historical Perspective

The use of nonuniform electric fields for causing useful separations of materials probably goes back to the nineteenth century, when Lowden[22] in 1891 patented an invention for the abstraction of metallic particles from used lubricating oils. As practiced, it probably involved what we would now term "mutual dielectrophoresis" of the polarizable metal particles. Here the metal particle produces a distortion of the electric field about it.

The resulting nonuniformity of the field evokes DEP, especially that of particle-to-particle type, hence the term *mutual DEP*. The agglomerated particles would then more readily precipitate. The Cottrell precipitator, patented in 1911, used corona effects to produce charging of the particles to be removed from smoke gases, followed by the subsequent electrophoresis of the particles to the outer electrode. While at the electrode occasional residual charges in the collected material help hold the deposit on the electrode by the action of DEP on the other already discharged particles. The importance of DEP in the Cottrell precipitator was not recognized until[1] recently. The Cottrell process was much modified and adpated to the removal of water emulsified in petroleum.[23,24] This process quite probably depends upon DEP, although this principle was not then well understood. Bates' process for petroleum refining[25] also used DEP. Müller[26] had suggested theoretical reasons why nonuniform electric fields should collect suspended molecules, but concluded that it was impractical. Pohl, while at the Naval Research Laboratory in 1943, first quantitatively defined and theoretically described dielectrophoresis, and established its usefulness for analyzing suspended pigments and other particles, but wartime restrictions delayed publication.[27] The phenomenon of DEP was rather fully defined and explored in a series of papers considering nonuniform field effects in nonbiological media.[27-40]

Pohl and Hawk's work in 1966[41] appears to be the first successful use of a purely physical means for producing a separation of live and dead cells. Since then DEP has been shown to be useful in a wide variety of biological applications from cell sorting, cell characterization, and analysis, to the determination that living cells emit radiofrequency ac fields and attract various foreign particles. The nonuniform electric field effects were found useful in handling all types of living cells including algae, yeasts, bacteria, mammalian blood cells, and the organelles such as sperm, chloroplasts, mitochondria, and plastids.[1] Smaller portions of cells and even viruses can be advantageously examined and handled using the techniques of DEP. It is important to emphasize that biological DEP is generally restricted to ac rather than dc fields.

The use of alternating (ac) fields rather than static (dc) electric fields here has arisen, of course, because we are interested in dielectrophoresis effects upon neutral particles, and not in electrophoresis effects upon charged particles, such as ions. In fact this is a major point of difference between the two techniques, dielectrophoresis and electrophoresis.

By means of DEP, investigations of biological particles such as cells or their parts have established that cells can be made to move concertedly, and in a manner directly dependent upon their individual electrical identities.

The behavior of biological particles in *uniform* electric fields appears to have been first studied in some depth by Muth,[42] who subjected emulsions of fat particles to high-frequency ac fields. He observed that the particles tended to assemble along lines in the field to form what has now become known as *"pearl chaining."* This is a result of a corollary phenomenon known as *mutual dielectrophoresis*, in which particles, although present in what might have been originally a uniform electric field, themselves distort the field by their presence. The resulting field nonuniformities produce attractive interactions by DEP among the particles. The successive interactions lead to the formation of "pearl chains," bunching, and agglomeration (see Figure 2). An explanation by Krasny-Ergen[43] was made in terms of describing the fat particles in terms of perfectly conducting spheres, which, of course, they are not. The analysis was nevertheless helpful. Liebesny[44] noted that erythrocytes formed pearl chains in high-frequency fields of rather uniform character. Muth's[42] studies on fatty emulsions were further investigated by Manegold.[45] The responses of a number of unicellular organisms by Heller and co-workers[46,47] using high-frequency electric fields (0.1–100 MHz) were noteworthy. They reported pearl-chain formation, cellular orientational effects, preferential ranges of such frequency responses, and cell alignment. The previously unexpected observation of seeing lone cells rotate rapidly was reported. Elongated cells such as *Euglena* were observed to respond in a novel manner. The application of low-frequency fields made the cells orient themselves along the field lines, while at high frequency the same cells would orient themselves at right angles to the field lines. Such complex behaviors offered challenging puzzles, and even evoked ancient arguments in terms of "vital forces" of mysterious character.

The occurrence of such mysterious orientational effects in cells exposed to external electric fields caused Heller and others[46,47] to suspect the

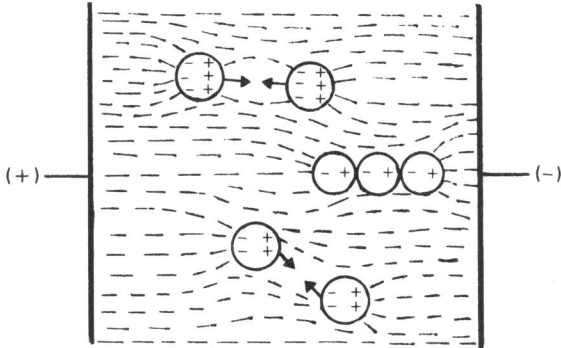

FIGURE 2. Agglomeration of particles in a dielectrophoratic field.

existing electromagnetic theory and its expression in terms of induced dipolar forces as being incapable of explaining cellular reorientation in electric fields. What further provoked puzzlement was the observation that such behavior had, until then, only been seen upon using living materials. The enigma was brilliantly solved in a series of studies by Fricke[48–50] and by Saito, Schwan, and Schwarz,[51] among others. The answers lay in the realization that cells were comprised of several differing regions of dielectrics, and that the cells were also nonspherical, i.e., were not isotropic.

Experimental studies by Füredi and Valentine[52] at various frequencies involved the bunching and orientation of microscopic particles made of polystyrene spheres, aluminum or carbon particles, ion exchange resin beads, ferric oxide or potato starch particles in water, saline solutions, or in castor oils. Orientation effects appeared to be absent in the salt solutions, but to be accentuated in castor oil, relative to that in water especially at high frequencies. Griffin and Stowell[53] confirmed and extended the studies of Teixera-Pinta et al.[47] on the bunching and orientational effects of *Euglena*. They observed that the cells did indeed align themselves along the field lines at frequencies below about 10 MHz, and realigned themselves to be at right angles to the field lines at frequencies up to about 100 MHz. But they discovered that at yet higher frequencies, above 100 MHz, another realignment parallel to the field lines occurred; see Figure 3. The critical frequency of the realignment was found to be dependent upon the conduc-

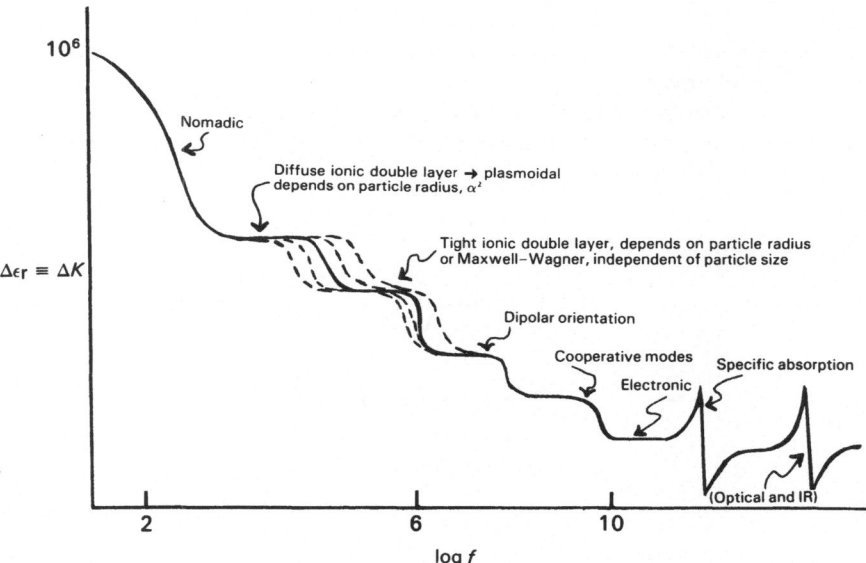

FIGURE 3. Orientation polarization represented by $K$, relative dielectric constant, as a function of the logarithm of frequency of the applied field.

tivity of the support medium. These remarkable results were later amplified and brilliantly analyzed by Saito *et al.*[51]

In the broad view of biological dielectrophoresis studies to date, we may note that DEP has two main ways to show itself: by translational motion and by rotational (orientational) motion. The translational response can be exhibited (a) by the motion of lone or clumped particles in a nonuniform electric field, or (b) by the mutual DEP of particles towards each other (resulting in bunching, or in pearl-chain formation) due to the field nonuniformities they make about themselves. The other observable effect of electric fields upon suspended neutral particles is that resulting in orientation, or even spinning of the particles. Interestingly, spinning effects can be observed even with particles such as yeast cells which appear to be closely spherical. In the following sections we shall discuss first the translational aspects of biological DEP, then the orientational aspects.

The equipment and experimental conditions needed for the DEP of biological suspensions can be extraordinarily simple. They can, of course, become quite complex, especially if one seeks to automate the procedures. Much useful and enlightening research can be done using simple equipment. For example, by using a microscope of modest magnification, a short pair of fine platinum wires mounted on a microscope slide, an inexpensive radiofrequency signal generator capable of supplying some 20–30 V rms, and an ac voltmeter, one can do DEP spectral comparisons and small-scale sorting of cells, organelles, and even some viruses. With care and patience, one can even examine single cells through the use of levitational DEP. In such simple experiments involving suspensions of cells or organelles, it is typical to use the following experimental range of conditions: Applied voltage, 0.01–30 V rms for electrode spacings of some 0.2–2 mm; suspension resistivities greater than 1000 $\Omega$ cm (i.e., suspension conductivities less than 1000 $\mu\Omega$/cm); frequencies of the applied field of 100 Hz to 1 GHz; and particle diameters of 0.2–100 $\mu$m. The results will be found to depend strongly upon the frequency of the applied field and upon the conductivity of the suspension.

## 4. Theory of Dielectrophoretic Force

An expression for the DEP force upon a rigid body in a fluid is derived here for the rather general case of dielectrics having nonzero conductivities in ac electric fields, in the frequency region in which magnetic effects may be neglected. These results were obtained by way of a calculation of the energy assuming the media to be homogeneous and isotropic. The calculations assume conservation of (electrical) energy and are therefore limited in validity when applied to cases where frictional (mechanical) losses dissipation has become an appreciable fraction of the average energy input. An improved basis would use conservation of momentum. The special case of a spherical body in a mildly divergent field is developed and discussed. (The equations in this section were derived in collaboration with Dr. Wm. L. McCubbin.)

One hundred years after the publication of Maxwell's equations, the theory of dielectrics is till a source of controversy.[1] The problem to which we address ourselves here concerns the correct expression for the dielectrophoretic force[27] on a conductive dielectric body in a conductive dielectric fluid, while subject to a nonuniform alternating electric field. A rather general expression for this force has already been obtained by Pohl and Crane,[55] but the particular form it takes in the case of a spherical body has been contested.[56] The purpose of the present chapter is to show that the cited criticisms are unfounded.

In order to do this we first derive the general force equation by a straightforward generalization of the energy method used by Stratton for perfectly insulating dielectrics.[57] We then obtain the particular expression for a spherical body and compare it with an expression found by Kallio and Jones[56] using an "effective dipole" approach. The difference in the two expressions are then analyzed and explained.

## 4.1. Derivation of the General Force Equation

We consider the method used by Stratton[57] to calculate the energy of a neutral dielectric body in a neutral dielectric fluid to be sound. Indeed, we believe that experience has shown that energy methods are inherently more reliable than calculations with reduced models insofar as the various contributions to the energy are of immediate intuitive comprehension. The replacing of an "actual" system by some "effective" one often carries the penalty that the directness of this connection with physical insight is lost. It will be shown below that precisely this situation has occurred in the present controversy.

Since some confusion has arisen in dielectric theory from careless definition of symbols, we have taken some care here to spell such matters out in some detail while adhering to standard usage. The electric field at any point will be described by the complex instantaneous field vector, $\tilde{E}(\mathbf{r}, t)$ (in the low-frequency range where the magnetic effects are negligible and "electrostatic" conditions may be said to prevail) where $\tilde{E}(\mathbf{r}, t) = E(\mathbf{r})e^{i\omega t}$. We shall require the amplitude, $E(\mathbf{r})$ to be a complex vector so that the phase of $\tilde{E}(\mathbf{r}, t)$ relative to other vectors may be written in exponential form. Thus

$$|E| = E_r + iE_i = |\mathbf{E}| e^{i\theta_E} \tag{A}$$

where

$$|E| = [E_r^2 + E_i^2]^{1/2} \quad \text{and} \quad \tan \theta_E = E_i/E_r \tag{B}$$

One of the simplest responses a dielectric medium can make to the impressed field is via electronic polarization (displacement of bound charges). The traditional vector used to describe the response is the displacement vector, $\tilde{D}(\mathbf{r}, t)$. Under most experimental conditions $\tilde{D}$ is linearly related to $\tilde{E}$, i.e., $\tilde{D} = \varepsilon\tilde{E}$, where $\varepsilon$ is real and is also a scalar if the medium is isotropic. In this case, $\tilde{D}$ and $\tilde{E}$ are in phase.

In addition, one may have the reorientation of permanent dipole entities and other relaxation processes that are essentially rearrangement processes not involving charge transport over macroscopic distances. The fact that such dipole orientation is usually thermally activated means that there is a delayed response resulting in a phase difference between $\tilde{D}$ and $\tilde{E}$. This phase lag is most conveniently accounted for by requiring $\varepsilon$ to be complex:

$$\tilde{D} = \varepsilon\tilde{E} = (\varepsilon' - i\varepsilon'')\tilde{E} = \tilde{E}|\varepsilon| e^{i\theta_E} \tag{C}$$

The imaginary part of the response is in phase with $\tilde{E}$ and for that reason can be considered as an ac conductivity. As is usual we may formally acknowledge this by writing $\varepsilon'' = \sigma_{e'}/\omega$, where the subscript emphasizes that dipole relaxation processes are responsible and the superscript signifies an in-phase contribution to the total in-phase conductivity defined below.

In addition to these two representative kinds of polarization processes (electronic and dipolar), we may have conduction by means of macroscopic charge transport. Again, because the latter conduction processes are often thermally activated, there will appear a phase lag of the conductive response to the applied field and we may write the conductivity due to charge transport $\sigma_c = \sigma_c' + i\sigma_c''$. In this notation the total in-phase conductivity (i.e., the observed conductivity) is

$$\sigma' = \sigma_c' + \sigma_\varepsilon' \tag{D}$$

It is, of course, recognized that there are polarization mechanisms with a fundamentally molecular basis in addition to the electronic and the dipolar orientation types referred to above. Indeed, the two other principal mechanisms, atomic[58] and nomadic polarization,[1,59,60] provide interesting examples of polarization mechanisms in which appreciable macroscopic charge transport may accompany their operation. There are in addition compounded mechanisms such as those involving ionic double layers on particles in media, etc., but we leave the detailed discussion of such cases to other[1] sources.

The total permittivity, $K$, can now be written as

$$K = \varepsilon - i\sigma_c/\omega \tag{E}$$

that is

$$K = \varepsilon' - (\sigma_c''/\omega) - i(\varepsilon'' + \sigma_c'/\omega) \tag{F}$$

with

$$K \equiv K' - iK'' \tag{1}$$

and the total displacement vector $\tilde{D}(\mathbf{r}, t)$ related to $\tilde{E}$ by

$$\tilde{D} = K\tilde{E} \tag{2}$$

In order to find the energy, $U$, in an field established in a fluid exhibiting both polarization and conduction, we must find a suitable generalization of the conventional electrostatic energy expression. The standard expression for the case of perfect dielectrics, where $E^0$ and $D^0$ are the field and displacement vectors, is

$$U = \tfrac{1}{2} \int_{\substack{\text{all} \\ \text{space}}} E^0 \cdot D^0 \, dv \tag{3a}$$

We assert that this generalization is

$$U = \tfrac{1}{2} \operatorname{Re} \int_{\substack{\text{all} \\ \text{space}}} \tilde{E} \cdot \tilde{D}^* \, dv = \tfrac{1}{2} \operatorname{Re} \int_{\substack{\text{all} \\ \text{space}}} \tilde{E}(\mathbf{r}, t) \cdot \tilde{D}(\mathbf{r}, t) \, dv \tag{3b}$$

where $\tilde{D}^*$ is the complex conjugate of $\tilde{D}$. More explicitly, the instantaneous energy is

$$U(t) = \tfrac{1}{2} \int_{\substack{\text{all} \\ \text{space}}} \operatorname{Re}\{\tilde{E}(\mathbf{r}, t) \cdot \tilde{D}^*(\mathbf{r}, t)\} \, dv \tag{3c}$$

where Re signifies the real part of the complex quantity in the bracket. The dynamical variable of interest to our problem is the average energy over a cycle,

$$\bar{U} = \tfrac{1}{2} \int_{\substack{\text{all} \\ \text{space}}} \operatorname{Re}\{E \cdot D^*\} \, dv \cos^2 \omega t$$

$$= \tfrac{1}{4} \int_{\substack{\text{all} \\ \text{space}}} \operatorname{Re}\{E \cdot D^*) \, dv \tag{4}$$

where $E$ and $D$ are the complex vector amplitudes of the applied field van the total displacement vectors, respectively.

So far we have been discussing the field in a homogeneous isotropic conductive dielectric fluid. Now let a rigid body be inserted into the fluid, and let any work necessary to restore the charge distribution at the electrodes be done so that the impressed field far from the body has its former value. [Alternatively, one may simply assume with Stratton[57] that the charge at the distant electrodes were held fixed.] The remaining difference of field energy with and without the body must represent the energy of the body in the field.

We shall use the subscripts 1 and 2 to denote the fluid medium and the body, respectively. Using $E_0$ and $D_0$ for the fluid in the absence of the body, we obtain from (4)

$$\bar{U}_0 = \tfrac{1}{4} \int_{v_1 + v_2} \mathrm{Re}\{E_0 \cdot D_0{}^*\} \, dv \tag{5}$$

Using $E$ and $D$ for any point after the introduction of the body, then the energy of the body is $\Delta \bar{U} = \bar{U} - \bar{U}_0$, or

$$\Delta \bar{U} = \tfrac{1}{4} \int_{v_1 + v_2} \mathrm{Re}\{E \cdot D^* - E_0 D_0\} \, dv \tag{6}$$

We now wish to show that Stratton's treatment[57] for the electrostatic (insulator) case carries through for the present case of a system of conductive dielectrics. The first step is to rewrite (6) as

$$\Delta \bar{U} = \tfrac{1}{4} \int_{v_1 + v_2} \mathrm{Re}\{E \cdot (D^* - D_0) + (E - E_0) \cdot D_0{}^*\} \, dv \tag{7}$$

In Stratton's treatment of nonconductive ideal dielectrics, i.e., where the vectors $E^0$ and $D^0$ are noncomplex and time independent, the following conditions hold at the boundary between $\varepsilon_2$ and $\varepsilon_1$:

$$\bar{n} \times [E^0_+ - E^0_-] = 0 \tag{8a}$$

$$\bar{n} \cdot [(D^0 - D^0_0)_+ - (D^0 - D^0_0)] = 0 \tag{8b}$$

where $\bar{n}$ is the outward normal vector to the boundary of the body and the subscripts $+$ and $-$ refer to the regions just outside and inside the boundary, respectively. Equation (8b) is true only if the boundary charge is zero.

This permits the use of the general vector result, [57] $\bar{P}$ and $\bar{Q}$ representing generalized vector quantities, [57]

$$\int_{v_1 + v_2} \bar{P} \cdot \bar{Q} \, dv = 0 \tag{9}$$

in which $\nabla \times \bar{P} = 0$ and $\nabla \cdot \bar{Q} = 0$ at all points.

In the present case of conductive dielectrics there may arise a real charge accumulated at the boundary due to the difference in conductivities of the two media. The simplest way of dealing with this is to show that (8b) with $D^0$ replaced by $D$ given the correct electrostatic expression when surface charge is present. Writing

$$\bar{n} \cdot [(D - D_0)_+ - (D - D_0)_-] = 0 \tag{10a}$$

implies

$$\bar{n} \cdot [(D_+ - D_-) - (D_{0+} - D_{0-})] = 0 \tag{10b}$$

i.e., that

$$\bar{n} \cdot [D_+ - D_-] = 0$$

From the definitions contained in (1) and (2), this implies

$$\bar{n} \cdot [(\varepsilon_1 E_+ - \varepsilon_2 E_-) - (i/\omega)(\sigma_{c1} E_+ - \sigma_{c2} E_-)] = 0 \tag{10c}$$

From the current continuity conditions, which with $\partial \rho_s(t)/\partial t = i\omega \rho_s$, is

$$i\omega \rho_s + \bar{n} \cdot [\sigma_{c1} E_+ - \sigma_{c2} E_-] = 0 \tag{10d}$$

this can be written as

$$\bar{n} \cdot \text{Re}\{\varepsilon_1 E_+ - \varepsilon_2 E_-\} - \rho_s = 0 \tag{10e}$$

From this one obtains

$$\bar{n} \cdot \text{Re}[D_+ - D_-] = \rho_s \tag{10f}$$

which is the simple result for the electrostatic limit case when a real surface density of charge exists. This result confirms the consistency of the use of $D$ for $D^0$ in Eqs. (8a) and (8b), so as to include the effect of surface charges such as may appear in the presence of conductive dielectrics. We now return to the expansion and development of Eq. (6).

As a consequence of theorem (9) and condition (10) the first integral in (7) may be set equal to zero, and we have

$$\Delta \bar{U} = \tfrac{1}{4} \int_{v_1} \text{Re}\{(E - E_0) \cdot D_0{}^*\} \, dv + \tfrac{1}{4} \int_{v_2} \text{Re}\{(E - E_0) \cdot D^*\} \, dv \qquad (11)$$

Now

$$\int_{v_1} \text{Re}\{E - E_0) \cdot D_0{}^*\} \, dv = \int_{v_1} \text{Re}\left\{\left[(E - E_0)\frac{K_1 K_1{}^*}{K_1}\right\} E_0{}^*\right] dv$$

$$= \int_{v_1} \text{Re}\left\{(D - D_0)\frac{K_1{}^*}{K_1} E_0{}^*\right\} dv \qquad (12)$$

Moreover, by theorem (9),

$$\int_{v_1 + v_2} \text{Re}[E_0{}^* \cdot (D - D_0)] \, dv = 0$$

giving

$$\int_{v_1} \text{Re}[E_0{}^* \cdot (D - D_0)] \, dv = - \int_{v_2} \text{Re}[E_0{}^*(D - D_0)] \, dv \qquad (13)$$

Combining (11), (12), and (13), we find

$$\Delta U = \tfrac{1}{4} \int_{v_2} \text{Re}[(E - E_0) \cdot D_0{}^*] \, dv - \tfrac{1}{4} \int_{v_2} \text{Re}\left[\frac{K_1{}^*}{K_1} E_0{}^*(D - D_0)\right] dv \qquad (14a)$$

Finally, using $D_0 = K_1 E_0$, $D = K_2 E$ inside the body

$$\Delta U = \tfrac{1}{4} \int_{v_2} \text{Re}\left[\frac{K_1{}^*}{K_1}(K_1 - K_2)EE_0{}^*\right] dv \qquad (14b)$$

This result is the generalization we sought in the case of conductive dielectrics in an ac field. It is a generalization of the result given by Stratton, among others, for the case of perfectly insulating dielectrics in a static field. It has the same form as the expression found by Pohl and Crane[55] by a somewhat different argument.

The required expression for the dielectrophoretic force is obtained[56] from $F = -\nabla(\Delta U)$, i.e.,

$$F = \tfrac{1}{4} \int_{\text{body}} \text{Re}\left[\frac{K_1{}^*}{K_1}(K_2 - K_1)\nabla E \cdot E_0{}^*\right] dv \qquad (15)$$

## 4.2. Force on a Conductive Sphere in a Conductive Fluid

It is a simple matter to rework the standard electrostatic problem of a sphere in a slightly divergent field using the requirement that the normal component of $D$ (rather than $D^0$) is continuous across the boundary. The resulting relation between $E$ inside the sphere and $E_0$, the field in the absence of the spherical body, as shown in standard texts for ideal dielectrics, but with complex dielectric constants,[57] and as justifiable by the procedure in Ref. 1, is

$$E = \frac{3K_1 E_0}{2K_1 + K_2} \tag{16}$$

If it is assumed that $E$, and hence $\nabla E_0{}^2$ vary but little over the diameter, $2a$, of the sphere, then from (15)

$$F = \frac{1}{4}\left(\frac{4}{3}\pi a^3\right) \mathrm{Re}\left[\frac{K_1{}^*}{K_1} \frac{(K_2 - K_1)3K_1}{K_2 + 2K_1} \nabla E_0 \cdot E_0{}^*\right]$$

or                                                                                                      (17)

$$F = \pi a^3 \, \mathrm{Re}\left[\frac{K_1{}^*(K_2 - K_1)}{K_2 + 2K_1}\right]\nabla E_0{}^2$$

This is to be compared with the result given by Kallio and Jones.[56] Combining their equations (1) and (D-5), their result in our notation, is[61]

$$F = 2\pi a^3 \, \mathrm{Re}\left[\frac{\varepsilon_1'(K_2 - K_1)}{K_2 + 2K_1}\right]\nabla E_0{}^2 \tag{18}$$

The factor 2 corresponds to the use of $\mathbf{E}_0$ as the rms field[62] rather than its amplitude as in (17).

The essence of the method by which (18) was derived consists in noting that the potential outside the sphere (in the conductive dielectric case) can be written

$$\phi_1(r, \theta) = \left[\left(\frac{K_2 - K_1}{K_2 + 2K_1}\right)\frac{a^3}{r^3} - 1\right]E_0 r \cos\theta \tag{19}$$

from which it is clear that the portion stemming from the polarization, $\theta_1'(r, \theta)$, is

$$\phi_1'(r, \theta) = \left(\frac{K_2 - K_1}{K_2 + 2K_1}\right)a^3 E_0 \frac{\cos\theta}{r^2} \tag{20}$$

Kallio and Jones then observe that the potential due to a "model" dipole of moment, $p_{\text{eff}}$, has the same formal dependence upon $r$, viz.,

$$\phi_d = \left(\frac{p_{\text{eff}}}{4\pi\varepsilon}\right)\frac{\cos\theta}{r^2} \tag{21}$$

The equating of (20) and (21) followed by the substitution $\varepsilon = \varepsilon_1'$ is then held to provide the effective moment, $p_{\text{eff}}$, required in the equation for the dielectrophoretic force,

$$F = (p_{\text{eff}} \cdot \nabla) E_0 \tag{22}$$

It will be clear by now that a different logic is involved. While (21) is indeed the potential of a classical static dipole, such a dipole is not necessarily an adequate model of a physical system with *time-dependent* conduction processes. That it is not adequate for the general problem of conductive dielectrics in ac fields seems to us to be established by the general bases upon which (15) was deduced, and from the rather systematic and symmetric way in which the permittivities are carried through to the final result. The contrasting of the two results, Eqs. (18) and (17) [or the more general Eq. (15)] seems to us to emphasize that which can arise in decoupling naturally complex product terms such as $\text{Re}[(K_1^*/K_1)(K_2 - K_1)]$ into separate model terms such as $\varepsilon_1'$ and $(K_2 - K_1)/(K_2 + 2K_1)$ as was done to reach Eq. (18).

We conclude that the theory for the energy and force relations of dielectrophoresis while simplistically assuming conservation of energy is most soundly developed along the lines pioneered by Stratton and followed here. Some experimental verification of a general sort for this theory concerning the case of a lossy spherical body in a lossy fluid was obtained by Crane and Pohl[63] on studying the dielectrophoretic collection rate of yeast cells, and on studying the dielectrophoretic force on single[1,61,62] yeast cells. Further experimental investigation of this interesting and fundamental problem is needed.

Since the response of matter to DEP is dependent upon its polarization, or rather to the degree of it, while in an inhomogeneous electric field, let us look more closely at the main mechanisms presently recognized for understanding polarization. There are four modes of polarization which occur on a molecular scale: (1) electronic, (2) atomic, (3) dipolar orientation, and (4) nomadic (giant, long-range, usually involving delocalized electronic orbitals). In addition to these four molecular modes, there occur several types of interfacial or interregional polarization of a more macroscopic character.

## 5. Polarization, the Electrical Distortion in Matter

### 5.1. Molecular Modes

The blocked or restricted motion of the charges in a body of matter appears to the experimenter as a polarization response. The continuous or semicontinuous motion of such charges through the sample appears as a conductive response. The two processes are linked over the entire frequency range by the Kramers–Kronig relations. This is well discussed in Fröhlich's book on dielectrics.[120]

#### 5.1.1. Electronic Polarization

The universal and common polarization response is that which arises as the elemental charges, the electrons and the nuclei, distort their relative positions as an external field is applied. The distortion of the electronic[18,19] distributions about the nuclei is usually quite small. One expects this upon noting that the fields within the atom (about $10^{11}$ V m$^{-1}$) are far larger than those usually applied by man (0 to $10^8$ V m$^{-1}$); a usual value for the shift of the charge centers in an atom subject to, say, 1 V m$^{-1}$ is about $10^{-8}$ Å, or $10^{-18}$ m. Electronic polarizability, although ubiquitous and universal, usually is not large. In most organic solids, the electronic polarization produces only a mild increment of the relative dielectric constant, $K$, of about 2–4. In inorganic solids, such as elemental silicon, a value of 12 is noted.

#### 5.1.2. Atomic Polarization

The so-called atomic polarization arises from the shifts of differently charged atoms with respct to each other. In salt, NaCl, for example, which is an ionic solid, an applied external field causes the positive sodium atoms to shift their positions slightly with respct to their negative counterparts. Normally such a contribution to the dielectric polarization of solids is low, about 1/7 that noted for the electronic polarization in organic materials. In rare cases, atomic polarization can be quite large. Witness, for example the perovskite minerals[18,19] such as $BaTiO_3$. It may, if prepared carefully, have a remarkably high relative dielectric constant of, say, 4000.

#### 5.1.3. Orientational (Dipolar) Polarization

As the name implies, this arises from the orientation of dipolar molecules or portions thereof. That is, since the various atoms comprising a molecule may carry somewhat differing net charges, there will be regions of positive and negative charge lying somewhat separated creating a "dipole." Such dipoles can respond to an external field by tending to reorient so as to minimize their potential energy. Among the familiar molecules which are dipoles are HCl, $H_2O$, amino acids, proteins, DNA, RNA, etc. The orientational polarization of a solid can be quite large. The same is true of some liquids. The relative dielectric constant of water, for example is 78. The greatest portion of this large value is due to the orientation of water molecules or to the migration of protons. The ability of such large groups to follow a rapidly alternating field such as supplied at radiofrequencies, is limited. We can understand the fact that the dielectric constant of water is only about 1.8 at frequencies above 20 GHz, but at rather lower frequencies is observed to be 78. The reader is referred to excellent reviews of the electronic, atomic, and orientational polarization of solids, liquids, and gases.[17-20]

### 5.1.4. Nomadic Polarization

A given field strength produces a polarization proportional to the number of dipoles per unit volume, and to the degree to which the charges in the material can be induced to move. Said another way, the wider the domain in which the changes can swing, the greater the effective dielectric constant of the material.

In the case of electronic or atomic polarization, the charges responding to the field may be numerous, but usually move only very short distances. The effective dielectric constant in such cases is correspondingly small. In the case of dipole orientation, the charges move rather larger distances upon application of the field as the dipoles reorient. The effective dielectric constant can accordingly be fairly large. In nomadic polarization, in contrast, the charges are free to move very long distances, and may journey many molecular segments away (hence the name "nomadic"). The effective dielectric constant can be then be huge. The resulting giant polarization can result in organic solids with dielectric constants exceeding 1000, even 100,000, whereas it is usual to observe dielectric constants on only about 2–38 in conventional organic solids.

These concepts can be, perhaps, better appreciated by considering the following simple equation:

$$P = Nqx \tag{23}$$

where $N$ is the number of dipoles per unit volume, $q$ is the magnitude of the charge at either end of the dipole, and $x$ is the distance of separation between the positive and negative charge centers comprising the dipole.

Nomadic polarization is a relatively recent discovery[1,59,60,121] that has opened the way to producing new materials with outstanding dielectric properties. It may be mentioned briefly in passing that one can now prepare organic polymers of high chemical and thermal stability which have both higher dielectric constants and lower dielectric losses than conventional materials such as the excellent perovskite solids. In addition, the nomadic[64,65] polarization solids respond without the hysteresis common to the perovskite materials.

The presence of nomadic polarization in biological materials can add substantially to the polarizability.[59,60] The roving carriers can be mobile electrons, mobile holes, or even mobile ions such as $H^+$, $K^+$, or $Na^+$.

As examples of these we have, in the electronic case, the cytochrome $c$ complexes, or the chlorophyll complexes.[66] As examples of the ionic type we have the various membrane-mobile ionophores on the one hand, and DNA on the other.[67]

Said another way, studying and understanding nomadic polarization in biological materials is often merely taking a look at the organic condition present in these materials, but from a very different viewpoint.

For comparison purposes we can look at some rough estimates of the incremental change in the relative dielectric constant, $\Delta K$, contributed by each of the above types of polarization. See Table 1.

## 5.2. Supramolecular Modes (Interfacial and Interregional Polarization)

It has been realized since the mid-1800s that the presence of material interfaces in matter provides discontinuities for the travel of changes, and provides opportunities therefore for the delays in their travel in response to an applied field. This the experimenter would view as an electrical polarization. One would then assign to the system as a whole an apparent dielectric constant containing the effects of these macroscopic polarizations, and in addition to those due to the molecular modes discussed above.

TABLE 1. Typical Contributions of Various Types of Molecular Polarization to the Relative Dielectric Constant, $\Delta K$

| Polarization type | $\Delta K$ |
|---|---|
| Electronic | 1–10 |
| Atomic (most organic compounds) | 1/7 of the electronic value |
| Atomic (inorganic solids) | 1–6,000 |
| Dipole orientation | 0–160 |
| Nomadic (hyperelectronic, organic polymers)[60] | 0–300,000 |
| Nomadic (hyperprotonic salts)[60] | 0–1,000,000 |

It is convenient to mentally separate the more macroscopic supramolecular polarization modes into those associated with the *bulk* properties of the several phases on the one hand, and into those more localized to the surfaces. We do not find the mythical two-dimensional conduction or polarization to be useful concepts today, for they lack both physical reality and predictive usefulness. Theories employing such ideas as surface conductivity quickly reduce to curve-fitting exercises. On the other hand the concept of the surface-associated ionic double layers has proven highly useful. Yet another concept, that of a plasma or a quasiplasma proves useful in dealing with biological systems. We are thinking here of a rigid or a semirigid structure within which ions of one charge are fixed, and between which mobile counterions can rove. A bacterial wall provides an example of this plasmoidal system of easily polarizable charges.

Broady speaking, disperse systems have a number of ways to exhibit enhanced polarizability.

### 5.2.1. Interfacial (Bulk–Bulk) Polarization

This is the well-known Maxwell–Wagner polarization.[1] It always arises when two media are in contact and have differing electrical properties. A strong field and a disparity of charges arises at the juncture, leading to the possibility of high effective polarizations. These are ubiquitous in biology.

### 5.2.2. Electrophoretic Dielectric Response[1,68]

Since suspended or colloidal particles usually bear an electric charge which is compensated only by a distributed atmosphere of counterions (gegenions) out in the support medium, the electrophoretic motion of the charges due to an applied field could give a dielectric response. It turns out upon detailed analysis that the response is "negative" in that the contribution is such as to *decrease* the overall polarization of the system.[1,68] Since this has not been observed to date in biological systems, we drop further discussion of it here.

### 5.2.3. Surface-Associated Interfacial Polarization[1]

The region of contact of two different materials, generally of differing chemical potentials or work functions will give rise to an interfacial potential and therefore a dipolar surface layer. In the event that one or both of the materials has a high dielectric constant or is perhaps easily dissociated, then the formation of ionic species at this surface dipole layer can occur.

This is the origin of ionic double layers. Obviously, the presence of easily adsorbed and easily dissociated impurities can play a big role.

The motion of the mobile charges associated with the ionic double layer can be conveniently regarded as those (a) closely bound and of low mobility or (b) loosely bound and of high mobility. The whole ionic double layer is often referred to as the Helmholz double layer. The inner or more tightly bound and the outer or more loosely layers are conventionally termed the Stern and the Gouy layers, respectively. Grimley and Mott[69] showed that analogous electrical double layers exist within solids near their surfaces. One example would be that of AgBr particles suspended in aqueous media. Here, on one (solid) side of the solid–liquid interface there is an ionic double layer of $Ag^+$ and $Br^-$ ions; on the aqueous side there is a double layer formed with aqueous ions. The aqueous double layer depends on the $pAg = \log_{10}[Ag^+]^{-1}$ in the solution. It is also modifiable by the presence of other adsorbable ions, especially if they are multivalent. Such "double–double" layers,[1] so to speak, are present in biological systems.

Among the most difficult systems to characterize from the electrical standpoint are aqueous suspensions. It is clear from experiment that the ionic double layers as well as the bulk–bulk properties play roles. The effective dielectric constant of about 10,000 can be obtained from what might at first seem to be a simple system of water (dielectric constant $\sim 80$) containing suspended polymer particles of dielectric constant about 2! Similar results have been obtained with glass spheres or with living cells suspended in aqueous media.[70–72]

Considerable stimulus to the understanding of these "anomalously" high polarizabilities of the suspensions was brought by Schwarz[73] and by Schwan et al.[68] using a model considering the inner ("bound") ions of the boundary layer. Their theory was tested against experiments on suspensions of polystyrene spheres in aqueous media, and found to fit well. Schwarz's model assumed spheres of radius $a$ having closely bound but mobile counterions of charge $e$, and mechanical mobility $\mu_b$ at a surface density $\rho_0$ held at the double layer of negligible thickness compared to the particle radius. Where the volume fraction of such spheres in the suspension is $p$, their effect upon the polarizability is to contribute a "static" dielectric increment of

$$\Delta K_s = \frac{9}{4} \frac{pe^2 a \rho_0}{1 + (p/2)^2 \varepsilon_0 kT} \tag{24}$$

The electrical dispersion of a simple Debye type was predicted having a critical frequency, $\nu_{crit}$, of

$$\nu_{crit} = \frac{1}{2\pi\tau} = (\mu_b kT/\pi a^2) \tag{25}$$

or

$$\tau_{\text{bound layer}} = a^2/(2\mu_b kT) \tag{26}$$

Note that the critical relaxation time is predicted to depend upon the square of the particle radius, as observable. This "bound-ion-layer" model of Schwarz was remarkably successful in several ways. The magnitude of the dielectric increment and its frequency shift with particle radius were reasonably in agreement with observation. Despite this welcome advance in the understandings of dispersions, the Schwarz theory deliberately excluded the possibility of "bound" counterion exchange with the surrounding medium. It also neglected any contribution to the particle polarizability that arises from the distortion of the outer or surrounding portions of the diffuse double layer.

The consequences of these restrictive assumptions have been reviewed by Dukhin[74,75] and others, who noted that the neglect of the outer regions leads to error in assessing the source of the polarization. The diffuse outer, and not the inner ionic double layer is according to Dukhin, the major source of polarization. The details of this comparison were recently reviewed.[1] So much for the effective electrical polarization of rigid spherical bodies surrounded by an ionic double layer, i.e., for a large class of colloidal particle suspensions.

### 5.2.4. Plasmoidal Polarization

In porous wet solids such as the ion exchange resins, or in the bacterial wall, and in gels it is apparent that some of the ions can be bound and be relatively fixed, while the counterions are relatively free and mobile. It turns out that the polarizabilities of such structures can be huge, especially at low frequencies.[1] The dielectric behavior of certain large polymeric electrolytes can also imitate this behavior.[76,77]

### 5.2.5. Polymeric Electrolytes

Biopolymers are both polar and ionic in character. They exhibit features characteristic of both polymeric solutes *and* of colloidal particles, giving complex behaviors difficult to describe simply. Enormous responses can occur at low frequencies. Minakata, Imai, and Oosawa[76] have studied theory and experiment for solutions of the polyelectrolyte, tetra-$N$-butylammonium polyacrylate ($Bu_4NPA$). They observed two low-frequency dielectric dispersion peaks, one at about 100,000 Hz, the other at about 1000 Hz. They suggest that the former is due to a bulk–bulk, Maxwell–Wagner process and that the later and slower process is associated with the displacement of counterions along the convoluted polymer chains.

Related studies on solutions of more rodlike poly-ions such as DNA and RNA were made by a number of authors.[67] The theory for such systems is well advanced.[78]

### 5.2.6. Overview of Dielectric Response Mechanisms

In Table 2 is shown a list of the numerous prominent types of electrical polarization mechanisms. Some estimate of the magnitude of the

*TABLE 2.* Types of Electrical Polarization in Biomater

| Type | Mechanism | Typical frequency response maxima (Hz) | Typical dielectric constant increment, $\Delta\varepsilon_r$, or $\Delta K$ |
|---|---|---|---|
| Electronic | Shift of electron orbitrals relative to nucleus | $10^{16}$ | 1–15 |
| Atomic | Shifts od atoms of one charge relative to others of differing charge | $10^{12}$–$10^{14}$ | 0.1–1 (4000 rare) |
| Dipole orientation | Twisting of preexisting group or molecular dipoles | $10^6$–$10^{11}$ | 1–38 |
| Nomadic | Long-range drift of electron, holes, or protons | $10^2$–$10^4$ | 100–$10^6$ |
| Maxwell–Wagner | Interfacial charge accumulations of charge drifting in bulk-to-bulk contents | $10^5$–$10^7$ | 1–$10^5$ |
| Electrophoretic | Drift of small ions in liquid media, relative to colloidal counterions | $10^5$ | Small, rare |
| Surface-associated | Distortion of ionic double layers, at a liquid–solid juncture | | |
| a. Schwarz | Response of ions of inner (tightly bound) Stern layer of Helmholtz double layer | $10^5$–$10^7$ | 1–1000 |
| b. Dukhin | Response of diffuse ions of outer (Gouy) layer of Helmholtz double layer | $10^2$–$10^5$ | 1–1,000,000 |
| c. Grimley–Mott | Electron–hole population shifts near interface within solids | $10^4$–$10^9$ | 1–100(?) |
| Plasmoidal | Ion drift in porous networks of relatively fixed counterions | $10^2$–$10^5$ | 1–1000 |

dielectric response is given to help the reader appreciate the extent of such responses. The response is expressed in terms of the increment in the relative dielectric constant which might typically be seen. The case of atomic polarization is a bit anomalous in that it is usually quite small in organic materials, but can be large in certain ionic solids such as $BaTiO_3$ or $NaNbO_3$.

Typical maximum ranges at which such mechanisms can follow the oscillations of an applied external ac field are also given in Table 2.

## 6. Orientational Dielectrophoresis

Whenever a body is subject to an electric field and the body is non-spherical, there exists the possibility of creating a torque. See Figure 4A. Even for spherical bodies, if the polarization is a tensor not parallel to the field there exists the possibility for a torque to arise. For example, early experiments by Griffin and Stowell[53] showed that *Euglena* (nonspherical) would align parallel to the applied field at low frequencies, or at very high

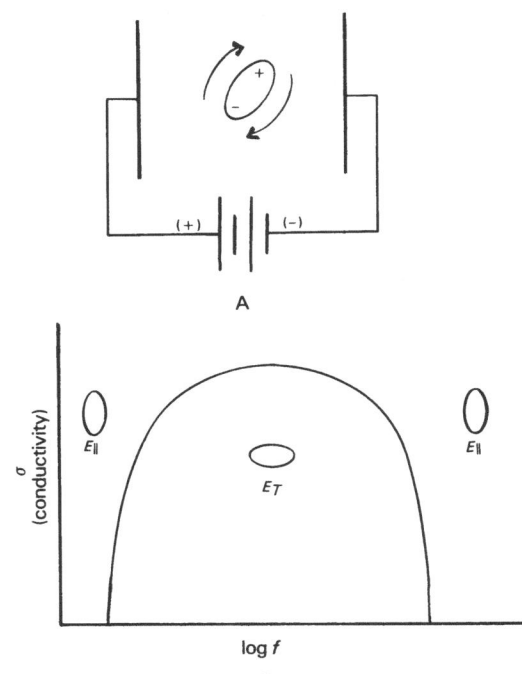

*FIGURE 4.* (A) Torque induced in nonspherical body by electric field. (B) Nonspherical body aligning parallel to low and high frequencies, but perpendicular at intermediate frequencies.

frequencies, but align perpendicular to the field at intermediate frequencies, as in Figure 4B. This feature is related to the ellipsoidal character of the cells, as analyzed later by Saito, Schwan, and Schwarz.[51] In fact they showed that for the case of an ellipsoidal object with three different radii of rotation, there can be even more frequency-rotation domains, and each of the three orientations can predominate at a particular frequency range.

Fomchenkov and Gavrilyuk[79] have developed the orientational DEP technique by using a pair of grid electrodes in concert with a nephthalometer to detect the changes in optical density of a suspension of cells as the field is applied. The response is found to be rather rapid, on the order of a minute or two.

## 7. Induced Cellular DEP

### 7.1. Experimental Considerations

There are a number of experimental aspects which affect biological DEP. One wants and indeed needs a nonuniform electric field to cause DEP. A variety of electrode shapes can be used, including pin–pin, wire–wire, pin–plate, and isomotive geometries. These are sketched in Figure 5. The isomotive electrode system is one designed to produce a constant DEP force over a finite region of a DEP chamber. It is particularly useful for analytical and in delicately comparative experiments. The pin–pin and wire–wire electrodes are relatively easy to construct and handle.

The electrode arrangements possible are many, but biological DEP, being done in aqueous media, imposes its own restriction on the dimensions. Much experimental experience as well as theoretical analysis shows

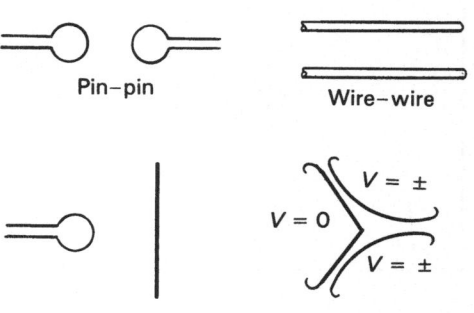

FIGURE 5. Various electrode shapes for producing dielectrophoretic fields.

that DEP effects in $H_2O$ have a very limited range. In practice, one should not expect to observe useful effects at distance greater than $1000 \mu m$ (1 mm) from any electrode when "safe" voltages are applied. By safe is meant that the power density is less than about 0.001–0.01 with cm$^{-3}$ for extended periods so as to avoid deleterious heating.

The voltages used in conventional DEP experiments normally do not exceed some 30 V rms, and 10 V rms is useful on the simpler wire–wire or pin–pin electrode systems.[1]

The electrode materials are preferably surfaced by a noble metal (Au, Pt). Stainless steel or carbon elctrodes give polarization troubles. Power supplies to give adequate voltages (0–30 V rms) over the range 10 Hz to 60 MHz are commercially available and need not be discussed further. The same remarks[1] apply to the voltmeters and conductance metere necessary.

DEP chambers can be very simple and still yield significant answers to the experimenter. The *isomotive* electrode geometry is useful in analytical and comparative experiments because it produces a DEP force independent of position over the working area of the chamber.[1,31,35,40] There are many other useful electrode shapes[1,81] that can be used depending on the application desired.

Biological DEP experiments can be successfully completed with relatively simple equipment. A DEP chamber can be constructed on a microscope slide.

The other equipment necessary, a microscope, voltmeter, conductance meter, and power supplies are widely available.

There are several physical variables which affect cellular DEP collection. The most important of these are applied voltage, particle concentration, solution conductivity, and frequency of the applied field. The DEP collection varies linearly with both voltage and concentration. There is a large departure from linearity with large voltages ($> \sim 30$ V rms) as heating and stirring effects disrupt collection. These two variables can be adjusted by the experimenter for optimum observations of the DEP collection.

DEP collection is a complicated function of conductivity and frequency. Figure 6A shows the DEP "spectrum" of a cell. As can be seen, the response as a function of frequency varies greatly. There a regions of minimal response and a number of peaks of strong response. As can be seen in Figure 6A these peaks decrease with increasing conductivity. The frequency at which these peaks occur is also seen to shift.

Since the frequency of the applied field and the conductivity of the suspension are so closely and complexly coupled, it is necessary to carefully monitor and control these two variables during an experiment. A spectrum of DEP collection vs. frequency has little meaning unless the conductivity of the suspension has been kept constant and has been specified. Matters

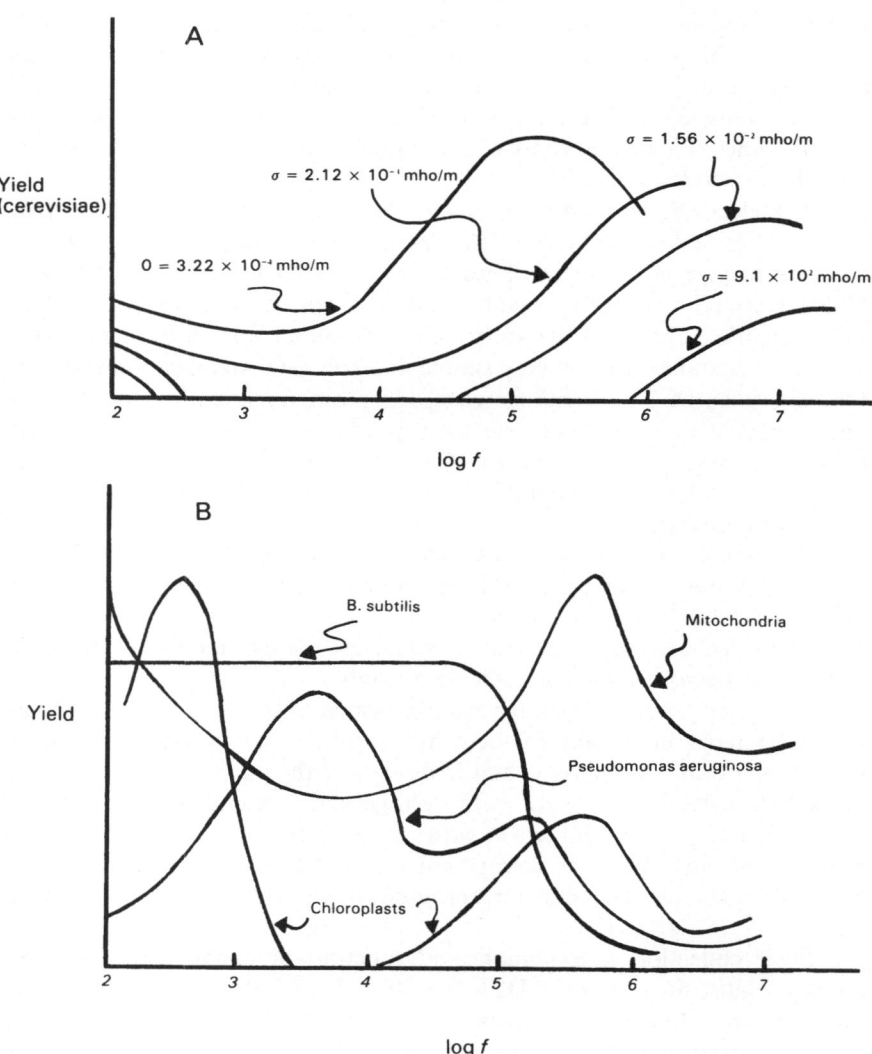

*FIGURE 6.* (A) DEP "spectrum" of a cell. (B) Different types of cells exhibiting unique DEP collection spectra.

can be simplified by working in media of high resistivity. We find that the DEP spectra are essentially identical if the resistivity of the medium is 100,000 to 1,000,000 $\Omega$ cm. [conductivity 1–10 $(\Omega \, m)^{-1}$].

Biological variables which affect DEP response are numerous. This area is only beginning to be investigated. There is still much to discover and understand.

An obvious observation from Figure 6B is that each type of cell has a DEP collection spectrum that is unique. Cells may be characterized by their spectra and separations made based on the differences in the spectra of two cell types.

Since the polarization of a cell depends on its physiological state, anything that affects its state can affect the DEP spectrum of that cell. Some early work[3,41] compared normal yeast cells with dead yeast that had been killed with heat. Such drastic treatment produced large differences in the DEP spectra. Separation of these two cells was relatively simple because of the large differences.

Studies of cells have shown large changes as the moves through its life cycle. DEP studies have shown differences in response with changing colony age.[1,3] These studies have shown the importance of using synchronized cells so that all cells are the same age.

Studies on the blood platelets of dogs with hemophilia, dogs that only transmitted the disease, and normal dogs showed differences in the DEP responses between the three.[1,4]

Other work has studied the effects of various chemicals (inhibitors, herbicides, etc.) on the DEP response of cell.[1,3]

The most subtle difference detected to date, perhaps, was in the work of Mason and Townsley.[10] They cultured yeast on two different growth media and were able to separate a mixture of these cells using DEP.

A number of experimental techniques have been developed for biological DEP work. We will describe several of the most useful techniques and give their advantages and limitations.

## 7.2. Batch Methods

The batch method is the easiest and was the first DEP technique developed. For biological DEP a batch chamber can be constructed on a microscope slide. The electrodes (two parallel wires for example) are placed in a pocket or inside a capillary tube that is mounted on the slide (Figure 7). The chamber is large enough to hold 1 or 2 drops of cell suspension and measurements of the number cells collected on the electrodes can be easily made 1 or 2 minutes after the field is applied. The chamber is then rinsed and dried and the process is repeated as often as desired. Batch methods are simple, low cost, and can accurately characterize cells, even

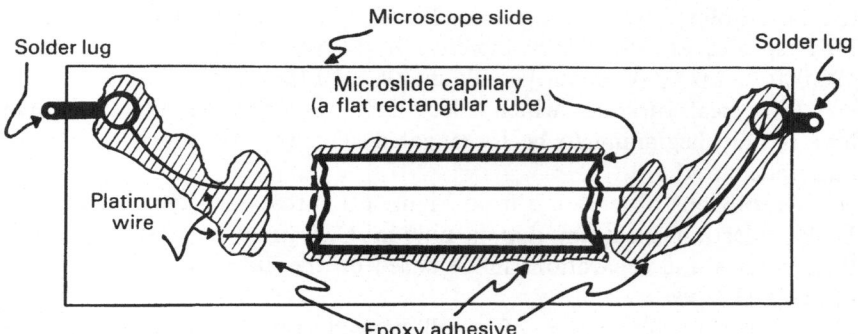

FIGURE 7.  Batch chamber for biological DEP.

detecting subtle differences in cell physiology.[1-6,10] Batch methods have some limitations which may or may not be significant in an experiment. The most common electrode configurations produce a DEP force that decreases rapidly with distance from the electrode surface. This makes separations difficult since positional differences in the DEP force can be greater than differences arising from physiological differences. Actual physical separations are almost impossible due to the small dimensions involved. Superior micropipetting techniques are required.[1,10] Since only cells collected on the electrodes are measured, the batch method cannot easily detect negative DEP where cells are repelled from the strong field regions. Batch chambers cannot measure this response quantitatively.

The semibatch technique evolved from the batch to provide a capability for separating cells. A current design is the "phonograph" chamber (Figure 8). The electrodes are grooved metal plates resembling

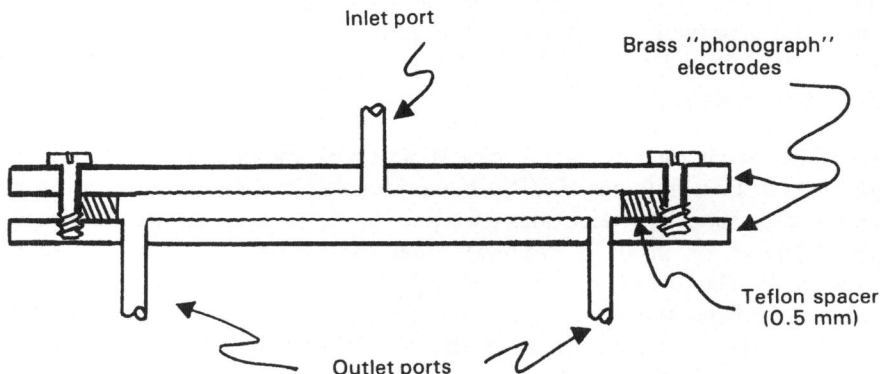

FIGURE 8.  Phonograph electrode DEP chamber.

phonograph records. A suspension of cells is pumped through the chamber and cells are held back by the DEP field. The cells are flushed out later when the field is turned off. This chamber is capable of handling and separating enormous numbers of cells. Characterizations of cells are not as easy to do in a semibatch chamber. Its main function is separation of cell types. A mixture of cells whose characteristic spectra are known is used and the chamber is "tuned" to collect one cell type preferentially over the other. The chamber then acts as a biological filter. Since the filtering action depends on positive DEP, the semibatch chamber has similar properties with respect to negative DEP as do batch chambers.

## 7.3. Continuous Methods

A method with many advantages is that of continuous stream DEP. Figure 9 shows a continuous flow chamber and electrode system. Cells are injected into the flowing stream so that they flow down a narrow region in the center of the chamber. This system[5] has many advantages. The electrodes approximate the isomotive shape to make delicate separations possible. The stream of cells can be moved either way by the field so both positive and negative DEP can be used. Separations are possible with the use of flow splitters. The system also is capable of being automated. This system cannot separate great numbers of cells as the semibatch one can, but it is much more versatile as it can separate and also quickly characterize cells.

## 7.4. Single-Cell Levitation

The most delicate measure of a cell's DEP response is made in single-cell DEP.[6] In the technique (shown in Figure 10), a single cell is levitated[7-9] between two electrodes. The DEP force balances the force of gravity in a manner similar to the Millikan oil-drop experiment. Single-cell DEP is a very delicate method of characterizing cells.[6] It can follow changes as a cell moves through its life-cycle. It is applicable as a rapid biopsy or cell analyzer. It can measure both positive and negative DEP, and is also capable of being automated.[82]

## 7.5. Micro-DEP

A new technique, micro-DEP uses the fact that growing cells (e.g., fetal, tumor, would-healing) oscillate electrically.[13-16] The cells emit non-uniform electric fields and can attract small polar particles by DEP. The technique consists of measuring the cells ability to collect tiny 2-$\mu$m-

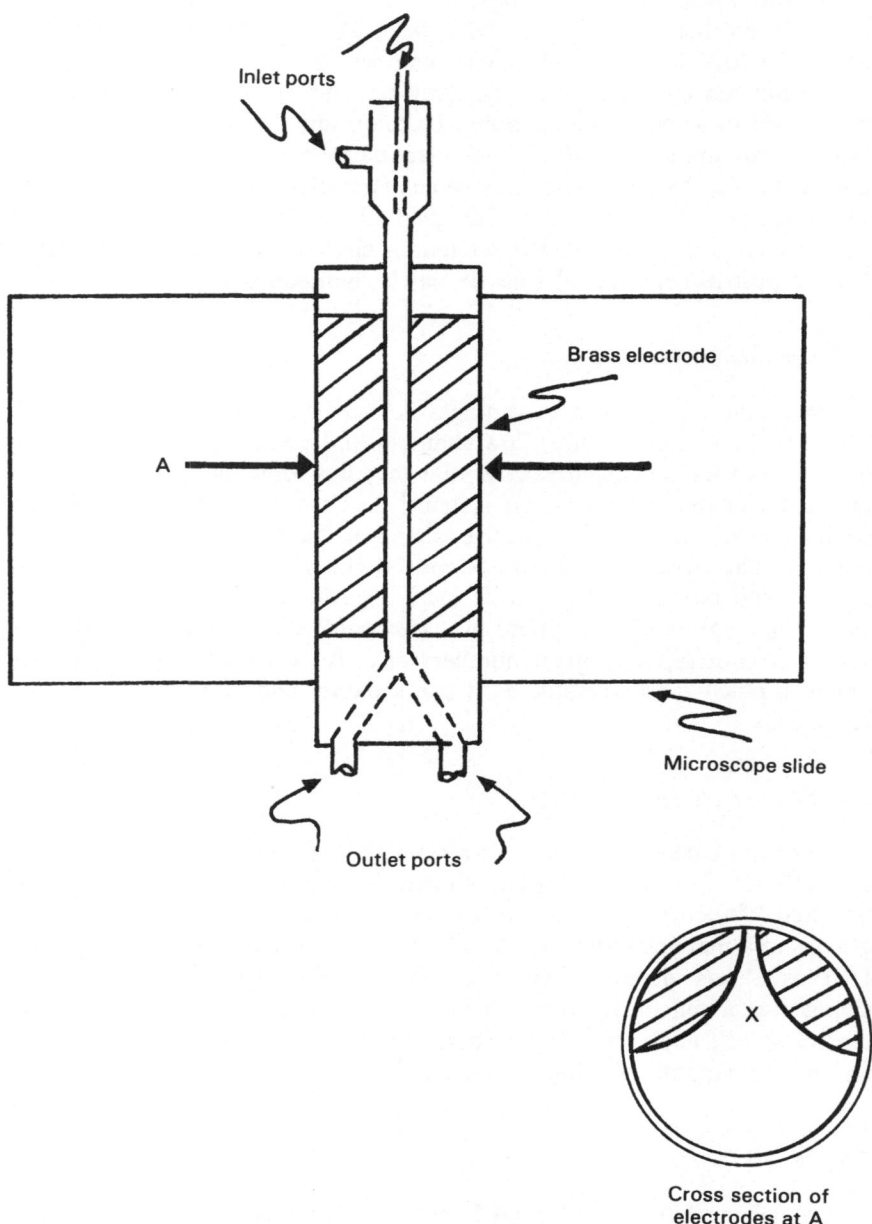

FIGURE 9.  Continuous DEP flow chamber and electrode system.

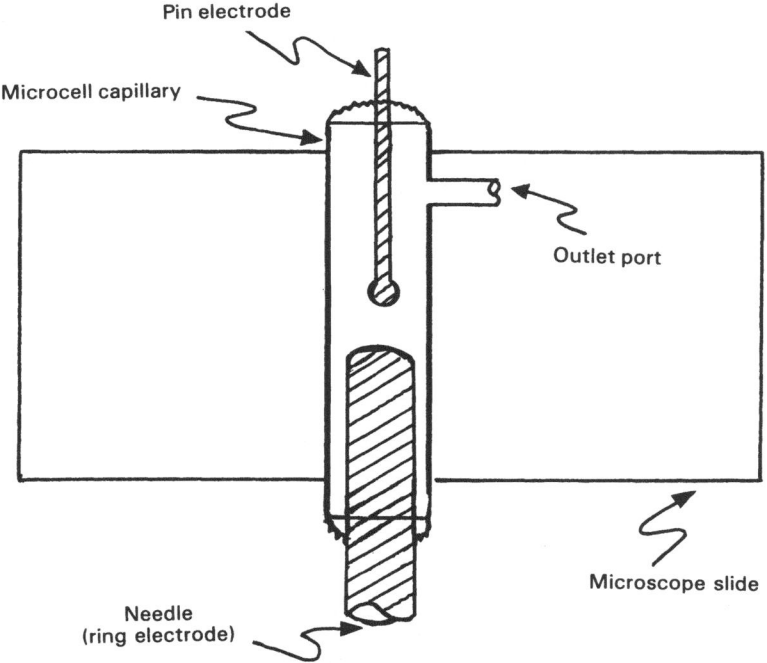

*FIGURE 10.* Single-cell levitation DEP cell.

diameter particles. The cells are compared as to the number of highly polarizable particles they collect ($BaTiO_3$), and as to the number of poorly polarizable particles they collect ($BaSO_4$). The difference (usually $2:1$ in favor of polar particles) is a measure of the DEP attraction. This technique is used to investigate the mechanisms of cell reproduction[15] and to investigate the nature of cancer cells.

## 8. The Use of DEP to Shape Tissue Models

DEP can be used to gather and orient living cells into new aggregates of preselected shape. The resultant shapes can then be fixed in form by using a thermosensitive[11] gel.

The clumping and aggregation of cells from suspensions in culture media to form masses is well known,[83,122] but the resultant aggregates except for rouleaux formation by erythrocytes, are normally rather random as to their organization. The aggregates so formed, moreover, may or may not be weakly adherent, a matter which depends strongly upon the fields, through dielectrophoretic action. It can be used to effect a more selective organization of polycellular masses, and how, by using a special matrix, even weakly adherent cells can be made to hold rather firmly together in large selectively shaped forms.

Since rapid collection of erstwhile single cells can be brought about by the use of dielectrophoresis, and the shape of the mass of cells so formed is controlled by the shape of the electrodes, by the frequency of the field, and by the shape of the cells, it appeared worthwhile to see if the shapes so formed could be made more permanent, so as to prepare desired structures of the cellular aggregates. To this end, the gel-forming properties of a relatively nontoxic block copolymer solution in water were used. A concentrated aqueous solution of the block copolymer of polyethylene oxide and polypropylene oxide (PEO-PPO) is quite fluid at 0°C to about 5°C, but sets reversibly to a rather stiff gel at about 30–40°C. The PEO-PPO polymer solutions are reported to be relatively nontoxic to most organisms (Pluronic resin F-127), Wyandotte Chemicals Corp.).

Suspensions of the desired microorganisms were prepared in the cold (0–25°C) polymer solution, which were then deionized by gently stirring the suspension with beads of a mixed-bed ion exchange resin (Rohm & Haas MB-3) until the specific resistivity exceeded $10^5\,\Omega$ cm. The cellular suspension was then decanted free of the resin beads. (It was found necessary to remove fines from the resin beads by repeated washings with distilled water before starting.) The cold fluid suspension of microorganisms in the PEO-PPO polymer was then subjected to a non-uniform electric field produced by a pair of parallel platinum wires 2 mm apart and each 425 $\mu$m in diameter, and upon which was impressed a voltage of 40 V rms at 600 kHz. The cells, as is typical of dielectrophoretic collection, collected rapidly and quite equally on *both electrodes*. When collection of the cells was essentially complete at the high value, the voltage was reduced to a "holding" value of 5 V rms and the preparation was warmed to 36–37°C. The matrix of cells and polymer set to a stiff gel. A "fleshy" material of low orderliness was prepared using the globular yeast cells (*Saccharomyces cerevisiae*).[82]

An analogous experiment using the rod-shaped *Bacillus subtilus* (circa 1 by 4 $\mu$m) cells provided a more tissuelike material, in that higher organization of the collected cell mass was apparent. It was observed that if these microorganisms are collected at a frequency of 2.5 MHz, they move endwise towards the electrodes and assemble in parallelized layers, having the small end towards the electrode. These same cells can be made to collect quite differently at a much lower frequency (about 1 kHz), where they move in sideways, collecting so as assemble flat against the wire. Once the collection is begun and cells are against the electrode, they may be made to stand erect or lie down alternately as the frequency is switched from high to low and back. This curious process may be repeated many times. The gelled matrix of such elongated cells can thus be prepared with either orientation. The complete process of collection and setting the collected matrix can take less than five minutes.

By choosing the frequency, voltage, electrode design, and suspension conductivity conditions, it is possible to prepare the collected cell masses in a variety of way. By using a cylindrical electrode geometry, the masses can be tube-shaped with the cells lying in monolayers or in thick multilayered structures. The cells, if they are elongated in shape can be deposited end-on or flat-down upon the central electrode wire. The end-on oriented variety of cellular matrices suggests that this may be the best form for subsequent post-treatment to produce unidirectionally functional layer structures, although such modification was not attempted here. Tubular masses of cells were easily obtained using a cylindrical electrode geometry.

The above experiments suggest that as the techniques for the use of dielectrophoresis to form tissue models improve, as better cellular organization is obtained, as improved cementing techniques are found, and as chemical or physical post-treatments for the matrices are developed, then a wide variety of interesting and perhaps functionally useful multicellular preparations can be fabricated from suitable single-celled organisms.

## 9. Applications of Orientational Dielectrophoresis

The several effects of externally applied fields upon living cells in suspension include electrophoresis, dielectrophoresis, and the induction of orientation, spinning and even fusion. The orientation of cells can be induced in either uniform or nonuniform fields. It is often advantageous to use nonuniform fields in this case so as to produce a concentration of the cells locally in which the orientation can be more easily detected. A particularly fruitful development along these lines has been made by Fomchenkov and Gavrilyuk.[79] They used optical detection of the light passing through a suspension of cells to which an electric field was applied. Jennings *et al.* had earlier used light scattering and dichroism to study bacterial orientation in electric fields.[87-90] The method of Fomchenkov and Gavrilyuk[86,90] employs a simple nephthalometer in conjunction with a DEP chamber having a grid of fine wires to affect the cells. Studies of the orientational dependence upon the applied frequency and various physiological factors were made with several microorganisms, including *Pseudomonas fluorescens*, *Escherichia coli B*, and *Chlorella vulgaris*. The forms of the spectra were found to depend upon the growth phase of the cultures. The addition of albumin to the suspensions was found to profoundly alter the orientational response spectrum, indicating that there had been appreciable surface adsorption causing alternation of the surface ionic double layers and hence of the DEP forces associated with such ionic double layers.

## 10. Natural rf Oscillations in Dividing Cells

### 10.1. Introduction

In 1968, H. Fröhlich[91] predicted that the chemical energy of cells could evoke high-frequency electrical oscillations. We have now observed such outputs from cells. We see them to be associated with cell division.

To observe such rf oscillations, we have used two very different techniques, each needing only relatively simple apparatus. The first technique is called micro-DEP and essentially requires only a microscope.[13-16] Here one observes the collection of various highly polarizable particles by a cell so as to examine the radio frequency (rf) field emitted from it. In the second technique, direct observation is made of the spinning of cells evoked by external rf fields. Both methods yields similar conclusions as to the nature, frequency, strength, and occurrence of the rf electrical oscillations of cells. We believe, therefore, that the presence of the postulated rf oscillations has been established beyond reasonable doubt. It now remains to study their meaning. Are they cause or effect, necessity or frill, in the life of cells? Where and how do they operate? What causes them? What controls, intracellular or intercellular, do they evoke or reflect? With what processes are they associated in the cellular life cycle?

In particular, the rf fields of cells can be examined by using the principles of DEP. If the cells generate electromagnetic fields, they may collect highly polarizable particles more readily than they do ones of low polarizability. Such a study can be done with simple means: a good microscope and a knowledge of what to look for.[13-16]

In the following sections we shall give a brief survey of the microdielectrophoresis and the spinning studies to date, and show how they demonstrate the existence of natural rf oscillations in cells. Next, we shall indicate the line of future research and make some suggestions as to the applications.

### 10.2. Micro-DEP (μ-DEP) by Living Cells

The μ-DEP experiments are relatively simple and direct. A typical experiment consists of mixing a suspension of cells with a suspension of smaller powder particles, then observing the number of particles associated with the cells after a short time. This rate (collection factor) is the compared for similarly sized particles of various degrees of electric polarizability (DK). The particles used are typically in the size range of 2 $\mu$m and are size-selected by appropriate settling and/or filtering procedures. In contrast, the mouse filbroblast cells used are rather larger, viz., ca. 2 $\mu$m $\times$ 5 $\mu$m for erythrocytes, 10–15 $\mu$m for murine "L" fibroblasts or ascites tumor cells, and 10–50 $\mu$m for fetal fibroblasts. The bacterial cells are typically about 2 $\mu$m $\times$ 5 $\mu$m (*B. cereus*), while the yeast cells (*S. cerevisiae*) are about 4–6 $\mu$m in diameter.

For particles of high polarizability one uses materials such as $BaTiO_3$, $SrTiO_3$, or $NaNbO_3$ or a polymer DP-1A with static $\varepsilon_r$ of about 2000, 400, 650, and 5000, respectively. For particles of relatively low polarizability (i.e., less than that of the suspending medium, $H_2O$, static $\varepsilon_r$ 80) numerous materials would serve. We have used $BaSO_4$, $Al_2O_3$, and $SiO_2$ (static $\varepsilon_r$ 11.5, 7, and 3.8, respectively).

At this point it is essential to apply the concept of the "effective dielectric constant" as developed both theoretically and in studies of DEP.[1] By this we mean the idea that one must consider the overall effect of *all* mobile changes in the system during ac measurements. Thus, the conductivity *and* the dielectric constant contribute to the effective dielectric constant at any particular frequency. Failure to carefully include *both* factors can lead to unwanted consequences and to misunderstandings. Without going into details (see, however, Ref. 41), we can put the point that for many cases of a body suspended in a fluid medium, the DEP force depends upon the difference of the effective dielectric constant of the body from that of the surrounding medium. The DEP force in a specific case depends, of course, upon the volume of the body and upon the gradient of the square of the field. For the moment, we need only concern ourselves with the dielectric factors. For simple cases the effective dielectric constant $|K_{\text{eff}}|$ depends upon the relative dielectric constant, the specific conductivity, and the frequency in the following way:

$$K_{\text{eff}} = [K_i^2 + (\sigma_i/2\pi\varepsilon_0 f)^2]^{1/2} \qquad (27)$$

where $\sigma_i$, $K_i$ are the specific conductivity and relative dielectric constant (GK), respectively, at frequency $f$.

The theoretical problem in real cases becomes quite complex but those details need not concern us further at this point. Suffice it is say for the moment that the *effective* dielectric constant (DK) of a liquid such as $H_2O$ can be changed at will over wide ranges by the addition of conductive salts. In this manner the relative polarizabilities and hence DEP forces upon bodies such as cells or crystals can be widely varied. Thus, at a given high frequency (1 MHz), although in pure water (DK $\sim$ 80) a pure (insulating) crystal of $BaTiO_3$ has a higher effective DK($\sim$ 2000), yet the adding of a trace of salt (e.g., 1 mM̄) increases the effective dielectric constant $|K_{eff}|$ of the water to over 13,000. Said another way, the erstwhile high DK material $BaTiO_3$ is attracted to the high field region by DEP in pure water, but is repelled even when in very dilute aqueous (salty) media at this frequency. It experiences positive DEP in the pure $H_2O$ and negative DEP in the salt solution. This is a most useful fact.

We have used this fact to examine the natural rf from cells.[13–16] When murine ascites cells, e.g., are in deionized M/4 sucrose, they show an extra gathering rate for the high DK ($BaTiO_3$ or $NaNbO_3$) over that for the low DK materials (e.g., $BaSO_4$ or $SiO_2$). The addition of 0.1 mM KCl or NaCl clearly suppresses this preference, which disappears in 1 mM solution, supporting the ideas given above.

Kinetic studies of the rate of accumulation of 2-$\mu$m particles by mouse "L" cell fibroblasts or by mouse ascites cells showed that only a few minutes mixing time suffice for comparative rate studies. A more-or-less steady state is reached in 10–15 min.

The natural rf oscillations of the cell might conceivably arise in smallish "domains" throughout the cell. If so, they might imitate the behaviors of known ferroelectric or ferromagnetic materials wherein an external "poling" pulse of a strong field could reorient the erstwhile randomly oriented domains to achieve a still higher overall polarization. Preliminary experiments in our labs indicate that a mild dc "poling" pulse does indeed increase the preference of mouse ascites tumor cells for the more polarizable powder particles as might be expected in this picture.

The frequency of the natural rf oscillations term of cell can be estimated from the salt effect. It will be recalled that the $\sigma$ term of $|K|$ contains the frequency $\omega$. Using this fact, and the known conductivity, $\sigma$, required to repress the $\mu$-DEP, we estimate a minimum frequency $f = 2\pi\omega$ as $\geqslant 5$ kHz. Later experiments using observations of cell-spin-resonance give a result of 1–30 kHz for the actual frequency, and confirm the $\mu$-DEP result.

An estimate of the field strength, $E_\mu$, appearing at the cell surface and due to the natural rf dipole can also be made from the properties of $BaTiO_3$. This is a ferroelectric solid. It has a hysteresis of polarization and requires a certain minimum $E$, $E_{crit.}$, to override its internal intrinsic polarization. That $BaTiO_3$ is attracted to and held by in excess of that amount expected for ordinary particles of similar size, density, etc., shows that the field due to the particles is greater than $E_{prior} \simeq 10$ V/cm. From this one can also estimate that the natural rf dipole of the cell is about $10^{-22}$ mCcm corresponding to a swinging of some $10^7$ electrons a distance of ca. 1 Å in the cell.

At this point one may ask how we can be sure that the natural dipole is an oscillating one. Might the natural dipole be a static one and cause the

$\mu$-DEP effects observed? No, because the presence of the conducting medium, $H_2O$, would quickly release ions to mask the static dipole.[14] It can be shown easily from dielectric theory that the natural cellular dipole must be one which oscillates at a rate faster than the masking rate (dielectric relaxation rate) of the aqueous medium in order for the dipole field to be sensed by the test particles (e.g., $BaTiO_3$ or $NaNbO_3$). As noted above, this frequency must at least be 5000 Hz. There cannot be a simple charge effect here due to just the differing DK's of the cell and the various particles. For example, it was suggested by one colleague that it was well known (Coehn[92] effect, 1898) that simple "triboelectric" charging of particles occurred upon separating materials of differing DK in a vacuum. Might not charges so generated result in the mutual attraction of the cells for the high DK particles? The answer is no, because of the presence of the conductive medium, $H_2O$, which would rapidly destroy such charge imbalance. Moreover, a simple experiment done by mixing aqueous suspensions of two powders, one of high DK, the other of low DK [e.g., $BaTiO_3$, DK $\sim 2000$; $BaSO_4$, DK $\sim 11.5$ in $H_2O$, DK $\sim 80$], showed normal behavior. The mixture settled at a rate intermediate to that of either pure particle suspension. Has a special "Coehn effect" existed, the mixture would be expected to agglomerate and settle more rapidly than either pure suspension.

In summary, the presence of natural rf oscillations by cells is confirmed (a) by the preference of (dividing) cells for accumulating particles of high DK over that for ones of low DK; (b) by the predicted suppression of this by salt effect; (c) by the predicted increase of this by "poling"; (d) as we shall see, by cellular spin resonance[95–99]; and (e) by direct detection.[123,124]

The biological aspects of $\mu$-DEP are of particular interest. It is observable in bacteria, yeast, and mammalian cells, and hence is probably universal. It is particularly evident in cells that are in the reproductive cycle, and is, therefore, most probably associated in the organizational process connected with mitosis and cell division. We have observed[93] using $\mu$-DEP a preference of cells for high DK particles in the following cases: (a) (synchronized) early phase *Bacillus cerius*,[93] (b) yeast (*Saccharomyces cerevisiae*),[13,14] (c) rapidly dividing fetal mouse fibroblasts,[14,93] (d) rapidly dividing mouse "L" fibroblasts, and (e) mouse ascites tumor cells.[13–16,93] On the other hand, normal nonpreferential pick-up of cells for either high DK or low DK particles was observed for non-dividing cells, e.g., confluent mouse "L" fibroblasts or the (enucleate) mouse erythrocytes. Preliminary work by Dr. Brian Goodwin, University of Sussex, England, found $BaTiO_3$ particles to be attracted to the rapidly dividing[94] frog blastula (Xenopus). Pattern formation was reported as predicted.[14,93]

## 11. Cellular Spin Resonance (CSR)

Can cells when placed in an external field, $E$, exhibit spin resonance, in analogy to that seen for ESR and NMR? Apparently so. Some time ago a phenomenon of cellular rotation during mild DEP was observed[3] to occur at specific, selected frequencies. Here, most of the yeast cells were held by a mild DEP force against a platinum electrode. As noted earlier, the cells exhibit mutual DEP and generally stack up upon each other, especially if a moderate to high concentration of cells is available. What was surprising and unexplained at that time was to occasionally see cells spin busily about an axis normal to the field lines. This was seen to occur at a rather sharp frequency for each particular cell that spun. As the frequency was changed, cells could be seen to stop spinning here and there in the field of view. Spinning could be seen to occur with cells at a wide range of frequencies anywhere in the field of view whether in contact with the electrode, against another cell, or floating freely briefly in the medium. As the frequency was changed, the spinning rate would vary, becoming a maximum in a narrow frequency range. A cell which had stopped as a result of a frequency change could often be seen to recommence spinning if the frequency were returned to the original value. In a young and growing culture, spin counts were high in the frequency range of $10^4$ Hz (*Saccharomyces cerevisiae*). In aged cultures such spin was absent or nearly so. We, therefore, associate spin with the reproductive state of the cells. Spin of cells induced by ac fields has also been observed by others,[95-99] but the specific frequency characteristics were apparently first reported by Crane[3] and Pohl.

There are, in theory, three main reasons why an applied electrical field could evoke cell-spinning:

1. Bipolar ion deposition by conduction onto the cell surface: This, however, is essentially a dc phenomenon discovered only[100] recently; it cannot account for the above observations seen with rf fields.
2. Asymmetric (tensorial) polarization: This can be shown to have an effective response only at the frequency at which the cell rotates. It is not, therefore, the cause of the observed spin (5 Hz) due to critical and narrow *high* ($10^5$ Hz) frequency responses.
3. Interparticle, delayed polarization, off-axis interactions. Here, in a fashion analogous to the action of the familiar shaded-pole motor, the action of the applied field combined with that of the phase-pole motor, the action of the applied field combined with that of the phase-delayed field arising from a nearby and actively polarizing particle produces a torque upon a given cell. This cell–cell interaction can be extensive and evoke massive rotational action of many

cells simultaneously. It is probably the major cause of most of the cellular spinning observed. In cannot, however account for the observed spinning of lone cells, whether they sit against an electrode or out in the fluid.

4. Natural *intrinsic* oscillating electric dipoles: This is the only known mechanism whereby one can account for all the above-described facts. We believe it provides a clear and logical explanation for them.

From the theory of such behavior we can predict, in a straightforward manner and using no adjustable parameters, the dipole strength and its field strength at the edge of a cell. In the case of yeast cells one obtains results that agree closely with those using the (quite unrelated phenomenologically) technique of micro-DEP. The two techniques support each other in these three aspects (i.e., dipole frequency, dipole strength, and surface field strength). The technique of CSR is presently an easy one to use and quantify.[123] It provides results offering great insight into cellular processes, especially those during cell division. We believe it has much potential.

## 12. Origins of the Natural rf Oscillations of Dividing Cells

The origin(s) of the observed natural radiofrequency oscillations of dividing cells are as yet not clear. They could arise from oscillating chemical reactions[101–103] coupling to physically mobile regions of ions so as to produce charge density waves. The study of oscillating chemical reactions is presently in an increasingly active state.[101–104] An interesting paper by Schmidt and Ortoleva[104] suggests how such oscillating reactions might be coupled to waves of electrical charge. The phenomenon of the *mechanically induced stimulation of electromagnetic radiation* (M.I.S.E.R.) from vibrated systems having ionic double layers may play a role.[105] A fine collection of reviews on oscillations in biological systems has appeared.[101,102]

There could also be a more directly physical (as opposed to chemical) origin. Fröhlich[91] showed in 1968 that an assembly of randomly oscillating similar dipoles[91] could be driven to operate in cooperatively condensed mode if the input power exceeded a certain minimum value. This and a series of stimulating subsequent papers opened a new avenue of cellular studies.[91,106,107,110]

Whatever the origin(s) of the natural rf cellular oscillations, it will be of much interest to learn if they are necessary or incidental, of cause or effect, in the living state.

As stated above we have good evidence that cells in the dividing state act like small radio sources, and emit radiofrequency (rf) electromagnetic radiation. These natural rf oscillations have been detected among bacterial, algal, yeast, avian, and mamalian cells. It would, therefore, appear that they are universal in living systems. Moreover, they are most readily detected from cells that are in the act of reproduction. We do not yet know precisely where in the cells life cycle they first appear and abate. We do not yet know with what chemical energy systems or sources they are associated. Are these natural rf oscillations necessity or frills in cellular life? Their ubiquity and hence probable ancient origin would bespeak "necessity."

If these natural rf oscillations are necessary and linked to cellular reproduction, what new control might we then have over cellular division? At the moment we judge these natural rf fields to be of very short range (microns) and to be linked more to intracellular rather than intercellular processes such as communication. This conservative view may be too cautious and needs examination. There are a number of critical questions which suggest themselves as ways to test the hypothesis that such natural rf oscillations are necessary in reproduction. Some of these are the following:

1. Does the electrical oscillation continue if the cell is made to stop in its reproductive cycle—as by use of agents such as colchicine, colcemid, and Nocodazole?

2. Is the electrical oscillation necessary (and/or observable) in all types of reproducing cells? Or do some cell types not need to oscillate to reproduce?

3. Is there a preferential orientation of the mitotic poles of cell groups? Is this linked with the electrical oscillations? Are they in phase? Is the orientation responsive to externally applied ac fields? Care must be exercised here against making snap judgments and due regard must be taken of the inability of electric fields (ac or dc) to penetrate media effectively unless the conductivity–frequency interrelations are favorable. Only high-frequency fields can penetrate media of even low conductivity, for example.

4. What is the strength, range, and frequency of the electrical oscillations associated with reproduction?

5. Is cellular growth affected by properly applied (see remarks above) external fields?

6. Does the critical frequency or range thereof of normal cells differ, from that of somatic-healing, embryonic, or tumor cells?

7. If the frequencies do differ, is this due to chemical processes under control of the gene Tu, following the line of though[108] of Ahuja and Anders?

8. Do these critical frequencies associated with reproduction processes (or their ranges) correspond to the frequency of the dielectric loss peaks of

normal cellular tissue? If so, does this imply a relation between the internal (reproductive)[111,112] oscillation of a cell and the electrical damping due to its surroundings—i.e., is there an electrical aspect to contact or density inhibition? To invasiveness?

9. During cell divisions from fertilized egg to blastula, is there evident dipole interaction due to the postulated[15,94] natural electrical rf oscillations? Or is there overall organization of the ac fields about cells?

10. What is the source of the reproductive rf oscillations? Is it (a) related to an oscillatory chemical reaction cycle, or (b) based upon a more physically derived phenomenon[14] such as that of the Fermi–Pasta–Ulam–Fröhlich type?

11. What is the relation of the natural rf oscillations to the phase of cell growth ($G_1$, $S$, $G_2$, or $M$)?

12. What energy system of the cells drives the oscillations?

To the extent that favorable answers are found to the questions posed above, our knowledge of the origins, meaning, and control of the natural rf oscillations connected with cell growth could mean better insights into disease control and somatic repair.

## 13. DEP-Guided Pulse Fusion of Cells

In nature, cell-to-cell fusion occurs in many situations, including muscle fiber formation, endo- and exocytosis, and bisexual reproduction. The achievement of high-yield *in vitro* fusion at will[113] opens remarkable possibilities for hybridization. If we are to apply the process of fusion to hybridization. then techniques of high efficiency are needed for forming two-cell fusions. For some purposes, such as doing studies with microelectrodes of the membrane properties of living cells, the formation of giant multinucleate cells is of advantage.

The fusion techniques which are based on the use of either Sendai virus or chemical agents[114] have shortcomings. The fusion kinetics in such cases cannot be controlled because of its unsynchronized nature. The yield, moreover, is very low in such processes, and they do not yet permit preselection of the multiplet number in fusion (i.e., 2, 3, 4, etc.). Furthermore, the chemical conditions required are questionable in that they are quite unnatural (e..g., high or low pH and high $Ca^{2+}$ concentrations). The use of such chemically extreme conditions must alter the membrane properties of the fused cells as compared to those of the original unfused cells.

The new electrical technique,[115,116,124,127] combining DEP under gentle ac inhomogeneous field conditions with a subsequent ultrashort dc pulse (ca. 30 $\mu$s) of about 1 V across the cell membrane is gentle, synchronous, and capable of high yield (up to 90% so far). The DEP is used to bring about juxtaposition of the desired cells. The dc pulse causes a brief but reversible puncture of the membrane. The conditions can readily be arranged to produce fusion and leave the fused aggregate viable. It is worth noting that the DEP + pulse-fusion technique does *not* strongly perturb the entire cell surface, but only a tiny contact region, typically 60 Å in diameter.

Electrical pulse-fusion can be done with a wide variety of cells, including bacterial, yeast, and higher plant cell protoplasts, as well as on erythrocytes and other mammalian cells.[115,126,127] Because of the great potential of this technique it deserves close attention. The DEP + pulse-fusion technique has been used to meld mesophyll cell protoplasts of *Avena sativa* (oat),[115] of *Petunia inflata*, and *Vicia faba* (broad bean).[115] The fused protoplasts proved to be viable as judged by vital stain (trypan blue), and could be developed into ex-plants in the case of *Petunia i.*

The use of high-voltage pulses to fuse cells in the presence of chemical agents has recently been reported.[117]

## Acknowledgments

This research was supported by the Pohl Cancer Research Laboratory, and by the National Science Foundation.

## References

1. H. A. Pohl, *Dielectrophoresis*, the *Behavior of Matter in Nonuniform Electric Fields*, Cambridge University Press, London (1978).
2. H. A. Pohl, "Dielectrophoresis: Applications to the Characterization and Separation of Cell," in *Methods of Cell Separation* (N. Catsimpoolas, ed.), Plenum Press, New York (1978), Vol. I, 1pp. 67–169
3. H. A. Pohl and J. S. Crane, *Biophys. J.* 11, 711 (1971).
4. J. E. Rhoads, H. A. Pohl, and R. G. Buckner, *J. Biol. Phys.* 4, 93 (1976).
5. H. A. Pohl and K. Kaler, *Cell Biophys.* 1, 15 (1979).
6. C. S. Chen and H. A. Pohl, *Trans. N. Y. Acad. Sci.* 238, 176–185 (1974).
7. J. S. Crane and H. A. Pohl, "Dielectric Properties of Single Yeast Cell," *J. Electrostatics* 5, 11 (1978).
8. J. S. Crane and H. A. Pohl, "Use of the Balanced-Cell Technique to Determine the Dielectric Properties of Single Yeast Cells," *J. Biol. Phys.* 5, 49–74 (1977).
9. K. Kaler and H. A. Pohl, "Dynamic Levitation of Living Individual Cells," QTRG Research Note No. 83, March (1979); and *J. Biol. Phys.* 8, 18–31 (1980); *IEEE-IAS Trans. on Industrial Applications*, IA-19, 1089–1093 (1983).
10. B. D. Mason and P. M. Townsley, *Can. J. Microbiol.* 17, 879 (1971).
11. Herbert A. Pohl, "Electrical Forming of Masses of Living Cells," *J. Coll. Interface Sci.* 39, 437 (1972).

12. M. Eisenstadt and I. H. Scheinberg, *Science* **176**, 1325 (1972).
13. H. A. Pohl, "Microdielectrophoresis of Dividing Cells," *Bioelectrochemistry* (H. Keyzer and F. Gutmann, eds.), Plenum Press, New York (1980), pp. 273–295.
14. H. A. Pohl, "Do Cells in the Reproductive State Exhibit a Fermi–Pasta–Ulam–Fröhlich Resonance and Emit Electromagnetic Radiation?" *Collective Phenomena*, **3**, 221–244 (1981).
15. H. A. Pohl, "Electrical Aspects of Cell Growth and Invasiveness," *J. Biol. Phys.* **7**, 1–16 (1979).
16. H. A. Pohl, "Oscillating Fields About Growing Cells," *Int. J. Quantum Chem.* **7**, 411–431 (1980).
17. C. Kittel, *Elementary Solid State Physics*, Wiley, New York (1973).
18. A. H. von Hippel, *Dielectrics and Waves*, Wiley, New York (1954), p. 39.
19. A. J. Deller, *Solid State Physics*, Macmillan, London (1962), p. 86.
20. J. A. Stratton, *Electromagnetic Theory*, McGraw-Hill, New York (1941), p. 141.
21. See especially Ref. 1, p. 535.
22. J. J. Lowden (to Henry G. Thomas), *U.S. Patent* No. 465, 822 (1891).
23. D. B. Dow, *U.S. Bureau of Mines Bulletin No. 250* (1926).
24. E. W. Stevens, *U.S. Patent* No. 1533711 (1925).
25. J. R. Bates, *U.S. Patent* 2,665,246 (1954).
26. F. H. Müller, *Wiss. Veröffentl. Siemens-Werken*, **17**, 20 (1938).
27. H. A. Pohl, *J. Appl. Phys.* **22**, 869 (1951).
28. H. A. Pohl, *J. Appl. Phys.* **29**, 1182 (1958).
29. H. A. Pohl and J. P. Schwar, *J. Appl. Phys.* **30**, 69 (1959).
30. H. A. Pohl, *J. Electrochem. Soc.* **107**, 386 (1960).
31. H. A. Pohl and C. E. Plymale, *J. Electrochem. Soc.* **107**, 383 (1960).
32. H. A. Pohl, *J. Appl. Phys.* **32**, 1784 (1961).
33. A. Lösche and H. Hultschig, *Koll. Z.* **141**, 177 (1955).
34. W. F. Pickard, *Prog. Dielectr.* **6**, 105 (1965).
35. H. A. Pohl, *J. Electrochem. Soc.* **115**, 155c (1968).
36. P. J. W. Debye, *Phys. Rev.* **91**, 210 (1953).
37. P. Debye, P. P. Debye, B. H. Eckstein, W. A. Barber, and G. J. Arquette, *J. Chem. Phys.* **22**, 152 (1954).
38. P. Debye, P. P. Debye, and B. H. Eckstein, *Phys. Rev.* **94**, 1412 (1954).
39. W. A. Barber, P. Debye, and B. H. Eckstein, *Phys. Rev.* **94**, 1412 (1954).
40. H. A. Pohl, *U.S. Parent* No. 3,162,592 (December 22, 1964).
41. H. A. Pohl and Ira Hawk, *Science* **152**, 647 (1966).
42. E. Muth, *Kolloid Z.* **41**, 97 (1927).
43. W. Krasny-Ergen, *Hochfreq. Elektr.* **48**, 126 (1936).
44. P. Liebesny, *Arch. Phys. Theo.* **19**, 736 (1939).
45. E. Manegold, *Kolloid-Z.* **111**, **111**, 11 (1950).
46. J. H. Keller, *Digest of 12th Ann. Conf. on Electronic Techniques in Medicine and Biology*, p. 56, IRE-AIEE-ISA (November 1959).
47. A. A. Teixera-Pinto, L. L. Nejelski, J. L. Cutler, and J. H. Heller, *Zxp. Cell. Res.* **20**, 548 (1960).
48. H. Fricke, *J. Gen. Physiol.* **9**, 137 (1925).
49. H. Fricke, *Phys. Rev.* **26**, 678 (1925).
50. H. Fricke, *J. Appl. Phys.* **24**, 644 (1953).
51. M. Saito, H. P. Schwan, and G. Schwarz, *Biophys. J.* **6**, 313 (1966).
52. A. A. Füredi and R. C. Valentine, *Biochim. Biophys. Acta* **56**, 33 (1962).
53. J. L. Griffin and R. E. Stowell, *Exp. Cell Res.* **44**, 684 (1966).
54. For example, the paper entitled "Effective Field of a Dipole in Polarizable Fluids," by E.

L. Pollock and B. J. Alder, *Phys. Rev. Lett.* **39**, 299 (1977); and the reply to it by R. L. Fulton, *Phys. Rev. A* **18**, 1318 (1978).

55. H. A. Pohl and J. S. Crane, *J. Theor. Biol.* **37**, 1 (1972).
56. G. A. Kallio and T. B. Jones, "Dielectrophoretic Levitation of Spheres and Shells," *Preprints of Symposium on Electrohydrodynamics*, Fort Collins, Colorado January 16–18 (1978); cf. T. B. Jones, *J. Electrostatics* **6**, 69 (1979).
57. J. A. Stratton, *Electromagnetic Theory*, McGraw-Hill, New York (1941), p. 112.
58. C. P. Smyth, *Dielectric Behavior and Structure*, McGraw-Hill, New York (1955).
59. M. Pollak and H. A. Pohl, *J. Chem. Phys.* **63**, 2980 (1975).
60. H. A. Pohl and M. Pollak, *J. Chem. Phys.* **66**, 4031 (1977).
61. J. S. Crane and H. A. Pohl, *J. Electrostatics* **5**, 11 (1978).
62. J. S. Crane and H. A. Pohl, *J. Biol. Phys.* **5**, 49 (1977).
63. J. S. Crane and H. A. Pohl, *J. Theor. Biol.* **37**, 15 (1972).
64. H. A. Pohl and J. R. Wyhoff, *J. Non-Cryst. Solids* **11**, 137 (1972).
65. J. R. Wyhof and H. A. Pohl, *J. Polymer Sci. Part A-2* **8**, 1741 (1970).
66. R. Pething, *Dielectric and Electronic Properties of Biological Materials*, Wiley, New York (1979).
67. S. Takashima, *Biopolymers* **5**, 899 (1967).
68. H. P. Schwan, G. Schwarz, J. Maczuk, and H. Pauly, *J. Phys. Chem.* **66**, 2626 (1962).
69. T. B. Grimley and N. F. Mott, *Discuss. Faraday Soc.* **1**, 3 (1947).
70. H. P. Schwan, *Adv. Biol. Med. Phys.* **4**, 147 (1957).
71. K. S. Cole, R. H. Cole, and H. J. Curtis, *J. Gen. Physiol.* **18**, 877 (1935); **19**, 609 (1936); **21**, 591 (1938).
72. H. Fricke and H. J. Curtis, *J. Phys. Chem.* **41**, 729 (1937).
73. G. J. Schwarz, *J. Phys. Chem.* **66**, 2636 (1962).
74. S. S. Dukhin and V. N. Shilov, *Colloid J. (USSR)* **31**, 564 (1970); **32**, 245 (1970); **32**, 90 (1970).
75. S. S. Dukhin and V. N. Shilov, *Dielectric Phenomena and the Double Layer in Disperse Systems and Polyelectrolytes*, Wiley, New York (1974).
76. A. Minakata, N. Imai, and F. Oosawa, *Biopolymers* **11**, 347 (1972).
77. A. Minakata and N. Imai, *Biopolymers* **11**, 329 (1972).
78. P. I. Meyer and W. E. Vaughan, *Biophys. Chem.* **12**, 329–339 (1980).
79. V. M. Fomchenkov and B. K. Gavrilyuk, *J. Biol. Phys.* **6**, 29 (1978).
80. H. A. Pohl, *Sci. Am.* **203**, 107 (1960).
81. H. A. Pohl and K. Pollock, *J. Electrostatics* **5**, 337 (1978).
82. H. A. Pohl, *U.S. Patent* No. 4,326,934, "Continuous Dielectrophoretic Cell Classification Method" (1982).
83. R. Cruickshank, J. P. Duguid, and R. H. A. Swain, *Medical Microbiology*, E. and S. Livingstone Ltd., London (1968), p. 122.
84. C. S. Chen, H. A. Pohl, J. S. Huebner, and L. J. Bruner, *J. Colloid Interface Sci.* **37**, 354 (1971).
85. I. P. Ting, K. Jolley, C. A. Beasley, and H. A. Pohl, *Biochem. Biophys. Acta.* **234**, 324 (1971).
86. H. A. Pohl and W. F. Pickard, *Dielectrophoretic and Electrophoretic Deposition*,Electrochemical Society, New York (1969).
87. B. A. Jennings and V. J. Morris, *J. Colloid Interface Sci.* **49**, 89 (1974).
88. V. J. Morris, P. J. Rudd, and B. R. Jennings, *J. Colloid Interface Sci.* **50**, 379 (1975).
89. P. J. Rudd, V. J. Morris, and B. R. Jennings, *J. Phys. D* **8**, 170 (1975).
90. N. M. Fomchenkov, V. N. Brezgunov, B. K. Gavrilyuk, V. V. Smolyaninov, and Z. F. Bunina, *J. Biol. Phys.* **7**, 45 (1979).
91. H. Fröhlich, *Int. J. Quantum Chem.* **2**, 641 (1968).

92. A. Coehn, *Ann. Phys.* **64**, 217 (1898).
93. H. A. Pohl, "Natural Electrical r.f. Oscillations From Cells," *J. Bioenerg. S. Biomembr.* **13**, 149–169 (1951).
94. Brian Goodwin, private communication.
95. J. H. Heller, *Digest of 12th Ann. Conf. of Electronic Techniques in Med. and Biol.*, p. 56, IRE-AIEE-ISA (November 1959).
96. A. A. Teixera-Pinto, L. L. Nejelski, J. L. Cutler, and J. H. Heller, *Exp. Cell. Res.* **20**, 548 (1960).
97. J. L. Griffin and R. E. Stowell, *Exp. Cell. Res.* **44**, 684 (1966).
98. A. W. Friend, E. D. Finch, and H. P. Schwan, *Science* **187**, 357 (1975).
99. M. Mischel and I. Lamprecht, *J. Biol. Phys.* **11**, 43–44 (1983).
100. Cf. Ref. 1, pp. 136–141.
101. J. E. Treherne, W. A. Foster, and P. K. Schofield, "Cellular Oscillators," in *J. Exper. Biol.* **81**, (review volume) 1979.
102. M. J. Berridge and P. E. Rapp, *J. Exper. Biol.* **81**, 217 (1979).
103. R. M. Noyes and R. J. Field, *Ann. Rev. PhSs. Chem.* **25**, 95 (1974).
104. S. Schmidt and P. Ortoleva, *J. Chem. Phys.* **71**, 1010 (1979).
105. H. A. Pohl, *Chem. Engr. Commun.* **4**, 237 (1980).
106. H. Fröhlich, *Cooperative Phenomena* **1**, 641 (1973).
107. H. Fröhlich, *IEEE Trans. Microwave Theory Techniques* **MTI-26**, 613 (1978).
108. M. R. Ahuja and F. Anders, "Cancer as a Problem of Gene Regulartion," in *Recent Advances in Cancer Research* (R. C. Gallo, ed.), (1977), Vol. I, pp. 103–117.
109. H. A. Pohl, "Quasi One-Dimensional Electronic Conduction and Nomadic Polarization in Polymers,'" *J. Biol. Phys.* **2**, 113–172 (1974).
110. H. Fröhlich, "Possibilities of Long and Short Range Electric Interactions of Biological Systems," *Neurosci. Res. Program Bull.* **15**, 67–73 (1977).
111. E. Zeuthen, "Artificial and Induced Periodicity in Cells," *Adv. Biol. Med. Phys.* **6**, 37–73 (1958).
112. M. G. Sargent, "Synchronous Culture of *B. subtilis* Obtained by Filtration with Glass Fibers," *J. Bacteriol.* **116**, 736–740 (1973).
113. G. Köhler and C. Milstein, *Nature (London)* **256**, 495 (1975).
114. R. H. Kennett, T. J. McKearn, and K. B. Bechtol, eds. *Monoclonal Antibodies*, "Hybridomas, A. New Dimension in Biological Analysis," Plenum Press, New York (1980).
115. U. Zimmermann, J. Vienken, and P. Scheurich, "Electric Field-Induced Fusion of Biological Cells," *Biophys. Structure Mechanism, Suppl.* **6**, 86 (1980).
116. U. Zimmermann, G. Pilwat, and H. A. Pohl, *J. Biol. Phys.* **10**, 43–50 (1982).
117. H. Weber, W. Förster, H. E. Jacob, and H. Berg, "Enhancement of Yeast Protoplast Fusion by Electric Field Effects," *Proc. Vth Internat. Symp. on Yeasts*, London, Ontario, Canada, July 1980.
118. F. Kaiser, "Limit Cycle Model for Brain Waves," *Biol. Cybernetics* **27**, 155 (1977).
119. H. Fröhlich, *Neurosci. Res. Bull.* **15**(4), 67–72 (1977).
120. H. Fröhlich, *Theory of Dielectrics*, Oxford Press, London (1949).
121. H. A. Pohl, P. S. Vijayakumar, L. Dunn, and W. T. Ford, "Stable Dielectrics with Giant Polarization," 1983 *Annual Report IEEE. Conf. on Electrical Insulation and Dielectric Phenomena*, p. 486.
122. S. Rowlands, in *Coherent Excitations in Biological Systems* (H. Fröhlich and F. Kremer, eds.), Springer-Verlag, New York (1983), pp. 145–161.
123. H. A. Pohl, "Natural Oscillating Fields of Cells," in *Coherent Excitations in Biological Systems* (H. Fröhlich and F. Kremer, eds.), Springer-Verlag, New York (1983), 199–210.
124. A. H. Jafary-Asi and C. W. Smith, Biological Dielectrics in Electric and Magnetic

Fields," *Annual Report 1983, IEEE Conf. on Elec. Insul. & Dielectric Phenomena*, pp. 350–355.

125. H. A. Pohl, "The Spinning of Suspended Particles in a Two-Pulsed, Three-Electrode Systems," *J. Biol. Phys.* **11**, 66–68 (1983).

126. H. A. Pohl, K. Pollock, and H. Rivera, "The Electrofusion of Cells," *Int. J. Quantum Chem., Quantum Biol. Symp.* **11**, 327–345 (1984).

127. H. Rivera, H. A. Pohl, S. Cherski, and M. Swicord, "Electrical Fusion of Cells and Nuclei in Vitro: A Preliminary Note," *J. Biol. Phys.* **11**, 63–65 (1983).

# 13

# Electrical Phenomena in Proteinoid Cells

## Aleksander T. Przybylski and Sidney W. Fox

*ABSTRACT:* Electrical phenomena in artificial cells are described. The constituent material of the cells, referred to as proteinoid or as thermal protein, have been extensively studied in the context of the origin of life, which led to the finding of excitability as one of the biofunctions. The activities found in proteinoid cells are such as to make them useful models for modern excitable cells as well as for protocells. For example, the proteinoid cells display double membrane, asymmetric permeability, membrane potentials, action potentials, and photoactivity.

## 1. Introduction

The cellular units described in this review are assembled from proteinoids, i.e., thermal copolymers of amino acids.

### 1.1. Proteinoid

The proteinoids, thermal copolyamino acids, represent a spectrum of polymers having compositions and structures like those of natural proteins and peptides. Since they are products of the synthetic laboratory, compositions not known in nature may be produced in abundance, and in sensitively graded products. These materials are referred to as artificial proteins, synthetic proteins, thermal proteins, or proteinoids. The last term is used because of the similarity to proteins. Since 1972, each semiannual

---

*Aleksander T. Przybylski and Sidney W. Fox* • Institute for Molecular and Cellular Evolution, University of Miami, 521 Anastasia Avenue, Coral Gables, Florida 33134.

index of *Chemical Abstracts* has listed these polymers as "proteins, thermal." They differ from protein in containing DL amino acids.

It is easily possible to include in a single polymer some proportion of each of the 20 amino acids common to modern protein, and to obtain a proteinoid that on hydrolysis gives an amino acid analysis typical of proteins.

The dominant linkage in the proteinoid is the peptide bond. The proteinoids rich in aspartic acid contain many aspartoylimide linkages, which, while known in protein, are not usual:

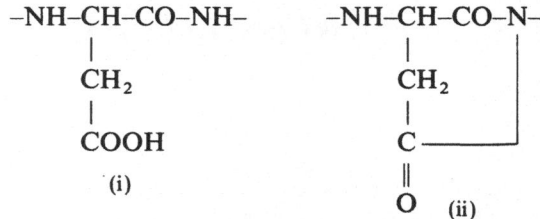

The aspartoylimide linkage may be easily opened by hydrolysis, giving α or β peptide links depending upon hydrolytic conditions.

Another quantitative difference of proteinoids from protein is the presence of branches through the side chains of the dicarboxylic amino acids. The branches are typically one to 12–15 residues in the main chain.[1]

The proteinoid polymer is highly nonrandom. For instance, during the thermal polymerization of glu, gly, and tyr only two tripeptides were formed: (pyro)glu-gly-tyr and (pyro) glu-tyr-gly.[2] Thirty-six tyr-containing tripeptides would be expected from random polymerization. The experimental finding was corroborated by others.[3]

Temperature, time, and added substances are the conditions that have been most explored in thermal polymerization.

## 1.2. Temperature

For convenience during a single working day, temperatures of 150–180°C are most often used for predominantly acid mixtures. Substantial polymerization then occurs in 3-8 hr.

Instead of temperature of 150°C or above, temperatures as low as 65°C can be employed for two weeks.[4] Intermediate temperatures and times can be employed. In the conduct of these reactions, temperature can be traded for time, or *vice versa*, in accord with classical physical chemistry. Overall casual comparisons suggest that the effects of activation energy barriers are absent or small for acidic or basic mixtures.

Various diluents increase the yield or facilitate the reaction at various temperatures. The additant that has been most studied, and which facilitates much, is phosphate, either as phosphoric acid,[5] polyphosphoric acid,[6] or various salts thereof.[7] Some constrained benefit is available from using phosphoric acid in an amino acid mixture from which aspartic and glutamic acids are absent.

The polymers range from single macromolecular types[8] to typically three or four major fractions having small differences between those fractions.[9]

### 1.3. Prosthetic Groups

Various prosthetic groups, e.g., heme or $Fe^{2+}$, can be incorporated by heating with the amino acid mixture. Although the chemical structure is not known in each case, such nonamino acid compounds are firmly held; they appear to be fixed by covalent bonds.

### 1.4. Hydrolyzability

The bonds between amino acids are mostly normal peptide bonds. Hydrolysis requires the same conditions that are used in hydrolysis of proteins.

When the thermal polyamino acid is an acidic type, complete recovery of amino acids may be observed, and the hydrolytic procedure may be quite the same as for proteins.

As the polymer contains increasing proportions of basic amino acid, especially lysine, above a minimum, hydrolysis requires stronger conditions and recovery of amino acids is incomplete. This is believed to be due to side-chain reactions between lysine and aspartic acid residues, for example. The production of spongy three-dimensional polymers has been noted.

### 1.5. Solubility

The various solubility classes of proteins have been mimicked. Thus, one may prepare proteinoids that behave like albumins, globulins, protamines and histones, gliadins, glutelins, or prolamines, etc.

### 1.6. Ionic Behavior

Many thermal polyamino acids may be salted-in or salted-out. Some are sticky, some are hygroscopic, some are neither.

## 1.7. History of Proteinoid Microspheres

The emergence of proteinoid microspheres as scientific objects is predominantly a history of experiments. In a seemingly logical progression, some scholars have deduced that components of cells as we know them arose first. That is to say, nucleic acids, proteins, and lipids arose and then the surrounding cell and its membrane assembled. The experiments have, however, indicated another sequence in evolution. Proteinoid cells arise easily, quickly, and abundantly. This was learned only by experiment. The further incorporation of this phenomenon into evolutionary theory visualizes the extensive evolution of such protocells into modern cells.[10]

An early experiment in the making of proteinoid and microspheres in geologically relevant conditions has been described as follows: "A mixture of dry amino acids, containing sufficient aspartic acid and glutamic acid, was placed in the depression of a piece of lava from the beds on the Kapoho field on the island of Hawaii. The reaction vessel, i.e., the piece of lava, was then placed in an oven at 170°C for several hours. The powder was thereby converted to a light amber-colored liquid so viscous that it remained *in situ*. The rock and polymer was then laved by hot 1% sodium chloride solution. The liquid, which was slightly turbid, now contained large number of microspherical units."[11]

The condensation of amino acids and chemical structure of the respective polymers, chromatograms of their hydrolysates, and morphological features of microspheres have been described.[12] The interaction of appropriate thermal copolyamino acids with hot or cold water proved to be a necessary condition for preparation of microspheres.[13] The proteinoid microspheres are spherical and usually uniform in diameter in the range from 0.5 to 7 $\mu$m. Factors controlling size of microspheres are: type of polymer, added substances, ratio of solid to liquid component in the mixture, presence and concentration of electrolytes in solution, temperature of solution, and rate of cooling.

By controlling the proportion of acidic and basic proteinoids in the mixture there can be produced gram-positive or gram-negative units.[14]

In a first stage of study of the microspheres the phase separation[15] and compartmentalization[16] were examined. The next stage summarized the behavioral[17] properties of these units and the metabolic characteristics.[18]

In the latest stage, catalytic properties related to internal chemistry structure was examined. The intracellular peptide synthesis was detailed[19,20] and the concept of origins of a protein synthesis cycle has been proposed.[21]

The most recent questions dealt with the origin of the genetic coding mechanism. Thermal lysine-rich polymers have been shown to catalyze the synthesis not only of peptide bonds, but also of internucleotide bonds.[22] An updated experimental proteinoid model has been advanced[23] as a result.

Of most note, spherules made from proteinoid and lecithin have been recently shown to possess electrical membrane phenomena.[24] Membrane, action, and oscillatory potentials were next recorded in the microspheres[25] made from proteinoid only. While many of the functions of the model protocells are much weaker than the corresponding functions in modern cells, e.g., the catalytic functions, the electrical behavior is quantitatively comparable. Consequently the units are models of modern excitable cells, as well as of excitable protocells, and are treated mainly as such in this chapter.

### 1.8. History of Electrical Excitation in Evolved Cells

Excitability is regarded as a main feature of living systems. Electrical excitation was earlier thought to be expressed only in nervous and muscle tissue. Later it turned out that similar phenomena are observed also in plants. Even more, reassembly experiments showed that algal protoplasmic drops display membrane potential and its changes resemble action potentials.[26,27] Excitability has been described in epithelia, plant cells, algae, protozoa, and ova.[28,29]

In the history of development of living systems, the membrane was the primary constituent element of the cell. What are the physical and chemical premises of an organization of the membrane and cell?

The only up-to-date suitable experimental model to help answer this question is the proteinoid cell[13] assembled from thermal proteins, or proteinoids. To date, no other experimental protein-related model of the first cell has been shown to have comparable properties.

In 1973 the first finding of excitability in a proteinoid cell was reported.[30] These data were further expanded.[24,25] Transmission electron microscope evidence of a double membrane in these cells,[13] selective permeability, and osmotic properties[31] as well as bilayer membranes made of proteinoids[32] provide an experimental background to understanding the origin of excitability.

## 2. Physical Properties of Microspheres

Among the physical properties of microspheres, their mechanical properties, stability to pH, to dehydration, and to external potassium ion are relevant in their bioelectrical responses.

## 2.1. Stability

One striking feature of the proteinoid microspheres is that they are long-lasting. When prepared aseptically, they do not disintegrate over many months, or even up to 6 years.[33] Similarly, their ability to maintain membrane potential and the generation of its electrical oscillation persists over at least many months.[34]

### 2.1.1. Dehydration

The microspheres can be dehydrated and lyophilized. When dissolved in water they possess the properties they had prior to dehydration.

### 2.1.2. Resistance to pH

The range of pH stability varies with the kinds of proteinoid of which the microspheres are composed. Microspheres composed of acidic polymers can typically tolerate only the 3.0–5.5 pH range, whereas those composed of basic and acidic polymers are not soluble at higher ranges such as pHs 8–9.[35]

The broadest range of 2.5–10.5 pH has been observed for microspheres of Anders' polymer made of copoly (asp, ser, glu, gly, ala, orn, lys, his, arg).[36] Heating of mixed amino acids with sea water salts yields acidic and basic proteinoids, simultaneously, which are stable, as microspheres, in alkaline solution.[37]

### 2.1.3. Resealing of the Membrane

The proteinoid can be obtained with various degrees of plasticity. While some of the membranes are semi-leaky, they reseal themselves after injury, for instance, after penetration by micropipette.

### 2.1.4. Stability to External ($K^+$) Concentration

Potassium concentration in the aqueous environment of microspheres is directly connected with their membrane electrical polarization. Across the pores within the membrane this ion evidently penetrates into the cell. The osmolar properties of the membrane determine the disruption of the entity. Thus, microspheres made of copoly (asp, glu) withstand 1 $M$ KCl but are dissolved in 3 $M$ KCl, whereas those made of copoly (asp, glu, his) withstand a 3 molar concentration of this salt.

## 3. Membrane

According to the accumulated knowledge, a membrane is a necessary component of the protocell, as it is of the modern cell. Membrane organization follows the phase separation stage. The presence of hydrophilic and hydrophobic residues, as well as electrically polar groups within the polymers, appear to be sufficient conditions for self-assembly of membraneous structures.

Compartmentalization and gradient (chemical, ionic, osmotic, electrical) across the membrane are two simultaneously present properties of the microspheres.

Because of the mutual orientation of membrane-making compounds during bilayer membrane formation, decrease in the bifacial energy takes place. This means that the micelle is more stable than separate constituent molecules. The corresponding entropy change $\Delta S$ is given by $\Delta S = d(\Delta F_i)/dT$, where $F_i$ is the standard free energy, and $T$ is the absolute temperature.[38]

### 3.1. Bilayer Proteinoid Membranes

Bimolecular lipid membranes[39] have become a useful model for studying the physicochemical properties of true membranes.

It turned out to be true that proteinoids, without any lipids, also form bimolecular membranes.[32] Despite the fact that black proteinoid membranes are not as long-lived in the ultrathin state as phospholipid membranes, they last long enough to be examined. Those rich in hydrocarbon-rich amino acid side chains mostly display properties characteristic for BLMs. The same polymers are among those that most readily combine with lecithin.

### 3.2. Spherical Proteinoid Membranes

A film type of membrane is the simplest form of membrane produced. It has been experimentally demonstrated that formation of spherical lipid membranes takes place under mechanical agitation of aqueous suspensions. This is true also in the case of proteinoid-only membranes. Transmission electron microscopic studies revealed that the thickness of a single layer in a membrane is around 500–5000 Å.

### 3.3. Tubular Proteinoid Membranes

Under conditions of dehydration and subsequent hydration, and especially in the presence of lysine-containing polymer, tubules are formed.

The formation is either as single tubules or outgrowing processes from the body of the cell, with subsequent branching.

## 4. The Effect of Light

The thermal polymers of amino acids reveal photo effects (Figure 1), which are almost certainly related to the photocatalytic activity that was reported in 1972 by Wood and Hardebeck.[40] This property is due to the presence in the polymer of flavin and pterin derivatives.[3,41] The pigments are formed during amino acid thermolysis. The chromophores appear to be covalently linked to thermal oligomers. One proposed chemical binding is

### 4.1. Spectral Characteristics of the Proteinoid

The uv spectrum of the proteinoids displays a broad absorbance. This is due, most likely, to selective absorption by the peptide bonds.[42] This region of absorption is parallel to that of some chromophores and, in particular, to pterin (Figure 2) which has been discovered in the thermal polyamino acids.[3,41,43]

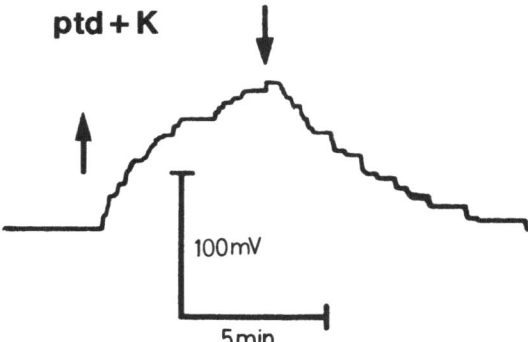

**ptd + K**

FIGURE 1. Photovoltaic potential of CRW XV-65 proline-rich proteinoid (10 mg/ml) in aqueous solution. Arrows pointing up indicate light (halogen lamp, 50 W) turned on; arrows pointing down indicate light turned off.

100 mV

5 min

## 4.2. The Effect of Light Intensity on Electrical Properties

The proteinoid and proteinoid microspheres display electrical phenomena when illuminated. The photovoltaic response of the proteinoid solution is proportional to light intensity, although not directly, whereas the photoelectrical discharges of the proteinoid-lecithin microspheres are greatest only during moderate illumination by white light of about 10 lux; they are not proportional. Strong light inhibits electrical discharges, in some cases fully. When microspheres are illuminated, they show electrical discharges and maintain a membrane potential, and during absence of light both discharges and potentials gradually diminish. After reillumination the membrane potential and electrical discharges reappear (Figure 3). The response in Figure 3 is enhanced by inclusion of chlorophyll.

The photosensitivity of proteinoids appears to explain the energy source for electrical phenomena in the proteinoid cells.

Data exist on chemical energy storage in the presence of cell-like proteinoid structures. For instance, a small conversion of ADP to ATP has been accomplished in aqueous suspension by use of cell-like structures aggregated from copoly (asp, glu, tyr). One explanation is that this occurs through formation of dopaquinone in the peptide structure during illumination.[44] Microspheres made of copoly (asp, glu, cys, leu, tyr) were suspended in water with $FeCl_2 \cdot 4H_2O$, $KH_2PO_4$, and ADP sodium salt in solution and illuminated for 14 hr by a 160 W Westron lamp. The chromatogram revealed a peak in the ATP region (Figure 4).

## 5. Electrical Phenomena

Physical factors resulting in membrane conductance (electrical discharges across membrane are determined by fast conductance changes)

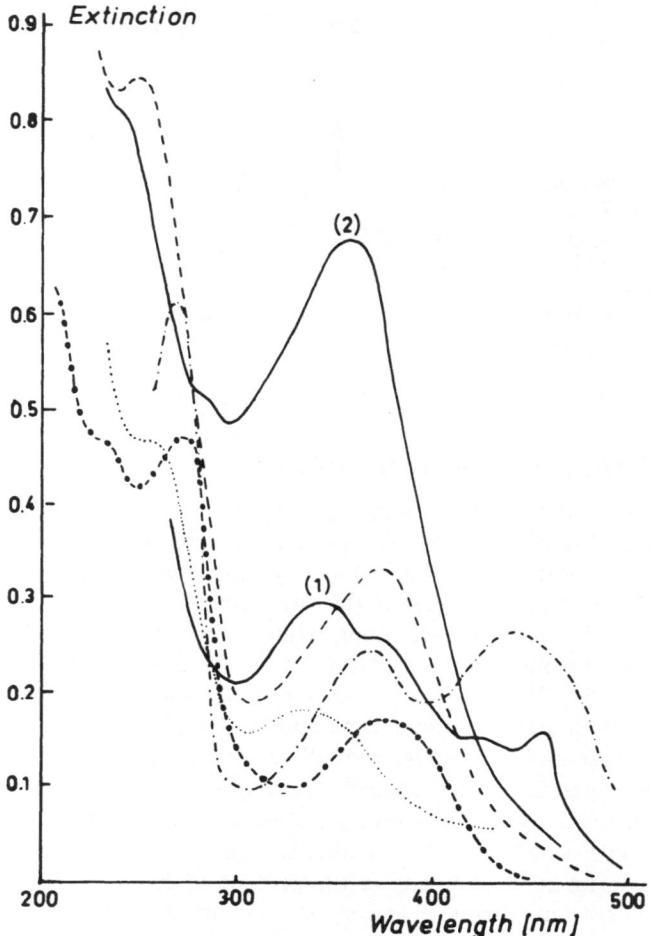

*FIGURE 2.* UV spectrum of thermal copoly (lys,ala,gly) (1) and its ether extraction product (2) compared with that of pterin (···), chrysopterin (–––), lactoflavin (–·–), and xanthopterin (–•–). After Ref. 45.

comprise a series of events from interphase surface charge via electric potential across the membrane, ion activity, and ion gradient on both sides of the membrane as well as of intrinsic membrane indices such as oriented dipoles at membrane surface, interlayer membrane dielectric constant, and intermembrane ion motility. If chromophores are present in the membrane, the electrical behavior of the membrane is defined by factors responsible for trapping of photons, formation of carriers, and charge separation. In this case the membrane is a reversible light energy transducer.

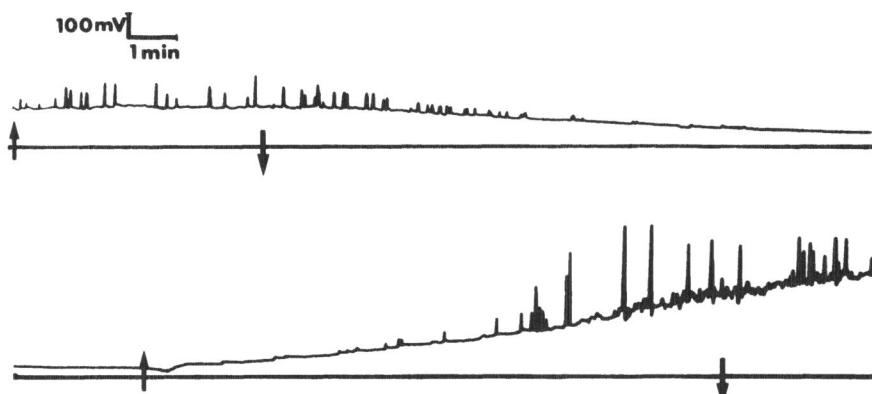

FIGURE 3. Effect of light on electric discharge of a microsphere composed of thermal copoly (asp,glu,arg) and lecithin, and entrapped crude chlorophyll. Arrows pointing up indicate light (halogen lamp, 50 W) turned on; arrows point down indicated light turned off. (Lower recording is continuation of upper recording.)

## 5.1. Electrical Resistance

Both theoretical premises and some experimental data allow us to try to find an explanation of the observed membrane polarization changes in artificial membranes. The first contributing factor which is involved is the membrane resistance. When the concentrations of electrolytes are the same across the membrane, the resistance is linear. Nonlinear resistance and rectifying properties of the membrane above this restriction are detected. If the lipid-like substances are characterized by high resistance, proteins display lower resistance, and their presence in membranes diminishes it con-

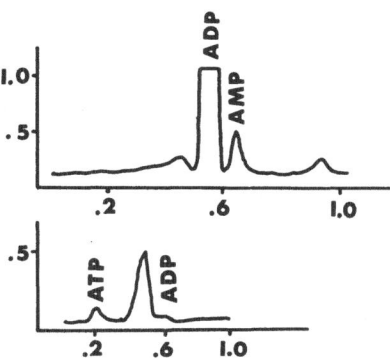

FIGURE 4. Chromatograms showing ATP produced in a suspension of copoly (asp,glu,cys,leu,tyr) microspheres in an aqueous solution of ADP and inorganic phosphate and with isobutyric acid $NH_4OH$:EDTA solvent (upper), and rechromatogram of ATP fraction with $LiCl_2$:acetic solvent (lower). Ordinate: CPM × $10^3$.

siderably. It has been found that an addition of natural protein (excitation-inducing material = EIM) reduces membrane resistance down to as low as $10^3 \, \Omega/cm^2$ (from an initial value of $10^8$–$10^{10} \, \Omega/cm^2$).[39]

## 5.2. Current–Voltage Dependence

The fact that acidic, basic, and neutral amino acids form proteins results in preservation of considerable charge of protein molecules, both in the case of natural proteins and artificially synthesized thermal proteinoids. This means that molecules of both kinds of substance possess similar charge-holding capacity and, consequently, physical and chemical reactivity as well as electrical properties. The voltage-dependent trans-bilayer orientation of melittin may be an example here.[45] When added in a concentration of 10–20 ng/ml to the *cis* side of an oxidized cholesterol black lipid membrane, melittin induced an asymmetric voltage-dependent conductance increase. The observed conductance increased during *trans*-negative potential, and decreased when the applied potential was *trans*-positive. In the conducting state peptide assumes a *trans*-bilayer position. The current–voltage dependence of BLM-only membrane within the 1 $M$ up to $1.10^{-3}$ $M$ concentration of NaCl is linear.

It has been found that some proteinoids and cells made of them display the negative resistance of current–voltage characteristics in the first quadrant of coordinates.

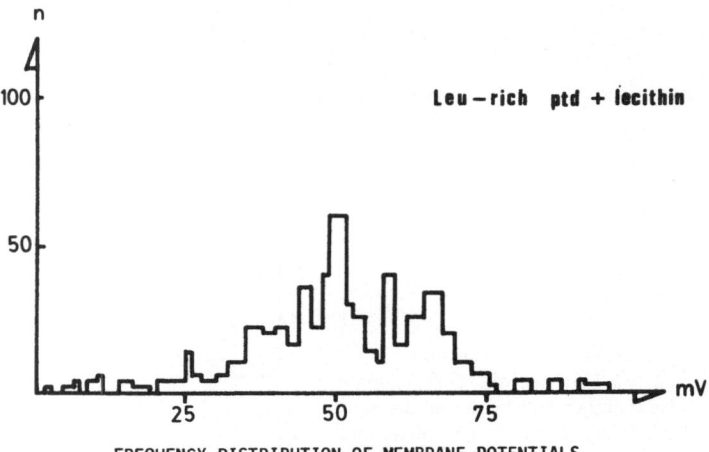

FREQUENCY DISTRIBUTION OF MEMBRANE POTENTIALS

*FIGURE 5.* Frequency distribution of membrane potentials recorded in leu-rich ptd-lecithin cells.

There have been obtained also data that demonstrate a region of excitability induced by electrical polarization. This results in the appearance of electrical discharges following the first polarization, and a passive hysteresis effect during electrical depolarization.

The data showed that, with the membrane voltage held constant, the current across the membrane displays nonlinear characteristics during stimulation. These findings seem to be crucial in explanation of the phenomenon of electrical behavior of the artificial proteinoid cell, as the negative resistance is a necessary characteristic for the generation of the oscillations observed.

## 5.3. Membrane Potential

Microspheres in suspension display low (1–5 mV) electrical polarization across the membrane. When suspended in aqueous solution with salts such as KCl, the membrane potential is higher. If lecithin and KCl with $K_2HPO_4$ buffer have been added to water and proteinoid, the resultant vesicles (after the mixture is heated) display membrane potentials as high as 50 mV or greater (Figure 5).

## 5.4. Electrical Discharges and Oscillations

If the strict conditions of preparation of BLM and their resistance are fulfilled, bilayer membranes, after addition of KCl and EIM from *Enterobacter cloacae*, display electrical properties such as membrane potential and current-induced and spontaneous discharges.[39]

Similar responses are observed in proteinoid cells (Figures 6 and 7).

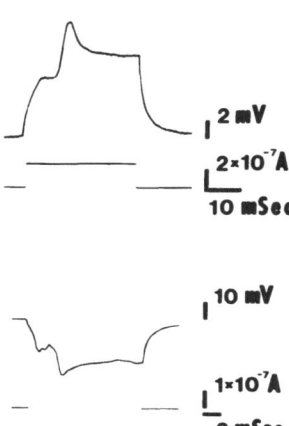

*FIGURE 6.* Current-induced changes of membrane polarization of the 2:2:1 proteinoid-lecithin cell.

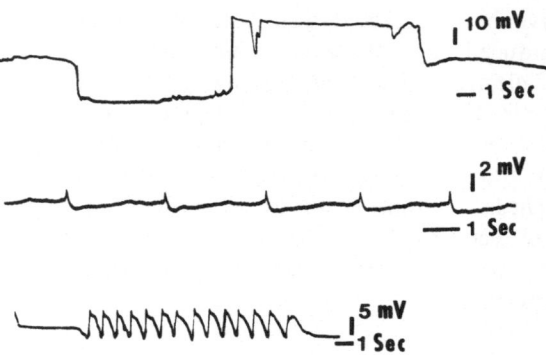

FIGURE 7. Membrane potential, spontaneous discharges, and oscillations of the 2:2:1 proteinoid-lecithin cell.

The similarity of the pattern of electrical discharges of the bilayer membrane and the spherical membrane of the proteinoid cell (Figure 8) indicate also a similar mechanism of impulse generation. Various patterns of electrical discharges observed in proteinoid cells are different in the shape and amplitude of the gradient. Some highly resemble spiking of a natural neuron (Figure 9). The repeatable examples of cells made of the same material display characteristic electrical discharges. However, similar patterns of electrical discharge were observed in spherules made of various proteinoids. This indicates that although the pattern may be primarily determined by chemical composition of the proteinoid cells, there are other contributing factors.

The examples illustrating electrical membrane phenomena in this chapter have not been observed in all experimental attempts. Each

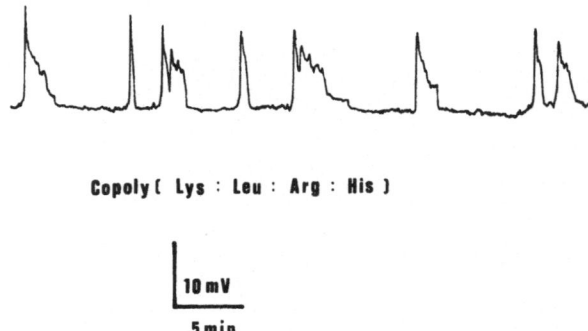

Copoly ( Lys : Leu : Arg : His )

10 mV

5 min

FIGURE 8. Pattern of spontaneous electrical discharges of the proteinoid cell made of copoly (lys,leu,arg,his).

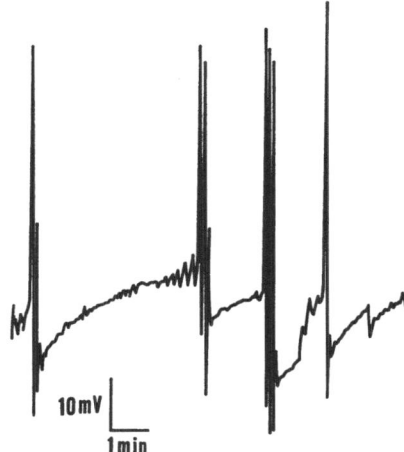

FIGURE 9. Pattern of spontaneous electrical discharges of the 2:2:1 proteinoid-
lecithin cell.

phenomenon has, however been observed several times; some are more
easily repeatable than others. The responses of microspheres from different
polyamino acids are undergoing numerical analyses.

## 5.5. Channels

A charge and ion transfer through the membrane and the kinetics of
this process seem to be similar, if not even common in planar, bilayer, and
spheroidal membranes. Four synthesized peptides, e.g., leu-ser-leu-gly,
having helical structure, were found to form ion channels across the lipid
membrane.[46] A synthetic peptide $K^+$ carrier[47] is further experimental
proof toward "simplification" of the mechanism underlying generation of
electrical impulses across the membrane.

This means that the molecular configuration of ion channel gating
may be realized by relatively simple amino acid residues, and by peptides
only. If so, the observed "channeling" phenomena both in BLM with the
presence of EIM and in proteinoid vesicles may be understood.

## 6. Some Electronic Properties

The above-mentioned example of mellitin[45] is a good illustration also
for some electronic features which can be studied in proteinlike compounds. Melittin is an amphipathic peptide of 26 residues with a
hydrophobic stretch of 19 amino acids followed by a cluster of four
positively charged residues at the COOH terminus.

When melittin was added in a concentration of 10–20 ng/ml to the *cis* side of an oxidized cholesterol membrane, the asymmetric voltage-dependent conductance of the resultant black lipid membrane was increased. The conductance increased during *trans*-negative potential and decreased when the voltage was *trans*-positive. These changes are completely reversible.

Analysis of melittin-induced conductance fluctuations suggests perturbations of the lipid bilayer structure, and formation of structural channels. Addition of pronase to the *trans*-side in the presence of a *trans*-negative membrane potential abolished the conductance as it did when added to the *cis* side. This implies that the conducting state of melittin requires a *trans*-membrane configuration. In a non-conducting absorbed state a hydrophobic loop, with the region from threonine-9 to proline-13 as the likely site of the turn, penetrates the lipid but does not extend across the membrane. In the conducting state the turn becomes extended as the peptide assumes a *trans*-bilayer position.

Energy calculations show that the membrane potential can supply sufficient energy for this transition. The membrane seems to be this structure at which energy transfer takes place through a low energy barrier, and where the uncontrolled diffusion of electrons is prevented by nonadiabatic behavior in the electron transfer process.[48] The polar electronic structure of lipids and proteins holding vibrating ions[49] is crucial here.

For understanding the function of excitable membranes, the finding[50] of the rotational correlation time of a methyl group in amino acid side-chain structure (at 37°C) is especially interesting. This time varies by two orders of magnitude, and in the case of methionine is equal to 800 ms. This can be considered as an experimental background of nonlinear behavior of a system composed of amino acids. Nonlinear pulses can be extremely stable during long lifetimes if geometrical (i.e., stereochemical) constraints of the molecule force the pulse (as a soliton) to move along the one-dimensional channel.[51,52] Other recently reported data indicate that the alpha-helical proteins are a perfect guide for charge transfer because each peptide group has a comparatively large dipole moment, and, due to this, the peptide groups of the alpha-helical protein molecules form repetitive potential wells for an external electron.[53]

## 7. Potential Applications

Since the proteinoid materials and results highly resemble those of natural proteins, many applications would be in modeling natural proteins, cells, and membranes in which we may change at will the composition of constituent amino acids and then monitor the behavior of the polymers. Additionally, the quantum chemical calculations of molecular orbitals of such polymers may be performed and the resultant picture may be easier to verify experimentally than in the case of natural peptides. The existing data on this subject on simulation of natural polypeptides and proteins[54,55] make such comparisons practical. Consequently, proteinoids may be very useful in the analysis of natural proteins, and the constructionistic methodology holds promise as a complementary approach to the classical analytical one.

Thermal polymers of amino acids are able to make spherical and planar membranes. This gives us both theoretical and practical tools for studying the composition and functions of natural membranes. Having such models, in which we can manipulate amino acid composition, we may produce synthetically membranes resembling natural ones, and in some aspects surpass them in desired properties.

Another heuristic value of the proteinoids would be to use membranes and compare them to known electronic devices such as, for instance, tunnel diode or unijunction transistor. Some electronic properties of membranes made of proteinoids highly resemble the respective characteristics of these electronic devices.[56] Here we closely approach the domain of the next step in computer technology. As yet predominantly, if not only, theoretical and conceptual attempts are described in the literature (e.g., Ref. 57) including the patent literature.[58–61]

The fact that proteinoid polymer contains covalently bonded chromophores[3,41,43] and is photoactive[62,63] opens the possibility of exploring its behavior during light stimulation as well as of considering this polymer as a candidate for material in photovoltaics.

Another potentiality is in the field of microencapsulation.[64,65] The ability to make microspheres, plus the fact that they have electrically charged double membranes, and also lack antigenicity, is especially interesting.

## 8. Conclusions and Prospects

The proteinoid model of the cell is based on the protein nature of biological cells. A large number of membrane phenomena have been catalogued. These include membrane potential, oscillatory rhythmic phenomena, action potentials, induced electrical discharge, hyperpolarization, asymmetric permeability, channeling phenomena, current–voltage characteristics with negative resistance, ion-dependent membrane potentials, homeostatic recovery, and photosensitivity. There exists a fundamental analogy between the electrical phenomena of the artificial proteinoid cell (protocell model) and evolved natural excitable cells.

The property of excitability is thus added to the lengthening list of biological properties observed in proteinoid microspheres. Because of the synthetic nature of the proteinoids, it is now possible to study in detail the relationship of structure to electrical biofunction.

## Acknowledgments

Results reported in this chapter are from grants provided by the National Foundation for Cancer Research and the National Aeronautics and Space Administration (NGR 10-007-008). Contribution No. 372 of the Institute for Molecular and Cellular Evolution.

## References

1. K. Harada and S. W. Fox, "Characterizations of Functional Groups of Acidic Thermal Polymers of α-Amino Acids," *BioSystems* 7, 222–229 (1975).
2. T. Nakashima, J. R. Jungck, S. W. Fox, E. Lederer, and B. C. Das, "A Test for Randomness in Peptides Isolated from a Thermal Polyamino Acid," *Int. J. Quant. Chem. Quant. Biol. Symp.* 4, 65–72 (1977).
3. J. Hartmann, M. C. Brand, and K. Dose, "Formation of Specific Amino Acid Sequences During Thermal Polymerization of Amino Acids," *BioSystems* 13, 141–147 (1981).
4. D. L. Rohlfing, "Thermal Polyamino Acids: Synthesis at Less than 100°C, *Science* 193, 68–70 (1976).
5. P. Neri, G. Antoni, F. Benvenuti, F. Cocola, and G. Gazzei, "Synthesis of Alpha, beta-poly [(2-hydroxy-ethyl)-DL-aspartamide], A New Plasma Expander," *J. Med. Chem.* 16, 893–897 (1973).
6. K. Harada and S. W. Fox, in *The Origins of Prebiological Systems* (S. W. Fox, ed.), Academic, New York (1965), pp. 289–298.
7. A. Vegotsky, "Thermal Copolymers of Amino Acids," Ph. D. dissertation, Florida State University (1961).
8. K. Dose and L. Zaki, "Hämoproteinoide mit Peroxidatischer und Katalatischer Activität," *Z. Naturforsch.* 26b 144–148 (1971).
9. S. W. Fox and T. Nakashima, "Fractionation and Characterization of an Amidated Thermal 1:1:1-Proteinoid," *Biochim. Biophys. Acta* 140, 155–167 (1967).
10. A. L. Lehninger, *Biochemistry*, 2nd ed., Worth and Co., New York (1975).
11. S. W. Fox, in *The Origins of Prebiological Systems* (S. W. Fox, ed.), Academic, New York (1965), pp. 361–372.
12. S. W. Fox, "A Theory of Macromolecular and Cellular Origins," *Nature* 205, 328–340 (1965).
13. S. W. Fox and K. Dose, *Molecular Evolution and the Origin of Life*, rev. ed., Dekker, New York (1977).
14. S. W. Fox and S. Yuyama, "Effects of the Gram Stain on Microspheres from Thermal Polyamino Acids," *J. Bacteriol.* 85, 279–283 (1963).
15. S. W. Fox, "The Evolutionary Significance of Phase-Separated Microsystems," *Origins Life* 7, 49–68 (1976).
16. S. Brooke and S. W. Fox, "Compartmentalization in Proteinoid Microspheres," *BioSystems* 9, 1–22 (1977).

17. S. W. Fox, "The Origins of Behavior in Macromolecules and Protocells," *Comp. Biochem. Physiol.* **67B**, 423–436 (1980).
18. S. W. Fox, "Metabolic Microspheres," *Naturwissenschaften* **67**, 378–383 (1980).
19. S. W. Fox and T. Nakashima, "The Assembly and Properties of Protobiological Structures: The Beginnings of Cellular Peptide Synthesis," *BioSystems* **12**, 155–166 (1980).
20. T. Nakashima and S. W. Fox, "Formation of Peptides from Amino Acids by Single or Multiple Additions of ATP to Suspensions of Nucleoproteinoid Microparticles," *BioSystems* **14**, 151–161 (1981).
21. S. W. Fox, "Origins of the Protein Synthesis Cycle," *Int. J. Quant. Chem. Quant. Biol. Symp.* **8**, 441–454 (1981).
22. J. R. Jungck and S. W. Fox, "Synthesis of Oligonucleotides by Proteinoid Microspheres Acting on ATP," *Naturwissenschaften* **60**, 425–427.
23. S. W. Fox, T. Nakashima, A. Przybylski, and R. M. Syren, "The Updated Experimental Proteinoid Model," *Int. J. Quant. Chem. Quant. Biol. Symp.* **9**, 195–204 (1982).
24. Y. Ishima, A. T. Przybylski, and S. W. Fox, "Electrical Membrane Phenomena in Spherules from Proteinoid and Lecithin," *BioSystems* **13**, 243–251 (1981).
25. A. T. Przybylski, W. P. Stratten, R. M. Syren, and S. W. Fox, "Membrane, Action, and Oscillatory Potentials in Simulated Protocells," *Naturwissenschaften* **69**, 561–563 (1982).
26. T. Takenaka, I. Inoue, Y. Ishima, and H. Horie, "Excitability of Surface Membrane of Protoplasmic Drop Produced from Protoplasm in Nitella," *Proc. Jpn. Acad.* **47**, 554–557 (1971).
27. I. Inoue, Y. Ishima, H. Horie, and T. Takenaka, "Properties of an Excitable Membrane Produced on the Surface of a Protoplasmic Drop in Nitella," *Proc. Jpn. Acad.* **47**, 549–553 (1971).
28. B. Gomperts, *The Plasma Membrane*, Academic Press, New York (1977).
29. J. L. Howland, *Cell Physiology*, Macmillan, New York (1973).
30. Y. Ishima and S. W. Fox, Abstract, *Third Ann. Mtg. Soc. Neuroscience* **17.10**, 172 (1973).
31. S. W. Fox, R. M. McCauley, P. O'B. Montgomery, T. Fukushima, K. Harada, and C. R. Windsor, in *Physical Principles of Biological Membranes* (F. Snell, J. Wolken, G. J. Iversen, and J. Lam, eds.), Gordon and Breach, New York (1969), pp. 417–430.
32. S. W. Fox, T. Adachi, W. Stillwell, Y. Ishima, and G. Baumann, in *Light Transducing Membranes: Structure, Function, Evolution* (D. W. Deamer, ed.), Academic Press, New York (1978), pp. 61–75.
33. L. Hsu, unpublished data.
34. A. T. Przybylski, in *Molecular Evolution and Protobiology* (K. Matsuno, K. Dose, K. Harada, and D. L. Rohlfing, eds.), Plenum Press, New York (1984), pp. 253–266.
35. S. W. Fox and S. Yuyama, "Abiotic Production of Primitive Protein and Formed Microparticles," *Ann. N. Y. Acad. Sci.* **108**, 487–494 (1963).
36. R. M. Syren, unpublished data.
37. W. D. Snyder and S. W. Fox, "A Model for the Origin of Stable Protocells in a Primitive Alkaline Ocean," *BioSystems* **7**, 222–229 (1975).
38. H. T. Tien, *Bilayer Lipid Membranes*, Marcel Dekker, New York (1974).
39. P. Mueller and D. O. Rudin, "Resting and Action Potentials in Experimental Bimolecular Lipid Membranes," *J. Theoret. Biol.* **18**, 222–258 (1968).
40. A. Wood and H. G. Hardebeck, "Light Enhanced Decarboxylations by Proteinoids," in *Molecular Evolution* (D. L. Rohlfing and A. I. Oparin, eds.), Plenum Press, New York (1972), pp. 233–245.
41. B. Heinz and W. Ried, "The Formation of Chromophores Through Amino Acid Thermolysis and their Possible Role as Prebiotic Photoreceptors," *BioSystems* **14**, 33–40 (1981).
42. G. H. Beaven and E. R. Holiday, "Ultraviolet Absorption Spectra of Proteins and Amino Acids," *Adv. Protein Chem.* **7**, 319–386 (1952).

43. B. Heinz and W. Ried, "Structure Elucidation of a Chromo Proteinoid," Abstract C2-17, 7th International Conference on the Origins of Life, Mainz, Germany (July 10–15, 1983).
44. S. W. Fox, T. Adachi, and W. Stillwell, in *Solar Energy: International Progress* (T. N. Veziroglu, ed.), Vol. 2, Pergamon Press, New York (1980), pp. 1056–1074.
45. C. Kempf, R. D. Klausner, J. N. Weinstein, J. Van Renswoude, M. Pincus, and R. Blumenthal, "Voltage-Dependent Trans-Bilayer Orientation of Melittin, *J. Biol. Chem.* **257**, 2469–2476 (1982).
46. S. J. Kennedy, R. W. Roeske, A. R. Freeman, A. M. Watanabe, and H. R. Besch, Jr., "Synthetic Peptides Form Ion Channels in Artificial Lipid Bilayer Membranes," *Science* **196**, 1341–1342 (1977).
47. R. J. Bradley, W. O. Romine, M. M. Long, T. Ohnishi, M. A. Jacobs, and D. W. Urry, "Synthetic Peptide $K^+$ Carrier with $Ca^{++}$ Inhibition," *Arch. Biochem.* **178**, 2, 468 (1977).
48. S. Larsson, "Electron Transfer in Biological Systems," *Int. J. Quantum Chem.: Quantum Biol. Symp.* **9**, 385–397 (1982).
49. V. Denner and F. Kaiser, "Phase Transition Behavior of a Greater Membrane Model, *Int. J. Quantum Chem.: Quantum Biol. Symp.* **9**, 41–57 (1982).
50. M. A. Keniry, R. L. Smith, H. S. Gutowsky, and E. Oldfield, in *Structure and Dynamics: Nucleic Acids and Proteins* (E. Clementi and R. H. Sarma, eds.), Adenine Press, New York (1983), pp. 435–450.
51. D. W. McLaughlin, in *Structure and Dynamics: Nucleic Acids and Proteins* (E. Clementi and R. H. Sarma, eds.), Adenine Press, New York (1983), pp. 55–60.
52. A. C. Scott, in *Structure and Dynamics: Nucleic Acids and Proteins* (E. Clementi and R. H. Sarma, eds.), Adenine Press, New York (1983), pp. 389–404.
53. A. S. Davydov, in *Structure and Dynamics: Nucleic Acids and Proteins* (E. Clementi and R. H. Sarma, eds.), Adenine Press, New York (1983), pp. 377–387.
54. M. Levitt, C. Sander, and P. S. Stern, "The Normal Modes of a Protein: Native Bovine Pancreatic Trypsin Inhibitor," *Int. J. Quantum Chem.: Quantum Biol. Symp.* **10**, 181–199 (1983).
55. M. Goodman, A. S. Verdini, N. S. Choi, and Y. Masuda, in *Topics in Stereochemistry* (E. L. Eliel and N. L. Allinger, eds.), Wiley-Interscience, New York (1970), Vol. 5, pp. 69–166.
56. A. T. Przybylski, "Material Model of the Neuron," Abstract Sanibel Symposia (March 1–15, 1984).
57. F. L. Carter, *Molecular Electronic Devices*, Dekker, New York (1982).
58. W. M. Biernat, "Molecular Storage Unit," United States Patent 3,119,099 (January 21, 1964).
59. F. S. Barnes, "Quantum State Memory," United States Patent 3,754,988 (August 28, 1973).
60. A. Aviram, "Organic Memory Device," United States Patent 3,833,894 (September 3, 1974).
61. C. Levinthal, "System for Storing and Retrieving Information at the Molecular Level," United States Patent 4,032,901 (June 28, 1977).
62. A. T. Przybylski and S. W. Fox, in *Solar Power Applications: Alternative Energy Sources IV* (T. N. Veziroglu, ed.), Ann Arbor Science, Ann Arbor, Michigan (1982), Vol. 3, pp. 95–102.
63. A. T. Przybylski, R. M. Syren, and S. W. Fox, in *Alternative Energy Sources V. Part B: Solar Applications* (T. N. Veziroglu, ed.), Elsevier Science Publishers B.V., Amsterdam (1983) pp. 367–377.
64. S. W. Fox and S. Brooke, in *Microencapsulation* (T. Kondo, ed.), Techno Books, Tokyo (1979), pp. 257–290.
65. A. T. Przybylski and S. W. Fox "Excitable Artificial Cells of Proteinoid," *Appl. Biochem. Biotechnol.* **10**, 301 (1984).

# 14

# Noise in Biomolecular Systems

## D. Vasilescu and H. Kranck

*ABSTRACT:* The principle of electrical spectrography and its measurement system is discussed. The phenomenon of noise in electrolytes and interfaces receives attention. Noise spectrography is found to have applications in some biomolecular systems, viz., DNA helix-to-coil transition, thermal transconformation, and "salt-free" premelting effects. Noise conductivity emission spectra of collagen solutions gave information on permanent dipole fluctuations and hydrodynamic properties of the system.

## 1. Introduction

Any sample of matter is able to produce electrical noise by the stochastic displacement of internal electrons or ions. We can also observe that in biological systems we are in the presence of water, biomolecules, and ions in perpetual fluctuation. All fundamental biomolecules such as nucleic acids, proteins, and lipids are surrounded by water and ions.

Counterions around charged groups (phosphates in nucleic acids) play an important part because they induce the essential properties of the polyelectrolyte. Ions in the double layers at the membrane–electrolyte interface are also in a particular category and they introduce specific properties of the interface. If a biomolecule possesses a permanent dipole, the fluctuations of the dipole can create a special type of noise.

In all these examples we can notice that ions are distributed in two groups: one in the bulk of the system (i.e., electrolyte Brownian motion) and one consisting of ions with special noise properties.

*D. Vasilescu and H. Kranck* • Laboratoire de Biophysique, Institut Polytechnique Méditerranéen, Université de Nice, Parc Valrose-06034, Nice, Cedex, France.

From the thermodynamic point of view the observation of this kind of noise—without any exterior perturbation—is a reversible process corresponding to the emission of an electric random signal.

Other phenomena are interesting from the noise point of view. They related to ion transport across membranes,[1-11] equilibrium and non-equilibrium kinetic systems,[12] nerve membrane noise,[13-20] and membrane current fluctuations from ionic channels ($Na^+$ channels and $K^+$ channels in axons) in stationary or nonstationary states.[21-24] Some of these studies have been described in extended reviews.[25-29]

Our study deals essentially with the ions in a liquid sample—at thermal equilibrium—as in electrolytes, in biomolecular counterionic atmospheres, and at the electrolyte–membrane interfaces.

The spontaneous fluctuations of permanent dipoles are also discussed in the case of the collagen molecule.

## 2. Principle of Noise Spectrography

When an electric noise signal $n(t)$ is delivered by the sample, it has to be detected, and its behavior has to be analyzed versus time $t$ or frequency $v$.

### 2.1. Basic Vocabulary[30,31]

If $n(t)$ is a continuous random variable, we define the following.
*The Mean Values:*

$$\bar{n} = \int_{-\infty}^{+\infty} np(n)\, dn \tag{1}$$

$$\overline{n^2} = \int_{-\infty}^{+\infty} n^2 p(n)\, dn \qquad \text{(mean square value)} \tag{2}$$

where $p(n)$ is the probability density function of $n$. In a stationary random process, the mean values are independent of time.
*The Temporal Mean Values:*

$$\langle n \rangle = \lim_{T \to \infty} \frac{1}{2T} \int_{-T}^{+T} n(t)\, dt \tag{3}$$

$$\langle n^2 \rangle = \lim_{T \to \infty} \frac{1}{2T} \int_{-T}^{+T} n^2(t)\, dt \qquad \text{(mean square value)} \tag{4}$$

*Ergodicity:*

$$\bar{n} = \langle n \rangle$$
$$\overline{n^2} = \langle n^2 \rangle \tag{5}$$

The ergodic processes are stationary.

*Autocorrelation Function:*

$$\mathcal{R}_n(\tau) = \langle n(t + \tau) n(t) \rangle, \qquad \tau \text{ is the correlation time}$$

When $\tau = 0$

$$\mathcal{R}_n(0) = \langle n(t) n(t) \rangle = \langle n^2 \rangle$$

*Power Spectrum (or Spectral Density):* The power spectrum is the Fourier transform of the autocorrelation function:

$$\mathcal{S}_n(\omega) = \hat{\mathcal{R}}_n(\tau) = \int_{-\infty}^{+\infty} \mathcal{R}_n(\tau) \exp(-i\omega\tau) \, d\tau \tag{6}$$

(with $\omega = 2\pi v$ and $i = \sqrt{-1}$) and the autocorrelation function is the inverse Fourier transform of $\mathcal{S}_n(\omega)$:

$$\mathcal{R}_n(\tau) = (2\pi)^{-1} \int_{-\infty}^{+\infty} \mathcal{S}_n(\omega) \exp(i\omega\tau) \, d\omega \tag{7}$$

If $\tau = 0$

$$\mathcal{R}_n(0) = \langle n^2 \rangle = (2\pi)^{-1} \int_{-\infty}^{+\infty} \mathcal{S}_n(\omega) \, d\omega \tag{8}$$

*Cross-Correlation Function:* If $n_A(t)$ and $n_B(t)$ are two ergodic random processes it is possible to define the cross-correlation function by

$$\mathcal{R}_{n_A n_B}(\tau) = \langle n_A(t + \tau) n_B(t) \rangle \tag{9}$$

If the $n_A$ and $n_B$ are statistically independent and if $\langle n_A \rangle = 0$ and $\langle n_B \rangle = 0$, then

$$\mathcal{R}_{n_A n_B}(\tau) \to 0 \qquad \text{for all } \tau \tag{10}$$

## 2.2. Noise Spectrograph Apparatus[32]

A schematic description of the cross-correlation spectrograph, built in our laboratory, is shown on Figure 1.

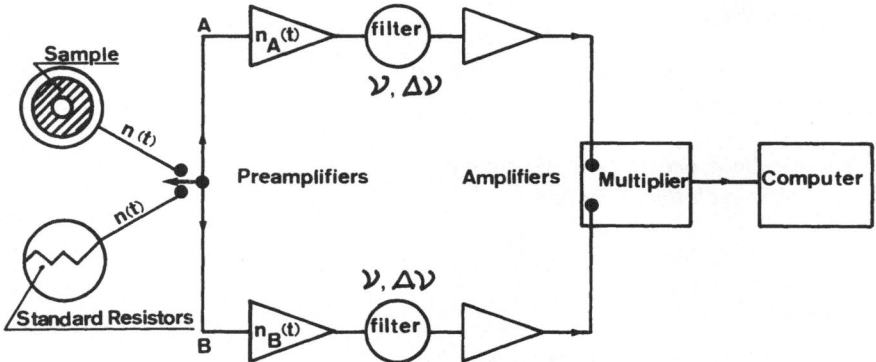

FIGURE 1.   Schematic diagram of the cross-correlation noise spectrograph.

In both $A$ and $B$ channels, the random voltage $n(t)$ delivered by the sample is amplified (frequency range 0–300 kHz), then filtered at a frequency $v$ (1 Hz $< v <$ 100 kHz) with a bandwidth $\Delta v = 0.05v$. Then the signal is detected through a multiplier–integrator built in our laboratory and transmitted to a computer.

Amplifiers $A$ and $B$ exhibit their intrinsic noncorrelated electronic noises [denoted by $n_A(t)$ and $n_B(t)$] and also independent of $n(t)$.

In each channel of the spectrograph, the total noise signal is

$$N_A = n(t) + n_A(t)$$
$$N_B = n(t) + n_B(t) \tag{11}$$

The cross-correlation function $\mathscr{R}_{N_A N_B}(\tau)$ of $N_A$ and $N_B$ noises is

$$\mathscr{R}_{N_A N_B}(\tau) = \mathscr{G}^2 \langle N_A(t+\tau)\, N_B(t) \rangle$$
$$= \mathscr{G}^2 [\mathscr{R}_n(\tau) + \overbrace{\mathscr{R}_{n_A n}(\tau) + \mathscr{R}_{n n_B}(\tau) + \mathscr{R}_{n_A n_B}(\tau)}^{L}] \tag{12}$$

where $\mathscr{G}$ is the total amplification gain in each channel.

Because $n$, $n_A$, and $n_B$ are noncorrelated noises, the last three terms, denoted $L$, are negligible; therefore the above relation becomes

$$\mathscr{R}_{N_A N_B}(\tau) = \mathscr{G}^2 \mathscr{R}_n(\tau) = \mathscr{G}^2 \langle n(t+\tau)\, n(t) \rangle \tag{13}$$

$\mathscr{R}_n(\tau)$ is the autocorrelation function of the noise produced by the studied sample. In our spectrograph, the multiplier–integrator computes this function for $\tau = 0$:

$$\mathscr{R}_n(0) = \mathscr{G}^2 \langle n(t)^2 \rangle \tag{14}$$

Hence, by a cross-correlation technique, it is possible to obtain the noise properties of the sample with high rejection of the parasitic noise of the device.

In our spectrograph, the mean square value of the sample noise $\langle n(t)^2 \rangle$ is compared with the noise produced by standard metal or metal layers resistors using an interpolation method.

The power spectrum of the thermal noise produced by an ideal resistor is

$$\mathscr{S}_n(\omega) = 2kTR_n \tag{15}$$

where $k$ is the Boltzmann constant, $T$ is the absolute temperature, and $R_n$ is the noise resistance. At a given temperature, this power spectrum is flat (i.e., frequency independent). This noise is also called "white noise."

The autocorrelation function $\mathscr{R}_n(\tau)$ is the inverse Fourier transform of the power spectrum, Eq. (7), and the noise voltage created by the resistor is

$$\langle n^2(t) \rangle = \mathscr{R}_n(0) = 4kTR_n\, \Delta v \tag{16}$$

This expression is the limiting value of $\mathscr{R}_n(\tau)$ (with $\tau = 0$) in a frequency range where $hv \ll kT$, i.e., when $v \ll 10^{13}$ Hz.

The mean square value of the voltage noise produced by a resistor at the frequency $v$, with $\Delta v$ band, is known as the Nyquist–Johnson theorem[33,34] [see also our previous paper[35]].

Finally, to achieve the best accuracy, the output of the multiplier–integrator is analyzed by a computer. A noise measurement is carried out on 200 sampling points. The sample equivalent noise resistance is obtained through a linear regression on at least four standard resistors.

## 3. Noise in Electrolytes and Interfaces

### 3.1. Dilute Aqueous 1.1 Electrolytes[35]

In an aqueous 1.1 electrolyte, at thermal equilibrium, we introduce the Landau length $l \equiv \mathscr{L}/\varepsilon_r' T$ where $\varepsilon_r'$ is the real part of the permittivity of water and $T$ is the absolute temperature. $\mathscr{L}$ is the universal constant[36]:

$$\mathscr{L} = \frac{e^2}{4\pi\varepsilon_0 k} = 1.669 \times 10^{-5}(\text{m K}) \tag{17}$$

with $k$ the Boltzmann constant, $e$ the electronic charge, and $\varepsilon_0$ the permittivity of free space.

If $\langle d \rangle$ is the mean distance between two ions in the electrolyte,

Coulombian interactions are not important if $l \ll \langle d \rangle$; i.e., the electrolyte may be considered as a plasma.[37]

At 25°C, with $\varepsilon'_r = 78.33$, $l = 7.15$ Å. At $10^{-2}$ mol/liter, $\langle d \rangle \simeq 44$ Å; thus for concentrations below $10^{-2}$ mol/liter the above holds.

For a dilute 1.1 aqueous electrolyte like NaCl or KCl, the electrical noise is the result of the Brownian movement of ions in water.

The solution of the ionic velocity Langevin equation $v(t)$ for one dimension, leads to an autocorrelation function

$$\mathscr{R}_v(\tau) = \frac{kT}{m} \exp(-\tau/mu) \tag{18}$$

where $m$ is the mass, $k$ the Boltzmann constant, $T$ the absolute temperature, and $u$ the mechanical mobility of the ion. The power spectrum of the process is

$$\mathscr{S}_v(\omega) = \frac{(2kT/m)\,\tau_r}{1 + \omega^2 \tau_r^2} \qquad \text{with} \qquad \tau_r = mu \tag{19}$$

For monovalent ions like Na$^+$, K$^+$, and Cl$^-$, the critical time $\tau_r$ has a value $\tau_r \simeq 10^{-13}$–$10^{-14}$ s at ambient temperature. Thus,

$$\omega^2 \tau_r^2 \ll 1 \qquad \text{and} \qquad \mathscr{S}_v(\omega) = \frac{2kT}{m}\,\tau_r = 2kTu \tag{20}$$

The spectral density of the process is constant at constant temperature, i.e., $\mathscr{S}_v(\omega)$ is independent of the noise frequency: the case of the so-called thermal noise.

One may also express the mean square value of the noise voltage produced by the electrolyte in terms of its resistance, Eq. (16). The resulting noise conductivity is

$$\sigma_n = CR_n^{-1}, \qquad C \text{ is a geometrical constant} \tag{21}$$

If we define the electrical noise mobility of the ions by

$$\mu_n = eu \tag{22}$$

where $u$ is the mechanical mobility and $e$ the electronic charge, this noise mobility is the equivalent of the classical mobility derived from the motion of an ion when an external electric field is applied. Thus

$$\sigma_n = \frac{N}{2} e(\mu_n^+ + \mu_n^-) \tag{23}$$

for a system of $N$ identical monovalent ions.

Noise measurements do not allow the separate evaluation of $\mu_n^+$ and $\mu_n^-$ but yield the mean mobility:

$$\langle \mu_n \rangle = \frac{\mu_n^+ + \mu_n^-}{2} \quad \text{and} \quad \sigma_n = Ne\langle \mu_n \rangle \tag{24}$$

Finally, it is helpful to relate the preceding results to the Kubo fluctuation–dissipation theorem,[38] in which formalism an alternating electric field is applied as a small perturbation. The response of the system leads to a generalized conductivity of the electrolyte:

$$\sigma(\omega) = \frac{Ne^2}{kT} \int_0^\infty \mathcal{R}_v(\tau) \exp(-i\omega\tau) \, d\tau \tag{25}$$

where $\mathcal{R}_v(\omega)$ is the autocorrelation function of the velocities.

Introducing the velocity power spectrum from Eq. (6) yields

$$\mathcal{S}_v(\omega) = \frac{2\langle v^2 \rangle \tau_r}{1 + \omega^2 \tau_r^2} \simeq 2\langle v^2 \rangle \tau_r \tag{26}$$

$$\sigma(\omega) = \frac{Ne^2}{kT} \times \frac{1}{2} \mathcal{S}_v(\omega) = \frac{Ne^2}{kT} \langle v^2 \rangle \tau_r \tag{27}$$

with $\langle v^2 \rangle = kT/m$ for one dimension and $\sigma(\omega) = Ne^2u = Ne\mu \simeq Ne\langle \mu_n \rangle$.

Therefore it appears that application of an alternating electric field to an electrolyte reveals an intrinsic property of its constituents ions, viz., their mechanical mobility expressed as their electrical mobilities.

For KCl and NaCl (concentration $< 10^{-2}$ mol/liter), the mean mobilities $\langle \mu_n \rangle$ evaluated by noise spectrography are identical with those obtained by conventional bridge methods.[35]

Concerning the effect of frequency, the noise emission spectra obtained are flat and there is no excess noise component varying with frequency $v$; the dilute electrolyte produces a white noise spectrum in accordance with theoretical considerations expressed above.

### 3.2. Noise at Electrode–Electrolyte Interface (Case of 1.1 Concentrated Electrolyte)

For concentrations higher than $10^{-2}$ mol/liter, the condition $\langle d \rangle \gg l$ is no longer verified and the electrolyte cannot be approximated any more by a plasma; Coulombic interactions become important. We have extended our previous measurements to concentrated aqueous KCl solutions ($10^{-2}$–2 mol/liter). Each solution has been studied at 25°C in the frequency range 500 Hz to 100 kHz.[39] From a technological point of view, the more concentrated the solutions are, the lower noise resistances they have; thus error measurements become very important. Figure 2 shows the variations of measured noise resistances $R_n$ of the cell versus frequency for different electrolyte concentrations. Each $R_n$ curve possesses an analog shape with an asymptotical value $R_\infty$ in the high-frequency domain; $R_\infty$ is close to $R_p$, the parallel resistance of the cell measured at 10 MHz with a conventional impedance bridge. The conductivities deduced from the $R_\infty$ values are in agreement with the table values found in Reference (40).

We can observe that in this case, the cell measured noise emission spectral density $\mathscr{S}_{mes}$ is not flat versus frequency. Therefore we may consider that $\mathscr{S}_{mes}$ is composed of two components: a constant noise spectral density due to the electrolyte itself $\mathscr{S}_s(\omega)$; an excess noise spectral density due to the electrode–electrolyte interfaces $\mathscr{S}_i(\omega)$. And

$$\mathscr{S}_{mes}(\omega) = \mathscr{S}_s(\omega) + \mathscr{S}_i(\omega) \tag{28}$$

We can deduce a similar equation for noise resistances:

$$R_{mes} = R_s + R_i \tag{29}$$

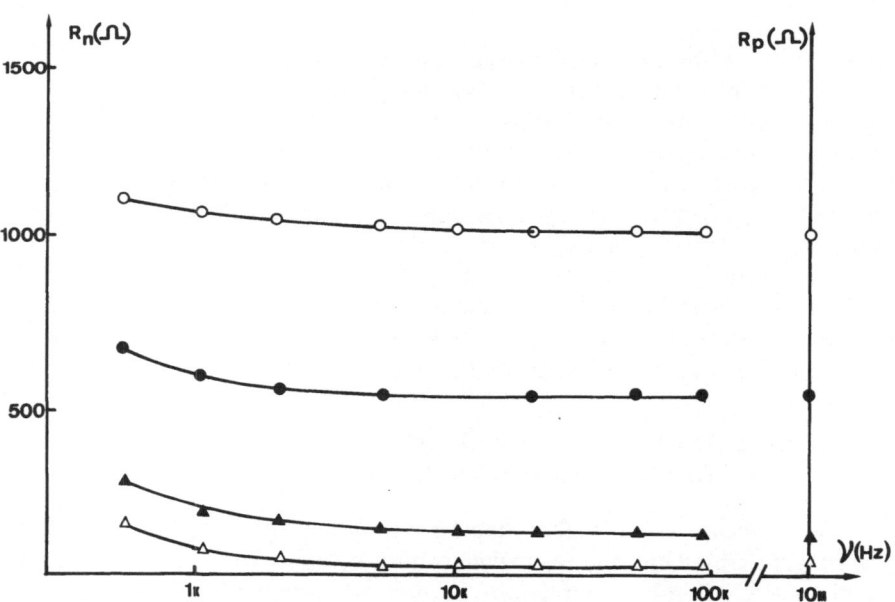

FIGURE 2.   Variation of noise resistance of the cell containing concentrated KCl versus the frequency $v$.  $\bigcirc$, 0.01 mol/liter;  $\bullet$, 0.02 mol/liter;  $\blacktriangle$, 0.1 mol/liter;  $\triangle$, 1 mol/liter. $R_p$ are the values of parallel resistance measured at 10 MHz with an impedance bridge.

The analysis of excess noise behavior has shown that its spectral density is

$$\mathscr{S}_i(\omega) = A/\omega^\alpha \tag{30}$$

where $\alpha$ is close to 1.

We can see in Figure 3 an example of noise resistance $R_{mes}$ variation versus the inverse of frequency, in the case of KCl 1 mol/liter. The line was obtained by means of a linear regression with a correlation coefficient of 0.99.

The limit value $R_s$ corresponding to infinite frequency is the asymptotic value $R_\infty$ (see above).

The origin of the excess noise can be attributed to the electrodes–electrolyte interface because there is no direct current through the cell measurement, and there is no electrode polarization; nevertheless there occurs an electrical double layer at the electrode–electrolyte interfaces.[41]

We also notice that in the absence of direct current, an electrochemical reaction able to produce the energy needed to keep up an electrical flow cannot exist. An electrochemical reaction is a sufficient but not a necessary condition for the emission of a $1/v$ noise.

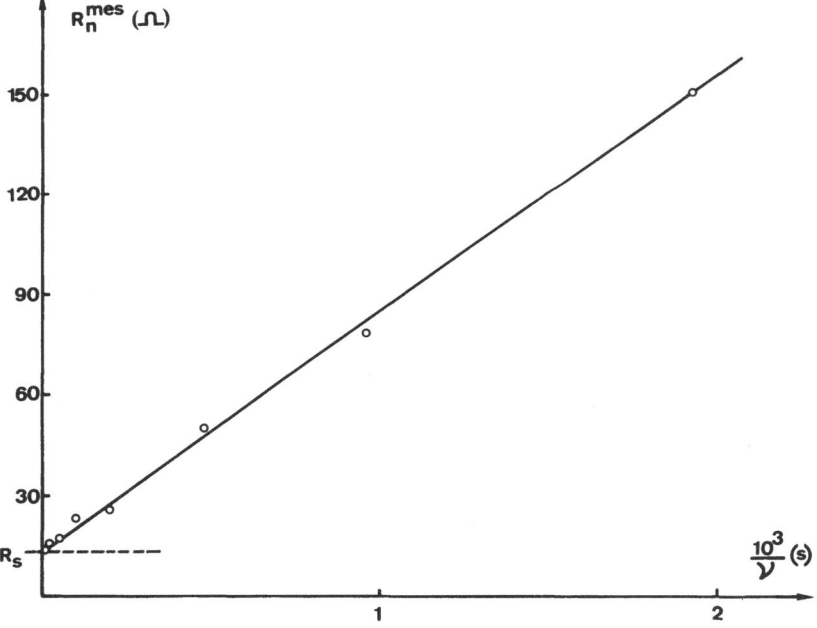

FIGURE 3. Variation of noise resistance of the cell containing KCl (1 mol/liter) versus $1/v$. $R_s$ is the noise resistance of the sample.

To conclude our discussion of noise measurements in electrolytes, in the absence of an applied electrical field, we can see that spectral density of electrolyte itself is constant versus frequency and corresponds to thermal white noise; the electrode–electrolyte interface spectral density is not a characteristic of concentrated electrolytes; it cannot be detected in the case of dilute electrolytes because

$$\mathscr{S}_i(\omega) \ll \mathscr{S}_s(\omega) \tag{31}$$

This is in agreement with the results described in Section 3.1.

### 3.3. Noise of the Synthetic Membrane–Electrolyte Interface

We present in this section some interesting results concerning the membrane–electrolyte interface properties in the case of a collodion membrane.[42] The membranes are collodion thick films ($\simeq 10 \, \mu m$) made from a 4% collodion solution in diethylether; these membranes contain fixed negative electrical charges inside. The electrolyte on both sides of the membrane is aqueous NaCl (concentration $2 \times 10^{-4}$–0.1 mol/liter), the active membrane surface is 1.2 cm$^2$, and the electrodes are made of platinized platinum. The power spectrum corresponding to the noise emitted by the total system is

$$\mathscr{S}_n(\omega) = 2kTR_e(Z) \tag{32}$$

where $R_e(Z)$ is the real part of the noise impedance under test. If $Z$ is represented by a parallel $(R, C)$ network, we can write

$$\mathscr{S}_n(\omega) = 2kTR/(1 + R^2C^2\omega^2) \tag{33}$$

With the Brophy and Webb method[43] it is possible to obtain by noise measurements the $R$ and $C$ values separately.[32]

In order to separate the different contributions creating the total noise, we have used Schwan's model.[44] This model, in the case of electrode polarization, is represented in Figure 4. The sample (cell without membrane) is represented by the parallel network $(R_x, C_x)$ and the electrode–electrolyte interfaces by the series $(R_{si}/2, 2C_{si})$ circuits. Thus, the noise resistance $R$ and the noise capacity $C$ are

$$R = (1 + R^2C^2\omega^2)\left(R_{si} + \frac{R_x}{1 + R_x^2C_x^2\omega^2}\right) \tag{34}$$

$$\frac{1}{\omega C} = \left(1 + \frac{1}{R^2C^2\omega^2}\right)\left[\frac{1}{C_{si}\omega} + \frac{1/\omega C_x}{1 + 1/(R_x C_x \omega)^2}\right] \tag{35}$$

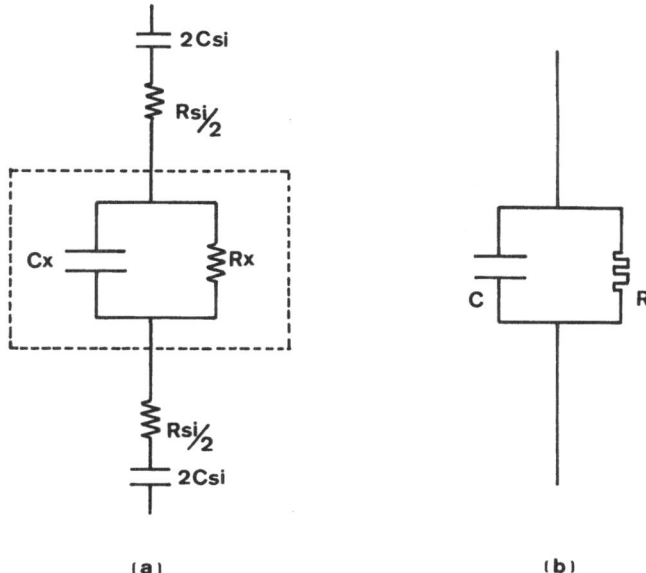

FIGURE 4. Schwan's model. (a) Complete model. (b) Device seen by the noise spectrograph.

This model was generalized in the following way: In a first step

- the membrane and its membrane–electrolyte interfaces are represented by the parallel network $(R_{xm}, C_{xm})$;
- the electrolyte and the membrane–electrolyte interfaces are represented by the series network $(R_{sie}/2, 2C_{sie})$.

In a second step

- the membrane alone is represented by the parallel network $(R_m, C_m)$;
- the membrane–electrolyte interfaces are represented by the series networks $(R_{sim}/2, 2C_{sim})$.

Figure 5 shows the variations of noise resistance $R_c$ and noise capacity $C_c$ of the cell, with membrane and without membrane, versus frequency (the electrolyte is NaCl $5 \times 10^{-3}$ mol/liter). We observe that $R_c$ and $C_c$ are decreasing functions when the frequency is increasing, in a different manner when considering the cell with membrane and without membrane. On the other hand, if we consider the behavior of the membrane elements itself, we can suppose that ion transport, in the absence of an applied field, is a simple diffusion process. Therefore the $R_m$ and $C_m$ values are independent of

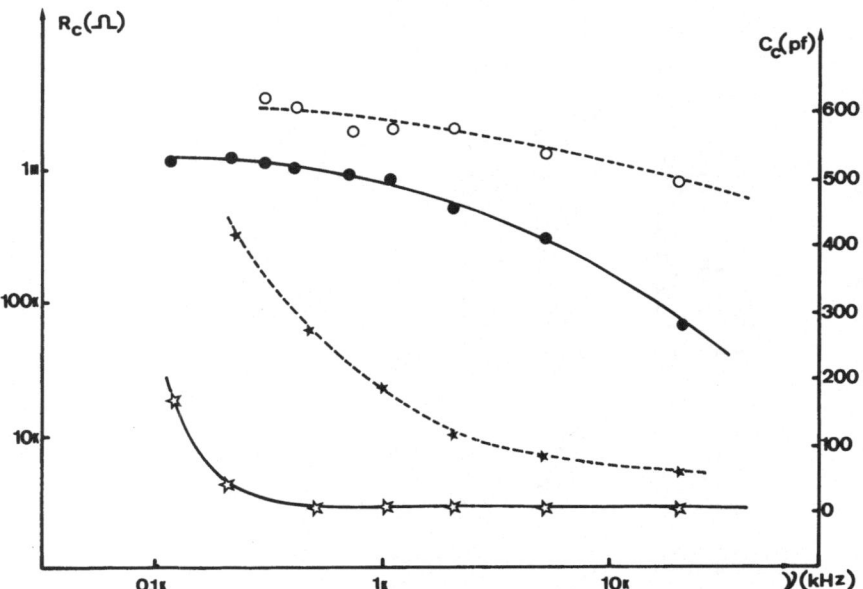

FIGURE 5. Variation of noise resistance $R_c$ and noise capacitance $C_c$ of the cell measurement versus the frequency $v$. The concentration of the electrolyte (NaCl) inside the cell is $5 \times 10^{-3}$ mol/liter. ●, $R_c$ in presence of membrane; ○, $C_c$ in presence of membrane; ★, $R_c$ without membrane; ☆, $C_c$ without membrane.

frequency in the studied range (100 Hz to 50 kHz); $1/v$ noise does not appear in this case.[4,45–47] Using bridge measurements in the high-frequency domain (0.5–5 MHz), we have noticed that $R_m$ and $C_m$ are very small with regard to the $R_{sim}$ and $C_{sim}$ interface elements; the noise interface elements $R_{sim}$ and $C_{sim}$ decrease considerably in the high-frequency domain and become negligible. So, as a first approximation, in our noise measurements, the network $(R_{xm}, C_{xm})$ represents essentially the contribution of the membrane–electrolyte interfaces. The behavior of these interfaces is well represented by the equations

$$R_{xm} = R_0/v^\alpha \quad \text{and} \quad C_{xm} = C_0/v^\beta \tag{36}$$

We have determined the $\alpha$ and $\beta$ exponents by varying the concentration of NaCl concentration. The results obtained are shown in Figure 6. We can notice the particular behavior of $\alpha$, which tends toward zero when the electrolyte concentration is zero. This reveals the disappearence of the membrane–electrolyte interface contribution in the case of very dilute electrolyte and the return to the thermal noise.

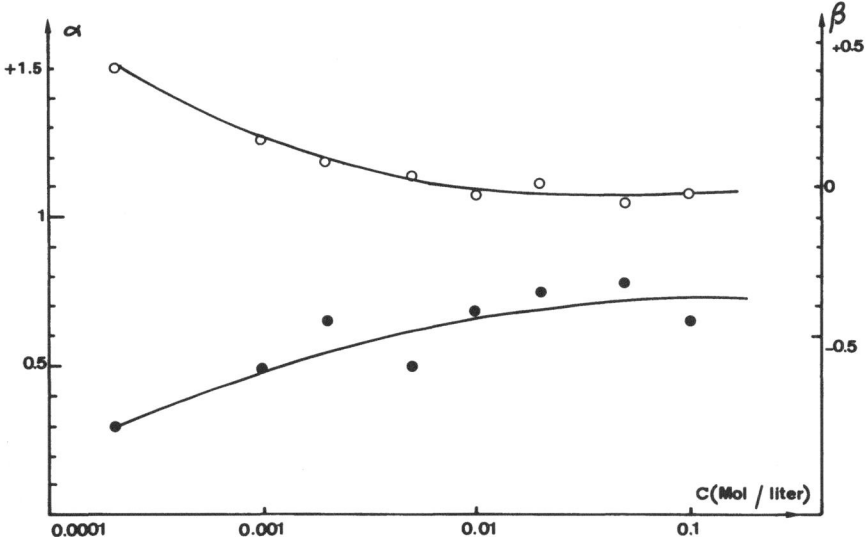

FIGURE 6.  Variation of exponent $\alpha$ (in $R_0/v^\alpha$) and exponent $\beta$ (in $C_0/v^\beta$) versus the electrolyte concentration. $\bullet$, $\alpha$; $\circ$, $\beta$.

We can also determine the power spectrum of the membrane–electrolyte interfaces:

$$\mathscr{S}_{nxm}(\omega) = 2kTR_e(Z_{xm}) = \frac{2kTR_{xm}}{1 + R_{xm}^2 C_{xm}^2 \omega^2}$$

$$= \frac{2kTR_0/\omega^\alpha}{1 + R_0^2 C_0^2 \omega^2/\omega^{2(\alpha+\beta)}} \tag{37}$$

In our case

$$R_{xm} C_{xm} \omega \gg 1$$

and

$$\mathscr{S}_{nxm}(\omega) \simeq \frac{2kT}{R_0 C_0^2 \omega^{(2-\alpha-2\beta)}} \tag{38}$$

Thus, we can write

$$\mathscr{S}_{nxm}(\omega) = S_0/v^\gamma \qquad \text{with} \quad \gamma = 2 - \alpha - 2\beta \tag{39}$$

This relation was experimentally verified. Figure 7 shows the variations of $\gamma$ versus electrolyte concentration. Our $\gamma$ values are in

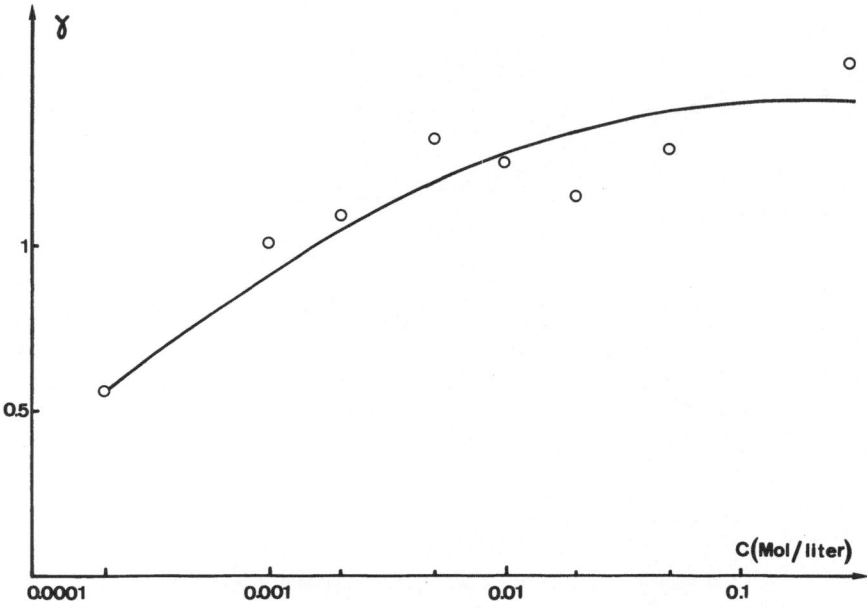

Variation of exponent $\gamma$ (in $S_0/\nu^\gamma$) versus the electrolyte concentration.

agreement with those of Green and Yafuso[1] for membranes in the presence of a weak electrical field. We also notice in Figure 7 that $\gamma \to 0$ when the concentration of electrolyte approaches zero; then the spectral density $\mathcal{S}_{nxm}$ becomes constant (independent of frequency) at infinite dilution of electrolyte—i.e., we reach the white noise.

Concerning the collodion membrane resistances, De Felice and Michalides[48] have found values up to 10 M$\Omega$/cm$^2$, in the presence of a direct current. If we extrapolate our results, using the relation $R_{xm} = R_0/\nu^\alpha$, for 1 Hz frequency, we find values in the range 0.5–100 M$\Omega$, in agreement with the De Felice and Michalides experiments. Gregor and Sollner[49] have found resistance values of about 10–100 $\Omega$ for collodion membranes by means of a Kohlrausch's bridge. This is a supplementary argument proving that the observed values of De Felice and Michalides[48] are the membrane–electrolyte interface $R_{xm}$. On the other hand, the $C_{xm}$ behavior different from $R_{xm}$ variation is related to the adopted model. A more complete treatment of the interface capacity due to the Gouy–Chapman double layer was proposed by Gur et al.[50,51]

Finally, the membrane–electrolyte interface noise generation can be divided into two parts:

a. A membrane–electrolyte interface noise resistance generator:

$$R_{xm} = R_0/v^\alpha$$

b. A modulation of the previous noise by the impedance:

$$Z_{xm} = R_{xm} - i/C_{xm}\omega$$

The (b) effect has also been observed by De Felice and Michalides.[48]

This is also the reason why in the Dorset and Fishman experiments[3] the membrane power spectrum measured at 1 kHz is below the corresponding thermal noise level.

## 4. Application to Some Biomolecular Systems

### 4.1. Ionic Atmosphere around DNA and Direct Visualization of the Thermal Transconformation (Helix → Coil Transition)

#### 4.1.1. DNA as a Polyelectrolyte

In order to simplify the model we present in Figure 8 the DNA in the double helical *B* form as a rodlike polyion. If the solvent is NaCl, the compensating counterions of negative phosphate sites are $Na^+$ and the coions of the electrolyte are $Cl^-$. It is possible to show that a correlation exists between the conformation of the macroion and some characteristic parameters of the electrolyte.[52]

In the double helical *B* form of DNA, the mean value of the helix arc between two neighbor phosphate sites is $H \simeq 7$ Å. The projection *b* of all the phosphate sites onto the double helix axis is 1.7 Å between consecutive projections (distance noted *b* on Figure 8).

The DNA polyelectrolyte behavior is expressed by the charge parameter $\xi$ (G. S. Manning[53]); see also the review, Reference 54.

$$\xi = \frac{l}{b} \tag{40}$$

where *l* is the Landau length.

In the Manning model, if $\xi > 1$, the DNA is thermodynamically unstable and the counterions will condense on it until $\xi$ reaches the value of unity. When $\xi = 1$, $l = b$; theoretically this situation corresponds to an extended DNA backone and *b* corresponds almost to the Landau length. In the temperature range 20–80°C, the Landau length does not vary

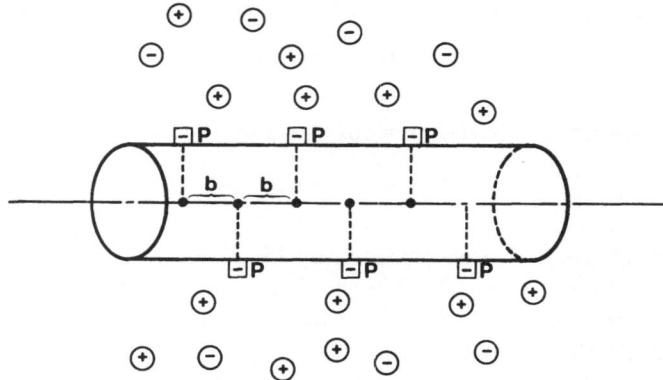

**FIGURE 8.** The rodlike DNA model. ⊟, phosphate sites; ⊕, counterions; ⊖, coions; $b$ is the distance between two neighbor phosphate sites projected onto the cylindrical axis.

significantly, because the $\varepsilon'_r T$ product of water is nearly constant (see Section 3.1). These conditions are those in which DNA solutions are studied, and $b$ is equal to the length $H$ of the arc of the helix calculated above.

Let $n_e$ and $n_s$ be the molar concentrations of phosphate sites and added salt, respectively. In his model Manning calculated the following expressions for $\gamma_{Na^+}$ and $\gamma_{Cl^-}$; the activity coefficients of the counterions and coions

$$
\gamma_{Na^+} = \frac{(\xi^{-1}\chi + 1)}{(\xi + 1)} \exp\left[ -\frac{1}{2} \frac{\xi^{-1}\chi}{(\xi^{-1}\chi + 2)} \right]
$$

$$
\gamma_{Cl^-} = \exp\left( -\frac{1}{2} \frac{\xi^{-1}\chi}{(\xi^{-1}\chi + 2)} \right)
$$

(41)

with $\chi = n_e / n_s$.

An experimental study of the activities $a_{Na^+}$ and $a_{Cl^-}$ for the helical ($h$) and coiled ($c$) forms allows us to test the validity of Manning's equations. Such measurements where carried out in our laboratory using glass-membrane or single-crystal electrodes, selectively sensitive to $Na^+$ or $Cl^-$.[55,56]

We can thus determine $\xi_h$ and $\xi_c$ corresponding to $\gamma_{Na^+}$ in the helical and coiled form of DNA and consequently evaluate the $b$ values.

In the case of DNA (0.3–0.8 g/liter) dissolved in $2 \times 10^{-3}$ mol/liter NaCl, $b_h = 1.7$ Å with $\xi_h = 4.2$ and $\xi_c = 1.2$ and $b_c \simeq 6$ Å. We notice that $b_c < 7$ Å; i.e., in the coil conformation the DNA backone is in an only partly unwould form.

The thermal transconformation or helix → coil transition is followed by an ejection of $Na^+$ counterions

$$\Delta a_{Na^+} = a^h_{Na^+} - a^c_{Na^+} \qquad (42)$$

($a^h$ and $a^c$ are the $Na^+$ activities for the helix and coil states of DNA).

DNA of different origins exhibit an extent of binding[57] $\Delta\psi = \Delta a_{Na^+}/n_e$ between 25% and 40% (with $n_e = 1$–$2 \times 10^{-3}$ mol/liter and Nacl concentrations 1–$2 \times 10^{-3}$ mol/liter).

### 4.1.2. Noise Measurements in DNA Solutions

DNA solutions in aqueous NaCl are employed as the dielectric of a cylindrical capacitor with gold electrodes; the voltage fluctuations are then studied as described above by noise spectrography. We report here some original results obtained in our laboratory.[58,63]

*4.1.2a. Spectral Density.* The equivalent noise resistance $R_n$ of DNA solutions was measured by comparison with calibrated white noise generators (wire resistors) in the frequency range 200 Hz to 50 kHz.

The noise resistance $R_n$ is related to the spectral density $\mathscr{S}_n(\omega)$ by

$$R_n = \frac{\mathscr{S}_n(\omega)}{2kT} \qquad (43)$$

if the emitted noise is white.

The spectral density $\mathscr{S}_n(\omega)$ is flat in the case of native and of thermally denatured DNA (see Figure 9). These results show that—within the 2%–3% precision of noise measurements—it is not possible to assign a permanent dipole moment to the DNA molecule (see Section 4.2). The variations of $R_n$—at the same temperature—observed before and after the helix → coil transition of DNA are related to counterion ejection phenomena described above and later in Section 4.1.2b.

Considering its molecular structure and distribution of fixed charges, no evidence can be found to assume that high molecular DNA could have a permanent dipole moment; the counterions are symmetrically distributed, and at thermodynamic equilibrium, in the absence of an applied electric field $E$, they do not contribute to create a dipole moment $\mu$.

Another effect was introduced by Oosawa[64] concerning the role of the mean square dipole $\langle \mu^2 \rangle$ due to the counterion fluctuation along the DNA polyion structure. The polarizability of the DNA along the double helix axis is given by

$$\alpha = \frac{\langle \mu^2 \rangle_{E=0}}{kT} \qquad (44)$$

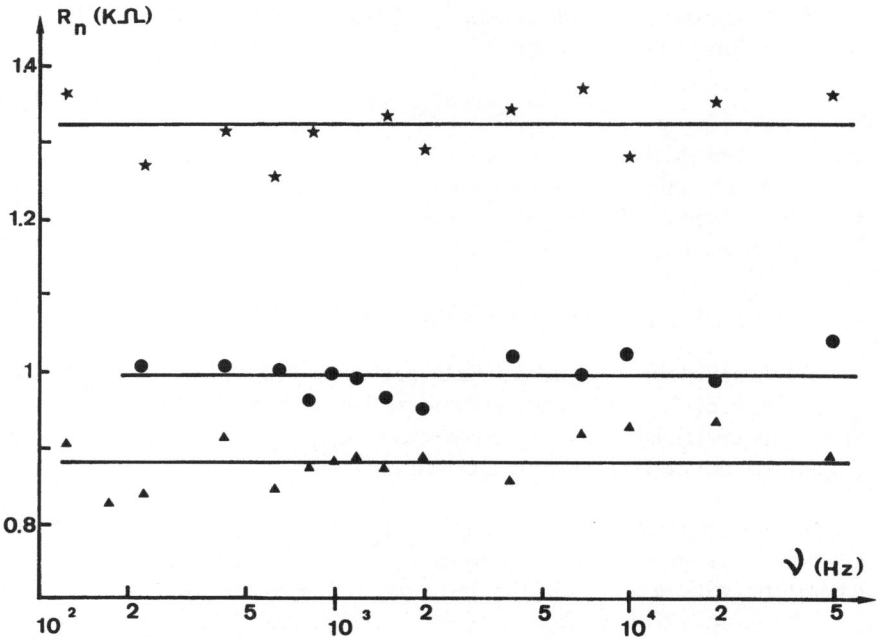

FIGURE 9.  Variation of noise resistance $R_n$ of a DNA solution in NaCl $4 \times 10^{-3}$ mol/liter, versus the frequency $v$. ★, native DNA (chicken erythrocyte at the concentration $C = 1$ g/liter at 25°C); ●, the same native DNA at 40°C; ▲, the same DNA after thermal transconformation (sample heated at 80°C) measured at 40°C.

When the fluctuation of bound counterions is expanded in a Fourier series, the resulting mean square dipole results as

$$\langle \mu^2 \rangle = e^2 L^2 \sum_K \left( \frac{L}{2\pi K} \right)^2 \langle C_K'^2 \rangle \qquad (45)$$

where $L$ is the length of the polyion, $K$ is the wave number,

$$C_K' = \frac{2}{L} \int_0^L \delta C_+ \sin \frac{2\pi K x}{L} \, dx \qquad (46)$$

$C_+$ is the average concentration of bound counterions, and $\delta C_+$ is the fluctuation of $C_+$ along the rodlike DNA, between $x$ and $x + dx$ positions.

The form of the $\langle \mu^2 \rangle$ equation shows that $\langle \mu^2 \rangle$ is proportional to the square of the wavelength $L/K$; i.e., fluctuations of the counterion atmosphere at long wavelengths (or low frequencies) make the main contribution to the mean square dipole moment.

This fact gives a possible explanation for the high value of the permittivity $\varepsilon$ observed in the low-frequency domain, where conventional dielectric measurement are carried on. Nevertheless, if this model introduces a possible dispersion phenomenon in the absence of an externally applied electric field $E$ (due to the various $K$ fluctuation modes of the counterions), this effect is too small to be observable by noise spectrography.

*4.1.2b. Thermal Transconformation.* The Arrhenius diagram in Figure 10 shows the variations of the noise resistance during the

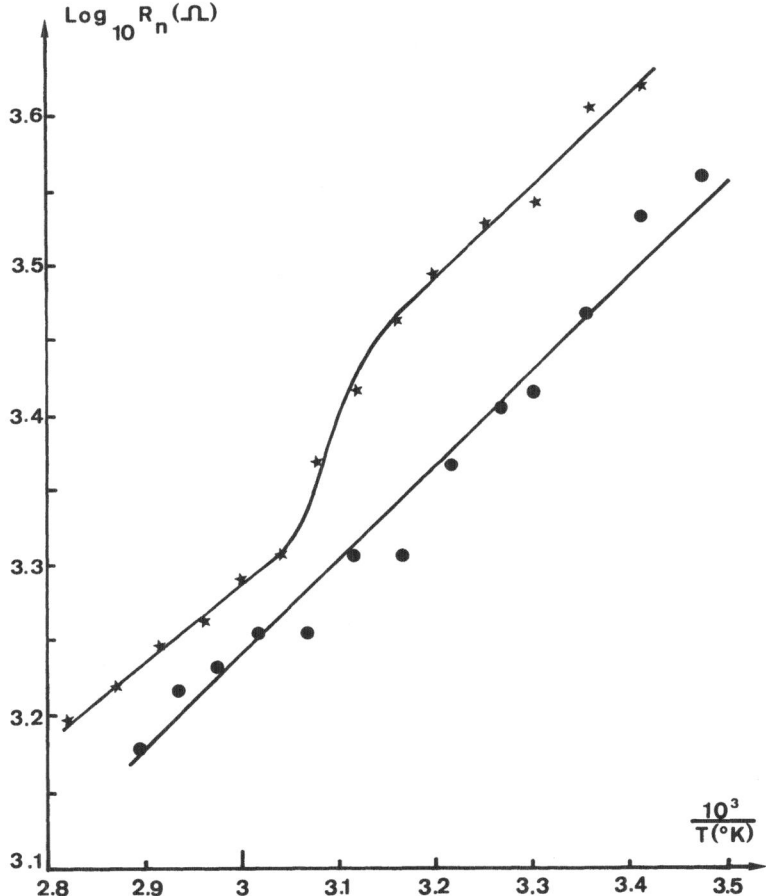

FIGURE 10. Arrhenius diagram of noise resistance for a solution of native DNA ($\star$) and thermally denatured ($\bullet$); the fixed noise frequency is $\nu = 1$ kHz.

helix → coil transition of a DNA solution. The breaks noted for a native DNA solution are characteristic of a phase transition.[65]

Within the regions preceding the melting point $T_m$ and following $T_m$, the $R_n(T)$ relation is of the form

$$R_n = R_0 \exp\left(\frac{E_G}{kT}\right) \tag{47}$$

The activation energies $E_G$ are, in first approximation, equal to the activation energy of a NaCl solution with a conductivity corresponding to that of DNA solution.

On the other hand, the increase in noise conductivity with $T$ for temperatures above the $T_m$ indicates that the number of free ions in solution increases also; this phenomenon corresponds to the ejection of a percentage $\alpha$ of the compensating $Na^+$ counterions from the phosphate sites, during the thermal transconformation.

The noise conductivity of a NaCl solution is

$$\sigma_n = (n_{Na^+} + n_{Cl^-})\, e\langle\mu_n\rangle \tag{48}$$

$\langle\mu_n\rangle$ is the mean noise mobility of the cations and anions since $n_{Na^+} = n_{Cl^-}$, we may write $\sigma_n = Ne\langle\mu_n\rangle$, with $N = n_{Na^+} + n_{Cl^-} =$ average number of free ions in the NaCl solution.

If the mean mobility $\langle\mu_n\rangle$ of the free ions in the DNA solution can be considered to be equal to that of the same ions in a NaCl solution of the same concentration, we may write

$$N = \frac{C}{e\langle\mu_n\rangle R_n} \tag{49}$$

where $C$ is a cell constant.

Figure 11 shows the variation of $N$ in two DNA solutions versus the temperature. Before and after the melting point $T_m$, we observe two horizontal plateaus; before and after the thermal transconformation, $N$ is constant.

The difference $\Delta N$ between the two flat parts corresponds to the increase in number of free ions in the DNA solution after the helix → coil transition. If the role of the coions is negligible (as predicted by the Manning model), the increase $\Delta N$ corresponds to an ejection of $Na^+$ counterions from the negative phosphate sites of DNA.

From Figure 11 we obtain

$\Delta N = \Delta n_{Na^+} = 0.363 \times 10^{24}$ ions/m³         for a calf thymus DNA solution

$\Delta N = \Delta n_{Na^+} = 0.463 \times 10^{24}$ ions/m³         for a chicken erythrocyte DNA solution

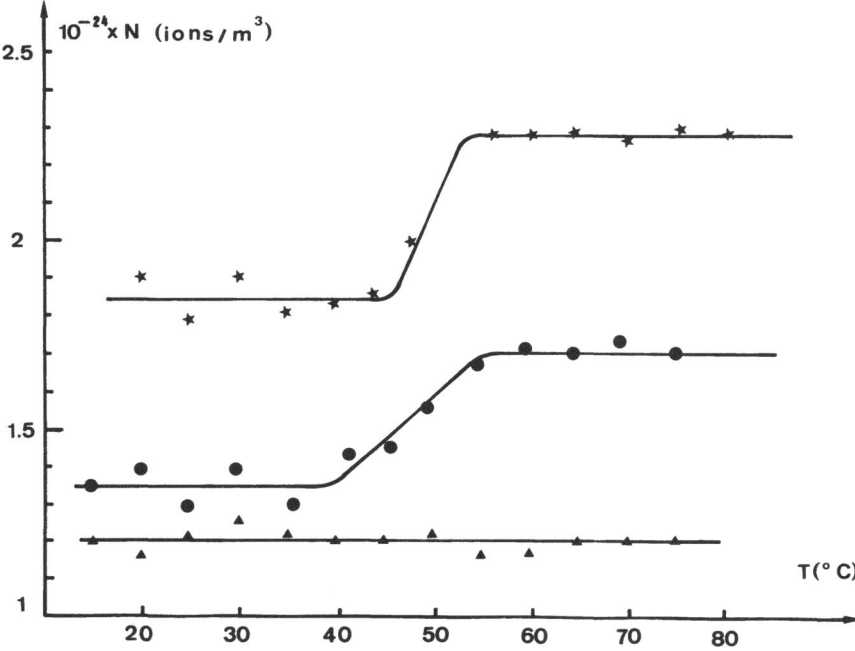

**FIGURE 11.** Variation of the number N of free ions in two solutions of DNA, as a function of temperature. ▲, NaCl $10^{-3}$ mol/liter; ●, DNA solution (calf thymus; $C = 0.6$ g/liter); ★, DNA solution (chicken erythrocyte; $C = 0.9$ g/liter).

The percentage $\Delta \psi$ of $Na^+$ ejected during the thermal transconformation is thus given by

$$\Delta \psi = 100 \frac{\Delta C_{Na^+}}{n_e} \tag{50}$$

with $C_{Na^+} = \Delta n_{Na^+}/N_0$, $N_0$ is Avogadro's number. In the case of our two solutions, we obtain the following values: $\Delta \psi = 33\%$ and $\Delta \psi = 27.5\%$. These values are in good agreement with those indicated for the extent of binding in Section 4.1.1.

*4.1.2c. Premelting Effect in DNA under "Salt-Free Conditions."* For the DNA solutions described above, the solvent contained an excess of counterions and thus the high conductivity of the electrolyte masks the observation of a premelting effect. Studies of DNA solutions under salt-free conditions, i.e., Na–DNA salt dissolved in pure distilled water, allow observation of this effect.

Figure 12 shows the Arrhenius diagram of noise conductivity $\sigma_n$ for DNA solutions under salt-free conditions. We observe a discontinuity at 23°C; this effect is not related to the released counterions because this temperature is below the DNA $T_m$. This premelting effect is related to the ionic sheath around the rodlike DNA. The Na$^+$ counterions in this sheath are less firmly bound to the polyion above the premelting point (23°C) than they are below that temperature.

This interpretation agrees with observations by Goswami and Das

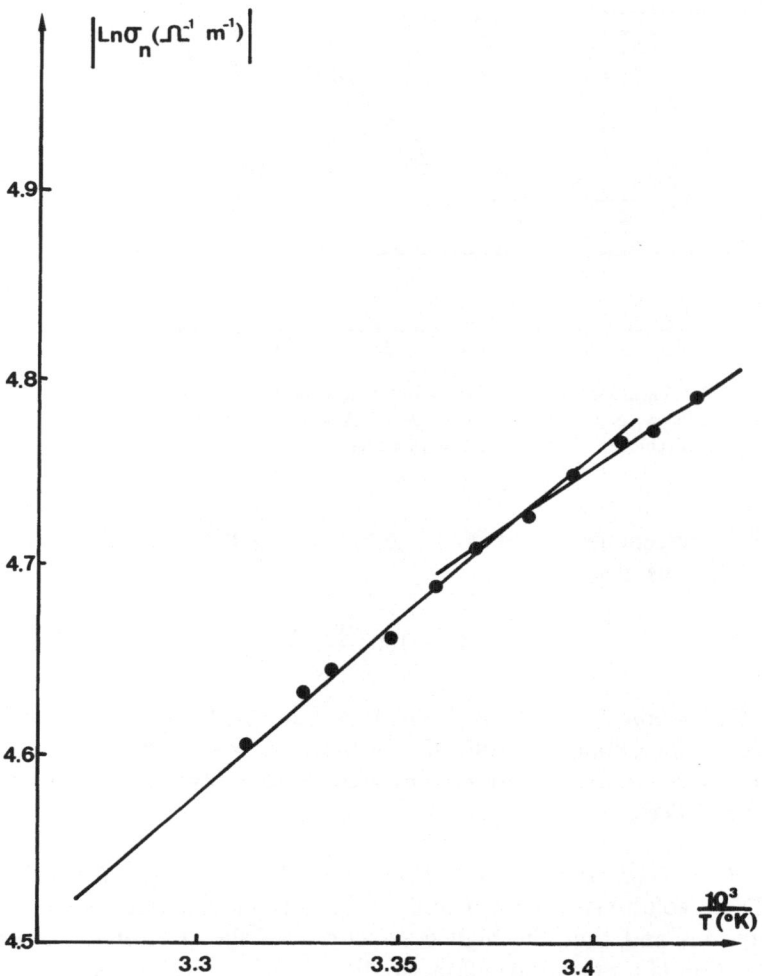

**FIGURE 12.** Arrhenius diagram of noise conductivity $\sigma_n$ for a calf thymus DNA solution ($C = 0.66$ g/liter) in "salt-free" conditions. The fixed noise analysis frequency is $\nu = 1$ kHz.

Gupta,[66] who report a dielectric relaxation in DNA salt-free solutions, between 100 Hz and 100 kHz, in which the relaxation times are different below and above 22.5°C.

The premelting effects observed in our noise measurements do not involve an electric dipole moment because of the absence of any applied field. The partial destabilization of the DNA counterionic sheath in the low-temperature region appears to be a characteristic of salt-free DNA solutions. The premelting process begins at the DNA periphery, without affecting the bases stacking or the coupling of the bases, and afterwards true melting takes place involving the helix → coil transconformation and ejection of the compensating counterions.

### 4.2. Demonstration of the Permanent Dipole Fluctuations of Collagen Molecules

### 4.2.1. Noise Conductivity Emission Spectra of Collagen Solutions

Collagen is the most important fibrous protein occurring in the animal kingdom. The collagen molecule—or tropocollagen—is a long cylindrical rod corresponding to a triple helical right-handed coiled-coil tertiary structure. Each of the three strands ($\alpha$ chain) is a left-handed helix containing about 1050 amino acid residues. The $\alpha$ polypeptide chain conformation is determined by the intervention of the regular tripeptidic unit ($Gly-R_2-R_3$) in which glycine takes a place in every third position; the two imino acids proline and hydroxyproline occur frequently in the $R_2$ and $R_3$ locations.[67]

The dipolar nature of amino acid intervening in molecular structure of collagen implies the existence of a permanent dipole $\mu$ orientated along the revolution axis of the cylindrical rod. By electrical birefringence measurements, Yoshioka and O'Konski[68] have predicted a value of about 15000 Debye for this important permanent dipole and an orientational relaxation time of 0.1–0.2 ms. These results are in agreement with the more recent birefringence experiments of Bernengo *et al.*[69,70] On the other hand, classical dielectric measurements on high conductive solutions (like collagen ones) and in the low-frequency range, are very difficult because they involve an important electrode polarizability and an induced dipole moment (due to the applied external electric field). The only known tentative dielectric experiments on collagen solutions are those of Hanss *et al.*[71] These authors have measured the dielectric dispersion in the high-frequency range from 20 kHz to 20 MHz and cannot see the relaxation due to the reorientation of collagen permanent dipoles situated in the low-frequency domain.

Actually, by noise spectrography measurements, electrode polarization effects are considerably eliminated and there is not an induced moment due to an external applied field.

We present here recently obtained results by measuring the emission noise spectrum—in the absence of applied electrical field—due to the molecular rotational Brownian motion of collagen molecules.[72]

Figure 13 shows the noise conductivity dispersion curve of a collagen solution (methanol + 5% water, $10^{-3}$ mol/liter HCl) prepared following the procedure proposed by Herbage.[73] A sigmoïdal conductivity curve is obtained when the solvent noise is substracted from the total measured noise conductivity.

To compare with classical dielectric relaxation studies, we have plotted the variation of the normalized noise conductivity $\sigma_{norm}$ of collagen solution versus the product $\omega\tau_N$ ($\tau_N$ is the critical time in noise measurements). Figure 14 shows that a very good agreement is obtained

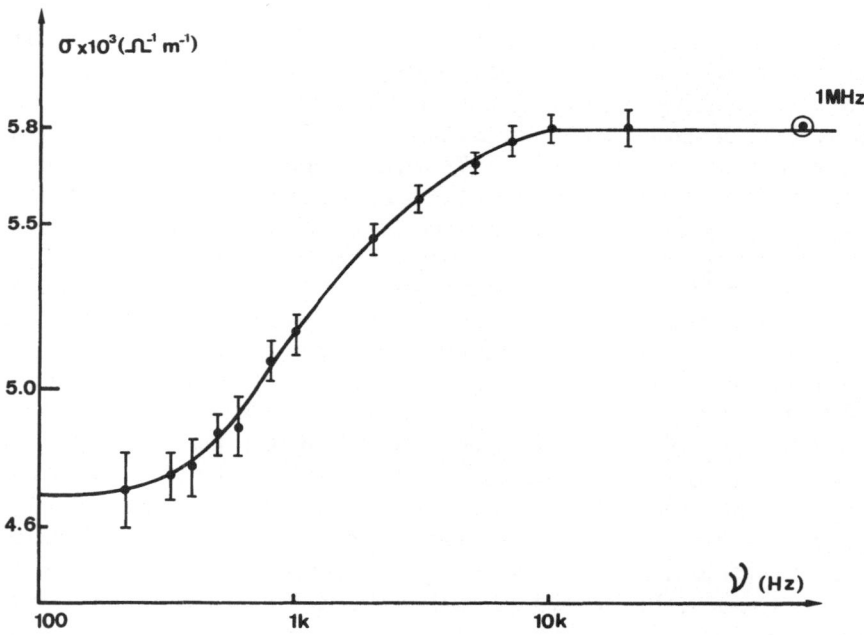

FIGURE 13. Noise conductivity of collagen (1 g/liter of collagen (1 g/liter in methanol + 5% water, $10^{-3}$ mol/liter HCl) versus frequency, at 25°C, after subtracting the noise conductivity of the solvent. Vertical bars correspond to the error measure. The 1-MHz data point was obtained by a bridge measurement.

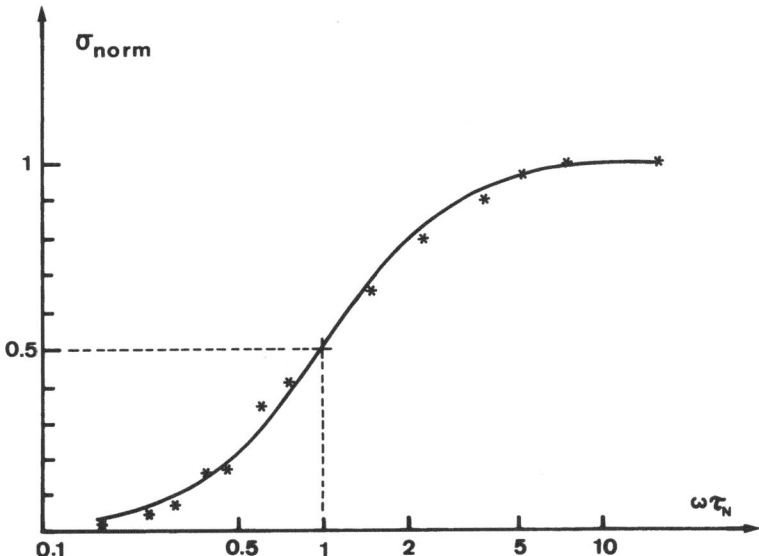

**FIGURE 14.** Normalized noise conductivity of collagen versus $\omega\tau_N$, at 25°C. The solid line represents the function $(1 + 1/\omega^2\tau_N^2)^{-1}$. The stars are the experimental values of $(\sigma - \sigma_s)/(\sigma_\infty - \sigma_s)$. $\sigma_s$ is determined at the lower-frequency plateau and $\sigma_\infty$ at the upper plateau.

between experimental data points with the first-order relaxation conductivity corresponding to the Debye equation

$$\frac{\sigma - \sigma_s}{\sigma_\infty - \sigma_s} = \left(1 + \frac{1}{\omega^2\tau_N^2}\right)^{-1} \tag{51}$$

where $\sigma_s$ and $\sigma_\infty$ are the conductivities for zero and infinite frequencies.

The measured noise critical time is $\tau_N \simeq 1.2 \pm 0.1 \times 10^{-4}$ s at 25°C and corresponds to a frequency $\nu = 1330$ Hz ($\omega\tau_N = 1$). Such an emission noise relaxation has been predicted in the case of polar molecules by several authors[74,75] and even measured experimentally for pure pentachlorodiphenyl by Le Bot *et al.*[76]

### 4.2.2. Power Spectrum of Permanent Dipolar Molecules

Let us consider a molecule with a permanent dipole moment $\boldsymbol{\mu}$; between time $t$ and $(t + \tau)$, the autocorrelation function of the moment is given by

$$\mathscr{R}_\mu(\tau) = \langle \boldsymbol{\mu}(t + \tau)\,\boldsymbol{\mu}(t) \rangle \qquad \text{with} \quad |\boldsymbol{\mu}| = \mu = \text{const} \tag{52}$$

In the plane rotator model developed by Scaife[77] (see Figure 15), $\mathcal{R}_\mu(\tau) = \mu^2 \langle \cos \Delta\theta \rangle$, it is possible to show that, when inertial effects are neglected (Debye approximation),

$$\mathcal{R}_\mu(\tau) = \mu^2 \exp(-\tau/\tau_r) \tag{53}$$

where $\tau_r$ is a critical time.

The power spectrum of the dipole moment

$$\mathcal{S}_\mu(\omega) = \frac{2\mu^2 \tau_r}{1 + \omega^2 \tau_r^2} \tag{54}$$

is similar to the $\mathcal{S}_v(\omega)$ expression, Eq. (26) (cf. Figure 15), but in this case the term $\omega^2 \tau_r^2$ is not negligible compared to unity. Otherwise, considering now a bistable system with permanent dipole moments fluctuating between two positions $+\mu$ and $-\mu$, the $\mathcal{R}_\mu(\tau)$ function is the same as in the plane

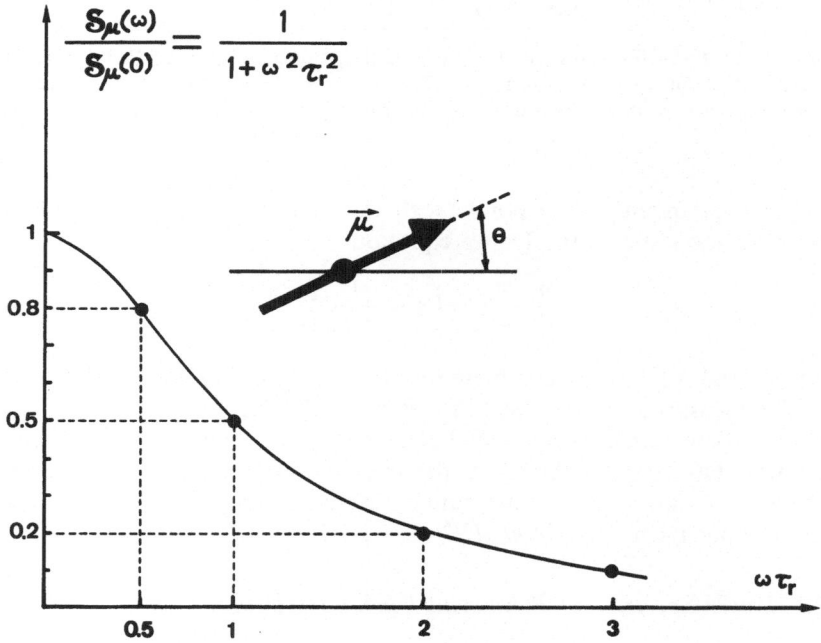

FIGURE 15. The rotator plane model of permanent dipole moment: variation of the normalized power spectrum of dipole moment orientation fluctuations

$$\frac{\mathcal{S}_\mu(\omega)}{\mathcal{S}_\mu(0)} = \frac{1}{1 + \omega^2 \tau_r^2}$$

rotator model where $\tau_r = 2\tau_e$ with $\tau_e$ the residence time in the $+\mu$ or $-\mu$ positions.[75]

The important fact is that the power spectrum $\mathscr{S}_\mu(\omega)$ is similar to the $\varepsilon'(\omega)$ Debye equation for the permittivity:

$$\varepsilon' - 1 = \frac{\varepsilon_s - 1}{1 + \omega^2 \tau^2} \tag{55}$$

### 4.2.3. Hydrodynamic Properties of Collagen Solutions:
### A Tentative Interpretation

If $D_\theta$ is the rotary diffusion coefficient of collagen molecule, we can introduce three different critical times:

$\tau_r = 1/D_\theta$     is the critical time appearing in the plane rotator phenomenological model.

$\tau_D = 1/2D_\theta$    is the dielectric relaxation time for a sphere, a prolate ellipsoid, and a cylinder.[78,79]

$\tau_B = 1/6D_\theta$    is the orientational relaxation time in electric birefringence measurements.

The theoretical $D_\theta$ value of collagen molecule approximated by a cylinder of radius $r$ and of length $2L$ may be obtained from a Perrin-like formula, demonstrated by Burgers:[80]

$$D_\theta = \frac{3kT}{8\pi\eta L^3}\left[\ln\left(\frac{2L}{r}\right) - 0.8\right] \tag{56}$$

in which $k$ is the Boltzmann constant, $T$ the absolute temperature, and $\eta$ the viscosity of the solvent.

A $D_\theta$ value of $1000 \pm 50 \text{ s}^{-1}$ has been determined from electric birefringence experiments, with collagen in water as solvent.[69,70] Thus, assuming a radius of $r \simeq 1$ nm, we can deduce a length of $2L \simeq 270$ nm for the collagen rod. In the case of our methanol–water collagen solutions, $\eta \simeq 0.65 \times 10^{-2}$ poise at 25°C and the calculated value of $D_\theta$ from Burgers formula is

$$D_\theta \simeq 1500 \text{ s}^{-1}$$

According to this $D_\theta$ value the three critical times described above can be calculated:

$$\tau_r = 6.70 \times 10^{-4} \text{ s}$$

$$\tau_D = 3.35 \times 10^{-4} \text{ s}$$

$$\tau_B = 1.15 \times 10^{-4} \text{ s}$$

We observe that the measured noise critical time $\tau_N \simeq 1.2 \times 10^{-4}$ s, is close to the electric birefringence $\tau_B$ value. This seems to be in agreement with our hypothesis of the noise conductivity spectra origin, i.e., the orientation fluctuations of collagen molecules bearing an electric permanent dipole in the absence of an electric applied field.

It is also noteworthy that $\tau_B$ is obtained in the birefringence technique after the suppression of an applied electric field; thus after a slight perturbation the molecules return to their spontaneous random fluctuations.

To demonstrate our assumption that $\tau_N$ is directly related to the hydrodynamical properties of the solution, other experiments have been carried out by varying the viscosity of the solvent: $\eta$ was decreased by heating the solution up to 40°C (see Figure 16). We observe that $\tau_N$ is shifted toward high frequencies. $\eta$ was increased by addition of glycerol at 25°C (see Figure 17). In this case, $\tau_N$ is shifted down to lower frequencies and the determination of the low-frequency plateau of sigmoïdal curve becomes impossible.

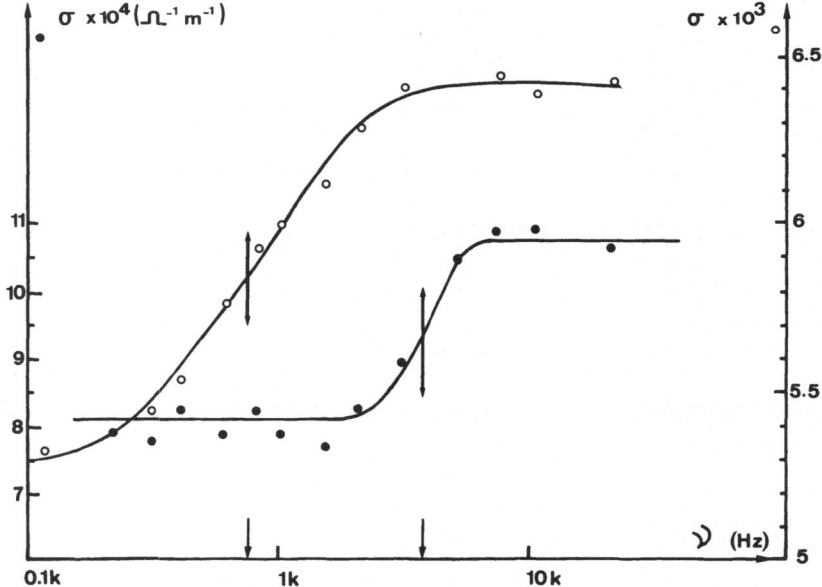

· *FIGURE 16.* Noise conductivity dispersion spectrum of collagen solution. Collagen concentration (2 g/liter in methanol + 5% water, $10^{-3}$ mol/liter HCl): ○, at 25°C; ●, at 40°C.

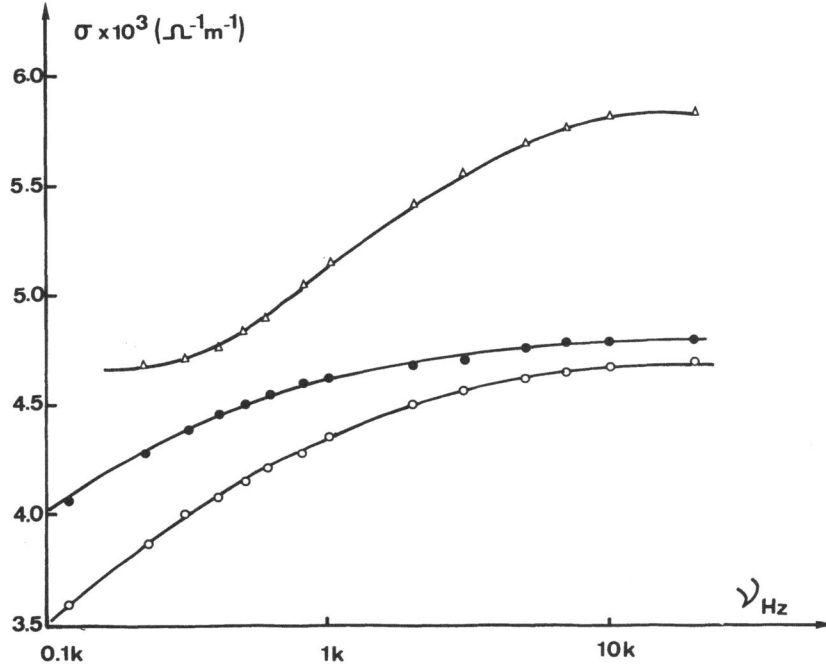

FIGURE 17. Noise conductivity dispersion of collagen solutions with glycerol added at 25°C. Collagen concentration (1 g/liter in methanol + 5% water, $10^{-3}$ mol/liter HCl). $\triangle$, 0% glycerol; $\bigcirc$, 10% glycerol; $\bullet$, 20% glycerol.

## 5. Conclusion

Noise spectrography is an efficient technique allowing observation of electrical phenomena in systems which are liable to be perturbed by the application of an external electric field. As a consequence, noise measurements are convenient for the study of electrochemical or bioelectrochemical systems such as electrolytes, biopolyelectrolytes, electrolyte–membrane interfaces, etc.

In the case of simple aqueous 1.1 electrolytes, we have discussed the origin of observed noise and the perturbation of the white noise power spectrum by interfacial effects. The electrolyte–electrode interface power spectrum corresponds to a $1/v$ noise.

The study of the membrane–electrolyte interfaces occurring in collodion membranes was focused on the corresponding complex power spectrum. The membrane–electrolyte interface power spectrum is not a simple $1/v$ noise but depends on electrolyte concentration. At infinite dilution of the electrolyte this power spectrum reverts to white noise.

For DNA solutions, the role of the counterionic atmosphere around the negative phosphate sites is critical. The general behavior of this ionic atmosphere may be readily studied by noise spectrography, and we have shown that

- DNA does not possess an inherent electric permanent dipole moment;
- the thermal transconformation affects directly the noise conductance because of the release of counterions from DNA phosphate sites;
- a premelting effect occurs in DNA solutions under salt-free conditions.

We have also demonstrated that it was possible to throw light on the orientation fluctuations of polar macromolecules like collagen, by measuring the noise emission conductivity $\sigma_n$ versus frequency. The critical time $\tau_N$ of orientation fluctuations of the collagen permanent dipoles cannot be assimilated to the classical dielectric relaxation time $\tau_D$, but the numerical value of $\tau_N$ is very near to the reorientation time $\tau_B$ measured by electrical birefringence.

From the theoretical point of view, no complete theories exist able to relate the microscopic orientation fluctuations of permanent dipoles to macroscopic dielectric properties of macromolecules in solution. A phenomenological understanding of the $\sigma_n$ shape is possible by considering the plane rotator model which possesses a sigmoidal power spectrum $\mathscr{S}_\mu(\omega)$ decreasing when the frequency is increased. On the hypothesis that, in the interval $(\omega, \omega + d\omega)$, the emitted noise by the fluctuating permanent dipoles corresponds to a noise resistance $R_n$ such as

$$\mathscr{S}_\mu(\omega) \propto R_n(\omega) \tag{57}$$

we can observe a noise conductivity $\sigma_n(\omega) \propto \mathscr{S}_\mu^{-1}(\omega)$; it is the shape we have found.

Besides, it is difficult to take into account such factors as dipole–dipole interactions, macromolecule–solvent interactions, and the local field. Thus, the interpretation of the noise conductivity dispersion amplitude seems difficult as far as the classical absorption dielectric relaxation is concerned.

*Acknowledgment*

The authors wish to acknowledge their gratitude to Professor F. Gutmann for the help given in many stimulating discussions.

# References

1. M. E. Green and M. Yafuso, "A Study of the Noise Generated During Ion Transport Across Membranes," *J. Phys. Chem.* **72**, 4072–4078 (1968).
2. M. E. Green, "Noise Spectra Across an Anion Membrane," *J. Phys. Chem.* **78**, 761–762 (1974).
3. D. L. Dorset and H. M. Fishman, "Excess Electrical Noise During Current Flow Through Porous Membranes Separating Ionic Solutions," *J. Membr. Biol.* **21**, 291–309 (1975).
4. P. G. Saffman and M. Delbrück, "Brownian Motion in Biological Membranes," *Proc. Natl. Acad. Sci. USA* **72**, 3111–3113 (1975).
5. B. Neumcke, "1/f Noise in Membranes," *Biophys. Struct. Mechanism* **4**, 179–199 (1978).
6. E. Frehland, "Current Noise Around Steady States in Discrete Transport Systems," *Biophys. Chem.* **8**, 255–265 (1978).
7. L. J. Bruner and J. E. Hall, "Autocorrelation Analysis of Hydrophobic Ion Current Noise in Lipid Bilayer Membranes," *Biophys. J.* **28**, 511–514 (1979).
8. E. Frehland, "Current Fluctuations in Discrete Transport Systems far from Equilibrium. Breakdown of the Fluctuation Dissipation Theorem," *Biophys. Chem.* **12**, 63–71 (1980).
9. P. C. Jordan, "Current Noise in Transport of Hydrophobic Ions Through Lipid Bilayer Membranes Including Diffusion Polarization in the Aqueous Phase," *Biophys. Chem.* **12**, 1–11 (1980).
10. Y. Fang, Q. Li, and M. E. Green, "Noise Spectra of Transport at an Anion Membrane–Solution Interface," *J. Colloïd Interface Sci.* **86**, 185–190 (1982).
11. Y. Fang, Q. Li, and M. E. Green, "Noise Spectra of Sodium and Hydrogen Ion Transport at a Cation Membrane–Solution Interface," *J. Colloïd Interface Sci.* **88**, 214–220 (1982).
12. Y. Chen, "Differentiation Between Equilibrium and Nonequilibrium Kinetic Systems by Noise Analysis," *Biophys. J.* **21**, 279–285 (1978).
13. A. A. Verveen and H. E. Derksen, "Fluctuation Phenomena in Nerve Membrane," *Proc. IEEE* **56**, 906–916 (1968).
14. D. J. M. Poussart, "Nerve Membrane Current Noise: Direct Measurements Under Voltage Clamp," *Proc. Natl. Acad. Sci. USA* **64**, 95–99 (1969).
15. D. J. M. Poussart, "Membrane Current Noise in Lobster Axon Under Voltage Clamp," *Biophys. J.* **11**, 211–234 (1971).
16. H. Lecar and R. Nossal, "Theory of Threshold Fluctuations in Nerves I. Relationships Between Electrical Noise and Fluctuations in Axon Firing," *Biophys. J.* **11**, 1048–1067 (1971); "II. Analysis of Various Sources of Membrane Noise," *Biophys. J.* **11**, 1068–1084 (1971).
17. E. Wanke, L. J. De Felice, and F. Conti, "Voltage Noise, Current Noise and Impedance in Space Clamped Squid Giant Axon," *Pflügers Arch.* **347**, 63–74 (1974).
18. H. Fishman, "Patch Voltage Clamp of Squid Axon Membrane," *J. Membr. Biol.* **24**, 265–277 (1975).
19. H. Fishman, D. J. M. Poussart, and L. E. Moore, "Noise Measurements in Squid Axon Membrane," *J. Membr. Biol.* **24**, 281–304 (1975).
20. J. R. Clay and M. F. Shlesinger, "Theoretical Model of the Ionic Mechanism of 1/f Noise in Nerve Membrane," *Biophys. J.* **16**, 121–136 (1976).
21. F. Conti, L. J. De Felice, and E. Wanke, "Potassium and Sodium Ion Current Noise in the Membrane of Squid Giant Axon," *J. Physiol.* **248**, 45–82 (1975).
22. H. M. Fishman, L. E. Moore, and D. J. M. Poussart, "Potassium-Ion Conduction Noise in Squid Axon Membrane," *J. Membr. Biol.* **24**, 305–328 (1975).
23. F. Conti, B. Neumcke, W. Nonner, and R. Stämpfli, "Conductance Fluctuations from the Inactivation Process of Sodium Channels in Myelinated Nerve Fibres," *J. Physiol.* **308**, 217–239 (1980).

24. F. J. Sigworth, "Interpreting Power Spectra from Nonstationnary Membrane Current Fluctuations," *Biophys. J.* **35**, 289–300 (1981).
25. A. A. Verveen and L. J. De Felice, "Membrane Noise," *Progr. Biophys. Mol. Biol.* **28**, 189–265 (1974).
26. P. Laüger, R. Benz, G. Stark, E. Bamberg, P. C. Jordan, A. Fahr, and W. Brock, "Relaxation Studies of Ion Transport Systems in Lipid Bilayer Membranes," *Quart. Rev. Biophys.* **14**, 513–598 (1981).
27. L. J. De Felice, *Introduction to Membrane Noise*, Plenum Press, New York (1981).
28. F. Conti and E. Wanke, "Channel Noise in Nerve Membranes and Lipid Bilayers," *Quart. Rev. Biophys.* **8**, 451–506 (1975).
29. Y. Chen, in *Advances in Chemical Physics* (I. Prigogine and S. A. Rice, eds.), Vol. XXXVII, pp. 67–97, Wiley, New York (1978).
30. A. Papoulis, *Probability, Random Variables and Stochastic Processes*, Mc-Graw Hill, New York (1965).
31. G. R. Cooper and D. C. McGillem, *Methods of Signal and System Analysis*, Holt, Rinehart and Winston, New York (1967).
32. H. Kranck, doctoral thesis, Nice University (1979).
33. J. B. Johnson, "Thermal Agitation of Electricity in Conductors," *Phys. Rev.* **32**, 97–109 (1928).
34. H. Nyquist, "Thermal Agitation of Electric Charge in Conductors," *Phys. Rev.* **32**, 110–113 (1928).
35. D. Vasilescu, M. Teboul, H. Kranck, and F. Gutmann, "Electrical Noise in Aqueous 1.1 Electrolytes," *Electrochim. Acta* **19**, 181–186 (1974).
36. D. Vasilescu, "Sur une notation rationalisée des paramètres caractéristiques d'un électrolyte 1.1 en solution," *J. Chim. Phys.* **7–8**, 1131–1132 (1974).
37. L. Landau and F. Lifchitz, *Physique Statistique*, Mir, Moscow (1976), p. 349.
38. R. Kubo, "Statistical-Mechanical Theory of Irreversible Processes." I. General Theory and Simple Applications to Magnetic and Conduction Problems, *J. Phys. Soc. Jpn* **12**, 570–586 (1957).
39. H. Kranck, D. Vasilescu, C. Bezot, G. Bossis, M. Teboul, and F. Gutmann, "Bruit électrique dans les électrolytes 1.1 aqueux. 2. Cas des fortes concentrations. Bruit d'interface," *Electrochim. Acta* **23**, 891–897 (1978).
40. Y-C. Chiu and R. M. Fuoss, "Conductance of the Alkali Halides." XII. Sodium and Potassium Chlorides in Water at 25°, *J. Phys. Chem.* **72**, 4123–4129 (1968).
41. J. O'M. Bockris and A. K. N. Reddy, *Modern Electrochemistry*, Vol. 2, Plenum Press, New York (1970).
42. H. Kranck, C. Bezot, M. Teboul, and D. Vasilescu, "Bruit électrique d'une membrane au Collodion—Effets d'interface," *Electrochim. Acta* **24**, 939–947 (1979).
43. J. J. Brophy and S. L. Webb, "Critical Fluctuations in Triglycine Sulfate," *Phys. Rev.* **128**, 584–588 (1962).
44. H. P. Schwan, "Alternating Current Electrode Polarization," *Biophysik* **3**, 181–201 (1966).
45. F. N. Hooge, "1/f Noise in the Conductance of Ions in Aqueous Solutions," *Phys. Lett.* **33**, 169–170 (1970).
46. F. N. Hooge, "Discussion of Recent Experiment in 1/f Noise," *Physica* **60**, 130–144 (1972).
47. A. Cyrot, doctoral thesis, Université Pierre et Marie Curie, Paris (1978).
48. L. J. De Felice and J. P. L. M. Michalides, "Electrical Noise from Synthetic Membranes," *J. Membr. Biol.* **9**, 261–290 (1972).
49. H. P. Gregor and K. Sollner, "Improved Methods of Preparation of "Permselective" Collodion Membranes Combining Extreme Ionic Selectivity with High Permeability," *J. Phys. Chem. Ithaca* **50**, 53–70 (1946).

50. Y. Gur, I. Ravina, and A. J. Babchin, "A Numerical Method for Solving a Generalised Poisson–Boltzmann Equation," *J. Colloïd Interface Sci.* **64**, 326–332 (1978).
51. Y. Gur, I. Ravina, and A. J. Babchin, "The Poisson–Boltzmann Equation Including Hydratation Forces," *J. Colloïd Interface Sci.* **64**, 333–341 (1978).
52. D. Vasilescu, H. Grassi, and M. A. Rix-Montel, in *Polyelectrolytes and their Applications* (A. Rembaum and E. Selegny, eds.), Reidel, Dordrecht (1975), Vol. 2, pp. 197–216.
53. G. S. Manning, "Limiting Laws and Counterion Condensation in Polyelectrolyte Solutions." I. "Colligative Properties," *J. Chem. Phys.* **51**, 924–933 (1969); II. "Self-Diffusion of the Small Ions," *J. Chem. Phys.* **51**, 934–938 (1969).
54. M. T. Record, C. F. Anderson, and T. M. Lohman, "Thermodynamic Analysis of Ion Effects on the Binding and Conformational Equilibria of Proteins and Nucleic Acids: The Roles of Ion Association or Release, Screening, and Ion Effects on Water Activity," *Quart. Rev. Biophys.* **11**, 103–178 (1978).
55. D. Vasilescu and M. A. Rix-Montel, "Mesure directe de l'éjection de cations Na$^+$ hors des sites phosphates lors de la dénaturation thermique du DNA," *Biochim. Biophys. Acta* **199**, 553–555 (1970).
56. M. A. Rix-Montel, H. Grassi, and D. Vasilescu, "Experimental Studies of Thermal Denaturation of the Na–DNA System with Respect to Manning's Model," *Biophys. Chem.* **2**, 278–289 (1974).
57. B. G. Archer, C. L. Craney, and H. Krakauer, "The Interaction of Na Ions with Synthetic Polynucleotides," *Biopolymers* **11**, 781–809 (1972).
58. D. Vasilescu, M. Teboul, R. Viani, and H. Grassi, "Étude expérimentale du bruit de fond dans des solutions de DNA," *C. R. Acad. Sci. Paris* **B266**, 1005–1008 (1968).
59. R. Viani, 3d cycle doctoral thesis, Nice University (1969).
60. B. Camous, 3d cycle doctoral thesis, Marseille University (1972).
61. D. Vasilescu, M. Teboul, H. Kranck, and B. Camous, "Showing up the Thermal Transconformation of Na–DNA in Solution by Noise Spectrography," *Biopolymers* **12**, 341–352 (1973).
62. H. Grassi, M. A. Rix-Montel, H. Kranck, and D. Vasilescu, "Premelting Effects in DNA Under Salt-Free Conditions," *Biopolymers* **14**, 2525–2535 (1975).
63. H. Kranck, C. Bezot, M. A. Rix-Montel, and D. Vasilescu, "Noise Conductance of DNA Under Salt-Free Conditions," *Biopolymers* **15**, 599–603 (1976).
64. F. Oosawa, "Counterion Fluctuation and Dielectric Dispersion in Linear Polyelectrolytes," *Biopolymers* **9**, 677–689 (1970).
65. J. Kumamoto, J. Raison, and J. Lyons, "Temperature "Breaks" in Arrhenius Plots: A Thermodynamic Consequence of a Phase Change," *J. Theor. Biol.* **31**, 47–51 (1971).
66. D. N. Goswami and N. N. Das Gupta, "On the Dielectric Polarization of DNA," *Biopolymers* **13**, 1549–1556 (1974).
67. G. N. Ramachandran and C. Ramakrishnan, in *Biochemistry of Collagen*, (G. N. Ramachandran and A. H. Reddi, eds.), Plenum Press, New York (1976), pp. 45–84.
68. K. Yoshioka and C. T. O'Konski, "Dipole Moment, Polarizability and Optical Anisotropy Factor in Collagen in Solution from Electric Birefringence," *Biopolymers* **4**, 499–507 (1966).
69. J. C. Bernengo, B. Roux, and D. Herbage, "Electrical Birefringence Study of Monodisperse Collagen Solutions," *Biopolymers* **13**, 641–647 (1974).
70. J. C. Bernengo, B. Roux, and D. Herbage, in *Electro-Optics and Dielectrics of Macromolecules and Colloids* (B. R. Jennings, ed.), Plenum Press, New York (1979), pp. 219–230.
71. M. Hanss, D. Herbage, and P. Comte, "Propriétés électriques du collagène en haute fréquence," *J. Chim. Phys.* **65**, 176–181 (1968).

72. H. Kranck, J. C. Bernengo, and D. Vasilescu, *Appl. Phys. Commun.* **2**(3), 189–202 (1982–83).
73. D. Herbage, doctoral thesis, Lyon (1972).
74. L. Davis, "Spontaneous Polarization Noise in Polar Dielectrics," *J. Appl. Phys.* **35**, 2004–2010 (1964).
75. Vera V. Daniel, *Dielectric Relaxation*, Academic, New York (1967), pp. 46–64.
76. J. Le Bot, E. Riaux, G. Grosvald, and R. Ollivier, "Analyse du spectre d'émission d'un diélectrique polaire," *J. Phys. (Paris)* **28**, 47–50 (1967).
77. B. K. P. Scaife, in *Complex Permittivity* (B. K. P. Scaiffe, compiler); The English University Press, London (1971), pp. 23–31.
78. F. Perrin, "Mouvement brownien d'un ellipsoide (I). Dispersion diélectrique pour des molécules ellipsoidales," *J. Phys. Radium* **5**, 497–511 (1934).
79. C. Tanford, *Physical Chemistry of Macromolecules*, Wiley, New York (1961), pp. 432–437.
80. J. M. Burgers, *Verhandel Koninkl. Ned. Akad. Wetenschap* **16**, 113–119 (1938).

# 15

# Cellular Spin Resonance (CSR)

## Hiram Rivera and Herbert A. Pohl

*ABSTRACT:* Small objects such as suspended live cells, organelles, tissue fragments, or even inanimate powder particles may be made to spin in an electromagnetic field. The spinning occurs in a resonant response to the applied frequency and reflects the dielectric properties (permittivity) of the suspension. There are three special cases: Spinning (1) in a static (dc) field, (2) in a simple oscillatory field, and (3) in a rotating field. The theory and examples for several interesting cases and their probable mechanisms are presented. The technique of cellular spin resonance (CSR) has several interesting applications. It sensitively detected alterations in surface properties due to a polyelectrolyte at concentrations of ca. 100 ppb. The CSR spectra of cells reflect their type and physiological state. Data to date indicate that live cells spin oppositely from dead ones at some frequency, even in mixtures. The dielectric properties of tiny particles can be readily determined.

## 1. Introduction

Living cells and other small particles in suspension can be observed to spin while in the presence of an alternating current (ac) electric field. The cellular spin rates phenomenon can be correlated with various characteristics such as cell age, culture age, health of the cell (normal versus tumor, etc), and cell type. The spinning of inanimate particles can be informative as to their dielectric properties. The present account describes the studies being done to reach an understanding of this exciting new technique.

*Hiram Rivera* • Physics Department, Oklahoma State University, Stillwater, Oklahoma 74078. *Herbert A. Pohl* • National Magnet Laboratory, Massachusetts Institute of Technology, Building NW-14, Cambridge, Massachusetts 02139.

## 2. Cellular Spinning

### 2.1. In a Static Field

Cells or other small particles may be observed to spin in an electric field that is either static, oscillatory, or rotating in direction.[1-6] The spinning of more or less spherical bodies in a *static* field has been known for some time. The theory and confirming experiments for that type due to surface charge deposition by ambipolar (bidirectional) current have been given.[4]

### 2.2. In a Simple Oscillatory Field

The spinning of cells while in a two-pole *oscillatory* field has been reported by a number of researchers.[1-12] The first account of *resonant* spinning was that of Pohl and Crane.[3] Their observations stemmed from the dielectrophoretic studies of baker's yeast, *Saccharomyces cerevisiae*. These studies were done while the cells were in between two parallel wire electrodes and subject to an ac field. The cells spun in a sharply resonant manner, in that each cell responded and spun only at a rather sharply defined specific applied frequency. Typically the cells were seen to spin at about 0.1–10 Hz while the applied field might be oscillating at, say 10–10,000,000 Hz. The cells were observed to spin rapidly either against one or the other of the electrodes or even out in the suspension. The frequency being applied could be adjusted so as to stop the spinning of some cells while starting others, and to slow the spinning rate or even to change the direction of rotation. Later studies found that the spin rate was proportional to the intensity of the applied field (Figure 1). The sharply resonant nature of the spinning response led to the use of the descriptive term "cellular spin resonance" (CSR). Since then, various investigators have quantitated the spin rate of budding yeast cells,[8] and the characteristic CSR of various cell lines, including human erythrocytes, Friend cells, and mesophyll protoplasts of *Avena sativa*.[11]

Cellular spin resonance in the simple oscillatory field can be broken down into two major types: one, a common type in which cells spin while in close proximity to another cell or other polarizable objects; and second, a rarer type in which cells spin while *alone* in suspension or against a smooth electrode. The observation of lone cells spinning out in the *middle* of the suspension is a rarer event than that of spinning alone against a smooth electrode or while interacting with other cells. Even so, there have been several research groups' reports observing and studying this event.[3,12-14] The fact that lone cell rotation is a rarer event can be attributed in part to dielectrophoresis.[4] Dielectrophoresis (DEP) is the

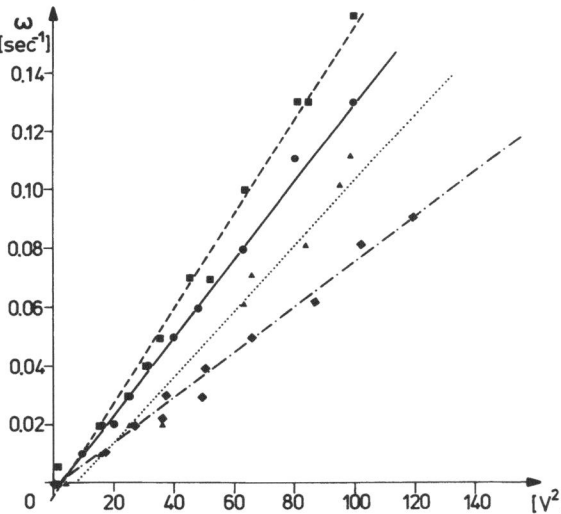

**FIGURE 1.** Spin rate of yeast *(Saccharomyces cerevisiae)* in a four-pole rotating electric field as a function of the square of the applied voltage on electrodes with a 1-mm gap. The frequency of the applied field was 60 kHz. Measurements of the spin rate $(w_c)$ for cells in various concentrations of sucrose in water. (Squares, circles, triangles, and diamonds designate data for 0, 100, 200, and 300 g sucrose per liter, respectively). The resistivity of the solutions was adjusted to 133 kΩ cm. The cells examined were from 10-day-old culture, and were classified as 98% viable by methylene blue stain test.

motion of bulk or particulate matter induced by the action of nonuniform electric fields. The movement of the particle will be towards the region of highest field intensity (positive dielectrophoresis) if the effective dielectric constant of the particle (or cell) exceeds that of the suspending medium. Conversely, the movement of the particle will be away from the region of higher intensity if the particle is of a lower dielectric constant of the medium. Normally, cells perturb the field and create a region of strong field intensity nearby, thereby attracting and linking other cells and forming "pearl chains." Because of this phenomenon of "mutual dielectrophoresis," any cell that is alone out in a suspension will tend to be attracted to other cells, thus preventing prolonged close examination of lone cell rotation. Also, there exist physical contraints such as thermal upsets, ionic injection, etc., causing field streaming.

There exists however, another reason for the rarity of observation of lone cell rotation in a two-pole field. This laboratory has recently done multiple field studies with *Saccharomyces cerevisiae* at dilute concentrations where only a few (2–10) cells were present in a *rotating* field provided by a four-pole electrode arrangement (Figure 2). Once a lone cell was observed

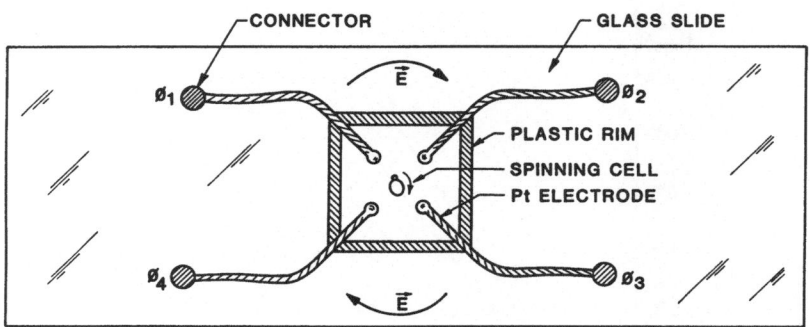

*FIGURE 2.* The four-electrode CSR chamber. The distance between opposed (i.e., N–S or E–W) tips is 1.2 mm. The inner least width of the well is 8.2 mm. It is 1.0 mm deep. The Pt tips are ca. 130 $\mu$m in diameter on 75-$\mu$m-diam Pt wire. All are mounted on a standard glass microscope slide.

to rotate freely in the rotating field, an ac field was then applied to only *two* poles while the remaining two were shorted to avoid field induction. Only a small percentage of the cells (ca. 1%) are observed to be capable of spinning in the two-pole field in this arrangement while out in suspension. Furthermore, the type of cells so far observed to be spinning while under these conditions was at the stage in its life just before the splitting of the mother cell into two daughter cells. The frequency of the applied field at which the cells were observed to rotate was between 10 and 100 kHz. This raises the question, is lone cell rotation linked to a particular stage of the life cycle?

What are the possible explanations for lone cell rotation in a simple oscillating field? The most compelling reason seems to be that of an internal dipolar oscillation within the cell.[15] This oscillation would be present only with live cells since upon the death of the cell spinning ceases. This seems to be supported by the fact that the cells spin at a much lower rate than that being applied by the external field. The cell rotates at somewhere between 0.1 and 30.0 Hz, while the external field oscillates at, say, 600 kHz. The presence of an internal dipolar field oscillation would interact with the externally applied field to provide a rotational torque and thus induce the spinning.

The cellular oscillations are not necessarily dipolar, but may oscillate as linear quadrupoles or higher multipoles. In view of the relatively weak character of the cellular oscillations it would also be expected that the externally applied oscillatory fields would serve to "pull" or change the frequency into resonance with that of the applied field. This would cause the CSR spectrum to be broadened by the external frequency pulling.

### 2.3. In a Rotating Field

Finally, there is the rotation of cells and other particles in *rotating* electric fields. If, for example, three or more electrodes are arranged in a ring and pulsed sequentially to produce a rotating electromagnetic field in the intervening space, this produces a polarization on a particle in the midregion. Moreover, this polarization takes a finite time to establish. The angular lag between the direction of the dipole thus created and the direction of the exterior rotating field now gives rise to a torque. The spin of the particle can thus be correlated with the field- and frequency-dependent dielectric properties, or permittivity. The CSR technique has potentially broad applications in minerological, as well as biophysical and medical problems.

Experimental evidence shows that there are correlations between the physiological state of the cell and its CSR spectrum. A comparison of the CSR of normal and cancerous fibroblasts was shown in an earlier paper.[16]

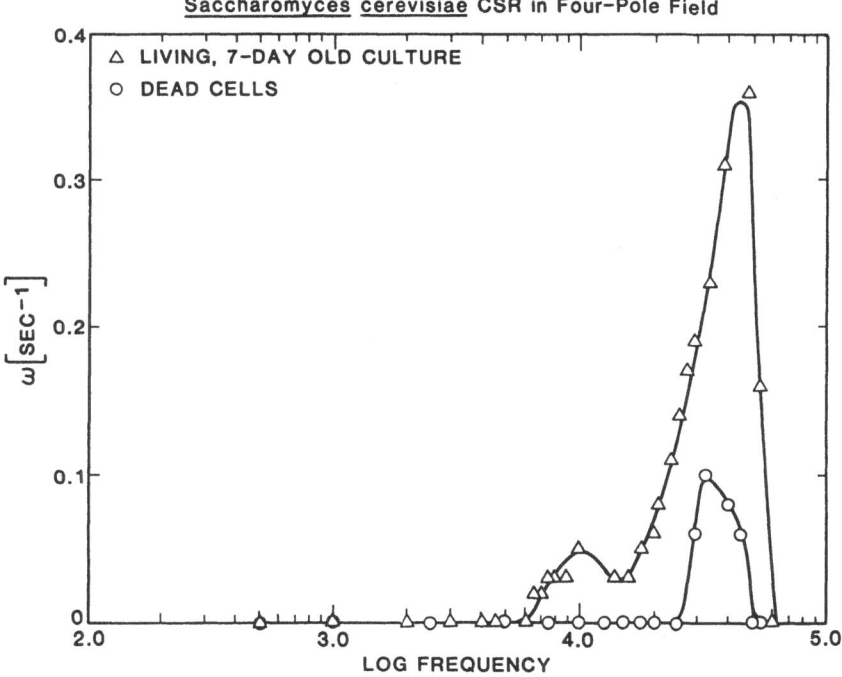

FIGURE 3. Spin rate spectra of living (triangles) and dead (circles) yeast *(Saccharomyces cerevisiae)*. The live cells were from a 7-day-old culture; the dead cells were heat-killed by exposure to 70°C for 3 min. The applied voltage was 10 V p–p, and the resistivities of the suspensions were 250 to 460 k$\Omega$ cm.

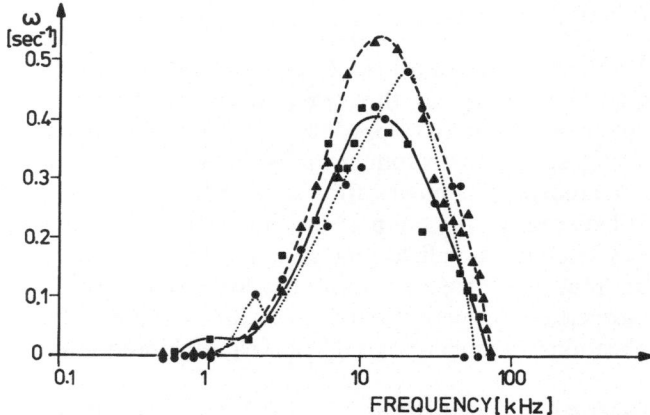

FIGURE 4. Dependence of the CSR spectrum of yeast *(Saccharomyces cerevisiae)* upon the colony age. (Circles, triangles, and squares refer to data for colony ages of 2, 6, and 8 days, respectively.) Note the shift of the 2- and the 20-kHz peaks to lower frequency as the colony age increases.

Figure 3 shows the difference, for example, between a live yeast cell and a dead one. It may turn out that the technique of CSR at a particular low frequency range will be useful as a "vital" test of cells. Each cell type has its own characteristic spectrum which identifies its state of being, in this case live or dead. Figure 4 illustrates the dependency on colony age. Figures 5–7 show several examples of CSR spectra for different cell lines.

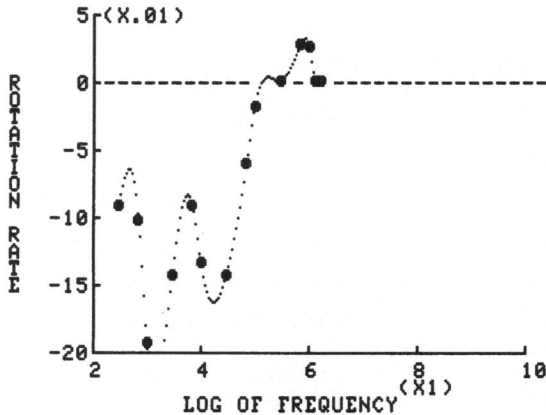

FIGURE 5. The CSR spectrum of 1-day-old culture of bovine kidney cells.

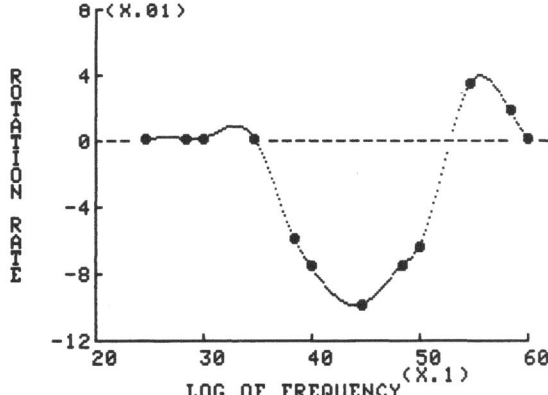

*FIGURE 6.* The CSR spectrum of CRFK (Crandall feline kidney) cells from a 4-day-old culture.

## 3. Particle Spinning

Inanimate particles can also spin while in the presence of ac electrical fields. The use of particles provides a model with which to test theories on spin resonance without having to be concerned with the ever-changing state of live cells. As can be seen from Figure 8, the conductivity of the solution is a critical factor in determining an accurate CSR. This is especially true at lower frequencies.

The general assumption made about particles is that they will spin in

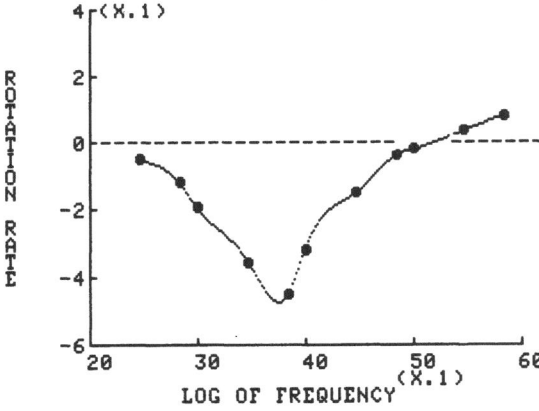

*FIGURE 7.* The CSR spectrum of green monkey (VERO) kidney cells from a 5-day-old culture.

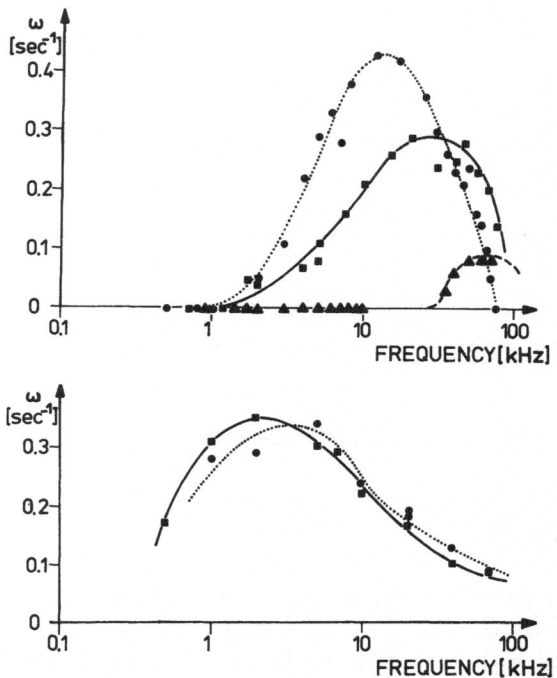

FIGURE 8. Effect of the conductivity of the suspension upon the CSR spectrum of single yeast *(Saccharomyces cerevisiae)* cells bearing a small bud and from a 6-day-old colony. Voltage 10 V p-p. Spinning was counterfield in direction. Dotted curve, cells in pure water, 2.4 μS/cm. Solid curve, cells in 8.9 μS/cm. Dashed curve, cells in 0.025 μS/cm. The conductivity was adjusted by adding NaCl. Note that the effect of increasing the conductivity is to shift the peaks to a higher frequency.

the direction of the ac field if they are more polarizable than the medium they are suspended in, and will spin against the field if they are less polarizable. In our case, if spinning with the field is clockwise, it is denoted by a ( + ) value, and if spinning counterclockwise it is denoted with a ( − ), that is, spinning against the rotation of applied field. As model particles of high polarizability we have used crystals of $BaTiO_3$ (ca. 2000) and of a low polarizability those of $BaSO_4$ (ca. 40). Figure 9 shows the CSR spectrum of $BaTiO_3$ while in pure water. It can be observed that the particles follow the field as the frequency increases until approximately 1000 kHz. Thereafter it crosses over to the negative region of spinning until about 600 kHz, where it again goes back to the positive side of the spectrum. $BaSO_4$ in pure water, conversely, is not observed to spin at any frequency. As seen by

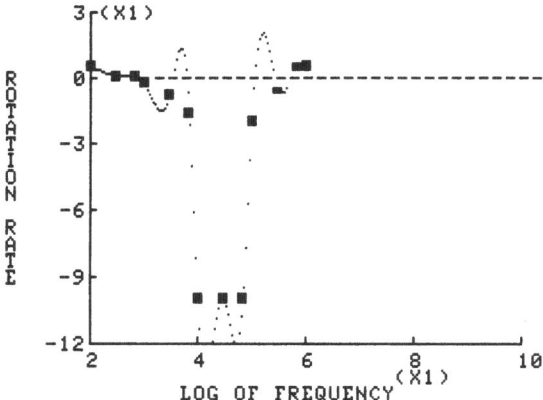

FIGURE 9. The CSR spectrum of barium titanate particles in high purity (5 μmho/cm) water.

Figures 10 and 11 the addition of Darvan No. 7, a polymeric polyelectrolyte based upon polyacrylic acid (manufactured by R. T. Vanderbilt Company, Inc.), is shown to alter the CSR spectrum of both the $BaTiO_3$ and $BaSO_4$. Almost incredibly low concentrations of it suffice to affect the CSR of suspended particles. It appears that concentrations as low as 100 parts per billion will substantially alter the CSR spectrum. This observation most readily points out two things: first, that the CSR spectrum technique

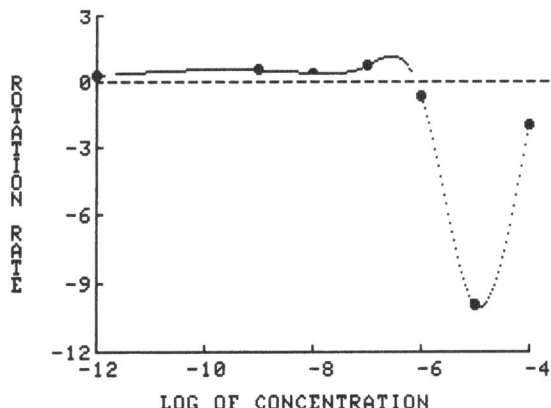

FIGURE 10. The CSR of barium titanate as affected by polyacrylate polyanions. The particle spin rate when driven by a rotating field at 600 kHz is seen to be affected even by very dilute solutions.

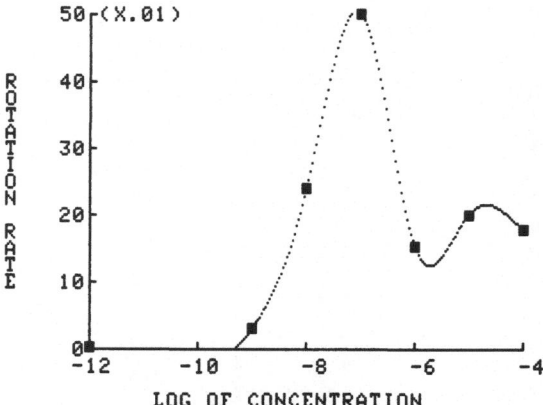

FIGURE 11. The CSR of barium sulfate as affected by polyacrylate polyanions. The particle spin rate when driven by a rotating field at 600 kHz is seen to be affected even by very dilute solutions.

is extremely sensitive and that it will detect small changes in the object being tested; and second, that the intrinsic properties of the models being tested can be altered at least as far as the CSR spectrum is concerned. We must conclude that the surface absorption of the polyelectrolyte, Darvan, appreciably modifies the exterior of the particles, causing a new set of parameters for the models to be set.

## 4. Theory

To facilitate the application and the understanding of CSR a simplified analysis of the theory is presented below. Briefly stated, it is found that the CSR spectrum gives spin rates proportional to the magnitude and sign of the effective (differential) polarization of the body in the suspending medium. From the observed CSR spectra, then, the size and course of the effective permittivity spectra of small bodies can be determined and the nature of the polarizabilities can be interpreted. In the following analysis, friction between the body and the floor of the chamber is neglected as a first approximation since we are dealing with tiny objects usually visible in Brownian motion.

*Torque on a Polarizable Sphere in a Rotating Field:*

$$\mathbf{T} = \boldsymbol{\mu} \times \mathbf{E} \tag{1}$$

$$T = \mu E_0 \sin \delta \tag{2}$$

where $\delta$ is the phase lag angle of the polarization behind the phase of the field $E_0(t)$. For a sphere the induced moment when in a medium is, after $t \to \infty$,

$$\mu = 2\pi a^3 \, \text{Re} \left[ \frac{\varepsilon_1^*(\varepsilon_2 - \varepsilon_1)}{\varepsilon_2 + 2\varepsilon_1} \right] \mathbf{E}_0 \tag{3}$$

where Re means the real part of the complex quantity in the bracket.

Assuming a simple Debye dielectric of one characteristic relation time, $\tau$, of the form

$$\varepsilon = \varepsilon' - j\varepsilon'' = \varepsilon_\infty + \frac{(\varepsilon_s - \varepsilon_\infty)}{1 + \omega^2\tau^2} \frac{-j(\varepsilon_s - \varepsilon_\infty)\,\omega\tau}{1 + \omega^2\tau^2} \tag{4}$$

where $\varepsilon$ is the absolute dielectric permittivity (complex), $\varepsilon = \varepsilon' - j\varepsilon''$, $\varepsilon'$ is the in-phase absolute dielectric constant, $\varepsilon''$ is the out-of-phase dielectric constant, $\varepsilon_\infty$ is the permittivity of "infinite" frequency, $j = \sqrt{-1}$, $\omega$ is the angular frequency of the applied field, $\mu$ is the induced moment, and $D$ is the dielectric replacement, we may write

$$D = \varepsilon E \tag{5}$$

$$D(t) = \varepsilon_\infty E(t) + (\varepsilon_s - \varepsilon_\infty) E(t) e^{-i\delta} = (\varepsilon' - j\varepsilon'') E(t)$$
$$= \varepsilon_\infty E(t) + (\varepsilon_s - \varepsilon_\infty) E(t)[\cos\delta - j\sin\delta] \tag{6}$$

From Eqs. (4), (5), and (6) we find

$$\sin\delta = \frac{\omega\tau}{(1 + \omega^2\tau^2)^{1/2}} \tag{7}$$

for the angle of the phase lag, $\delta$.

In a rotating field as from a four-electrode system with potentials $V_x = V_0 \sin\omega t$ and $V_y = V_0 \cos\omega t$ applied to the $x$ and $y$ electrode pairs, the magnitude of the maximum field, $E_0$, and the potential differences remain constant in the midregion of the symmetric electrodes.

Combining Eqs. (2), (3), and (7), we obtain as an expression for the torque, and using $K = \varepsilon/\varepsilon_0$,

$$T = \mu E_0 \sin\delta = 2\pi a^3 \, \text{Re} \left[ \frac{K_1^*(K_2 - K_1)}{K_2 + 2K_1} \right] \frac{\omega\tau}{(1 + \omega^2\tau^2)^{1/2}} \tag{8}$$

where $\varepsilon_0$ is the permittivity of free space. For a sphere slowly rotating in a fluid medium we may compute the frictional drag from Stokes' formula.[17]

The electrical torque and the hydrodynamic drag will be equal in steady state. Hence, we can write

$$T_{\text{hydro}} + T_{\text{el}} = 0$$

$$8\pi a^3 \eta \omega_c = 2\pi a^3 \varepsilon_0 \left[ \frac{\omega\tau}{(1 + \omega^2\tau^2)^{1/2}} \right] \text{Re} \left[ \frac{K_1^*(K_2 - K_1)}{K_2 + 2K_1} \right] E_0^2 \qquad (9\text{a})$$

or

$$\omega_c = \frac{\varepsilon_0}{4\eta} \left\{ \frac{\omega\tau E_0^2}{(1 + \omega^2\tau^2)^{1/2}} \right\} \text{Re} \left[ \frac{K_1^*(K^2 - K_1)}{K_2 + 2K_1} \right]$$

$$\omega_c = \frac{\varepsilon_0}{4\eta} \left[ \frac{\omega\tau E_0^2}{(1 + \omega^2\tau^2)^{1/2}} \right] K_{\text{eff}} \qquad (9\text{b})$$

where $\eta$ is the viscosity of the medium, $\omega_c$ is the rotational (angular speed) of the spherical body, $a$ is the radius, $\omega$ is the angular frequency of the applied field, $\tau$ is the characteristic relaxation time of the body dielectric, and $K_1$ and $K_2$ are the complex constants of the medium and sphere, respectively,

$$K_i = K_i' - jK_i'' = K_i' - j\frac{\sigma_i}{\varepsilon_0 \omega}$$

$$K_{\text{eff}} = \text{Re} \left[ \frac{K_1^*(K_2 - K_1)}{K_2 + 2K_1} \right] \qquad (10)$$

For the special case of a sphere of insulating character in a conductive fluid

$$\mu = 4\pi a^3 \varepsilon_0 K_1' \left( \frac{\sigma_2 - \sigma_1}{\sigma_2 + 2\sigma_1} \right) \qquad (11\text{a})$$

and if $\sigma_2 \approx 0$ then

$$\mu \cong -2\pi a^3 \varepsilon_0 K_1' \qquad (11\text{b})$$

and the rotational speed (in a direction opposite to the rotation of the field) is

$$\omega_c = \frac{\varepsilon_0 K_1'}{4\eta} \frac{\omega\tau E_0^2}{(1 + \omega^2\tau^2)^{1/2}} \qquad (12)$$

In this case, $\tau$ can be evaluated in advance. We can expect the relaxation time of the insulating sphere in the conducting medium to approximate $\tau = RC$, where $R \cong \rho_1/2a$ and $C = 4\pi\varepsilon_0 a$ or $\tau \approx 4\pi\varepsilon_0 \rho_1$.

## 5. Conclusions

We conclude that the simplified theoretical analysis predicts the rate of cellular rotation, $\omega_c$, to be proportional to the square of the applied field intensity (as observed in Figure 1); to be inversely proportional to the viscosity; and to be proportional to the field frequency, to the relaxation of the (presumed Debye-type) cell, and to the "effective polarizability," $K_{eff}$. We expect, and, so far, find, $\omega_c$ to reflect the magnitude and sign of $K_{eff}$.

For example, it is known from earlier studies of the DEP of yeast that $K_{eff}$ is generally negative in the region 500 Hz to 70 kHz for live cells, and positive for dead ones. This agrees with the observed sign of $\omega_c$. A plot of $\omega_c/\omega_E$ versus $\omega_E$ can be expected to provide a convenient method for obtaining dielectric spectra for single cells, to give relative values of $K_{eff}$ as a function of the applied frequency.

## Acknowledgments

The authors acknowledge with thanks their support by the Pohl Cancer Research Laboratory, Inc., and Mrs. Jill F. Dotson for her generous contribution of cell cultures and valuable advice.

## References

1. A. A. Teixera-Pinto, L. L. Nejelski, J. L. Cutler, and J. H. Heller, *Exp. Cell Res.* **20**, 548 (1960).
2. A. A. Furedi and R. C. Valentine, *Biochem. Biophys. Acta* **56**, 33 (1962).
3. H. A. Pohl and J. S. Crane, "Dielectrophoresis of Cells," *J. Biophys.* **11**, 711–727 (1971).
4. H. A. Pohl, *Dielectrophoresis, The Behavior of Matter in Non-Uniform Electric Fields,* Cambridge University Press, Cambridge (1978).
5. Maja Mischel, Arthur Voss, and H. A. Pohl, "Cellular Spin Resonance in Rotating Electric Fields," *J. Biol. Phys.* **10**, 223–226 (1983).
6. H. A. Pohl, "Cellular Spin Resonance, A New Method For Determining The Dielectric Properties of Living Cells," *Int. J. Quantum Chem.* **10**, 161–174 (1983).
7. H. A. Pohl, "Cellular Spin Resonance in Pulsed Rotating Electric Fields," *J. Biol. Phys.* **11**, 59–62 (1983).
8. M. Mischel and I. Lamprecht, *Z. Naturforsch.* **35c**, 1111 (1980).
9. H. A. Pohl, "The Spinning of Suspended Particles in a Two-Pulsed, Three Electrode System," *J. Biol. Phys.* **11**, 66–68 (1983).
10. H. A. Pohl, "Cellular Spin Resonance, CSR," *J. Theor. Biol. Phys.* **93**, 207–213 (1981).
11. U. Zimmermann, J. Vienken, and G. Pilwat, *Z. Naturforsch.* **36c**, 173 (1981).
12. H. A. Pohl and T. Braden, "Cellular Spin Resonance of Aging Yeast and of Mouse Sarcoma Cells," *J. Biol. Phys.* **10**, 17–30 (1982).
13. C. S. Chen, "On The Nature and Origins of Biological Dielectrophoresis," Ph. D. thesis, Oklahoma State University, Stillwater, Oklahoma 74078 (1973).

14. M. Mischel and I. Lamprecht, "Rotation of Cells in Nonuniform Rotating Alternating Fields," *J. Biol. Phys.* **11**, 43 (1983).
15. H. A. Pohl, "Natural Cellular Electrical Resonances," *Int. J. Quantum Chem., Quantum Biol. Sump.* **9**, 399–409 (1982).
16. M. Mischel and H. A. Pohl, "Cellular Spin Resonance: Theory and Experiment," *J. Biol. Phys.* **11**, 98–102 (1983).
17. H. Lamb, *Hydrodynamics 3rd* ed., Cambridge University Press, Cambridge (1906), p. 546.

# 16

# Dielectrophoretic Cell Sorting

## J. Kent Pollock and Herbert A. Pohl

ABSTRACT: Living cells can be sorted with the aid of dielectrophoresis (DEP). The method depends upon the natural differences in the effective dielectric constants of the several types (species, physiological states, etc.) being handled. The use of additive stains or chemical modifiers or other tagging agents is not usually required. In the apparatus to be described, streams of cells can be subjected to DEP forces and their deflection into or out of the region of most intense electric field can be measured to provide dielectric relaxation spectra and comparisons of cell types or cell abnormalities. Moreover, since an actual physical force is exerted upon the cells, and a real displacement results, such an apparatus is readily arranged to provide continuous cell sorting base upon the intrinsic dielectric properties of the passing cells. Since there is much need for means to sort and to characterize living cells, we have focused upon how DEP should be applied to these desired ends. Our aim has been to devise a dual-purpose instrument capable of sorting cells and of characterizing them by their unique dielectric polarizability spectra. Early models succeeded in providing manual operation indicating an ability to make exquisitely delicate distinctions between cells, and to make actual separations, once the DEP spectra were determined. More recently, we have begun the redesign to obtain automation of the spectrometer mode of operation. Early tests show it to function successfully so as to provide simple spectra of a given cell type introduced into the instrument. The theory will be presented for the operation, and for the interpretation of the DEP spectra of cells. The experimental results for several cellular systems will also be presented.

## 1. Introduction

Dielectrophoresis (or DEP) has been used for some years to study the dielectric response of cells and cell organelles.[1] DEP has been used to characterize cells[2-8] and to separate them.[9-12]

_J. Kent Pollock_ • Pohl Cancer Research Laboratory, 515 Harned Avenue, Stillwater, Oklahoma 74075. _Herbert A. Pohl_ • National Magnet Laboratory, Massachusetts Institute of Technology, Building NW-14, Cambridge, Massachusetts 02139.

The first successful batch separations[9-11] aroused interest in a continuous stream sorter that could handle much larger volumes of cells and produce easily collectible fractions of the separated suspension. A stream-centered continuous DEP sorter was developed that was also capable of characterizing cells by their unique dielectric polarizability spectra.

This dual-purpose cell sorter/analyzer has recently been redesigned to automate the spectrometer mode of operation. Early tests show that it readily provides a spectrum of a given cell type much more quickly than the manual spectrometer used previously.

## 2. Theory

DEP can easily be illustrated with Figure 1. Figure 1a shows a neutral particle and a charged particle in a uniform electric field. The neutral par-

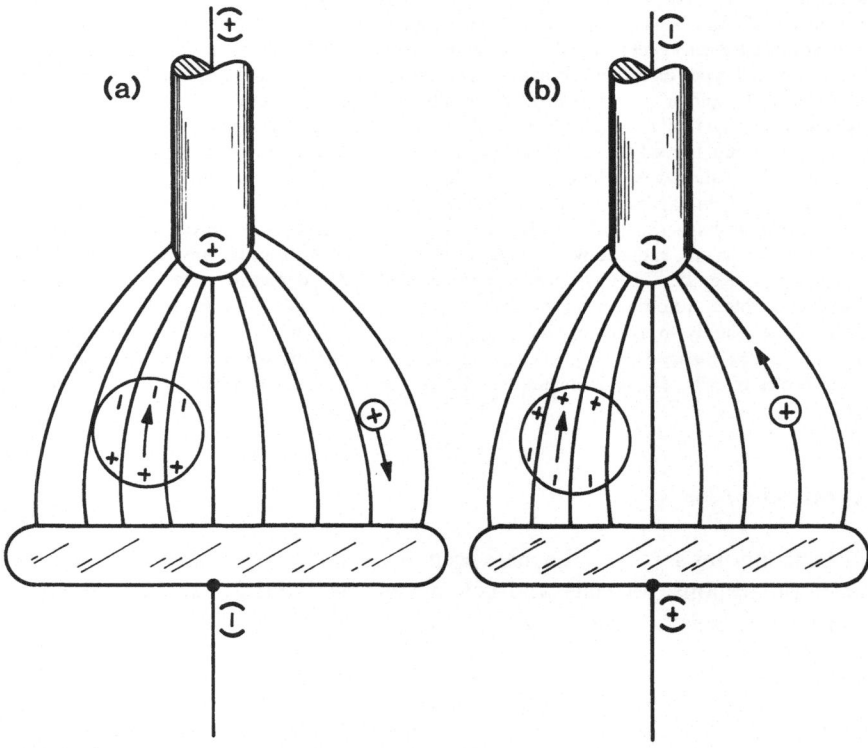

**FIGURE 1.** Diagram comparing the behaviors of neutral and charged bodies in a non-uniform electric field.

ticle polarizes, with one side becoming slightly more positive and the other slightly more negative. All charges experience a force equal to the charge times the local field, $F = q \times E$. Since the field is uniform, the neutral particle will experience no net force. The charged particle will experience a translational force, viz., electrophoresis, whose direction depends on the field direction.

Figure 1b shows the same particles in a nonuniform electric field. The charged particle will again exhibit electrophoresis. The neutral particle will again be polarized. However, in the nonuniform field, the neutral particle will experience a net force (DEP) toward the strong field region. The DEP force arises because the local field varies over the dimension of the particle, so that the force on one side of the particle is greater than the force on the other.

Another aspect of DEP can be seen in Figure 1b. If the field is reversed, the neutral particle will still be attracted to the strong field region. The charged particle will reverse its motion. By applying an ac field, only DEP

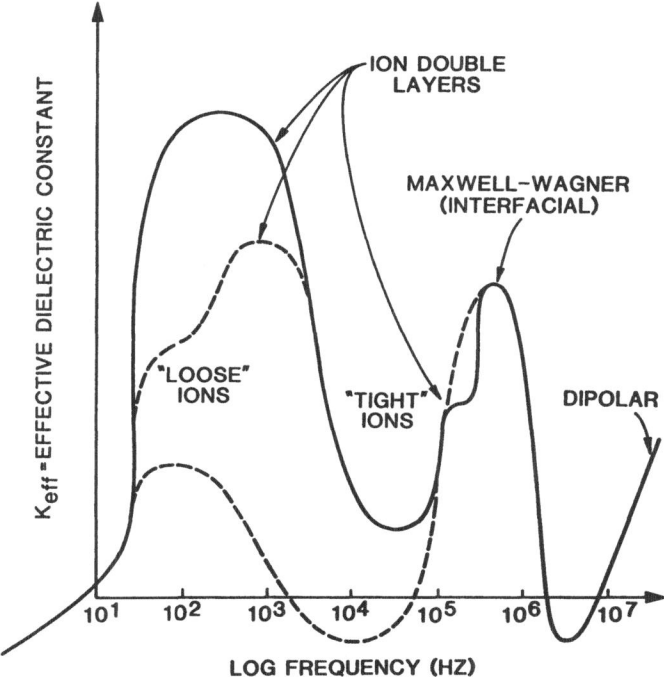

FIGURE 2. Diagram of typical spectral origins in biological dielectrophoresis. The effective dielectric constant of a living cell in an aqueous medium of very low conductivity is observed to display a variation with the applied frequency such as shown.

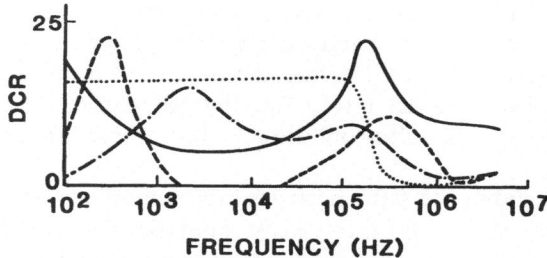

FIGURE 3. Diagram of the yield spectrum, or dielectrophoretic collection rate onto a wire–wire electrode of various cells as affected by the frequency of the applied field. The ordinate as shown is given in relative values as the average length of the pearl chains of cells gathered onto the wire after a definite period (say, 2 min) of field at a given maximum value.

can be observed as the neutral particle moves steadily toward the strong field region and the charged particle "shudders" about its original position.

Particles not only move toward the strong field region but can also move away from it. This negative DEP arises from the interaction of the particle and its supporting medium. In actual practice, particles are suspended in air or liquid and the field affects both the particle and the suspending medium. In such a case, the DEP force experienced by the particle depends on the differences between the polarizability of the particle and the polarizability of the medium. If the particle is in a liquid medium more polar than itself, the DEP force will be away from the strong field region and the particle will experience negative DEP.

The DEP force on a neutral particle, such as a cell, has been derived assuming approximate conservation of energy by Pohl[1] as

$$F = 2\pi a^3 \, \text{Re} \left[ \frac{\varepsilon_1^*(\varepsilon_2 - \varepsilon_1)}{\varepsilon_2 + 2\varepsilon_1} \right] \nabla(E_0^2) \tag{1}$$

where $\varepsilon_2$ is the complex dielectric constant of the cell, $\varepsilon_1$ is the complex

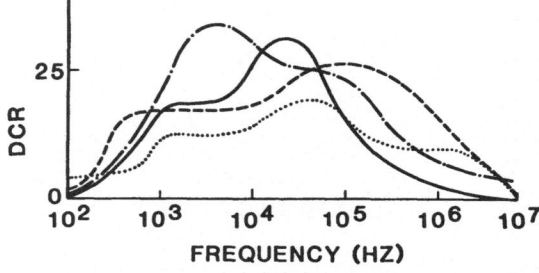

FIGURE 4. Diagram of the yield spectrum or dielectrophoretic collection rate for various cells as affected by frequency of the applied field. Ordinate in relative values, as in Fig. 3.

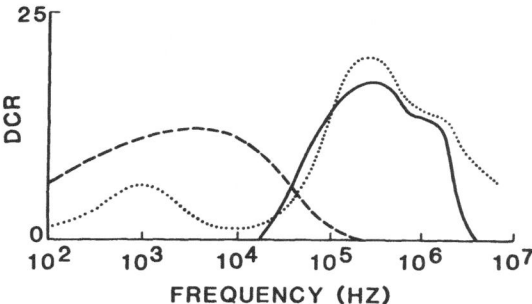

FIGURE 5. Diagram of the yield spectrum for various cells as affected by the frequency of the applied field. Ordinate in relative values, as in Fig. 3.

dielectric constant of the water, and $a$ is the cell radius. Re stands for the real part of the complex quantity in the bracket. Sauer,[13] assuming an exact conservation of momentum in the field, derived the expression

$$F = -\frac{3}{4} V \left( \frac{b^*}{3+b^*} \right) + \left( \frac{b}{3+b} \right) \frac{1}{4} (\varepsilon_1^* + \varepsilon_1) \nabla(E_0^2) \qquad (2)$$

where $V$ is the cell volume and $b = (\varepsilon_2 - \varepsilon_1)/\varepsilon_1$. Sauer's expression is now considered a superior solution; however, the discrepancy with Eq. (1) is small except in the case with high specific wattage.

As can be seen from these equations, the DEP force is a quite complex relation involving the complex dielectric constants of both cell and supporting medium.

The effective dielectric constant of the cell can itself vary dramatically with frequency. Figure 2 shows the typical polarization mechanisms present in most cells. Broadly, the low-frequency range is dominated by

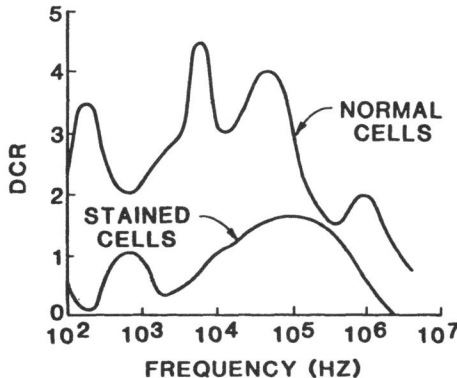

FIGURE 6. Diagram of the yield spectrum or dielectrophoretic collection rate for the alga, *Chlorella vulgaris*, stained (lower curve) and unstained cells (upper curve) compared.

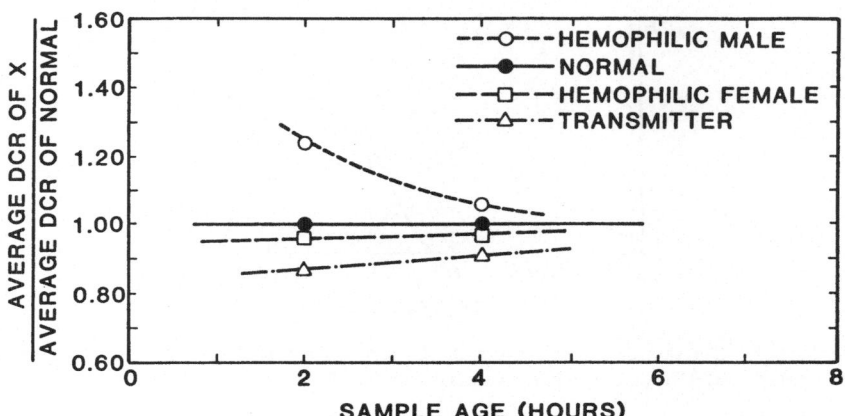

FIGURE 7. The average dielectrophoretic collection rate at 1.0 MHz of various types of canine (beagle) thrombocytes relative to that of normal thrombocytes as a function of sample age, showing the effect of various types of hemophilic cells.

surface properties while the higher-frequency responses involve the polarization of the cell interior. The loosely bound ion model of Dukhin[14] and the plasmoidal model of Einolf and Carstensen[15–17] have the lowest characteristic frequency of $10^2$–$10^3$ Hz. The tightly bound ionic double layer model of Schwarz[13] has a characteristic frequency range of $10^4$–$10^5$ Hz. Above $10^5$ Hz, the interior cell polarization is seen with the familiar Maxwell–Wagner or interfacial polarization being dominant.

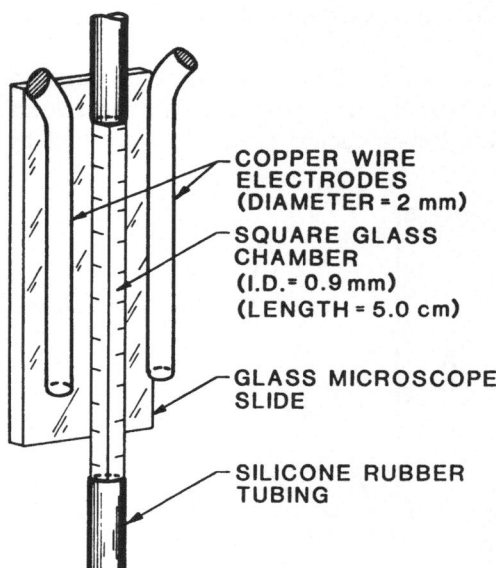

FIGURE 8. Exploded view of the continuous flow chamber for the dielectrophoretic analysis and sorting of microorganisms.

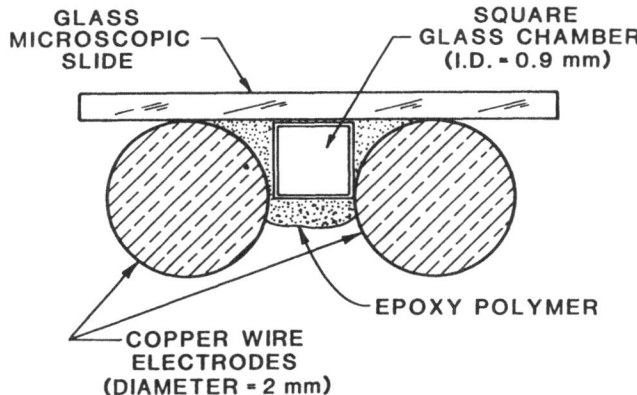

*FIGURE 9.* Cross-sectional view of the continuous flow chamber showing the square capillary in cross section.

The interaction of the highly variable effective dielectric constant of the cells with the supporting medium results in the spectra seen in Figures 3–7. It is apparent that each species has a unique spectrum and that there are regions where negative DEP (cells repelled from the intense field region and thus no collection) occurs. It can also be seen that there occur frequencies at which one cell type can be collected in preference to another. Such batch separations have, in fact, been performed by Pohl and Hawk[9] and by Mason and Townsley.[11]

## 3. Sorter Design

Subsequent work in this laboratory[8,12] has been on the development of a continuous, stream-centered DEP cell sorter/analyzer. There are certain problems inherent in sorting cells that are not encountered in sorting minerals, for example.

Living cells are delicate and must be suspended in water. To avoid heating effects in the water, the voltage must be kept low. Even with low voltage, there are still stirring effects from thermal gradients, charge injection, etc. that can interfere with the sorting. Some cells tend to adhere to surfaces, so a sorter would have to avoid letting the cells come into contact with any surfaces. For delicate separations, the force experienced by the cell must not vary with the cell's position in the sorter chamber. Figures 8 and 9 show a stream-centered DEP sorter flow chamber. This chamber approximates the desired isomotive electrode shape[1] to ensure equal force on the cells throughout the chamber. The cells are injected into the center

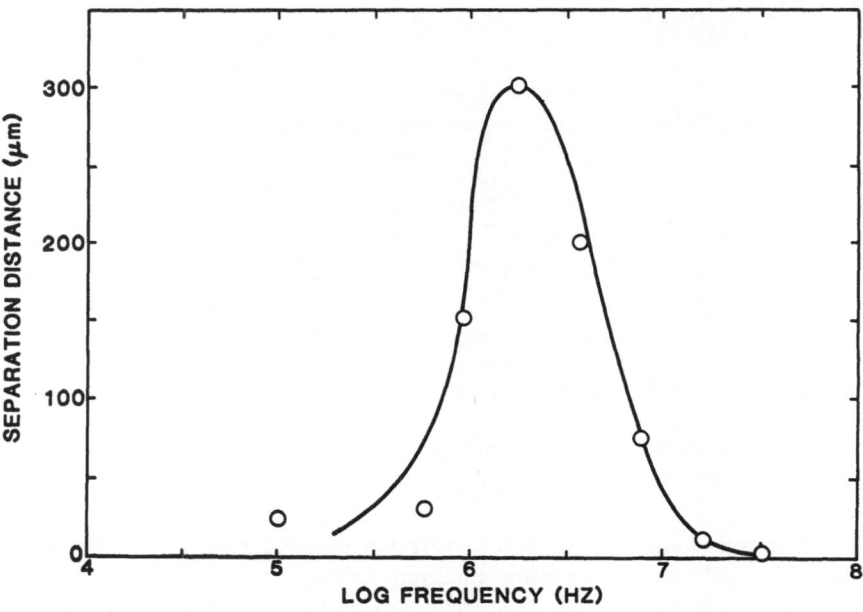

FIGURE 10.   Dielectrophoretic cell sorting: *Saccharomyces cerevisiae* vs. *Netrium digitus.*

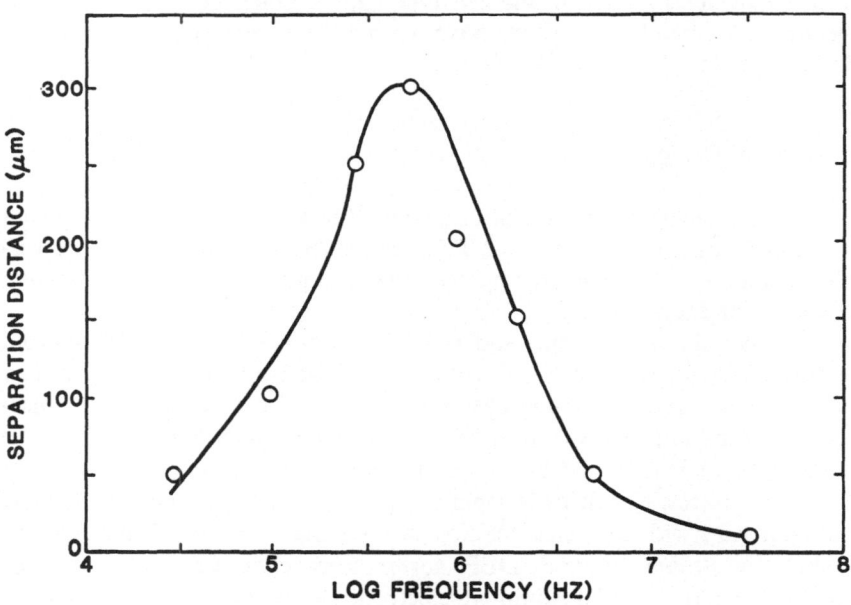

FIGURE 11.   Dielectrophoretic cell sorting: The algae *Ankistrodesmus* vs. *Staurastrum.*

of the flow chamber and are deflected as they flow down past the electrodes. The stream-centered design prevents the cells from touching surfaces that they might adhere to and also allows both positive and negative DEP to occur.

## 4. Results

This apparatus has been used[12] to separate a yeast *(Saccharomyces cerevisiae)* and an alga *(Netrium digitus)*, shown in Figure 10, and also to separate two algae *(Ankistrodesmus falcatus* and *Staurastrum gracile)*, Figure 11. In both cases the cell stream split into two streams. The figures show the separation distance between the two streams as a function of frequency. Separation was confirmed visually as the cells were shaped differently and easily distinguished.

The sorter has been automated under computer control which allows spectra to be obtained much more quickly and with more detail. Figures 12 and 13 show spectra of live and dead yeast, respectively, obtained from the automated sorter while operating in the analyzer mode. The spectra show both positive and negative deflections and can be used to select conditions to separate a mixture of live and dead cells.

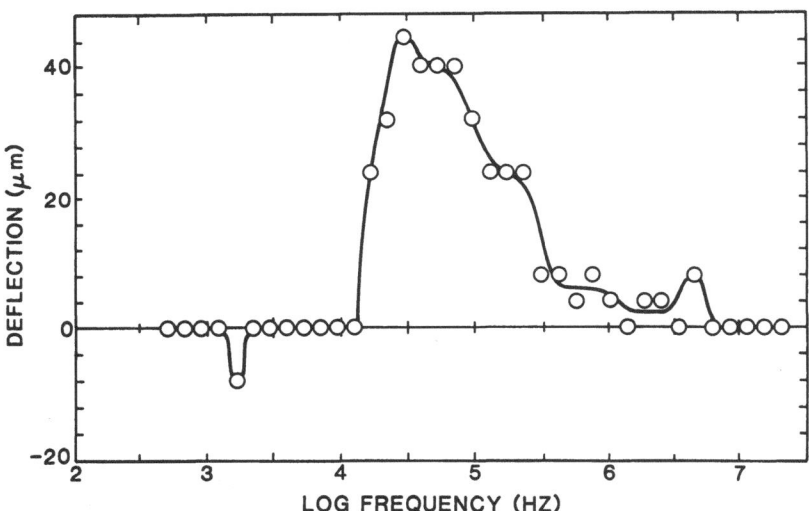

FIGURE 12. The lateral deflection, in microns after 2.5 cm travel of the live yeast, *Saccharomyces cerevisiae* in the automated DEP spectrum analyzer. Resistivity, 300 kΩ cm. Voltage, 2.5 V rms.

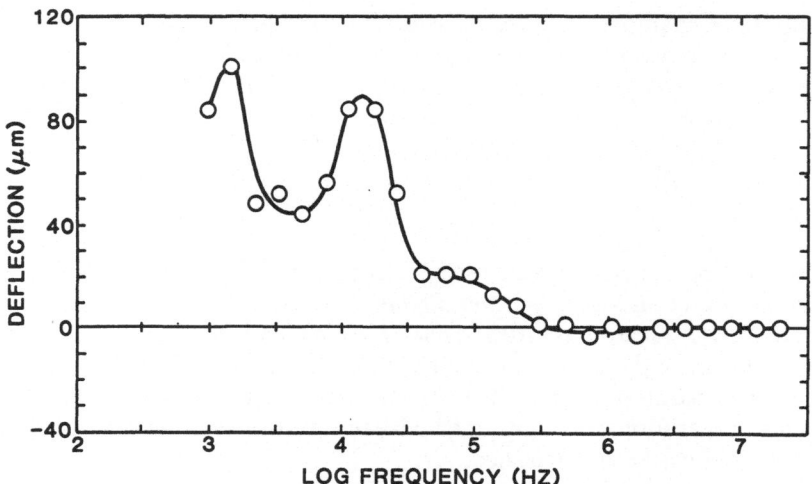

FIGURE 13. The lateral deflection of dead yeast in the automated DEP spectrum analyzer; conditions as in Fig. 12 above, but at 120 kΩ cm.

## 5. Conclusion

A DEP cell sorter/analyzer has been developed[8,12] which uses the polarizability of cells to characterize and sort them. All types of cells (algae, bacteria, mamalian) can be handled easily by DEP. This technique gives the biologist a new tool to both characterize and handle cells.

## References

1. H. A. Pohl, *Dielectrophoresis, The Behavior of Matter in Nonuniform Electric Fields*, Cambridge University Press, Cambridge (1978).
2. H. A. Pohl and J. S. Crane, *Biophys. J.* **11**, 711 (1971).
3. I. P. Ting, K. Jolley, C. A. Beasley, and H. A. Pohl, *Biochem. Biophys. Acta* **234**, 324 (1971).
4. C. S. Chen and H. A. Pohl, *Trans. N.Y. Acid. Sci.* **238**, 176 (1974).
5. J. E. Rhoads, H. A. Pohl, and R. G. Buckner, *J. Biol. Phys.* **4**, 93 (1976).
6. J. S. Crane and H. A. Pohl, *J. Biol. Phys.* **5**, 49 (1977).
7. J. S. Crane and H. A. Pohl, *J. Electrostatics* **5**, 11 (1978).
8. H. A. Pohl, K. Kaler, and J. K. Pollock, "Continuous Positive and Negative Dielectrophoresis of Microorganisms," *J. Biol. Phys.* **9**, 67–85 (1981).
9. H. A. Pohl and I. Hawk, *Science* **152**, 647 (1966).
10. J. S. Crane and H. A. Pohl, *J. Electrochem. Soc.* **115**, 584–588 (1968).
11. B. D. Mason and P. M. Townssley, *Can J. Microbiol.* **17**, 879 (1971).
12. H. A. Pohl and K. Kaler, *Cell Biophys.* **1**, 15 (1979).

13. F. A. Sauer, *Coherent Exitations in Biological Systems* (H. Fröhlich and F. Kremer, eds.), Springer-Verlag, New York (1983), pp. 134–144.
14. S. S. Dukhin and V. N. Shilov, *Dielectric Phenomena and the Double Layer in Disperse Systems and Polyelectrolytes*, Wiley, New York (1974).
15. C. W. Einolf and E. L. Carstensen, *J. Phys. Chem.* **75**, 1091 (1971).
16. C. W. Einolf and E. L. Carstensen, *Biophys. J.* **13**, 8 (1973).
17. E. L. Carstensen and R. E. Marquis, "Dielectric and Electrochemical Properties of Bacterial Cells", in *Spores VI* (P. Gerhart, R. Costilow, and H. L. Sadoff, eds.), American Society for Microbiology, Washington D. C. (1975), p. 563.
18. G. T. Schwarz, *J. Phys. Chem.* **66**, 2636 (1962).

# 17

# A Qualitative, Molecular Model of the Nerve Impulse

## Conductive Properties of Unsaturated Lyotropic Liquid Crystals

*Paavo K. J. Kinnunen and Jorma A. Virtanen*

*ABSTRACT:* An approach to the molecular mechanism of nerve impulse is described. Basically, we propose an electronic conduction band to exist in properly arranged ethylenic double bonds of unsaturated nerve membrane lipids. Electron–electron interaction in the conduction band is brought about by a pair of holes residing on a charge-transfer band of cholesterol and phospholipid carbonyls. Rectification of signal and driving force for propagation is provided by transmembrane ion fluxes which generate a lateral field along the axis of the nerve fiber. Transmembrane ion currents are controlled by Na-channel proteins, which in turn are regulated by a transient phase transition in the membrane lipids. In the latter process a crucial role is played by phosphatidylserine. Under resting potential this lipid in the outer surface of the membrane is deprotonated and in the liquid crystalline state. Following a depolarizing pulse an electrostatically triggered phase transition takes place due to protonation of phosphatidylserine with subsequent phase separation of a crystalline membrane lipid domain and opening of Na channels.

## 1. Introduction

In spite of extensive research the molecular mechanisms underlying the excitability of biological membranes are still by and large unresolved. The success of the Hodgkin–Huxley equations in describing the action potential

*Paavo K. J. Kinnunen and Jorma A. Virtanen* • Department of Medical Chemistry and Department of Chemistry, University of Helsinki, Siltavuorenpenger 10, SF-00170 Helsinki 17, Finland.

associated temporal changes in transmembrane ion conductances has led to the wide acceptance of their depolarization–repolarization model of the impulse transmission.[1,2] These equations represent an empirical description of the ion currents further determined by altered molecular organization in the membrane. The Hodgkin–Huxley (HH) model can also be presented with two first-order differential equations.[3] Several monographs and reviews are available which summarize the experimental findings and some of the excisting models in detail.[4–7] Alternatives for the HH model include the dipole model of Goldman.[8–10] He suggested the conformation of nerve membrane phospholipid polar head groups to control the transmembrane ion flow. Cope among others has presented evidence for a phase transition in axons during excitation.[11] According to Cope cells should be considered as ion exchangers whose association sites have a profound preference for sodium and potassium ions. During excitation this preference would be reduced by a conformational change in the proteins. Cope has also forwarded evidence for superconductivity phenomena being associated with nerve processes.[12,13]

According to the HH model of nerve impulse, transmembrane channel proteins should control the permeability of the nerve membrane for $Na^+$ and $K^+$. Recent reports describe the purification and reconstitution of the $Na^+$ channel protein.[14–18] Molecular details of this protein are still to be established. Further studies are likely to provide valuable information for pursuing the details of nerve excitation. The nerve membrane ion channels are regulated by the so-called gating currents, predicted by Hodgkin and Huxley and measured by Armstrong and Bezanilla.[19,20] Hodgkin and Huxley pointed out that the mechanisms responsible for the activation of the ion channels involve a negatively charged particle or a dipole. The total number of electronic charges causing the gating currents by moving within the membrane should equal six. The quantity of $Na^+$ entering an axon exceeds by approximately three-fold the ion flux expected on the basis of the voltage change. The reasons for this discrepancy are unclear. There is also considerable between species variation in the amount of $Na^+$ entry into the axons. Extracellular $Na^+$ is not mandatory for excitation but can be replaced by a number of cations.[21–23] Therefore, although the ion currents and gradients across a neuronal membrane definitely are involved in the impulse transmission, they are mere reflections of more specific changes in the membrane.

Plenty of data indicate the presence of mobile electrons in the nerve membrane. Calvin *et al.* observed the reduction of spin labels upon contact with nerve preparations.[24] Another spin probe TEMPO* is used as a model anesthetic.[25] The structure of TEMPO has very little in common

---

* TEMPO, 2, 2′, 6, 6′-tetramethylpiperidine-1-oxyl.

with the other local anaesthetics, cocaine, lidocaine, procaine, and the like. Most probably the mechanism of action of the latter group of compounds involves interference with protein–lipid interactions as well as binding to nerve membrane phospholipids,[26–30] phosphatidylserine in particular.[28] The mechanism of action of tetrodotoxin (TTX) seems to involve galactolipids.[31] In most cases the results by Huneeus-Cox *et al.* can be explained by oxidation of the nerve membrane by the anaesthetic compounds.[32] For instance, chloroform increases the capacitance of planar lipid bilayers and reduces the gating currents in axons.[33,34]

The high although diffuse electron density in the outer half of nerve membrane bilayer has been detected by x-ray diffraction.[35] It was concluded to be due to an unusually high content of protein. An additional contribution could be due to the presence of auxiliary electrons. The impulse associated magnetic fields indicate the propagation of electric charges in the nerve.[36] These nerve impulse associated magnetic fields as well as the electric and magnetic field induced changes in cells are difficult to reconcile with the HH model.[36–40]

Here a new model for nerve impulse is described. The basic concept is the possibility for electronic conduction along properly arranged ethylenic double bonds of unsaturated nerve membrane lipids.[31] This electron band has been named by us a semiconjugated molecular orbital system, SMOS. in addition to providing the conduction band in the nerve membrane, SMOS is considered to be involved in the maintenance of the anisotropy of the liquid crystalline membrane. The nerve impulse is proposed to consist of a pair of electrons* propagating in SMOS. The spins and momenta of these electrons are opposite. The electron pair is stabilized by an interaction with a charge transfer band of cholesterol (acceptor) and phospholipid carbonyls (donor) formed in the membrane surface. Destabilization of the lattice due to charge movements and dissociation of the charge-transfer couple leads to a transient phase change from the resting, induced liquid crystalline state to crystalline state of the membrane phospholipids with subsequent activation of the ion channels. Flow of ions through the membrane generates a lateral field along the nerve fiber (capacitative current) which drives the signal further.

It is to be stressed that the present model does not contradict the basis for the HH model. It is an attempt to describe the molecular events preceding the opening of the transmembrane ion channels.

---

* These are represented by those involved in the stabilization of SMOS. See Section 3.

## 2. Ordering in the Nerve Membrane

The ultrastructure of neuronal cells has been studied employing a variety of techniques. The structural details of a native axon are, however, still far from being elucidated.

Electron diffraction patterns have been obtained of myelin and rat central nervous system membranes after osmium tetroxide and formalin fixation.[41-44] Strongly diffractive crystalline material present in an $OsO_4$ fixed specimen was interpreted as precipitated dye as the observed spotted pattern was much weaker in formalin treated tissue samples. $OsO_4$ easily reacts with ethylenic double bonds. An alternative and more likely explanation therefore is that regular, electron dense crystalline structures are formed due to reaction of $OsO_4$ with a highly organized lattice of unsaturated membrane lipids.

The use of fluorescent probes,[45-51] 2-toluidinyl-naphthalene-6-sulfonate (TNS) in particular, when applied to isolated axons has revealed the high degree of anisotropy of axonal membranes. Part of the TNS-dye is bound in the membrane to a highly organized hydrophobic lattice. The Kosower Z value* of these high affinity sites appears to be of the order of 75, corresponding to absolute ethanol.[49] A six times higher number of more polar low affinity sites for TNS are present in an axon. However, only the nonpolar high affinity sites signal the excitation and propagation of an impulse. During depolarization of the membrane potential there is a transient decrease in the emission intensity of the high affinity site bound TNS molecules with a shift in the emission maximum towards longer wavelengths. In the resting state the emission spectrum of the high affinity site bound TNS is similar to the spectrum of anhydrous crystals of the dye. The light emitted by the membrane bound dye is nearly completely polarized. The axis of polarization shows the absorption and emission dipoles to parallel the longitudinal axis of the axon.

## 3. Electronic Conduction in Liquid Crystalline Membranes. Role of Unsaturated Lipids

One class of organic conductors are polyenes in which conductivity is due to intramolecular overlapping $\pi$ orbitals. The different modes of soliton conduction have been reviewed by MacDiarmid and Heeger.[52] The practical use of polyenes is limited by thermally activated chain rotation, which interrupts the conjugation. Stacked, planar delocalized

---

* Z values provide an empirical scale of solvent polarities, as defined by Kosower (J. Am. Chem. Soc. 80, 3253–3260 (1958)).

$\pi$ orbitals containing polymers provide an alternative studdied with charge-transfer polymers and also found in some radical ion salts.[53] These conductors have, like semiconductors in general, positive activation energy, and unlike in metallic conductors their conductivity increases with increasing temperature.

We propose the unsaturated lipids of biological membranes to provide the cells with an electronic conduction band. A striking feature of the unsaturated lipids in biomembranes is the very constant location of ethylenic *cis*-double bond in monounsaturated acyl chains between carbon atoms 9 and 10. A noteworthy exception is nervonic acid (*cis*-15-24: 1), which is abundant in central nervous system membranes. A microviscosity barrier has been observed in bilayers of dioleoylphosphatidylcholine at the depth of the oleic acid double bonds.[54] Phospholipids belong to lyotropic liquid crystals and possess remarkable short- and long-range order. In a cell membrane lattice the local concentration of ester carbonyls and acyl chain ethylenic double bonds as well as the local concentration of cholesterol $C=C$ bonds is very high.

Conjugated systems consist of $sp^2$-hybridized atoms connected with $\sigma$ bonds. Conjugation is an intramolecular phenomenon. If molecular association is due to or accompanied with intermolecular conjugation we have employed the term semiconjugated molecular orbital system, SMOS.[31] The basic unit of SMOS in a nerve membrane is proposed to consist of one nervonic and two oleic acid side chains of membrane lipids.[31] Good orbital overlap requires the ethylenic double bonds to be properly oriented. The degree of orbital overlap is further regulated by the conformation of the membrane lipids and subsequently depends on changes produced by electrostatically or thermally induced phase transitions in the membrane.

Simple molecular orbital theory allows some general conclusions to be made. If four separate ethylenic double bonds approach each other symmetrically with $p$ orbitals perpendicular to the plane new orbitals and energy levels are formed. Both highest occupied "molecular" orbital (HOMO) and lowest unoccupied "molecular" orbital (LUMO) are illustrated schematically in Figure 1. Two important features emerge.

    i. Both HOMO and LUMO electron densities are highest at the ends of the double bond system.
    ii. If HOMO is antibonding then LUMO is bonding between the different double bonds.

Both (i) and (ii) are commensurate with good conduction. Their validity is not limited to four ethylenic bonds but can be applied to a helically arranged system consisting of several double bonds. An example of the latter can be found in dolichol.[31]

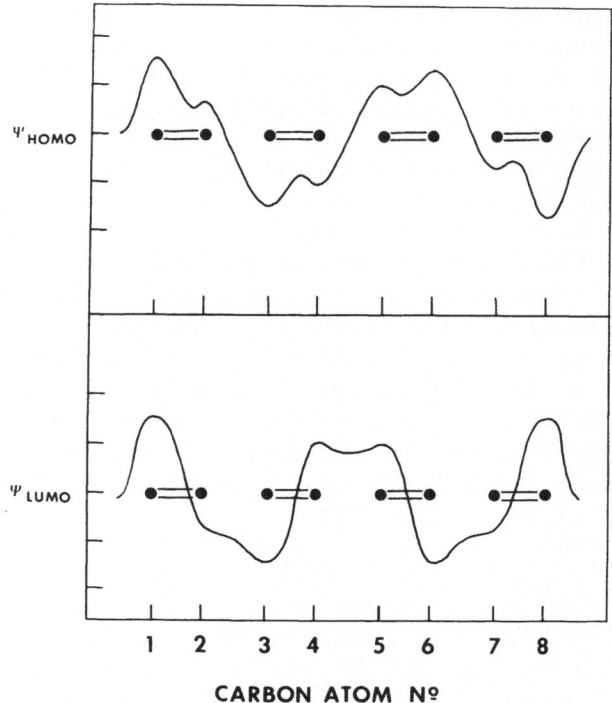

**CARBON ATOM Nº**

*FIGURE 1.* Schematic illustration of the HOMO and LUMO of a supermolecule consisting of four strongly interacting ethylenic double bonds. See text for details.

According to Winstein, the cyclic double bond systems may exhibit homoaromaticity.[55] These can be treated with the Woodward–Hoffmann* rules.[56] For a system consisting of three double bonds we get two more rules.

    iii. Ground state interaction and reaction is allowed when all contacts between the double bonds are suprafacial (i.e., on the same side, denoted by $\langle {}_{\pi}2_s + {}_{\pi}2_s + {}_{\pi}2_s \rangle$). One double bond has suprafacial contacts and the two others have antarafacial contacts (i.e., on the opposite sides, denoted by $\langle {}_{\pi}2_s + {}_{\pi}2_a + {}_{\pi}2_a \rangle$).

    iv. Interaction and subsequent reaction is allowed between two ground state double bonds and one excited double bond if two

---

* Initially, these rules were put forward by Woodward and Hoffmann for the analysis of pericyclic reactions. Briefly, vicinal orbitals with the same symmetry stabilize the interaction between approaching molecules. Likewise, conservation of orbital symmetry in a reaction favors the formation of a chemical bond. For basics of fundamental symmetry principles in orbital theory, see Ref. 56.

bonds      are      suprafacially      and      one      is      antarafacially
($\langle _\pi 2_s + _\pi 2_s + _\pi 2_a \rangle$)   or,   all   double   bonds   are   antarafacially
($\langle _\pi 2_a + _\pi 2_a + _\pi 2_a \rangle$) arranged.
An excited double bond can be substituted by a double bond having one or two extra electrons.

These rules can be used to estimate the stability of double bond systems
found in living cells. Accordingly, the phospholipid ester bond carbonyls
and the double bond of cholesterol could interact in a manner illustrated
schematically in Figure 2a. Organization is of the $\langle _\pi 2_s + _\pi 2_s + _\pi 2_s \rangle$ type
and a benzene-like stabilized structure is formed. The tight, helically
arranged double bond system of the type $\langle _\pi 2_s + _\pi 2_s + _\pi 2_a \rangle$ proposed to
exist in dolichol and in conducting nerve membrane is illustrated in
Figure 2b. Cyclic structure is stabilized by one extra electron. Although
allowed by symmetry the reaction to an anionic cyclohexane derivative is
energetically impossible.

Taking into account the above we can inspect the conductivity of a
linear array of a loosely associated double bond system. From (i) it follows
that exchange of electrons and electronic interactions with surroundings
should take place at the ends of the helically arranged double bonds. In
order to have high conductivity this is to be expected. From (ii) it follows

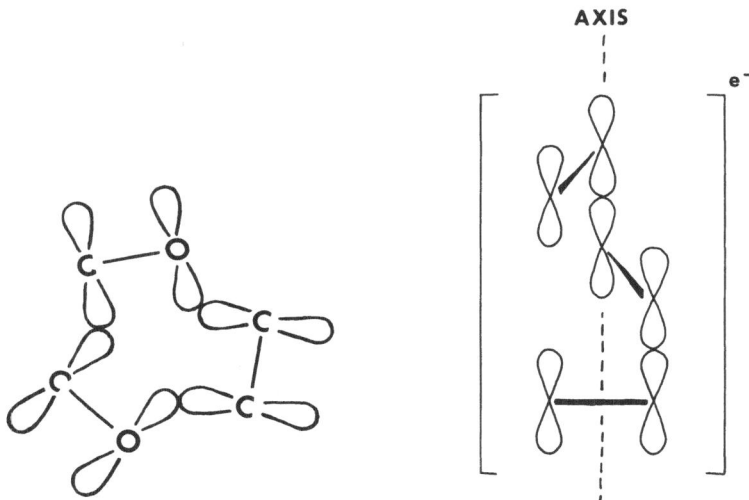

FIGURE 2.   (Left) Illustration of the *p*-orbitals of phospholipid carbonyls interacting with
the cholesterol double bond. All orbitals lie in the same plane (plane of the page). (Right) The
basic, three ethylenic double bond unit in a helical array in SMOS, stabilized by an extra electron. The upper- and lowermost double bonds have suprafacial whereas the double bond in
the middle has antarafacial contacts.

that an extra electron on LUMO increases the association of the double bonds. Two extra electrons further enhance the interaction. Association in turn lowers the energies of all populated orbitals. Once formed this system cannot decompose easily to separate double bonds and double bonds in radical anion state, nor polymerize to a polyene radical anion or dianion. Combining rules (i) and (ii) it may be concluded that if a double bond system has one extra electron or an extra pair of electrons the system may grow at one end and disappear at the other, forming a soliton. If the electron density is proper these solitons are likely to possess phase coherence. Coherent electron transfer can also occur in periodically alternating occupied and unoccupied states of the double bond units of optimal size. Stabilization due to association favors the formation of conducting LUMO electron pairs.

Two basic requirements for superconductivity are the pairing of electrons and phase coherence.[57] Both of these can be satisfied in highly anisotropic arrays of loosely associated double bonds which are not directly chemically bonded as further association due to extra electrons would then be impossible. These criteria seem to be fulfilled by proper liquid crystalline state of matter.

## 4. Crystalline → Liquid Crystalline Phase Transition of Phospholipid Membranes

The crystalline → liquid crystalline phase transition of hydrated phospholipids produces dramatic changes in the organization and physical properties of these lyotropics. The phase behavior of phospholipids has been intensively studied and several excellent reviews are available (see refs. 58–73, for instance). Detailed information on the conformational, orientational, and organizational changes in phospholipids undergoing phase transitions is lacking. Hydration organizes these lipids by a favorable change in solvent entropy (hydrophobic effect). Most significant stabilization of the membrane lattice is, however, derived from London–van der Waals dispersion forces and Madelung potential.

The feasibility of phospholipid phase transition phenomena in the propagation of nerve impulse as well as in information transfer in biological systems in general has been noted earlier. The phase transition of model and biomembranes exhibits considerable hysteresis.[74,75] The possible role of fast and slow hysteresis phenomena in biomembranes for information storage was put forward by Träuble.[74]

The available data on physical changes in isolated axons upon excitation are summarized in Table 1. Comparison of these data with results from studies on phospholipid model membranes reveals a very close

*TABLE 1.* Comparison of Phospholipid Liquid Crystalline → Crystalline (Cooling) Transition with Nerve Membrane During Excitation

|  | Nerve | Phospolipid model membrane |
|---|---|---|
| Enthalpy | Exothermic[a(76,77)] | Exothermic[(78)] |
| ANS-fluorescence | Decreased[b(49)] | Decreased[(74)] |
| IR-spectrum, $C=O$ stretch at $1750^{-cm}$ | Shift to higher frequencies[c(79)] | Shift to higher frequencies[d(80)] |
| Fluidity[e] | Decreased[(81)] | Decreased[(82,83)] |
| Permeability for $Na^+$ | Increased[f(84)] | Maximal at Tm[(85)] |
| Order by birefringence | Increased[(86)] | Increased[(87)] |
| Light scattering | Increased[(86)] | Increased[(88)] |

[a] Followed by heat absorption.
[b] Determined with labeled axons.
[c] Assigned to phospholipid carbonyls.
[d] Measured with dipalmitoylphosphatidylcholine.
[e] Determined by the ratio of pyrene monomer to excimer fluorescence intensities.
[f] Hodgkin–Huxley model.

relationship between these two phenomena. Also shown in Table 1 are changes in phospholipid membranes undergoing a transition from liquid crystalline into crystalline state (cooling transition). Several of the findings on nerve membranes were already assigned to phospholipids. Comparison of these data lends strong support to the concept that at least part of the resting nerve membrane phospholipids exist in the liquid crystalline state and that during the passage of an impulse part of these liquid crystalline phospholipids enter transiently into the crystalline state. In accordance with the increased permeability of liposomes held at their phase transition temperature, this transient phase change of the nerve membrane should contribute to the observed transmembrane ion currents during excitation. Distearoylphosphatidylcholine bilayers exhibit typical ion channels when at the phase transition temperature.[(89)] Antonov *et al.* concluded that owing to this property of phosphatidylcholine no absolute requirement in biological membranes for proteineous ion channels would exist.

Factors controlling the state of the membrane lipids are of importance. During excitation the nerve membrane appears to be transiently in the crystalline state. This could be due to the high content of saturated fatty acids in the nerve membrane lipids. The effects of nerve membrane proteins should, however, also be considered. Yet, it must be possible under resting potential to have some regions of the nerve membrane in liquid crystalline state, as indicated by the results in Table 1. In order to distinguish between liquid crystalline phospholipid model membrane and the state of the participating lipids of nerve membrane under resting potential we call this state an *electrostatically maintained liquid crystalline state*. It is to be stressed that it is necessary to have only a part of the membrane lipids under potential control, i.e., during excitation in crystalline and under resting potential in the induced liquid crystalline state.

The likely factors contributing to the maintenance of the induced liquid crystalline state are as follows: (i) SMOS, which presumably requires, for π-orbital overlap, that the participating lipids be in a conformation where the glycerol backbone lies parallel to the membrane surface and both acyl chains start perpendicular to the membrane-water interface[90]; (ii) membrane potential, which exerts pressure on the membrane due to Coulombic attraction between the charged ion layers on the inner (negative) and outer (positive) surfaces of the membrane. In this respect the peculiar lipid composition of nerve membrane is of importance. The acyl chains of nerve membrane lipids consist of both unsaturated and saturated species.[91,92] The phase transition temperature of fully hydrated dipalmitoyl- and distearoylphosphatidylcholine are at 41 and 56°C, respectively, whereas that for dioleoylphosphatidylcholine is at $-22$°C.[78] The crystalline → liquid crystalline transition of dipalmitoylphosphatidylcholine is accompanied by thinning of the membrane by approximately 5 Å.[93,94] Lateral expansion compensates for the reduction in membrane thickness and produces the overall increase in free volume which occurs at the transition; (iii) in addition to the above, the resting membrane potential maintains a higher than equilibrium concentration of $Na^+$ in the electrical double layer of the outer axon surface. Subsequently, this causes acidic phospholipids to be deprotonated and in liquid crystalline state (see Section 7). Owing to the high content of saturated lipids, however, if the membrane potential is removed (depolarization) the membrane spontaneously undergoes a transition to the crystalline state with subsequent changes in the lipid conformation.

Hysteresis of axon membrane phase transition can explain the transmembrane negative resistance reported by Moore.[95] At higher potentials under current control the membrane remains in the induced liquid crystalline state. With superimposed depolarizing pulses the membrane reaches a critical potential value allowing the membrane to go to the crystalline state. Both crystalline → induced liquid crystalline and induced liquid crystalline → crystalline transitions take place at their corresponding critical threshold potentials, which differ due to the hysteresis of the membrane lipid phase behavior. The abrupt changes and hysteresis in membrane potential during heating–cooling cycles of an axon, as reported by Inoue et al.,[96] are explained in a similar manner.

Several local anesthetics lower the transition temperature of phospholipids.[97,98] In these cases the anesthetic potency correlates with their ability to lower the transition temperature. Lee concluded the Na channel to be active in crystalline bilayers only.[98] Fluidization of the membrane by local anaesthetics would result in the inactivation of the ion channels. An alternative, although closely related, view is allowed according to the present model. Lowering of the nerve membrane phase transition temperature ("apparent fluidization") by these drugs prevents the membrane from transiently entering the phase transition region and thus blocks the impulse conduction.

(iv) The fourth factor involved in the regulation of the phase behavior is the counterion-induced transition of acidic phospholipids (phosphatidylserine in axons) from the crystalline to liquid crystalline state (see below).

## 5. Salt-Induced Conformational Changes in Phosphatidylserine

The anomalous behavior of the acidic phospholipids phosphatidylserine and -glycerol in response to electrolytes has been observed. In contrast to DLVO theory, screening of the phospholipid negative charge by counterions results in an expansion of the lattice.[90,99,100] Using monolayers and liposomes of phosphatidylglycerol as substrates for phospholipases A1 and A2 we have obtained evidence for a NaCl-induced continuous phase change in this phospholipid.[90] According to our results phosphatidylglycerol at pH 7.4 in the absence of electrolytes is in a condensed, crystalline state. In the presence of salt, the conformation of the lipid corresponds to that in the liquid crystalline state. Phosphatidylserine can be expected to behave in a similar manner.

Although phosphatidylserine is in general asymmetrically distributed in cell membranes with the bulk of this lipid in the cytoplasmic leaflet of the bilayer, some phosphatidylserine appears to reside in the outer lipid monolayer of the axonal membrane. Furthermore, this phosphatidylserine is involved in the nerve action potential.[101] Treatment of an axon with extracellular serine decarboxylase converts phosphatidylserine to -ethanolamine, which results in a decrease in the action potential spike height. Catalysis of the reversed reaction by this enzyme in the presence of excess L-serine converts phosphatidylethanolamine to -serine. This produces an average of 28% increase in the action potential amplitude. It is worth noticing that several anaesthetic compounds have been shown to bind phosphatidylserine *in vitro*.[28] The role of phosphatidylserine phase behavior in the nerve action potential will be discussed in somewhat more detail in Section 7.

## 6. Charge-Transfer Complex of Phospholipid and Cholesterol

Cholesterol is ubiquitous for nearly all mammalian membranes and is generally present in a 1:1 stoichiometry with phospholipids.[102] The structural and functional roles of cholesterol in biological membranes are not known. Several of the properties of this lipid are anomalous. For instance, it has an extinction coefficient of $20 \times 10^3$ with a maximum in the uv absorption at 235 nm.[103]

Cholesterol complexes with phospholipids. Preference for phosphatidylcholine has been reported but not confirmed.[104–107] In mixed monolayer films of cholesterol and phospholipids, at an air/water interface, cholesterol has a pronounced condensing effect. The area per molecule in the mixed film is less than the area per molecule in films of each of the compounds alone.[108–110] The nature of this interaction is uncertain. The hydroxyl group of cholesterol has been suggested to be hydrogen bonded to the carbonyl oxygens of the phospholipid acyl ester bonds.[111–115] Studies with diether phospholipids argue against such a role for the phospholipid carbonyl moieties.[116] On the other hand the phospholipid carbonyls do have a marked effect on the permeability of lipid bilayers as revealed by studies with liposomes consisting of either diether or diester phospholipids and cholesterol.[117] A cholesterol induced conformational change in phospholipids has been used to explain the changes in Raman $C=O$ stretch which occurs upon the sterol binding to dipalmitoylphosphatidylcholine.[118] Hydrogen bonding of cholesterol hydroxyl to phospholipid phosphate oxygen has been proposed as well.[119,120] It is significant that in studies on the association of cholesterol with lipids the conclusions made are based on the assumption that the only interacting site in cholesterol is provided by the hydroxyl function. Yet, the experimental data are compatible with a model where the cholesterol hydroxyl is hydrogen bonded to the phosphate oxygen and the phospholipid ester bond carbonyls form a charge-transfer complex with the cholesterol $C=C$ bond.* Binding of chloroform, iodine, and barbiturates to phosphatidylcholine has been demonstrated.[121–123] Formation of charge-transfer complexes between phosphatidylcholine and a variety of electron acceptors including 2,4-dinitrophenol has been reported.[124–126] A number of investigators have reported on the conductivity of both dry and hydrated phospholipid films.[127–135] Hydration, unsaturation, and cholesterol (or the presence of other suitable electron acceptors) reduce the resistance and the activation energy for conduction.[130,134,135]

---

* The possibility of phosphatidylcholine (donor)–cholesterol (acceptor) charge-transfer complex formation was initially suggested by Gutmann [*Bioelectrochemistry* (H. Keyzer and F. Gutmann, eds.), Plenum Press, New York (1980), pp. 159–169].

Cholesterol reduces the permeability of phospholipid bilayers to solutes such as $Na^+$, $Cl^-$, and glucose.[136] Introduction of cholesterol into diether phosphatidylcholine membranes does not significantly affect their permeability for $Na^+$.[117] It was concluded that the ester carbonyls are strongly involved in the interaction between cholesterol and phospholipid. Here we want to point out the possible analogy between the structure of ionophores of the crown ether type and the surface of a phospholipid membrane. Accordingly, the binding of $K^+$ to phospholipid membranes could involve coordination of the cation to the carbonyls in a manner similar to the binding of $K^+$ to, say, valinomycin.[137] Thus, cholesterol would reduce the permeability of phospholipid membranes mostly due to its interaction with the phospholipid carbonyls involved in the cholesterol–phospholipid charge-transfer complex. Segregation of different phospholipids into separate domains with defined stoichiometry and arrangement including membrane proteins provides the cell with specific functional areas and sites in its membranes, e.g., sites with differing permeabilities for $Na^+$, $K^+$, and $Ca^{++}$.

The importance of phospholipid carbonyls for the transmembrane ion conductance is further exemplified by the observation that the ester carbonyls contribute significantly to the capacitance of phospholipid bilayers with a reciprocal relationship in the 0.1–50-Hz range.[138] The noise spectrum of an axon is dominated by capacitance and is of the $f^{-1}$ type in a range covering the above frequencies.[139–142]

Cholesterol abolishes the thermotropic chain melting transition of phospholipids.[143–144] In the present model gating currents are proposed to represent the transfer of an electron pair from the cholesterol–phospholipid charge-transfer band into SMOS, which in turn causes the dissociation of the complex. This is compatible with the effects of pressure on nerve action potential.[25,145] For isolated axons held at elevated pressures the action potential and the time-to-peak-current are prolonged while the conductance amplitude remains practically unaffected. It was concluded that elevated pressures primarily influence gating processes. According to McConnell *et al.* the complex formed would in this case be nonionic owing to increased volume caused by reduced Coulombic attraction.[146] On the other hand, the free volume increase upon crystalline → liquid crystalline transition will delay the relaxation of the membrane back to the resting state while at elevated pressures. Hysteresis can be expected owing to nonequilibrium thermodynamics.

Upon dissociation of the phospholipid–cholesterol charge-transfer complex the liberated phospholipid molecule should adopt the thermodynamically favorable crystalline state predicted from the discharge of the counterions from phosphatidylserine as well as from the high content of saturated lipids in the nerve membrane. During passage of an action potential the absorption of an axon at 280 nm is increased, while there is a decrease in the absorption at 245 and 260 nm.[147] These deflections in the uv spectrum deviate temporally and the increase at 280 nm precedes by 0.5 ms the changes at shorter wavelengths. Part of the above findings could reflect changes in the state of the charge-transfer band. Likewise, uv radiation has been shown to block the gating currents.[148] The measurements on the binding of TNS to axons, among other observations made, clearly show the high degree of anisotropy of the nerve membrane and specifically indicate an uniaxial ordering. Phase transition of phospholipids is a cooperative process. Therefore, phase transition of nerve membrane phospholipids could in part be a transition in one dimension and involve those phospholipids dissociated from cholesterol during action potential while organized in linear arrays parallel to the longitudinal axis of the fiber.[149]

## 7. A Qualitative, Molecular Model of the Nerve Impulse

In the resting state there is a potential of approximately $-80$ mV across the membrane with positive outside and negative inside. As was discussed above, this resting potential keeps the membrane in an induced liquid crystalline state. SMOS is in a superconducting state and the charge-transfer band of phospholipid and cholesterol exists in the glycerol backbone region of the membrane. Phosphatidylserines in the outer leaflet of the bilayer are screened by $Na^+$, which keeps these acidic phospholipids in a liquid crystalline state.*,†

The first event after a depolarizing pulse is the discharge of membrane

---

* The electrostatic analysis put forward by Träuble for acidic phospholipids can be applied to nerve membrane phosphatidylserine as well [in *Structure of Biological Membranes* (S. Abrahamsson and I. Pascher, eds.), Nobel Foundation Symposium, Plenum Press, New York (1977) Vol. 34, pp. 509–550]. As noted by Träuble, DLVO-theory as such does not suffice in describing the properties of acidic phospholipids. For the nerve membrane this results in the following. The resting potential keeps higher than equilibrium concentration of $Na^+$ in the double layer of the exterior membrane lipid monolayer. This corresponds to increased ionic strength and leads to deprotonation of phosphatidylserine. Owing to Coulombic repulsion the negatively charged PS molecules are mixed in liquid crystalline state in the membrane. Upon depolarization protonation of PS occurs as a consequence of surge of $Na^+$ from the membrane surface. This is followed by sequestering of PS into a separate crystalline phase which contains $Na^+$ channels in active form. Notably, low extracellular pH should also reduce the gating currents as it enhances the electron affinity of the membrane surface CT-band electron acceptors. More thorough analysis of these events and their connection to the operation of $Na^+$ channel will be presented elsewhere.
† Note also the article by Antonov *et al.* in *Biophysics* **27**, 862–867 (1982).

capacitance. During this there is a surge of $Na^+$ from the outer membrane surface which leaves the phosphatidylserines without counterions, and therefore these lipids can enter, after the dissociation of the phospholipid–cholesterol charge-transfer couple, into crystalline state. If the charge-transfer band is not intact the counterions return and neutralize the acidic phosphatidylserines without the initiation of an impulse.

Capacitance discharge is followed by the asymmetric gating currents. In our model they represent the removal of a pair of electrons from the (by rigorous formalism) superionic[150] charge-transfer band to SMOS. Changes in the CT-band can be illustrated as follows:

$$A^= D^{++} A^= D^{++} A^= D^{++} \rightarrow A^= D^{++} A^{\infty\infty} D^{++} A^= D^{++}$$

The $2e^-$ leaves the CT band and enters SMOS. Upon removal of the electron pair from the CT band the CT couple dissociates and the phospholipid starts to undergo a phase transition from induced liquid crystalline state to crystalline state preconditioned by the unscreened, protonated phosphatidylserines. Concomitantly with the initiation of phospholipid phase transition the energy level of the holes in the CT band is elevated and the removed $2e^-$ cannot return to $A^{\infty\infty}$.

Phase transition of membrane phospholipids now starts and eventually leads to the activation of the $Na^+$ channels. The transmembrane ion flow generates lateral field in the membrane. Owing to this field the auxiliary $2e^-$ in SMOS will move towards the resting membrane away from the region undergoing phase transition. In this manner, the $2e^-$ on SMOS and the vacant sites on $A^{\infty\infty}$ become separated, the latter being bound to and in part initiating the phase transition. Yet, the $2e^-$ is coupled to $D^{++}$, which as $A^{\infty\infty} D^{++}$ propagates in the membrane. Thus, the generated, existing impulse consists of

i. $2e^-$ on SMOS and originating from $A^=$ of the cholesterol–phospholipid CT band.
ii. $A^{\infty\infty} D^{++}$ propagating in the CT band.
iii. Phospholipids undergoing phase transition from induced liquid crystalline state to crystalline state. These phospholipids consist of phosphatidylserine and those which have been in $A^= D^{++}$, i.e., represented by $D^{++}$ in $A^{\infty\infty} D^{++}$.
iv. Activated ion channels in crystalline phospholipid lattice.

As is evident from the above we are concerning only the events taking place in the outer lipid monolayer of the axon membrane bilayer. Although the cytoplasmic leaflet under physiological conditions is involved, in our model we are assuming that most of the electronic activity resides in the outer half of the membrane bilayer. This assumption is supported by our preliminary experiments, where we found extremely high conductivities in proper phospholipid monolayers at a solid/water interface.

The present model for nerve impulse resembles closely the exciton mechanism of high-temperature superconductivity, as put forward by Little and Ginzburg.[151–157] Little's polymer consists of a polyene spine with polarizable dye side chains, the latter forming the exciton band.[151] Ginzburg proposes high-temperature superconductivity to be found in thin metallic films placed between highly dielectric layers.[153,154]

## Acknowledgments

The comments by Drs. Heikki Seppä, Jorma Vuorinen, and Antti-Pekka Tulkki from the University of Technology, Otaniemi, Finland, and those by Dr. Petri Vainio from the Department of Medical Chemistry, University of Helsinki, are appreciated. The basic concepts have been presented in the Soviet–Finnish Symposium on Nuclear and Plasma Membranes, Helsinki, Finland, March 18–19 1982, and in the Scandinavian Workshop on Plasma Lipoproteins, Punkaharju/Helsinki, Finland, June 22–29, 1982, as well as in lectures delivered by PKJK while a visiting lecturer of the Soviet Academy of Sciences, Moscow, USSR, March 12–19, 1983. Financial support was provided by the Finnish State Medical Research Council and the University of Helsinki.

## References

1. A. L. Hodgkin, "The Ionic Basis of Nervous Conduction," *Science* **145**, 1148–1154 (1964).
2. A. F. Huxley, "Excitation and Conduction in Nerve: Quantitative Analysis," *Science* **145**, 1154–1159 (1964).
3. J. L. Hindmarsh and R. M. Rose, "A Model of the Nerve Impulse Using Two First-Order Differential Equations," *Nature* **296**, 162–164 (1982).
4. A. L. Hodgkin, *The Conduction of the Nervous Impulse*, C. C. Thomas, Springfield, Massachusetts (1964).
5. B. Hille, in *Progress in Biophysics and Molecular Biology* (J.A.V. Butler and D. Noble, eds.), Vol. 21, Pergamon, New York (1970), pp. 3–34.
6. A. C. Scott, "The Electrophysics of Nerve Fiber," *Rev. Mod. Phys.* **47**, 487–533 (1975).
7. I. Tasaki, *Physiology and Electrochemistry of Nerve Fibers*, Academic, New York (1982).
8. D. E. Goldman, "A Molecular Structural Basis for the Excitation Properties of Axons," *Biophys. J.* **4**, 167–188 (1964).
9. L. Y. Wei, "Role of Surface Dipoles on Axon Membrane," *Science* **163**, 280–282 (1969).
10. S. P. Almeida, J. D. Bond, and T. C. Ward, "The Dipole Model and Phase Transitions in Biological Membranes," *Biophys. J.* **11**, 995–1001 (1971).

11. F. W. Cope, in *Bioelectrochemistry* (H. Keyzer and F. Gutmann, eds.), Plenum Press, New York (1980), pp. 297–329.

12. F. W. Cope, "Evidence from Activation Energies for Superconductive Tunnelling in Biological Systems at Physiological Temperatures," *Physiol. Chem. Phys.* 3, 403–410 (1971).

13. F. W. Cope, "Enhancement by High Electric Fields of Superconduction in Organic and Biological Solids at Room Temperature and a Role in Nerve Conduction," *Physiol. Chem. Phys.* 6, 405–410, (1974).

14. R. Henderson and J. H. Wang, "Solubilization of a Specific Tetrodotoxin-Binding Component from Garfish Olfactory Nerve Membrane," *Biochemistry* 11, 4565–4569 (1972).

15. W. S. Agnew, S. R. Levinson, J. S. Brabson, and M. A. Raftery, "Purification of the Tetrodotoxin-Binding Component Associated with the Voltage-Sensitive Sodium Channel from *Electrophorus electricus* Electroplax Membranes," *Proc. Natl. Acad. Sci. USA* 75, 2606–2610 (1978).

16. J. A. Talvenheimo, M. M. Tamkun, and W. A. Catterall, "Reconstitution of Neurotoxin-Stimulated Sodium Transport by the Voltage-Sensitive Sodium Channel purified from Rat Brain," *J. Biol. Chem.* 257, 11868–11871 (1982).

17. M. Condrescu and R. Villegas, "Ion Selectivity of the Nerve Membrane Sodium Channel Incorporated into Liposomes," *Biochim. Biophys. Acta* 688, 660–666 (1982).

18. B. K. Krueger, J. F. Worley, and R. J. French, "Single Sodium Channels from Rat Brain Incorporated into Planar Lipid Bilayer Membranes," *Nature* 303, 172–175 (1983).

19. A. L. Hodgkin and A. F. Huxley, "A Quantitative Description of Membrane Current and its Application to Conduction and Excitation in Nerve," *J. Physiol.* 117, 500–544 (1952).

20. C. M. Armstrong and F. Bezanilla, "Currents Related to Movement of the Gating Particles of the Sodium Channels," *Nature* 242, 459–461 (1973).

21. R. Lorente de Nó, F. Vidal, and L.M.H. Larramendi, "Restoration of Sodium-Deficient Frog Nerve Fibers by Onium Ions," *Nature* 179, 737–738 (1957).

22. J. W. Moore, "Temperature and Drug Effects on Squid Axon Membrane Ion Conductances," *Fed. Proc.* 17, 113 (1958).

23. I. Tasaki, I. Singer, and A. Watanabe, "Excitation of Internally Perfused Squid Giant Axons in Sodium-Free Media," *Proc. Natl. Acad. Sci.* 54, 763–769 (1965).

24. M. Calvin, H. H. Wang, G. Entine, D. Gill, P. Ferruti, M. A. Harpold, and M. P. Klein, "Biradical Spin Labeling for Nerve Membranes," *Proc. Natl. Acad. Sci* 63, 1–8 (1969).

25. J. J. Kendig and E. N. Cohen, "Pressure Antagonism to Nerve Conduction Block by Anesthetic Agents," *Anesthesiology* 47, 6–10 (1977).

26. C. R. Badger and G. M. Helmkamp, "Modulation of Phospholipid Transfer Protein Activity. Inhibition by Local Anesthetics," *Biochim. Biophys. Acta* 692, 33–40 (1982).

27. H. S. Hendrickson and M. C. E. van Dam-Mieras, "Local Anesthetic Inhibition of Pancreatic Phospholipase $A_2$ Action on Lecithin Monolayers," *J. Lipid Res.* 17, 399–405 (1976).

28. M. B. Feinstein and M. Paimre, "Specific Reaction of Local Anesthetics with Phosphodiester Groups," *Biochim. Biophys. Acta* 115, 33–45 (1965).

29. J. L. Browning and H. Akutsu, "Local Anesthetics and Divalent Cations Have the Same Effect on the Headgroups of Phosphatidylcholine and Phosphatidylethanolamine," *Biochim. Biophys. Acta* 684, 172–178 (1982).

30. Y. Boulanger, S. Schreier, and I.C.P. Smith, "Molecular Details of Anaesthetic–Lipid Interaction as Seen by Deuterium and Phosphorus-31 Nuclear Magnetic Resonance," *Biochemistry* 20, 6824–6830 (1981).

31. J. A. Virtanen and P. K. J. Kinnunen, *A Qualitative, Molecular Model of the Nerve Impulse. Conductive Properties of Unsaturated Lipids*, University of Helsinki Offset Press (1981).

32. F. Huneeus-Cox, H. L. Fernandez, and B. H. Smith, "Effects of Redox and Sulfhydryl Reagents on the Bioelectric Properties of the Giant Axon of the Squid, *Biophys. J.* **6**, 675–689 (1966).
33. J. Reyes and R. Latorre, "Effect of the Anesthetics Benzyl Alcohol and Chloroform on Bilayers Made from Monolayers," *Biophys. J.* **28**, 259–280 (1979).
34. J. M. Fernándes, F. Bezanilla, and R. E. Taylor, "Effect of Chloroform on Charge Movement in the Nerve Membrane," *Nature* **297**, 150–152 (1982).
35. J. C. Nelander and A. E. Blaurock, "Disorder in Nerve Myelin: Phasing the Higher Order Reflections by Means of the Diffuse Scatter," *J. Mol. Biol.* **118**, 497–532 (1978).
36. J. P. Wikswo, J. P. Barack, and J. A. Freeman, "Magnetic Field of a Nerve Impulse: First Measurements," *Science* **208**, 53–55 (1980).
37. R. A. Luben, C. D. Cain, M. C.-Y. Chen, D. M. Rosen, and W. R. Adey, "Effects of Electromagnetic Stimuli on Bone and Bone Cells in Vitro: Inhibition of Responses to Parathyroid Hormone by Low-Energy Low-Frequency Fields," *Proc. Natl. Acad. Sci. USA* **79**, 4180–4184 (1982).
38. N. Zisapel and M. Laudon, "Dopamine Release Induced by Electrical Field Stimulation of Rat Hypothalamus in Vitro. Inhibition by Melatonin," *Biochem. Biophys. Res. Commun.* **104**, 1610–1616 (1982).
39. R. Dixey and G. Rein, "³H-Noradrenaline Release Potentiated in a Clonal Nerve Cell Line by Low-Intensity Pulsed Magnetic Fields," *Nature* **296**, 253–256 (1982).
40. R. Plonsey, "The Nature of Sources of Bioelectric and Biomagnetic Fields," *Biophys. J.* **39**, 309–312 (1982).
41. H. Fernández-Moran, "Diffraction of Electrons by Structures Resembling Myelin Lamellae," *Exp. Cell Res.* **2**, 673–679 (1951).
42. J. H. Matheja, "Electron Diffraction of Membranes," *Biophysik* **7**, 163–168 (1971).
43. R. S. Khare and R. K. Mishra, "Electron Diffraction Studies on the Effect of Certain Drugs on Brain Cell Membranes," *Stud. Biophys.* **38**, 205–209 (1973).
44. S. W. Hui, "Electron Diffraction Studies on Membranes," *Biochim. Biophys. Acta* **472**, 345–371.
45. I. Tasaki, A. Watanabe, R. Sandlin, and L. Carnay, "Changes in Fluorescence, Turbidity, and Birefringence Associated with Nerve Excitation," *Proc. Natl. Acad. Sci. USA* **61**, 883–888 (1968).
46. I. Tasaki, A. Watanabe, and M. Hallett, "Properties of Squid Axon Membrane as Revealed by a Hydrophobic Probe 2-*p*-toluidinylnaphatalene-6-sulfonate," *Proc. Natl. Acad. Sci. USA* **68**, 938–941 (1971).
47. I. Tasaki, A. Watanabe, and M. Hallett, "Fluorescence of Squid Axon Membráne Labelled with Hydrophobic Probes," *J. Membrane Biol.* **8**, 109–132 (1972).
48. I. Tasaki, E. Carbone, K. Sisco, and I. Singer, "Spectral Analysis of Extrinsic Fluorescence of the Nerve Membrane Labeled with Aminonaphatelene Derivatives," *Biochim. Biophys. Acta* **323**, 220–233 (1973).
49. I. Tasaki, "Energy Transduction in the Nerve Membrane and Studies of Excitation Processes with Extrinsic Fluorescent Probes," *Ann. NY Acad. Sci.* **227**, 247–267 (1974).
50. L. B. Cohen, B. M. Salzberg, H. J. Davila, W. N. Ross, D. Landowne, A. S. Waggoner, and C. H. Wang, "Changes in Axon Fluorescence During Activity: Molecular Probes of Membrane Potential," *J. Membrane Biol.* **19**, 1–36 (1974).
51. L. B. Cohen and B. M. Salzberg, "Optical Measurement of Membrane Potential," *Rev. Physiol. Biochem. Pharmacol.* **83**, 35–88 (1978).
52. A. G. MacDiarmid and A. J. Heeger, in *Molecular Electronic Devices* (F. L. Carter, ed.), Marcel Dekker, New York (1983), pp. 259–271.
53. K. C. Kao and W. Hwang, *Electrical Transport in Solids*, Pergamon Press, New York (1981).

54. K. R. Thulborn, F. E. Treloar, and W. H. Sawyer, "A Microviscosity Barrier in the Lipid Bilayer due to the Presense of Phospholipids Containing Unsaturated Acyl Chains," *Biochem. Biophys. Res. Commun.* **81**, 42–49 (1978).
55. S. Winstein, "Nonclassical Ions and Homoaromaticity," *Qt. Rev.* **23**, 141–176 (1969).
56. R. B. Woodward and R. Hoffmann, *The Conservation of Orbital Symmetry*, Verlag Chemie, Academic, New York (1970), pp. 101–107.
57. A. C. Rose-Innes and E. H. Rhoderick, *Introduction to Superconductivity*, Pergamon Press, New York (1978).
58. D. Chapman, "Phase Transitions and Fluidity Characteristics of Lipids and Cell Membranes," *Qt. Rev. Biophys.* **8**, 185–235 (1975).
59. A. G. Lee, "Functional Properties of Biological Membranes: A Physical–Chemical Approach," *Progr. Mol. Biol. Biophys.* **29**, 5–56 (1975).
60. A. G. Lee, "Lipid Phase Transitions and Phase Diagrams II. Lipid Phase Transitions," *Biochim. Biophys. Acta* **472**, 237–281 (1977).
61. A. G. Lee, "Lipid Phase Transitions and Phase Diagrams II. Mixtures Involving Lipids," *Biochim. Biophys. Acta* **472**, 285–344 (1977).
62. J. F. Nagle, "Theory of the Main Lipid Bilayer Phase Transition," *Ann. Rev. Phys. Chem.* **31**, 157–195 (1980).
63. D. Marsh, "Statistical Mechanics of the Fluidity of Phospholipid Bilayers and Membranes," *J. Membrane Biol.* **18**, 145–162 (1974).
64. S. Marcelja, "Chain Ordering in Liquid Crystals II. Structure of Bilayer Membranes," *Biochim. Biophys. Acta* **367**, 165–176 (1974).
65. H. L. Scott, "Some Models for Lipid Bilayer and Biomembrane Phase Transitions," *J. Chem. Phys.* **62**, 1347–1353 (1975).
66. F. W. Wiegel, "An Exactly Solvable Two-Dimensional Biomembrane Model," *J. Stat. Phys.* **13**, 515–530 (1975).
67. R. E. Jacobs, B. Hudson, and H. C. Andersen, "A Theory of the Chain Melting Phase Transition of Aqueous Phospholipid Dispersions," *Proc. Natl. Acad. Sci. USA* **72**, 3993–3997 (1975).
68. H. M. Zacharis, "A Linear Function for the Melting Behaviour of Lipids," *Chem. Phys. Lipids* **18**, 221–231 (1977).
69. R. G. Priest, "Semiphenological Model for the Lipid Bilayer Phase Transition: Finite Chains in Three Dimensions," *Chem. Phys.* **66**, 722–725 (1977).
70. M. I. Kanehisa and T. Y. Tsong, Cluster Model of Lipid Phase Transitions with Application to Passive Permeation of Molecules and Structure Relaxations in Lipid Bilayers," *J. Am. Chem. Soc.* **100**, 424–432 (1978).
71. S. Doniach, "Thermodynamic Fluctuations in Phospholipid Bilayers," *J. Chem. Phys.* **68**, 4912–4916 (1978).
72. S. Marcelja and J. Wolfe, "Properties of Bilayer Membranes in the Phase Transition or Phase Separation Region," *Biochim. Biophys. Acta* **557**, 24–31 (1979).
73. K. A. Dill and R. S. Cantor, in *Physics of Amphiphiles. Micelles, Vesicles and Microemulsions* (V. Degiorgio and M. Corti, eds.), Elsevier, Amsterdam (1985), pp. 376–393.
74. H. Träuble, "Phasenumwandlungen in Lipiden. Mögliche Schaltprozesse in biologischen Membranen," *Naturwissenschaften* **58**, 277–284 (1971).
75. H. Träuble and H. Eibl, "Electrostatic Effects on Lipid Phase Transitions: Membrane Structure and Ionic Environment," *Proc. Natl. Acad. Sci. USA* **71**, 214–219 (1974).
76. J. V. Howarth, R. D. Keynes, and J. M. Ritchie, "The Origin of Initial Heat Associated with a Single Impulse in Mammalian Nonmyelinated Nerve Fibers," *J. Physiol.* **194**, 745–793 (1968).
77. A. Fraser and A. H. Frey, "Electromagnetic Emission at Micron Wavelengths from Active Fibers," *Biophys. J.* **8**, 731–734 (1968).

78. B. D. Ladbrooke and D. Chapman, "Thermal Analysis of Lipids, Proteins and Biological Membranes. A Review and Summary of Some Recent Studies," *Chem. Phys. Lipids* **3**, 304–367 (1969).

79. M. H. Sherebrin, B.A.E. MacClement, and A. J. Franko, "Electric Field-Induced Shifts in the Infrared Spectrum of Conducting Nerve Axons," *Biophys. J.* **12**, 977–989 (1972).

80. D. G. Cameron, J. K. Kauppinen, H. H. Casal, and H. H. Mantsch, "The Thermotropic Behaviour of Diacyl Phosphatidylcholines: A Fourier Transform Infrared Study" (manuscript).

81. D. Georgescauld and H. Dulochier, "Transient Fluorescent Signals from Pyrene Labeled Pike Nerves During Action Potential. Possible Implications for Membrane Fluidity Changes," *Biochem. Biophys. Res. Commun.* **85**, 1186–1191 (1978).

82. H. J. Pownall and L. C. Smith, "Viscosity of the Hydrocarbon Region of Micelles. Measurement by Excimer Fluorescence," *J. Am. Chem. Soc.* **95**, 3136–3140 (1973).

83. A. K. Soutar, H. J. Pownall, A. S. Hu, and L. C. Smith, "Phase Transitions in Bilamellar Vesicles. Measurements by Pyrene Excimer Fluorescence and Effect on Transacylation by Lecithin: Cholesterol Acyltransferase," *Biochemistry* **13**, 2828–2836 (1974).

84. A. L. Hodgkin and A. F. Huxley, "Currents Carried by Sodium and Potassium Ions Through the Membrane of the Giant Axon of Loligo," *J. Physiol.* **116**, 449–472 (1952).

85. D. Papahadjopoulos, K. Jacobson, S. Nir, and T. Isac, "Phase Transitions in Phospholipid Vesicles. Fluorescence Polarization and Permeability Measurements Concerning the Effect of Temperature and Cholesterol," *Biochim. Biophys. Acta* **311**, 330–348 (1973).

86. L. B. Cohen, R. D. Keynes, and B. Hille, "Light Scattering and Birefringence Changes During Nerve Activity," *Nature* **218**, 438–441 (1968).

87. M. J. Janiak, D. M. Small, and G. G. Shipley, "Interactions of Cholesterol Esters with Phospholipids: Cholesterol Myristate and Dimyristoyl Lecithin," *J. Lipid Res.* **20**, 183–199 (1979).

88. P. N. Yi and R. C. MacDonald, "Temperature Dependence of Optical Properties of Aqueous Dispersions of Phosphatidylcholine," *Chem. Phys. Lipids* **11**, 114–134 (1973).

89. V. F. Antonov, V. V. Petrov, A. A. Molnar, D. A. Predvoditelev, and A. S. Ivanov, "The Appearance of Single-ion Channels in Unmodified Lipid Bilayer Membranes at the Phase Transition Temperature," *Nature* **283**, 585–586 (1980).

90. P. K. J. Kinnunen, T. Thurén, P. Vainio, and J. A. Virtanen, in *Physics of Amphiphiles. Micelles, Vesicles and Microemulsions* (V. Degiorgio and M. Corti, eds.), Elsevier, Amsterdam (1984), pp. 687–701.

91. J. S. O'Brien and G. Rouser, "The Fatty Acid Composition of Brain Sphingolipids: Sphingomyelin, Ceramide, Cerebroside, and Cerebroside Sulfate," *J. Lipid Res.* **5**, 339–342 (1964).

92. J. S. O'Brien and E. L. Sampson, "Fatty Acid and Fatty Aldehyde Composition of the Major Brain Lipids in Normal Human Gray Matter, White Matter, and Myelin," *J. Lipid Res.* **6**, 545–551 (1965).

93. D. Chapman, R. M. Williams, and B. D. Ladbrooke, "Physical Studies of Phospholipids. VI. Thermotropic and Lyotropic Mesomorphism of some 1,2-diacyl-phosphatidylcholines (Lecithins)," *Chem. Phys. Lipids* **1**, 445–475 (1971).

94. H. Träuble and D. H. Haynes, "The Volume Change in Lipid Bilayer Lamellae at the Crystalline–Liquid Crystalline Phase Transition," *Chem. Phys. Lipids* **7**, 324–355 (1971).

95. J. W. Moore, "Excitation of the Squid Axon Membrane in Isoosmotic Potassium Chloride, *Nature* **183**, 265–266 (1959).

96. I. Inoue, Y. Kobatake, and I. Tasaki, "Excitability, Instability and Phase Transitions in Squid Axon Membrane Under Internal Perfusion with Dilute Salt Solutions," *Biochim. Biophys. Acta* **307**, 471–477 (1973).

97. P. Connor, B. S. Mangat, and L. S. Rao, "The Labilization of Lecithin Liposomes by

Steroidal Anaesthetics: A Correlation with Anaesthetic Activity," *J. Pharm. Pharmacol.* **26**, 120P (1974).

98. A. G. Lee, "Model for the Action of Local Anaesthetics," *Nature* **262**, 545–548 (1976).
99. J. F. Tocanne, P. H. J. Th. Ververgaert, A. J. Verkleij, and L. L. M. van Deenen, "A Monolayer and Freeze-Etching Study of Charged Phospholipids. I. Effects of Ions and pH on the Ionic Properties of Phosphatidylglycerol and Lysylphosphatidylglycerol," *Chem. Phys. Lipids* **12**, 201–219 (1974).
100. H. Hauser, M. C. Phillips, and M. D. Barratt, "Differences in the Interaction of Inorganic and Organic (Hydrophobic) Cations with Phosphatidylserine Membranes," *Biochim. Biophys. Acta* **413**, 341–353 (1975).
101. A. M. Cook, E. Low, and M. Ishijimi, "Effect of Phosphatidyl Serine Decarboxylase on Neural Excitation," *Nature* **239**, 150–151 (1972).
102. R. A. Demel and B. de Kruyff, "The Function of Sterols in MEMBRANES," *Biochim. Biophys.* Acta **457**, 109–132 (1976).
103. J. E. Bell and C. Hall, in *Spectroscopy in Biochemistry* (J. E. Bell, ed.), CRC Press, Boca, Raton, Fla. (1981), Vol. 1, pp. 3–36.
104. P. W. M. van Dijck, B. de Kruyff, L. L. M. van Deenen, J. de Gier, and R. A. Demel, "The Preference of Cholesterol for Phosphatidylcholine in Mixed Phosphatidylcholine–Phosphatidylethanolamine Bilayers," *Biochim. Biophys. Acta* **455**, 576–587 (1976).
105. R. A. Demel, J. W. C. M. Jansen, P. W. M. van Dijck, and L. L. M. van Deenen, "The Preferential Interaction of Cholesterol with Different Classes of Phospholipids," *Biochim. Biophys. Acta* **465**, 1–10 (1977).
106. W. I. Calhoun and G. G. Shipley, "Sphingomyelin–Lecithin Bilayers and Their Interaction with Cholesterol," *Biochemistry* **18**, 1717–1722 (1979).
107. A. Blume, "Thermotropic Behaviour of Phosphatidylethanolamine–Cholesterol and Phosphatidylethanolamine–Phosphatidylcholine–cholesterol Mixtures," *Biochemistry* **19**, 4908–4913 (1980).
108. D. Chapman, N. F. Owens, and D. A. Walker, "Physical Studies of Phospholipids. II. Monolayer Studies of Some Synthetic 2, 3-diacyl-DL-phosphatidylethanolamines and Phosphatidylcholines Containing *Trans* Double Bonds," *Biochim. Biophys. Acta* **120**, 148–155 (1966).
109. J. Tinoco and D. J. McIntosh, "Interactions Between Cholesterol and Lecithin in Monolayers at an Air–Water Interface," *Chem. Phys. Lipids* **4**, 72–84 (1970).
110. R. A. Demel, W. S. M. Geurts van Kessel, and L. L. M. van Deenen, "The Properties of Polyunsaturated Lecithins in Monolayers and Liposomes and the Interaction of these Lecithins with Cholesterol," *Biochim. Biophys. Acta* **266**, 26–40 (1972).
111. H. Brockerhoff, "Model of Interaction of Polar Lipids, Cholesterol, and Proteins in Biological Membranes," *Lipids* **9**, 645–650 (1974).
112. C.-H. Huang, "Roles of Carbonyl Oxygens at the Bilayer Interface in Phospholipid–Sterol Interaction," *Nature* **259**, 242–244 (1976).
113. N. Chatterjee and H. Brockerhoff, "Evidence for Stereospecific Phospholipid–Cholesterol Interaction in Lipid Bilayers," *Biochim. Biophys. Acta* **511**, 116–119 (1978).
114. P. L. Yeagle, W. C. Hutton, C.-H. Huang, and R. B. Martin, "Headgroup Conformation and Lipid–Cholesterol Association in Phosphatidylcholine Vesicles: A $^{31}P\langle^{1}H\rangle$ Nuclear Overhauser Effect Study," *Proc. Natl. Acad. Sci. USA* **72**, 3477–3481 (1975).
115. P. L. Yeagle and R. B. Martin, "Hydrogen-Bonding of the Ester Carbonyls in Phosphatidylcholine Bilayers," *Bio Biophys. Res. Commun.* **69**, 775–780 (1976).
116. S. Clejan, R. Bittman, P. W. Deroo, Y. A. Isaacson, and A. F. Rosenthal, "Permability Properties of Sterol-Containing Liposomes from Analogues of Phosphatidylcholine Lacking Acyl Groups, *Biochemistry* **18**, 2118–2125 (1979).
117. F. T. Schwarz and F. Paltauf, "Influence of the Ester Carbonyl Oxygens of Lecithin on

the Permeability properties of Mixed Lecithin–Cholesterol Bilayers," *Biochemistry* **16**, 4335–4339 (1977).

118. E. Bicknell-Brown and K. G. Brown, "Raman Studies of Lipid Interactions at the Bilayer Interface: Phosphatidylcholine–Cholesterol," *Biochem. Biophys. Res. Commun.* **94**, 638–645 (1980).

119. A. Darke, E. G. Finer, A. G. Flock, and M. C. Phillips, "Nuclear Magnetic Resonance Study of Lecithin–Cholesterol Interactions," *J. Mol. Biol.* **63**, 265–279 (1972).

120. S. P. Verma and D. F. H. Wallach, "Effects of Cholesterol on the Infrared Dichroism of Phosphatide Multilayers," *Biochim. Biophys. Acta* **330**, 122–131 (1973).

121. M. Okazaki and I. Hara, "Solubility of Phosphatidylcholine in Choloroform. Formation of Hydrogen Bonding Between Phosphatidylcholine and Chloroform," *Chem. Phys. Lipids* **17**, 28–37 (1976).

122. G. L. Jendrasiak, "The Interaction of Iodine with Lecithin Micelles," *Chem. Phys. Lipids* **4**, 85–95 (1970).

123. A. G. Lee, "Interactions Between Phospholipids and Barbiturates," *Biochim. Biophys. Acta* **455**, 102–108 (1976).

124. B. Rosenberg and H. C. Pant, "The Semiconducting Rectifier Behaviour of a Biomolecular Lipid Membrane," *Chem. Phys. Lipids* **4**, 203–207 (1970).

125. B. Bhowmik and G. L. Jendrasiak, "Charge Transfer Complexes of Lipids with Iodine," *Nature* **215**, 842–843 (1967).

126. B. Rosenberg and B. B. Bhowmik, "Donor–Acceptor Complexes and the Semiconductivity of Lipids," *Chem. Phys. Lipids* **3**, 109–124 (1969).

127. B. Rosenberg and G. L. Jendrasiak, "Semiconductive Properties of Lipids and Their Possible Relationship to Lipid Bilayer Conductivity," *Chem. Phys. Lipids* **2**, 47–54 (1968).

128. G. L. Jendrasiak, "Effect of Iodine on the Electrical Resistance of Lipid Bilayer Membranes," *Chem. Phys. Lipids* **3**, 98–101 (1969).

129. L. Y. Wei and B. Y. Woo, "Electronic Conduction in Lipid Films with Metal Contacts," *Biophys. J.* **13**, 877–889 (1973).

130. G. L. Jendrasiak and J. H. Hasty, "The Electrical Conductivity of Hydrated Phospholipids," *Biochim. Biophys. Acta* **348**, 45–54 (1974).

131. I. Lundström and M. Stenberg, "Charge Injection and Charge Storage in Lipid Multilayers," *Chem. Phys. Lipids* **12**, 287–302 (1974).

132. G. L. Jendrasiak and J. C. Mendible, "The Effect of Phase Transition on the Hydration and Electrical Conductivity of Phospholipids," *Biochim. Biophys. Acta* **424**, 133–148 (1976).

133. G. L. Jendrasiak and J. C. Mendible, "The Phospholipid Head-Group Orientation: Effect on Hydration and Electrical Conductivity," *Biochim. Biophys. Acta* **424**, 149–158 (1976).

134. A. A. Shulyndin, "The Lateral Resistance of Phospholipid Bilayer Membranes of Different Structures," *Izv. Akad. Nauk USSR* (Ser. Biol., in Russian) **3**, 456–459 (1980).

135. A. A. Shulyndin, "The Effect of Cholesterol on the Lateral Resistance of Phospholipid Membranes," *Izv. Akad. Nauk USSR* (Ser. Biol., in Russian) **1**, 154–156 (1981).

136. D. Papahadjopoulos, S. Nir, and S. Ohki, "Permeability Properties of Phospholipid Membranes: Effect of Cholesterol and Temperature," *Biochim. Biophys. Acta* **266**, 561–586 (1971).

137. M. M. Shemyakin, Yu. A. Ovchinnikov, V. T. Ivanov, V. K. Antonov, E. I. Vinogradova, A. M. Shkrob, G. G. Malenkov, A. V. Evstratov, I. A. Laine, E. I. Melnik, and I. D. Ryadova, "Cyclodepepsipedtides as Chemical Tools for Studying Ionic Transport Through Membranes," *J. Membrane Biol.* **1**, 402–430 (1960).

138. R. G. Ashcroft, H. G. L. Coster, and J. R. Smith, "The Molecular Organization of Bimolecular Lipid Membranes. The Dielectric Structure of the Hydrophilic/Hydrophobic Interface," *Biochim. Biophys. Acta* **643**, 191–204 (1981).

139. R. Pethig, *Dielectric and Electronic Properties of Biological Materials*, Wiley, New York (1979).
140. A. A. Verveen and L. J. DeFelice, "Membrane Noise," *Prog. Biophys. Molec. Biol.* **28**, 189–265 (1974).
141. F. Conti, L. J. DeFelice, and E. Wanke, "Potassium and Sodium Ion Current Noise in the Membrane of the Squid Giant Axon," *J. Physiol.* **248**, 45–82 (1975).
142. F. Conti and E. Wanke, "Channel Noise in Nerve Membranes and Lipid Bilayers," *Quart. Rev. Biophys.* **8**, 451–506 (1975).
143. B. D. Ladbrooke, R. M. Williams, and D. Chapman, "Studies on Lecithin–Cholesterol–Water Interaction by Differential Scanning Calorimetry and X-Ray Diffraction," *Biochim. Biophys. Acta* **150**, 333–340 (1968).
144. S. Mabrey, P. L. Mateo, and J. M. Sturtevant, "High Sensitivity Scanning Calorimetric Study of Mixtures of Cholesterol with Dimyristoyl- and Dipalmitoylphosphatidylcholine, *Biochemistry* **17**, 2464–2468 (1978).
145. J. V. Henderson and D. L. Gilbert, "Slowing of Ionic Currents in the Voltage-Clamped Squid Axon by Helium Pressure, *Nature* **258**, 351–352 (1975).
146. H. M. McConnell, B. M. Hoffman, and R. M. Metzger, "Charge Transfer in Molecular Crystals," *Proc. Natl. Acad. Sci. USA* **53**, 46–50 (1965).
147. P. O. Makarov and M. V. Krasovitskaya, "Investigation of the Molecular Mechanism of Neurodynamics by the Method of Cytospectrophotometry," *Biofizika* **15**, 492–496 (1970).
148. J. M. Fox, B. Neumke, W. Nonner, and R. Stämpfli, "Block of Gating Currents by Ultraviolet Radiation in the Membrane of Myelinated Nerve," *Pflugers Arch.* **364**, 143–145 (1976).
149. M. K. Jain, F. Ramirez, T. M. McCaffrey, P. V. Ioannou, J. F. Marecek, and J. Leunissen-Bijvelt, "Phosphatidylcholesterol Bilayers. A Model for Phospholipid–Cholesterol Interaction," *Biophys. Biochim. Acta* **600**, 678–688 (1980).
150. P. J. Strebel and Z. G. Soos, "Theory of Charge Transfer in Aromatic Donor–Acceptor Crystals," *J. Chem. Phys.* **53**, 4077–4090 (1970).
151. W. A. Little, "Possibility of Synthetizing an Organic Superconductor," *Phys. Rev.* **134**, 1416–1424 (1964).
152. W. A. Little, "The Exciton Mechanism in Superconductivity," *J. Polymer Sci. (Part C)* **29**, 17–26 (1970).
153. V. L. Ginzburg, "The Problem of High Temperature Superconductivity," *Contemp. Phys.* **9**, 355–374 (1968).
154. V. L. Ginzburg, "The Problem of High-Temperature Superconductivity. *II*," *Sov. Phys. Usp.* **13**, 335–352 (1970).
155. V. L. Ginzburg, "High Temperature Superconductivity," *J. Polymer Sci. (Part C)* **29**, 3–16 (1970).
156. V. L. Ginsburg and D. A. Kirzhnits, "On the Problem of High Temperature Superconductivity," *Phys. Rep. (Sec. C)* **4**, 343–356 (1972).
157. V. L. Ginzburg and D. A. Kirzhnits, eds., *High Temperature Superconductivity*, Plenum Press, New York (1982).

# 18

# Conformation and Electronic Aspects of Chlorpromazine in Solution

## Phoebe K. Dea and Hendrik Keyzer

*ABSTRACT:* High-resolution proton and carbon-13 nuclear magnetic resonance spectroscopy studies have been carried out on chlorpromazine base and hydrochloride, and related phenothiazine analogs. Complete unambiguous assignments were possible after the analysis of long-range spin–spin coupling, application of selective heteronuclear and homonuclear decoupling, and ultraviolet irradiation experiments. Negative stain electron micrographs of chlorpromazine hydrochloride showed aggregation of the micelles in water which were shown by NMR analysis to have *gauche–trans* side-chain conformation in water and chloroform. Selective proton line broadening was observed in chlorpromazine hydrochloride micelles upon ultraviolet irradiation. These results were consistent with a staircase-like stack of the aggregate in which the ring nitrogen of one molecule interacts with the aromatic ring of the next; the side chains of contiguous molecules orienting in the same direction. In water, the phenothiazine portions of two stacks attract each other so that the side chains extend towards the outside of the micelle. A rationale is adduced for the *gauche–trans* conformation of the side chain based on the mutual repulsion of the aromatic protons adjacent to the ring nitrogen and the side-chain protons. In chloroform, aggregates are completed by joining two stacks of chlorpromazine hydrochloride via the lyophobic quaternary amines. A possible orientation for the iodine molecule in the chlorpromazine hydrochloride is discussed.

## 1. Introduction

Molecular organization is a feature of biological activities and of many biochemical processes, especially in events demanding specificity. Alexander and Trim[1] showed in the 1940s that surface-active molecules, particularly

*Phoebe K. Dea and Hendrik Keyzer* • Department of Chemistry and Biochemistry, California State University, Los Angeles, 5151 State University Drive, Los Angeles, California 90032.

those capable of forming micelles, could effectively elicit biological activities at the cellular level, often at the biomembrane. Many drug molecules, including local anesthetics, antihistamines, and tranquilizers are surface active; the majority are amphiphiles[2]—that is, they contain within the same molecule nonpolar and polar groups. Chlorpromazine hydrochloride is a good example of such an amphiphile with a critical aqueous micelle concentration[3] of 0.7%.

Psychotropic compounds such as the phenothiazine derivatives possess disparate lipophilic and hydrophilic sites leading to the well-known "phenalkylamine" pattern.[4,5] In aqueous environments the nonpolar portion is hydrophobic, and the polar part hydrophilic. The type and magnitude of the effects of these compounds in the living system depend on the electronic nature of the side chain and ring substituents.[4,5,6] Quantitative changes depend on the electron withdrawing power of the aromatic ring substituents, particularly at the 2-position in the phenothiazine series. Qualitative effects are directed by the structure and molecular weight of the side chain.

The side chain can also affect electronic events of the tricyclic ring system. Electron spin resonance experiments allowed Fenner[7] to suggest that the influence of the side chain of phenothiazine on the formation of free radicals showed a correlation between the redox activity of the phenothiazine nucleus and dynamic aspects of stereochemistry. For example, there was a difference in the formation of cationic free radicals between promazine and alimemazine. The latter has a branched side chain, and forms cationic free radicals only under irradiation. It differs from promazine in pharmacodynamic properties, reported[8] to result in considerably shifted ion exchange equilibria. A kinetic study of the oxidation of dopamine by dialkylaminoalkyl phenothiazine cationic free radicals showed[9] that a strong correlation existed between side-chain structures and oxidation rates; phenothiazine free radicals with two carbon side chains had faster rates than those with three carbon side chains, albeit both were very rapid at physiological pH.

Ionic surfactants are necessarily associated with counterions which may modify their properties significantly.[10] Studies[11] have indicated that optimum physiological effects of chlorpromazine hydrochloride (CPZ · HCl) are regulated by pH constraints. Subsequent physicochemical studies[12,13] have shown that the environment of the anionic counterion of chlorpromazine could have a profound effect on the behavior of the molecule. Such environmental changes could affect the conformation of the molecule, and therefore its interaction with receptors or molecular competitors *in vivo*.

## 1.1. On the Structure of Chlorpromazine

McDowell[14] has published an x-ray structure of chlorpromazine base which shows a phenothiazine nucleus folded about the SN axis with an angle between the best planes of the benzene rings of 139.4°. See Figure 1. The contraction of the CS bond (1.75 Å), the CSC angle (97.3°), and the folding of the phenothiazine nucleus are characteristic of these molecules, and are explained by assuming *d*-orbital interaction in the bonding of the S atom. The phenothiazine moiety is thought to be a system of conjugated double bonds with an extensive pool of delocalized electrons as well as lone pairs of electrons on the N and S atoms. Structural features of phenothiazines are closely similar. See Table 1.

Coubeils and Pullman[21] calculated, by means of a molecular orbital method, that the conformation of the side chain attached to the ring nitrogen depended on the folding of the phenothiazine ring systems along the S–N (ring) axis. These data were in agreement with available x-ray

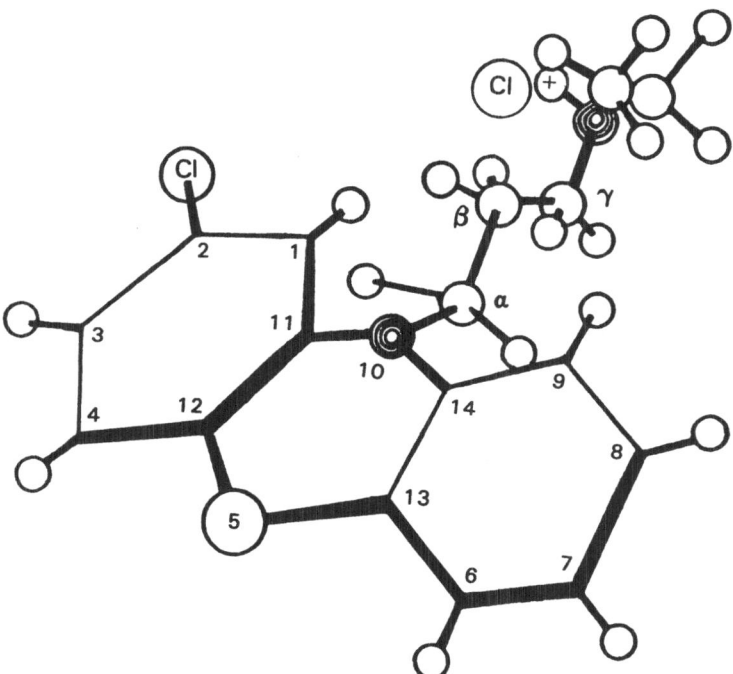

*FIGURE 1.* A model of chlorpromazine hydrochloride with *gauche–trans* side-chain conformation. (In this model one $C_\alpha$ proton and two $C_\gamma$ protons are equidistant from H-9, while the
other $C_\alpha$ proton and the two $C_\beta$ protons are equidistant from H-1.)

TABLE 1. Some Structural Features of Phenothiazines

| Compound | Geometric formula | CSC angle (deg) | Dihedral angle (deg) | Bond distances (Å) | | References |
|---|---|---|---|---|---|---|
| | | | | CS | CN | |
| Phenothiazine | | 99.6 100.9 | 153.3 185.5 | 1.770 1.762 | 1.406 | 15, 16, 17, 18 |
| Chlorpromazine | | 97.3 | 139.4 | 1.75 | 1.41 | 19 |
| Thiethylperazine | | 99.0 | 139.0 | 1.78 | 1.425 | 20 |

crystallographic information on chlorpromazine, thiethylperazine, diethazine, and mopazine. Kier,[22] using extended Hückel calculations, predicted the preferred conformation of chlorpromazine (and haloperidol) to be for an extended side chain. Tollenaere, Moereels, and Koch[23] established an extremely flexible nature for neuroleptic tricyclic drugs. Reboul and Cristau[24] defined 16 conformational parameters that permitted description of the topology of a series of polycyclic psychotropic amines of known crystallographic structure. They then examined[25] nine neuroleptics, three antihistamines, three anti-Parkinson agents, and four antidepressants in terms of these parameters, and found that no preferred conformation existed for the neuroleptics or the antidepressants. Galy *et al.*[26] examined multiconformational aspects of some phenothiazine derivatives, and concluded that different biological activities of the same compound may be associated with conformational mixtures. The nonplanarity of the rings in solution was claimed to be[27] preserved for chlorpromazine, compatible with a boat formation.

Molecular motion of the side chain in chlorpromazine base and hydrochloride have been studied[28] in solution and found to be very rapid. In chlorpromazine base and the hydrochloride rapid methyl group rotation does occur,[29] with a slightly higher activation barrier for the base. The difference was thought[29] to be due to the fact that the $CH_3-N-CH_3$ bond angles about the terminal N are not exactly tetrahedral, i.e., 109.9° in the base,[19] and 110.8° in the hydrochloride.[30] This phenomenon is found to be general for similar amines.[31,32] A number of workers[20,33-37] have used nuclear magnetic resonance (NMR) extensively in attempts to determine the structure of phenothiazine derivatives in solution. NMR studies of $CPZ \cdot HCl$ should yield information about the configuration of the ring nitrogen bonds, the orientation of the side chain, and the disposition of the methyl groups. However, considerable differences in results and their interpretation have been reported.

## 1.2. Electronic Aspects of Chlorpromazine

The phenothiazines are known[38,39] to be electron donors with ionization potentials of about 7.0–7.4 eV as determined from charge transfer spectra. These studies were prompted by the suggestion[40] that the behavior of chlorpromazine might well depend on its solid state electronic properties. Many charge transfer studies were undertaken[41] of the phenothiazines in the solid state and solution, showing that they were capable of forming intermediate to strong complexes. They are known[42] to form exciplexes (excitation activated complexes), generally rather weak adducts produced in a one-quantum process involving a virtual photon, as in the case of the phenothiazines and melanine.[43] Similar adducts are the indole exciplexes,[44] or adducts of nucleosides with nucleotides,[44] or those of amines with hydrocarbons.[45] After the supply of excitation energy has ceased, the exciplex may decay by a variety of radiative and/or non-radiative processes, and sometimes form the precursor of a chemical reaction, as in the rather slow reaction between 6-hydroxydopamine and chlorpromazine.[46] However, the charge transfer complex may also dissociate directly into ion pairs.[47] Such ion radicals form even in moderately polar solvents,[48] say $\varepsilon \approx 6$. Since the free energy difference between a free and solvated ion pair usually exceeds $kT$,[41] it appears that dissociation occurs via a nonrelaxed state of the ion pair.[47] However, some dissociation may also result from the fully relaxed exciplex in view of the positive entropy contribution arising from desolvation.

Association is a feature of chlorpromazine base and related derivatives in the supercooled state.[6] Such associations may be preceded by the formation of intramolecular complexation,[49] i.e., different regions of the same molecular species acting as donor and acceptor.[50] It is interesting to note that in this context the solid state structure of chlorpromazine base is not symmetrical,[19] presumably due to van der Waals or perhaps weak charge transfer interaction between the chlorine ring substituent and the side-chain amine group.[18] Donor and acceptor properties can be modified by chemical substitution, and are sensitive to steric hindrance. In a way, intramolecular charge transfer reproduces the effect of delocalized electrons.[51]

In this study we have determined by means of $^1H$ and $^{13}C$ NMR spectroscopy the conformation and some electronic aspects of CPZ · HCl and CPZ base in a variety of solvents and in the presence of electron acceptors. The NMR experiments were complemented by negative stain electron microscopy.

## 2. Experimental

### 2.1. Materials

Chlorpromazine hydrochloride was a gift from Smith, Kline, and French Co. 10-Methyl phenothiazine was obtained from Eastman Organic, and 2-chlorophenothiazine from Aldrich Co. These compounds were used without further purification. Phenothiazine obtained from Matheson, Coleman, and Bell was vacuum sublimed twice immediately prior to use. CPZ base was prepared by titrating the hydrochloride with 1 $N$ NaOH solution. An ether solution of the base was washed repeatedly with water, and then dried with anhydrous $Na_2SO_4$. The ether was evaporated, and the base stored under vacuum in the dark. The deuterated solvents were all from Wilmad Glass Co. and 1,4-dioxane from Fisher Scientific.

## 2.2. $^1H$ NMR Measurements

The $^1$H NMR spectra were recorded on a Bruker WM-500 spectrometer* operating at a field strength of 11.74 Tesla. Preliminary and comparative experiments were performed on a Bruker WP-60 spectrometer. Dioxane was used as internal standard. All chemical shifts reported are in $\delta$ (ppm) downfield from TMS; $\delta_{TMS} = \delta_{dioxane} - 3.52$ ppm.

## 2.3. $^{13}C$ NMR Measurements

$^{13}$C NMR spectra were obtained on a Bruker HX-90e spectrometer. Longitudinal relaxation ($T_1$) measurements were performed with inversion recovery methods by using a ($180°-\tau-90°$) pulse sequence. The pulse width corresponding to a $90°$ flip angle is $10 \mu s$. $T_1$ values were determined from sets of 5–8 spectra by computer fitting a plot of intensity as a function of $\tau$.

## 2.4. Electron Microscopy

Dispersions for microscopy were prepared by dissolving 10 mg CPZ·HCl in 1 ml of water and adding to it an equal volume of 2% aqueous solution of potassium phosphotungstate. A drop of the mixture was placed on a mica–carbon-coated grid, drained, dried, and examined with a Philips 201C transmission electron microscope. Operating voltage was 80 kV. Photographic images were taken on 35 mm film. Magnification ranged from 150,000 to 200,000 in the final print.

## 3. Results and Discussion

### 3.1. Assignments of $^1H$ Spectra

The proton NMR spectra of three related compounds, phenothiazine, 10-methylphenothiazine, and 2-chlorophenothiazine were obtained and analyzed for chemical shift comparison. Phenothiazine and 10-methylphenothiazine exhibit $C_{2v}$ symmetry, which greatly simplified the spectral assignment. Our assignments on these two compounds are in agreement with those of Fronza and co-workers,[52] who reported the

* The Southern California Regional NMR Facility located at the California Institute of Technology, Pasadena, California, supported by NSF grant No. CHE 79-16324.

chemical shifts of several phenothiazines including phenothiazine and 10-methylphenothiazine in $CDCl_3$ and DMSO-$d_6$. The chemical shift values are summarized in Table 2. The numbering of the atoms is shown in Figure 1.

The spectrum of the aromatic protons of CPZ · HCl at 10 mg/ml in $D_2O$ is shown in Figure 2. The assignment of the benzenoid protons was accomplished by analyzing the long-range coupling constants and by selective homonuclear decoupling experiments. The only singlet present in the atomic region at 6.58 ppm was readily assigned to the H-1 proton. The two well-resolved triplets at 6.78 and 7.04 ppm were assigned to the H-7 and H-8 protons, respectively, by comparison with related phenothiazine compounds. Selective decoupling experiments were the most useful for assigning the protons coupled to H-7 and H-8. Irradiating the H-8 proton at 7.04 ppm caused the H-7 triplet to become a doublet and resulted in the collapse of the doublet at 6.70 ppm to a singlet. The resonance at 6.70 ppm can therefore be assigned to the H-9 proton. Similarly, irradiation of the triplet centered at 6.78 ppm enabled the assignment of the resonance at 6.90 ppm to the H-6 proton. This assignment was confirmed by the presence of a 1.5-Hz long-range coupling observed under high-resolution conditions between the H-6 and H-8 protons. The remaining two doublets in the aromatic region correspond to the H-3 and H-4 protons, which are directly coupled with identical coupling constants of 8.5 Hz. Irradiation of one of these protons collapsed the other into a singlet. The specific assignment of these two protons was achieved through the observation of a long-range coupling of 1.5 Hz between H-3 and H-1.

The assignments of the chemical shifts of CPZ · HCl in DMSO-$d_6$ and CPZ · HCl and CPZ base in $CDCl_3$ were carried out in a similar manner (Table 2). Table 3 lists the coupling constants observed for the aromatic protons of CPZ · HCl.

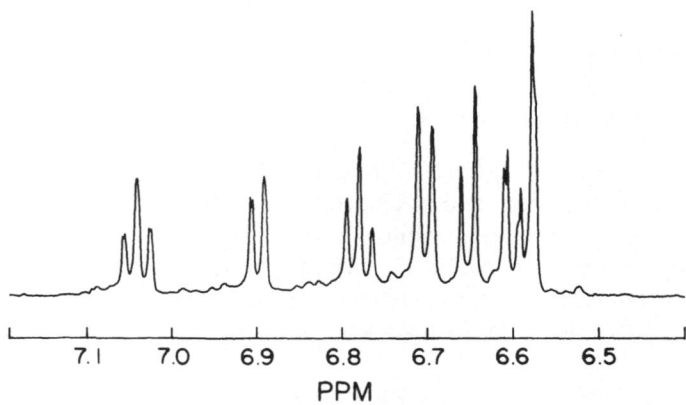

*FIGURE 2.* Spectrum of aromatic protons of chlorpromazine hydrochloride in $D_2O$.

*TABLE 2.* Summary of Proton Chemical Shifts

| Compound | Chemical shift (ppm) | | | | | | | | | | | |
|---|---|---|---|---|---|---|---|---|---|---|---|---|
| | H-1 | H-2 | H-3 | H-4 | H-6 | H-7 | H-8 | H-9 | H-$\alpha$ | H-$\beta$ | H-$\gamma$ | N(CH$_3$)$_2$ |
| Phenothiazine in CD$_3$CN | 6.67 | 7.01 | 6.80 | 6.95 | 6.95 | 6.80 | 7.01 | 6.67 | — | — | — | — |
| 10-Methyl phenothiazine in CD$_3$CN | 6.91 | 7.21 | 6.94 | 7.14 | 7.14 | 6.94 | 7.21 | 6.91 | — | — | — | — |
| 2-Chloro phenothiazine in DMSO-$d_6$ | 7.83 | — | 7.91 | 8.05 | 8.05 | 7.92 | 8.13 | 7.80 | — | — | — | — |
| CPZ·HCl in D$_2$O | 6.58 | — | 6.60 | 6.66 | 6.90 | 6.78 | 7.04 | 6.70 | 3.58 | 1.81 | 2.84 | 2.46 |
| CPZ·HCl in CDCl$_3$ | 6.70 | — | 6.79 | 6.92 | 7.02 | 6.83 | 7.06 | 6.76 | 3.89 | 2.21 | 2.92 | 2.49 |
| CPZ·HCl in DMSO-$d_6$ | 7.12 | — | 7.01 | 7.09 | 7.19 | 7.00 | 7.24 | 7.18 | 3.97 | 2.05 | 3.13 | 2.70 |
| CPZ·HCl/I$_2$ in DMSO-$d_6$ | 7.09 | — | 7.01 | 7.06 | 7.17 | 6.99 | 7.23 | 7.18 | 3.97 | 2.03 | 3.14 | 2.74 2.75 |
| CPZ base in CDCl$_3$ | 6.73 | — | 6.73 | 6.86 | 6.96 | 6.90 | 6.99 | 6.71 | 3.72 | 1.77 | 2.23 | 2.05 |

*TABLE 3.* List of Coupling Constants for the Aromatic Protons in CPZ · HCl

| Solvent | Coupling constant (Hz) | | | | | |
|---------|---------|---------|---------|---------|---------|---------|
|         | $J_{13}$ | $J_{34}$ | $J_{67}$ | $J_{68}$ | $J_{78}$ | $J_{89}$ |
| $D_2O$  | 1.5 | 8.5 | 8.0 | 1.5 | 8.0 | 8.0 |
| DMSO-$d_6$ | 2.0 | 8.3 | 7.3 | 1.3 | 8.0 | 8.0 |
| CDCl$_3$ | 2.0 | 8.1 | 8.1 | 1.5 | 8.1 | 8.1 |

## 3.2. Solvent Effects

The chemical shifts of 10 mg/ml of CPZ · HCl were measured in three different solvents: $D_2O$, DMSO-$d_6$, and CDCl$_3$. Large upfield shifts for all the CPZ · HCl proton resonances were observed when the solvent was $D_2O$ or CDCl$_3$, compared to those observed in DMSO-$d_6$ (Table 2). Note the similarities in the chemical shifts of CPZ · HCl in $D_2O$ and in CDCl$_3$.

The conformation of the *N,N*-dimethylaminopropyl side chain in CPZ · HCl was found to be solvent dependent. The percent *trans* conformer calculated from the parameter $N = |J + J'|$ for each $CH_2CH_2$ fragment following the approach of Abraham et al.[53] are summarized in Table 4. The population of *trans* conformer for the $C_\alpha$–$C_\beta$ fragment was found to decrease drastically in water and in chloroform solutions, although the $C_\beta$–$C_\gamma$ fragment was predominantly *trans* in all three solvents. The side

*TABLE 4.* Conformational Analysis of the *N,N*-Dimethylaminopropyl Side Chain

|          | Fragment | $N = |J + J'|$ (Hz) | $\Delta E$ (kcal/mole) | % trans |
|----------|----------|---------|---------|---------|
| CPZ · HCl | $C_\alpha$–$C_\beta$ | 12.0 | −1.24 | 6 |
| in $D_2O$ | $C_\beta$–$C_\gamma$ | 16.2 | 1.21 | 80 |
| CPZ · HCl | $C_\alpha$–$C_\beta$ | 14.3 | 0.32 | 46 |
| in DMSO-$d_6$ | $C_\beta$–$C_\gamma$ | 15.8 | 0.98 | 72 |
| CPZ · HCl | $C_\alpha$–$C_\beta$ | 12.5 | −0.64 | 14 |
| in CDCl$_3$ | $C_\beta$–$C_\gamma$ | 16.2 | 1.21 | 80 |
| CPZ base | $C_\alpha$–$C_\beta$ | 14.0 | −0.23 | 43 |
| in CDCl$_3$ | $C_\beta$–$C_\gamma$ | 14.0 | −0.23 | 43 |
| CPZ base | $C_\alpha$–$C_\beta$ | 12.0 | −1.24 | 6 |
| in DMSO-$d_6$ | $C_\beta$–$C_\gamma$ | 16.0 | 1.09 | 76 |

chain of CPZ base in $CDCl_3$ adopted a conformation similar to that of CPZ·HCl in DMSO-$d_6$, i.e., *trans–trans*. These results are in agreement with the work by Fronza and co-workers.[20,34] In $D_2O$ the side chain of the ultrasonicated CPZ base was *gauche* for $C_\alpha$–$C_\beta$ and *trans* for $C_\beta$–$C_\gamma$, essentially the same as for the salt in water.

Some information on the structure of the aggregate may be obtained by examining the spin-lattice relaxation time ($T_1$) of the various carbons in chlorpromazine. These are listed in Table 5. The $T_1$ values for CPZ·HCl in various solvents decreased in the following order: $CDCl_3$, DMSO-$d_6$, $D_2O$. In $CDCl_3$, the CPZ base had longer $T_1$ values than CPZ·HCl, indicating that the aggregation was weaker in the case of the free base.

## 3.3. Irradiation

In the presence of the cationic free radicals formed upon irradiation of CPZ·HCl in $D_2O$ and in $CDCl_3$ with uv light, we observed a gradual broadening of the proton resonances, with the exception of the protons at positions 4 and 6, whose doublets remained relatively sharp. The aromatic region of CPZ·HCl under these conditions is shown in Figure 3. The

*TABLE 5.* Carbon-13 $T_1$ Values for 1.0 $M$ CPZ·HCl in Various Solvents

| Carbon number | $T_1$ (s) | | | | |
| | $D_2O$ | DMSO-$d_6$ | | $CDCl_3$ | |
| | CPZ·HCl | CPZ base | CPZ·HCl | CPZ base | CPZ·HCl |
|---|---|---|---|---|---|
| C-1 | 0.07 | 0.20 | 0.07 | 1.20 | 0.92 |
| C-2 | 1.54 | 4.30 | 3.52 | 19.43 | 17.90 |
| C-3 | 0.06 | 0.20 | 0.20 | 1.00 | 0.83 |
| C-4 | 0.07 | 0.17 | 0.17 | 1.15 | 0.93 |
| C-6 | 0.07 | 0.22 | 0.19 | 1.20 | 0.84 |
| C-7 | 0.06 | 0.21 | 0.07 | 0.93 | 0.84 |
| C-8 | 0.07 | 0.18 | 0.18 | 1.20 | 0.84 |
| C-9 | 0.07 | 0.22 | 0.07 | 1.20 | 0.92 |
| C-11 | 1.59 | 5.65 | 4.31 | 24.95 | 18.00 |
| C-12 | 2.40 | 5.01 | 5.26 | 25.60 | 25.70 |
| C-13 | 2.40 | 6.20 | 5.08 | 31.85 | 21.10 |
| C-14 | 2.07 | 4.98 | 4.22 | 22.43 | 16.69 |
| $C_\alpha$ | 0.10 | 0.20 | 0.07 | 1.03 | 0.51 |
| $C_\beta$ | 0.08 | 0.19 | 0.15 | 0.88 | 0.36 |
| $C_\gamma$ | 0.06 | 0.09 | 0.17 | 0.73 | 0.42 |
| $CH_3$ | 0.61 | 0.55 | 0.48 | 1.01 | 0.98 |

relationships most widely used to describe the relaxation of a nucleus by an unpaired electron are the Solomon–Bloembergen equations[54] given below:

$$\frac{1}{T_{1,M}} = \frac{2}{15} \frac{\gamma_I^2 g^2 S(S+1) \beta^2}{r^6} \left( \frac{3\tau_{c,1}}{1+\omega_I^2 \tau_{c,1}^2} + \frac{7\tau_{c,2}}{1+\omega_S^2 \tau_{c,2}^2} \right)$$
$$+ \frac{2}{3} \left( \frac{A}{\hbar} \right)^2 S(S+1) \left( \frac{\tau_{e,2}}{1+\omega_S^2 \tau_{e,2}^2} \right)$$

and

$$\frac{1}{T_{2,M}} = \frac{1}{15} \frac{\gamma_I^2 g^2 S(S+1) \beta^2}{r^6} \left( 4\tau_{c,1} + \frac{3\tau_{c,1}}{1+\omega_I^2 \tau_{c,1}^2} + \frac{13\tau_{c,2}}{1+\omega_S^2 \tau_{c,2}^2} \right)$$
$$+ \frac{1}{3} \left( \frac{A}{\hbar} \right)^2 S(S+1) \left( \tau_{e,1} + \frac{\tau_{e,2}}{1+\omega_S^2 \tau_{e,2}^2} \right)$$

where $S$ is the electron spin quantum number for the paramagnetic nuclei, $g$ is the $g$ factor of the electron, $\beta$ is the Bohr magneton of the electron, $\gamma$ is the gyromagnetic ratio of the nuclei interacting with the paramagnetic nuclei, $\omega_I$ is the nuclei resonance frequency, $\omega_S$ is the electron resonance frequency, $\tau_c$ and $\tau_e$ are the dipolar and scalar correlation times, respectively, and $A/\hbar$ is the isotropic hyperfine interaction constant.

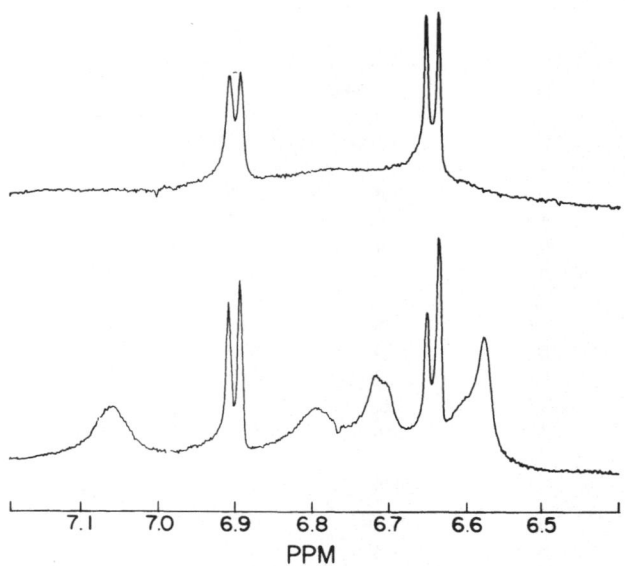

FIGURE 3. Aromatic region of irradiated chlorpromazine hydrochloride in $D_2O$. Bottom: Brief exposure; Top: Long exposure (approx. 20 min).

Since the relaxation times depend to the sixth power on the distance of the nuclei from the paramagnetic center, the free radical site must be further away from the H-4 and H-6 protons. This would eliminate the sulfur atom as a possible free radical site, leaving the nitrogen as the most likely candidate. Furthermore, it is not possible for the tricyclic ring to assume a conformation such that the protons at positions 4 and 6 are further away from the nitrogen free radical site than their immediate neighboring protons at positions 3 and 7, suggesting that intermolecular interaction must be important.

Irradiation of CPZ base in DMSO-$d_6$ and CDCl$_3$ did not yield broadening of the aromatic proton signals, indicating that free radical aggregates did not exist for the drug in the base form in these solvents. CPZ·HCl in DMSO-$d_6$ behaves like the base.

CPZ base was present in D$_2$O mostly as microcrystallites, as indicated by the necessity of ultrasonication to obtain a sufficient signal-to-noise ratio for NMR measurements. In the solid, then, the base side chain appears to be C$_\alpha$–C$_\beta$ *gauche*, C$_\beta$–C$_\gamma$ *trans*, as for CPZ·HCl in D$_2$O. Ragg and co-workers[34] maintain that all promazines exhibit the above side-chain conformation in aqueous solution, differing in their view from others,[35,36,55-57] and that the tricyclic system stabilizes the observed gauche trans form.

Figure 4 shows several conformations designated quasiequatorial (essentially equivalent to *intra*), i.e., the ring N–C bond inside the interplanar fold, coplanar, and quasiaxial (essentially equivalent to *extra*). Several workers[7] claimed a preference for an *extra* or quasiaxial CPZ·HCl conformation. Fronza and colleagues also claimed[34] that the *intra* ring N-substituent bond is favored for phenothiazine but not for phenothiazine derivatives. In solids the orientation of the N–C$_\alpha$ bond was found[20] to be closer to intra than extra, and probably closer to a flatter set of ring N bonds.

In the quasiaxial state, no *a priori* preference for C$_\alpha$–C$_\beta$ *gauche* exists. The quasiequatorial case with the C$_\alpha$ protons mirrored in the SNC plane yields propyl protons on either side of that plane equidistant to H$_1$ and H$_9$,

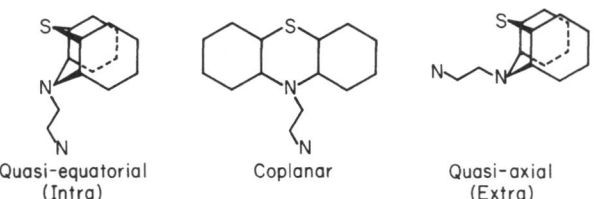

Quasi-equatorial       Coplanar       Quasi-axial
(Intra)                               (Extra)

*FIGURE 4.* Several conformations of chlorpromazine.

i.e., one $C_\alpha$ and both $C_\beta$ protons from $H_1$, and one $C_\alpha$ and both $C_\gamma$ protons from $H_9$. 180° rotation of $C_\beta$ protons about the $C_\alpha–C_\beta$ bond yields another conformer in which one $C_\alpha$ and both $C_\beta$ protons are equidistant from $H_9$, requiring the $C_\alpha$ proton and both $C_\gamma$ protons to be equidistant from $H_1$. This conformation has the rationale of a kind of symmetry. The interplanar fold of the tricyclic system provides a constraint which stabilizes the *gauche–trans* side-chain conformation. This structure is shown by the model in Figure 1.

### 3.4. Conformation in CPZ · HCl Micelles

A micellar model[58] was proposed for phenothiazine drugs consisting of 10–12 monomers vertically stacked with chain alternating in opposite directions. In this model the sulfur atom of (the) tricyclic ring is superimposed on the ring nitrogen of the adjacent molecules. Examination of a variety of molecular models reveals that the only way for the H-4 and H-6 NMR signals to remain sharp is for the tricyclic rings of the CPZ · HCl molecules to overlap in a staircase-like stack, with the ring N of one molecule interacting with the aromatic ring of the contiguous molecule via a charge transfer mechanism. See Figure 5. The interplanar distance is likely to be in the neighborhood of about 3.3 Å, the distance most frequently encountered[59] for charge transfer complexes. Moreover, the side chains are on the same side of the stack. An intermolecular distance of about 3 Å would allow considerable latitude of movement of the side chains and possible pyramidal ring N inversions. The quasiequatorial conformation is probably the preferred state in the aqueous micelle with solvation at the ionic end of the side chain. Quasiequatorial oriented chains would also allow CPZ · HCl to form aggregates with less steric hindrance.

It must be expected that there is a tendency for the ionic side chains to

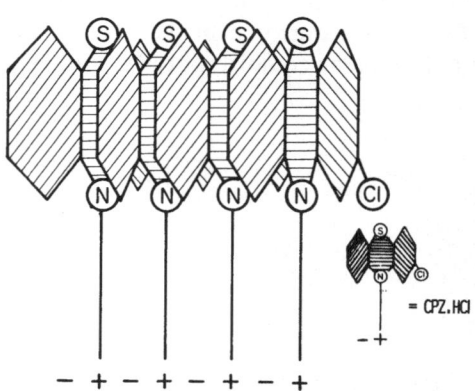

= CPZ.HCl

FIGURE 5. View of staircase-like stack of chlorpromazine hydrochloride molecules.

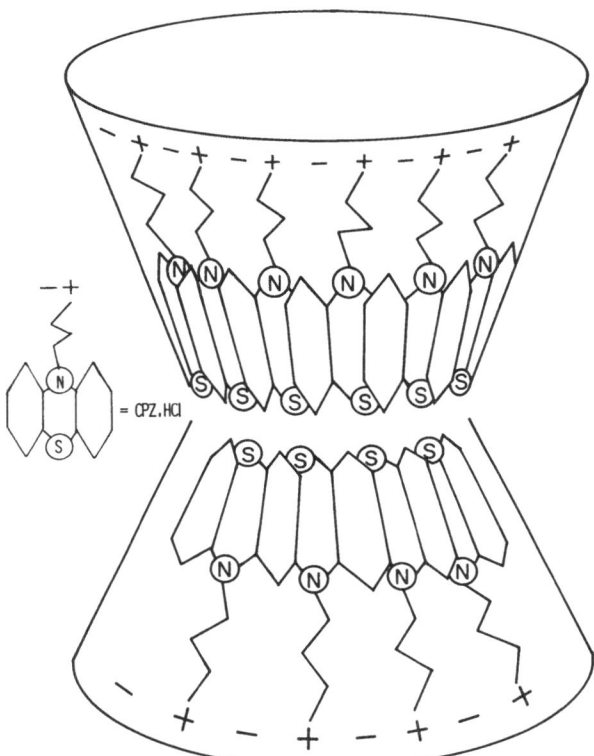

FIGURE 6. Suggested three-dimensional arrangement of chlorpromazine hydrochloride molecules in aqueous micelles: Two staircase-like stacks interacting via the sulfur atoms of the phenothiazine moieties.

repel one another in aqueous solution. The stacks may interact hydrophobically via the phenothiazines to complete the micelle, i.e., with the ionic chains facing outward. See Figure 6. The electron microscopy data bear out the notion of different size aggregates. See Figure 7. If CPZ · HCl is given the planar dimensions of approximately 10 by 10 Å, aggregates smaller than 20 molecules in number appear to be possible. At the other end of the scale are aggregates containing in excess of 500 molecules. Even though CPZ · HCl here was above the critical micelle concentration, it is well known[60] that aggregates can exist at concentrations orders of magnitude less than that for critical micelles.

In the extremely lipophilic chloroform the ionic tails of the CPZ · HCl molecules with their negative counterions tend to crowd together in the micelle, while the phenothiazines overlap in a manner similar to that in water. Like the aggregate in water, the CPZ · HCl array in $CDCl_3$ can also

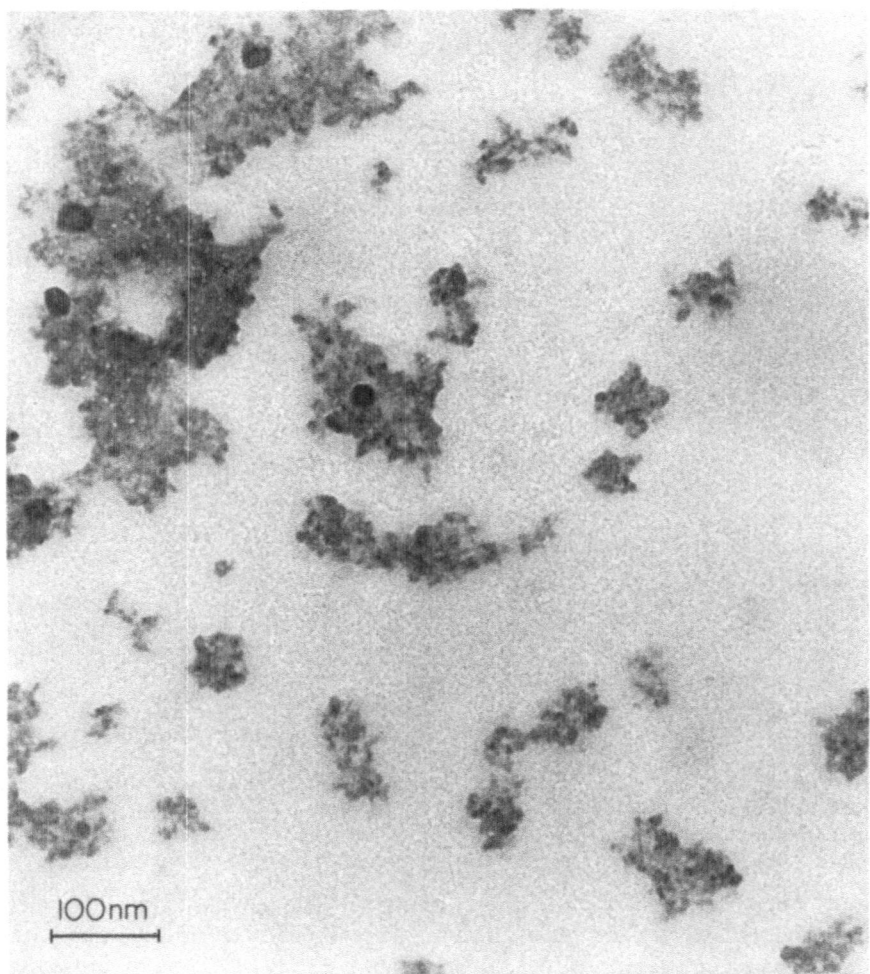

*FIGURE 7.* Electron micrograph of chlorpromazine hydrochloride aggregates in water.

form a staircase stack. The micelles are completed via clustering of the ionic species lyophobically of a pair of stacks. See Figure 8. In this model, the outer portion of the tricycles, the overlapping phenothiazines, form the exterior of the micelles solvated by chloroform molecules.

In addition to being quasiequatorial, the conformation of the side chain of CPZ · HCl is quite similar in $D_2O$ and $CDCl_3$, being 80% *trans* at the $C_\beta$–$C_\gamma$ and roughly 10% at the $C_\alpha$–$C_\beta$ bonds. The conformation of the side chain in dimethylsulfoxide is quite different. Since CPZ base in $CDCl_3$

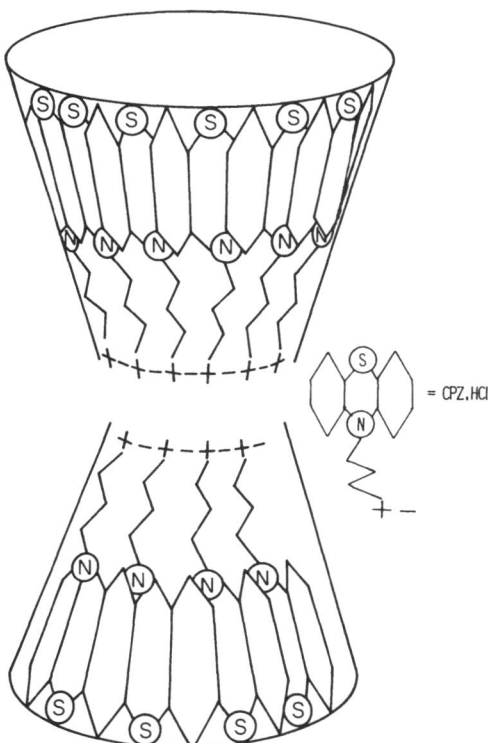

FIGURE 8. A suggested three-dimensional arrangement of chlorpromazine hydrochloride in chloroform. Two staircaselike stacks interacting via the quaternary groups of side chains.

does not aggregate, just like CPZ · HCl in DMSO, one might expect no side-chain conformation preferences, as was indeed observed. Conformational preferences have been found for imipramine hydrochloride in $CDCl_3$ solution but none for imipramine base.[61] Barbe and Chauvet-Monges[53] also saw conformational preferences for CPZ · HCl in solution.

A micellar structure results in a considerable contribution towards the free energy of the system. Consider $n$ electrons distributed over $N$ molecular sites, $nN$. The number of ways $W$ in which similar systems may be assembled is given by

$$W = N!/(N-n)! \, n!$$

If one electron is added, the free energy contribution due to the entropy change $\Delta S$ involved is

$$-T \, \Delta S = kT \ln N$$

where $\Delta S$ is related to $W$ by the well-known Boltzmann relation

$$\Delta S = k \ln W$$

At 300 K, this free energy gain is about 0.055 log $N$ eV. In a monolayer of even 20 molecules, the energy gain is about 0.072 eV.

### 3.5. Electronic Disposition. Interactions of CPZ · HCl with an Electron Acceptor

In the presence of the electron acceptor $I_2$ (1:1 molar ratio), we observed some changes in the chemical shift of the CPZ · HCl tricyclic ring, particularly the H-1 proton, and the methyl protons of the side chain. These results are summarized in Table 1. The most dramatic effect is that in the presence of $I_2$, the methyl resonance is split due to coupling with the amine proton. The observed splitting of 4.4 Hz is independent of the strength of the magnetic field, and is of the same order of magnitude as observed[13] for $SO_2$ adducts of several heterocyclic compounds containing $N$-methyl groups. The doublet collapsed to a singlet when $D_2O$ was added, and the proton peak at about 9.5 ppm disappeared.

In the complex with $I_2$, actually[62] $I_2Cl^-$, the side chain of CPZ · HCl in DMSO is most likely in the quasiaxial conformation. However, the $I_2Cl^-$ molecule interacts with the ring nitrogen, causing an NMR shift in the H-1, H-4, and H-6 atoms. The polarized $I_2Cl^-$ molecule must be situated somewhat more towards the benzene ring containing the Cl substituent to account for the NMR shifts observed in the H-1 and H-4 proton signals. A competition probably exists between the substituent Cl and the chloride ion for the $I_2$ molecule. However, the iodine sequesters the chloride ion to the extent that the quaternary proton splits the methyl resonance signal via the quaternary nitrogen.[13]

Self-complexing may occur,[50] with different regions of the same molecular species acting as donor and acceptor. These effects are important in solution, where they may lead to association. In this context, it is important to stress that in the micellar aggregate of CPZ · HCl intermolecular charge transfer complexing is a way of aiding electron delocalization, just as the addition of a foreign electron acceptor does. Complete electron transfer would result in a wholly ionic compound and not in a charge transfer complex.

For high stability, charge transfer should be incomplete[63] as observed in the CPZ · HCl aggregate. If, as Fenner[7] and Fronza and co-workers[20] suggest, the quasiaxial state of CPZ · HCl is the more stable, the quasiequatorial conformation of this molecule represents a state of excitation supplying an energy possibly sufficient to maintain an exciplex aggregate. In such an exciplex, an electron is taken not to the bottom of the conduction band but into an excited state energy level with the energy gap, below the bottom of the conduction band. Thus, the electron affinity is increased and the probability of complexation is correspondingly enhanced. If the electron is raised not from the Fermi level of the donor but from a higher, excited state energy level, the conjugation potential is reduced. Both processes may occur simultaneously. Excitation leads to a considerable change in the magnitude and direction of the dipole moment.[64] The light-irradiated CPZ · HCl in $D_2O$ and $CDCl_3$ then belongs to the class of light-activated[65] charge transfer self-complexes.[41]

The clearest single feature of an exciplex is the reversibility of its equilibrium with the isolated components.[66] Since energy differences involved in such excitation processes are generally quite small, even of the order of $kT$, such exciplexes can be produced via a variety of energy supplying mechanisms, e.g., electrochemical phenomena at membrane surfaces. A corrolary of aqueous micellar assemblages of such particles is that counterions are attracted by the hydrophilic regions to become electrocatalytically active.[67] Incorporation of an acceptor into a cationic micelle can increase the rate of reaction involving hydrated electrons by about 60 times compared with an anionic micelle.[68] This has profound implications for electron transfer reactions at biological membranes in the presence of CPZ · HCl.

## References

1. A. E. Alexander and A. R. Trim, *Proc. R. Soc. London Ser. B* **113**, 220 (1946).
2. A. T. Florence, *Adv. Colloid Interface Sci.* **2**, 115 (1968).
3. A. T. Florence, in *Micellization, Solubilization and Microemulsions* (K. L. Mittal, ed.), Plenum Press, New York (1977), Vol. 1.
4. A. W. Nineham, *Pharm. Bull.* **1150** (1962).
5. F. Gutmann and H. Keyzer, *Rev. d'Aggressologie* **10**, 27 (1968).
6. H. Keyzer, Ph.D. thesis, University of New South Wales, Sydney, Australia (1966).
7. H. Fenner, in *Phenothiazines and Structurally Related Drugs* (I. S. Forrest, C. J. Carr and E. Usdin, eds.), Raven Press, New York (1974), p. 5.
8. V. T. Gorshov *et al.*, *Zh. Fiz. Khim.* **51**, 2680 (1977).
9. M. R. Gasco and M. E. Carbotti, *J. Pharm. Sci.* **86**(5), 612 (1979).
10. K. L. Mittal and P. Mukerjee, in *Micellization, Solubilization and Microemulsions* (K. L. Mittal, ed.), Plenum Press, New York (1977), Vol. 1, p. 1.
11. C. M. Gooley, H. Keyzer, and F. Setchell, *Nature* **223**, 80 (1969); H. Keyzer, C. Lowe, W. Plumtree, and F. Gutmann, in *The Fourth International Conference on Phenothiazines and Related Drugs* (H. Eckert and F. Gutmann, eds.), Plenum Press, New York (1980).
12. S. Chan, C. M. Gooley, and H. Keyzer, *Tetrahedron Lett.* **13**, 1193 (1975).
13. D. Beltran, S. Chan, and H. Keyzer, in *Bioelectrochemistry* (H. Keyzer and F. Gutmann, eds.), Raven Press, New York (1980).

14. J. J. H. McDowell, in *Phenothiazines and Structurally Related Drugs* (I. S. Forrest, C. J. Carr, and E. Usdin, eds.), Raven Press, New York (1974), p. 33.
15. N. M. Cullinane and W. Thees, *Trans. Faraday Soc.* **36**, 507 (1940).
16. R. G. Wood, C. H. McCale, and G. Williams, *Phil. Mag.* **31**, 507 (1941).
17. J. D. Bell, J. F. Blount, O. V. Briscoe, and H. C. Freeman, *Chem. Commun.* 1656 (1968).
18. A. Feinberg and S. H. Snyder, *Proc. Natl. Acad. Sci. U.S.A.* **72**, 1899 (1975).
19. J. J. H. McDowell, *Acta Crystallogr.* **B25**, 2175 (1969).
20. G. Fronza, E. Ragg, and R. Mondelli, *Actual. Chim. Therap.* **8**, 245 (1981).
21. J. L. Coubeils and B. Pullman, *Theor. Chim. Acta (Berl.)* **24**, 35 (1972).
22. L. B. Kier, *J. Theor. Biol.* **40**, 211 (1973).
23. J. P. Tollenaere, H. Moereels, and M. J. Koch, *Eur. J. Med. Chem.* **12**, 199 (1977).
24. J.-P. Reboul and B. Christau, *Eur. J. Med. Chem.* **12**, 71 (1977).
25. J.-P. Reboul and B. Christau, *Eur. J. Med. Chem.* **12**, 76 (1977).
26. A.-M. Galy, C. Levayer, J.-P. Galy, and J. Barbe, in *Phenothiazines and Structurally Related Drugs* (E. Usdin, H. Eckert, and I. S. Forrest, eds.), Elsevier/North Holland, New York (1980), p. 21.
27. L. S. Isbrandt, R. K. Jensen, and L. Petrakis, *J. Magn. Reson.* **12**, 143 (1973).
28. D. W. Larsen and J. Y. Corey, *J. Am. Chem. Soc.* **99**, 1740 (1977); R. J. Abraham, L. J. Kricka, and A. Ledwith, *J. Chem. Soc. Perkin Trans.* **2**, 1648 (1974).
29. B. A. Soltz, J. Y. Corey, and D. W. Larsen, *J. Phys. Chem.* **83**, 2162 (1979).
30. P. Marsau, *C. R. Acad. Sci. Paris Ser. C* **274**, 1806 (1972).
31. M. L. Post, O. Kennard, and A. S. Horn, *Acta Crystallogr.* **31B**, 1008 (1975).
32. M. L. Post and A. S. Horn, *Acta Crystallogr.* **33B**, 2590 (1977).
33. J. Barbe and A. M. Chauvet-Monges, *C. R. Acad. Sci. Paris Ser. C* **279**, 935 (1974).
34. E. Ragg, S. Fronza, and R. Mondelli, *J. Chem. Soc. Perkin Trans. II* **12**, 1587 (1982).
35. J. Barbe and A. Blanc, *C. R. Acad. Sci. Paris Ser. C* **282**, 117 (1976).
36. J. Barbe and A. Blanc, *C. R. Acad. Sci. Paris Ser. C* **282**, 299 (1976).
37. J. Barbe, A. Blanc, and A. Chauvet-Monges, *C. R. Acad. Sci. Paris Ser. C* **284**, 109 (1977).
38. A. Fulton and L. E. Lyons, *Aust. J. Chem.* **21**, 873 (1968).
39. T. Kitagawa, *J. Molec. Spectrosc.* **26**, 1 (1968).
40. G. Karreman, I. Isenberg, and A. Szent-Gyorgi, *Science* **130**, 1191 (1959).
41. F. Gutmann, H. Keyzer, and L. E. Lyons, *Organic Semiconductors, Part B*, Malabar, Florida (1983).
42. M. Gordon and W. R. Ware, *The Exciplex*, Academic, New York (1975); *Molecular Association* (R. Foster, ed.), Academic, New York (1979), p. 2; P. Frochlich and E. L. Wehry, "The Study of Excited State Complexes (Exciplexes)," in *Modern Fluorescence Spectroscopy* (E. L. Wehry, ed.), Plenum Press, New York (1976), Vol. 2, p. 319.
43. F. Gutmann and H. Keyzer, *Electrochim. Acta* **13**, 693 (1968); I. S. Forrest *et al.*, *Rev. Agressologie (Paris)* **2**, 147 (1966).
44. M. A. Slifkin, *Charge Transfer Interactions of Biomolecules*, Academic, London (1971), p. 251.
45. N. Mataga, *Bussei Kenkyu* **18**, A37 (1973).
46. R. I. Kukhtim *et al.*, *Opt Spektrosk.* **35**, 845 (1973); V. V. Slobodyanik *et al.*, *Dopov. Akad. Nauk. Ukr., RSR, A* **35**, 1033 (1973).
47. M. Itoh *et al.*, *Bull. Chem. Soc. Jpn* **47**, 1078 (1974).
48. Y. Taniguchi and N. Mataga, *Chem. Phys. Lett.* **13**, 596 (1972); Y. Taniguchi *et al.*, *Bull. Chem. Soc. Jpn* **45**, 764 (1972); G. Bokestein and M. Buck, *Rec. Rev. Chim. Pays-Bas.* **92**, 1095 (1973).
49. R. S. Mulliken and W. B. Person, *Molecular Complexes*, Wiley, New York (1969); M. A. Slifkin and R. H. Walmsley, *Photochem. Photobiol.* **13**, 57 (1971).
50. J. Ferrari *et al.*, *J. Am. Chem. Soc.* **95**, 948 (1973); F. Gutmann and H. Keyzer, *J. Chem.*

*Phys.* **46**, 1969 (1967); T. Tamaki, *Bull. Chem. Soc. Jpn* **46**, 2527 (1973); *J. Am. Chem. Soc.* **97**, 3209 (1975); S. C. Abbi and D. M. Hanson, *J. Chem. Phys.* **60**, 319 (1974); A. K. Prokofev, *Russ. Chem. Rev.* **45**, 519 (1976); L. G. Schroff and A. J. A. van der Weerdt, *Tetrahedron Lett.* **18**, 1649 (1973); I. V. Turovskii *et al., Dokl. Akad. Nauk. SSSR Phys. Chem.* **223**, 746 (1975).

51. K. Mutai *et al., Chem. Lett.* 1047 (1977); M. L. Olsen and K. R. Sundberg, *J. Chem. Phys.* **69**, 5400 (1978); J. R. C. Applequist and K.-K. Fung, *J. Am. Chem. Soc.* **94**, 2952 (1972); *Acc. Chem. Res.* **10**, 79 (1977); S. L. Mair, *Acta Crystallogr.* **A34**, 66 (1978).

52. G. Fronza, R. Mondelli, G. Scapini, G. Ronsisvalle, and F. Vittorio, *J. Magn. Reson.* **23**, 437 (1976).

53. R. J. Abraham, L. J. Kricka, and A. Lewith, *J. Chem. Soc. Perkin Trans.* **2**, 1648 (1974).

54. I. Solomon and N. Bloembergen, *J. Chem. Phys.* **25**, 261 (1956); R. A. Dwek, *Nuclear Magnetic Resonance in Biochemistry: Applications to Enzyme Systems*, Oxford University Press, London (1973).

55. J. J. Kaufman and W. S. Koski, in *Drug Design* (E. J. Ariens, ed.), Academic, New York (1975), Vol. IV.

56. J. J. H. McDowell, *Acta Crystallogr.* **B33**, 771 (1977); **B36**, 2178 (1980).

57. P. Marsau and J. Gauthier, *Acta Crystallogr.* **B29**, 992 (1973); J. R. Rodgers, A. S. Horn, and O. Kennard, *J. Pharm. Pharmacol.* **28**, 246 (1976).

58. D. Attwood, A. T. Florence, and J. M. N. Gillan, *J. Pharm. Sci.* **63**, 988 (1974); D. Attwood and J. Gibson, *J. Pharm. Pharmacol.* **30**, 176 (1978).

59. J. C. A. Boeyens and I. H. Herbstein, *J. Phys. Chem.* **69**, 2160 (1965).

60. B. L. Lindman and H. Wennerström in *Topics in Current Chemistry* (F. L. Boshke, ed.), Springer-Verlag, Berlin (1980), Vol. 87.

61. L. J. Kricka and A. Ledwith, *J. Chem. Soc. Perkin Trans.* **2**, 16 (1974).

62. H. Keyzer, R. F. Landel, and A. Rembaum, "Insoluble Polymeric Quaternary Trihalogen Salt Coated Substrates," U.S. Patent 3,898,336 (August, 1975); "Polymeric Organic Halogen Salts," U.S. Patent 3,778,746 (December, 1973).

63. B. D. Silverman, *Phys. Rev. B* **16**, 5153 (1977).

64. W. Robertson *et al., J. Chem. Phys.* **35**, 464 (1961); N. Tyutuckov *et al., Theor. Chim. Acta (Berlin)* **20**, 385 (1971).

65. M. Masumar *et al., Chem. Phys. Lett.* **59**, 188 (1978); W. Lachish *et al., ibid.* **65**, 574 (1979); W. Lachish and D. J. Williams, *ibid.* **72**, 225 (1980).

66. H. Knibbe *et al., Ber. Bunsen Ges. Phys. Chem.* **73**, 839 (1969); S. Yomosa, *J. Phys. Soc. Jpn* **36**, 1655 (1974).

67. C. A. Bunton, *Progr. Solid State Chem.* **8**, 167 (1973).

68. H. J. Frank *et al., Ber. Bunsen Ges. Phys. Chem.* **80**, 547 (1976); M. Gratzer *et al., ibid.* **79**, 475 (1975).

# 19

# Electrochemistry of Drug Interactions and Incompatibilities

## G. M. Eckert, F. Gutmann, and H. Keyzer

*ABSTRACT:* Molecular associations involving drugs are of considerable pharmacological importance. Incompatibilities and the binding of "active" drugs to "inert" formulation components are characterized by such associations and may result in rendering otherwise carefully planned therapeutic regimens ineffective. The "large cation–large anion" incompatibility is considered. Association between two drugs or a drug and tissue component *in vivo* may greatly modify the distribution and therefore the therapeutic effect of the drug. In spite of complications caused by tissue binding and metabolic modification the therapeutic interaction between some pairs of drugs is suitable for study by physicochemical *in vitro* techniques. Further details with respect to the interactions of the anticoagulant heparin are given. The study of the mechanism of drug action is greatly assisted by the *in vitro* study of simplified model systems. The chlorpromazine/iodine system is discussed. Electrochemical techniques, such as conductimetric titration, voltammetry, and potentiometry, are indicated when there is a decrease (as in cation–anion association) or increase (as in the dissociation of a charge transfer complex) in the concentration of charge carriers. Other physicochemical techniques are complementary with these electrochemical techniques. Biological and clinical studies are essential before deciding the practical therapeutic significance of an interaction demonstrated by physicochemical means.

## 1. Scope and Limitations

In the complicated situation involved in the effect of a drug administered to the living subject, the action of the drug may be greatly modified by other drugs.

*G. M. Eckert* • Clinical Pharmacologist, The St. George Hospital, Kogarah, N.S.W. 2217, Australia.   *F. Gutmann* • School of Chemistry, Macquarie University, North Ryde, N.S.W. 2113, Australia.   *H. Keyzer* • Department of Chemistry and Biochemistry, California State University, 5151 State University Drive, Los Angeles, California 90032.

A drug–drug interaction (or simply drug interaction) occurs "when the overall biological response of two or more drugs is markedly different from the simple sum of the effects of each component given singly."[1]

It is important to note that "drug interaction" by this definition is a biological effect. A drug interaction may occur without an *in vitro* physiochemical interaction between the two drugs, and conversely a physicochemical interaction does not necessarily imply a biological interaction.

The interaction between drugs and tissue components is also of practical pharmacological significance and it is convenient to consider some aspects of this.

The term "drug incompatibility" is applied to unfavorable reactions which occur between two or more drugs before the administration of the drugs.

The above definitions notwithstanding, the word "interactions" has of course, many meanings. No attempt will be made here at further clarification.

## 1.1. Significance of Drug Interactions

### 1.2.1. Therapeutic Importance

Clinically, drug interactions are important because of their significance in therapeutics.

Incompatibility, such as that between anionic and cationic antiseptics discussed below, may inactivate the agent(s) involved.

Study of drug interactions which occur after administration of the drugs to the patient is more difficult.

With the development of potent, more specific drugs during the post-World War II period, the fashion for mixing drugs in small-scale mixtures, dispensed individually by the pharmacist, was replaced by industrially manufactured, quality controlled, single (or sometimes fixed dose combination) drug formulations resulting in a decreasing interest in compatibilities. Along with this development was the pious hope, at least among medical academics, that polypharmacy ("shot-gun therapy") would disappear in favor of definite diagnoses and specific therapy. This hope, however, has not been realized. The common use of more than one drug for most patients in both hospitals and the community has been documented.[2,3] In a survey carried out by one of the authors (G.E.) the average number of drugs prescribed concurrently per hospitalized patient was 6.8. Polypharmacy in practice is not dead. Instead of the drugs being mixed in the bottle before administration they are now mixed in the patient. It should be noted that during the same period, there has been an increase in the number and use of highly potent drugs rather than simple placebos, greatly increasing the effects of interaction. *In vitro* techniques have a limited, but significant, role in the elucidation of this complex problem. Many adverse drug interactions are known,[4,5] but not all interactions are disadvantageous. The anticoagulant heparin is rendered safer in clinical use by the ability of protamine sulfate to reverse its action *in vivo*; the interaction of penicillin and probenecid at the renal tubular excretion mechanism enables prolonged duration of action of the antibiotic. Procaine penicillin, a sparingly water soluble association complex between the procaine cation and the penicillin anion, is used therapeutically by intramuscular injection to allow a sustained action of the penicillin, which is rapidly excreted if administered as the soluble sodium salt.[6] Association complexes between drugs and proteins have been applied to the formulation of "sustained release" preparations for therapeutic use.[7]

### 1.1.2. Mechanisms of Drug Disposition and Action

A study of drug interactions may also be of value in the understanding of the mechanisms by which drugs act (that is, the forces between the drug and its "receptor site"), how they are attached to or transported across cellular structures such as cell membranes, and also in the understanding of pathogenesis of disease states.

As Albert[8] pointed out, the association between anion dyes and cationic drugs, used as the basis for methods of colorimetric analysis, "is a model for the uptake of drug cations by receptors and for carrier-aided passage of agents through semipermeable membranes."

In some cases, it is convenient to study a model system, such as chlorpromazine-iodine, to elucidate particular fundamental properties of a given therapeutic agent.

### 1.1.3. Chemical Analysis

Precipitation and extraction of association complexes and the applications of spectral changes in charge transfer reactions have been extensively applied to the chemical analysis of drugs. Although these reactions are the same or similar to those discussed here in the context of therapeutics, application to chemical analysis will be considered to be beyond the scope of this review.

### 1.2. Limitations of In Vitro Investigation

*In vitro* methods follow the reaction between two or more molecular species in an arbitrarily simplified system. The conditions under which the drugs act in the patient are much more complicated. It is convenient to discuss the processes involved in the therapeutic actions in three stages, namely pharmaceutical, pharmacokinetic, and pharmacodynamic.

### 1.2.1. The Pharmaceutical Stage

This is all the events involved in the preparation from the drug, of a formulation suitable for administration.

The problems of drug incompatibility and binding of "active" drug to "inert" formulation components are significant here.

### 1.2.2. The Pharmacokinetic Stage

This is the stage between administration of the drug and its arrival at its site of action, and then its subsequent removal. Absorption, distribution, metabolic transformation, and excretion are involved. That is, pharmacokinetics deals with the movement of the drug into, around, and out of the body.

Drugs are commonly modified by metabolism, and, in fact, it may be the drug metabolite which is biologically active rather than the molecular species of the drug itself. This clearly limits the application of *in vitro* methods.

Distribution of the drug, which may be critical to its ultimate action, may be greatly modified by binding to tissue components such as plasma protein and its transport across cell membranes and barriers such as the blood brain barrier.

### 1.2.3. The Pharmacodynamic Stage

This is the effect of the drug at its site of action, which may be its specific receptor site. Interactions at the pharmacodynamic stage do not lend themselves readily to direct investigation by physicochemical means. Thus, the application of physicochemical methods to the study of drug interactions is limited because these interactions do not always, or even usually, occur between two uncharged molecular species in a system approximating to a simple aqueous solution. Pharmacokinetic and pharmacodynamic considerations greatly complicate the situation.

As usual great care must be taken in extrapolating laboratory results to the clinical situation. As was said by a pharmacologist speaking at the 1929 meeting of the Section of Pharmacology and Therapeutics of the American Medical Association, "There is no short cut from chemical laboratory to clinic, except one that passes too close to the morgue."[2] Confirmation of the clinical significance of chemical facts must always be obtained by clinical means.

### 1.3. Advantages of the In Vitro Investigation of Drug Interactions and Incompatibilities

Certain drug combinations are suitable for investigation by *in vitro* physical methods.

Physicochemical methods will confirm or deny a straightforward reaction between two drugs, and this information may be of value because some of the medical documentation is of a "deductive" (or "it may be expected that...") nature, which is recorded and repeated as observed fact. The alleged interaction between heparin and lidocaine hydrochloride, discussed later, in Section 7, is an example of this.

Physicochemical reactions which are involved in incompatibilities and binding of active drug to formulation components are clearly suitable for study by physicochemical means.

The investigation of cationic and antibacterial agents by anionic agents, discussed later, in Section 3, illustrates this.

A few drugs act in the blood in an essentially unchanged condition. The anticoagulant heparin (although bound to plasma protein) acts in this way by what appears to be an electrochemical mechanism.[9,10] Further, the anticoagulant action is reversed clinically by cationic agents, such as protamine sulfate, which neutralize the negative charge of the anionic heparin. The reaction involved in the inactivation of heparin can be studied by electrochemical techniques.

The study of the binding of drugs to tissue components (for example, chlorpromazine to melanin) by electrochemical techniques may provide information of fundamental therapeutic significance. The study by physicochemical methods of simplified model systems may be of value in supporting or denying postulated pharmacodynamic mechanisms.

### 1.4. Interactions Investigated by Electrochemical Techniques

Some charge transfer reactions and ion associations are accompanied by observable changes of electric properties such as conductivity and capacitance and may therefore be followed by electrochemical techniques.

### 1.4.1. Charge Transfer Reactions

Szent-György,[11-13] among others, has drawn attention to the importance of charge transfer in biological systems. For example, porphyria-inducing drugs such as barbituates, carbamides, and amides have been shown[14] to act at the electron exchange center of the coenzyme Q. Such reactions, which involve donors as well as acceptors of roughly the same energy, are difficult to detect because there will be no color change, no obvious charge transfer band in the absorption spectrum, no electron spin resonance signal, nor change in dipole moment. However, if one carries out these interactions *in vitro* as electrode reactions which occur within an "active space" or within an electrolytic double layer, significant changes do occur and may be detected by electrochemical methods.[15]

Charge transfer complexes of heparin are of special interest because it has been shown that the antithrombin III–heparin cofactor, a major inhibitor of thrombin, as well as other serine proteases are inhibited by complex formation.

The pharmacological activity of charge transfer complexes has been reviewed.[11–13,16] The use of conductometric titration to study the charge transfer reactions of chlorpromazine is indicated in Section 1.4.2.

### 1.4.2. Ion Association

The inactivation of cationic antimicrobial agents by anionic agents is well known (see Section 3). Since the reaction between ionic agents is accompanied by changes in concentration of free charge carriers, the application of electrochemical techniques is indicated. The molecular forces involved in this type of reaction have been considered and reviewed by Nancollas,[17] Davies,[18] Diamond,[19] Higuchi,[20] Frieser,[21] Szwarc,[22] and others. These ionic associations possibly provide a simplified model for what has been referred to as the "hydrophobic bond,"[23] while the metachromasia which accompanies many of these complex formations, when one of the species is colored, is suggestive of charge transfer interaction.

Free energy of solution has been suggested as the significant factor,[24] while polarity, thermodynamics of Coulombic attraction, molecular size, molecular shape, charge distribution, and charge shielding have all been considered. The forces involved with substrate–enzyme complex formation, at drug receptor sites and in antigen–antibody reactions, probably involve a similar range of molecular forces.

Examples will now be considered of the application of electrochemical techniques to the interactions of chlorpromazine and other phenothiazine derivatives, heparin, and antibiotics.

## 2. Methodologies

### 2.1. Introduction

To detect an interaction between two drugs or a drug and a tissue component, change of some physical property which changes when the molecular structures associate or dissociate must be measured. An exhaustive list of such properties cannot be presented because any measurable property may have an application. Conductivity, spectra, viscosity, and surface tension, among others, have been applied.

For charge transfer complexes, if the permittivity of the solvent is high enough, the complex may dissociate, giving rise to ions capable of carrying current and thus affect the conductance of the solution. A conductance maximum is seen in the titration curve. In the absence of dissociation, complexation often may be ascertained from the formation of a spectroscopically observable new absorption band,[25] either in the visible or in the uv, but this does not hold if the complex dissociates to any appreciable degree. Thus, conductimetric titration[26] supplements spectroscopy in that formation of ions, usually not discernible spectroscopically, is readily followed conductimetrically.

Similarly, if association of ionic species occurs, the interaction is accompanied by a decrease in the concentration of free charge carriers, giving rise to a conductance minimum in the titration curve.

### 2.2. Conductivity Titration

A variant of the general Job[27] method introduced by Gutmann and Keyzer[26,28-30] may be used: a solution of a potential electron donor in a solvent is titrated against a potential electron acceptor in the same solvent. Departures from linearity indicate an interaction and the relative concentrations at a conductance maximum or minimum give the stoichiometry of the adduct formed. For the biologically active substances described in the following sections, the titrations, unless otherwise stated, are carried out in unbuffered aqueous solution. The addition of buffers introduces additional variables without enhancing the accuracy of the system as a biological mode. Slifkin[31] has reported that buffers can themselves take part in charge transfer interactions, and thus their presence is liable to cause additional complications.

In the aqueous solutions of ionic drugs used in many of the investigations the background conductivity is high and an instrument of sufficient accuracy is required. We have satisfactorily used throughout a Wayne-Kerr Conductance Bridge model B-224 operating at 10,000 rad (viz., 1592 Hz) and yielding an accuracy of $+0.1\%$ or better.

## 2.3. Voltammetry

Voltammetric study of charge transfer complexes is a recent development.[15] The technique has been applied to drug interactions:[32] in this study the general arrangement was conventional. The working electrode and the counterelectrode were Pt/Pt. Potentials were measured against a standard calomel electrode mounted in close proximity to the working electrode and the currents measured in terms of the potential differences developed across a voltage dropping resistor of a value always below 1% of the equivalent cell resistance, so as to minimize linearization of the voltammograms. All voltammograms were taken at 37°C.

Experimental conditions should be kept as simple as possible. The authors[32] did not use supporting electrolyte or deaerated solutions. The use of a potentiostat obviated the corrections for the voltage drop of the solution.

At this stage in the development of the subject quantitative and detailed evaluation of the voltammograms is probably not meaningful.

## 2.4. Potentiometry

The formation of a charge transfer complex may also be indicated from potentiometry: the electrode potential of an "active" electrode, usually Au or Pt, is measured against a reference electrode potential, say a saturated calomel electrode. Donor is then titrated against an acceptor solution, or vice versa, and a maximum or a minimum in the potential vs. titrant concentration curve indicates the complex stochiometry.[33] The technique has been applied to study the interactions between E. coli and antibiotics such as streptomycin and kanamycin, using a three-compartment cell.[34]

## 2.5. Other Physical Techniques

Other physical techniques that have been used to follow the formation of charge transfer complexes include NMR,[35] ESR,[25] surface tension,[36] viscosity,[37] nuclear quadrupole resonance,[38] refractive index,[39,40] dipole moments,[39,40] or polarography[15] and spectroscopy,[41] especially in the uv and ir. Electron scanning microscopy[42] has been employed to study suspenoids formed in some of the interactions. As noted above, this variety of physical methods is mutually supplementary.

## 2.6. Biological and Clinical Techniques

Biological and clinical investigations are necessary to substantiate or otherwise the clinical significance of an interaction which has been demonstrated by *in vitro* physical techniques. For example, microbiological culture techniques have been used for suggested antibiotic interactions,[43] and blood coagulation studies with heparin interactions,[44] while clinical epidemiological monitoring procedures[45] have more general application.

## 3. Incompatibility between Anionic and Cationic Antibacterial Agents and Other Anionic and Cationic Drugs

Cationic surface active agents such as benzalkonium chloride and cetyltrimethylammonium bromide (Figure 1) have use as antibacterial agents, i.e., "disinfectants" and "antiseptics," and have the advantage that the antimicrobial action is combined with a cleaning action associated with their detergent activity.

Domagk,[46] as early as 1935, soon after the introduction of the cationic disinfectants, noted their inactivation by soap and other anionic detergents (Figure 2). The inhibition of the synthetic cationic detergents by anionic compounds, including phospholipids and alkylsulfates, was studied, using microbiological techniques.[47-50]

Valko and DuBois[51] made the interesting observation that bacteria apparently "killed" by cationic quaternary ammonium detergent germicides may be revived by the addition of a high molecular weight anion, provided that the application of the latter occurs within a period of 10–30 min.

The mechanism of action and incompatibilities of the nondetergent cationic acridine series of antibacterials was studied in detail.[52]

Although the reaction between "cationic" and "anionic" agents is frequently accompanied by visible precipitation, therapeutic incompatibility was observed, in some cases, without any visible change.[53] Barker,[53] reporting to the British Pharmaceutical Conference, stated "the opinion was expressed that, if generalization was at all possible, it could be said that anionic compounds were incompatible with cationics and that the incompatibility might not always be visible."

Because of the unreliability of visible change as a guide to incompatibility, a "general rule" was evolved to the effect that "cationic drugs are incompatible with anionic drugs, whilst cationic drugs are compatible with nonionic agents."[54] Convenient as this may have been, especially for surface active agents, it leads to false conclusions through applications which were not verified experimentally. There is no simple relationship between

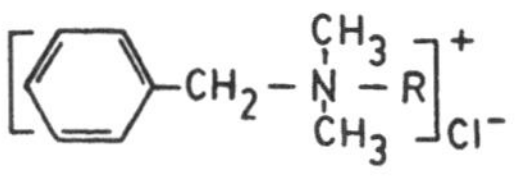

*FIGURE 1.* Structure of a typical cationic detergent antibacterial agent. A mixture of alkyldimethylbenzylammonium chlorides of the general formula in which R represents a mixture of the alkyls from $C_8H_{37}$ to $C_{18}H_{37}$.

FIGURE 2. Structure of a typical anionic detergent. Sodium lauryl sulfate is a mixture of sodium normal primary alkyl sulfates, consisting chiefly of sodium dodecyl sulfate.

$$\left[ C_{12}H_{25}OSO_3 \right] Na^+$$

the molecular weight of anion/cation pair and the incompatibility of the salts of that anion and cation; the uncritical extension from detergents to other drugs is particularly unwarranted.

If "cation–anion" incompatibility occurs, a change in concentration of free charge carrier could be expected to accompany the reaction which could therefore be followed by conductimetry. Gutmann and Keyzer[26] applied conductimetric techniques to demonstrate a charge transfer interaction between chlorpromazine and iodine. At about the same time, but independently, Rodgers[55] used similar techniques to follow pharmaceutical interactions between drugs involving removal of charge carrier species from solution. Harris[43,56] extended the pharmaceutical application of conductometric techniques. Harris called the false "general rule" into question, observing, "it is generally considered, often without confirmation, that two large organic ions of opposite charge cannot co-exist in solution." Analytical chemists who regularly apply ion association to extraction and other techniques are well aware of the complicated nature of the interactions between "large cations" and "large anions". The mechanisms involved are more subtle than a spurious general reaction; association may occur between ions and neutral species and may fail to occur between examples which obey the so-called "general rule." Irrespective of the mechanism or physicochemical principles involved, many examples of association are accompanied by changes in concentration of free charge carriers and are therefore suitable for investigation by electrochemical methods.

## 4. Interactions of Chlorpromazine

### 4.1. The Significance of Chlorpromazine

Chlorpromazine (CPZ) is of considerable interest because of its pronounced pharmacological activity. It is highly psychotropic[6] and is notable for its excellent electron donor properties. The drug is always administered as the hydrochloride because the free base is insoluble in water; however, their electronic ring properties are similar. Pure zone refined CPZ free base has a melting point of 56.5°C; it tends to supercool forming a glass. Under uv irradiation, or in sunlight, both the free base and the hydrochloride tend to go over into the free radical, formed by donation of an electron from the ring nitrogen. This free radical is quite stable, especially in acid solution, where it can form spontaneously, yielding an oily tar. The statement in the Merck Index and similar statements in literature that it forms "an oily liquid" refers to the free (ion) radical and not to the compound as such. Any results obtained in this oily tar thus refer to the free radical and not to pure CPZ·HCl.

The physiological activity of the drug is quite extensive; it decreases the release of acetylcholine, tends to expand biological membranes, inhibits amine receptor activation, affects the membrane enzyme activity, tends to inhibit ventricular activity, is anticholinergic, antiemetic, affects Ca changes in tissues, affects glycogen in the liver, binds to microsomes, affects $O_2$ uptake, acts as a mild local anesthetic, etc. Its main application, however, is in psychiatry, where it still remains the drug of choice for many forms of schizophrenia. For a full discussion of the drug including its side-effects, the reader is referred to the Proceedings of the 3rd[59] and 4th[60] NIMH sponsored Conferences on Phenothiazine and Related Drugs.

## 4.2. Other Phenothiazine Drugs

A considerable number of phenothiazine derivatives, varying in the side chain attached to the nitrogen and a substituent in the ring, have been introduced into clinical practice. The relationship between the chemical structure and the pharmacological activity of these drugs has been reviewed.[61]

A number of phenothiazine drug structures are shown in Figure 3. Of the drugs tabulated in this figure, promethazine is an antihistamine and antiemetic while the others are tranquilizers and antiemetics.

## 4.3. Chlorpromazine–Iodine

### 4.3.1. Significance of the Chlorpromazine–Iodine Reaction

Chlorpromazine plus iodine is not a combination that is likely to occur in a pharmaceutical formulation nor is the interaction of any significance in therapeutics. The study of the reaction is of biological importance, however, because of the information provided with respect to the electronic properties of the phenothiazine derivatives, and these properties in turn are likely to have significance in understanding the underlying mechanisms of the drug. That is, $CPZ–I_2$ is a model system.

Model systems using a purely physiochemical approach have greatly aided the investigation of pharmacological problems especially in testing alternative postulated mechanisms. "Even though the model system is merely a method for visualizing a problem in simpler molecular terms and

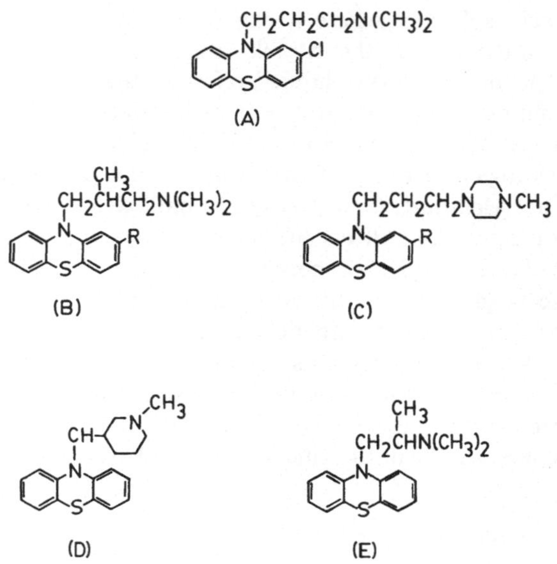

FIGURE 3. Structures of some phenothiazine derivatives. A: chlorpromazine. B: trimeprazine (R = H), methotrimeprazine (R = OCH$_3$). C: prochlorperazine (R = Cl), trifluoperazine (R = CF$_3$). D: mepazine. E: promethazine.

is not an attempt to reproduce physiological conditions, it can serve the purpose of eliminating those mechanisms which violate the principles obtained from these studies."[62]

Iodine, for which the electron acceptor properties are well studied, is a suitable reagent with which to assess the electron donor properties of the phenothiazine derivatives.

### 4.3.2. Conductimetric Titrations and Spectrophotometric Study of the Chlorpromazine–Iodine Reaction

Conductimetric titration curves[63,64] of CPZ · HCl in acetonitrile with iodine and its converse are displayed in Figure 4 and exhibit a maximum at 1:2 ratio.

Freshly prepared CPZ base titrated with iodine yields a relatively sharp peak at a 1:1 ratio at 20.5°C (Figure 5A). Exposing the CPZ solution to daylight for a prolonged period and then titrating it at 20°C with iodine solution yields an irregularly shaped curve (Figure 5B) with a distinct maximum at a 2:3 ratio and a faint shoulder on the higher iodine content side. Repeating this titration at 34°C (Figure 5C) shows an increase in the peak height of the curve in addition to an increased

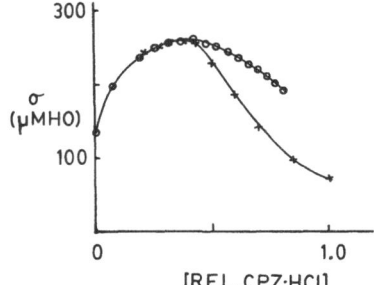

FIGURE 4. Conductimetric titrations at 20°C of chlorpromazine hydrochloride (CPZ·HCl) with iodine, both dissolved in acetonitrile. Initial concentrations: CPZ·HCl and $I_3$, $1.24 \times 10^{-3}$ $M$. $\sigma$ means conductance, REL. signifies relative.

shoulder. Repeating this titration at 50°C (Figure 5D) promotes the trend in the last titration to such an extent that the shoulder has now become the maximum and the ratio of CPZ:$I_2$ is now largely 1:2. Some time later (two months), a sample of CPZ of the same base, which when freshly prepared yielded the 1:1 maximum, was dissolved in acetonitrile and titrated with iodine solution and clearly yields a maximum at a 2:3 ratio, as does the converse titration (Figures 5E and 5F).

Titrating CPZ base with $I_2$ in dimethyl sulfoxide also yields a maximum at a 2:3 proportion.

The spectra of CPZ·HCl and CPZ·HCl/I are compared in Figures 6a and 6b.[63] Spectra of CPZ·HCl in Nujol mulls and KBr disks at 18 and 60°C (mp CPZ is 56.5°C) are identical.

CPZ, like many other phenothiazine derivatives with a 10-substituted side chain, supercools readily and for prolonged periods, but the infrared spectrum of liquid CPZ is identical above and below the melting point, while differing considerably from the solid base. In fact there are more similarities between liquid CPZ and the complex CPZ:$I_2$ than between the liquid and solid CPZ. The spectra of CPZ in the solid and liquid state, as well as its complex with iodine, are displayed in Figures 7a and 7b.

FIGURE 5. Conductimetric titrations of chlorpromazine base with iodine, both in acetonitrile solution. A, Titration of freshly prepared, recrystallized, and immediately used chlorpromazine at 20°C. B, Titration at 20°C of chlorpromazine which has been allowed to age in acetonitrile solution. C, Same as for B but titrated at 34°C. D, Same as for B but titrated at 50°C. E, Titration at 20°C of chlorpromazine aged in the solid state. F, Converse of E. Initial concentrations: A, CPZ, $3.15 \times 10^{-3}$ $M$; $I_2$, $3.14 \times 10^{-3}$ $M$. B, C, and D: CPZ, $3.22 \times 10^{-3}$ $M$; $I_2$, $3.14 \times 10^{-3}$ $M$. E and F: CPZ, $3.00 \times 10^{-3}$ $M$; $I_2$, $3.1 \times 10^{-3}$ $M$. $\sigma$ and REL. as for Figure 4.

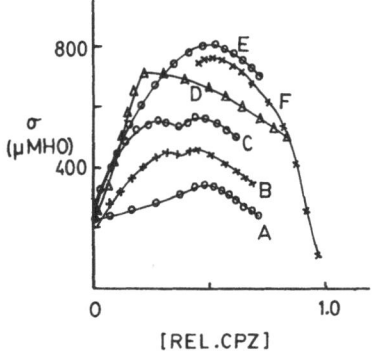

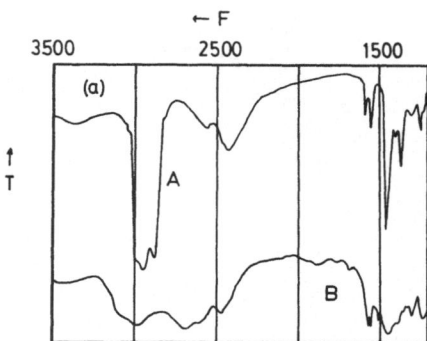

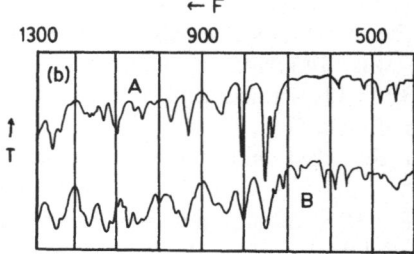

FIGURE 6. (a) Spectra of chlorpromazine hydrochloride and of its iodine complex in the frequency range $F = 3500-1200$ cm$^{-1}$. Curve marked A is the spectrum of chlorpromazine hydrochloride in a paraffin mull. Curve marked B refers to the iodine complex of chlorpromazine hydrochloride. (b) Same for Figure 6a but in the range $1300-400$ cm$^{-1}$. $T$ is transmittance in arbitrary units.

In solution, in which favorable orientations between donor and acceptor are more probable than in the solid state, CPZ complexes with $I_2$ not only in the ratio 2:3, as they do in the solid state, but also in different proportions.

CPZ, when freshly prepared, recrystallized, and immediately titrated with $I_2$ gives a conductivity peak at 1:1. Aged CPZ either in acetonitrile solution or the solid state leads to a 2:3 ratio with $I_2$, whereas on heating the solution an $I_2$ complex of a 1:2 composition is obtained.

Hence at least three different complex species are likely to exist between $I_2$ and phenothiazine and its derivatives.

The 2:3 species is thought to be a definite new entity, i.e., one complex involving two phenothiazine molecules and three $I_2$ molecules, rather than a 50/50 mixture of 1:1 and 1:2 complexes; the ESR data show the strongest signal at the 2:3 ratio, which in turn is twice as strong as the signal for the 1:1 mixture and certainly stronger than the signal for the 1:2 ratio. It is not unlikely that the true state of affairs involves a mixture of these three different complex species, with, at room temperature, the 2:3 complex being the most stable, and dative in the ground state.

From the stability of the phenothiazine:iodine and CPZ:$I_2$ complexes in their electrical properties and in view of the fact that the stable form for

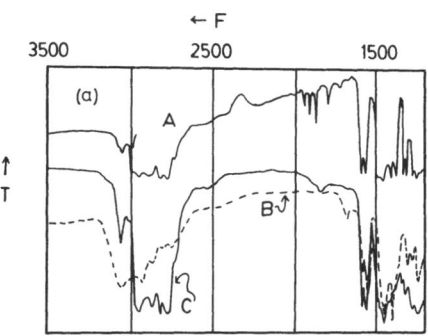

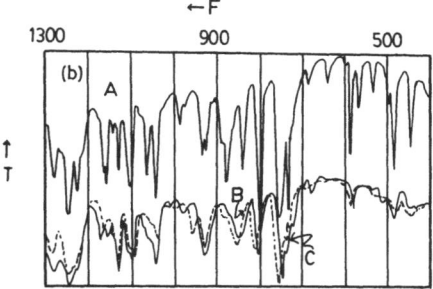

*FIGURE 7.* (a) Spectra of chlorpromazine and of its iodine complex in the frequency range $F = 3500–1200\ cm^{-1}$. Curve marked A refers to crystalline chlorpromazine at 20°C. Curve marked B refers to the iodine complex of chlorpromazine base at 20°C. Curve marked C refers to chlorpromazine in the liquid state at 60°C. (b) Same as for Figure 7a but in the range $1300–400\ cm^{-1}$. $T$ is transmittance in arbitrary units.

both occurs for the 2:3 ratio it is likely that the donor activity of CPZ resides in the phenothiazine nucleus and not in the side chain. Should some donor activity exist in the side chain of CPZ then it would be expected to occur at the nitrogen atom. This side-chain nitrogen atom can be isolated and blocked from any further donor activity by salt formation.

Even so the CPZ:HCl molecule complexes with $I_2$ in a 1:2 proportion in acetonitrile, similar to the high-temperature form of CPZ base. Further, CPZ base complexes with chloranil, as does phenothiazine, in a 1:1 ratio.

The above reasoning rules out the formation of any truly ionic or covalent bonds between the phenothiazine nucleus and $I_2$. Thus no new infrared absorption band associated with such bands may be produced.

One of the most significant differences between the spectra of CPZ and $CPZ:I_2$ is the appearance of a new, weak, but distinct band at about $1700\ cm^{-1}$ (see Figure 7a). This new band also appears in the spectrum of the $CPZ:HCl:I_2$ complex (Figure 7a). This new band also appears in the spectrum of the $CPZ:HCl:I_2$ complex (Figure 6a) and must therefore be due to some vibration concerning the phenothiazine nucleus. The new band increases in intensity and shifts to higher frequency, i.e., from 1680 to

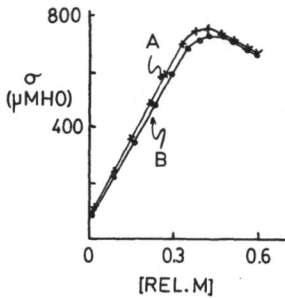

FIGURE 8. Phoreograms in acetonitrile solution of meprazine derivatives with iodine: curve A, trimeprazine; curve B, methotrimeprazine. [REL. M] is relative meprazine concentration.

1720 cm$^{-1}$, as the molar iodine content is increased from 1:0.5 to 1:1.5. This absorption is not inconsistent then with a C–N stretching vibration in which double-bond character increases as suggested above.

### 4.3.3. Relationship between Donor Strength and Pharmacological Activity in the Phenothiazine Derivative

Charge transfer complexes of iodine with chlorpromazine (CPZ), methotrimeprazine (MTZ), trifluoperazine (TFP), trimeprazine (TMZ), prochlorperazine (PCP), thioproperazine (TPP), and promethazine (PMZ) have been studied[65] by conductometric and spectral methods.

All the conductivity titrations showed a well-developed maximum at the 2/3 stoichiometry for phenothiazine derivative:iodine complexes. The curves obtained with the meprazine and perazine derivatives are shown in Figures 8 and 9, respectively.

All the iodine complexes exhibit a new band at approximately 1700–1650 cm$^{-1}$, which, in some cases, appears as a doublet.

The spectral changes following upon complexing with iodine are even more pronounced in the case of bromine as the acceptor. The infrared spectra of the aging Br$_2$ complexes undergo changes indicative of chemical

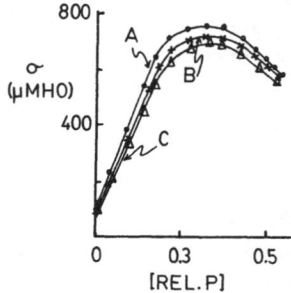

FIGURE 9. Phoreograms in acetonitrile solution of perazine derivatives with iodine: curve A, prochlorperazine; curve B, trifluoperazine; curve C, methotrimeprazine. [REL. P] is relative phenothiazine derivative.

reaction, e.g., the free bases show a new, broad absorption band at about $3400 \text{ cm}^{-1}$ due probably to ammonium ion formation. This is further confirmed by the evolution of HBr, probably the by-product of the ring bromination. Such a reaction does not take place in the $I_2$ complexes, as has already been demonstrated[63] in the case of $CPZ:I_2$ and $CPZ \cdot HCl:I_2$, most likely because $I_2$ is a weaker acceptor than $Br_2$.

Maximum conductivity, $\sigma_p$, may be used as a measure of electron donor or acceptor strength.

The current is carried by ions which are produced[25,26,66] by charge transfer complex formation and consequent dissociation into ions in a solvent. Formation depends on the ionization potential of the donor and on the electron affinity of the acceptor as well as on their mutual orientations. Once the complex is formed it may dissociate into ions, and this involves the interaction of the complex constituents with the solvent. If, however, in a particular series of complexes, only substituents in the donor are changed, then the total number of ions formed will be a measure of the ionization potential of the donors.

The total number of ions is measured[25,26,66] by the maximum conductivity, $\sigma_p$, at the stoichiometry of the charge transfer complex in solution. A large $\sigma_p$ reflects a smaller ionization potential of the donor molecule, other conditions being kept constant, and thus a greater complexing ability. Increasing the electron withdrawing power of a substituent in the donor increases the ionization potential of the donor, and should, therefore, decrease the complexing ability of the donor.

The phenothiazines, arranged in order of maximum conductivity, are given in Table 1.[65] The correlation between the estimate of charge transfer activity and specific physiological effects is shown[65] in Table 2.

Since it is known that CPZ and its derivatives can act as an electron acceptor, it would be expected that the weaker donor activity of a

TABLE 1. Maximum Conductivity of Phenothiazine Derivative/$I_2$ Complexes in Acetonitrile

| Compound | Abbreviations | Maximum conductivity $\sigma_p$ ($\mu$mho) |
|----------|---------------|---------------------------------------------|
| Promethazine | PMZ | 725 |
| Chlorpromazine | CPZ | 785 |
| Trimeprazine | TMZ | 755 |
| Methotrimeprazine | MTZ | 740 |
| Prochlorperazine | PCP | 752 |
| Trifluoperazine | TFP | 720 |
| Thioproperazine | TPP | 690 |

TABLE 2. Comparison of Pharmacological Activity with the Donor Properties of the Phenothiazine Derivatives[a] Following the Sequence for Decreasing Electron Donating Abilities: PMZ ≥ CPZ ≥ TMZ ≥ PCP; TMZ > MTZ; PCP > TFP > TPP. Abbreviations are listed in Table 1.

| Physiological activity | Comparison with donor strength[b] | The bracketed number refers to | Comment |
|---|---|---|---|
| Hypothermic | CPZ (100)[c] <br> TMZ (59) > MTZ (30) <br> PCP (40) > TPP (3) | The comparative body temperature drop induced by a fixed dose of thiazine derivative. | Proportional to donor strength. |
| Antihistaminic | PMZ ($10^4$) ≥ CPZ ($10^2$) <br> TMZ ($1.2 \times 10^4$) > MTZ ($10^4$) <br> PCP (50) > TPP (250) | The number of lethal doses of histamine against which a fixed quantity of PH derivative will give protection. | Proportional to donor strength. |
| Antiemetic | CPZ (100) <br> PCP (=35) > TFP (6) <br> > TPP (0.5) | The comparative dose required to abolish emesis induced by a fixed quantity of apomorphine. | Inversely proportional to donor strength. |
| Antiadrenaline | PMZ (=2000) ≥ CPZ (100) <br> TMZ (500) > MTZ (100) <br> PCP (100 = 600) > TPP (300) | The comparative dose required to offset the effect of a lethal dose of adrenaline. | Inversely proportional to donor strength. |

[a] All these data have been taken from May and Baker Ltd. (Dagenham) Pharmaceutical Publications, which contain a digestion of the relevant literature and references for each compounds cited here. The publications follow: "Phenergan," (PMZ), Ed. 6 (1964); "Largactil," (CPZ), Ed. 7 (1963); "Stemetil," (PCP), Ed. 2 (1962); "Vallergan," (TMZ), Ed. 6 (1965); "Veractil," (MTZ), Ed. 2 (1960); "Terfluzin," (TFP), Ed. 4 (1966); "Majeptil," (TPP), Ed. 3 (1963).

[b] The sequence given above this table is adhered to.

[c] CPZ is arbitrarily taken as the standard and is given the value of 100, while the other derivatives are given values according to the CPZ standard.

phenothiazine derivative would cause it to be a relatively stronger electron acceptor. This would explain the inverted sequence observed for adrenaline antagonism and antiemetic activity of the phenothiazine drugs.

Postulating also an acceptor role for the thiazine derivatives rather cuts across the conventional view that their drug action is due to their striking donor properties, but this notion widens the scope of present theories explaining the influences of ring substitution in these compounds with respect to physiological charge transfer activity.

## 4.4. Interactions between Chlorpromazine and Neural Transmitters

It has been suggested[15] that the pharmacological activity of chlorpromazine and related phenothiazine drugs is due to an electrode reaction involving the formation of a surface charge transfer complex with a neurotransmitter on the synaptic membrane. Other results[15] indicate that phenothiazines are bound to the receptor site on the synaptic membrane by a combination of charge transfer and electrostatic forces with hydrophobic bonding also being involved.

These possibilities make a study of the physicochemical reaction between chlorpromazine (and other phenothiazine derivatives) and neural transmitters of special interest.

The electrochemical and spectral studies of the reaction between phenothiazine derivatives and acetylcholine, serotonin, and dopamine derivatives confirmed that charge transfer reactions occur.[67]

## 4.5. Chlorpromazine–Melanin Interactions

### 4.5.1. The Melanins

The melanins (from the Greek "melas," meaning black), are naturally occurring pigments which are polymers derived from oxidation of tyrosine and dopa. There is considerable species variation in this group of pigments, which is widely distributed. They are found in melanocytes in the basal layer of the human epidermis, giving the characteristic color toning of the skin; in the eye, melanin is found in the retina, ciliary body, and choroid. Another variant of melanin is found in "squid ink."

Melanin is amphoteric; it may act as a donor but more usually as an acceptor.[68] The detailed structures and molecular weights of the melanins derived from the various sources such as "squid ink" or human hair are unknown at present.

Natural melanin, at physiological temperatures, however, carries considerable spontaneous surface charges and it behaves as an electret.[69] The magnitude of the effect depends on hydration; any specimen of natural melanin is not electrically neutral but carries a net electrical charge. To what extent these observations are due to sample preparation is not clear, but, if inherent, then the physiological properties of melanin must be greatly affected.

The behavior of melanin is dependent on hydration. Wet melanin switches from a high to a low resistivity state upon the application of a small electric field. This has been attributed to an emulsion inversion.[70,71] Wet melanin is said to be an emulsion of electron-rich domains in an electron-poor medium. Many chemical reactions, such as redox reactions or complexation with donors, could do likewise.

### 4.5.2. Biological Relationship between Chlorpromazine and Melanin

Chlorpromazine and other phenothiazine tranquilizers preferentially accumulate at sites of high melanin content such as hair and certain glandular tissues, and it has been suggested that this is based on an interaction possibly by formation of a charge transfer complex.[72] In such an adduct chlorpromazine would function as an electron donor[73] and melanin as an electron acceptor.[74]

"Melanosis" manifested by skin pigmentation, ocular opacities, and pigmented retinopathy have been reported as side effects of long-term therapy.[75,76] These observations make the potential interaction of chlorpromazine with melanin of practical and clinical significance.

### 4.5.3. Conductimetric Titration of Chlorpromazine Against Melanin

The species variability, unknown molecular weights, and poor solubility in most solvents of the melanins present difficulties in the choice and preparation of the material for titration. In the conductimetric study cited[66] melanin was prepared according to Piatelli and Nicolaus[77] substituting frozen Japanese squid (cuttlefish) for the Mediteranean sepia species used by the Italian authors. A sample of human melanin from Japanese hair was made available by Bolt,[78] who used a new and mild procedure for extraction and purification. The natural melanins varied in their solubility in suitable solvents. Melanin derived from human liver melanoma appeared to be the least polymeric and most readily soluble. For squid melanin DMSO was used as the "solvent," although some insoluble residue was left. DMSO, however, is not a truly inert solvent, but interacts with melanin, particularly at high temperatures. To minimize this reaction all tests were carried out in cold saturated solutions of melanin in DMSO, or suitable dilutions thereof. Triethanolamine completely dissolved human hair melanin and was employed in a series of tests. This solvent also appeared to react with the pigment, because the conductivity of the respective solution was considerably below that of the pure solvents.

*In vitro* formation of a charge transfer complex between chlorpromazine as the electron donor and melanin, derived from squid or human hair as the electron acceptor, was demonstrated by an increase in conductivity in a mixture of solutions of the components of a number of solvents. If triethanolamine or diethylacetamide were used as solvents, this increase was demonstrable by simple mixture of the components but could be further increased by uv irradiation. In other solvents, such as dimethylsulfoxide, significant increases of conductivity were demonstrable only after uv irradiation.

As the molecular weights of the melanin preparations derived from squid or human hair are unknown at present, the stoichiometry of the complex cannot be calculated.

### 4.5.4. Biological Significance of Chlorpromazine–Melanin Interaction

Potts[74] discussed the formation of chlorpromazine–melanin charge transfer complexes and concluded that this was a distinct possibility, but, at that time, experimental proof was still lacking. Support for the charge transfer complex formation between chlorpromazine and melanin has been obtained by using additional and supplementary techniques.[66] More recently, this complex has been identified in chlorpromazine-eye melanin.[79,80]

Additional support for the formation of a charge transfer complex has been derived from an EPR study[66]; the signal due to melanin alone was reduced by the addition of chlorpromazine and decreased further on exposure to visible light.

It is of interest that *in vivo* experiments in which chlorpromazine hydrochloride is administered to mice as the adduct of electron acceptors implicate the free radical of the drug in hypothermia and sedation.[81]

The melanin–chlorpromazine complex may also play a role in reducing the toxicity of chlorpromazine in mice. The $LD_{50}$ in black mice is more than twice that of chlorpromazine in albino mice.[82]

Melanin has been shown to form adducts with chloroquine, paraquat,[83,84] and clindamycin.[85] The significance of drug interaction with melanin has been discussed.[86]

### 4.6. Chlorpromazine–Heparin Interaction[32]

Heparin, as described in Section 6, is an anionic, mucopolysaccharide, which is known[32] to complex with, for example, other di- and polyamines. A spectral shift of 90 Å has been reported for its adduct with the cationic toluidine blue. Cationic dyes tend to be electron donors, and thus it is in this reaction that heparin appears to exhibit an electron acceptor character. An interaction between sodium heparin and chlorpromazine hydrochloride in aqueous solution is shown from the results of conductimetric titration (Figure 10).[32] Both conductivity and capacitance curves exhibit a minimum. The true stoichiometry for the interaction, as explained in Section 6, cannot be established from these results. The conductivity minimum is comparable with that shown by the cationic and anionic detergents in Section 3 rather than the conductivity maximum shown by chlor-promazine–iodine. Since the capacitance, too, exhibits a minimum close to the same composition of titrants as that in the minimum in the conduc-tivity curve, the resulting entity is less polar than its components. In view of the low concentration employed, these capacitance changes cannot be due to a bulk effect but must be due to electrode reactions proceeding within the electrode double layers; the permittivity of the bulk of the solution is necessarily close to that of water. However, the two electrode double layer capacitances, which electrically are in series with the bulk capacitance, are determined by electrode reactions occurring at the elec-trode surface in the double layer. The capacitance changes are seen to be quite large; the minimum value of 4.3 nF is 60% of the 7.1 nF of pure chlorpromazine hydrochloride solution and 22% of the 19.7 nF of the heparin solution. The reaction product thus appears to be adsorbed at the electrode(s) where its different polarity causes the capacitance changes observed. The capacitance minimum may also be due to the inner Helmholtz plane being pushed further away from the electrode with the adsorption of the bulky molecular adduct.

Conductivity minima which have been reported for the interactions of quaternary ammonium compounds have been ascribed to associations which lower the critical micelle concentration and thus reduce the concen-tration of the substituted ammonium ions. Such an effect is considered to be somewhat unlikely in the present case because the chlorpromazine con-

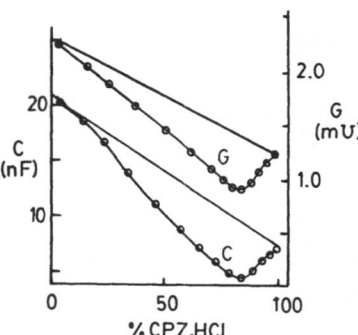

FIGURE 10. Phoreograms of the heparin–chlor-promazine HCl interaction 200 units/ml of heparin $10^{-3}M$ chlorpromazine hydrochloride both in aqueous solution; Gold electrodes; 17°C. The values of the conductance $G$, in millimho, refer to the right-hand ordinate scale, while capacitance values, $C$, in nF, are scaled on the left. The phoreogram as well as the capacitance plot show well-defined minima. The linear base lines, which would be followed in the absence of a charge carrier interaction, are also indicated.[32]

centration was kept to less than one third of its known critical micelle concentration.

The spectra show a displacement of the chlorpromazine peak at 3080–3100 Å. While this slight bathochromic shift could be interpreted as evidence for the formation of a contact charge transfer complex, by itself it is not thought to be sufficient evidence. If heparin as an electron acceptor is forming a charge transfer complex with the strong electron donor chlorpromazine, a conductivity maximum should result. The opposite occurs. The role of hydrophobic bonding in the association should be considered.

Whatever the mechanism, a pharmaceutical incompatibility between sodium heparin and chlorpromazine hydrochloride solutions is confirmed. The clinical significance of an interaction between the two drugs after administration has not been demonstrated. The type of electrode reaction demonstrated in the model *in vitro* system may occur *in vivo*, because a biological membrane has been shown to be capable of acting as an electrode and it known that heparin is taken up by cell membranes. Binding of chlorpromazine to other mucopolysaccharides may be of significance in the pharmacokinetic disposition of the drug.

## 4.7. Chlorpromazine–Phenytoin Interaction[32]

The incompatibility between chlorpromazine hydrochloride and phenytoin sodium solution (Figure 11) is confirmed and elucidated by conductimetric titrations (Figure 12). A sharp and significant change in slopes is seen to occur at 1:1 stoichiometry. A conductivity minimum is observed, and this is again accompanied by a capacitance minimum at the same point. Again there is seen to occur an electrode reaction as discussed in the previous reaction. This is supported by the results of cyclic voltammetry (Figure 13), which shows a new peak, absent from the voltammograms, at +250 and −300 MV. With chlorpromazine hydrochloride, only the well-known CPZ free radical peak at +700 MV is observed. With the combination, a large subsidiary loop at high anodic voltages, which is probably due to filming, is seen (that is, to a relatively stable reaction product, due to anodic oxidation, being deposited on the electrode surface and only removed at much lower positive voltages). The peaks of anodic and cathodic sweep directions are widely separated; the separation voltage difference itself does not appear to follow any obvious relation to the scan rate.

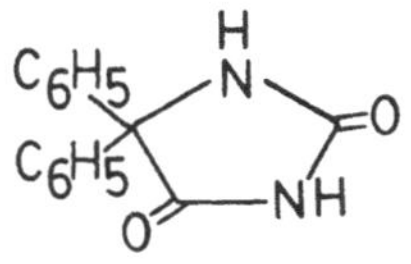

*FIGURE 11.* Structure of phenytoin. Note: The sodium salt was used in the conductivity and voltammetry studies described.

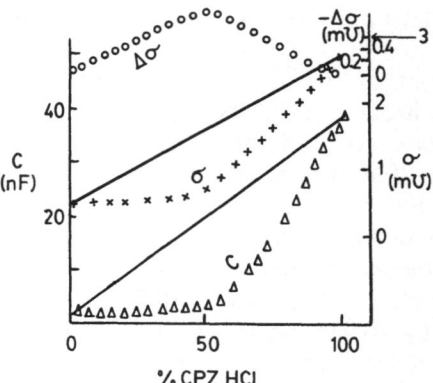

FIGURE 12. Titration phoreogram of the phenytoin Na chlorpromazine–HCl interaction. Both reagents were $10^{-3} M$ aqueous solutions at 17°C, gold electrodes were used. The conductance $\sigma$, as well as the conductance differences $\Delta\sigma$, are scaled in millimho on the right and the capacitances $C$, in nF, on the left. The linear base lines valid for no carrier interaction are also shown. The $\Delta\sigma$ conductance plot at top gives the absolute values of the conductance differences from the base lines; It is seen to peak at a 1:1 ratio.[32]

Although the pharmaceutical incompatibility is confirmed it is not possible on these results alone to comment on any clinically significant interaction between the antiepileptic phenytoin and the psychotropic chlorpromazine.

## 5. Interaction between Beta-Lactam Antibiotics and Aminoglycoside Antibiotics

### 5.1. Beta-Lactam Antibiotics

The beta-lactam antibiotics, featuring the beta-lactam ring (Figure 14), include

a. the penicillins, including benzylpenicillin, carbenicillin, ticarcillin, cloxacillin, flucloxacillin, ampicillin, and amoxycillin;
b. the cephalosporins, including cephalothin and cephamandole.

The beta-lactam antibiotics are anionic, unstable to hydrolysis, and act by inhibiting the synthesis of the bacterial cell wall.

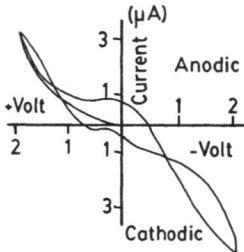

FIGURE 13. Cyclic voltammogram of the phenytoin Na:CPZ·HCl system. Both reagents were in aqueous solution at 37°C, Pt/Pt electrodes. Potentials vs. SCE. Sweep rate 100 mV/sec. The peaks at about $-300$ mV and at about $+250$ mV do not occur in the voltammograms of the components. The peak at about $+700$ mV is due to the chlorpromazine free radical.

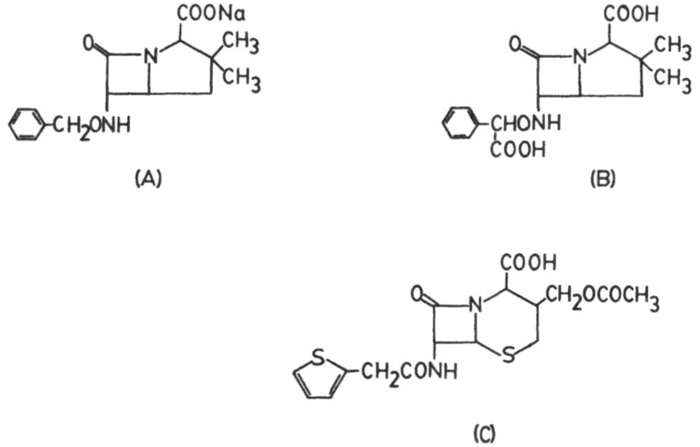

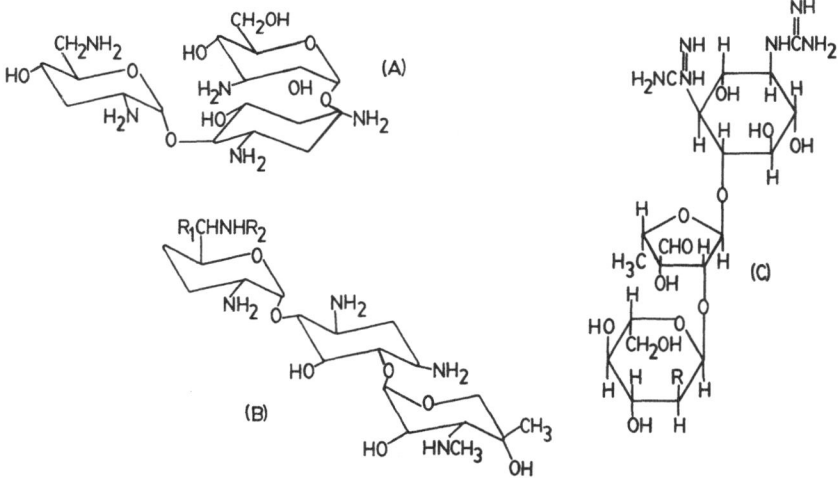

**FIGURE 14.** Structures of beta-lactam antibiotics examples. A: benzyl penicillin sodium. B: carbenicillin. C: cephalotin.

## 5.2. Aminoglycoside Antibiotics

The aminoglycoside antibiotics, including streptomycin, kanamycin, gentamicin, tobramycin, amikacin, and netilmycin, are, as the group name implies, glycosides of amino sugars and sugar derivatives. See Figure 15.

The aminoglycosides act by inhibiting protein synthesis by interaction with the 30$s$ ribosomal subunit. In the form used in therapeutics, namely, the salts of the amine with acid, the aminoglycoside antibiotics are cationic.

**FIGURE 15.** Structures of some aminoglycoside antibiotics. A: tobramycin. B: gentamicin; $c_1$, $R_1 = R_2 = CH_3$; $c_2$, $R_1 = CH_3$, $R_2 = N$; $c_3$, $R_1 = R_2 = N$. C: streptomycin, $R = CH_3NH$.

### 5.3. Clinical Use of Combinations of Beta-Lactam and Aminoglycoside Antibiotics

Combinations of a beta-lactam and an aminoglycoside antibiotic are commonly administered in life-threatening sepsis, especially in patients whose resistance to disease has been lowered by malnutrition or impaired immunological status.[87] Because the two groups of antibiotics act by different mechanisms and have different spectra of antibacterial activity, a combination of one of each type of antibiotics is said to have a broad range of activity. Although empirical antibiotic therapy is inconsistent with many of the accepted principles of good management of infected patients, many seriously ill patients (such as the neutropenic patient with leukemia) must be treated as soon as there is evidence of sepsis. Otherwise, death is likely to occur before the results of bacterial culture are available.

### 5.4. Problems with Combinations of Beta-Lactam and Aminoglycoside Antibiotics

Although many of the possible combinations have been used in clinical practice, the relative advantages and disadvantages of each combination have not been adequately elucidated. There is even considerable confusion about the *in vivo* or *in vitro* inactivation or synergy with the gentamicin/carbenicillin combination[88]—a combination which has one of the longest records of clinical use in this series of combinations. Similarly, the relative nephrotoxicity of the various possible combinations is not clear.[89]

### 5.5. Electrochemical Investigation of the Beta-Lactam–Aminoglycoside Antibiotic Interaction

Since the beta-lactam antibiotics are anionic and electron donors and the aminoglycosides are cationic and electron acceptors, aspects of the suspected interaction can be studied by electrochemical means. Conductometric titration has demonstrated a well-defined interaction between carbenicillin and tobramycin and carbenicillin and gentamicin, while cephalothin and gentamicin and cephalothin and tobramycin showed negligible interaction. Iodine reacted with carbenicillin but not with cephalothin. At the time of writing, studies with these and other combinations were continuing to confirm or reject physicochemical reactions and to establish the stoichiometry involved. The information provided will be fundamental with respect to these antibiotics and the clinical significance will be determined by considering the chemical facts in combination with microbiological and clinical studies.

## 6. Interactions between Heparin and Antibiotics

### 6.1. Heparin

Heparin, which occurs naturally in mast cells, is a sulfate mucopolysaccharide.[44,90] Although the complicated molecular form in which heparin occurs has a molecular weight between 750,000 and 1,000,000, the commercial heparin which is used therapeutically is a mixture, with molecular weights between 6,000 and 20,000.[44]

There is some controversy with respect to the nature of the repeating unit of the polysaccharide; it is commonly stated[44] to be a disaccharide of glucosamine and a hexuronic acid (either $\alpha$-L-iduronic acid or $\beta$-D-glucuronic acid) linked 1:4 with three sulfate groups per disaccharide unit. Alternatively, a tetrasaccharide unit with seven ionic sites (five sulfates and two carboxylates) has been postulated.[91] It is agreed, however, that the polysaccharide is highly ionic even at pH well below physiological pH. Two suggested structures for the repeating unit of heparin are indicated in Figure 16.

The situation is complicated by species differences and variability related to the manufacturing process.

Therapeutically, heparin is most commonly used as an anticoagulant. The effect in preventing the coagulation of blood depends on the presence in the blood of a cofactor which is an $\alpha_2$-globulin known as antithrom-

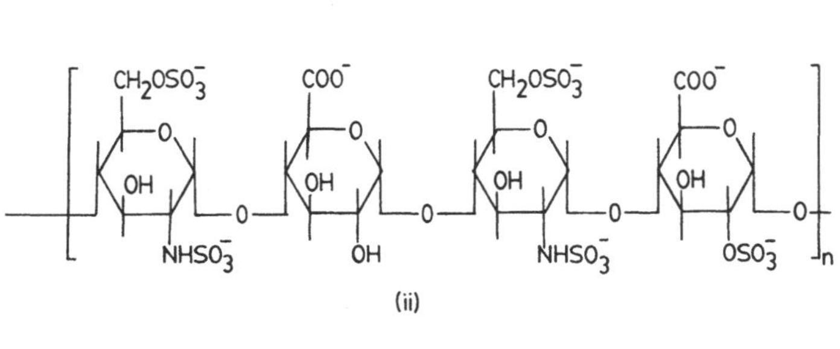

FIGURE 16. Postulated structures for the repeating unit of heparin. (i) Reference 44. (ii) Reference 91.

bin III. Antithrombin III heparin cofactor is a major physiological inhibitor of thrombin and of the activated forms of certain other coagulation factors. Heparin has other effects such as plasma lipid clearing, complexation with and detoxification of histamine, and anti-inflammatory action,[92] although it is not commonly used for these purposes.

Blood coagulation is associated with electrochemical events and the anticoagulant action of heparin is lost if the anionic charge is neutralized.

## 6.2. Heparin–Cation Interactions

The anticoagulant effect of heparin can be rapidly neutralized *in vivo* with protamine sulfate, a strongly cationic substance which combines with and inactivates the anionic heparin. This reaction, which is used therapeutically,[6] can be followed by conductometric titration.[32,93] Conductometric titration has also been used to investigate the interaction with chlorpromazine hydrochloride.[32] The photochemistry of heparin–acridine orange complexes has been studied.[94]

A chemical assay for heparin based on the interaction with the polycations known as ionenes has been described.[91]

The possible interaction between heparin and the cationic aminoglycoside antibiotics would appear suitable for investigation by electrochemical methods.

## 6.3. Heparin–Antibiotic Interactions

### 6.3.1. Background to Method and Results of Conductivity Titrations of Heparin against Aminoglycoside and Penicillin Antibiotics

Certain aminoglycoside antibiotics, including gentamicin and streptomycin, have been observed to interact with the acidic mucopolysaccharide heparin, condroitin sulfate, and hylauronic acid[95,96] with reduction of antibacterial activity. The interaction of aminoglycosides with the acidic mucopolysaccharides has been suggested to result in the accumulation of the drug in the kidney and in the inner ear[96] and to partly account for the well-documented nephrotoxicity[97] and ototoxicity of this group of antibiotics.

In view of the above observations it seems appropriate to follow the reaction between heparin and antibiotics[42] by conductometric titration, using the usual method. $0.5 \times 10^{-3}$ molar solutions of antibiotic were used because these are well-characterized chemical entities and the concentrations are reasonably close to the practical therapeutic range. Heparin, however, is not well characterized and its concentrations are usually indicated in terms of a "unit of activity" which is defined in terms of biological activity.[6] It was found that solutions containing between 30 and 40 units/ml were satisfactory for the purpose of these titrations. Where possible, analytical grade unpreserved antibiotic was used, but the commercial preserved products were always also used in order to obtain conditions more closely allied to the therapeutic situation.

### 6.3.2. Discussion of Heparin–Antibiotic Conductivity Titrations

The results suggest a positive interaction between heparin and the aminoglycoside antibiotics. There appears to be no interaction between heparin and penicillin or cloxacillin; within the precision of the measurement, a perfect straight line results, linking the conductance of the two components.

The Janz–Danyluk[98] plots for tobramycin with preservative indicate that, in the millimolar concentration region here employed, tobramycin in aqueous solution is highly associated, forming triple ions and higher aggregates. These, then, appear to associate further with heparin, forming highly complex adducts which are sufficiently large to form colloidal suspensoids. Some of them, at least, dissociate, giving rise to the conductance maxima which appear in the phoreograms. At least part of the alternating current involved in the conductance as measured is likely to be due to such carriers.

In these carriers, the antibiotic is quite firmly held in higher aggregates, and not likely to be available for any bactericidal action; thus, complexation with heparin should considerably lower biological activity, as confirmed in microbiology tests.

No stoichiometry can be assigned to the heparin interactions because of the uncertainty of the exact composition of commercial solutions of this drug.

A colloidal reaction product appears to form at the volume ratio corresponding to the conductance peaks in the case of the aminoglycoside–heparin mixture. The suspensoid readily redissolves on agitation; the turbid reaction mixture clarifies on shaking.

Scanning electron microscopy, using an instrument with a resolution of about 350 nm, failed to resolve the individual particles, which thus appear to form a true colloid. Higher resolution using electron transmission microscopy with very thin, dipped layers showed many particles of dimensions of about 100 nm.[42] The colloidal interaction between mucopolysaccharides and aminoglycosides was studied by Deguchi *et al.*,[96] who observed that the stability of the complexes depended on the ionic strength.

Differences in surface activity of antibiotics are of special interest, in view of the nephrotoxicity of the aminoglycosides, which has been postulated to be due to their binding to renal lysosome membranes.[95] The capacitance changes, as well as the conductance changes observed, would indicate that the adsorption behavior of the aminoglycosides differs in a fundamental way from that of the nonaminoglycosides, such as penicillin or cloxacillin. Kunin[95] has suggested that heparin may form "loose complexes that readily dissociate" with antibiotics. However, he invokes purely electrostatic binding to tissue sites, while our findings suggest that it involves, at least in part, preferential adsorption.

The nature of the actual interaction cannot be established from conductimetry alone. In a discussion of the reaction between heparin and polycations, Rembaum and Haack[91] noted the associated liberation of sodium chloride, and inferred that "changes in chain conformations must occur" during the reaction. They suggest that the mechanism of the reaction involves an "isoelectric jump," on which there has been a considerable amount of experimental and theoretical work, in order to elucidate the reaction thermodynamics. A similar mechanism may well be operative in the heparin–aminoglycoside interactions.

However, these highly complex molecular species are likely to possess also electron-donating regions; it has recently been shown that conjugative interactions between aromatic rings and oxo groups result in acceptor behavior in the orthogonal configuration, with donor behavior in the planar arrangement. The behavior depends largely on the relative energy of the interacting orbitals.[42]

Although the clinical significance of the interaction has not been established, the antagonism of the antimicrobial activity of the aminoglycoside antibiotics by heparin has been demonstrated by bacterial culture techniques. Kunin[95] reported that the reduction in antimicrobial activity is lowest for streptomycin while gentamicin and kanamycin exhibit greater but similar lowering activity. This observation was confirmed in the microbiological tests carried out in conjunction with the electrochemical investigations.[42] Here it was found that tobramycin was inactivated more than gentamicin, gentamicin was slightly more inactivated than kanamycin, while streptomycin was least affected. When this sequence was compared with the deviation in conductivity at the point of greatest departure in the conductance curve, it was found that tobramycin produced the greatest effect and streptomycin the least. Thus, there is some qualitative agreement between the microbiological testing and the results of conductometric titrations.

## 7. Heparin–Lidocaine Noninteraction

Cationic lidocaine (lignocaine) hydrochloride is commonly assumed to be incompatible with anionic sodium heparin.[44] The possibility of interaction between these two drugs is significant because both drugs are routinely administered together in certain cardiovascular conditions. The presumed interaction has, however, been shown not to occur. Conductivity titration of heparin sodium against lidocaine hydrochloride yielded a nearly perfect straight line.[32,93] The capacitances, too, remained substantially constant throughout the titration.

Clinical investigations on hemodialysis patients receiving both drugs did not detect any interaction.[99]

Because the two drugs may be indicated concurrently, it is important to confirm or deny significant interaction. As usual in the clinical situation, an immediate definite decision is required—"what to do for this patient now." To complicate the regimen or to deny optimum drug therapy because of a spurious interaction is to do considerable harm and to do this harm repetitively.

## 8. Metachromasia

### 8.1. Definition

Metachromasia,[100] from "meta," a change in the kind of, and "chroma," color, refers to the qualitative change which occurs in the color of certain dyes when they interact with other substances. The term was originally applied by Paul Ehrlich to the phenomenon of color change when a dye is adsorbed onto a substrate. Thus, cartilege and other mucopolysaccharide-containing tissues are stained red by toluidine blue. Similarly, Hartley[101] observed that when bromophenol blue solution is added to cetrimide solution the color changes from purplish blue to clear blue.

### 8.2. Applications of Metachromasia

The phenomenon has been studied with respect to histochemical staining. It has also been applied to adsorption indicators in titration[102–105] and to the detection of micellar aggregations.[105]

Metachromasia is of potential value in model systems for pharmacological phenomena because the strengths of the forces involved are probably in the same range as those involved in certain biological molecular processes.[11] The metachromatic properties of heparin have received considerable attention in this context.[90]

### 8.3. Mechanism of Metachromasia

Metachromasia has been attributed to the formation of ion pairs[106,107] of micelles.[107–109] Lewis and co-workers[110] have shown that

534                                    G. M. Eckert, F. Gutmann, and H. Keyzer

the ion pair phenomenon does not in itself alter the spectral absorption of associated ions; higher-ordered ion clusters are necessary before significant shifts are observed. The influence of molecular size of the counterion, hydration energy, and hydrophobic bonding in micelles have been considered. Szent-Györgyi[11] has discussed the continuity of ion pair association with charge transfer complex formation and the relationship of these with the phenomenon of metachromasia.

### 8.4. Electrochemical Investigations of Metachromasia

Because of the possibility of ion association and charge transfer the use of conductometric titration is suggested. Preliminary unpublished conductometric studies by the authors on the reaction between bromophenol blue and quaternary ammonium and arsonium salts failed, within the sensitivity of the instrument used, to detect any conductivity changes. The reaction is, however, accompanied by metachromasia and can be followed spectrophotometrically, showing again the complementary nature of the physical methods available.

### References

1. J. Thomas, "Drug Interactions," in *Drugs/Actions and Uses. Selected Review Articles,* No. 1, New Ethicals, Sydney (1969).
2. A. Melville and C. Johnson, *Cured to Death. The Effects of Prescription Drugs,* Angus and Robertson, London (1982).
3. W. H. W. Inman, ed., *Monitoring for Drug Safety,* MTP Press, Lancaster (1980).
4. E. Martin, ed., *Hazards of Medication,* Lippincott, New York (1971).
5. *Australian Pharmaceutical Handbook and Formulary,* 13th edition, The Pharmaceutical Society of Australia, Canberra (1983).
6. *Martindale, The Extra-Pharmacopoeia,* 28th edition, Pharmaceutical Press, London (1982).
7. Japan Atomic Energy Research Institute, *Sustained Release Drug Complexes,* Chemical Abstracts, 97, 44339z (1982) from Jpn. Kokai Tokyo Koho JP, 82 56, 421 (Cl. A61K9/00), 5 April 1982, Appl. 80/132, 670, 24 September 1980; 4 pp.
8. A. Albert, *Selective Toxicity,* 4th Ed., Methuen, London (1968).
9. E. Chargaff and K. B. Olsen, "Studies on the Chemistry of Blood Coagulation VI. Studies on the Action of Heparin and Other Anticoagulants. The Influence of Protamine on the Anticoagulant Effect in Vivo," *J. Biol. Chem.* 122, 153–167 (1937).
10. A. Ur, "Changes in the Electrical Impedance of Blood During Coagulation," *Nature* 226, 269–270 (1970).
11. A. Szent-Györgyi, *Introduction to a Submolecular Biology,* Academic, New York (1960).
12. A. Szent-Györgyi, *Bioelectronics,* Academic, New York (1968).
13. A. Szent-Györgyi, *The Living State and Cancer,* Marcel Dekker, New York (1978).
14. M. L. Cowger, R. E. Labbe, and M. Sewell, *Arch. Biochem. Biophys.* 101, 96 (1963).
15. J. P. Farges and F. Gutmann, "Charge-Transfer Complexes in Electrochemistry," in

*Modern Aspects of Electrochemistry*, No. 12 (J. O'M. Bockris and B. E. Conway, eds.), Plenum Press, New York (1977), pp. 267–314.

16. C. Tamara, *Bull. Chem. Soc. Jpn* **46**, 2388 (1973).

17. G. H. Nancollas, "Thermodynamics of Ion Association in Aqueous Solution," *Quart. Rev.* **14**, 402–426 (1960).

18. C. W. Davies, *Ion Association*, Butterworths, London (1962).

19. R. M. Diamond, "The Aqueous Solution Behaviour of Large Univalent Ions. A New Type of Ion Pairing," *J. Phys. Chem.* **67**, 2513–2517 (1963).

20. T. Higuchi, A. Michaelis, and J. H. Rytting, "Role of Solvating Agents in Promoting Ion Pair Extraction," *Anal. Chem.* **43**(2), 287–289 (1971).

21. H. Frieser, "Relevance of Solubility Parameter in Ion Association Extraction Systems," *Anal. Chem.* **41**(10), 1354 (1969).

22. M. Szwarc, "Ions and Ion Pairs," *Acc. Chem. Res.* **2**, 87–96 (1969).

23. P. Mukerjee, K. J. Mysels, and P. Kanauan, "Counterion Specificity in the Formation of Ionic Micelles—Size, Hydration, Hydrophobic Bonding Effects," *J. Phys. Chem.* **71**, 4166–4175 (1967).

24. N. A. Gibson and D. C. Weatherburn, "The Distribution of Salts and Large Cations Between Water and Organic Solvents," *Anal. Chim. Acta* **58**, 159–165 (1972).

25. F. Gutmann and L. E. Lyons, *Organic Semiconductors*, 2nd Ed., revised by F. Gutmann and H. Keyzer, Krieger, Malabar, Florida (1983).

26. F. Gutmann and H. Keyzer, "Conductivity Titrations of Charge-Transfer Complexes in Solution—II," *Electrochim. Acta* **11**, 555–568, 1163–1169 (1966).

27. P. Job, *C. R. Acad. Sci. (Paris)* **180**, 928 (1925).

28. F. Gutmann and H. Keyzer, "Charge-Transfer Complexes of Chlorpromazine in Solution: A Conductimetric Study," *Electrochim. Acta* **12**, 1255–1262 (1967).

29. F. Gutmann and H. Keyzer, *Electrochim. Acta* **13**, 693 (1968).

30. F. Gutmann, "Conductimetric Titrations of Charge Transfer Complexes in Solution: A Review," *J. Sci. Indus. Res.* **26**, 19–28 (1967).

31. M. A. Slifkin, *Charge Transfer Interactions in Biomolecules*, Academic, London (1971), pp. 42, 47.

32. G. M. Eckert, J. P. Farges, and F. Gutmann, "Exploratory Studies of Some Drug Charge Transfer Interactions," *J. Biol. Phys.* **6**, 161–175 (1978).

33. M. E. Starzak, *J. Biol. Phys.* **2**, 57 (1974); M. D. Ryan and G. S. Wilson, *Anal. Chem.* **47**, 885–890 (1975); P. Groll and F. Grass, *Electrochim. Acta* **16**, 31 (1971).

34. C. Boitre, "Membrane System for Determining the Interactions Between Biologically Active Substances and Living Microorganisms," *Boll. Lab. Chim. Prov.* **22**(6), 1013–1023 (1971).

35. R. Foster and C. A. Fyfe, "Electron–Donor–Acceptor Formation by Compounds of Biological Interest II," *Biochim. Biophys. Acta* **112**, 490–495 (1966).

36. I. Blei, "Complex Formation Between Chlorpromazine and Adenosine Triphosphate," *Arch. Biochem. Biophys.* **109**, 321–324 (1965).

37. L. S. C. Wan, "Interaction of Salicyclic Acid with Quaternar Ammonium Compounds," *J. Pharm. Sci.* **57**, 1903–1906 (1968).

38. R. A. Bennett and H. O. Hooper, *J. Chem. Phys.* **47**, 4855 (1967).

39. R. Sahai, V. Singh, and M. Chudhan, "Dielectric Studies of Molecular Complexes of DDT with Some Compounds of Biological Interest," *Monatsh. Chem.* **112**(10), 1129 )1981).

40. R. Sahia, V. Singh, and R. K. Verma, "Role of Dielectric Constant on Stoichiometry of Some Iodine Molecular Complexes and Their Transformation into Ion Pairs," *Indian J. Chem. Sec. A* **20A**(10), 1017 (1981).

41. G. Briegleb, *Elektron Donator Akzeptor Komplexe*, Springer, Berlin (1961).

42. G. M. Eckert, F. Gutmann, and M. Kabos, "The Electrochemical Interactions of Heparin and the Aminoglycoside Antibiotics," *J. Biol. Phys.* **10**, 51–63 (1982).
43. W. A. Harris, "The Interaction of Neomycin and Other Aminoglycosides with Pectin and Amaranth," *Australian J. Pharm.* **52** (1971).
44. A. Gallus and G. Engel, *Heparin*, The Society of Hospital Pharmacists of Australia (1978).
45. F. H. Gross and W. H. W. Inman (eds.), *Drug Monitoring*, Academic, London (1977).
46. G. Domagk, "Eine neue Klasse von Desinfectionsmittel," *Deutsche Medizinische Wochenschrift* **61**, 829–832 (1935).
47. Z. Baker, R. W. Harrison, and B. F. Miller, "Inhibition by Phospholipids of the Action of Synthetic Detergents on Bacteria," *J. Exp. Med.* **74**, 621–637 (1941).
48. I. Michaels, "The Use of Inhibitory Agents in Investigations of Antibacterial Activity," *Pharm. J.* **(1950)**, 263–264, 302–303.
49. C. A. Lawrence, "Inactivation of the Germicidal Action of Quaternary Ammonium Compounds," *J. Am. Pharm. Assoc. Sci. Ed.* **37**, 57–61 (1948).
50. W. Nixon and M. W. Cheetham, "Surface Active Agents. A Note on Some Visual Incompatibilities," *Pharm. J.* **(1950)**, 46.
51. E. I. Valko and A. S. DuBois, "The Antibacterial Action of Surface Active Cations," *J. Bacteriol.* **47**, 15–25 (1943).
52. A. Albert, S. D. Rubbo, R. J. Goldacre, M. E. Davey, and J. D. Stone, The Influence of Chemical Constitution of Antibacterial Activity. Part II: A General Survey of the Acridine Series," *J. Exp. Pathol.* **26**, 160–192 (1945).
53. H. E. R. Barker, "Cationic–Anionic Incompatibility and Ointments Containing Cation Active Antiseptics," *Austral. J. Pharm.* **(1948)**, 801–807.
54. The Pharmaceutical Society of Australia, *The Australian Pharmaceutical Formulary*, 8th ed., (1955).
55. D. H. Rodgers, "Conductimetric Study of the Interaction of Hexylresorcinol and Amaranth with Quaternary Ammonium Compounds," *J. Pharm. Sci.* **54**, 459–460 (1965).
56. W. A. Harris, "A Preliminary Conductimetric Study of Some Cation–Anion Interactions," *Austral. J. Pharm.* **(1968)**, S87–S90.
57. G. M. Eckert and C. Griffiths, "Some Difficulties Associated with the Dispensing of Surface Active Agents," *Austral. J. Pharm.* **(1952)**, 839–840.
58. P. P. DeLuca and H. B. Kosternabauder, "Interaction of Preservatives with Macromolecules IV. Binding of Quaternary Ammonium Compounds by Non-Ionic Agents," *J. Am. Pharm. Assoc. Sci. Ed.* **49**(7), 430–437 (1960).
59. *Proceedings of the 3rd NIMH Sponsored Conference on Phenothiazines and Related Drugs* (I. S. Forrest, C. J. Carr, and E. Usdin, eds.), Raven Press, New York (1974).
60. *Proceedings of the 4th NIMH Sponsored Conference on Phenothiazines and Related Drugs*, Zurich, 1979, Elsevier, Holland (1980).
61. F. Goodman and C. Gilman, *The Pharmacological Basis of Therapeutics*, Fifth Edition, MacMillan, New York (1975).
62. C. Boitre, M. Marchetti, C. Del Vecchio, G. Lionetti, and A. Memoli, "A Physicochemical Model for the Mechanism of Action of Antihistamines and Cortisol," *J. Med. Chem.* **12**(5), 832–836 (1969).
63. F. Gutmann and H. Keyzer, "Study of Phenothiazine and Chlorpromazine–Iodine Complexes," *J. Chem. Phys.* **46**(5), 1968–1974 (1967).
64. A. Brau, J. P. Farges, and F. Gutmann, "Electrochemical Studies of Phenothiazine–Iodine Charge-Transfer Complexes," *Electrochim. Acta* **17**, 1803–1811 (1972).
65. F. Gutmann and H. Keyzer, "Charge Transfer Complexes of Thiazine Derivatives and Their Possible Physiological Significance," *Agressologie* **9**, 27–35 (1968).
66. I. S. Forrest, F. Gutmann, and H. Keyzer, "In Vitro Interaction of Chlorpromazine and Melanin," *Agressologie* **7**, 147–156 (1966).

67. F. Gutmann, L. C. Smith, and M. A. Slifkin, "Charge Transfer Interactions of Chlorpromazine with Neutral Transmitters," in *The Phenothiazines and Structurally Related Drugs* (I. S. Forrest, C. J. Car, and E. Usdin, eds.), Raven Press, New York (1974).

68. S. Lukiewicz, K. Reszka, and Z. Matuszak, *Bioelectrochem. Bioenerg.* 7(1), 153–165 (1980).

69. R. Capeletti and T. R. Crippa, "Electret State and Hydrated Structure of Melanin," *Bioelectrochem. Bioenerg.* 8(5), 555 (1981).

70. F. W. Cope, "Inversion of Emulsions of Aggregated Electrons as a Possible Mechanism for Electrical Switching in Wet Melanin and in Amorphous Semiconductors. A Manifestation of Cooperative Electron Interaction," *Physiol. Chem. Phys.* 9(6), 543 (1977).

71. F. W. Cope, "Critical Exponent Analysis of Activation Energies of Nonlinear Arrhenius Plots as a Test for Cooperative Interactions in Amorphous Semiconductors and in Biological Systems," *Physiol. Chem. Phys.* 9(4, 5), 329, 443 (1977).

72. F. M. Forrest, I. S. Forrest, and Roisin, "Biochemical and Post-Mortem Studies on a Patient Treated with Chlorpromazine," *Agressologie* 4(3), 259–265 (1963).

73. I. S. Forrest, F. M. Forrest, and M. Berger, "Free Radicals as Metabolites of Drugs Derived from Phenothiazines," *Biochim. Biophys. Acta* 29, 441–442 (1958).

74. A. M. Potts, "The Reaction of Uveal Pigment in Vitro with Polycyclic Compounds," *Invest. Ophthal.* 3, 405–416 (1964).

75. A. C. Greiner and G. A. Nicolson, "Pigmentation in Viscera Associated with Prolonged Chlorpromazine Therapy," *Can. Med. Ass. J.* 91, 627–635 (1964).

76. A. C. Greiner and K. Berry, "Skin Pigmentation and Corneal and Lens Opacities with Prolonged Chlorpromazine Therapy," *Can. Med. Ass.* 90, 363–365 (1964).

77. M. Piatelli and R. A. Nicoulaus, "The Structures of Melanins and Melanogenesis I," *Tetrahedron* 15, 66–75 (1961).

78. A. Bolt, private communication.

79. I. A. Menon, "A Qualitative Study of the Melanin for Blue and Brown Human Eyes," *Exp. Eye Res.* 34(4), 531 (1982).

80. I. A. Menon, E. V. Gan, and H. F. Haberman, "Electron Transfer Properties of Melanin and Melanoproteins," *Pigm. Cells* (1972), pt. 2, 297–309.

81. H. Keyzer, C. Lowe, W. Plumtree, and F. Gutmann, *Phenothiazines and Structurally Related Drugs*, *Proceedings 4th International Phenothiazine Conference, Zurich, 1979.* Elsevier, Amsterdam (1980).

82. M. Zarach-Krutysza, "Susceptibility of Albino and Black Mice to Some Drugs with Affinity to Melanins," *Acta Pol. Pharm.* 38(4), 513 (1981).

83. H. Tjaelve, M. Nilsson, and B. Larsson, "Studies on the Binding of Chlorpromazine and Chloroquine to Melanin 'in vivo'," *Biochem. Pharmacol.* 30(13), 1845 (1981).

84. B. Larsson and H. Tjaelve, "Studies on the Mechanism of Binding to Melanin," *Biochem. Pharmacol.* 28(7), 1181 (1979).

85. M. Barza, A. Kane, and J. Baum, "Marked Differences Between Pigmented and Albino Rabbits: The Concentration of Melanin in Iris and Choroid Retina," *J. Inf. Dis.* 139(2), 203 (1979).

86. M. M. Salazar, "The Significance of the Interpretation of Drugs with Melanin," *Diss. Abst. Int. B.* 37(8), 3908 (1977).

87. J. Klatersky and M. J. Staquet (eds.), *Combination Antibiotic Therapy in the Compromised Host*, Raven Press, New York (1982).

88. L. J. Riff and G. G. Jackson, "Laboratory and Clinical Conditions for Gentamicin Inactivation by Carbenicillin," *Ann. Int. Med.* 130, 887–891 (1972).

89. M. Mork-Hansen and K. Kaaber, "Nephrotoxicity in Combined Cephalothin and Gentamicin Therapy," *Acta. Med. Scand.* 201, 463–467 (1977).

90. L. B. Jaques, "Heparin and Related Polyelectrolytes," in *Polyelectrolytes and Their Applications* (A. Rembaum and E. Selegny, eds.), Reidel, Dordrecht (1975), pp. 146–161.

91. A. Rembaum, personal communication.

92. T. F. Dougherty and D. A. Dolowitz, "Physiologic Actions of Heparin not Related to Blood Clotting," in *Chemical Dynamics*, Papers in honor of Henry Eyring (J. O. Hirschfelder and D. Henderson, eds.), Wiley, Interscience, New York (1971).

93. G. M. Eckert and F. Gutmann, "The Electrochemistry of Some Drug Interactions," *Electroanal. Chem. Interfacial. Electrochem.* **62**, 267–272 (1975).

94. J. M. Menter, R. E. Hurst, and S. S. West, "Photochemistry of Heparin–Acridine Orange Complexes in Solution: Photochemical Changes Occurring in the Dye and Polymer on Fluorescence Fading," *Photochem. Photobiol.* **29**, 473–478 (1979).

95. C. M. Kunin, "Binding of Antibiotics to Tissue Homogenates," *J. Infect. Dis.* **121**, 55–64 (1970).

96. T. Deguchi, A. Ishii, and M. Tanaka, "Binding of Aminoglycoside Antibiotics to Acidic Mucopolysaccharides," *J. Antibiotics* **XXXI**, 150–155 (1978).

97. G. B. Appel and H. C. Neu, "The Nephrotoxicity of Antimicrobial Agents. Parts I, II and III," *New Engl. J. Med.* **296**, 663–670, 722–728, 784–787 (1977).

98. G. J. Janz and S. S. Danyluk, *Electrolytes* (B. Pearse, ed.), Pergamon Press, Oxford (1962), p. 255.

99. P. C. Farrell, M. F. O'Rourke, and E. Turnbull, "Precise Anticoagulation During Coronary Care Management," Abstract, *Am. Soc. Artif. Int. Org.* **8**, 71 (1979).

100. A. S. MacNalty (ed.), *The British Medical Dictionary*, The Caxton Publishing Co., London (1961).

101. G. S. Hartley, "The Effect of Long-Chain Salts on Indicators: The Valence-Type of Indicators and the Protein Error," *Trans. Faraday Soc.* **30**, 444 (1934).

102. N. W. Tschoegl, "Analysis of Water Soluble Synthetic Soaps," *Rev. Pure Appl. Chem.* **4**(3), 171–206 (1954).

103. E. L. Colichman, "Photocolorimetric Method for Determination of Quaternary Ammonium Salts," *Anal. Chem.* **19**, 430 (1947).

104. C. W. Ballard, J. Isaacs, and P. G. W. Scott, "The Photometric Determination of Quaternary Ammonium Salts and of Certain Amines by Compound Formation with Indicators," *J. Pharm. Pharmac.* **6**, 971 (1954).

105. M. R. J. Salton and A. E. Alexander, "Estimation of Soaps and Ionised Detergents," *Research* **2**, 247 (1949).

106. E. L. Colichman, "Spectral Study of Long Chain Quaternary Ammonium Salts in Bromophenol Blue Solutions," *J. Am. Chem. Soc.* **73**, 3385 (1951).

107. E. L. Colichman, "Conductance of Long Chain Quaternary Salts in Bromophenolate Blue Solutions," *J. Am. Chem. Soc.* **72**, 1834 (1950).

108. P. Mukerjee, "Use of Ionic Dyes in the Analysis of Ionic Surfactants and Other Ionic Organic Compounds," *Analyt. Chem.* **28**, 870–873 (1956).

109. H. B. Klevens, "Analysis of Colloidal Electrolytes by Dye Titration," *Analyt. Chem.* **22**(9), 1142–1145 (1950).

110. S. E. Sheppard and A. L. Geddes, "Effect of Solvents Upon the Absorption Spectra of Dyes. V. Water as Solvent: Quantitative Examination of the Dimerization Hypothesis," *J. Am. Chem. Soc.* **69**, 2003–2009 (1944).

# 20

# Muscular Contraction

## Mohammad Amin

*ABSTRACT:* It is currently believed that the actomyosin ATPase in muscle converts chemical energy released in ATP hydrolysis directly into the mechanical form and that similar processes are at work for all other biodynamic phenomena such as cytoplasmic streaming, chromosomal movement during mitosis, and the movement of cilia and flagella. A critical appraisal of the available evidence does not provide unequivocal support for this hypothesis. An alternative mode of energy transduction is suggested which involves the generation of proton motive force across the length of filamentous and tubular macromolecular assemblies when these are interposed between an ATPase and a region of neutral pH or another enzyme which catalyzes an $H^+$-consuming reaction at about the same rate as these are produced in ATP hydrolysis. Assuming that such macromolecular assemblies have high protic conductance and also bear negative fixed charges it has been shown that the proton flux down its chemical potential gradient generates a more or less surface localized electric field which drives the diffuse cationic layer in the opposite direction. The associated hydrodynamic flow can be harnessed for any given biodynamic process by an appropriate spatial organization of the macromolecular assemblies. This idea has been worked out quantitatively for striated muscle in which ATP is hydrolyzed in the region of overlap of the thick and the thin filaments. The latter are, respectively, anchored in the structural complexes of the $M$ line and the $Z$ disk both containing the enzyme CPK which catalyzes the proton-consuming Lohmann reaction. It follows that $H^+$ flux along the thick and the thin filaments would be directed, respectively, to the $M$ line and the $Z$ disk and cause the diffuse cationic layers around these filaments to move towards the I band and the H band, respectively. The viscous drag couple exerted on the myofilaments by the movement of the diffuse cationic layer in the form of three-dimensional loops is estimated to be sufficient to account for the observed muscular tension. The analysis of the power output of the muscle in terms of a viscoelectric equivalent circuit yields Hill's equation in the first approximation. The problem of unexplained energy is resolved in terms of the flux of $H^+$ present in myosin heads at the beginning of contraction, while the extra expenditure of ATP during recovery serves to recharge the heads with $H^+$. The crucial result for bioenergetics is that ATP hydrolysis releases the greater part of the free energy in the form of proton motive force in muscular contraction, suggesting a unified mode of energy transduction in living systems.

*Mohammad Amin* • School of Life Sciences, Jawaharlal Nehru University, New Mehrauli Road, New Delhi-110067, India.

## 1. Introduction

Aloysis Galvani's experiments on the twitching of the muscles of frog legs,[1] performed in the 1780s, are a landmark in the multidisciplinary field of bioelectrochemistry. The process of the contraction of the muscles is currently believed to be based on a chemomechanical transduction process in which ions have an insignificant role to play. Bioelectrochemistry has come to a dead end at the threshold of the contractile machinery. This is all the more strange since electrochemistry possesses, in the electrokinetic phenomena, the potentialities to deal also with the dynamic phenomena at the cellular level. After the introduction of the cross-bridge model of muscular contraction[2,3], Peter Mitchell's chemiosmotic theory, which demands transmembrane proton-motive force as an essential intermediate for the transduction of the energy released in the electron transport chains of mitochondrial and thylakoid membranes into the chemical form, has revolutionized the field of bioenergetics. While the chemiosmotic theory is being fruitfully applied to the study of transport processes in cell membranes, it has had little impact on the research on muscles. Since contractility is a major bioenergetic process in animals, one must ask whether the conversion of the chemical energy of ATP hydrolysis in muscle into the mechanical form may not involve the generation of the proton-motive force as an intermediate step. The current view that the chemical energy is directly transformed into the mechanical form demands a critical reappraisal of the cross-bridge model of muscular contraction. If the model is not found to be inviolable, then it would become feasible to approach the problem in a manner that removes the conceptual discrepancies.

Section 2.1 gives a brief survey of the evidence supporting the cross-bridge model. The conclusion is that there is at present no conclusive evidence available to prove that the model is correct. On the other hand there are a few experimental findings that are difficult to reconcile with the model. This survey is neither thorough nor comprehensive and is meant merely to justify a new approach suggested by the author,[4-6] based on the assumption of high proton conductance of the myofilaments in the contractile apparatus of the muscles. The proton-motive force is generated across these filaments by ATP hydrolysis at the cross-bridges and proticity flow is to the structures at which the myofilaments are anchored and where $H^+$ are consumed in the Lohmann reaction. The highly anisotropic ionic distribution at the surfaces of the myofilaments causes the cations ($K^+$) of the diffuse layer to move in the direction opposite to the flux of $H^+$. Hydrodynamic flow in three-dimensional loops along the myofilaments is thought to generate viscous force couples which bring about the sliding of the two sets of myofilaments relative to each other. The quantitative results obtained from this theory are in good agreement with experimental results.

This approach resolves the difficulty created by the hypothesis of direct energy transduction from the chemical to mechanical form. As in the case of membrane transport energized by $H^+$ ATPases, in muscles also there is first a buildup of a proton-motive force across the lengths of the myofilaments with the electroosmotic flow of the cationic layer converting it to the mechanical output.

The proton flux mediated electro-osmosis or in short proto-osmosis is the perimembrane and transfilament analog of Mitchell's (transmembrane) chemiosmotic theory and could be operating at the cellular and organismal levels in diverse dynamical processes (see "Transport in Plants," Chapter 21 in this volume). If the phenomenon is experimentally verified in muscles and in other systems, the dynamic aspects of cell biology would become amenable to quantitative treatment using the principles of electrochemistry.

## 2. The Mechanism of Muscular Contraction

### 2.1. The Cross-Bridge Model

From the studies of the structure of striated muscle fibers,[2] it became evident that the change in the length of the fibers was due to a relative sliding of two sets of interdigitating filaments without any change in the lengths of the filaments themselves. The thick filaments are stabilized into a regular hexagonal arrangement at their centers by a criss-cross of the *M*-line proteins. The thin filaments emanate from the *Z* disk on both of its sides and penetrate the lattice spaces between the thick filaments in a ratio of one to two. The degree of penetration (Figure 1) increases as a muscle shortens and the distance between two adjacent *Z* disks becomes shorter. The structural units extending from one *Z* disk to an adjacent one are called sarcomeres and their lengthwise alignment forms a myofibril whose total length changes, additively, as the sarcomeres shorten. Sliding of the thin into the thick filaments requires the presence of some physical factor

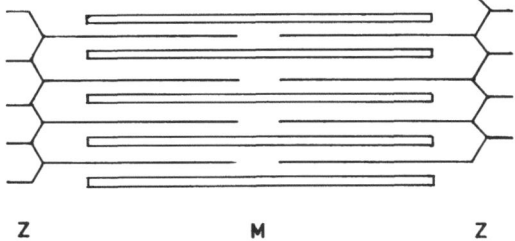

FIGURE 1. A contractile unit showing the thin filaments emanating from the *Z* disks and the centrally located thick filaments.

Z      M      Z

which can generate and transmit force from one set of the filaments to the other. Furthermore, it should be possible for this force to be generated in all sarcomeres of a myofibril and in all myofibrils of a fiber at the same time to enable the fiber to contract and pull up a load attached to the tendon. The question how the synchronization of the contractile activity is achieved was resolved in terms of electrical signals which propagate along the membranes of the transverse tubules.[7] The system of the transverse tubules, or the $T$ system, penetrates and ramifies in the muscle cells to reach every sarcomere and makes junctions with another tubular system, the sarcoplasmic reticulum (SR),[8] which surrounds the sarcomeres. The electrical signal at the membranes of the $T$ system causes the release of $Ca^{2+}$ from SR, which in turn triggers the process of contraction.[9] As to the nature of the physical (force-generating) factor, the idea of a direct mechanical coupling between the sets of myofilaments was supported by the observation that the magnitude of the force was close to zero at the length of the sarcomeres for which one could expect zero overlap between the filaments and increased linearly as the degree of overlap was increased up to about two thirds of the maximum overlap.[10,11] The molecular entities operating as independent force generators were identified as the regular projections, the cross-bridges, occurring in the thick filaments with the required length and flexibility to enable them to physically contact the thin filaments around them.[2,3] A cyclical interaction of the cross-bridges with the actins of the thin filaments, involving an attachment of the two, a structural deformation of the cross-bridge–actin complex (causing a stretching of an elastic element within the complex), and a detachment of the cross-bridges from the actins, would lead to the generation of a tensile force between the myofilaments. The thick filaments have been shown to be made up of an antiparallel arrangement of myosin molecules so that the tail of the myosins points to the center of the filament and a flexible part [subfragment 2 (S2) of the heavy meromyosin (HMM)] carrying two globular heads [subfragment 1 (S1) of HMM] points outward.[12] The cross-bridges $(S2 + 2 \times S1)$ have, therefore, opposite orientation on the opposing sides with reference to the $M$ line as is required for their proposed activity as the force generators. One could also show that in the rigor state, cross-bridges make strong links with the actins.[13] From the biochemical side the evidence that S1 is the sole protein component capable of hydrolyzing ATP strengthens the model in a very basic manner. It is, therefore, not surprising that the majority of physiologists consider the problem of muscular contraction to have been resolved in principle.

If one looks more closely at the situation, however, one cannot escape the feeling that the cross-bridge model has shifted the problem of finding the mechanism of contraction from the level of the myofilaments to the properties of the cross-bridges. These have to perform the (Maxwellian) demonic task of approaching the actins, forming a bond of low potential energy, proceeding to form the bonds, successively, of relatively lower potential energies (instead of making the bond of the lowest energy directly), getting detached, and repeating the same performance over and over. As regards the details of the models, in relation to the behavior of the cross-bridges, the evidence is weak or negative. One would have expected the rigor links as the final state of the cross-bridge–actin interaction, i.e., the end of the power stroke. From several studies it appears that the rigor link is probably never formed in contracting muscles.[14] Against the expectation, that the cross-bridge–actin complex would run back the different intermediate states of the interaction, if the muscle is stretched in the rigor state, no other state than the rigor complex was detectable.[15] The model envisages global structural alterations in S1 or S2 with or without the actins, caused by the hydrolysis of ATP. No such changes have been detected, in spite of great efforts and sophisticated techniques.[14] At the same time one cannot say that the evidence has disqualified the cross-bridge model, since there is always the possibility of the techniques not being sensitive enough to detect the (expected) events.

A number of pieces of evidence weaken the case of the cross-bridge model. The movement of the cross-bridges occurs, during contraction, not only in the region of the overlap between the myofilaments but also in the overlap-free region[16] and continues even after the contraction is over. Thus the movement of the cross-bridges can occur even without the involvement of the actin and without a concomitant generation of force.

Biochemically, quite a few problems remain unresolved for the cross-bridge model. One would have expected that ATP hydrolysis would occur during the power stroke, i.e., when S1 is in contact with the actin, so that the energy released in the process of ATP hydrolysis can be instantaneously used for conversion into mechanical work, Strangely enough the hydrolysis occurs when the two moieties are physically separate from each other; upon the binding of ATP to S1 there is rapid dissociation of the rigor link and hydrolysis takes place when the two have just dissociated.[17] The splitting of ATP by S1 is not accompanied by any significant release of free energy. Instead, the greater part of the free energy is released during the dissociation of the products of hydrolysis from the hydrolytic site of S1. This process is accelerated if S1 can interact with actins, as happens upon the release of $Ca^{2+}$ by SR. It has been found difficult to relate the steps of the dissociation of the products to the proposed multistep power stroke of the cross-bridges. From the known rates of dissociation of the products it is concluded that these processes are too slow to occur during the very short time intervals available to the cross-bridge–actin complex when the muscle shortens at a fair velocity. Stretching of the activated muscle fibers[18] leads to the generation of tensions greater than the isometric tension, but the amounts of ATP consumed are aften only 5% of the amount used in isometric contraction for an equal time interval.

According to the cross-bridge model the generation of force must be accompanied by a synchronous expenditure of ATP. There is, however, a large body of evidence indicating a temporal dissociation of the energy output by the muscle and the input of chemical energy.[19] It has been found that the work output and heat generation during a given time interval always exceeds the enthalpy change accompanying the splitting of phosphocreatine (PCr) during the same interval. The discrepancy has been observed under all possible experimental conditions, being most pronounced at the beginning of the contraction process when the amount of the unexplained energy can be more than the amount that can be accounted for by all the chemical reactions occurring in the muscle. After the contraction is over, there seems to be an overexpenditure of metabolic energy, which also raises the question about the mode of its utilization or storage. The net balance between the input of chemical energy and the liberation of energy (heat + work) by the muscles is established only after long periods of recovery.

From the above survey of the experimental evidence it should be apparent that the cross-bridge model cannot be considered as inviolable, and there is sufficient reason to look for some other viable models which can resolve all puzzling aspects of muscular contraction in a consistent manner.

### 2.2. The Search for a Biodynamic Principle

The structural organization of the contractile apparatus as described in Section 2.1 relates to the striated muscle fibers and is not universal for all contractile cells and tissues. In the smooth muscles[20] the types of filaments and their organization is fundamentally different. In a large number of organisms, myosin molecules are not organized in the form of filaments, and yet some kind of contractility and motility is displayed by them.[21] That the anchoring of myosin molecules is not essential even in the striated fibers was demonstrated by Oplatka *et al.*[22] Dispersed myosin can generate tension in striated fibers in which the filamentous myosin was neutralized. Furthermore, active streaming along actin threads is observed when enzymically active water-soluble myosin fragments are provided with ATP. Streaming of cytosol is also not confined to actin filaments. Microtubules, microfilaments, neurofilaments, and possibly other cytoplasmic fibers are associated with the genesis of a variety of motile and streaming processes.[23,24] If it is recognized that it is possible to cause a relative movement of two filamentous, parallel structures by generating a hydrodynamic eddy flow between the two, one may say that the fundamental dynamic process in the cells is a hydrodynamic flow and motility and contractility are derived from it. On the other hand, two arguments disfavor the idea of conformational changes as the primary process behind biodynamic effects. (i) Conformational changes will have to be invoked not only for myosin–F-actin system but also for dynein–microtubule system and for all other cases where ATPases operate in conjunction with some filaments. (ii) Myosin does not show any global conformational changes either alone or in the actomyosin complex.[14]

If a basic biodynamic principle exists it must have an intimate link with the basic bioenergetic phenomena, especially those related to the synthesis and hydrolysis of ATP. The reversible ATPase of the mitochondrial membrane synthesizes ATP if a proton-motive force is generated across the mitochondrial membrane, and, conversely, generates a proton-motive force by hydrolyzing ATP so long as $n_H F\Delta p \geqslant -\Delta G_{ATP}$, where $n_H$ is the number of $H^+$ translocated across the membrane for each ATP molecule synthesized, $F\Delta p$ is the electrochemical potential difference of $H^+$ across the membrane, and $-\Delta G_{ATP}$ is the free energy of ATP hydrolysis.[25] Thus the reversible ATPase of the mitochondrial membrane acts as a proton pump. The hydrolysis of ATP is associated with the obligatory production of at least one proton since the oxygen atom of the water molecule accepts an electron from one of the hydrogen atoms when the water molecule is attacked by the extremely reactive reactive phosphoryl group of the ATP. In the case of the reversible ATPase of the mitochondrial membrane and in other $H^+$-ATPases associated with cell membranes[26] the proton-motive force, $\Delta p$, generated in the process of ATP hydrolysis is used for diverse activities, mainly for transporting substances against their electrochemical gradients. The question arises now about the fate of the $H^+$ liberated during ATP hydrolysis by an ATPase which is not located in or at a membrane. Does it get neutralized the moment it is produced or could it perform some still unknown function? If we assume that the ATPase is associated with a macromolecular assembly such as a filament or a microtubule the proton can either move into the surrounding aqueous phase or possibly move along the macromolecular assembly to a more distant location. In the first case the local pH would fall rapidly and may lead to a decrease in the rate of hydrolysis. By the same token the second possibility will only be of interest if the protons can move along the filaments much more rapidly than they can diffuse in water in a hydrated form. Onsager and others[25,27,28] have proposed that the hydrogen bonded chains in, and the layers of structured water around, the biological macromolecules provide the necessary conditions for a rapid flow of protons. The high conductance of filamentous macromolecular assemblies for the flow of proticity would decrease the chance of $H^+$ escaping into the aqueous medium and ensure the maintenance of optimal pH for the ATPase.

For a proton flux to occur from the site of ATP hydrolysis to a distant location along a filament two more conditions must be fulfilled. The first is that there has to be another chemical process running synchronously with the hydrolysis of ATP at the point of arrival of the protons, i.e., at the other end of the filament. This process could be, trivially, the solvation of $H^+$ as $H^+$ get pushed out of the proton wire at the terminal point and a subsequent neutralization in the aqueous phase by some basic groups. However, one may also conceive some second enzymic reaction to be taking place at the other end of the filament at the same rate as the rate of ATP hydrolysis. The second essential condition is that electroneutrality should be maintained in all parts of the system. This means that an equally rapid flux of ions in the medium around the filaments has to occur to unbuild the diffusion potential which would be generated across the length of the filaments by the (vectorial) flux of $H^+$.

Biological macromolecules usually contain an excess of acidic groups over basic groups, and consequently a higher concentration of cations than that of anions obtains in the cells. It is therefore safe to assume that the proton conducting filament will be surrounded by a diffuse layer of cations, predominantly the ubiquitous $K^+$. Being relatively less electropositive $K^+$ is not expected to enter the inner Helmholtz plane[29] and thus have a high mobility $u_K$ for movements tangentially along the surface of the filament. Given the physiological concentrations of $K^+$ of around 100 mM a rather condensed diffuse layer with a Debye length of about 1 nm would form at the surface. The surface concentration of $K^+$, $c_K$, would further depend on the surface charge density $\sigma^-$ at the surface of the filament. Assuming the surface conductance, $g_K = Fu_K c_K$, to be sufficiently high, the repercussions of the proton flux along the filaments may be analyzed as follows. As a first approximation the filament is thought to be segregated, together with the diffuse layer of an extension of 1 nm, from the rest of the medium by an imaginary cylindrical surface. The cationic layer in the enclosed volume would be referred to as the K layer. The concentration of anions in the K layer will be assumed, in accordance with the Gouy–Chapman theory, to be negligible. As the protons move along the filament, the electric field built up by their flux has a large axial component $E_z$ and would tend to stop further charge displacement in the same direction, i.e., tend to block a continued flux of $H^+$. However, $E_z$ would act also on the $K^+$. The maximum value of $E_z$ in the K layer would be at the outer Helmholtz plane, and its magnitude would decrease gradually with the radial distance from the surface. Thus $K^+$ will move with different velocities in the direction opposite to the initial movement of $H^+$ under the action of $E_z$. If the surface conductance of the $K^+$ at OHP is high the outer regions of the K layer would be shielded from the effect of $E_z$ to some extent. Macromolecules, however, do not have smooth surfaces, and a fine distinction, within the K layer, between the OHP and other regions may not be justified. Nonetheless the effect of $E_z$ can be assumed to be fully shielded in the region outside of the imaginary cylindrical surface. The K layer will thus fully take over the function of providing the charge compensating flux upon itself. In other words a diffusion potential gradient generated by a proton flux along the filament has, for the anisotropic ionic distribution around the filament, the sole effect of causing the K layer to execute a charge-compensating movement in the opposite direction. The movement of the K layer is electro-osmotic, i.e., coupled to the flux of water, and will bring about a streaming of the electrolyte around the filament. Thus the flow of proticity along macromolecular assemblies can generate the streaming of the cytosol within the cells. With specific reference to the hydrolysis of ATP there seems to be a possible mode of utilizing the power output of the proton-motive force and proton flux by harnessing the electro-osmotic flow for some purpose. It may be noted that this streaming would not only solve the problem of maintaining the local pH around the ATPase at the initial value, but would also serve to enhance the rate of the reaction by driving the substrates to the enzyme in tune with the rate of their breakdown. While the rate of the forward reaction will be enhanced the microstreaming will drive the products away from the reaction site and reduce the rate of the reverse reaction. This may prove to be useful for understanding the phenomenal rates of enzymic reactions in biological systems and the tight coupling between the diverse reactions occurring synchronously at different sites within a cell.

Can the phenomenon of $H^+$-flux mediated electro-osmosis provide the biodynamic principle which we have been looking for? It is clear that cellular transport processes such as axonal transport and cytoplasmic streaming could directly result from proto-osmosis if the ATPases inject protons into the filamentous systems. Ciliary movements could possibly result from a periodic proto-osmotic insurge of fluid in the cilia. A relative (contractile) movement of two sets of filaments could be caused by a proto-osmotic loop flow between the pairs of the filaments by means of the associated visous drag couple. But the question of the relevance of the effect can be solved only if it can be shown that the magnitude of the effect is sufficient, quantitatively, to explain at least one biodynamic phenomenon. The best-studied system in this context, from structural and biochemical aspects, is that of muscles. The necessary quantitative data for the comparision of theoretical result with experiments is available, but it has to be first confirmed that the conditions for the proposed electro-osmotic flow obtain in muscles.

In the contractile apparatus of striated muscles it is relatively easy to identify and locate the source, the conductors, and the sink of protons. The ATPase is located in the cross-bridges; creatine phosphokinase, CPK, the enzyme catalyzing the (proton-consuming) Lohmann reaction, in the structural complexes of the $Z$ disk and the $M$ line,[30] while the thin and the thick filaments may serve as proton conductors for the flux of $H^+$ from the overlap region to the $Z$ disk and $M$ line, respectively. The alternating arrangement of $M$ lines and the $Z$ disks constitutes a linear array of electrodes, and their spatial segregation introduces the vectorial character or directionality in the fluxes of $H^+$ along the thick and the thin filaments to these electrodes. This segregation is important for obtaining vectorial effects (contractile force) from scalar causes (chemical reactions such as the ATP hydrolysis).[31,32] This matter is not fully clear for the cross-bridge model,[33] especially when we recall the problems related to the stretching of muscle fibers.[18]

In an isometric contraction, the rate of ATP hydrolysis is $10^{-6}$ $M/s$ per gram wet weight of muscle,[18] implying that the pH in the overlap region would fall to $pH = 2.5$ within a second. We also know that the Lohmann reaction is perfectly synchronized with the rate of ATP hydrolysis so that the level of ATP does not fall during normal activity of a muscle. Thus from a purely biochemical viewpoint it is imperative that a flux of $H^+$ must occur from the overlap region to the $M$ line and to the $Z$ disk since in the absence of such a flux the activity of the myosin ATPase would drop rapidly and no contraction could occur for more than a fraction of a second. Even if the Lohmann reaction is switched off by inhibitors and buffers are used instead to neutralize $H^+$, the restricted volume of the overlap region cannot provide sufficient buffering capacity and a proton efflux from this region will have to take place. In this situation it is difficult to say whether the neutralization of $H^+$ will take place in the regions around the $M$ line and the $Z$ disk or whether the $H^+$ may move along some surface paths to the outside medium. However, within the sarcomere their path will remain along the myofilaments, if the contention of their high protic conductance is correct.

## 2.3. The Proto-Osmotic Mechanism of Muscular Contraction

The rate-limiting step in the hydrolysis of ATP by the myosin ATPase, M, (located in S1) is the transition of the intermediate complex, $M^{xx}.ADP.P$ to $M.ADP.P$.[17] This transition occurs at a rapid rate if S1 can interact with the actins (A) of the neighboring thin filaments in the form $AM^{xx}.ADP.P \rightarrow AM.ADP.P$. As against the small fraction (10%–20%) available from the hydrolysis of ATP itself ($M.ATP \rightarrow M^{xx}.ADP.P$), the above transition entails the liberation of more than 50% of the free energy. To understand the nature of the high-energy state $M^{xx}.ADP.P$, the results of experiments on the proton-early-burst[34] are of great possible significance. When ATP is added to myosin or HMM, the first one or two molecules get hydrolyzed extremely rapidly. The proton-burst is found to be consistently less in magnitude than the phosphate-burst. The magnitude of the early burst of $H^+$ also decreases with decreasing ionic strength. The obvious conclusion that can be drawn is that the $H^+$ remain associated with S1 and do not readily go into the aqueous phase and that the cooperation of the ions of the electrolyte is required for their diffusion away from the reaction site. The protonation of S1 would also explain the fall in the rate of hydrolysis after the early phase. If we identify $M^{xx}$ as the protonated state of the enzyme, the enhancement in the rate of hydrolysis brought about by the presence of actins in the medium would be due to the transfer of $H^+$ from S1 to A. In the muscle the interaction of S1 with the actins of the thin filaments is dependent on the concentration of $Ca^{2+}$, and, therefore, the enhancement of the rate of ATP hydrolysis is under neuronal control as would be expected. From the above, all that is required for continued hydrolysis of ATP at a rapid rate is the transfer of $H^+$ from S1 to actins, and $Ca^{2+}$ would merely serve to provide short-duration contacts between the two moieties. Upon stimulation of the muscle the first contacts of myosin heads with the actins will deliver more than one $H^+$ from S1 to the actins. This extra amount of $H^+$ would correspond to the $H^+$ which had resulted from the hydrolysis of ATP molecules in the recovery period and were held at S1. Their flux will be used later on to explain the problem of the "unexplained energy" mentioned in Section 2.1.

There is no point in speculating about the happenings at the molecular level which lead to the liberation of $H^+$ from the two S1 of the crossbridges when these contact the actins. During their release the protons at the heads have two possible paths open to them, one via the S2 to the thick filament and $M$ line and the other along the thin filament to the $Z$ disk. During the contact the protons will take that route for which the difference $\mu_H$ (filament) $- \mu_H$(CB) is less, where $\mu_H$ (filament) stands for the electrochemical potential of $H^+$ at either the thick or the thin filament at the position of the cross-bridge and $\mu_H$(CB) for the same quantity at the crossbridge. Apart from the conductances of the filaments, the above given differences in $\mu_H$ will depend on the rates of the $H^+$ consuming Lohmann reactions at the $M$ line and the $Z$ disk. Assuming, in order to avoid unnecessary complications, that the net fluxes to the $M$ line via all thick filaments and to the $Z$ disk via all thin filaments are equal, we proceed to follow the movement of the $K$ layer.

The movement of the $K$ layer, Figure 2, along the thick filaments will be from the $M$ line towards the tips of the thick filaments. From the tips $K^+$ will enter into the I bands, i.e., the space between the tips and the $Z$ disk. The charge-compensating movement of the K layer along the thin filaments is from the $Z$ disk towards the $M$ line, and here too $K^+$ will enter the space between the ends of the thin filaments and the $M$ line, i.e., the

FIGURE 2. Proton and $K^+$ fluxes for a pair of thin filaments and a thick filament. Only one cross-bridge has been shown. In the lower part of the diagram the loop flow of the K layer has been indicated.

H band. Both in the I band and the H band the $K^+$ will get recruited at the electrical double layers of the thin and the thick filaments respectively, as the $K^+$ of the K layer at these filaments are forced into the overlap region by the diffusion potential generated by the proton flux. Thus $K^+$ move in three-dimensional loops, the movement being, for the major part of the loops, along the myofilaments. As $K^+$ are kept close to the filaments in the form of a condensed laterally mobile layer by the negative fixed charges of the myofilaments, their lateral movement is expected to exert a viscous drag on the filaments. The directions of the viscous forces on the filaments are such as to cause their movements deeper into the interfilamental spaces. Taken together the hydrodynamic flows along the thick and the thin filaments constitute a viscous force couple which will cause the sarcomere to shorten against a load. These ideas will be developed in an analytical form in the subsequent sections and, whenever possible, quantitative comparisions will be made with the expermental data.

## 3. The $H^+/K^+$ Circuit in Sarcomere

### 3.1. $H^+/K^+$ Counterflux at Filaments

The equations for the fluxes of $H^+$ and $K^+$ and the expression for the diffusion potential will be derived now for a filament on the basis of the assumptions discussed in Section 2.2. Across the length $l$ of the proton-conducting filament a chemical potential difference $FV_H = -RT \ln c_H(l)/c_H(0)$ is assumed to be maintained by two synchronized chemical reactions which inject and take out $H^+$ at $z = 0$ and $z = 1$ respectively. The K layer has the radial extension from $r = a$ to $r = a + d$ with the thickness $d$ corresponding roughly to the Debye length. The concentration of $K^+$ in the K layer will be assumed to be homogeneous and their mobility assumed constant in the $z$ and $r$ variables. The electric field component in the $z$ direction, $E_z = -dV/dz$, will be assumed to be equal for all $r$ in the region of the K layer and zero outside. Writing $g_H$ for the protic conductance of the filament and $g_K$ for the surface conductance of the K layer, and assuming both to be independent of $z$, the flux equations are

$$J_H = g_H(V_H - V) \tag{1}$$

$$J_K = g_K(V_K - V) \tag{2}$$

Here $V_K$ is the Nernst potential difference for $K^+$, i.e., $V_K = -(RT/F)\ln c_K(l)/c_K(0)$. The expression for the diffusion potential difference $V$ follows from the condition of the conservation of electroneutrality applied to the restricted volume confined by the surface $r = a + d$, which gives $J_H = -J_K$. Hence

$$V = \frac{g_H}{g_H + g_K} V_H + \frac{g_K}{g_H + g_K} V_K \tag{3}$$

Eliminating $V$ from Eq. (1) we get

$$J_H = -J_K = [g_H g_K/(g_H + g_K)](V_H - V_K) \tag{4}$$

### 3.2. The Proton Circuit

In order to deal with the flux of $H^+$ in a sarcomere it suffices to consider a pair of thin filaments of a length $l_a$ and a thick filament interposed between them (Fig. 2). The proton flux to the $M$ line via the thick filament will be taken as equal to the flux of $H^+$ via the pair of the thin filament to the $Z$ disk. This means that starting with $V_K = 0$, no imbalance in the $K^+$ concentration will be created in the sarcomere and $V_K$ can be ignored. Protons are injected in the thin filaments by the cross-bridges (CB) at different sites along their lengths. A small resistance $r_{CB}$ is introduced for the path of $H^+$ from the CB to the actins. The resistances $r_H = g_H^{-1}/l_a$ and $r_K = g_K^{-1}/l_a$ are introduced for the pair of the thin filaments; i.e., the $g_H$ and $g_K$ are twice the value of the individual conductances of the thin filaments. For the discussion of the circuit the thin filament pair may be considered now as a single filament.

The injection of protons in the thin filaments is due to the contacts of CB at discrete point $z_i$ in the overlap region each causing a small flux $J_i$. Writing $r_H + r_K$ as $R$, Eq. (4) goes over to

$$J_i = V_H/[(l - z_i) R + r_{CB}] \tag{5}$$

where $2l$ is the length of the sarcomere and $l - z_i$ is the distance of the CB at $z_i$ from the $Z$ disk containing the enzyme CPK. $V_H$ has therefore the value $V(M^{**}) - V(CPK)$. The net flux of $H^+$ from all CB in the overlap region extending from $z_0$ to $z_1$ is the sum over all $J_i$. More conveniently $J_H$ can be obtained by the integral

$$J_H = \int_{z_0}^{z_1} \frac{V_H dz}{[(l - z) R + r_{CB}]} = \frac{V_H}{R} \ln \frac{Rl_a + r_{CB}}{R(l_a - \theta) + r_{CB}} = \frac{f(\theta)}{R} V_H \qquad (6)$$

where $\theta$ is the overlap between the two filaments and the overlap function $f(\theta)$ is defined by the identity in Eq. (6). If $r_{CB} \ll Rl_a$ then

$$f(\theta) = -\ln(1 - \theta/l_a) = \theta/l_a + (\theta/l_a)^2/2 + (\theta/l_a)^3/3 + \cdots$$
$$+ (\theta/l_a)^n/n + \cdots \qquad (7)$$

It can be seen that $f(\theta)$ does not vary appreciably around the slack length. Within the range of the applicability of the expansion $(-1 < \theta/l_a < +1)$ and without including the possibility of the thin filaments crossing over the region of the thick filament devoid of CB (the pseudo-H-band) $V_H J_H$ gives the power output of the proton circuit (heat generation in the static case considered here) as a function of overlap.

The same treatment can be extended to the thick filaments with (possibly) different values for $r_{CB}$ (pertaining to S2) and $R$.

The most crucial quantity for the proto-osmotic theory of muscular contraction is the value of $V_H$. According to our interpretation of the low magnitude of proton early burst the heads, S1, act as small capacitors of $H^+$. The accumulation of $H^+$ in S1 may be in the form of _$NH_2$ to _$NH_3^+$ conversions in S1. The chemical potential of $H^+$ in S1 is raised by repetitive hydrolysis of ATP when S1 are not in contact with the G-actins. To eliminate migration of $H^+$ across S2 to the thick filaments, it is postulated that in relaxed muscle there is a break in the hydrogen-bonded system between the hydrolytic sites and the thick filaments. In order to estimate the difference in the chemical potential of $H^+$, $\Delta\mu_H$ between S1 and the $M$ line and the $Z$ disc we make use of the fact that about 60% of the free energy of ATP hydrolysis, $\Delta G_{ATP}$, is available during the steps of the dissociation of products from the ATPase. If this fraction of $\Delta G_{ATP}$ is converted into $\Delta\mu_H$, then $\Delta\mu_H$ will be equal to 30 kJ/mol, corresponding to $V_H = 0.3$ V. For quantitative evaluations a lower value, $V_H = 0.2$ V will be used in the following sections.

## 4. Fenn Effect and Hill's Equation

### 4.1. Energy Output of Muscles

Fenn[35] observed that the sum of heat and work output was greater if a muscle is allowed to shorten against a small load than the amount of heat which is produced in isometric contraction over the same period of time[35] and that the component of heat production by itself was also greater. This effect can be explained in terms of the viscoelectric equivalent circuit of Figure 3, in which the dependence of $g_K$ on the movement of a filament has been schematically represented. A linear dependence $g_K(v) = g_0 + sv$ will be assumed for the increase in the surface conductance of the K layer as the filaments move codirectionally with a velocity $v$, over the value $g_0$. The output of the circuit is

$$\dot{E}(v) = V_H J_H + Pv = V_H^2 f(\theta) \frac{g_H(g_0 + sv)}{g_H + (g_0 + sv)} + Pv \qquad (8)$$

The first term in Eq. (8) can be expanded as

$$\dot{Q}(v) = V_H^2 f(\theta) T_H(0) \{ g_0 + T_H(0) sv - [T_H^2(0)/g_H](sv)^2 + O(3) \} \qquad (9)$$

where $T_H(0) = g_H/(g_H + g_0)$ and $O(3)$ are terms in the expansion of an order of third and higher powers in $sv$. The rate of heat liberation for isometric contraction is

$$\dot{Q}(0) = V_H^2 f(\theta) g_0 T_H(0) \qquad (10)$$

Equation (9) predicts a rise in the rate of heat liberation with increasing velocity of shortening for small values of $sv$ in accordance with the findings of Fenn. For higher speeds of shortening the quadratic term will become effective and cause a decline in the rate of heat production with increasing speed. This is analytically in agreement with the recent and more accurate observations of Hill.[36] It may be noted that for $v < 0$–that is, for stretching

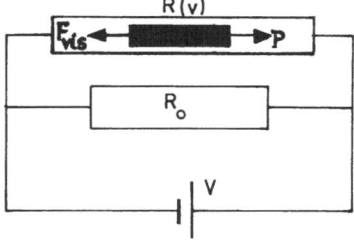

FIGURE 3. The viscoelectric equivalent circuit of muscle. A myofilament has been shown in black. The rectangular boxes represent the resistances $(sv)^{-1}$ and $g_0^{-1}$.

of a muscle with a load $P$ exceeding that load $(P_0)$ for which the contraction is isometric—the rate of heat production falls below $Q(0)$. This result is in agreement with the observations of Curtin and Davies.[18]

### 4.2. Force–Velocity Relation

The rate of energy released in excess of $\dot{Q}(0)$ by the miniature contractile unit of Fig. 3 while shortening against a load $P$, $P < P_0$, is $\Delta\dot{E}(v) = \dot{E}(v) - \dot{Q}(0)$. Since $\Delta\dot{E}(v)$ vanishes for $P = P_0$ one can express it as a power series in terms of the difference $P_0 - P$:

$$\Delta\dot{E}[v(P)] = c_1(P_0 - P) + c_2(P_0 - P)^2 + \cdots \tag{11}$$

An approximate force–velocity relation, $v = v(P)$, can be obtained by equating $\Delta\dot{E}[v(P)]$ of Eq. (11) with $\Delta\dot{E}(v)$ as given by Eqs. (8) and (10) and retaining terms only up to the second order in $v$ and $P_0 - P$:

$$v = \frac{a + P - [(a+P)^2 - 4asT_H(0)\{c_1(P_0 - P) + c_2(P_0 - P)^2\}/g_H]^{1/2}}{2asT_H(0)/g_H} \tag{12}$$

An even more approximate relation is obtained by keeping the linear terms only:

$$v = \frac{c_1(P_0 - P)}{a + P} \tag{13}$$

where $a = f(\theta) V_H^2 T_H(0)^2 s$.

In order to generalize Eqs. (8)–(13) to a myofibril, the unit of Fig. 3 can be reflected at the $z = 0$ plane and a sarcomere constructed by a parallel arrangement of several such units with the appropriate lattice symmetry. A myofibril containing $N$ sarcomeres will have a shortening velocity $2Nv$. The conductances $g_0$ and $g_H$ can be now interpreted as referring to all of the thin and the thick filaments in a half sarcomere and $P_0$ and $P$ as the load applied to the myofibril. Thus the only change required to be made is to substitute $v/2N$ for $v$ in all of the relevant equations and interpret $v$ as the velocity of shortening of muscle. Writing $2Nc_1$ as $b$ in Eq. (13); we get

$$v = \frac{b(P_0 - P)}{a + P} \tag{14}$$

which is Hill's (1938) force–velocity relation.[37]

In the following a set of compatible values for $g_H$, $g_0$, and $s$ will be determined using the measured value of $a = 4.8N$ [18] with the assumption that only 10% of the total power output of the proton circuit is dissipated as heat in the proton conductors. The estimates of the values will be obtained for contraction of muscle around the slack length and $f(\theta)$ will be taken as unity. The assumption of low heat dissipation in the proton conductors is compatible with our basic assumption of the high proton conductance of the myofilaments. Thus

$$\frac{J_H(V_H - V)}{J_K V} = \frac{g_0}{g_H} = 1/9 \tag{15}$$

The net flux of $H^+$ along the thick and the thin filaments in a half sarcomere ($M$ to $Z$ line) of a unit cross-sectional area is about $1.0 \times 10^{-10}$ mol s$^{-1}$ during an isometric contraction. This estimation is based on the observed rate of ATP hydrolysis $(1.0 \times 10^{-6}$ mol/g/s)[18] and a slack length of the half sarcomere of about $1.0 \times 10^{-6}$ m. This flux does not include the flux of $H^+$ stored at the CB before the onset of contraction. In a tetanus of 10 s duration the ratio of the unexplained energy to the net amount of energy liberated by the muscle is about 30%,[19] so that the actual $J_H$ is about $1.3 \times 10^{-10} \times F$ (ampere). Half of this current flows along the thin filaments with the resistance: $g_H^{-1} + g_0^{-1} = (10/9) \cdot g_H^{-1}$. Taking $V_H = 0.2$ V, we have $g_H = 3.5 \times 10^{-4}$ mho.

The parameter $s$ can be determined using the relation $a = V^2 s$, given after Eq. (13). With $V = 0.9 V_H$ $s = 148 N V^{-2}$.

The above estimates are mainly intended to give the idea of the orders of magnitudes of the quantities. It is seen from Eq. (8) that $sv/(g_H + g_0)$ must be less than unity for the expression to converge rapidly. With the above given estimates it is true only for small velocities of shortening $(v \leqslant 2$ cm s$^{-1})$. Presumably both $V_H$ and $g_H$ have been underestimated above by about 10% or more and the value of $s$ would be less than that given above.

### 4.3. Isometric Tension

For the viscous force between a myofilament and the K layer we will use the same relation as that which gives the viscous force between two layers of a viscous fluid[38]:

$$F_{vis} = \eta A \, dv/dr \tag{16}$$

where $A$ is the area of contact between the adjacent layers and $\eta$ the viscosity of the fluid. For the filaments this relation is modified as

$$F_{vis} = \eta_s A(v_K - v)/d \tag{17}$$

where $\eta_s$ is the viscosity in the K layer, $A = 2\pi r_f l_f$ ($r_f$ and $l_f$ being the radius and the length of the filament), $v_K$ the velocity of the K layer, and $v$ that of the filament. It will be assumed that the velocity of water in the K layer is the same as that of the K$^+$. Since $u_s$, the mobility of K$^+$ in the K layer, will be inversely proportional to $\eta_s$ we have $\eta_s u_s = u_0 \eta_0$ ($u_0 = 7.5 \times 10^{-4}$ cm$^2$ V$^{-1}$ s$^{-1}$, $\eta_0 = 0.01$ Poise). For the thin filaments $r_f = 3.5 \times 10^{-7}$ cm and $l_f = 1.0 \times 10^{-4}$ cm. In isometric case $v = 0$ and $V = 0.18$ V. The number of thin filaments in a half sarcomere segment of an area of cross section of 1 cm$^2$ is estimated as $10^{11}$, so that $A = 22$ cm$^2$. From the above given data the viscous drag on the thin filaments,

$$F = A u_0 \eta_0 (V/l_f)/d = 29N \tag{18}$$

For the thick filaments the number of filaments is half of the number of the thin filaments but the radius is twice as large. Therefore both arms of the viscous drag couple will exert the same force on the two sets of the interdigitated myofilaments and cause them to slide into each other.

The question that remains yet to be resolved is whether the layers of water in between the filaments will not slip for large values of $P$. One must be able to show that there is always a velocity gradient $dv/dr$ of sufficient magnitude at any given surface surrounding, say, a thick filament so that the collective viscous force calculated for such surfaces around all of the thick filaments, $\eta_0 A_s dv/dr$ exceeds or equals the value $P_0$, which is around $20N$.[18] The weakest link would be at the surface of zero velocity surrounding the thick filament, shown in cross section in Figure 4 as a wavy line. This problem may be analyzed as follows. In the absence of a pressure gradient, the general equations for the velocity and velocity gradient in a cylindrically symmetric hydrodynamic flow are given by[38]

$$dv/dr = C/r$$
$$v(r) = C \ln r + D \tag{19}$$

Taking the velocity at distance $d$ ($d = 10^{-7}$ cm) away from the surface of the thick filament ($r_f = 7 \times 10^{-7}$ cm) as $v_K = u_0 V/1 = 1.7$ cm s$^{-1}$ and

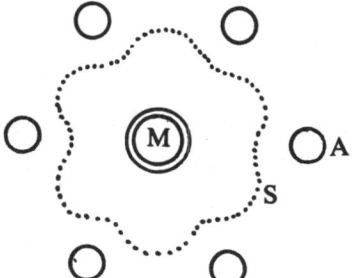

FIGURE 4. Cross-sectional view of a thick filament (M) surrounded by six thin filaments (A). The wavy line (S) represents the surface of zero velocity.

as zero at a distance of $16 \times 10^{-7}$ cm, corresponding roughly to the distance of the surface of zero velocity $(S)$ from the axis $(r = 0)$ of the thick filament one gets $v(r) = 2 \ln r / 16 \times 10^{-7}$ and $dv/dr = 2/r$, viz. $(8 \times 10^{-7} \text{ cm} \leqslant r \leqslant 16 \times 10^{-7} \text{ cm})$. With the area of contact, for $\theta = l_M = 0.8 \times 10^{-4}$ cm, as 40 cm$^2$ and $\eta_0 = 0.01$ Poise, the viscous force for a half sarcomere and unit cross-sectional area is found to be $5N$. Though of the right order of magnitude the value is one fourth of the required magnitude. The maximum chance of underestimation is the value of $A_S$ then $v_K$ and possibly also $\eta_0$. Even if one tries to bring up the value of the viscous force at $S$ it is clear that the possibility of slipping is a critical factor in the operation of the viscous force model. At the same time it may have some bearing on the "give" effect which is observed when the muscle is loaded with loads exceeding $P_0$.[3]

## 5. Discussion

The agreement of the quantitative results obtained from the proto-osmotic mechanism with the observed experimental value in terms of orders of magnitudes is rather encouraging, especially in view of the fact that most of the data concerning the proto-osmotic effect and viscoelectric coupling are lacking at present. The hydrodynamic aspects of the approach are quite complex. The movements of the two heads of a cross-bridge as it approaches and comes under the influence of the hydrodynamic flow along the thin filament may provide a clue as to the necessity of two heads per cross-bridge, since a single head would tend to bounce away from the thin filament. It is also of considerable significance that around 40% of the free energy of ATP is available for the dissociation of S1 from the actins. If electrostatic forces are involved in the interaction of the cross-bridges with the actins the separation of the CB from the actins and their liberation from the flow of the fluid around the actins may require an input of energy.

The solution to the great puzzle of the unexplained energy in terms of the H$^+$ stored at the cross-bridges also helps to explain the fast response of a muscle. The continued movement of the cross-bridges in the overlap free zone and after the end of the contraction is explicable in terms of the inertial hydrodynamic flow associated with proto-osmosis. This approach also does not require any long duration contact of a cross-bridge with the actins. The dissociation of ADP and P can occur at any time after the actual contact is made.

The explanation of muscular force in terms of a viscous force couple explains, in a straightforward manner, the development of a higher tension during stretching than during contraction, since the viscous force depends on the relative velocity of the movement of the filaments versus the

K layers around them. At the same time the energy expenditure is less because of the reduced $g_K(v)$ and hence $J_H$. The observed effect is therefore just the opposite of the Fenn effect.

With the help of the viscoelectric couplings:

$$F_{vis} \sim v_K - v$$
$$v_K \sim V/\eta_s, \qquad \eta_s \sim 1/v$$
$$\dot{v} \sim F_{vis} - P \tag{20}$$
$$V \sim 1/(1 + v)$$

the mechanical transients of muscle[3] can be qualitatively explained. This will be illustrated for the length jump transients in tension. In a length jump ($\Delta l < 0$) experiment the fully stimulated muscle is allowed to shorten a little and the time course of the tension is followed. Initially $F_{vis} = P_0$, $v = 0$, and $V$ is large so that $v_K$ would tend to be large if $\eta_s$ were small. During the length jump the K layer, whose movement was initially hindered by the fixed filaments, gains momentum so that when the filaments are again immobilized it exerts a strong viscous drag on them. This corresponds to the early rapid phase of the tension recovery. During the length jump, however, the diffusion potential is also unbuilt to some extent, so that $v_K$ becomes smaller than it was initially and remains small till $V$ is recharged by the retarded movement of the K layer after the filaments have been immobilized at the end of the length jump. This situation explains the plateau in the tension recovery curve after the first phase of rapid recovery and the gradual rise in tension to the initial isometric tension.

From the encouraging results which have been obtained above by the applications of electrochemistry and hydrodynamics it may be concluded that the approach merits further exploration both theoretically and experimentally. The values of $g_H$, $g_0$, and $s$ and the magnitude of the proton-motive force have to be determined by experiments especially designed for this purpose. In its present form the approach is liable to suffer from the limited competence of the author. However, it is hoped that the approach has been developed here to the extent that it is convincing enough to be subjected to experimental scrutiny.

There are two possible applications of the approach: The concept of high proton mobility implies that $H^+$ will be the first among the ions to react to a change in the electrochemical potential in the different compartments of a muscle fiber. Thus $H^+$ are expected to move into the T-system upon the arrival of the propagated action potentials and thence into the sarcoplasmic reticulum via the junctions between the two.[8] Their interaction with calsequestrin in the terminal cisternae may effect an enhancement in the activity of $Ca^{2+}$ and these may also enhance the permeability of the $Ca^{2+}$ channels in the SR membrane. This interpretation is in line with the findings of Somlyo *et al.*[39] and Shoshan *et al.*[40] and may be useful in elucidating the mechanism of excitation contraction coupling.

Activation of the contractile machinary by means of the interaction of $Ca^{2+}$ with Troponins (TN) is amenable to a new interpretation in the framework of the proto-osmotic approach. It is suggested that the above interaction serves to establish the hydrogen-bonded systems of F-actin. Each TN overlaps two G-actins and is thus in a position to interfere with their mutual linkages and break the protic circuit of F-actin in the absence of $Ca^{2+}$. The interaction of $Ca^{2+}$ with TN-C and of the latter with TN-I and TN-T may restore the normal linkages between the affected G-actins and "switch on" the circuit of $H^+$ along the F-actin. All other modes of activation are also suggested to be by means of the establishment of proton conduction paths in the myofilaments, from S1 to the *M* line and the *Z* disc.

It is evident that the present approach is more versatile, from the structural point of view, to cope with muscle types in which the organization and the types of myofilaments differ from the striated muscle. It is hoped that the proto-osmotic approach will find applications in the diverse motile contractile and transport activities in living systems. The involvement of proton-motive force in ciliary movements has already been established.[41] Among the several effects of altering the pH of the medium on the mechanical properties of muscle given by Roos and Boron[42] two interesting observations may be mentioned. In 1880 W. H. Gaskell could bring the frog ventricle to resume beating for a short time by increasing the pH of the perfusate. Isaac Newton is reported to have observed that when he put a drop of vinegar on pieces of eel heart, their rhythmic contractions were abolished. This qualifies Newton to be credited with having observed the first manifestation of proton-motive force in muscle.

## References

1. D. M. Needham, *Machina Carnis*, Cambridge University Press, Cambridge (1971).
2. H. E. Huxley, "The Mechanism of Muscular Contraction," *Science* **164**, 1356–1366 (1969).
3. A. F. Huxley, "Muscular Contraction," *J. Physiol.* **243**, 1–43 (1974).
4. M. Amin, "$H^+$-flux Mediated Electroosmosis: III Mechanism of Muscle Contraction," *Natl. Acad. Sci. Lett.* **5** (8), 277–279 (1982).
5. M. Amin, "Theory of Muscle Contraction: I Isometric Contraction," *J. Biol. Phys.* **11** 91–97 (1984).
6. M. Amin, "Theory of Muscle Contraction: II Isotonic Contraction, *J. Biol. Phys.* **11**, 123–126 (1984).
7. L. L. Constantin, "Contractile Activation in Skeletal Muscle," *Prog. Biophys. Molec. Biol.* **29**, 197–224 (1975).

8. C. Franzini-Armstrong and L. D. Peachey, "Striated Muscle—Contractile and Control Mechanisms," *J. Cell Biol.* **91** (3), 166s–186s (1981).

9. S. Ebashi, "Regulation of Muscle Contraction," *Proc. R. Soc. London Ser. B* **207**, 259–286 (1980).

10. R. W. Ramsey and S. F. Street, "The Isometric Length–Tension Diagram of Isolated Skeletal Muscle Fibers of the Frog," *J. Cell. Comp. Physiol.* **15**, 11–34 (1940).

11. A. M. Gordon, A. F. Huxley, and F. J. Julian, "The Variation in Isometric Tension with Sarcomere Length in Vertebrate Muscle Fibers, *J. Physiol.* **184**, 170–192 (1966).

12. H. Ishikawa, in *Cell and Muscle Motility* (R. M. Dowben and J. W. Shay, eds.), Plenum Press, New York (1983), Vol. 4, pp. 1–136.

13. J. W. S. Pringle, "The Contractile Mechanism of Insect Fibriller Muscle," *Prog. Biophys. Molec. Biol.* **17**, 1–60 (1967).

14. S. Highsmith and R. Cooke, in *Cell and Muscle Motility* (R. M. Dowben and J. W. Shay, eds.), Plenum Press, New York (1983), Vol. 4, pp. 207–238.

15. R. Cooke, "Stress Does Not Alter the Conformation of a Domain of the Myosin Cross Bridge in Rigor Muscle Fiber," *Nature* **294**, 570 (1981).

16. H. E. Huxley, "Structural Changes in the Actin- and Myosin-Containing Filaments During Contraction, *Cold Spring Harbor Symp. Quant. Biol.* **37**, 361–376 (1973).

17. E. W. Taylor, "Mechanism of Actomyosin ATPase and the Problem of Muscle Contraction,"*Crit. Rev. Biochem.* **6**, 103–164 (1979).

18. N. A. Curtin and R. E. Davies, "Chemical and Mechanical Changes During Stretching of Activated Frog Skeletal Muscle," *Cold Spring Harbor Symp. Quant. Biol.* **37**, 619–626 (1973).

19. N. A. Curtin and R. C. Woledge, "Energy Changes and Muscular Contraction," *Physiol. Rev.* **58** (3), 691–761 (1978).

20. G. Gabella, in *Smooth Muscle* (E. Bülbring, A. F. Brading, A. W. Jones, and T. Tomita, eds.), Edwards Arnold, London (1981), pp. 1–46.

21. K. R. Porter, in *Cell Motility*; Cold Spring Harbor Conference on Cell Proliferation, Vol. 3 Book A; "Motility, Muscle and Non-muscle Cells," Cold Spring Harbor Laboratory (1976), pp. 1–28.

22. A. Oplatka, J. Borejdo, H. Gadasi, R. Tirosh, N. Liron, and E. Reisler, in *Proteins of the Contractile Systems*, Proceedings of the Ninth FEBS Meeting Vol. 31 (E. N. A. Biro ed.), North-Holland/American Elsevier, New York (1975), pp. 41–46.

23. R. D. Goldman and D. M. Kneipe, "Functions of Cytoplasmic Fibers in Non-muscle Cell Motility," *Cold Spring Harbor Symp. Quant. Biol.* **37**, 523–534 (1973).

24. B. Grafstein, in *Handbook of Physiology. The Nervous system*, American Physiological Society, Bethesda, Maryland (1977), Sec. 1, Vol. 1, pp. 691–717.

25. D. B. Kell, "On the Functional Proton Current Pathway of Electron Transport Phosphorylation, An Electrodic View, *Biochim. Biophys. Acta* **549**, 55–99 (1979).

26. I. C. West, *The Biochemistry of Membrane Transport*, Chapman and Hall, London (1983).

27. L. Onsager, in *Physical Principles of Biological Membranes* (F. Snell, J. Wolken, G. Iverson, and J. Lam, eds.), Gordon & Breach, New York (1970), pp. 137–141.

28. J. F. Nagle and S. Tristram-Nagle, "Hydrogen Bonded Chain Mechanisms for Proton Conduction and Proton Pumping, *J. Membrane Biol.* **74**, 1–14 (1983).

29. J. O'M. Brockris and A. K. N. Reddy, *Modern Electrochemistry*, Vols. 1 and 2, Plenum Press, New York (1970).

30. T. Wallimann, D. C. Turner, and H. M. Eppenberger, in *Proteins of the Contractile Systems*, Proceedings of the Ninth FEBS Meeting, Vol. 31 (E. N. A. Biro, ed.) (1975), pp. 119–124.

31. P. Glansdorf and I. Prigogine, *Thermodynamic Theory of Structure, Stability and Fluctuations* Wiley-Interscience, New York (1974).

32. A. Katchalsky and P. F. Curran, *Non-Equilibrium Thermodynamics in Biophysics*, Harvard University Press, Cambridge, Massachusetts (1964).
33. C. W. F. McClare, "Chemical Machines, Maxwell's Demon and Living organisms," *J. Theor. Biol.* **30**, 1–34 (1971).
34. J. F. Koretz, T. Hunt, and E. W. Taylor, "Studies on Mechanism of Myosin and Actomyosin ATPase," *Cold Spring Harbor Symp. Quant. Biol.* **37**, 179–184 (1973).
35. W. O. Fenn, "The Relation Between the Work Performed and Energy Liberated in Muscular Contraction," *J. Physiol. (London)* **58**, 373–395 (1924).
36. A. V. Hill, "The Effect of Load on the Heat of Shortening of Muscle, *Proc. R. Soc. London Ser. B* **159**, 297–318 (1964).
37. A. V. Hill, "The Heat of Shortening and the Dynamic Constants of Muscle, *Proc. R. Soc. London Ser. B* **126**, 136–195 (1938).
38. G. Joos, *Lehrbuch der Theoretischen Physik*, Akademische Verlagsgesellschaft, Frankfurt (1959).
39. A. P. Somlyo, A. V. Somlyo, H. Gonzalez-Serratos, H. Schuman, and G. McClellan, in *Muscle Contraction: Its Regulatory Mechanisms* (S. Ebashi, K. Maruyama and M. Endo, eds.), Springer Verlag, Berlin, (1980), pp. 421–433.
40. V. Shoshan, D. H. MacLennan, and D. S. Wood, "A Proton Gradient Controls a Calcium-Release Channel in Sarcoplasmic Reticulum, *Proc. Natl. Acad. Sci. U.S.A.* **78**, 4828–4832 (1981).
41. S. Khan and R. M. Macnab, "Proton Chemical Potential, Proton Electrical Potential and Bacterial Motility," *J. Mol. Biol.* **138**, 599–614 (1980).
42. A. Roos and W. F. Boron, "Intracellular pH," *Physiol. Rev.* **61**(2), 296–434 (1981).

# 21

# Transport in Plants

## Mohammad Amin

ABSTRACT: An electrokinetic phenomenon has been deduced to occur at surfaces bearing fixed negative charges and possessing high conductance for protons. Anisotropic distribution of ions close to the surface has the consequence that a vectorial flux of $H^+$ laterally along the surface causes a charge compensating flux of the diffuse cationic layer in the opposite direction. If such a surface forms the internal lining of a capillary connecting two regions at different pH then the proton flux in the direction of its chemical gradient causes an electroosmotic flow of water and solutes in the opposite direction. In plants the transport network is spanned by capillary-like conducting systems bearing negative fixed charges at their internal surfaces while photosynthesis and nitrate reduction in the leaves and respiration in the other tissues generate $H^+$-chemical potential difference between these tissues. The assumption of high proton mobility in the transport pathways of plants not only solves the problem of pH regulation in these tissues when considered individually but also provides a source of energy for long- and medium-distance transport via the mechanism of $H^+$-flux mediated electroosmosis. The collective proton motive force generated by the cells of a sink tissue causes a flux of $H^+$ to the leaf mesophyll cells and drives the photosynthates to the sinks. Uptake of water and turgor regulation in cells emerge as active processes coupled to $H^+/K^+$ antiport via $K^+$-selective pores when the latter are considered as a limiting case of small capillaries with a radius of a hydrated potassium ion. In these pores the efflux of $H^+$ is driven by $H^+$ ATPase and the condition of the preservation of electrical neutrality within the pore requires $K^+$ to be propelled inwards along with the water molecules within the pore. In the xylem the upward movement of the xylary sap arises from the efflux of $H^+$, generated by the respiratory activity of the tissues of the shoot and root stele via the radial symplasmic pathways crossing the endodermis to the soil. Ascent of sap is an active transport process and can occur without transpiration; the latter merely increases the rate of water flow by generating a pressure gradient which is codirectional with the electro-osmotic flow. This approach to transport in plants is supported by experimental evidence, permits a unified treatment of long-distance transport and cellular transport in a consistent manner, and is in full accord with established principles of bioenergetics.

*Mohammad Amin* • School of Life Sciences, Jawaharlal Nehru University, New Mehrauli Road, New Delhi-110067, India.

## 1. Introduction

An electrokinetic phenomenon, related to electro-osmosis, has been theoretically deduced to occur under very stringent conditions at the surfaces of macromolecular assemblies, particularly the cell membranes. This phenomenon may be described in simple terms as follows. If a rapid proton flux occurs by a specific mechanism along the surface of a macromolecular assembly bearing an excess of negative fixed charges, then the cationic layer adjacent to the proton conducting surface executes a charge compensating movement in the direction opposite to the proton flux. The difference from electro-osmosis lies in the mode of generation of the electric field and its localization in the region close to the surface. This phenomenon has been provisionally named proto-osmosis.

Assuming that proto-osmosis may occur at the surfaces of the capillaries forming the transport network of plants a novel (and heretical) approach to the mechanism of transport in plants emerges, which explains, at the moment only qualitatively, most of its puzzling aspects in a consistent and unified manner. A very rudimentary treatment has been possible at the moment. In the absence of essential information, a detailed discussion would have been premature. In Section 2 the phenomenon of transport in plants is briefly discussed to provide readers of diverse backgrounds an access to the topic and to bring out the inadequacies of our present understanding. Leaves, in which photosynthesis and nitrate reduction are the dominating metabolic processes, are identified as $H^+$ consuming tissues and a flux of $H^+$ is suggested to be directed towards them from the heterotrophic tissues in which growth processes lead to the generation of an excess of acidic groups. In Section 3 the structural and functional prerequisites for proto-osmosis are described for the case of capillaries and the hydrodynamic consequences of proto-osmosis have been worked out. In Section 4 turgor* regulation in plant cells is discussed in terms of the $K^+$-antiport mechanism discussed in Section 3, as a limiting case of proto-osmosis in $K^+$-selective pores in the cell membranes of plant cells. Active $H^+$ efflux along with proto-osmosis in the pathways of the transport network of plants have been then used in Section 5 for the transport of $K^+$, water and photosynthates to the different tissues of a plant.

---

* **turgor**\ 'tər-gər, -ˌgȯ(ə)r \n [LL, turgidity, swelling, fr. L *turgēre*] : the normal state of turgidity and tension in living cells; *esp*: the distension of the protoplasmic layer and wall of a plant cell by the fluid contents (Webster's New Collegiate Dictionary, 1977).

## 2. Transport Phenomena in Plants

### 2.1. The Transport Network

Higher terrestrial plants have an elaborate transport network comprising two long-distance pathways, the xylem and the phloem, which run parallel to each other from the fine roots to the leaf veins and the radial, medium distance, transport system formed by the rays.[1,2] Water and inorganic nutrients are transported from the roots to the leaves and other tissues in the xylem conduits, vessels, and tracheids, while the photosynthates reach the fruits, the buds, and the roots via the conducting elements of the phloem, the sieve tubes. The radial and tangential distribution of substances is facilitated by the presence of narrow, tubular, connections called the plasmodesmata, between adjacent cells of the xylem and the phloem parenchyma. As there is no membrane barrier at such locations, the cells interconnected by plasmodesmata form a symplasm. Most of the cells in plant tissues are organized in symplasms indicating the importance of plasmodesmal connections for intercellular transport. Transport in the rays is also symplasmic excepting the terminal junctions with the xylem conduits and the sieve tubes. At these locations specialized cells mediate the exchange of substances by means of a large number of pits in the intervening walls. The sieve tubes also constitute a symplasm but here the transverse end walls have relatively large pores, presumably to facilitate the movement of the viscous phloem sap. However, these sieve plate pores often appear to be surrounded by endoplasmic reticulum or traversed by filamentous proteins.[3] Among the two types of xylem conduits, both of which are dead cells when functionally mature, the vessels come closer to forming capillaries since their transverse septa are perforated. The tracheids have pits separating them from each other since the primary wall is continuous at the pits. Nonetheless, the high density of pits in the side walls of both the vessels and the tracheids points to their major involvement in the exchange of water and solutes between them and the surrounding xylem parenchyma.

### 2.2. Ascent of Sap

The problem of elucidating the mechanisms of transport in the various parts of the transport network has been treated in the past in a compartmentalized manner. As regards the question how considerable volumes of water with dissolved nutrients reach the leaves, the cohesion theory[4-7] postulates that the evaporation of water from the surface of leaves (transpiration) provides the driving force. As the radius of the xylem conduits is too large for the capillary action to be of much help[7] negative pressure (tension) is required both for holding water columns and for moving them upward at the observed rate. For a 100-m-tall tree the necessary tension is calculated to be about $-2 \times 10^6$ Pa, which is, to give a better idea, $-20$ atmosphere! Cohesive forces between water molecules in pure water have been calculated to be strong enough to sustain much higher tensions than those expected to obtain in plants.[6,7] However, the xylary sap contains dissolved gases and cavitation occurs very frequently even in medium size plants and herbs.[8,9] If negative pressures were maintained by transpiration, then, once a gas phase is generated by embolism, all water should evaporate and only gases and water vapor should be found in the conduits. Not only does this not happen, very surprisingly the cavitated columns rejoin under favorable conditions. Many small herbs which cavitate freely during the day are found to guttate in the morning.[9] In trees there is no certainty whether intact water columns exist in the new xylem conduits up to late summer, after excessive cavitation has occurred. The old xylem conduits exist in the cavitated state, with small water columns alternating with much larger air gaps (Jamin's chains).[10] There is a diurnal and a much greater seasonal variation in the content of water in all parts of plants,[4,10] and it is difficult to understand this in the framework of the cohesion theory. This theory was assumed to be supported by the observation of shrinkage of stem diameter around noon corresponding to a high rate of transpiration. Recent studies[11,12] have shown that the shrinkage occurs earlier in the upper parts of the stem than in the lower and that it is primarily the bark which shrinks and less the wood. The pressures recorded with a pressure bomb are also cited in support of the occurrence of high tensions in the xylem. This technique assumes that twigs behave as an elastic system so that the positive pressures required to cause exudation of sap from the cut end are taken as equal in magnitude to the negative pressure in the conduits, when the twig is a part of the intact plant. The values vary markedly with the degree of cavitation in the twigs[4] and such observations as the tension in a 0.1-m wheat plant $(-1.2 \times 10^6$ Pa), which is insignificantly different from that recorded in a 80-m-tall redwood tree $(-1.55 \times 10^6$ Pa),[6] make it very doubtful that the assumption behind the technique is correct.

Even the contention that transpiration provides the sole driving force for sap flow is against the facts. This was recognized by Dixon as a result of his experiments on twigs which were fully submerged in water to stop transpiration. There was a persistent uptake of water by the shoots. Similar experiments, carried out more recently,[13,14] have shown the water uptake by the twigs and leaves to be coupled to growth processes supported by respiration. Since growth occurs in intact plants also under conditions of high humidity when transpiration is negligible, the sap must be held and moved upwards by other agencies than the pressure difference generated by transpiration. Many inhibitors, eg., KCN, reduce the uptake of water by plants, and anaerobiosis such as caused by flooding the root system, can induce wilting.[6] That transpiration cannot be the essential factor in water uptake and transport is also clearly implied in the systematic observations of Gibbs,[15] who found a marked increase in the water content of wood in maple and other diffuseporous species after the leaf fall in autumn. Thus the problem of ascent of sap cannot be solved until the component of the driving force due to metabolic activity has been identified.

## 2.3. Transport of Photosynthates

Translocation of photosynthates from the leaves (sources) to the roots and other sinks is an even more perplexing phenomenon. While the phloem sap is generally much more viscous than the xylary sap, the radii of the conducting elements are in the range of 5–35 $\mu$m, which is an order of magnitude less than that of the xylem conduits.[3] The hydraulic conductivity of the sieve tubes is further reduced by the sieve plates having pores of radii of about 1 $\mu$m. In view of the low hydraulic conductivity the pressure gradient of about 10 kPa due to gravity is too meager to drive the sap.[7] The velocity of the sap flow as determined by various methods shows wide variations (0.3–300 cm h$^{-1}$) and bears no linear relationship with the dry weight transfer per unit time.[16] Tracer studies[17,18] reveal a fast component of the sap flow, i.e., of velocities greater than 1 m h$^{-1}$. Furthermore, water, potassium ions, and inorganic phosphate ions seem to leak out of the tubes and not to follow the sap movement in general. The most puzzling aspect of translocation process is that the flow in one and the same sieve tube can occur in two opposite directions (bidirectional flow).[18–20]

Since the concentration of sucrose and other photosynthates in the sieve tubes is higher in the region of sources than of the sinks, Münch[21] proposed that a turgor pressure difference between the two would be generated by a relatively greater imbibition of water by the sieve tubes in the source region. There are many difficulties in accepting this pressure flow hypothesis.[22,23] Conceptually, the idea of free inhibition of water in the source region is difficult to reconcile with the low water potential in the leaves. Water deficit in the xylem does not always lead to a decrease in the rate of translocation of photosynthates.[23] From the anatomical point of view, the presence of the sieve plates, even if the pores are not occluded, is not the best measure nature could have adopted to promote pressure flow. Indeed there is an obvious reason to hinder pressure flow since, as far as the chemical potential of water is concerned, the direction of pressure flow in the sieve tubes should be the same as in the xylem conduits, i.e., towards the leaves, as both conducting systems begin and end in the same regions of the roots and the leaves.

The metabolic activity of the sinks as well as of the sources determines the rate of translocation at any given time.[24] While this is to be expected in any case the crucial involvement of $K^+$ in phloem transport[25–27] has not been understood; $K^+$ deficiency reduces transport in a very critical manner. The possibility of the requirement of $K^+$ in phloem loading has been suggested by Malek and Baker.[28] As the considerations related to the sieve tubes alone seem to be of little help in finding the driving agency for the transport of photosynthates, in the next section we will look into the possible role of the sinks and the sources in the transport process.

## 2.4. Coupling of the Long-Distance Transport to Cellular Metabolism

It is difficult to conceive that the plant tissues merely take up materials as and when these are transported to them by the operation of such agencies as a pressure gradient. Whether it is water and minerals for the leaves or photosynthates for the sinks, the receiving tissue should be able to control the rate of supply according to its demand. The best way of achieving this is for the tissue to control the mechanism of transport itself, just as an isolated cell regulates its supplies from the medium by the active uptake processes operating in its membrane.

Active transport processes in cellular membranes are well understood in terms of Mitchell's chemiosmotic theory.[29–31] The idea that, directly or indirectly, active transport is energized by the electrochemical gradient of protons which is generated by $H^+$-ATPase, and that the solute molecules are taken up by symport and antiport processes with $H^+$ as the working ion, has been fully confirmed in the case of plant cells as well.[32] The electrochemical gradient of $H^+$, $-\Delta\mu_H$, which, divided by the Faraday constant $F$, is called the proton motive force or pmf (pmf $\equiv -\Delta\mu_H/F$), is generated by the active efflux of $H^+$ brought about by the hydrolysis of ATP at the internal membrane surface. In the context of the above paragraph we may now pose the question: If instead of a cell in a nutrient medium we take the case of a tissue which is connected by a (as yet unspecified) channel to a source of nutrients at a distance from it, then how could an active extrusion of $H^+$ by the cells of the tissue be utilized to drive the nutrients to the tissue? The $H^+$ extruded by the "sink" tissue (collectively) must then be able to operate at the site of the "source." In the case of the translocation of photosynthates in plants the connecting channel is the symplasmic pathway of the sieve tubes. Thus we may look for a mechanism of long-distance perimembrane transport with $H^+$ as the working ion.

This approach, while reasonable from the point of view of bringing about a unification of cellular and long-distance transport, can be further justified. Giaquinta[33] has given an account of experimental findings in support of an $H^+$ efflux mediated uptake of sucrose by the sieve tube–companion cell complex in the leaf veins. Artificially imposed alkalinity and low $K^+$ levels have a drastic negative effect on phloem loading. Surely the companion cells would be operating their $H^+$ pumps for the active uptake of sucrose, but this alone cannot explain the extraordinary rates of sucrose uptake by these cells, the dependence of the rate on the demand (size and rate of respiration) of the sink, and the curious situation that the sucrose taken up by these cells is passed on further into the sieve tube system. Clearly then the flux of $H^+$ extruded by the companion cells in the leaf veins is mainly coming from the sink tissue.

There are other reasons which make the above suggested flux of $H^+$ from the sinks to the source via the symplasmic pathway of the sieve tubes a realistic possibility. Smith and Raven[34] analyzed the problem of maintaining neutral pH in the cytoplasms of the heterotrophic and the autotrophic cells of plants. In the former (sink) cells, growth processes supported by respiration with hexose as energy source lead to the production of an excess of acidic groups over basic groups. Although the vacuoles can serve to some extent as reservoirs of $H^+$, a net efflux of $H^+$ from the cell to the external medium is considered to be the only long-term solution. As the Donnan free space (DFS) of the cell walls is also of restricted capacity, the $H^+$ efflux into the apoplasm constitutes a large expenditure of metabolic energy. In the case of the sink tissues of plants, however, there is the possibility of the removal of $H^+$ by the symplasmic pathway provided by the sieve tubes. At the other end of this pathway we have the autotrophic tissues of the leaf mesophyll. These face a grave problem of pH regulation of the opposite nature. Both photosynthesis and nitrate reduction produce an excess of $OH^-$ groups. Either $OH^-$ have to be pumped out or $H^+$ taken up to keep cytoplasmic pH neutral. Isolated leaves are known to make the medium around them alkaline[34] just as the roots do the opposite. The problem of both types of tissues can be easily solved in an energetically feasible manner if $H^+$ are assumed to move from the sinks to the sources via sieve tubes. In this manner the problem of pH regulation in plants is transformed into a potential solution of the transport problem. The sinks and the sources constitute two giant electrodes where $H^+$ and $OH^-$ are produced in different metabolic processes. Between the two tissues a pmf is thereby generated which is capable of delivering a power output $= J_H \cdot \text{pmf}$, if the proton circuit can be completed to permit the flux $J_H$ of $H^+$ to occur from the sink to the source. The clue as to how to complete the circuit is already available from the transmembrane transport in which $H^+$ flux is charge compensated by anions and cations which move in and out of the cell. We have merely to investigate which of the two possibilities is realized in the long-distance perimembrane transport.

## 3. Proto-Osmosis in Organismal Capillaries

### 3.1. Construction of an Organismal Capillary

An organismal capillary (OC) will be defined as a capillary having the following two properties:

1. The internal surface of the capillary bears excess negative fixed charges when in contact with an electrolyte at neutral pH.
2. The internal surface of the capillary is a good conductor of proticity.

The terms "proticity" and "conductor of proticity" have come into common usage in biology since the advent of Mitchell's chemiosmotic theory. Here it would suffice to say that proticity consists in a movement of protons (and not $H_3O^+$, etc.), and its flow is possible in hydrogen-bonded crystals such as ice and in hydrogen-bonded chains both inside and at the surfaces of macromolecular assemblies.[36-38] The process of $H^+$ transport in these proton conductors consists of a forward proton hop followed by a turning of the bonding defect. These two processes must strictly alternate and an accumulation of $H^+$ cannot occur along the proton conductor. $H^+$ may enter the conductor and get ejected at the other end of the proton conductor due to two chemical reactions which produce and consume $H^+$ at the ends of the conductor or by an imposed pH difference. It is essential for the operation of an OC that this injection–ejection mechanism is adopted so that the density $\sigma^-$ of the negative fixed charges remains unaffected by the flow of proticity.

A possible realization of an OC is to construct a symplast, shown in Figure 1, consisting of two cells, $M_0$ and $M_1$, joined by a plasmodesma without a desmotubule as the OC. The bathing media, $B_0$ and $B_1$, may contain a 1 mM KCl solution and are kept at a low and relatively high pH, respectively. The symplast contains a 100 mM solution at a pH of 7. The cell membranes are assumed to be permeable to $K^+$, $Cl^-$, and water. The whole internal surface of the symplast is assumed to be a continuous proton conducting system. Protons are supposed to enter at the external surface of $M_0$ at a proton deficient defect[38] and cross into the internal surface of $M_0$. At the other side of the symplast $H^+$ cross through the membrane of $M_1$ and go into $B_1$ via an excess proton defect. In this manner a flux of $H^+$ can occur from $B_0$ to $B_1$ down the pmf $[=2.3\,(RT/F)\,\Delta\mathrm{pH}]$ if some arrangement is made to complete the circuit within the symplast since $B_0$ and $B_1$ are not physically linked otherwise.

The internal surface of the OC may be conceived, to get a geometrical picture, as covered by a few layers of structured water while the polar residues of the phospholipids protrude further out, in the bulk phase of the

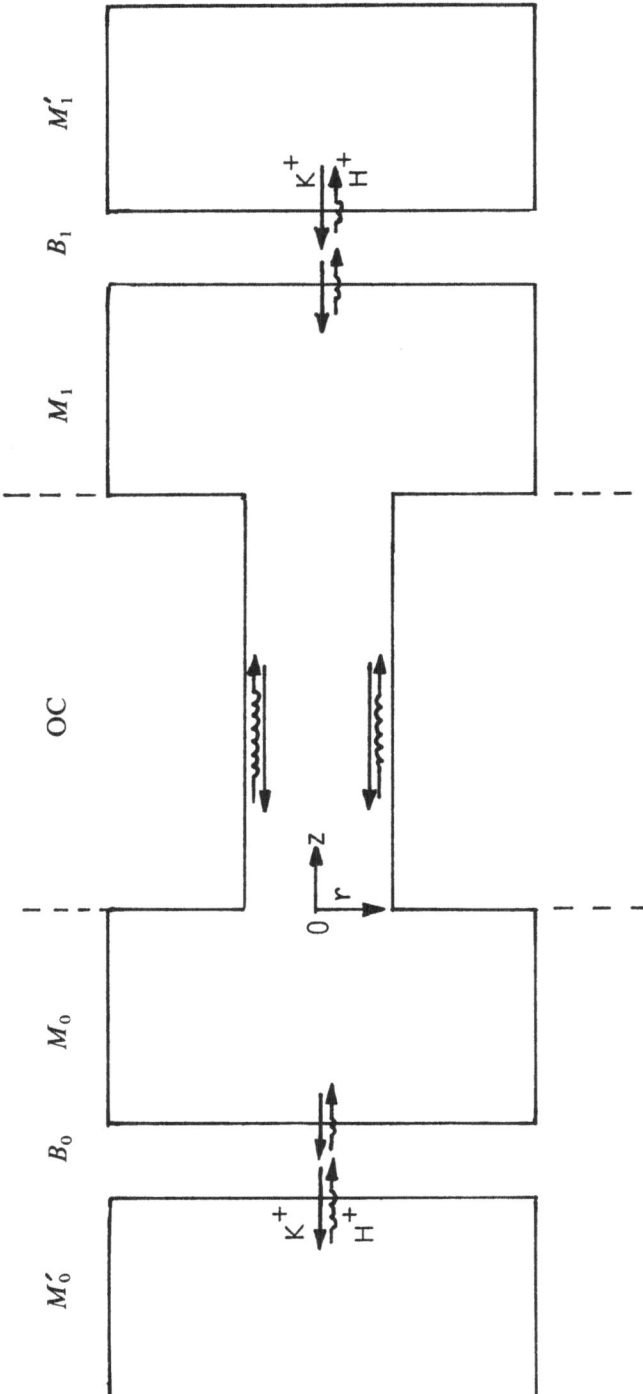

FIGURE 1. A basic transport unit that has been successively constructed beginning with the bathing media, $B_0$ and $B_1$, and a bicellular symplast with a plasmodesma. See text for further details.

lumen. The quantitative description in the next section is based on this picture in order to arrive at the essential conclusions in as simple a manner as the author could manage. But it should not be taken to mean that this realization of the surface is necessarily true at the molecular level.

### 3.2. The $H^+/K^+$ Counterflux

In the OC of radius $a$ and extending from $z = 0$ to $z = 1$, the proton conducting layers of structured water have a thickness $\varepsilon$ and occupy the region $R_H : a - \varepsilon < r \leqslant a$ with an area of cross section $A_H$. $\sigma^-$ is located at the cylindrical surface $r = a - \varepsilon$. The counterions, $K^+$, are assumed to be localized in a region $R_K : (a - \varepsilon) - d \leqslant r < a - \varepsilon$, of cross-sectional area $A_K$, where $d$ is the thickness of the screening $K^+$ layer. $Cl^-$ will be excluded from $R_K$ so that the net charge in $R_K$ due to $K^+$ is equal to $2\pi a l \sigma^-$ while in the region $R_L : r < (a - \varepsilon) - d$ the concentrations of $K^+$ and $Cl^-$ are equal.

A steady flux of $H^+$, $J_H$, across the OC, en route from $B_0$ to $B_1$, can occur only if it is charge balanced by either a flux of $Cl^-$ in the same $+z$ direction or a counterflux of $K^+$ in the $-z$ direction or an appropriate flux of both $K^+$ and $Cl^-$. Owing to the anisotropic ionic distribution in the OC a forward flux of $H^+$ in $R_H$ generates an electric field which has a large $E_z$ component in $R_H$ and $R_K$ and a lesser magnitude in $R_L$. Furthermore, the screening $K^+$ layer would further reduce $E_z$ in $R_L$ by moving laterally under the influence of the electric field. This shielding effect will depend on the concentration of $K^+$ in $R_K$ and hence on $\sigma^{-}$ [39] as well as on the mobility $u_K$ of the $K^+$ ions in the region $R_K$. Assuming, for simplicity, that the shielding is perfect, the condition of electroneutrality requires that the flux of $K^+$, $J_K$ (restricted to $K^+$ in $R_K$) must be equal in magnitude and opposite in direction to $J_H$.

The conductance of the proton-conducting $R_H$ region, $g_H$, can be taken as independent of $z$. Writing $\Delta V = V(l) - V(0)$ for the diffusion potential across $R_H$ we have

$$J_H = g_H(\Delta V_H - \Delta V) \tag{1}$$

where

$$\Delta V_H = -(RT/F) \ln[c_H(l)/c_H(0)] \tag{2}$$

which would be roughly equal to $-2.3(RT/F)[\text{pH}(B_l) - \text{pH}(B_0)]$ if the conductance of the rest of the surface of $M_0$ and $M_1$ is much greater than $g_H$. Henceforth the involvement of $M_0$ and $M_1$, which were introduced to provide the injection–ejection mechanism of $H^+$ in the OC, will be omitted and it will be assumed that $H^+$ get properly introduced in the OC from the bathing medium directly. The concentration of KCl on the two sides of the OC will be taken as variable, but $\Delta V_H$ will be kept fixed.

For $J_K$ the Nernst–Planck equation may be written as

$$J_K = -u_K c_K A_K F[(RT/F) \, d \ln c_K/dz + dV/dz]$$ (3)

If $E_z$ is not too large (as, for example, would be the case if the OC were an aquous pore in a membrane, which is discussed in Section 3.4), the concentration of $K^+$ in $R_K$ will be determined by $\sigma^-$ and therefore independent of $z$ apart from a narrow region at $z = 0$ and $z = 1$. The integral over $z$ may thus be split into three parts, from $z = 0$ to $\Delta z$, from $\Delta z$ to $l - \Delta z$, and from $l - \Delta z$ to $l$. The integration of the diffusion potential gradient gives

$$\int_0^1 c_K(z) \frac{d \ln}{dz} c_K(z) \, dz = c_K(\Delta z)[\ln c_K(\Delta z) - \ln c_K(0)] + 0$$

$$+ c_K(l - \Delta z)[\ln c_K(l) - \ln c_K(l - \Delta z)]$$

$$= c_K \ln c_K(l)/c_K(0)$$

since $c_K(\Delta z) = c_K(l - \Delta z) = c_K$. In the integration of $c_K(z) \, dV/dz$, the contribution of the terminal segments becomes neglible for $\Delta z \ll 1$. For the steady state flux of $K^+$ we obtain then

$$J_K = g_K(\Delta V_K - \Delta V)$$ (4)

where

$$\Delta V_K = -(RT/F) \ln c_K(l)/c_K(0)$$ (5)

and

$$g_K = u_K c_K F A_K/l$$ (6)

In Eq. (3) and onwards the dependence of $u_K$ and $c_K$ on the radial variable $r$ has been ignored and these values have been considered as averaged over $r$. The $r$ dependence of these quantities will be discussed in the next Section.

From $J_K = -J_K$ it follows that

$$V = (g_H \Delta V_H + g_K \Delta V_K)/(g_H + g_K)$$ (7)

This is the usual Planck–Henderson expression for the diffusion potential. In the present situation it is containing the contribution of the protons and $K^+$ only. Eliminating $\Delta V$ from Eqs. (1) and (4) we have

$$J_K = g_H g_K(\Delta V_K - \Delta V_H)/(g_H + g_K)$$ (8)

### 3.3. Proto-Osmosis

Due to the viscous coupling between $K^+$ and water molecules, there will be a flux of water and salt in the lumen. Whether a pressure gradient $[p(0) - p(l)]/l$ is generated by such a flux or it is imposed across the OC, it will in turn also bring about a flux of $K^+$ and generate a streaming potential difference $\Delta V_s$. Furthermore, the protons in $R_H$ will also react to the flux of $K^+$ due to pressure flow via $\Delta V_s$. Therefore all the three fluxes, $J_H, J_K$, and the flux of water, $J_W$, are interlinked.

To solve this, rather intricate problem a subdivision of $R_K$ is undertaken into an outer layer $R_K^0$ close to the surface: $r = a - \varepsilon = a_0$, and an inner layer $R_K^i$, with the radial extensions of $a_0 - d_0 \leqslant r < a_0$ and $a_0 - d \leqslant r < (a_0 - d_0)$, respectively. It is then assumed that the $K^+$ in $R_K^i$ can move laterally due to a flow of water in the lumen but those in $R_K^0$ will not. This may be justified by noting that the viscosity $\eta$ would be higher in $R_K^0$ and the interaction of $K^+$ with the fixed charges will be stronger. Both $K^+$ layers are, however, assumed to move under the influence of the electrochemical gradient of $K^+$. The mobilities of $K^+$ in the two regions will be denoted by $u_K^0$ and $u_K^i(u_K^i > u_K^0)$, and the average concentrations by $c_K^0$ and $c_K^i(c_K^0 > c_K^i)$, respectively.

Neglecting the fluxes of $Cl^-$ and $K^+$ in $R_l$ due to $\Delta V_s$ (which again acts maximally at $R_K$ and $R_H$) the equations of the three fluxes are written as

$$J_H = g_H(\Delta V_H - \Delta \bar{V}) \tag{9}$$

$$J_K = g_K(\Delta V_K - \Delta \bar{V}) + \alpha F \Delta p \tag{10}$$

$$J_W = g_W \Delta p + \beta(\Delta V_K - \Delta \bar{V}) \tag{11}$$

Here $\Delta p = p(0) - p(l)$, and $\Delta \bar{V}$ includes the contributions both of the streaming potential and the diffusion potential difference across the OC. $\Delta \bar{V}$ and the coupling constants $\alpha$ and $\beta$ have now to be determined.

The general solution for the velocity of a viscous fluid of viscosity $\eta$ in a capillary of radius $a_0$ is given by[40]

$$v_W(r) = -(\Delta p/4\eta l) r^2 + C \ln r + D \tag{12}$$

$C = 0$ because the velocity must remain finite at $r = 0$. At $r = a_0$ there is the flow of water associated with the movement of the $K^+$ of $R_K^0$ so that one can put $v_W(a_0) = s v_K^0$ where $s \leqslant 1$ and $v_K^0$ is the velocity of $K^+$ in $R_K^0$ due to the electrochemical gradient $(\Delta V_K - \Delta \bar{V})/l$. Hence

$$v_W(r) = (\Delta p/4\eta l)(a_0^2 - r^2) + s v_K^0 \tag{13}$$

The flux of water is obtained by integrating the above equation from $r = 0$ to $r = a_0$, over the area of the OC,

$$J_W = (\pi a_0^4 / 8\eta l)\, \Delta p + \pi a_0^2 s v_K^0 \qquad (14)$$

Since $v_K^0 = J_K^0 / F c_K^0 A_K^0$, $J_K^0 = g_K^0(\Delta V_K - \Delta\bar{V})$, and $g_K^0 = u_K^0 c_K^0 A_K^0 F / l$,

$$J_W = (\pi a_0^4 / 8\eta l)\, \Delta p + (\pi s a_0^2 u_K^0 / l)(\Delta V_K - \Delta\bar{V}) \qquad (15)$$

Comparing Eq. (15) with Eq. (11) we get

$$g_W = \pi a_0^4 / 8\eta l \qquad (16)$$

and

$$\beta = \pi s a_0^2 u_K^0 / l \qquad (17)$$

In order to evaluate $\alpha$, the flux of $K^+$ in $R_K^i$ solely due to $\Delta p$ has to be determined. The velocity of $K^+$ in the inner region can be ascertained, up to a constant of proportionality, $t$, from Eq. (13) with $v_K^0 = 0$, as the velocity of water at $r = a_0 - d_0$. Therefore,

$$v_K^i = t v_W(a_0 - d_0) = (2 a_0 d_0 - d_0^2)(t / 4\eta l)\, \Delta p \qquad (18)$$

Though $t < 1$, its value is expected to be less than that of the parameter $s$ since the latter refers to the region $R_K^0$ where the viscosity $\eta$ is greater.[41] Now

$$J_K(\Delta p) = J_K^i = A_K^i c_K^i v_K^i \qquad (18a)$$

Using Eq. (18) and comparing the result with Eq. (10) we get,

$$\alpha = t(2 a_0 d_0 - d_0^2) A_K^i c_K^i / 4\eta l \qquad (19)$$

Since $J_W$ is electroneutral, $J_H = -J_K$ and Eqs. (9) and (10) together determine the net potential difference:

$$\Delta\bar{V} = (g_H\, \Delta V_H + g_K\, \Delta V_K + \alpha F\, \Delta p)/(g_H + g_K) = \Delta V + \Delta V_s \qquad (20)$$

Writing $T_H$ for the transference number of $H^+$, $[T_H = g_H/(g_H + g_K)]$ and eliminating $\Delta\bar{V}$ from Eqs. (9)–(11) the final equations for proto-osmosis are written as

$$J_H = g_K T_H(\Delta V_H - \Delta V_K) - \alpha T_H F\, \Delta p \qquad (21)$$

$$J_K = -J_H \qquad (22)$$

$$J_W = \beta F T_H(\Delta V_H - \Delta V_K) + (g_W - \alpha F \beta T_H / g_H)\, \Delta p \qquad (23)$$

The results obtained in the last two sections will now be discussed and summarized. In Section 2.4 a question was raised whether it is possible to have a mechanism by which a pmf between two cells can be used to drive nutrients towards the cell which extrudes $H^+$ actively. Using the concept of proticity we have arrived at the conclusion that it is possible to achieve this if the path is an organismal capillary. While the process of injection and ejection of $H^+$ in relation to the cells of plant tissues will be discussed in Section 5, the events within the OC have been worked out in this section. It has been shown, that the flow of proticity in one direction induces a flux of $K^+$ at the internal surface of the OC in the opposite direction. In an ideal case, which corresponds to a high surface conductance of the $K^+$ layer, the anions on the two sides of the OC stay where they are, merely exchanging the counterions, $H^+$ and $K^+$, at the sides where the $H^+$ and $K^+$ conductors terminate.

The flow of the $K^+$ layer at the internal surface of the OC drives a flux of water and an equal number of $K^+$ and $Cl^-$ in the lumen (i.e., an electroneutral flux) in the opposite direction to the original $H^+$ flux. A net flow of water can ensue even against a pressure gradient till the pressure difference is smaller than $\Delta p(\max)$, which can be obtained from Eq. (23) by putting $J_w(\Delta p(\max)) = 0$. The peripheral $K^+$ layer and the adjacent layers of water move counterdirectionally to the flow of water in the OC caused by an adverse $\Delta p$ up to a value of $r$, $r_0$, as given by the condition $v_w(r_0) = 0$, Eq. (13). For all values of $\Delta p > p_{bd}$, where $p_{bd}$ is obtained by solving Eq. (13) for $v_w(r=0) = 0$, there is a bidirectional flow in the capillary. The value of $r$ at which the velocity vector changes its direction can also be determined using Eq. (13).

Because of the coupling between the fluxes it is possible to drive all three with any one of the forces. It should be noted that it is only when there is a balance of all the electrical forces that a normal flow, given by the Poiseuille equation,[40] is obtained. When $\Delta V_K \neq \Delta \bar{V}$, a reduction in the hydrodynamic conductivity is understandable in terms of the existence of the streaming potential $\Delta V_s$, Eq. (20), and its retarding influence on the movement of the $K^+$ layer. However, as the conductances of $R_H$ and $R_K$ are high in the OC, the streaming potential is expected to be nearly zero, since it is inversely proportional to the sum of these conductances, Eq. (20).

## 3.4. $K^+$ Antiport by Proto-Osmosis

Proton-motive force mediated $K^+$ antiport is a major transport process in the membranes of plant cells. $K^+$ exchanged for $H^+$ serve to maintain cytoplasmic pH neutral and promote the activities of a large number of enzymes.[42] $K^+$ antiport plays a vital role in the uptake of water by the roots,[43,44] in turgor regulation of stomatal guard cells,[45] and in phloem loading.[27] Having analyzed the case of macroscopic organismal capillaries with the conclusion that a $H^+/K^+$ counterflux occurs in them, it is reasonable to investigate the possibility that $K^+$ antiport in the membranes works in a similar manner. If we take the limit of $R_L$ going to zero in the OC described in Section 3.1, and choose $a_0$ equal to the radius of a hydrated $K^+$ ion, we obtain a $K^+$ selective pore with a lining with proton conducting property.

$H^+$-ATPases are implicated in most of the well-studied symport and antiport processes in plant cells[46] acting as electrogenic proton pumps. Using these as an $H^+$-injecting device operating close to or as a part of the $K^+$ selective pore, one can also fulfill the injection condition imposed on an OC. The ejected protons may (1) reenter the cytoplasm, (2) get neutralized at the external membrane surface, or (3) move on in another OC connected to the cell surface with a proton conductor. Our present interest lies in the fact that $K^+$ would get actively transported into the cell during the outward movement of $H^+$ through the pore. This implies that the condition of the preservation of electroneutrality is being applied within the pore and thus the inward movement of $K^+$ could be called electroperistaltic transport.

As the length $l$ of the OC takes small values the electric field $E_z$ becomes increasingly large, and for a membrane of 10 nm thickness, a potential difference of 100 mV gives $E_z$ of the order of $10^5$ V/cm. Thus one has to consider the possibility that $c_K = c_K(E_z, z)$ in the Nernst–Planck equation, Eq. (3). Depletion of ions from $Na^+$ and $K^+$ pores in the neuronal membrane has been considered as a possibility.[47] In the present context the treatment of Section 3.2 would remain unaltered if the depletion is assumed to be confined to the middle region of the pore, while $K^+$ in the outer regions of the pore-forming proteins remain associated with the excess of negative fixed groups. It may be recalled that a concentration gradient in the central region was not assumed for the derivation of Eq. (4). Furthermore, even if the $K^+$, which may occur in the central region at $E_z = 0$ in association with some negatively charged groups, get removed at high $E_z$, the conductance $g_K'$ of the pore does not vanish. The $c_K'(E_z, z)$ would in that situation refer to the time-averaged value of the $K^+$ in transit through the central region of the pore. Thus the basic equations of the $H^+/K^+$ counterflux, Eqs. (7) and (8), can be retained if allowance is made for the fact that $g_K' = g_K'(\bar{V}')$. The potential dependence of $g_K'$ would also be assumed to take care of the changes in $K^+$ mobility $u_K'$ in case the pore undergoes any conformational changes at some large value of $\Delta \bar{V}'$.

In the case of the $K^+$ selective pore, we are left only with the regions $R_H$ and $R_K^0$ so that the coupling between $K^+$ flux and water flux cannot be the same as postulated in Section 3.3. For the pore we may provide a simple coupling by supposing that all water molecules within the pore move into the cell for each $K^+$ crossing the full length of the pore. Thus if $\lambda$ moles of water cross per mole of $K^+$, $J_W = \lambda J_K/F$.

In order to study the behavior of the $K^+$ pores it is assumed that all other transport processes in the membrane apart from the $K^+$ antiport are switched off. The equations for the fluxes are

$$J_H = g'_H(\Delta V_H - \Delta \bar{V}') \tag{24}$$

$$J_K = g'_K(\Delta V_K - \Delta \bar{V}') - \alpha' \Delta \mu_W \tag{25}$$

$$J_W = -g'_W \Delta \mu_W + \beta'(\Delta V_K - \Delta \bar{V}') \tag{26}$$

To simplify the analysis, the cell is first allowed to equilibrate its water chemical potential $\Delta \mu_W = \mu_W(1) - \mu_W(0)$ occurring in Eq. (26)[7] and achieve a turgor pressure $p_0$ above the atmospheric pressure corresponding to its osmotic potential. As the $H^+$ pump is switched on, the flux of water into the cell will raise its turgor pressure above $p_0$. This pressure difference $p - p_0$ will be written as $\Delta p$, and in the above equations (25) and (26) $\Delta \mu_W$ will be replaced by $\bar{V}_W \Delta p$, where $\bar{V}_W$ is the partial molal volume of water.[7]

Proceeding on lines similar to those in Section 3.3 the coupling constants are found as $\beta' = \lambda g'_K/F$ and $\alpha' = Fg'_W/\lambda$. The diffusion potential, including the streaming potential term is, for the pore, given by

$$\Delta \bar{V}' = (g'_H \Delta V_H + g'_K \Delta V_K + F'g_W \bar{V}_W/\lambda)/(g'_H + g'_K) \tag{27}$$

and the three fluxes as

$$J_H = -J_K = g'_K T'_H(\Delta V_H - \Delta V_K) - FT'_H g'_W(\bar{V}_W/\lambda) \Delta p \tag{28}$$

$$J_W = (\lambda/F) J_K \tag{29}$$

The physiological content of these equations will be taken up in Section 4.1, where the turgor regulation in stomatal guard cells has been discussed.

### 3.5. A Basic Unit of Plant Transport

In Figure 1 the bicellular symplast was considered to be interposed between two bathing media at different pH. $M_0$ and $M_1$ were used for providing the injection–ejection mechanism, but nothing further was mentioned about their involvement in the context of the transport through the OC. This description will now be completed by the inclusion of a cell $M'_0$ which is equipped with the $K^+$ antiport device, and the basic mechanism of plant transport will be elucidated in the framework of the present approach.

The bathing medium $B_0$ will now be identified with the cell wall phase between $M_0$ and $M'_0$. Composed of macromolecules with a substantial excess of negative fixed charges[2] $B_0$ can be looked upon as a proton conductor on lines similar to those suggested for membrane surface. The bathing medium $B_1$ is at pH of 7 and contains, apart from a dilute KCl solution, a neutral solute $S$. The cell membranes of $M_0$ and $M_1$ will be assumed to have $K^+$ selective pores with the proton-conducting property but without a $H^+$-ATPase. The neutral solute $S$ is assumed to be able to pass through the $K^+$ selective pores of all cell membranes in the system. The concentrations of the solutes are adjusted in such a manner that the osmotic potential (including the matric potential[7]) is the same in all regions at the beginning of our *Gedanken experiment* to be described below.

The proton extrusion pump of $M'_0$ is switched on. Taking a low value of $V_K$ and the initial (also low) turgor pressure $p_0$, Eq. (28) predicts a large proton flux from $M'_0$ to $B_1$. (The coefficients in the equation as well as the chemical potential differences refer to the path and the end points between $M'_0$ and $B_1$.) $H^+$ will cross the apoplasmic gap $B_0$ to enter the inner surface of $M_0$ via $K^+$ pores, move along the continuous internal surface of the symplast, and reach $B_1$ through the $K^+$ pores of $M_1$. The same path will be followed by $K^+$ in the reverse direction, up to the cytoplasm of $M'_0$. As a consequence of the $J_K$, water and $S$ will move into the symplast and water with a fraction of $S$ which entered $M_1$ will reach the cell $M'_0$. In effect $M'_0$ has been able to suck in water and imbibe $S$ from $B_1$ using the symplast as a straw. The aspects concerning the role played by the OC will be discussed in later sections for the different types of OC occurring in the different parts of the transport network. It suffices here to remark that their job is essentially to enhance water and solute flux against a pressure gradient which may exist across the symplast. The desmotubules lower the hydraulic conductivity of plasmodesmata while the surface mediated flow (with the $K^+$ layer) can continue. If devoid of the desmotubules the plasmodesmata will bring about a (bidirectional) convection flow within the symplast and improve the transfer of $S$ from $M_1$ to $M_0$.

The influx of $K^+$ in $M'_0$ will continue till $c_{el}(\Delta V_H - \Delta V_K) = c_{pr}\Delta p$, where $c_{el}$ and $c_{pr}$ are the coefficients of the respective terms in Eq. (28). The cell has then a high concentration of $K^+$ and a high turgor. The difference $c_{el}(\Delta V_H - \Delta V_K) - c_{pr}\Delta p$ is thus a measure of the thirst of a cell. The three quantities $V_H$, $V_K$, and the turgor pressure $p$ of a cell can be defined with respect to some standard set of values, such as $pH = 7.1$ mM KCl, and 1 atm, and used for the purpose of determining the cell's behavior in respect of exchange of materials with any given medium connected to it by an organismal capillary. This would certainly be an incomplete description, since it ignores the contribution of all ionic transport processes other than those of $H^+$ and $K^+$. It will nonetheless be useful as a fair approximation to the actual situation since $K^+$ transport is by far the most prominent among other ions. Though the conductances $g'_K$, $g'_W$, and $g'_H$ were used for the $K^+$-selective pore with an $H^+$-ATPase associated with it, the same set of equations, viz., Eqs. (27)–(29), is applicable to the case when the $K^+$ pores without $K^+$ antiport device are included in the description of the membrane transport. $K^+$ flux in these channels would be due to the difference $\Delta V_K - \Delta V_m$, where $\Delta V_m$ is the potential difference across the membrane. The membrane potential $\Delta V_m$ also displays an electrogenic term,[2] but that need not be identical to $\Delta \bar{V}'$, given by Eq. (27), in view of the fact that there are several other antiport and symport processes taking place within the cell, and most of them are known to be mediated by the electrogenic proton pumps.

## 4. Turgor Regulation

### 4.1. Stomatal Guard Cells

Transport processes in plants are intimately related with the aspect of turgor regulation of the cells and symplasms of the diverse plant tissues. In this section the aim is to show that the process of turgor regulation can be understood to a large extent with the $K^+$ antiport mechanism discussed in Section 3.4, by relating it to the processes of pH regulation in the tissues (Section 2.4). This will lead to the recognition of a few basic principles which can then be used to explain the exchange of substances between the tissues and the uptake of water and nutrients by plants.

Turgor regulation in a cell capable of photosynthesis has been best studied in stomatal guard cells. Having functional chloroplasts, these cells are expected to experience the problem of pH regulation the moment the rate of generation of $H^+$ (coupled to respiration) lags behind compared to the generation of $OH^-$ groups (Section 2.4). This is bound to have an adverse effect on $K^+$ antiport since $H^+$ will now have to be pumped out against an adverse electrochemical gradient. The guard cells are maximally exposed to water loss and also have, relatively, the maximal $CO_2$ tension, i.e., the highest acidity (carbonic acid) in their apoplasm. Not being a part of a symplasm, the guard cells succumb to the excessive strains imposed on them due to their location resulting in the loss of $K^+$ and water.[45] What appears to be most interesting is the gradual fall in the pH of their vacuoles concomitant with the loss of $K^+$ to the subsidiary cells.[2] Obviously, the complex of subsidiary cells would have taken up $K^+$ by the $K^+$-antiport at their membranes. One could interpret the loss of water and $K^+$ by the guard cells as a process of passive leakage, with the charge of $K^+$ balanced by the divalent malate ions or some other anion, but these possibilities are less favored[48] than a coupling of the $K^+$ flux with $H^+$ counterflux. The fact that the pH of the guard cell does fall raises the possibility that the processes—uptake of $K^+$ by the subsidiary cells by $K^+$-antiport and the loss of $K^+$ with a concomitant fall in pH of the vacuoles of the guard cells—are more intimately coupled than could be explained by the passive leakage hypothesis. In other words, the subsidiary cells might have actively injected $H^+$ into the vacuoles of the guard cells and extracted $K^+$ by their $K^+$-antiport system acting across the intervening apoplasm and membranes.

This leads us to the last component in the scheme of Figure 1, namely, the addition of a cell $M'_1$ separated by an apoplasmic gap (in place of $B_1$) from $M_1$. $M'_1$ will be given a low value of $V_H$, which means that its $H^+$ pumps are not fully active for some reason. Now if the differences in the values of $V_H$, $V_K$, and turgor pressure $p$ between $M'_0$ and $M'_1$ favor $K^+$ flux from $M'_1$ to $M_0$, the latter will extract $K^+$ water and neutral solutes from $M'_1$. In other words, a reverse proto-osmosis will occur upon a cell with nonfunctioning proton pumps by another with well functioning proton pumps. The latter will treat the former as a source of $K^+$ water and $S$. In the case of stomatal guard cells the gap between them and the subsidiary cells is small and a fast exchange of substances is possible. The rise in the turgor of the subsidiary cell due to the water extracted from the guard cells brings about the necessary imbalance in the turgor interrelationship of the complex to account for the stomatal closure at the observed fast rates. However, the guard cells have been used here merely to bring out the concept of reverse proto-osmosis, and the possibility will be used later for the sink–source relationship even when the cells or symplasts are at greater distances from each other.

One more point which emerges from the above analysis of the guard cells is the close correlation of the turgor pressure $p$ with the $K^+$ concentration in a cell. Though a simple linear relationship between the two may not exist, one can say that a cell with a high value of $V_K$ will also have a high $p$ value. We may thus define a quantity $V(HK) = V_H - V_K$, called the proto-osmotic potential, for every cell and symplasm in plants and formulate the following rule: If the proto-osmotic potential of one cell is higher than that of another and both are connected by an organismal capillary, then the cell at the higher $V(HK)$ value will extract $K^+$ water and solutes from the other.

## 4.2. Leaf Mesophyll and Sink Tissues

As discussed in Section 2.4, there is a large production of $OH^-$ groups in the mesophyll cells, creating for them the problem of pH regulation. Though these cells are not expected to behave as the guard cells or, in a way, are saved from a similar fate by the timely closure of the stomatal aperture, the activity of their $H^+$ pumps is bound to decrease or become gradually less effective during the day. At the same time the leaves keep on obtaining $K^+$ from the xylem along with the transpiration stream so that their $V_K$ remains high. Thus in general one can assign a low value of $V(HK)$ to the mesophyll.

The sink tissues on the other hand have a continuous overproduction of $H^+$ in their cytoplasm. Along with the ATP energized $H^+$ pumps these would tend to extrude $H^+$ to the apoplasm via the labyrinth of surface paths between the tonoplast and plasmalemma (which cannot be denied the proton conduction property) as long as there is a region of low $V_H$ (high pH) accessible to them. However, the $H^+$ extrusion, whichever way it may occur, will be reduced when the rate of respiration is low or when the cell is generally impoverished. In both respects, the tissues of the bark are better off than those of the stele as these are closer to the source of nutrients and oxygen. We may therefore assign higher values of $V(HK)$ to the phloem parenchyma than to the xylem parenchyma. The time-dependent variation of the $V(HK)$ values between the tissues of the leaves and those of sinks are given in the next section.

## 5. Long-Distance Transport in Plants

### 5.1. Organismal Capillaries in the Plant Transport Systems

The basic transport unit of Fig. 1 will now be generalized to encompass the diverse medium- and long-distance pathways and then used for describing transport in the network by assigning $V(HK)$ values to the different tissues.

In the basic transport unit of Fig. 1, $M_0$ and $M_1$ were required to be permeable to $K^+$ and water and to enable $H^+$ to enter them and move along their internal surfaces by the specific transport mechanism described in Section 3.1. In Section 3.5, the apoplasm between $M_0'$ and $M_0$ was also assumed to show proton conduction. Thus two xylem conduits with their walls in contact with each other constitute, together with the pits between them as the organismal capillaries, a variation of the basic transport unit. The wall phase without the pits may be considered as having an appreciably lower conductance for $H^+$, $K^+$, and water, while the pits form a number of parallel arranged organismal capillaries. The conduction system of the xylem is then of the type $M_0-n'OC-M_1-n''OC-\cdots M_x$, or a multiplet of (generalized) transport units, which will be abbreviated as xTU. Other instances of transport units with parallel arranged organismal capillaries are $K^+$ pores in membranes, sieve plate pores, and several plasmodesmata between two cells. For each transport pathway all the variations in its transport units must be taken into account for evaluating the effective conductance.

As described in Section 2.1, the xylem conduits communicate not only with each other but also with the tissues around them. A number of lateral transport systems abut against the axially arranged xylem conduits, which will not be described here. However, the possibility of a lateral movement of the xylary sap will be utilized below for the qualitative description of the transport of water in plants.

The transport system of the sieve tubes is also a multiplet of generalized transport units (pTU) with the sieve plate pores as organismal capillaries. Though the sieve tubes are thin as such, the dependence of the hydraulic conductivity, $g_W$, on the fourth power of the radius, makes the distinction between them and the pores essential. The effects of occlusion on $g_W$ are difficult to calculate but the general features of the sieve tube system permit one to surmise that it has been made to cope with adverse hydrostatic gradients of variable but often large magnitudes.

As in the case of the xylem conduits, the sieve tubes communicate with the cells and tissues around them. Lateral transport systems, including the pathways for the exchange of substances between them and the xylem conduits,[49,50] may be visualized as lateral branches of the long distance pathways (xTU and pTU). These offshoots (lTUs) link all regions of the cross section of the root and the shoot with the xTU and pTU.

### 5.2. Transport in Xylem

The main result of Section 3.3 was that a flux of water occurs from a cell $M_1'$ to $M_0'$ when the difference in the proto-osmotic potential $\Delta V_H - \Delta V_K = \Delta V(HK)$ between the two favors a flux of $H^+$ from $M_0'$ to $M_1'$. Equation (23) may be more conveniently written as

$$J_W = -C_{el}\,\Delta V(HK) + C_{pr}\,\Delta p \tag{30}$$

where $C_{el}$ and $C_{pr}$ are, respectively, the coefficients of the electric potential difference and pressure difference in Eq. (23). If the transport unit of Fig. 1 is kept vertically, with $M_0'$ at a height $\Delta h$ above $M_1'$, the effective pressure difference across the unit is $\Delta p + \rho g\,\Delta h$, where $\rho$ is the density of the fluid in the transport unit. Therefore, the cell $M_0'$ can lift solutes and water towards itself, proto-osmotically, up to a height

$$\Delta h = [C_{el} \Delta V(HK) - C_{pr} \Delta p]/C_{pr} \rho g \qquad (31)$$

If the transport unit is placed horizontally then $M_0'$ can develop a pressure difference

$$\Delta p = (C_{el}/C_{pr}) \Delta V(HK) \qquad (32)$$

In general, given any source of water $S(i)$ at a given set of values, $p$, $V_H$, and $V_K$ and at a difference of height $h(i)$, then a plant cell $(l)$ connected to this source of water by a generalized transport system, will cause a water influx

$$J_W(S(i)) = -C_{el}(i, l) \Delta V(HK) + C_{pr}(i, l)[\rho g \Delta h(i, l) + \Delta p] \qquad (33)$$

where $C_{el}$ and $C_{pr}$ refer to the electrical and hydraulic conductances of the generalized transport pathway separating the cell from the source of water. The cell $(l)$ can belong to any tissue of the plant and need not be a leaf mesophyll cell. $S(i)$ can be other cells of the plant or the ground water. The total influx into the cell is the sum of $J_W(S(i))$ extending over all the plant cells and ground water.

In the case of the leaves, water influx is raised by the lowering of their turgor pressure by transpiration. This larger influx compensates for the loss of water by transpiration, the efflux of phloem sap, and is utilized for the growth of the leaf cell itself. the flux of water from the ground is favored by the low $V(HK)$ value of ground water but disfavored by $\Delta h$ and low $C_{el}$ value for xTU. The cells of the leaf mesophyll would therefore be expected to imbibe water from the sink tissues located near them and having high $V_K$ and $p$ (turgor pressure) values. Thus, strictly speaking, water uptake by the leaves involves all transport channels in plants and not only the xylem. The ascent of sap is comparable to the operation of a series of low-powered pumps, raising water in stages from a low-lying large reservoir to a high altitude, with the help of small water reservoirs all along the path. The tissues at all heights act as the small reservoirs and the lateral pathways lTU serve to connect these tissues with the major transport pathway of the xylem (xTU). In principle Eq. (33) can be used to describe the transport of water in plants by putting the time-dependent values of $V(HK)$ of all cells in the equation and summing over the whole tissue of a plant.

Variations in the value of $V(HK)$ of the leaves are expected due to the problem of pH regulation in the mesophyll cells discussed in Section 2.4. Around midday $V(HK)$ of the leaves would fall and reduce the rate of uptake of water by these cells. However, the $K^+$-antiport system of the leaves is not expected to suffer all too much since the loss of water to the atmosphere is reduced around this time by the closure of the stomata by the more sensitive guard cells.

One mode of transport not covered by Eq. (33) is that via the Jamin chains. It is possible[51] that between two water columns of these chains there is a thin film of electrolyte adhering to the walls of the xylem conduits by electrostatic forces of the electrical double layers. $H^+/K^+$ counterflux can occur here as well, though the amount of water transported in this manner would be small in comparison to proto-osmotic flow in the filled columns. Nevertheless, as more and more intact columns cavitate during the day over the summer months, this mode of supply would gain importance. It is, presumably, this mode of transport by which trees replenish their water content in autumn after the leaf fall. Cavitation is caused by low pressures generated in xTU by the active water uptake by the leaf cells at a rate exceeding the rate of refilling of the xTU by the sources $[S(i)]$.

If a plant is decapitated at the ground level a large part of the proton flux to the soil is cut off, but that due to the sink tissues of the root system continues to flow. As the casparian strip blocks the apoplasmic pathway from the stele to the cortex, the $H^+/K^+$ counterflux must occur via the radial symplasm. This situation corresponds to the case of the horizontally placed transport unit described earlier in this section. $H^+$ extruded into the stelar apoplasm move into the radial symplasm, cross the endodermis, and move out to the soil via the apoplasm of the cortex. $K^+$ and water follow the reverse path to reach the xylem conduits. The pressure developing in the xylem is given by Eq. (32). This pressure may be identified with root pressure,[6] since it shows the dependence on the rate of respiration and sensitivity to poisons through $V(HK)$. This approach not only explains the need for the casparian strip but also predicts that if the plasmodesmata in the radial symplasm are provided with desmotubules, higher root pressures will result. The extreme sensitivity of the rate of uptake of water on $K^+$ concentration (up to 1 mM) is also explicable due to the coupling of water influx with $K^+$ influx.

## 5.3. Transport in Phloem

The sink tissue of and around the phloem, $R(P)$, may be assigned a relatively higher $V(HK)$ than the average $V(HK)$ values of mature leaves and the sink tissues of and around the xylem, $R(X)$. This may be justified by noting that these have a shorter distance from the source of nutrients (sieve tubes) and also a good supply of oxygen. As the leaves produce excess $OH^-$ groups during the processes of nitrate reduction and photosynthesis, a $\Delta V(HK)$ is generated between the leaves and $R(P)$, causing the latter to execute reverse proto-osmosis on the mesophyll. As a result, $K^+$ water and photosynthates, $S$, flow into the sieve tubes via the companion cells in the leaf veins.

The flux of water and solutes, $S$, this time to a sink cell ($l$) will be again given by Eq. (33), the sources on which the summation has to be carried out in order to get the total influx being all the mesophyll cells in the shoot. For the flow of $S$ in the sieve tubes the values of the coefficients $C_{pr}$ and $C_{el}$ would be for the system pTU. Because of the dependence of these coefficients on the inverse of the distance between the source and sink, only the sinks close to a leaf can draw $S$ towards them. However, the amount of $S$ pulled by such a region of sinks, $R(P_0)$, into pTU is more than it can consume and a larger amount is left into the sieve tubes, say up to a path length pTU(0). In the process of drawing $K^+$, water, and $S$ to them the cells of $R(P_0)$ acquire high turgor pressure and a high level of $K^+$. The sink tissues lying close to $R(P_0)$ but further away from the leaf, $R(P_1)$, can thus execute reverse proto-osmosis on $R(P_0)$. $H^+$ flux will therefore occur from the thirsty cells of $R(P_1)$ to the satiated cells of the region $R(P_0)$, as a consequence of which the sap in the path length pTU(0) will be moved to pTU(1), closer to $R(P_1)$. This process will continue along pTU till the last region of sink at the end of the chain has obtained its share. In a steady state, $H^+$ from the closest sink will move into the leaves, from the next region $R(P_1)$, to $R(P_0)$, and so on till the end of pTU. This constitutes in end effect a constant flux of $H^+$ to the leaf commensurate with the rate of generation of $OH^-$ in the volume of the leaf. The relative contribution of a sink to this flux $J_H$, and conversely, the supply of $S$ to this sink will depend on the rate of respiration in the sink tissue and its closeness to the leaf. If no photosynthesis and nitrate reduction is taking place in the leaves the $H^+$ extruded by the respiring cells of the sinks must be secreted out into the soil and a flux of $K^+$ and water into the sinks will occur. During $OH^-$ production in the leaves the major part of the active efflux of $H^+$ by $R(P)$ is directed towards the leaves. As regards the flux of $H^+$ from the xylem parenchyma $R(X)$, one part, which has been discussed in Section 5.2, is directed to the soil and served to bring about the uptake of water and nutrients from the soil, and the other part presumably moves into the sieve tubes. This latter flux is expected from the relatively lower pH of the xylary sap, lower hydrostatic pressure, and a lower $V_K$ value. This flux would drive $K^+$, water, and $S$ into xTU and nourish the tissues of the region $R(X)$. Part of water and $K^+$ thus entering the xylem would join the flux of water to the leaves and raise the level of $K^+$ in the leaves, which is essential to lower their $V(HK)$ value, both for efficient phloem loading and the transport of the sap by proto-osmosis. $K^+$ can be thought to circulate in the total transport network of plants in the loops formed by axial and lateral transport systems,[49,50] driving the photosynthates in the sieve tubes like conveyor belts (i.e., the $K^+$ layer).[52]

## 6. Discussion

A large number of problems related to transport in plants were posed in Section 2. It cannot be said, by any means, that the phenomenon of plant transport has been explained here by invoking a variation of electro-osmosis, since no quantitative results have been obtained. However, even within the limited scope of the treatment, the gross features of transport in plants can be qualitatively understood in terms of the new approach. The ad hoc distinction between the mechanisms of long-distance transport and cellular transport is lifted and a unified and consistent description of transport processes at all levels is achieved if proto-osmosis is assumed to operate in the transport network of plants. $H^+$ remains, as in the case of cellular transport, the working ion with the pmf as the driving force for long- and medium-distance transport. The difficulties of pH regulation in the tissues of the sinks and the sources were removed, and what would constitute a drain on the energy resources of the cells of these tissues becomes a source of power output, $J_H \times$ pmf, for driving medium- and long-distance transport.

Plants appear to manifest proto-osmosis at a macroscopic scale because of the spatial separation of the $H^+$-consuming (leaf) and $H^+$-producing (sink) tissues. Unfortunately the relevant data on the rates of these processes are not available in a form to make it possible to check the results of the theory with experimental observations. A quantitative comparison of theory with experiments has been attempted for the case of muscles (in this volume) in which proto-osmosis is suggested to operate at a microscopic level. Apart from the system of the sieve tubes, axons could be suitable objects for the experimental verification of the phenomenon. Bowling[53] has measured transplate potentials of a few millivolts, but more electrophysiological studies are required. It should be mentioned that electro-osmosis has been often considered as a mechanism for translocation, and Fensom's theory comes very close to the present approach.[54,55] The basic difference between the two approaches is in the source and the localization of the electric field, which brings about the movement of the $K^+$. Proto-osmosis is more efficient than electro-osmosis brought about by an electric field across the whole lumen, since in the former case little energy is dissipated in the current flow in the lumen. Equation (7) for the diffusion potential predicts that the transplate potentials in the sieve tubes will be greater at low $K^+$ level on account of reduced $g_K$. However, the field measurable with electrodes is expected to be much lesser than the field acting on the $K^+$ layer at the surface. In this respect the techniques of Pohl[56] and Jaffe and Nuccitelli[57] may be more suitable. It is clear that the experimental findings of the above-mentioned authors could be explained in terms of the flow of proticity, and this author would like to think that the externally recorded electric fields provide experimental support for proto-osmosis as a widespread phenomenon in all active cells.[58]

An encroachment is made on the field of active transport in Section 3.4 by suggesting proto-osmosis as the mechanism of $K^+$-antiport merely on the basis of a limit process. $K^+$-antiport seems to be the dominant transport process caused by electrogenic proton pumps in higher plants. Furthermore, the flux of $K^+$ is strongly coupled to cell turgor. I was particularly encouraged to suggest active uptake of water coupled to $K^+$ influx because of the findings of MacRobbie[45] that the rise and fall in the turgor of the guard cells cannot be accounted for by simultaneous changes in their osmotic potential. If the $K^+$-antiport is driven by a reversible $H^+$-ATPase, the proposed mechanism contains within it the possibility of driving ATP synthesis by a reversal of $K^+$ flux, since a pmf of reverse polarity (to that generated by ATP hydrolysis) would be generated due to $J_H = -J_K$. This aspect could be of wider interest in active transport in general.

The treatment of a possible electrokinetic phenomenon, and its application to plant transport processes in the rudimentary form as presented here, reflects the limitations of the author. From the viewpoint of electrochemistry the treatment is not rigorous and needs improvement. Plant physiologists will find the treatment much too limited in scope and deficient in many respects. In partial defence I would like to point out that the topic is extremely vast, and restricting the treatment to $H^+$, $K^+$, water, and neutral solutes appeared to be the best possible way to express the basic scheme in clear and simple terms. Even in this limited scope, however, the approach yields encouraging results and opens up possibilities of fruitful applications in other areas of plant science. Being deeply interwoven with cellular metabolism, the concept of proton flux mediated transport may provide the missing mechanism for the long-distance couplings of metabolic processes. As an interesting application to problems of plant ecology, mention may be made that the effect of acid rain is, possibly, a simple concequence of the blockage of translocation of photosynthates due to the neutralization of the $OH^-$ groups in the leaves. Thus in spite of its shortcomings, this approach from a multidisciplinary standpoint, I hope, will prove to be useful for the study of plants.

## References

1. J. Moorby, *Transport Systems in Plants*, Longman, London, (1981).
2. U. Lüttge and N. Higinbotham, *Transport in Plants*, Springer-Verlag, New York (1979).
3. M. V. Parthasarathy, in *Transport in Plants I, Phloem Transport, Encyclopedia of Plant Physiology*, New Series, Springer-Verlag, Berlin (1975), Vol. 1, pp. 3–38.

4. M. H. Zimmermann, *Xylem Structure and the Ascent of Sap,* Springer-Verlag, Berlin (1983).
5. M. H. Zimmermann and J. A. Milburn, in *Physiological Plant Ecology II, Encyclopedia of Plant Physiology,* New Series, Springer-Verlag, Berlin (1982), Vol. 12B, pp. 135–152.
6. J. A. Milburn, *Water Flow in Plants,* Longman, London (1979).
7. P. S. Nobel, *Introduction to Biophysical Plant Physiology,* W. H. Freeman, San Francisco (1974).
8. J. A. Milburn, "Cavitation Studies on Whole Ricinus Plants by Acoustic Detection," *Planta* **112**, 333–342 (1973).
9. J. A. Milburn and M. E. McLaughlin, "Studies of Cavitation in Isolated Vascular Bundles and the Whole Leaves of Plantago Major L," *New Phytol.* **73**, 861–871 (1974).
10. G. Haberlandt, *Physiological Plant Anatomy,* Macmillan, London (1914).
11. R. C. Dobbs and D. R. M. Scott, "Distribution of Diurnal Fluctuations in Stem Circumference of Douglas Fir," *Can. J. For. Res.* **I**, 80–83 (1971).
12. B. Klepper, V. D. Browning, and H. M. Taylor, "Stem Diameter in Relation to Plant Water Status," *Plant Physiol.* **48**, 683–685 (1971).
13. P. E. Weatherley, in *The Water Relations of Plants,* (J. A. Rutter and D. Whitehead eds.), Blachwell, Oxford (1963), pp. 85–100.
14. J. A. Rutter, *Transpiration,* Oxford University Press, London (1972).
15. R. D. Gibbs, in *The Physiology of Forest Trees,* (K. V. Thimann ed.), Ronald, New York (1958), pp. 43–69.
16. M. J. P. Canny, in *Transport in Plants I, Phloem Transport, Encyclopedia of Plant Physiology,* New Series, Springer-Verlag, Berlin (1975), pp. 139–153.
17. A. J. Peel, in *Transport in Plants I, Phloem Transport, Encyclopedia of Plant Physiology,* New Series, Springer-Verlag, Berlin (1975), pp. 171–195.
18. D. S. Fensom, in *Transport in Plants I, Phloem Transport, Encyclopedia of Plant Physiology,* New Series, Springer-Verlag, Berlin (1975), pp. 223–244.
19. L. C. Ho and A. J. Peel, "Investigation of Bidirectional Movement of Tracers in Sieve Tubes of Salix verminalis L.," *Ann. Botany N. S.* **33**, 833–844 (1969).
20. W. Eschrich, in *Transport in Plants I, Phloem Transport†  Encyclopedia of Plant Physiology,* New Series, Springer-Verlag, Berlin (1975), pp. 245–255.
21. J. A. Milburn, in *Transport in Plants I, Phloem Transport, Encyclopedia of Plant Physiology,* New Series, Springer-Verlag, Berlin (1975), pp. 328–353.
22. I. F. Wardlaw, "Phloem Transport, Physical, Chemical or Impossible," *Ann. Rev. Plant Physiol.* **25**, 515–539 (1974).
23. J. Andrew, C. Smith, and J. A. Milburn, "Water Stress and Phloem Loading," *Ber. Deutsch. Bot. Ges.* **93**, 269–280 (1980).
24. D. R. Geiger and S. A. Sovonick, in *Transport in Plants I, Phloem Transport, Encyclopedia of Plant Physiology,* New Series, Springer-Verlag, Berlin (1975), pp. 256–288.
25. C. E. Hartt, "Effect of Potassium Deficiency upon Translocation of $^{14}C$ on Attached Blades and Entire Plant of Sugar Cane," *Plant Physiol.* **44**, 1461–1469 (1969).
26. S. Amir and L. Reinhold, "Interaction Between K-Deficiency and Light in $^{14}C$-Sucrose Translocation in Bean Plants," *Physiol. Plantarum* **24**, 226–231 (1971).
27. K. Mengel, "Effect of Potassium on Assimilate Conduction to Storage Tissue," *Ber. Deutsch. Bot. Ges.* **93**, 353–362 (1980).
28. F. Malek and D. A. Baker, "Proton Co-transport of Sugars in Phloem Loading," *Planta* **135**, 297–299 (1977).
29. P. Mitchell, "Keilin's Respiratory Chain Concept and its Chemiosmotic Consequences," *Science* **206**, 1148–1159 (1979).
30. D. G. Nicholls, *Bioenergetics; An Introduction to Chemiosmotic Theory,* Academic, London (1982).

31. I. C. West, *The Biochemistry of Membrane Transport*, Chapman and Hall, London (1983).
32. J. A. Raven and F. A. Smith, in *Plant Membrane Transport: Current Conceptual Issues* (R. M. Spanswick, W. J. Lucas and J. Dainty, eds.), Elsevier/North-Holland Biomedical Press, Amsterdam (1979), pp. 161–178.
33. R. Giaquinta, "Mechanism and Control of Phloem Loading of Sucrose," *Ber. Deutsch. Bot. Ges.* **93**, 269–280 (1980).
34. F. A. Smith and J. A. Raven, in *Transport in Plants II. Part A: Cells, Encyclopedia of Plant Physiology*, New Series, Springer-Verlag, Berlin (1976), Vol. 2, pp. 317–346.
35. L. Onsager, in *Physical Principles of Biological Membranes* (F. Snell, J. Wolken, G. Iverson, and J. Lam, eds.), Gordon & Breach, New York (1970), pp. 137–141.
36. L. Onsager, in *Physics and Chemistry of Ice*, (E. Whalley, S. J. Jones, and L. W. Gold, eds.), Royal Society, Ottawa (1973), pp. 7–12.
37. D. B. Kell, "On the Functional Proton Current Pathway of Electron Transport Phosphorylation; An Electrodic View," *Biochim. Biophys. Acta* **549**, 55–99 (1979).
38. J. F. Nagle and S. Tristram Nagle, "Hydrogen Bonded Chain Mechanisms for Proton Conduction and Proton Pumping," *J. Memb. Biol.* **74**, 1–14 (1983).
39. J. O'M. Bockris and A. K. N. Reddy, *Modern Electrochemistry*, Plenum Press, New York (1973).
40. G. Joos, *Lehrbuch der Theoretischen Physik*, Akademische Verlagsgesellschaft, Frankfurt (1959).
41. D. A. Haydon, in *Recent Progress in Surface Science*, (J. F. Danielli, K. G. A. Pankhurst, and A. C. Riddiford, eds.), Vol. 1, Academic, New York (1964), pp. 94–158.
42. H. J. Evans and R. A. Wildes, in *Potassium in Biochemistry and Physiology*, International Potash Institute, Berne (1971), pp. 13–39.
43. W. D. Jeschke, in *Plant Membrane Transport: Current Conceptual Issues* (R. M. Spanswick, W. J. Lucas, and J. Dainty, eds.), pp. 17–32, Elsevier/North-Holland Biomedical Press, Amsterdam (1979).
44. M. T. Marré, G. Romani, M. Cocucci, M. M. Moloney, and E. Marré, in *Plasmalemma and Tonoplast: Their Functions in the Plant Cell* (D. Marmé, E. Marré, and R. Hertel, eds.), Elsevier Biomedical Press, Amsterdam (1982), pp. 3–14.
45. E. A. C. MacRobbie, in *Plant Membrane Transport; Current Conceptual Issues* (R. M. Spanswick, W. J. Lucas, and J. Dainty, eds.), Elsevier/North-Holland Biomedical Press, Amsterdam (1979), pp. 97–112.
46. R. M. Spanswick, "Electrogenic Ion Pumps," *Ann. Rev. Plant Physiol.* **32**, 267–289 (1981).
47. M. Amin, "An Approach to the Steady State Current–Voltage Relationship of Excitable Membrane Based on Adsorption Phenomena," *Biophys. J.* **14**, 1–7 (1974).
48. T. C. Hsiao, in *Transport in Plants II Part B, Encyclopedia of Plant Physiology*, New Series Vol. 2, Springer-Verlag Berlin (1976), pp. 195–221.
49. W. Höll, in *Transport in Plants I, Phloem Transport, Encyclopedia of Plant Physiology*, New Series, Springer-Verlag, Berlin (1975), pp. 432–450.
50. J. S. Pate, in *Transport in Plants, Phloem Transport, Encyclopedia of Plant Physiology*, New Series, Springer-Verlag, Berlin (1975), pp. 451–473.
51. M. Amin, "Ascent of Sap in Plants Means of Electrical Double Layers," *J. Biol. Phys.* **10**, 103–109 (1982).
52. M. Amin, "A Mechanism of Translocation Based on High Proton Mobility and $K^+$ Counterflux at Negatively Charged Surfaces in Sieve Elements," *J. Biol. Phys.* **11**, 111–116 (1984).
53. D. J. F. Bowling, "Evidence for the Electro-osmotic Theory of Transport in the Phloem," *Biochim. Biophys. Acta* **183**, 230–232 (1969).
54. D. S. Fensom, "The Bioelectric Potentials of Plants and Their Functional Significance. I An Electrokinetic Theory of Transport," *Can. J. Botany* **35**, 573–582 (1957).

55. D. C. Spanner, in *Transport in Plants, I. Phloem Transport, Encyclopedia of Plant Physiology*, New Series, Springer-Verlag, Berlin (1975), pp. 301–327.

56. H. A. Pohl, in *Bioelectrochemistry* (H. Keyzer and F. Gutmann, eds.), Plenum Press, New York (1980), pp. 273–296.

57. L. F. Jaffe and R. Nuccitelli, "Electrical Controls of Development," *Ann. Rev. Biophys. Bioeng.* **6**, 445–476 (1977).

58. M. Amin, "Theory and Applications of a New Phenomenon of Electro-osmosis Based on Rapid $H^+$ Jumps: I. Rate of Enzymatic Reactions," *Natl. Acad. Sci. Lett.* **5**(6), 203–204 (1982).

# 22

# Electrochemical Methods for the Prevention of Microbial Fouling

## H. P. Dhar

*ABSTRACT:* Microbial fouling on metal surfaces is undesirable in many practical situations. The chemical and mechanical methods of fouling prevention include the use of biocidal chemicals, toxic paints, brushes, ultrasonic vibrations, uv light, etc. An electrochemical method would be desirable in many situations, as the method works directly at the metal–electrolyte interface. The electrochemical methods fall into two categories: anodic and cathodic. The anodic methods depend on the generation of a biocide like $Cl_2$ on the surface of the metal, or on the anodic production of toxic metal ions. The cathodic methods comprise the production of $H_2$ and $H_2O_2$ on the metal surface. In the former approach, water is electrolyzed at a current density in excess of $1$ mA cm$^{-2}$, which encourages the formation of calcareous deposits in seawater. In the latter approach, $H_2O_2$ is produced on the metal from the electrochemical reduction of dissolved $O_2$ in the electrolyte. A current density of $5$–$20$ $\mu$A cm$^{-2}$ has been found to be effective in decreasing bacterial content by $2$–$3$ orders of magnitude. The generation of $H_2O_2$ on the metal surface is accompanied by a slight change of pH towards the alkaline side. Thus, there is less danger of the formation of calcareous deposits for the cathodic $H_2O_2$ method for the prevention of microbial fouling.

## 1. Introduction

Microbial fouling is caused by the accumulation of microorganisms, mainly bacteria and to a leser extent protozoa and diatoms, and their metabolic products at interfaces. Such an interface can be a solid/liquid, a liquid/gas, or a gas/solid interface. In the present contribution attention will be focussed only on the solid/liquid interface where electrochemical methods

---

*H. P. Dhar*  •  Energy Research Corporation, 3 Great Pasture Road, Danbury, Connecticut 06810.

can be easily applied for the prevention of microbial fouling in particular, and macrobial fouling (or biofouling) in general.

At the beginning of the macrobial stage of fouling, diatoms, protozoa, and fungi appear in increasing abundance, and eventually a complex assemblage of microorganisms, their metabolic products, and water bearing detritus is formed.[1,2] The more serious macrobial fouling starts through the larval settlement of the sessile organisms such as barnacles, bryozoa, oysters, mollusks, etc.

Through the work of an increasing number of investigators, evidence is accumulating that the formation of a macrobial fouling layer proceeds in stages. The first stage is the adsorption of dissolved organic compounds in seawater,[3,4] and other macromolecules, ions, and colloids which may serve as nutrients for the settling microorganisms.[5] Corpe[6] observed that motile rod-shaped bacteria are the first organisms to colonize the "conditioned" substrates. Gerchakov et al.[7,1] and Marszalek et al.,[8] studying the fouling conditions in subtropical seawater, showed that the glass and metal substrates exhibit a characteristic succession of periphytic microorganisms. Rod-shaped microorganisms were the first organisms to appear on glass and metal samples in as little as four hours. Up to two weeks, microorganisms were essentially bacteria with other microorganisms occurring in relatively insignificant number. Within a five-week period, a complex two-tier microbial fouling layer was produced.

A microbial film, besides being the first step of the formation of a more serious macrobial fouling, is the cause of several unwanted phenomena. Such a film can reduce water flow in boiler pipe lines[9,10] caused by the capability of such a film in increasing the fluid frictional resistance. The latter aspect has been studied by Picologlou et al.[11] and McCoy et al.[12] A microbial film can also reduce the efficiency of the passage of heat through the otherwise conducting walls of a heat exchanger. Heat conductivity of a microbial fouling layer is essentially the same as the conductivity of a stagnant layer of water.[13] With entrapped particles of low thermal conductivity, heat conductivity of a fouling layer decreases further. Thus the presence of a microbial layer can be critical to systems such as an Ocean Thermal Energy Conversion (OTEC) power plant which operates at a low theoretical Carnot efficiency. A microbial fouling layer can contribute to the corrosion of a metallic substratum.[14–18] One or more of the following factors may be responsible[19] for corrosion by a microbial film: (1) influence on the rate of cathodic or anodic corrosion reaction; (2) change in the surface metal film resistance by metabolism or products of metabolism; (3) creation of a corrosive environment; and (4) establishment of a barrier by growth and multiplication so as to create electrolytic concentration cells on the metal surface.

Characklis[20,21] reviewed the literature on the growth of biofilm and its effect on frictional resistance. Extensive recent work on the characteristics of microbial film has also been carried out by Characklis and co-workers.[11,13]

The literature on the prevention of biofouling has been reviewed by Benson *et al.*[22] and by Mitchell and Benson.[23] However, electrochemical methods for fouling prevention received less attention. In many situations, an electrochemical method for fouling prevention would be desirable, and this review is an attempt to summarize the electrochemical methods available to achieve this objective.

## 2. Approaches to Biofouling Prevention

### 2.1. Chemical and Mechanical Methods

Microorganisms are known to exhibit a wide range of tolerance to biocidal chemicals, with bacteria showing the highest resistance.[22] With invertebrates, fouling starts with the larval settlement, and this stage is believed to be the most susceptible to biocidal chemicals.

Among the means used to remove a fouling layer, the use of chlorine and chlorine compounds[13,24] is widely practiced. Chlorine is needed to the extent of 50 ppm to remove an established fouling layer, but then the removal may not be complete, and the restart of a fouling process is rapid.

The use of paints containing a soluble toxic compound like salts of heavy metals (e.g., Cu, Hg, and As) or organic compounds of tin and lead is common. Toxic material may diffuse out of the matrix or the entire coating gradually may erode exposing a fresh toxic surface. Protein deterioration, enzyme inactivation, and degenerative effects are the ways a toxine would work on a fouling layer to destroy or to prevent its formation. The value of a biocidal chemical needed to control fouling in an open system like that in seawater situations would be prohibitive, and many of these treatments in the open system are of environmental concern.

Mechanical methods involving rubber balls, ultrasonics, uv light, and streams of air bubbles are often-tried methods[22] for removal of a fouling layer.

### 2.2. Electrochemical Methods

The electrochemical methods generally fall into two categories: anodic and cathodic. The use of electricity to control or to prevent fouling has been suggested for a long time.

### 2.2.1. Historical

Humphry Davey in 1825 developed a method[25] for protecting a copper-bottomed ship from fouling by connecting the ship to strips of Sn, Zn, or Fe through a metallic filament. Copper, being a weakly positive metal, is rendered negative by the more positive metals like Fe when in contact with an electrolyte like seawater. It is stated[25] that within limits, no marine deposition occurred, while only a very small amount of surface copper was dissolved into the sea.

A German patent was issued to Edison[26] in 1890. He set up a field between a ship's hull and electrodes hung overside. The ship was made negative in the electric circuit, and the method supposes that a fouling organism would avoid an electric field. There are no reports of the practical use of Edison's method of fouling prevention.

### 2.2.2. Anodic Methods

An anodic method for fouling prevention would depend upon either the anodic production of a biocide, using an external source of chloride electrolyte, or the anodic production of toxic metal ions.

Delius et al.[27] received a U.S. patent for biofouling prevention for their process in which an electrode is alternatively connected to anodic and cathodic potentials. During the anodic action chlorine evolves and prevents fouling. Bennett et al.[28] developed an anodic method in which oxygen is evolved at the anode, having a coating of ruthenium dioxide. The action is presumably due to an acidic condition generated during oxygen evolution. At higher current densities, e.g., 1 mA cm$^{-2}$, chlorine generation is likely in a chloride electrolyte.

Another anodic method depended on the dissolution of toxic metals like Cu and Ag. Green[29] got a patent for an antifouling process in which he used Cu anodes to prevent fouling. In this process, minute bubbles of chlorine evolved and copper ions were released at the anode during the electrolysis in the saline solution.

Silver cations generated by passing low-intensity direct current of magnitude 2–5 $\mu$A cm$^{-2}$ through pure silver electrodes have been found to be very much antibacterial in inhibition and killing oral bacteria,[30] and in the treatment of orthopedic infections.[31] Spadaro et al.[32] have shown that electrochemically generated silver ion was effective at a concentration of 5 $\mu$g ml$^{-1}$. Figure 1 plots data obtained by Spadaro et al.[32] showing the effect of current level and time on the viability of two kinds of bacteria. Various concentration levels of Ag$^+$ ion in the solution are given in the figure caption. Berger et al.[33] have shown that the bactericidal concentration of silver was 10–100 times less than that of silver sulfadiazine.

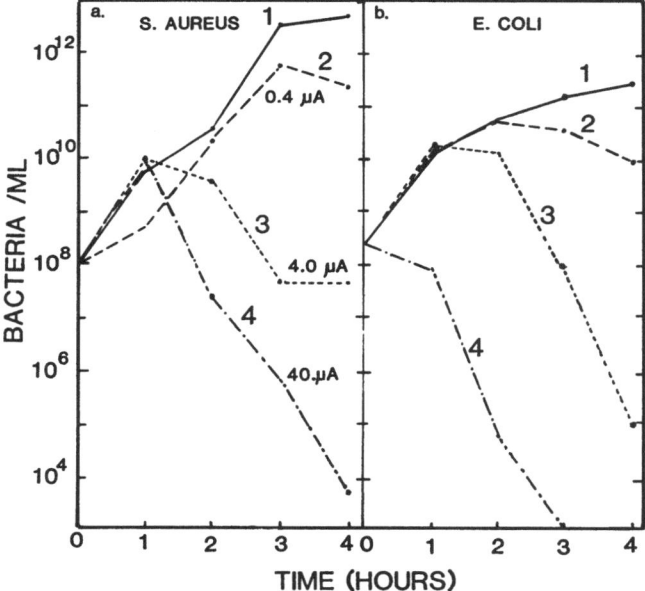

*FIGURE 1.* Inhibition of *Staphylococcus aureus* and *Escherichia coli* at the silver anode chamber as a function of time and current level. (a) 1: control Ag wire (0.7 $\mu$g Ag cm$^{-3}$); 2: 0.4 $\mu$A, 4.3 $\mu$g Ag cm$^{-3}$; 3: 4.0 $\mu$A, 7.0 $\mu$g Ag cm$^{-3}$; 4: 40 $\mu$A, 79 $\mu$g Ag cm$^{-3}$; (b) 1: control (see a); 2: 0.4 $\mu$A, 4.9 $\mu$g Ag cm$^{-3}$; 3: 4 $\mu$A, 5.8 $\mu$g Ag cm$^{-3}$; 4: 40 $\mu$A, 92 $\mu$g Ag cm$^{-3}$. Reference: Spadaro *et al.*[32]

Gerchakov *et al.*[1] have shown that toxic metals like Cu and Cu:Ni alloys also foul up in seawater in a relatively short period of time (e.g., three weeks). An initial growth of an oxide coating seemed to aid in the fouling process.

Baboian *et al.*[34] examined the anodic reaction of electrolysis of seawater as a means of preventing fouling. They used current density in excess of 50 mA cm$^{-2}$.

### 2.2.3. Cathodic Methods

*2.2.3a. Hydrogen.* Hydrogen is produced in the alkaline medium according to the reaction

$$2H_2O + 2e \rightarrow H_2 + 2OH^-$$

Several investigators examined this process as a means of removing or preventing biofouling layers.

Tschaikowski[35] obtained a British patent for developing a method to prevent rusting and fouling of ship's bottoms by continuous formation of molecular hydrogen and caustic soda formed by the electrolysis of seawater. The author mentions a current density of 6.5 $\mu$A cm$^{-2}$ in the electrochemical cell formed by graphite anode and the underwater part of the vessel, cathode, and at a cell potential of 4.0 V.*

Castle[36] found that a cathode made from a steel plate could be prevented from fouling in seawater when a current density in excess of 1 mA cm$^{-2}$ was used. He also found that the cathodic products of electrolysis in solution within short distances from the electrode was unfavorable for fouling organisms; however, the existence of a continuous electric field did not keep the free-swimming larvae of fouling organisms away from electrodes immersed in the sea, and the attached foulers like barnacles tolerated a current density of 1 A cm$^{-2}$ for 15 min.

Littauer and Jennings[37] examined the cathodic electrolysis of seawater as a means of fouling prevention at various current densities, and obtained partial protection. However, at current densities greater than 1 mA cm$^{-2}$, calcareous deposits formed on the electrode.

Stoner[38] developed an electrochemical method for removing adsorbed pathogenic bacteria on carbon and graphite surfaces through the application of alternating potential in the range of $\pm 5$ V and at a current density of 1–20 mA cm$^{-2}$. Any aqueous medium could be treated in this way.

*2.2.3b. Hydrogen Peroxide. The Method.* The present authors[39–42] have developed a method for the prevention of microbial fouling. The method is based on the *in situ* generation of hydrogen peroxide on a cathodic electrode surface utilizing dissolved oxygen which is present to the extent $3 \times 10^{-4}$ mol l$^{-1}$ in water. The method has been tested under laboratory conditions with a limited number of bacterial species in saline electrolytes including seawater.

Hydrogen peroxide has bactericidal properties, and is produced on a cathode surface as one of the reduction products of oxygen (another product is $H_2O$ or $OH^-$). In an alkaline medium the following reaction occurs:

$$O_2 + 2H_2O + 2e \rightarrow H_2O_2 + 2OH^-$$

The optimal potential for the production of $H_2O_2$ was calculated[40] to be 0.2 to $-0.6$ V (NHE).

*Initial Experiments.* The effectiveness of $H_2O_2$ as a bactericide,

---

* This seems to be erroneous. At a current density of 6.5 $\mu$A cm$^{-2}$, the prevailing reaction would be $O_2$ reduction.

generated on an electrode surface, was initially tested in a thin divided cell (dimensions $3 \times 2 \times 0.02$ cm) made of transparent and conducting tin oxide coated glass. The thin cell was placed on the platform of a microscope fitted with the dark field and phase contrast viewing optics. As one of the signs that a bactericidal chemical was produced at the cathode compartment, the bacteria were observed to lose motility with the application of electric potentials in the region where $H_2O_2$ generation was expected. The current density was approximately 5 $\mu$A cm$^{-2}$. The presence of $H_2O_2$ in the cathode compartment and $Cl_2$ in the anode compartment was detected by chemical tests. Trivial calculation shows that at 10% current efficiency of the experimentally obtained value of 5 $\mu$A cm$^{-2}$, the region of 0.02 cm near the surface has an $H_2O_2$ concentration of around $5 \times 10^{-6}$ mol l$^{-1}$. The latter value was determined in separate viability experiments with $H_2O_2$ to be in the lethal range.

In addition to direct observations, assessment of bacterial viability in the electrode compartments was made by diluting and culturing aliquots from each compartment. Such experiments revealed that bacterial concentration in both the cathode and anode compartments decreased in an exponential fashion with the time of application of potential.

*Adsorption of Bacteria on Electrodes.* Further testing of the method was carried out while bacteria were allowed to adsorb on polarized electrodes made from tin oxide coated glass and titanium metal in saline solutions and seawater. The electrodes were kept polarized in a divided cell (volume 50 ml) for a period of time in an electrolyte containing bacteria. At the end of the polarization period, electrodes were removed and, adsorbed bacteria fixed on the surface and counted by taking scanning electron micrographs. Figure 2 shows the number of adsorbed bacteria from artificial seawater containing $2 \times 10^7$ bacteria ml$^{-1}$ on tin oxide coated glass cathode and anode as a function of potential. The region of potential is consistent with the theoretical values calculated for generating $H_2O_2$. A blank sample of electrode placed in the same electrochemical cell would adsorb approximately $5 \times 10^{10}$ bacteria m$^{-2}$. A comparison of this latter number with those in Figure 2 reveals that a diminution of two orders of magnitude was achieved on the cathode surface. On the anode surface, a similar order of diminution was observed (Figure 1), and is due to the generation $Cl_2$ on the surface.

Figure 3 shows adsorption data of bacteria on a Ti cathode in natural seawater. The corresponding cathodic current, ranging from 1 to 20 $\mu$A cm$^{-2}$, is also shown in the plot as a function of potential.

Polarizing potentials could be applied in the form of pulses, and still gave rise to a diminution in the amount of adsorbed bacteria. Figure 4 demonstrates such results for tin oxide glass electrodes in 3% saline solution.

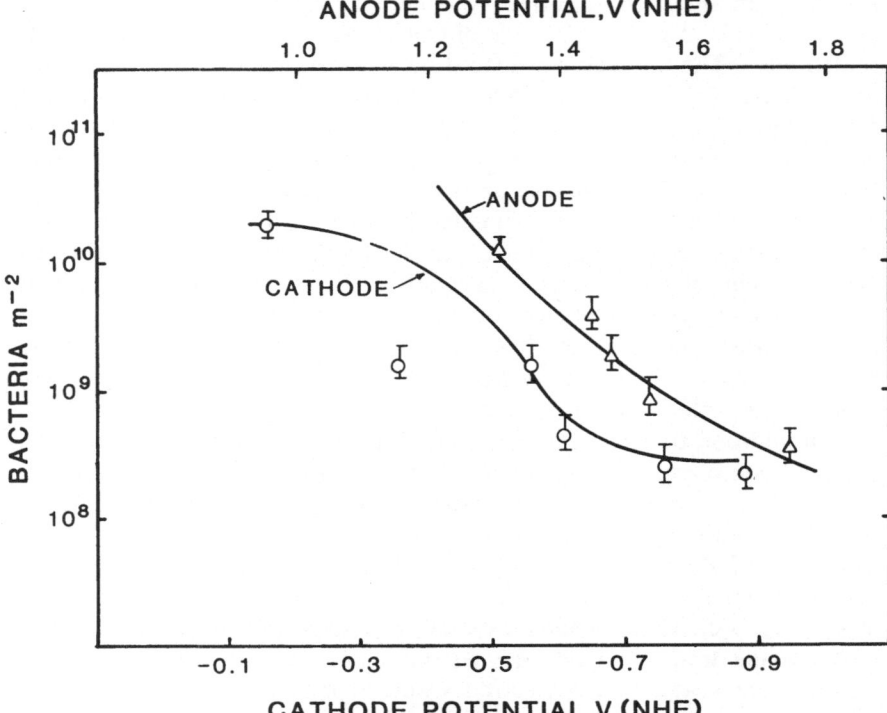

FIGURE 2. Adsorbed *Pseudomona atlantica* on tin oxide glass cathode and anode versus steady electrode potentials. The electrolyte, artificial seawater, contained approximately $2 \times 10^7$ bacteria $ml^{-1}$. Immersion time in the electrolyte at each potential, 3 h. Reference: Dhar *et al.*[42]

The developed method described above has been tested (H. Dhar, unpublished data) under various flow conditions of up to 100 cm s$^{-1}$ in the laboratory. The results indicated that the data obtained in the static system could be reproduced under flow conditions with titanium and stainless steel electrodes.

*Interpretation of the Observed Effect.* The observed diminution in the adsorbed bacteria on the electrode surface, besides being due to the bactericidal effect of $H_2O_2$, may also be in part due to (1) changes in surface energy of the electrode, (2) depletion of oxygen near the polarized surface, (3) change of pH near the surface, and (4) van der Waals forces between bacteria and the electrode surface.

The contributions from the first two effects were shown to be negligible.[40] The mechanism of pH effect might arise from the effect of pH

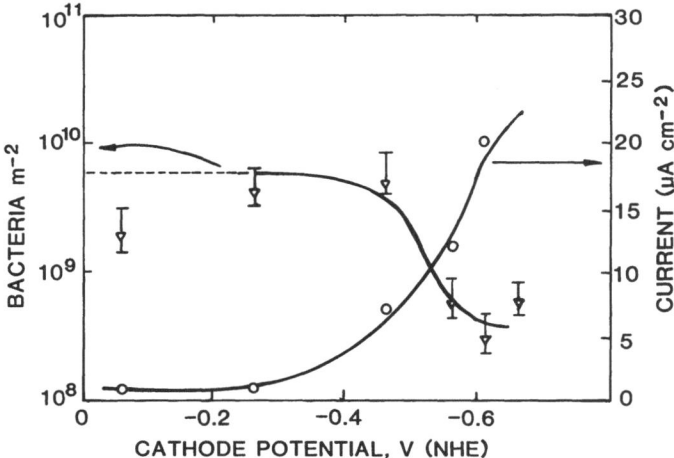

FIGURE 3. Adsorbed *Vibrio anguillarum* on Ti cathode versus steady electrode potentials. The electrolyte, natural seawater, contained approximately $2 \times 10^7$ bacteria ml$^{-1}$. Immersion time at each potential, 3 h. Reference: Dhar *et al.*[42]

on the double layer surrounding the bacteria, and thus upon their interaction with the electric field provided by the electrode.

The cathodic reduction of $O_2$ to $H_2O_2$ or to $OH^{-1}$ is accompanied by a change in pH in the vicinity of the electrode surface, causing an increase in bulk pH. Correspondingly at the anode, the likely reactions are the evolution of $O_2$, which would consume $OH^-$, or the evolution of $Cl_2$, which would react with water to produce HCl and HOCl. Hence, the cathode compartment will tend to become alkaline and the anode compartment acidic. At the anode, therefore, the negative charge that exists on bacteria[43] in solution will be decreased and attraction to the positive electrode lessened. On the cathode side,* the pH is increased, and it may be inferred that increasing $OH^-$ adsorption will increase the bacterial charges, thus decreasing adsorption of charged bacteria on the cathode.

Gordon *et al.*[44] studied adsorption of bacteria on polarized Cu and Pt metal surfaces in seawater, and attributed their adsorption results, in general, to pH changes at the interface. A copper surface was polarized cathodically from $-0.12$ to $-0.24$ V (NHE), and anodically from $-0.01$ to $0.07$ V (NHE). Platinum was polarized cathodically from $0.40$ to $0.22$ V (NHE), and anodically from $0.60$ to $1.30$ V (NHE). While anodic polarization was seen to decrease bacterial content on both metals, the cathodic polarization produced mixed results for different kinds of bacteria.

---

* In this statement it is assumed that charge on the anode is positive.

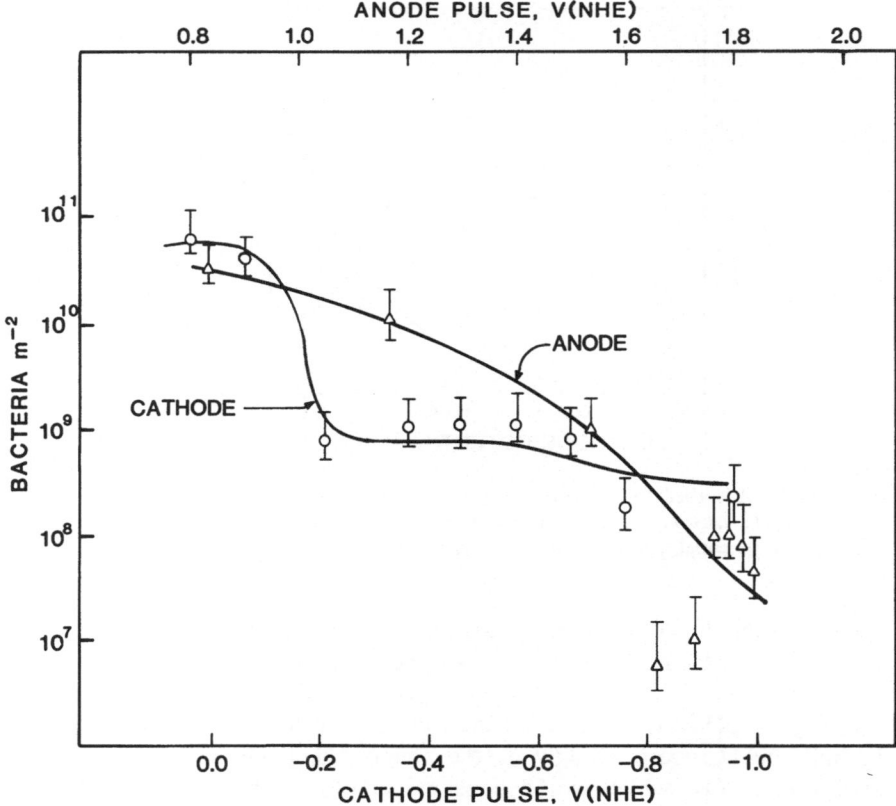

FIGURE 4. Adsorbed *Vibrio anguillarum* on tin oxide glass cathode and anode versus pulse height potentials. The electrolyte, 3% saline solution, contained approximately $1 \times 10^7$ bacteria ml$^{-1}$. Pulse height: 10 s on; 10 s off. Rest potential for off pulse 0.24 V. Immersion time in the electrolyte at each potential, 6 h. Reference: Dhar et al.[41]

The authors measured the pH changes as a function of polarization potential at a distance of 50 $\mu$m from the interface. Figures 5 and 6 show the change of pH as a function of overpotential (measured from the rest potential in each case) for Cu and Pt, respectively. At the Cu–seawater interface pH was seen to vary from 9.5 at $-1.03$ V (NHE) to 7 at 0.15 V (NHE), and at the Pt–seawater interface pH was observed to vary from 9 at $-0.54$ V (NHE) to 7.8 at 1.3 V (NHE).

The effect of the long-range intermolecular forces between the electrode and bacteria has been considered by Rutter and Vincent[45] in terms of the DLVO theory.[46,47] The total free energy of interaction comprises

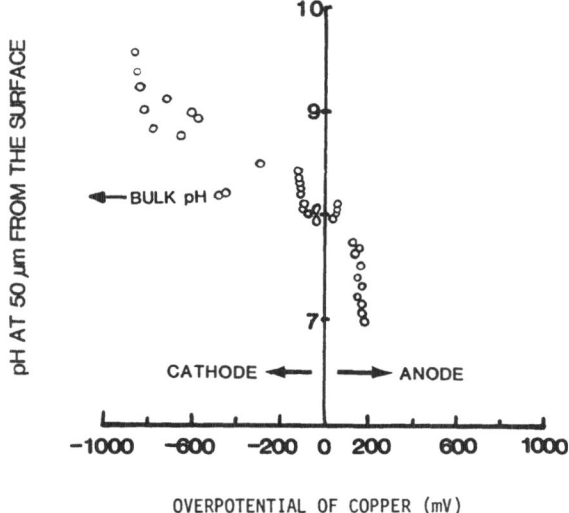

FIGURE 5. Interfacial pH at 50 μm from a copper surface in seawater as a function of over-potential [rest potential at −0.039 V (NHE)]. Reference: Gordon *et al.*[44]

two terms: $G_A$, due to van der Waals forces, and $G_E$, due to overlap of electrical double layers associated with charge groups present on the particle and the electrode. The term $G_A$ is always attractive, irrespective of the charges on the particle and electrode. At high electrolyte concentrations, corresponding to the salinity of seawater (ca. 0.5 mol l$^{-1}$), and for particle and surface of the same charge, $G_E$ becomes a small repulsive term, leading

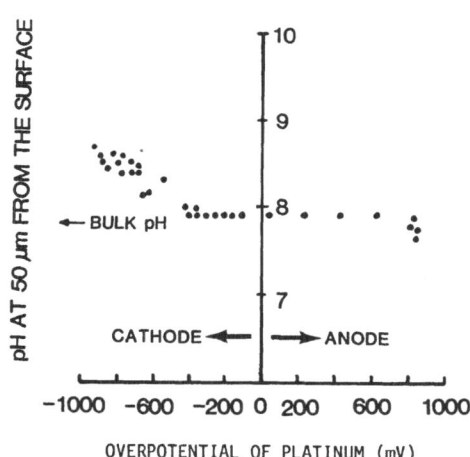

FIGURE 6. Interfacial pH at 50 μm from a platinum surface in seawater as a function of overpotential [rest potential at 0.465 V (NHE)]. Reference: Gordon *et al.*[44]

to a net attractive $(G_A + G_E)$. In cases where the surface and the particle are of opposite charges, the net effect of the two terms, $G_A$ and $G_E$, is a strong attraction, decreasing only with increasing bulk electrolyte concentration.

According to the above arguments, from a consideration of van der Waals forces, attractive interactions between bacteria and electrodes are expected irrespective of the nature of charges on them. The observations made above with a number of bacterial species suggest the dominant nature of the bactericidal action of the generated species on the surface.

### 2.2.4. Pros and Cons of an Electrochemical Approach

An electrochemical method for fouling prevention could be effectively applied as the method works at the metal–electrolyte interface where the prevention is needed. Thus, an excess use of a chemical which is always necessary in the chemical mode of fouling prevention could be avoided in the electrochemical method. The method could be ideal for situations where biofouling has to be avoided at the earliest stage, for example, in an OTEC power plant. In some systems, it would be possible to prevent fouling and electrochemical corrosion simultaneously by a proper manipulation of the cathodic potential.

The cathodic electrochemical method suffers from the possibility of encouraging precipitation of $Mg(OH)_2$ and $CaCO_3$ on the surface, and the anodic method from dissolution of the anode. A proper choice of electrode potentials and electrode materials is necessary for the success of an electrochemical method. Theoretical considerations suggest that, assuming $k_{sp} = 1.2 \times 10^{-11}$ mol$^3$ l$^{-3}$ for $Mg(OH)_2$ and concentration of $Mg^{+2}$ to be 0.52 mol l$^{-1}$ in seawater, $Mg(OH)_2$ should precipitate at a pH of 9.2. Precipitation of $CaCO_3$ should occur at a slightly lower pH. A combination of cathodic and anodic processes may be suitable to counteract any effect of pH changes which might adversely affect any operation.

## References

1. S. M. Gerchakov, F. J. Roth, B. Sallman, L. R. Udey, and D. S. Marszalek, "Observation on Microfouling Applicable to OTEC Systems," *Proc. Ocean Thermal Energy Conversion Biofouling and Corrosion Symposium*, Seattle, Washington (1977), pp. 63–75.
2. J. M. Sieburth, *Microbial Seascapes*, University Park Press, Baltimore, Maryland (1975).
3. R. E. Baier, "Influence of the Initial Surface Condition of Materials on Bioadhesion," *Proc. 3rd International Congress on Marine Corrosion and Fouling*, Northwestern University Press, Evanston, Illinois (1973), pp. 15–48.
4. G. I. Loeb and R. A. Neihof, "Marine Conditioning Films," *Adv. Chem. Ser.* **145**, 319–335 (1975).

5. K. C. Marshall and G. Bitton, "Microbial Adhesion in Perspective," *Adhesion of Microorganisms on Surfaces* (G. Bitton and K. C. Marshall, eds.), Wiley, New York (1980), pp. 1–5.

6. W. A. Corpe, "Primary Bacterial Films and Marine Microfouling," *Proc. 4th International Congress on Marine Corrosion and Fouling*, Antibes, France (1977), pp. 97–100.

7. S. M. Gerchakov, D. S. Marszalek, F. J. Roth, and L. R. Udey, "Succession of Periphytic Microorganisms on Metal and Glass Surfaces in Natural Seawater," *Proc. 4th International Congress on Marine Corrosion and Fouling*, Antibes, France (1976), pp. 203–211.

8. D. S. Marszalek, S. M. Gerchakov, and L. R. Udey, "Influence of Substrate Composition on Marine Microfouling," *Appl. Environ. Microbiol.* **38**, 987–995 (1979).

9. W. A. Corpe, "Ecology of Microbial Attachment and Growth on Solid Surfaces," *Proc. Symp. Microbiol. Power Plant Thermal Effluents*, Iowa City, Iowa (1978), pp. 57–65.

10. L. Seifert and W. Kruger, "Unusual High Frictional Factor in a Long Water Supply Line," *Ver. Dtsch. Ing. Z.* **92**, 189–191 (1950).

11. B. F. Picologlou, N. Zelver, and W. G. Characklis, "Biofilm Growth and Hydraulic Performance," *J. Hydraul. Div. Am. Soc. CIV. Eng.* **106**, 733–746 (1980).

12. W. F. McCoy, J. D. Bryers, J. Robbins, and J. W. Costerton, "Observations of Fouling Biofilm Formation," *Can. J. Microbiol.* **27**, 910–917 (1981).

13. W. G. Characklis, "Biofilm Development and Destruction in Turbulent Flow," *Ozone: Science and Engineering* **1**, 167–181 (1979).

14. F. L. LaQue, "Behavior of Metal and Alloys in Seawater," *Corrosion Handbook* (H. H. Uhlig, ed.), Wiley, New York (1948), pp. 383–430.

15. A. J. Costello, "The Corrosion of Metals by Microorganisms: A Literature Survey," *Int. Biodetn. Bull.* **5**, 191–218 (1969).

16. W. P. Iversion, "Biological Corrosion," *Advances in Corrosion Science*, Vol. 2, (M. G. Fontana and R. W. Staehle, eds.), Plenum Press, New York (1972).

17. K. D. Efird, "The Interrelation of Corrosion and Fouling of Metals in Seawater," *Mater. Perform.* **15**, 16–25 (1976).

18. S. M. Gerchakov, "Biofouling and Effects of Organic Compounds and Microorganisms on Corrosion Processes," *Symp. Microbiol. Power Plant Thermal Effluents*, Iowa City, Iowa (1978), pp. 67–72.

19. R. F. Hadley, "Corrosion by Microorganisms in Aqueous and Soil Environments," *Corrosion Handbook* (H. H. Uhlig, ed.), Wiley, New York (1948), pp. 466–481.

20. W. G. Characklis, "Attached Microbial Growths I. Attachment and Growth," *Water Res.* **7**, 1113–1127 (1973).

21. W. G. Characklis, "Attached Microbial Growths II. Frictional Resistance due to Microbial Slimes," *Water Res.* **7**, 1249–1259 (1973).

22. P. H. Benson, D. L. Bringing, and D. W. Perrin, "Marine Fouling and its Prevention," *Marine Technol.* **Jan**, 30–37 (1973).

23. R. Mitchell and P. Benson, "Control of Marine Biofouling in Heat Exchanger Systems," *MTS J.* **15**(4), 11–21 (1981).

24. J. A. Fava and D. L. Thomas, "Use of Chlorine to Control OTEC Biofouling," *Ocean Eng.* **5**, 269–288 (1978).

25. P. M. Lauren, "Sir Humphry Davy's Battle with the Sea," *Chemistry* **50**, 14–17 (1977).

26. H. Kühl, *Schiffbau* **37**, 224–226 (1936).

27. G. Delius and C. P. Tatro, "Process for Protecting Ships from Barnacles," U.S. Patent 1,021,734 (1912).

28. J. E. Bennett and J. E. Elliott, "Anodically Polarized Surface for Biofouling and Scale Control," U.S. Patent 4,256,556 (1978).

29. W. G. Green, "Anti-fouling, Barnacles, Algae Eliminator," U.S. Patent 3,241,512 (1966).

30. E. A. Thibodeau, S. L. Handelman, and R. E. Marquis, "Inhibition and Killing of Oral

Bacteria by Silver Generated with Low Intensity Direct Current," *J. Dental Res.* **57**, 922–926 (1978).

31. R. O. Becker and J. A. Spadaro, "Treatment of Orthopedic Infections with Electrically Generated Silver Ions," *J. Bone Joint Surgery* **60-A**, 871–881 (1978).

32. J. A. Spadaro, T. J. Berger, S. D. Barranco, S. E. Chapin, and R. O. Becker, "Antibacterial Effects of Silver Electrodes with Direct Current," *Antimicrob. Agents Chemother.* **6**, 637–642 (1974).

33. T. J. Berger, J. A. Spadaro, R. Bierman, S. E. Chapin, and R. O. Becker, "Antifungal Properties of Electrically Generated Metallic Ions," *Antimicrob. Agents Chemother.* **10**, 856–860 (1976).

34. R. Baboian, G. S. Haynes, B. S. Ryskiewich, and B. J. Freedman, "Biofouling Prevention of Flat Surfaces Using in situ Electrolysis of Seawater," *Mater. Perform.* **19**(10), 42–46 (1980).

35. T. L. Tschaikowski, "A Process to Prevent the Rusting and Fouling of Ship's Bottoms Made of or Coated with Iron or with Steel," British patent 3338 (1903).

36. E. S. Castle, "Electrical Control of Marine Fouling," *Ind. Eng. Chem.* **43**, 901–904 (1951).

37. E. Littauer and D. M. Jennings, "Prevention of Marine Fouling by Electrical Currents," *Proc. International Congress on Marine Corrosion and Fouling*, 2nd, Athens (1968), pp. 527–536.

38. G. E. Stoner, "Electrochemical Inactivation of Pathogens," U.S. Patent 3,725,226 (1973).

39. H. P. Dhar, J. O'M. Bockris, and D. H. Lewis, "Electrochemical Inactivation of Marine Bacteria," *J. Electrochem. Soc.* **128**, 229–231 (1981).

40. H. P. Dhar, D. H. Lewis, and J. O'M. Bockris, "Electrochemical Diminution of Surface Bacterial Concentration," *Can. J. Microbiol.* **27**, 998–1010 (1981).

41. H. P. Dhar, J. O'M. Bockris, and D. H. Lewis, "Cathodic Electrochemical Process for Preventing or Retarding Microbial and Calcareous Fouling," U.S. Patent No. 4,440,611 (1984).

42. H. P. Dhar, D. W. Howell, and J. O'M. Bockris, "The Use of in situ Electrochemical Reduction of Oxygen in the Diminution of Adsorbed Bacteria on Metals in Seawater," *J. Electrochem. Soc.* **129**, 2178–2182 (1982).

43. S. C. Daniels, "Mechanisms Involved in Sorption to Solid Surfaces," *Adsorption to Microorganisms to Surfaces* (G. Bitton and K. C. Marshall, eds.), Wiley, New York (1980), pp. 7–58.

44. A. S. Gordon, S. M. Gerchakov, and L. R. Udey, "The Effect of Polarization on the Attachment of Marine Bacteria to Copper and Platinum Surface," *Can. J. Microbiol.* **27**, 698–703 (1981).

45. P. R. Rutter and B. Vincent, "The Adhesion of Microorganisms to Surfaces: Physicochemical Aspects," *Microbial Adhesion to Surfaces* (R. C. W. Berkeley, J. M. Lynch, J. Melling, P. R. Rutter, and B. Vincent, eds.), Ellis Horwood Ltd., Chichester (1980), pp. 79–92.

46. B. V. Derjaguin and L. D. Landau, *Acta Phys. Chim. U.S.S.R.* **14**, 633 (1941).

47. J. W. Verwey and J. Th. G. Overbeek, *The Theory of Stability of Lyophilic Colloids*, Elsevier, Amsterdam (1948).

# Author Index

# Subject Index